CONCORDIA UNIVERSITY
3 4211 00106 3794

Y0-BEB-376

THE PROTEINS

CHEMISTRY, BIOLOGICAL ACTIVITY, AND METHODS

VOLUME II, PART B

The Proteins

CHEMISTRY, BIOLOGICAL ACTIVITY, AND METHODS

Edited by

HANS NEURATH
Department of Biochemistry
University of Washington
Seattle, Washington

KENNETH BAILEY
Department of Biochemistry
University of Cambridge
Cambridge, England

VOLUME II, PART B

ACADEMIC PRESS INC., PUBLISHERS
NEW YORK, 1954

Library of Congress Catalog Card Number: 52-13366

PRINTED IN THE UNITED STATES OF AMERICA

44060

CONTRIBUTORS TO VOLUME II, PART B

KENNETH BAILEY, *Department of Biochemistry, University of Cambridge, Cambridge, England.*

WILLIAM C. BOYD, *Professor of Immunochemistry, Boston University School of Medicine, Boston, Massachusetts.*

N. MICHAEL GREEN, *University of Sheffield, Department of Biochemistry, Sheffield, England.*

WALTER L. HUGHES, *Johns Hopkins University, McCollum-Pratt Institute, Baltimore, Maryland.*

J. C. KENDREW, *Cavendish Laboratory, Cambridge, England.*

HANS NEURATH, *Department of Biochemistry, University of Washington, Seattle, Washington.*

H. TARVER, *School of Medicine, University of California, Berkeley, California.*

ACKNOWLEDGMENT

The Editors wish to acknowledge the invaluable aid of Dr. Martha Sinai in preparing the Subject Indexes for Volumes I and II.

HANS NEURATH
KENNETH BAILEY

CONTENTS

Page

CONTRIBUTORS TO VOLUME II, PART B v

21. Interstitial Proteins: The Proteins of Blood Plasma and Lymph BY WALTER L. HUGHES. 663

I. Origin and General Properties of Plasma Proteins. 664
II. Maintenance of Vascular Volume. 673
III. Proteins Functioning as Transporters of Metabolites 698
IV. Protective (Anti-Infection) Functions—Antibodies 708
V. Clotting Components. 725
VI. Enzymes 732
VII. Hormones. 733
VIII. Isolated Components of Unknown Function 733
IX. Appendix—Inclusive Systems of Plasma Fractionation 741
X. General References. 753

22. The Proteins of Immune Reactions BY WILLIAM C. BOYD 756

I. Immunity. 756
II. Antigens 757
III. Antibodies. 782
IV. Antibody—Antigen Combination. 814
V. Complement and Complement Fixation. 839

23. Structure Proteins. I BY J. C. KENDREW 845

I. General Introduction 846
II. Silk Fibroin 849
III. The Keratin Group. 859
IV. The Collagen Group 909
V. Miscellaneous Structure Proteins. 946

24. Structure Proteins. II. Muscle BY KENNETH BAILEY 951

I. Introduction. 952
II. The Structure Proteins of Skeletal Muscle. 957
III. Particulate Components. 1002
IV. Extractability of Muscle Proteins. 1005
V. Estimation of Muscle Proteins. 1010
VI. Amino Acid Composition of the Structure Proteins and Its Significance. 1018
VII. X-Ray and Electron Microscope Studies of Muscle and Muscle Proteins 1024
VIII. Models of Muscle Contraction. 1040
IX. Biological Activities Associated with the Sarcoplasm 1050
X. Appendix and Conclusions. 1051

Page

25. Proteolytic Enzymes BY N. MICHAEL GREEN AND HANS NEURATH 1057

I. Introduction 1058
II. Methods of Preparation 1060
III. Physicochemical Properties 1070
IV. Chemical Composition 1082
V. Stability 1086
VI. Enzymatic Activity 1095
VII. Inhibition 1144
VIII. Action of Proteolytic Enzymes on Proteins 1171

26. Peptide and Protein Synthesis. Protein Turnover BY H. TARVER 1199

I. Introduction 1200
II. Synthesis of and Interaction between Peptide Bonds 1201
III. Synthesis of Protein and Incorporation of Isotopic Amino Acids *in Vitro* 1224
IV. Incorporation of Amino Acids *in Vivo*—Turnover 1259
V. General Conclusions 1291
VI. Appendix 1292

Author Index for **Volume II** 1297

Subject Index for **Volume II** 1353

CONTENTS OF VOLUME II, PART A

12. Nucleoproteins and Viruses BY R. MARKHAM AND J. D. SMITH
13. The Oxidizing Enzymes BY THOMAS P. SINGER AND EDNA B. KEARNEY
14. Respiratory Proteins BY FELIX HAUROWITZ AND RICHARD L. HARDIN
15. Toxic Proteins BY W. E. VAN HEYNINGEN
16. Milk Proteins BY THOMAS L. MCMEEKIN
17. Egg Proteins BY ROBERT C. WARNER
18. Seed Proteins BY SVEN BROHULT AND EVALD SANDEGREN
19. Proteins and Protein Metabolism in Plants BY F. C. STEWARD AND J. F. THOMPSON
20. Protein Hormones BY CHOH HAO LI

CONTENTS OF VOLUME I (TWO PARTS)

PART A

1. **The Isolation of Proteins** BY JOHN FULLER TAYLOR
2. **The General Chemistry of Amino Acids and Peptides** BY P. DESNUELLE
3. **The Amino Acid Composition of Proteins** BY G. R. TRISTRAM
4. **The Structure and Configuration of Amino Acids, Peptides and Proteins** BY BARBARA W. LOW
5. **Optical Properties of Proteins** BY PAUL DOTY AND E. PETER GEIDUSCHEK
6. **Electrochemical Properties of the Proteins and Amino Acids** BY ROBERT A. ALBERTY

PART B

7. **The Size, Shape and Hydration of Protein Molecules** BY JOHN T. EDSALL
8. **Protein Interaction** BY IRVING M. KLOTZ
9. **Protein Denaturation** BY FRANK W. PUTNAM
10. **The Chemical Modification of Proteins** BY FRANK W. PUTNAM
11. **The Relation of Chemical Structure to the Biological Activity of the Proteins** BY R. R. PORTER

Author Index

Subject Index

CHAPTER 21

Interstitial Proteins: The Proteins of Blood Plasma and Lymph[1,2]

By WALTER L. HUGHES

Page

I. Origin and General Properties of Plasma Proteins 664
1. Limits of Free Diffusion 664
2. Description of Blood, Blood Cells and Plasma 665
3. Requirements for Circulatory Function 667
4. Distribution in Plasma, Extravascular Fluid and Cells 669
a. Reservoirs of Plasma Proteins 669
b. Evidence for Plasma Proteins within Cells 670
c. Evidence for Cellular Proteins in Plasma 672
5. Metabolism of Plasma Proteins 672
II. Maintenance of Vascular Volume 673
1. Physiology of Peripheral Circulation 673
2. Capillary Porosities and Renal Function 674
3. Osmotic Properties of the Plasma Proteins 676
4. Serum Albumins 677
a. Definitions 677
b. Heterogeneity 678
c. Crystallization 680

(1) This chapter holds no claims to completeness having been developed around my particular interests. However, it is hoped that the gain in perception and in critical appraisal so achieved will offset the otherwise flagrant omissions. Fortunately, some proteins which might naturally fall in this chapter have been covered in other chapters and so will be mentioned here only in passing. These include hemocyanin (Chap. 14), and fibrin (Chap. 20). It is unfortunate, from my point of view, that omission of fibrin could not justify complete omission of the clotting process since this appears hopelessly complex at the present time. Instead, a brief description has been included centering around fibrinogen and prothrombin with references to more extended treatments. The discussion of immunity has been restricted largely to the nonimmune properties of antibodies, a more complete discussion being found in Chap. 22.

(2) While taking full responsibility for the views here expressed, I would like to express my appreciation to the many friends with whom I have spent hours of stimulating discussion, and whose ideas must naturally appear throughout this chapter. My particular thanks go to David Gitlin, Henry Isliker, John Pappenheimer and the editors.

Page
d. Physicochemical Properties 681
e. Stability 682
f. Composition 686
g. Reactivities 688
h. Physiological Functions 697
III. Proteins Functioning as Transporters of Metabolites 698
1. Iron-Binding Globulin, Transferrin 698
2. Zinc and Copper Components of Plasma. Ceruloplasmin 700
3. Lipoproteins 701
a. Cenapse Acide 704
b. β_1-Lipoprotein (X-Protein) 704
c. α-Lipoproteins 706
d. Gofman's Components—Lipide Transport 707
IV. Protective (Anti-Infection) Functions—Antibodies 708
1. Definition of Antibodies 708
2. Heterogeneity 710
3. Purification 715
4. Physical and Chemical Properties 720
V. Clotting Components 725
1. Fibrinogen 725
2. Prothrombin 730
3. Plasminogen 731
VI. Enzymes 732
VII. Hormones 733
VIII. Isolated Components of Unknown Function 733
1. Carbohydrate-Containing Albumin (McMeekin) 734
2. Seromucoid (Rimington) 735
3. Acid Glycoprotein (M1 of Winzler and Mehl) 735
4. Fetuin 737
5. Electrophoretic Components—a Tally Sheet 739
IX. Appendix—Inclusive Systems of Plasma Fractionation 741
1. Method 6 745
2. Method 9 749
X. General References 753

I. Origin and General Properties of Plasma Proteins

1. Limits of Free Diffusion

The all-important purpose of a circulatory system is obviously the convection of metabolites. Simple calculation will show that diffusion processes cannot supply sufficient metabolite at distances greater than a few millimeters. Thus the diffusion distance for O_2 at normal atmospheric tension at the rate of consumption of typical mammalian tissue has been calculated as 0.2 mm.[3,4]

(3) O. Warburg, *Biochem. Z.* **142,** 317 (1923).
(4) R. W. Gerard, *Am. J. Physiol.* **82,** 381 (1927).

2. Description of Blood, Blood Cells and Plasma

The convecting system evolved in the higher animals developed first through a hemocoele (as in crustacea)—a blood reservoir from which the fluid is pumped through a continuously branching system to the tissues from whence it drains back to the blood cavity, and eventually to a closed vascular system in the higher mollusks (octopus) and the vertebrates.[5,6] Throughout this evolution blood has consisted of a variety of "formed elements" (cells and smaller particles such as the blood platelets) and a large variety of molecules ranging from salts through sugars, amino acids, and other metabolites to the large molecules, the plasma proteins.

The withdrawal of blood, for the purpose of obtaining plasma proteins, requires special precautions if rapid changes following withdrawal are to be avoided. Blood is a self-sealing fluid, and the clotting process accomplishing this is activated by a chemical or physical stimulation of certain components of blood or of the injured tissues. In the case of vertebrate blood such changes may be delayed by preventing contact of the blood with injured cells or with "wettable" surfaces (by insertion of a tubular needle directly into the vascular system and by the use of vessels, tubing, etc., coated with paraffin or other nonwetting material). At the present time this may be most conveniently accomplished by the use of tubing, vessels, and even needles made of plastic[7] or by "siliconing" glassware[8] and coating needles and other metal parts with silicone oil.

Alternatively, blood coagulation may be prevented by stopping one of the chain of events leading to the formation of a clot. This has proven a successful method when study of the blood-clotting process itself was not intended. For this purpose workers have added citrate,[9,10] oxalate, or the sodium salt of ethylenediaminetetra-acetic acid[11] as complexing agents for calcium. These interrupt the clotting process by preventing the conversion of prothrombin to thrombin (a process involving calcium ions). Clotting may also be prevented by drawing the blood through a column of cation-exchange resin on the sodium cycle so as to exchange sodium for the blood calcium.[12] Heparin may also be used. This can act at a later stage in the clotting process to prevent the action of thrombin on fibrinogen.[13]

(5) C. L. Prosser, Comparative Animal Physiology, Chap. 15, W. B. Saunders, Phila., 1950.
(6) M. Florkin, Biochemical Evolution, Academic Press, New York, 1949.
(7) C. Walter, *Surg. Forum, Proc. 36th Congr. Am. Coll. Surgeons. 1950* (*Pub. 1951*), p. 483.
(8) E. G. Rochow, An Introduction to the Chemistry of the Silicones, Wiley & Sons, New York, 1951.
(9) G. A. Pekelharing, *Beitr. wissensch. Med.* **1,** 433 (1891).
(10) R. Lewisohn, *Med. Record* **87,** 141 (1915).
(11) G. Schwarzenbach, *Helv. Chim. Acta* **30,** 1798 (1947); F. Proescher, *Proc. Soc. Exptl. Biol. Med.* **76,** 619 (1951).
(12) A. Steinberg, *Proc. Soc. Exptl. Biol. Med.* **56,** 124 (1944).
(13) However, *in vivo,* it appears to prevent prothrombin conversion. D. S. Riggs, *New Engl. J. Med.* **242,** 179, 216 (1950); M. Burstein, *Compt. rend. soc. biol.* **144,** 750, 1338 (1952).

If "defibrinated" blood is desired, the blood may be drawn without any precautions and stirred (whipped) while clotting takes place. This prevents gross occlusion of the cells by the clot which contracts to small shreds.[14]

Plasma, the noncellular portion of blood, may be obtained by simple settling (used with horse blood) or by centrifugation of the cellular elements.

The various types of cells in human blood possess quite different densities permitting their separation in a density-gradient column.[15-17] On a larger scale this may be accomplished by differential centrifugation. This process has been made more efficient by the design of special centrifuges[18] and by use of aggregating reagents, such as dextran,[19] for the erythrocytes. The blood platelets have been separated by differential centrifugation and by adsorption on ion-exchange resins.[20]

Blood plasma so obtained from fasting animals is a straw-colored fluid, which visually appears stable, if sterile. However, continuous changes take place on storage, not only among the clotting components, as mentioned above, but also among the lipoproteins, complement, etc.[21] Some changes may be due to enzymes present and some to the natural instability of the protein components. Certainly, these changes may be minimized by storage in the cold, although at 0°C. considerable precipitation of fibrinogen or of cold-insoluble globulin may occur. For this reason the storage temperature of blood has frequently been specified as 4°C. Blood plasma may be stored frozen or dried (by "lyophilization" —sublimation of water under vacuum after freezing the plasma) with marked improvement in stability for many components. However, some of the lipoproteins are badly damaged by this technic (see page 705).

The constancy in composition of blood plasma led to Claude Bernard's concept of the *milieu interieur* which bathed all the cells and buffered them from an adverse external environment. The constancy to which Bernard referred was that of the circulating electrolytes (Table I), which is, in fact, unusually invariant for a "biological constant." Proteins (including lipoproteins) constitute the bulk of the remaining plasma components and these show much greater variation, which can be measured by physicochemical methods against this constant ionic background.

(14) This process of some historical interest was described by Hewson, who first isolated fibrin: W. Hewson, An Experimental Inquiry into the Properties of the Blood, T. Codell, London, 1771.
(15) B. L. Vallee, W. L. Hughes, Jr., and J. G. Gibson, 2d, *Blood* **1,** 82 (1947).
(16) J. W. Ferrebee and Q. M. Geiman, *J. Infectious Diseases* **78,** 173 (1946).
(17) D. W. Fawcett and B. L. Vallee, *J. Lab. Clin. Med.* **39,** 354 (1952).
(18) E. J. Cohn, personal communication.
(19) E. S. Buckley, Jr., and J. G. Gibson, 2d, *Proc. Univ. Lab. Phys. Chem. Related Med. and Public Health* **1,** 45 (1950).
(20) J. L. Tullis, personal communication.
(21) L. E. Krejci, L. Sweeney, and E. B. Sanigar, *J. Biol. Chem.* **158,** 693 (1945).

Thus density measurements can be used to estimate total protein, although they will err if the lipide content of the plasma varies. Actually the density method largely ignores the lipides since these have densities close to that of water. Therefore the measurements give estimates for protein agreeing well with nitrogen analysis if one assumes all of the proteins to contain 16% N.[22]

TABLE I
Ionic Composition of Blood Plasma

Cations, meq./l.		Anions, meq./l.	
Na^+	142	Cl^-	103
K^+	5	HCO_3^-	27
Ca^{++}	5	$HPO_4^=$	2
Mg^{++}	3	$SO_4^=$	1
		Organic acids	6
		Protein	16
	155		155

Alternatively, refractive-index measurements may be used. Since the refractive-index increment of lipide is relatively close to that of protein, this method agrees well with the estimation of protein as total non-dialyzable solids.[22a]

3. Requirements for Circulatory Function

The proteins of mammalian plasma show related physicochemical properties, all being negatively charged at physiological pH and having molecular weights varying from 40,000 to 150,000 (a few larger—see Table VI). This relative homogeneity would appear related to function in the blood stream. These molecular sizes are similar to the "pore" sizes of the reticulo-endothelium (these pores being apparently designed so that relatively small variation in membrane structure may permit wide variation in the amount of protein passed). Only traces of protein pass through the membranes of the general capillary bed (including the renal glomerulus) and larger amounts through the membranes of the liver capillaries.[23] Certainly, limited diffusibility would appear desirable

(22) D. D. Van Slyke, A. Hiller, R. A. Phillips, P. B. Hamilton, V. P. Dole, R. M. Archibald and H. A. Eder, *J. Biol. Chem.* **183,** 331 (1950).

(22*a*) S. H. Armstrong, Jr., M. J. E. Budka, K. C. Morrison, and M. Hasson, *J. Am. Chem. Soc.* **69,** 1747 (1947).

(23) E. M. Landis, *Physiol. Revs.* **14,** 404 (1934).

for the maintenance of a high concentration of these substances in the blood. However, perhaps the more interesting question is why the membranes permit any passage at all. Is this an imperfection of nature, a compromise in the interests of high permeability of certain small metabolites, or is such protein permeability necessary as an aid to their function? In the case of some proteins, such as the hormones, the latter interpretation would seem correct. Also, in the case of proteins whose function is transport of metabolites or protection against infection, permeability might seem desirable. It will be interesting in the following pages to look for further functional correlation of the relative permeability for different plasma proteins.

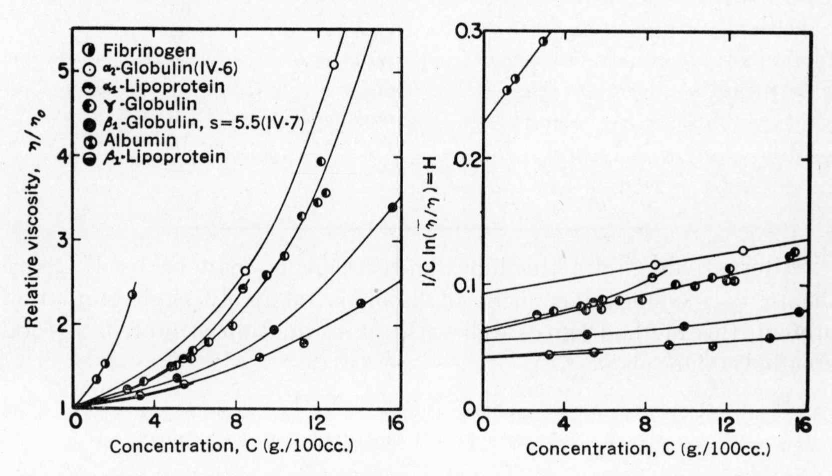

FIG. 1. Viscosity of various plasma proteins at 37°C. (From Oncley *et al.*)[61]

In the interests of osmotic efficiency (see sec. II) all plasma proteins at physiological pH are negatively charged and also they are all closely related in size (Table VI). These factors may be further related to hydrodynamic efficiency. Certainly, protein–protein interaction, which is accentuated when the proteins bear opposite electric charges, would decrease osmotic pressures and increase blood viscosity. This relative weakness of interaction between plasma proteins has been a boon to the investigator interested in the isolation of individual components.

The plasma proteins, excepting fibrinogen, also show similar and relatively small intrinsic viscosities (see Fig. 1). This, too would be a desirable feature in the circulation since it would decrease the work of the heart. Fibrinogen is, of course, necessarily asymmetric (and hence viscous) since it is a precursor to the network of the blood clot.

4. Distribution in Plasma, Extravascular Fluid, and Cells

a. Reservoirs of Plasma Proteins

While the proteins of plasma include a number of readily identifiable components of dimensions such as to favor their confinement to the vascular system, absolute retention is not achieved. Thus it has been shown by tracer technics that approximately 50 per cent of small (relative to plasma content) infusions of serum albumin disappear from the circulation within the first day.[24–26] A considerable portion of this "lost" protein may be explained as being contained in the intercellular spaces due to leakage from the capillaries.[26a] The volume of such spaces is generally considered to be about 1/5 of the total volume of the mammalian organism[27] and hence is approximately 3 × the plasma volume. The protein content of this extravascular fluid must vary widely, the average value lying somewhere between zero and 2–4 per cent—the protein content of lymph.[28] Thus, there would appear sufficient latitude to explain the protein loss following transfusion.

However, lymph represents a concentrated (with respect to protein) extravascular fluid, and the very small permeability of capillaries to plasma proteins coupled with practically complete equivalence between plasma colloid osmotic pressure and average capillary hydrostatic pressure (above tissue pressure)[29] suggests that the extravascular interstitial fluid must contain far less, on the average, than the 1 per cent plasma protein which the above argument would demand. Drinker has pointed out that a quiescent limb does not normally become edematous despite lack of lymph flow.[28] This again suggests that leakage of protein must be extremely small. These arguments apply to the extravascular fluid of skeletal muscle and the appendages. In

(24) A. M. Seligman, *J. Clin. Invest.* **23,** 720 (1944); K. Sterling, *ibid.* **30,** 1228 (1951).

(25) M. P. Deichmiller, F. J. Dixon, and P. H. Maurer [(*Federation Proc.* **12,** 386 (1953)] measured half-lives of several homologous serum albumins (I^{131} tagged): Man 15 days, dog 8 days, rabbit 6 days, and mouse 1.2 days.

(26) This effect was probably first demonstrated in humans by Janeway and Heyl, who followed by Heidelberger's quantitative precipitin technic the disappearance of bovine serum albumin from the blood stream and found 50 per cent disappearance during the first day followed by a much slower rate thereafter.

(26*a*) The dynamic equilibrium between vascular and extra vascular protein has been demonstrated with homologous antibody by D. Gitlin and C. A. Janeway [*Science* **118,** 301 (1953)].

(27) J. L. Gamble, Jr., and J. S. Robertson, *Am. J. Physiol.* **171,** 659 (1952); M. Gaudino, I. L. Schwartz, and M. F. Levitt, *Proc. Soc. Exptl. Biol. Med.* **68,** 507 (1948).

(28) C. K. Drinker and J. M. Yoffey, Lymphatics, Lymph, and Lymphoid Tissue, Harvard Univ. Press, Cambridge, 1941.

(29) E. M. Landis, *Am. J. Physiol.* **82,** 217 (1927); **83,** 528 (1928).

keeping with them, perfusion of the isolated hind limb[30] shows no loss of plasma protein to extravascular space.[31]

Therefore, the large protein losses described above would not appear to be generally distributed, and some extravascular "reservoir" of plasma proteins approaching blood plasma concentrations must exist in some of the internal organs (e.g., the liver) where capillary permeability is known to be very large. (High permeability for the capillaries of the portal system would seem necessary because of the low hydrostatic pressure—see sec. II.)

b. Evidence for Plasma Proteins within Cells

Another possible reservoir of plasma proteins lies within cells. Recent immunochemical studies have demonstrated the rapid appearance of heterologous proteins within cells and particularly within cell nuclei following their injection. Homologous plasma proteins within cells and cell nuclei have also been observed.

An "immunohistochemical method" developed by Coons for this purpose consists in staining alcohol-fixed tissue slices with an antibody to which has been coupled a fluorescent dye. After suitable washing procedures to remove unspecifically adsorbed antibody, the location of the protein against which the antibody is directed may be observed microscopically under ultraviolet light.[32] In this way Coons has been able to observe the rapid penetration (i.e., within 10 minutes) of proteins such as ovalbumin and serum albumin within a variety of cells following their intravenous injection into mice.[33] Haurowitz has also demonstrated the penetration following injection of foreign proteins labeled with radioactive iodine into liver cells of the rabbit. In this way he found radioactivity in the fractions designated: microsomes, the mitochondria, and the nuclei. However, the activity could be separated from the nucleoprotein itself.[34]

Such penetration within cells might be assumed to be the first stage in the immune response of the animal similar to the previously observed

(30) J. R. Pappenheimer, E. M. Renkin, and L. M. Borrero, *Am. J. Physiol.* **167,** 13 (1951).

(31) J. R. Pappenheimer, personal communication. Pappenheimer measures osmotic-pressure changes and therefore might not detect changes in extravascular protein which was not osmotically active. The presence of such "bound" plasma protein is suggested by the occurrence of plasma proteins together with interstitial connective tissue throughout the body in histological sections stained immuno-chemically.[37] If this is not an artifact in the preparation of the histological specimen, further modification of the picture of capillary permeability may be required.

(32) A. H. Coons and M. H. Kaplan, *J. Exptl. Med.* **91,** 1 (1950).

(33) A. H. Coons, E. H. Leduc, and M. H. Kaplan, *J. Exptl. Med.* **93,** 173 (1951).

(34) C. F. Crampton and F. Haurowitz, *Science* **111,** 300 (1950); F. Haurowitz and C. F. Crampton, *J. Immunol.* **68,** 73 (1951).

phagocytosis of particulate antigens.[35] However, the distribution observed by Coons was much more widespread than the cells usually implicated in this mechanism.[36] Using Coons's technic, Gitlin *et al.* have studied the distribution of homologous plasma proteins within cells[37] and have found a similar distribution of serum albumin, γ-globulin, and β-lipoprotein in human tissues. Furthermore, that this represents a true penetration for γ-globulin was indicated by injecting this into a child who genetically was unable to synthesize this protein. The child's tissues, which before injection did not stain for γ-globulin, showed a normal staining pattern after injection.

It is not known whether plasma proteins can reversibly enter and again leave the cell, although heterologous proteins have been shown to disappear from cells at about the same time as disappearance from the circulation should occur.[33] A quantitative approach to these experiments has not been possible, so that the magnitude of the effects is unknown. However, it was noted that only traces of plasma proteins could be detected within muscle cells.[37] Therefore, any large cellular depots must again reside within the body cavity.

The significance of intracellular plasma proteins may lie in their close relation to protein metabolism. Schoenheimer and Rittenberg *et al.* have demonstrated the dynamic equilibrium of plasma proteins[38] with tissue proteins, and Whipple and Madden have shown their importance as amino acid reservoirs in the fasting animal.[39] However, it is not clear whether the plasma proteins are transported intact to the point of their incorporation or are first broken down to the constituent amino acids (see Tarver, Chap. 26). A choice cannot be made from known physical parameters, for while the concentration of most combined amino acids (as proteins) is 10 to 100 times as great as the free concentration in plasma, the capillary permeability to the free amino acid is probably 100-fold that of the protein;[40] and while evidence for cellular uptake of proteins is slowly accumulating, the evidence for uptake of amino acids is much more striking.[41]

(35) A. Bowin and A. Delaunay, Phagocytose et Infections, Herman & Co., Paris, 1947.

(36) For the effect of altered physicochemical properties on protein distribution within cells (phagocytosis?) compare: D. Gitlin, *Proc. Soc. Exptl. Biol. Med.* **74,** 138 (1950).

(37) D. Gitlin, B. H. Landing, and A. Whipple, *J. Exptl. Med.* **97,** 163 (1953).

(38) R. Schoenheimer, S. Ratner, D. Rittenberg, and M. Heidelberger, *J. Biol. Chem.* **144,** 545 (1942).

(39) G. H. Whipple and S. C. Madden, *Medicine* **23,** 215 (1944).

(40) J. R. Pappenheimer, *Physiol. Revs.* **33,** 387 (1953).

(41) H. N. Christensen, T. R. Riggs, and N. E. Ray, *J. Biol. Chem.* **194,** 41 (1952).

c. Evidence for Cellular Proteins in Plasma

The presence of stray proteins in plasma as leakage from tissues might be expected either from dying cells or as a result of a slight permeability of the cell membranes. Substances whose presence in plasma might be so explained are the hydrolytic enzymes such as amylase, phosphatases, and esterases. The presence of catalase, occasionally considered to be a plasma enzyme, has been traced to hemolysis during blood collection, since its concentration is proportional to that of hemoglobin in the plasma.[42] The presence of peptidases in serum has been similarly ascribed to erythrocyte destruction.[42a] Recently, ovalbumin has been detected in chicken plasma, highest in laying hens. However, the concentration never exceeded 0.1 per cent of the total protein.[43]

In pathological conditions, changes in the plasma proteins are frequently observed which are proving increasingly useful in diagnosis. These changes can either involve an altered level of a normal plasma constituent (e.g., decreased albumin concentration in nephrosis) or the appearance of one or more new components, normally not detectable in plasma. An interesting example of the latter case is McCarty's crystalline "C reactive component"[44] (a protein occurring during the acute phase of certain infections). Further discussion of pathological changes is considered outside the scope of this chapter, and the reader is referred to the excellent review by Gutman.[44a]

5. Metabolism of Plasma Proteins

Metabolism of the plasma proteins is considered at greater length in Chap. 26 (Tarver), but it is interesting to point out in passing that the major component of plasma, serum albumin, appears to act as a protein reserve in the fasting animal. The turnover rate for human serum albumin is nevertheless quite slow (7 per cent/day),[25] indicating the relative stability of this protein toward catabolism or exchange phenomena. Nevertheless, the stability *in vitro* of albumin under physiological conditions is so great[45,46] that disappearance cannot be a function of "wearing out" in the circulation. Some sites of formation of certain

(42) G. E. Perlmann and F. Lipmann, *Arch. Biochem.* **7**, 159 (1945).
(42a) E. L. Smith, G. E. Cartwright, F. H. Tyler, and M. M. Wintrobe, *J. Biol. Chem.* **185**, 59 (1950).
(43) M. E. Marshall and H. F. Deutsch, *J. Biol. Chem.* **189**, 1 (1951).
(44) M. McCarty, *J. Exptl. Med.* **85**, 491 (1947).
(44a) A. B. Gutman, *Advances in Protein Chem.* **4**, 156 (1948).
(45) G. Scatchard, S. T. Gibson, L. M. Woodruff, A. C. Batchelder, and A. Brown, *J. Clin. Invest.* **23**, 445 (1944).
(46) P. D. Boyer, G. A. Ballou, and J. M. Luck, *J. Biol. Chem.* **162**, 199 (1946).

plasma proteins have been found; serum albumin is formed in liver slices,[47] immune globulin in plasma cells.[48,48a]

II. Maintenance of Vascular Volume

1. Physiology of Peripheral Circulation

In the simpler circulatory systems, as exemplified by the crustacea, blood upon leaving the heart passes through smaller and smaller vessels in a continuously ramifying network, the "vascular tree," until it is eventually discharged into intercellular spaces from which it seeps back to the body coelom and thence by veins to the heart. The osmotic relationships which must exist are thus obscured by active transport phenomena across cell surfaces.

In the closed vascular system of the vertebrate, on the other hand, the laws of hydrodynamics and diffusion suffice to explain the phenomena. To prevent their collapse, pressure within the blood vessels must be greater than in the surrounding tissue spaces. Since there is only slight net fluid flow through the permeable capillary walls, the hydrostatic pressure must be balanced by a chemical potential (osmotic pressure).[49] In fact, the *mean* hydrostatic pressure drop across the capillary wall equals, within experimental error, the colloid osmotic pressure.[50,51] The ionic and small-molecule composition of lymph is identical with that of plasma, when allowance is made for Donnan equilibria across the capillary membrane.[52] Consequently, active secretion by capillaries is unimpor-

(47) T. Peters and C. B. Anfinsen, *J. Biol. Chem.* **186,** 805 (1950).

(48) A. Fagraeus, Antibody Production in Relation to the Development of Plasma Cells, Esselts Aktiebolag, Stockholm, 1948.

(48a) L. L. Miller, W. F. Bale, and C. G. Bly [*J. Exptl. Med.* **99,** 125, 133 (1954)] have shown the incorporation of C_{14} (ϵ-lysine) in all of the plasma proteins, except γ-globulin, by perfused liver and in γ-globulins by the perfused eviscerated carcass. The perfused carcass also incorporated small amounts of activity into the α and β globulins. However, this may have represented merely binding of metabolic products of lysine by the lipoproteins, rather than protein synthesis.

(49) E. H. Starling, *J. Physiol.* **19,** 312 (1896).

(50) A. Krogh, E. M. Landis, and A. H. Turner, *J. Clin. Invest.* **11,** 63 (1932).

(51) An interesting difference arises in the octopus, which in spite of a high hydrostatic pressure[28] shows a low colloid osmotic pressure.[51a] It would be interesting to know how the octopus has resolved this difficulty, whether by a large pressure drop between the heart and the permeable capillary, or by a much larger amount of lymph flow, or still otherwise.

(51a) Aortal pressure = 25–80 mm. Hg.[28] Colloid osmotic pressure = 2.7 mm. P. Meyer, *Compt. rend. soc. biol.* **120,** 305 (1935).

(52) E. Muntwyler, R. C. Mellors, F. R. Mautz, and G. H. Mangun, *J. Biol. Chem.* **134,** 389 (1940).

tant or non-existent. (The possibility of special secretory mechanisms for certain trace components cannot, of course, be excluded.)

Since there is a continuous pressure drop along the length of the capillary, at the arterial end the hydrostatic pressure exceeds osmotic pressure and fluid flows out through the capillary wall, while at the venous end fluid is drawn into the capillary from the interstitial space. Fluid not reabsorbed is collected by the lymphatic system and emptied into a vein. This lymph contains the plasma proteins which have leaked out, now concentrated by water reabsorption. Nevertheless, filtration appears to be relatively unimportant as a physiological process; the rate of flow even at the extreme arterial end is slow enough that for most substances diffusion is the important consideration.[40] That is, filtration (motion of solute molecules with the stream) is unimportant compared to Brownian movement. (This is, of course, fortunate at the venous end of the capillary where the metabolite must diffuse against the current.) Another corollary of this is the fact that capillaries must be geometrically so distributed as to permit adequate diffusion to every cell, as indicated in sec. I. (Krogh has shown them to be of the order of 0.02 mm. apart.)[53]

Thus far we have dealt with capillary pressure drops across the capillary wall. Absolute capillary pressure must of course exceed atmospheric pressure by this drop (colloid osmotic pressure) plus tissue pressure. Tissue pressure may be considered as the elastic resistance to expansion of the tissue. Consequently, if the colloid osmotic pressure decreases (hypoproteinemia) while the hydrostatic capillary pressure remains constant, the tissue must swell (edema). Conversely, in therapy the injection of plasma protein raises the colloid osmotic pressure, and fluid flows back into the blood stream, from whence it is excreted through the kidney, relieving the edema. (The mechanism of the diuresis produced by elevating the colloid osmotic pressure of plasma is not so obvious, since by decreasing the chemical potential of water, glomerular filtration should be decreased.)

2. Capillary Porosities and Renal Function

Membranes may be considered as molecular sieves permitting the separation of molecules on the basis of size. This has been well demonstrated for artificial membranes[54] although completely uniform pore size has not been attained.

Viewed as a sieve, the vascular membrane would appear to have different pore sizes in different organs, varying from the smallest pores

(53) A. Krogh, *J. Physiol* **52,** 409 (1919).
(54) W. J. Elford and J. D. Ferry, *Brit. J. Exptl. Path.* **16,** 1 (1935).

in the kidney glomeruli to the largest in the liver capillaries.[23,55] Extensive data are available on the renal glomerulus, based on kidney function studies and on the beautiful experiments of Richards *et al.* on isolated glomeruli. It is generally agreed that proteins larger than albumin do not pass the glomerular membrane. However, among molecules of sizes similar to albumin some discrepancies appear.

Although Richards was unable to detect any protein in the ultrafiltrate from the glomerulus,[56] albuminuria is a common occurrence in kidney disease. Although this might be attributed to an altered glomerular permeability, over-all kidney function studies are complicated by the reabsorption of molecules from the glomerular filtrate by the renal tubules. In kidney disease, when glomerular abnormalities are not present, albuminuria could be attributed to a lack of tubular reabsorption. Therefore altered glomerular permeability appears an unnecessary hypothesis.

We might ask whether the sensitivity of Richards' test for protein was adequate. Clearance studies on hemoglobin excretion indicate that once the threshold is passed (about 100 mg. per cent—a measure of tubular reabsorption capacity), the urinary concentration bears a direct relation to other small molecules which are not reabsorbed. From these data, the concentration of hemoglobin in the glomerular filtrate has been calculated to be 3 per cent of that in plasma.[57]

Albumin has the same molecular weight as hemoglobin, although hemoglobin may dissociate into half molecules. However, albumin is definitely more asymmetric, with an axial ratio of 4:1 vs. 1:1 for hemoglobin, and it is not certain what dimension of the molecule should be used as a measure of size when discussing membrane permeability. Since Brownian motion precludes the possibility of orientation in the stream flowing through the pore, it would seem likely that effective size might prove more a function of the length of the molecule than of its volume, and if the path through the pore were tortuous, a rigid rod would be still further impeded.

Therefore, the concentration of albumin in the glomerular filtrate might be estimated at <60 mg. per cent (i.e., <2 per cent of the plasma level). (Dock measured 20 mg. per cent in perfusates of chilled rabbit kidneys.[57a]) This concentration might well be below the sensitivity of Richards' test.[56]

The permeability of the renal glomerulus to molecules as large as

(55) A recent electronmicrograph of the kidney appears to confirm the physiological evidence. B. V. Hall, *Federation Proc.* **12,** 467 (1953).
(56) P. A. Bott and A. N. Richards, *J. Biol. Chem.* **141,** 291 (1941).
(57) C. L. Yuile, *Physiol. Revs.* **22,** 19 (1942).
(57*a*) W. Dock, *New England J. Med.* **227,** 633 (1942).

albumin may merely point to the importance of a renal mechanism for the excretion of large metabolic fragments of proteins and other structural elements of cells. Owing to their extremely low concentration, little is known about the proteinaceous components of normal urine.[58]

3. Osmotic Properties of the Plasma Proteins

The osmotic efficiencies of several plasma proteins have been studied by Scatchard *et al.*[59–61] (Fig. 2). Serum albumin is responsible for ¾ of the colloid osmotic pressure of human plasma, although it represents approximately only ½ of the plasma proteins. This high efficiency is the result of its smaller size (65,000 vs. 90,000 for the average molecular weight of plasma proteins).

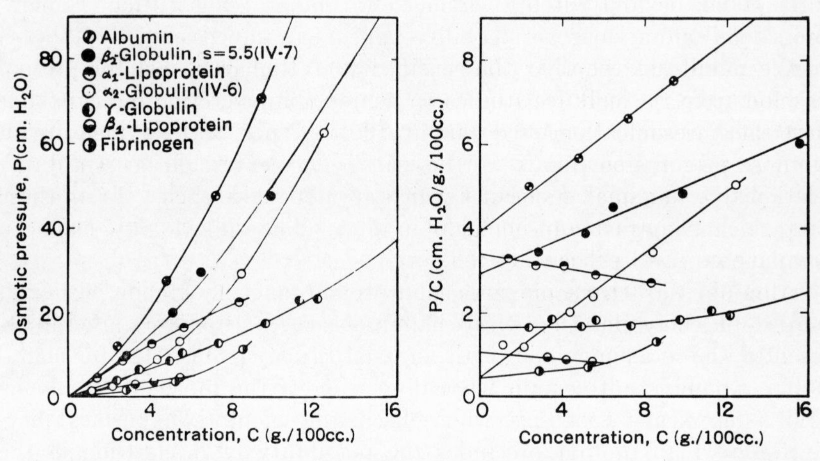

Fig. 2. Osmotic pressure of various plasma proteins at 37°C. (From Oncley *et al.*)[61]

Under physiological conditions approximately ⅓ of the colloid osmotic pressure is contributed by the Donnan effect.[59] A proper evaluation of this effect cannot be obtained from titration curves because of ion-binding. It is most readily derived from a complete study of the ion distribution in osmotic measurements where it can be obtained directly from the distribution of free (not bound) ions. Thus Scatchard *et al.* have obtained it from chloride-activity measurements on both sides

(58) A very large mucoprotein (mol. wt. = 7×10^6) which obviously cannot pass the glomerular membrane is present in amounts of 0.02 g./l. of normal human urine. I. Tamm and F. L. Horsfall, Jr., *J. Exptl. Med.* **95,** 171 (1952).

(59) G. Scatchard, A. C. Batchelder, and A. Brown, *J. Clin. Invest.* **23,** 458 (1944).

(60) G. Scatchard, A. C. Batchelder, and A. Brown, *J. Am. Chem. Soc.* **68,** 2320 (1946).

(61) J. L. Oncley, G. Scatchard, and A. Brown, *J. Phys. Chem.* **51,** 184 (1947).

of the membrane, or by measurements of sodium chloride distribution, assuming that sodium is not bound.[60]

The effect of albumin administration on plasma volume has been studied by Heyl, Gibson, and Janeway who report that 1 g. of albumin increases plasma volume by 17 ml.[62] in striking agreement with the predicted value of 18 ml. (at 25 mm. Hg pressure) calculated from osmotic-pressure measurements.[59]

4. Serum Albumins

a. Definitions

The mammalian serum albumins, which as indicated above provide most of the colloid osmotic pressure of plasma, have been variously defined as our knowledge has increased, each new redefinition having been generally in the direction of excluding some related protein found to differ from the bulk of the serum albumins in some particular. Thus the earliest classifications on the basis of solubility defined serum albumins as those proteins soluble in 0.5 saturated ammonium sulfate or soluble upon dialysis against distilled water saturated with carbon dioxide.[63,64] That such albumins contained appreciable quantities of quite different components was realized at that time by chemical studies of the composition of subfractions obtained from them by crystallization[65] or other means. Electrophoretic analysis, as developed by Tiselius,[66] further resolved the components and made possible a new definition of human serum albumin as the major component in electrophoretic analysis at pH 7.7 or 8.6 and having mobilities of -5.2 and -6.0 Tiselius units in phosphate $\Gamma/2$ 0.2, pH 7.7 and veronal $\Gamma/2$ 0.1, pH 8.6, respectively.[67] (The veronal buffer would appear to be more critical as it better resolves any α_1-globulin present.)[67] Bovine albumin and probably those of other species can be similarly defined.

Analysis for carbohydrate appears to be a simple chemical test for albumin purity, since from such electrophoretically homogeneous (i.e., 98 per cent or better) human or bovine albumin the majority of the protein may be crystallized free of carbohydrate, and the carbohydrate

(62) J. T. Heyl, J. G. Gibson 2d, and C. A. Janeway, *J. Clin. Invest.* **22,** 763 (1943).
(63) P. Panum, *Virchow's Arch. pathol. Anat. Physiol.* **4,** 419 (1852); Kauder, *Arch. exptl. Path. Pharmakol.* **26**; P. S. Denis, Memoir sur le Sang, Paris, 1859.
(64) For further historical data see E. J. Cohn, *Physiol. Revs.* **5,** 349 (1925).
(65) A. Gürber, *Sitzber. physik. med. Ges., 1894, Würzburg,* p. 143 (1895).
(66) A. Tiselius, *Trans. Faraday Soc.* **33,** 524 (1937).
(67) S. H. Armstrong, Jr., M. J. E. Budka, and K. C. Morrison, *J. Am. Chem. Soc.* **69,** 416 (1947).

present before crystallization can reasonably be allocated to the small amounts of α-globulins present by assigning them a carbohydrate content of 10 per cent or less.[68,69]

b. *Heterogeneity*

However, albumins so defined still do not represent a homogeneous protein; nor do they become so upon repeated recrystallization from ethanol–water mixtures.[68] Heterogeneity may be detected by electrophoretic analysis at more critical pH values. Thus the most highly purified albumins have continued to show two or more components at pH 4.0 as first observed by Luetscher.[70] They also show reversible boundary spreading at pH 4.6.[71] The latter test is more informative. The pH 4.0 measurements, although quite reproducible, are more difficult to interpret, since they are carried out in a region where the stability of albumin is questionable and, furthermore, the patterns obtained are particularly sensitive to the ionic environment.

Serum albumins, repeatedly recrystallized, were also found to be heterogeneous when tested for constancy of solubility with excess saturating body in buffered ethanol–water systems.[72] As an exception, McMeekin reported a fraction of horse serum albumin crystallized as the sulfate at pH 4.0, which showed constant solubility when recrystallized from ammonium sulfate and then equilibrated with the same solvent.[73] Unfortunately, he did not extend his studies over a wide enough range of saturating body to see how rigorously this criterion was obeyed. Kendall[74] has also reported a crystalline complex of human serum albumin with fatty acids. This showed remarkably constant solubility. However, his material must have been heterogeneous with regard to fatty acid content (i.e., mixed stearic, oleic, linoleic, etc.) and the albumin itself was presumably heterogeneous as regards its sulfhydryl content, as will be discussed below. Therefore, it seems probable that Kendall was studying a solid solution in equilibrium with a solution of similar composition. Northrop *et al.* have observed a similar phenomenon with pepsin.[75]

(68) E. J. Cohn, W. L. Hughes, Jr., and J. H. Weare, *J. Am. Chem. Soc.* **69,** 1753 (1947).

(69) By definition are thus excluded previously described "carbohydrate-containing albumins." (These will be discussed under glycoproteins, sec. VIII.)

(70) J. A. Luetscher, *J. Am. Chem. Soc.* **61,** 2888 (1939).

(71) R. L. Baldwin, P. M. Laughton, and R. A. Alberty, *J. Phys. Chem.* **55,** 111 (1951).

(72) W. L. Hughes, unpublished observations.

(73) T. L. McMeekin, *J. Am. Chem. Soc.* **61,** 2884 (1939).

(74) F. E. Kendall, *J. Biol. Chem.* **138,** 97 (1941).

(75) R. M. Herriot, V. Desreux, and J. H. Northrop, *J. Gen. Physiol.* **24,** 213 (1940).

Simple chemical evidence of heterogeneity lies in the sulfhydryl content of bovine and human serum albumins. The author has always found considerably less than one thiol group per molecule of human or bovine albumin unless the albumin had been fractionated with the aid of mercury.[76] However, material so fractionated has still appeared heterogeneous by solubility criteria and by electrophoretic analysis at acid pH.

A small degree of heterogeneity is also found upon ultracentrifugal analysis in 0.1 ionic strength buffers. This is usually evinced as a slight asymmetry of the schlieren diagram. However, in "bad" preparations a distinct shoulder may be observed moving ahead of the main peak with a sedimentation constant attributable to an albumin dimer. The amount of this component is a function of the age and past history of the albumin sample. Storage in the "dry" state at room temperature appears to be particularly deleterious. However, relative to the other measurements of heterogeneity, ultracentrifugal evidence would appear of minor importance, since the amount of heterogeneity is usually less than 10 per cent and in a good preparation is probably less than 5 per cent.

As to the ultimate basis of albumin heterogeneity, several postulates might be kept in mind. First, it might be assumed that the biosynthetic mechanism is not precise enough for exact duplication of molecules of the complexity of serum albumins. However, such a postulate appears unlikely from what is known already of biosynthetic specificity. A second postulate, of consequences similar to the first, would be that the albumin molecule may undergo slight alterations during its life in the circulation. Thus a population of young and old molecules would be heterogeneous. In terms of structure this may be exemplified by the fractional sulfhydryl content of serum albumin. The sulfhydryl is known to be a very labile group subject to oxidative or coupling reactions under physiological conditions. However, to extend such reasoning to other groupings is difficult at the present state of our knowledge. Oxidative alterations in the phenolic group of the tyrosyl residues of this protein have not been reported. Hydrolytic changes which might appear as changes in net charge (hydrolysis of amides) do not occur in large degree, as judged by electrophoretic homogeneity. The small amount of reversible boundary spreading observed by Alberty at the isoelectric point could be explained, he feels, by a unit variation in the net charge, and yet there are 86 hydrolyzable amide linkages.[71]

The heterogeneity observed electrophoretically and by solubility studies may be due to tightly bound charged impurities. Serum albumins bind many anions with extreme avidity although not in simple stoichiometric amounts (see sec. 4*g*. and Chap. 8). Thus purified albumins

(76) W. L. Hughes, Jr., *J. Am. Chem. Soc.* **69,** 1836, (1947).

usually contain small amounts of the higher fatty acids in amounts of a fraction of a mole of each per mole of protein.[68] Whether such heterogeneity can explain solubility behavior is uncertain. The author has found but slight evidence for fractionation in terms of fatty acid content under the conditions of his solubility tests (ethanol–water systems). On the other hand, Kendall reports considerable fractionation of fatty acid content during crystallization from ammonium sulfate.[74]

A final postulate regarding albumin heterogeneity may be that there is a variation among individual animals of the same species. This would make the albumins, now available in large amounts obtained from pooled plasmas, more heterogeneous than those described by the early workers, which were frequently obtained from a single animal. Immunological evidence for individual variations in proteins or protein conjugates is available, being the explanation of blood types. Such variation is presumably also the cause of failure of tissue grafts from one individual to another.

In spite of the evidence for heterogeneity cited above, the properties of serum albumin studied by most workers can be described in terms of a single protein component. Therefore in the following paragraphs serum albumin will be treated as if it were homogeneous unless specifically stated to the contrary.

c. *Crystallization*

Serum albumins are generally purified by crystallization procedures, although a remarkably pure product has been obtained by ethanol fractionation.[77] Salting-out procedures for crystallizing horse[65,73,78,79] and human[74,80] serum albumins have been described, bovine serum albumin thus far not having yielded to such methods.[81] The salts used have included ammonium sulfate, magnesium sulfate, sodium sulfate, and sodium dihydrogen phosphate. Of these, ammonium sulfate is by far the most convenient because of its greater solubility, magnesium sulfate and sodium sulfate being too insoluble to precipitate albumin in the cold. Sodium dihydrogen phosphate provides a useful buffering

(77) E. J. Cohn, L. E. Strong, W. L. Hughes, Jr., D. J. Mulford, J. N. Ashworth, M. Melin, and H. L. Taylor, *J. Am. Chem. Soc.* **68**, 459 (1946).

(78) S. P. L. Sørensen, *Compt. rend. trav. lab. Carlsberg* **18**, No. 5, 1 (1930).

(79) L. F. Hewitt, *Biochem. J.* **30**, 2229 (1936).

(80) M. E. Adair and G. L. Taylor, *Nature* **135**, 307 (1935).

(81) The author has observed crystallization in bovine albumin solutions concentrated by vacuum distillation of the supernatant after precipitation of the globulins by Cohn's method 6 (precipitates I through IV-4). In this case the albumin was salted-out by simultaneous concentration of the acetate buffer added in fractionation.

action. Crystallization at low ionic strengths with the aid of organic precipitants has been used for human and bovine albumins.[68,74] Ethanol has been the reagent of choice, but methanol would appear equally useful, and at least upon occasion acetone may be used.[82] When sufficiently purified, usually by previous crystallization, bovine and human serum albumin (or at least a fraction of them) may be crystallized from water.[68] Horse albumin has also been crystallized as its sulfate at pH 4 from water.[73]

In general, serum albumins crystallize most readily at a pH slightly alkaline to their isoelectric points (0.2–0.5 pH units) from concentrated solutions containing 10 per cent or more of protein. A variety of small molecules which are tightly bound to albumin facilitates crystallization from ethanol–water systems. These include the higher alcohols from pentanol to decanol, toluene, chloroform, and sodium oleate. Fatty acid salts also appear to be involved in Kendall's method for the crystallization of human albumin from ammonium sulfate.[74] Lewin has further shown that a variety of ions and small molecules may be bound to serum albumin without hindering the ability of the albumins to crystallize.[82] The mechanism of action of these agents is unknown. However, it is suggestive that sodium oleate and decanol have been found to decrease markedly the dielectric increment of serum albumin.[83] This indicates a change in the charge distribution, which must be one of the orienting forces for crystallization.

Both human and bovine serum albumins crystallize as dimer units when linked with a mercury atom.[84] The amount of mercuric salt added must be carefully adjusted to equivalence with the sulfhydryl content if optimal yields are to be obtained. The crystals form from aqueous ethanol solutions under conditions comparable to those for crystallizing the monomer, excepting that the dimer solubility is appreciably lower, so that they can be readily recrystallized from water. In this way the mercaptalbumin may be readily separated from the thiol-free protein, two or three recrystallizations usually sufficing to give the pure dimer as judged by ultracentrifugal analysis (Fig. 5) or content of mercury.

d. Physicochemical Properties

The physicochemical properties of serum albumin have been frequently mentioned in Vol. I of this text and will be only summarized here. In general, they can be explained in terms of an elongated ellipsoid of

(82) J. Lewin, *J. Am. Chem. Soc.* **73,** 3906–11 (1951).

(83) H. Dintzis, Doctor's Dissertation, Harvard University, Cambridge, Mass., 1952.

(84) Details of these crystallization procedures are in preparation (Hughes and Dintzis).

revolution with a molecular weight (anhydrous) of 65,000, a length of 150 A., and a diameter of 38 A. (axial ratio of 4:1). Assuming an ellipsoid of revolution, these dimensions best coordinate the various physicochemical data (i.e., unit cell dimensions of the crystal, sedimentation and diffusion constants, viscosity, and length from double-refraction of flow). However, the assumption of an ellipsoid of revolution may be a poor approximation to the true shape, and even small deviations from this (e.g., a cylinder or a molecule with holes in it) will require quite different dimensions, as Schulman has pointed out for fibrinogen.[90]

Even the molecular weight, which is capable of direct measurement, is a matter of some uncertainty. It will be noticed that the value of 65,000 is appreciably smaller than the previously accepted value of 69,000.[68] This new value is based primarily on the recent x-ray crystallographic data of Low,[85] who calculated 65,200 and 65,600 for two types of human serum albumin crystals and was confirmed in the ultracentrifuge by Creeth[86] and by Koenig and Perrings,[87] who calculated molecular weights of 65,400 and 64,500, respectively.

Regarding the earlier data leading to a mean value of 69,000, methods giving an average molecular weight (such as osmotic pressure and light scattering) will all overestimate the true molecular weight if heavier components, as discussed above, are present. Previous sedimentation constants have also varied owing to technical difficulties, of which the most serious has been correct measurement of temperature.[88] Recently Waugh has discovered a temperature change in the rotor upon acceleration, due to expansion of the metal.[89]

Various physicochemical evidence in the past has indicated a molecular weight of serum albumin much smaller than 65,000.[91,92] However, further investigation has always revealed either faulty experimental technic or incorrect interpretation of the data. Now that such methods are bolstered by analytical composition indicating that human albumin contains only one thiol and one tryptophanyl per 65,000, and bovine one thiol and one polypeptide chain (one free α-amino and one free α-carboxyl group),[93,94] the evidence would appear fairly conclusive.

(85) B. W. Low, *J. Am. Chem. Soc.* **74,** 4830 (1952).
(86) J. M. Creeth, *Biochem. J.* **51,** 10 (1952).
(87) V. L. Koenigs and J. D. Perring, *Arch. Biochem. and Biophys.* **41,** 367 (1952).
(88) S. Shulman, *Arch. Biochem. and Biophys.* **44,** 230 (1953).
(89) D. F. Waugh and D. F. Yphantis, *Rev. Sci. Instruments* **23,** 609 (1952).
(90) See Ref. 310.
(91) T. Svedberg, *Nature* **139,** 1051 (1937).
(92) G. Weber, *Biochem. J.* **51,** 155 (1952); *Discussions Faraday Soc.* **13,** 33 (1953).
(93) H. Van Vunakis and E. Brand, Abst. 119th meeting Am. Chem. Soc., p. 28c (1951).
(94) P. Desnuelle, M. Rovery, and C. Fabre, *Compt. rend.* **233,** 987 (1951).

Thus it would seem better to use albumin to test the physicochemical methodology, rather than vice versa.

Isoelectric serum albumin contains approximately one hundred pairs of positive and negative charges. While the word pair implies a very special charge distribution, its choice is suggested by the apparent high symmetry in charge distribution, resulting in a net charge asymmetry which can be represented by a dipole vector of a positive and a negative charge separated by 150 Å (700 Debye units) and inclined at an angle of 30° with the long axis of the molecule. The variation in magnitude of this dielectric increment from preparation to preparation has now been explained in terms of the fatty acid content of the albumin. Removal of fatty acid (by passage through anion-exchange columns on the acetate cycle) increased the dielectric increment, and subsequent addition of sodium oleate reduced it to the original value. The most marked effects were noticed with the first mole of fatty acid added per mole of protein. More surprisingly, perhaps, decanol had a similar effect.[83]

The solubility of serum albumin is higher than that of the other major plasma components in practically all solvent systems, so that separations are usually based on precipitating the other components while leaving the albumin in solution. Albumin is generally considered to be water-soluble, this being one of the early definitions of the albumin fraction. However, fractions of serum albumins can be crystallized from distilled water and then prove to be relatively insoluble in this solvent.[68,73] This is particularly true of the mercury albumin dimer.

Albumin solubility may be conveniently investigated in alcohol–water systems where the alcohol concentration, ionic strength, pH, temperature, and specific ion effects may be independently varied. Since constant solubility with variation of saturating body cannot be achieved, solubility may only be expressed in relative terms as a function of changing the parameters. In this way the effect of several variables on the solubility can be studied and reproducible values obtained for any given albumin preparation. The effects of ionic strength and ethanol concentration are thus illustrated in Fig. 3 over a limited pH range.[95] Albumin possesses a marked solubility minimum at the isoelectric point, the solubility increasing manyfold as the charge on the protein increases. Albumin solubility may increase with rising temperature as in alcohol–water systems, be largely independent of temperature as in the presence of polyelectrolytes,[96] or decrease with rising temperature as in the presence of zinc ions.[97]

(95) D. Mittelman, *in* E. J. Cohn, Blood Cells and Plasma Proteins, Academic Press, New York, 1953, p. 24.

(96) H. Morawetz and W. L. Hughes, *J. Phys. Chem.* **56**, 64 (1952).

(97) A. Weber, personal communication.

Albumin shows a tendency to interact with other proteins so that its solubility in natural systems may be quite different from that in the isolated state. Thus it is frequently carried down in globulin fractions under conditions where it should be soluble. In protein fractions of liver

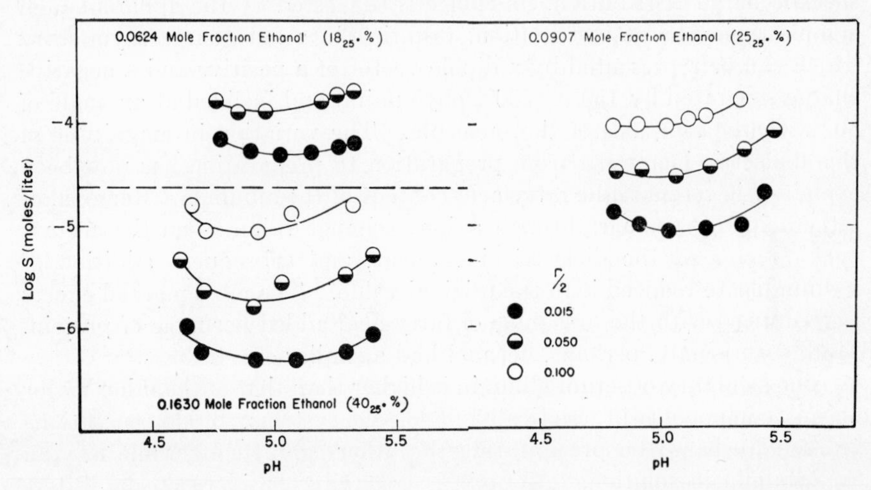

FIG. 3. Solubility of crystallized human serum albumin in ethanol–water mixtures at −5°C.[95]

cells, combination was so firm that only immunochemical methods could detect the albumin present.[98] Fortunately, albumin is present in plasma in much larger amounts than any other component, so that interaction usually does not greatly diminish yields.

e. Stability

Any discussion of protein solubility leads naturally into a discussion of protein stability, since the effect of time on solubility measurements as well as the extremes in variation of pH, temperature, and solvent composition which can be tolerated must be known. Serum albumin is one of the most stable of the plasma proteins. In the cold it appears stable from pH 10 to pH 4. It may be stable at even greater acidities; however, it shows progressive degrees of aggregation in the ultracentrifuge.[99] This may be the fibril formation described by Waugh.[100]

At elevated temperatures serum albumin in concentrated aqueous

(98) E. J. Cohn, Enzymes and Enzyme Systems, Harvard Univ. Press, Cambridge, Mass., 1951.
(99) H. A. Saroff, personal communication.
(100) D. F. Waugh, *J. Am. Chem. Soc.* **68**, 247 (1946).

solution shows a stability maximum between pH 6.5 and 7.0.[45] The stability may be further increased by the addition of many anions.[45,46,101] Sodium caprylate, 0.04 *M*, has proven particularly useful in this regard, making it possible to heat ("pasteurize") albumin for several hours at 60°C.[101] Under such conditions the stability of albumin relative to many other components of plasma is further increased, rendering possible the purification of albumins by selective denaturation of some of the globulins.[68] Under these conditions the rate of denaturation approximately doubles for each 2° rise in temperature,[101] appearing kinetically similar to the heat denaturation of other proteins.

In concentrated urea solutions serum albumin immediately undergoes changes associated with protein denaturation: increase in optical rotation and increase in viscosity[102–104] (see also Chap. 9). These are followed by further changes in the viscosity, finally resulting in a gel. The first rapid changes appear to be quite reversible. The secondary changes have now been interpreted as a disulfide interchange reaction resulting in the formation of intermolecular disulfide bridges and thus a three-dimensional network. Experimental evidence indicates that sulfhydryl groups are essential for the secondary changes.[105] The mechanism may be formulated thus:

$$R{-}SH + \left|\begin{matrix} S \\ S \end{matrix}\right\rangle R_1 \rightarrow R{-}S{-}S{-}R_1{-}SH$$

Where R and R_1 represent two protein molecules. The new sulfhydryl group formed may then react again and a chain reaction is started. Serum albumin dissolved in glacial acetic acid also gels over the course of some days, indicating that a similar mechanism may be involved.

Perhaps of greater interest is the amount of reversibility achieved in the first rapid reaction in urea. Albumin recovered from this treatment has been shown to have approximately the same solubility, electrophoretic mobility, viscosity, anion-combining ability, and immunological properties as the starting material.[104,106] Apparently the structure of serum albumin is largely determined by the peptide backbone and the

(101) G. Scatchard, L. E. Strong, W. L. Hughes, Jr., J. N. Ashworth, and A. H. Sparrow, *J. Clin. Invest.* **24,** 671 (1945).
(102) R. B. Simpson and W. Kauzmann, *J. Am. Chem. Soc.* **75,** 5139 (1953).
(103) W. Kauzmann, *in* McElroy and Glass, The Mechanism of Enzyme Action, Johns Hopkins Univ. Press, Baltimore, 1954.
(104) H. Neurath, G. R. Cooper, and J. O. Erickson, *J. Biol. Chem.* **142,** 249 (1942).
(105) C. Huggins, D. F. Tapley, and E. V. Jensen, *Nature* **167,** 592 (1951).
(106) J. O. Erickson and H. Neurath, *J. Exptl. Med.* **78,** 1 (1943).

disulfide cross-links, and, provided these links are unaltered, considerable amounts of configurational changes may occur reversibly. Further examples of this will appear in the following discussion.

f. Composition

Serum albumin appears to be composed exclusively of amino acid residues. By definition (see sec. II-4*a*) no carbohydrate (giving tests by methods involving furfuraldehyde formation) is present,[68] and nothing resembling a prosthetic group has ever been reported. The absorption spectrum can be interpreted in terms of the aromatic amino acid content.[107–109] The elementary composition has been reported by Brand.[110] Complete amino acid analyses[110–113] by several workers are summarized in Table II. Recoveries of up to 98 per cent of the protein give further evidence for the presence of only amino acids in this protein.

Particular uncertainty dwells upon the tryptophan determination, since all hydrolytic methods appear to give considerable destruction of this amino acid.[114] However, color reactions on the intact protein confirm the fact that bovine serum albumin contains more tryptophan than human. Tryptophan analyses were regularly carried out by the early workers during the fractionation of horse serum albumin, obtaining values which indicated one or two groupings per albumin molecule. However, inconsistencies concerning the effect of protein fractionation on the tryptophan content appear in their work.[115] Certainly a reliable method of tryptophan analysis is badly needed, and would provide an additional criterion for protein homogeneity.

The sulfur distribution in serum albumin should also be further investigated. Brand found in bovine serum albumin 42 sulfurs per 69,000 g. of protein (this becomes 40 per 65,000 g.) and *even* numbers of cysteine and methionine residues[110] (Table II). However, there is only

(107) D. Gitlin, *J. Immunol.* **62,** 437 (1949).
(108) Frequently, an absorption maximum at 405 mμ appears due to contamination by hemin.
(109) There is a very small amount of absorption in the range 340–400 mμ which cannot reasonably be assigned to aromatic amino acids or Tyndall scattering.
(110) E. Brand, *Ann. N. Y. Acad. Sci.* **47,** 187 (1946).
(111) E. Brand, B. Kassel, and L. J. Saidel, *J. Clin. Invest.* **23,** 437 (1944).
(112) S. Moore and W. H. Stein, *J. Biol. Chem.* **178,** 53 (1949).
(113) D. Shemin, *J. Biol. Chem.* **159,** 439 (1945).
(114) J. R. Spies and D. C. Chambers, *Ind. Eng. Chem., Anal. Ed.* **21,** 1249 (1949).
(115) Thus Hewitt[79] found that the tryptophan content of horse albumin decreased to 0.3 per cent upon isolation of the carbohydrate-free fraction. Whereas McMeekin[73] found a tryptophan content of 0.5 per cent for the carbohydrate-free fraction, which was not altered by further fractionation.

TABLE II
COMPOSITION OF SERUM ALBUMINS[a]
(Residues per 65,000 grams)

	Bovine	Human	Comparative analyses	
			Bovine	Equine
Alanine	46	—		
Glycine	17	14	(16)	
Valine	33	43	(36)	
Leucine	61	54	(68)	(50)
Isoleucine	13	8	(14)	
Proline	27	29	(32)	
Phenylalanine	26	31	(24)	
Tyrosine	18	17	(20)	(17)
Tryptophan	2	1		(1–2)
Serine	26	20		
Threonine	32	25		
Cysteine	1	1		
Cystine/2	34	34	34	34
Methionine	4	6	4	0
Arginine	22	23	(23)	(21)
Histidine	17	15	(16)	(18)
Lysine	55	55	55	
Aspartic acid	50	48	50	
Glutamic acid	75	75	75	
Amide N	(36)	(41)		
Total S	40	41	40	37

Nitrogen content of bovine serum albumin: measured = 16.07
calculated from above composition = 16.2[b]

[a] Data are largely from Tristram's Chap. 3 but corrected for a molecular weight of 65,000. However, cysteine data have been corrected in the light of my own evidence (see text), and the probably more precise estimates of Shemin[113] by isotope dilution for glycine, lysine, aspartic and glutamic acids have been substituted for Stein and Moore's and Brand's values. The values in columns 3 and 4 were included for direct comparison with column 2 having been carried out for a given amino acid by a single laboratory.[111–113]

[b] i.e. number of nitrogen atoms × 14 ÷ 65,000 × 100.

one cysteine in serum albumin,[116] hence, *if* the total sulfur content is correct, there must be an *odd* number of methionine residues or an additional unknown type of sulfur linkage must exist.

Serum albumin may contain a single polypeptide chain since only one terminal α-amino group (aspartyl) exists in human, equine, porcine, and bovine serum albumin.[93]

(116) W. L. Hughes, *Cold Spring Harbor Symposia Quant. Biol.* **14,** 79 (1950).

g. Reactivities

The reactivities of serum albumin, as of proteins in general, are so numerous that discussion will be limited to those particularly significant as regards to structure and those more or less peculiar to serum albumin. A more general discussion will be found in Chap. 10.

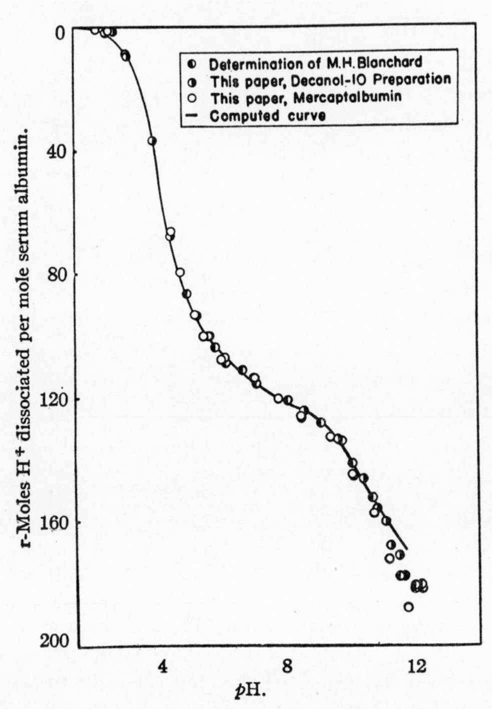

FIG. 4. Titration curve of human serum albumin. (From Tanford.)[117]

Hydrogen-ion equilibria will be considered first since they involve practically all of the reactive groupings of serum albumin and must be considered as competitive reactions in most of the other interactions to be discussed later (for further details see Chap. 6, sec. II and Chap. 8, sec. VI-1). The titration curve has been recently redetermined by Tanford[117] (Fig. 4), substantially in agreement with earlier data.

The points in Fig. 4 are experimental; the curve is theoretical, calculated from the amino acid composition with the assignment of appropriate intrinsic pK's (Table VII, Chap. 8, page 791) to each type of residue and allowing for the effect of protein charge on the equilibrium (excepting in the acid range) after the manner of Scatchard.[118]

(117) C. Tanford, *J. Am. Chem. Soc.* **72,** 441 (1950).
(118) G. Scatchard, *Ann. N. Y. Acad. Sci.* **51,** 660 (1949).

The intrinsic pK of the tyrosyl was estimated by a spectrophotometric titration. The intrinsic pK's of the carboxyl, imidazole, and amino groups were then estimated from the experimental data (Fig. 4) by curve-fitting with the help of measurements of the effect of temperature on the equilibria. The heats of ionization of carboxyl, imidazole, and amino groups are sufficiently different (Table VI, Chap. 6 and Table VII, Chap. 8) to allow the estimation of the contribution of each to any portion of the titration curve.

It will be noted that the calculated intrinsic constants correspond fairly closely to those for model compounds (Table VI, Chap. 6). However, the differences may have some significance. Thus the carboxyl groups definitely seem to be more acidic, and this weakened proton affinity appears to be a generally weakened cation affinity, being reflected in zinc binding and calcium binding,[119] which is much weaker per carboxyl grouping for isoelectric albumin than for simple carboxylate compounds. It is interesting to speculate whether this weak cation affinity is a corollary of the strong anion affinity of this protein (see below).

Another interesting feature of the titration curve is the very steep portion in the acid range, corresponding to titration of carboxyl groups as though no charge effects were present. Part of this effect has been explained by anion binding;[117] however, to attribute it completely to anion binding requires the assumption of the unreasonably large value of 89 chlorides bound at pH 2. Therefore it would seem more reasonable to explain it partly by structural changes in the molecule—an uncoiling which would decrease the electrical potential. The decreased sedimentation constant observed in acid solutions[120] could be similarly interpreted in terms of increased asymmetry. Such uncoiling must be quite reversible since there is no hysteresis in the titration curve upon neutralizing an albumin sample which has been brought to any acidity down to pH 2.0.[117]

The thiol grouping would seem a logical starting point when discussing the specific reactivities of serum albumin: first, because it is the most reactive grouping (in the sense of organic chemistry); secondly, because its nature and location permit a dimerization reaction between mercaptalbumin molecules which can be followed by the usual physicochemical techniques (Figs. 5–7); and thirdly, because its singular presence makes unnecessary the statistical considerations which complicate studies of the other groupings.

The reaction of this grouping with mercuric ions to form crystallizable dimers, as described above, permitted separation of the mercaptalbumin from the remaining albumin. However, excepting the thiol group (usually present on ⅔ of the albumin molecules), no other differences

(119) F. R. N. Gurd and W. L. Hughes, unpublished observations.
(120) H. Gutfreund, *Discussions Faraday Soc.* **13** (1952).

have been observed in these two fractions, suggesting that the remainder may have lost their sulfhydryl by oxidative or coupling reactions. This view is strengthened by the discovery that a thiol group may be liberated in this fraction by thioglycolate under conditions where the mercaptalbumin fraction does not form additional thiols.[121]

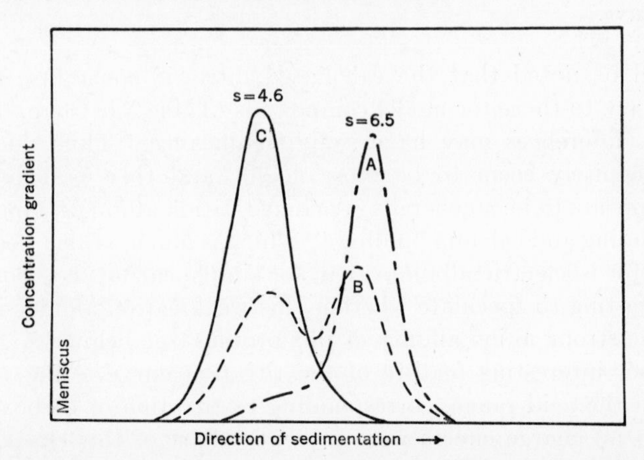

FIG. 5. Sedimentation diagrams of crystallized mercury serum albumin at pH 7. A fresh solution of the crystal sediments at the rate of an albumin dimer ($s = 6.5$ – curve A). After standing one week, some dissociation has taken place (curve B). Addition of excess mercuric chloride causes immediate dissociation and curve C is obtained with a sedimentation constant ($s = 4.6$) typical of serum albumin.[116]

While no other metal ions have been found to dimerize albumin like mercury, a similar reaction occurs with a "bivalent" mercurial:

$$\mathrm{HOHgCH_2{-}CH}\begin{matrix}\diagup \mathrm{O{-}CH_2} \diagdown \\ \diagdown \mathrm{CH_2{-}O} \diagup\end{matrix}\mathrm{HC{-}CH_2HgOH}$$

whereupon reaction proceeds manyfold faster.[122] Both dimers may be dissociated by reagents competing with albumin for the mercury (thiols, CN^-, and I^- being particularly effective) and by reagents competing with the mercury for the albumin (Ag^+, Hg^{++}). The dissociating effect of excess mercuric salts (Fig. 6) is explained by the following reaction scheme if it is assumed that the first reaction proceeds much further to completion than the second:

$$\begin{array}{c}\mathrm{AlbSH + HgCl_2 \leftrightharpoons AlbS{-}HgCl + H^+ + Cl^-} \\ + \\ \mathrm{AlbSH} \\ \upharpoonleft\downharpoonright \\ \mathrm{(AlbS)_2Hg + H^+ + Cl^-}\end{array}$$

(121) M. Hunter, personal communication.
(122) R. Straessle, *J. Am. Chem. Soc.* **73**, 504 (1951).

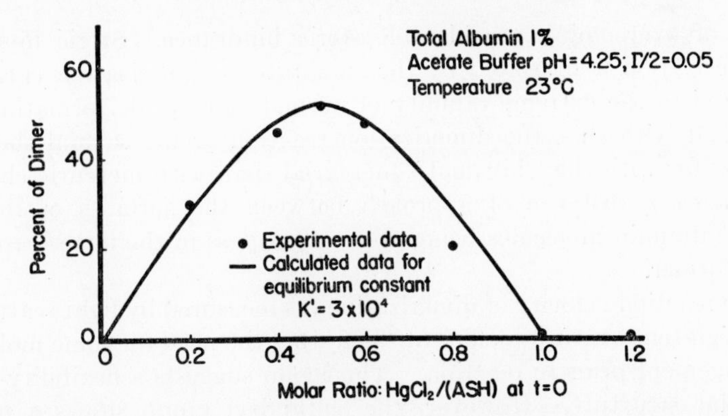

FIG. 6. Per cent of mercaptalbumin dimer at equilibrium as a function of $HgCl_2$/Alb-SH ratio. Experimental data from light-scattering measurements.[123]

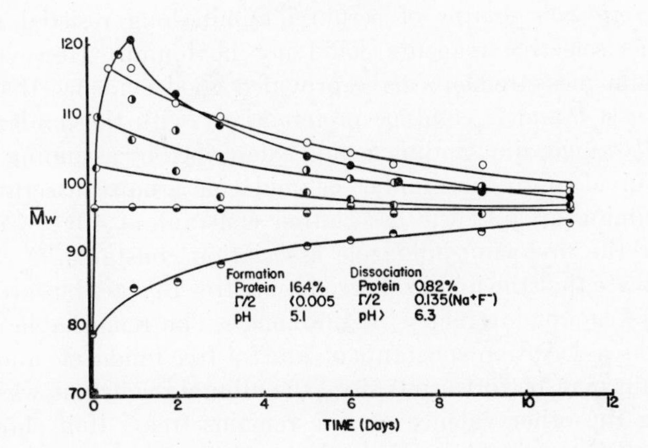

FIG. 7. Formation and dissociation of the mercuric dimer of serum albumin as determined by light-scattering measurements at room temperature. Serum albumin and the stoichiometric amount of $HgCl_2$ were mixed at zero time under conditions given for *Formation*. At successive intervals samples were diluted to the conditions given for *Dissociation*. Following dilution, immediate measurement of the turbidity provided the rising curve *Association* and then continued measurement of the diluted samples produced the converging curves showing association or dissociation towards equilibrium. (Courtesy, H. Edelhoch.)

The specificity of this reaction can be understood if it is remembered that the affinity of mercury for a thiol ion is about 10^{10} times greater than for any other grouping found in albumin.[124] Thus dimerization is favored over some much weaker intramolecular bridge even to the

(123) H. Edelhoch, E. Katchalski, R. Maybury, W. L. Hughes, and J. T. Edsall, *J. Am. Chem. Soc.*, **75**, 5058 (1953).

(124) The affinity of Hg^{++} for RS^- appears similar to its affinity for CN^- ($K_1 = 10^{18}$), J. Bjerrum, *Chem. Revs.* **46**, 381 (1950).

extent of overcoming considerable steric hindrance. Steric hindrance to dimerization is indicated by the slow rate of dimerization (Fig. 7)[123] compared to the extreme rapidity of normal mercaptide formation. In agreement with this, the dimerization reaction occurs several thousand times faster with the "bivalent" mercurial than with mercuric chloride, the necessary distance of approach between the surfaces of the two mercaptalbumin molecules being about 10 A. less in the latter case than in the former.

The reaction velocity of dimerization (as measured by light scattering) has a high temperature coefficient suggesting the need for some molecular rearrangement prior to reaction. This again suggests a flexibility in the molecular structure. However, the sulfhydryl group shows a normal reactivity toward methyl mercury iodide, giving the same equilibrium constant as cysteine.[116] This indicates that the sulfhydryl cannot be bound in some internal linkage.

The imidazole groups of serum albumin long resisted a definitive attack, as selective reagents could not be found. However, dialysis-equilibrium measurements have provided good evidence that zinc ions, between pH 5 and 7, combine preferentially with the imidazole groups. In this pH range, zinc binding could be described by assuming 15 identical sites (equivalent to the number of imidazole groups in serum albumin) and assuming an intrinsic association constant of $10^{2.82}$ (compared to $10^{2.76}$ for the first zinc–imidazole association constant).[125] These data thus indicate that the imidazole groups are free to react but are so situated that pairs cannot interact with zinc ions. The remarkable correspondence of the association constants of zinc for free imidazole and its residue in albumin may be fortuitous since the albumin constant was calculated assuming the other valence of zinc remains free. If it should also be linked to the protein, the values calculated for the zinc–imidazole linkage will be too high. This would allow zinc–imidazole interaction, like proton–imidazole interaction, to be weaker in albumin than in model substances. Cupric ions and perhaps those of some other metals also react preferentially with the imidazole groups,[126–128] but no other specific reactions with the imidazole groups have been reported. The imidazole groups react simultaneously with the tyrosyl residues during iodination (see next page).

(125) F. R. N. Gurd and D. S. Goodman, *J. Am. Chem. Soc.* **74,** 670 (1952).
(126) C. Tanford, *J. Am. Chem. Soc.* **74,** 211 (1952).
(127) H. A. Saroff and H. J. Mark, *J. Am. Chem. Soc.* **75,** 1420 (1953).
(128) I. M. Klotz has shown some competition by the sulfhydryl group for copper. I. M. Klotz, J. M. Urquhart, and H. A. Fiess, *J. Am. Chem. Soc.* **74,** 5537 (1952).

The 55 *epsilon amino groups* of lysine represent the largest class of cationic groups in serum albumin. They can be completely converted, by reaction with *O*-methylisourea, to homoarginyl groups[129] or with acetic anhydride to acetamino groups.[130] The guanidinated derivative is crystallizable and appears quite similar to the starting material with regard to immunological properties[131] and anion binding.[132] However, the derivative is markedly less soluble than serum albumin being insoluble at pH 5 and low ionic strengths (below 0.01 Γ/2). Kinetic studies with *O*-methylisourea[129] indicate that not all lysyl groups are equivalent in their reactivity; however, such variation is not seen in the titration curve (Fig. 4) which could be fitted by assuming a single p*K* for lysyl residues. Structural alterations, caused by the high charge on the protein when titrating amino groups, may have obliterated these differences.

Tyrosyl residues have been titrated spectrophotometrically as indicated above. Unlike egg albumin,[133] no hysteresis is observed upon reversing this titration. Furthermore, the intrinsic ionization constant appears quite close to appropriately substituted phenols and might be taken as evidence against hydrogen bonding. However, the absorption maximum of serum albumin at 279 mμ is definitely shifted from the maximum in tyrosine and its peptides (275 mμ), while treatment with dilute acetic acid shifts it toward the tyrosine maximum.[107] Therefore, the tyrosyl residues may be hydrogen-bonded in some reversible fashion. (For further discussion see Chap. 5.)

The tyrosyl residues react quite specifically with iodine in potassium iodide (I_3^-), provided one stops the reaction short of stoichiometric equivalence, as some other grouping, probably histidyl, also reacts, but more slowly.[134] Some monoiodotyrosyl residues may be formed initially, but the main product is diiodotyrosyl residues[134,135] which have the expected p*K*.[117] Partially iodinated samples crystallize isomorphously with the starting material.[134] Several workers have iodinated and crystallized horse serum albumin.[136–138]

(129) W. L. Hughes, Jr., H. A. Saroff, and A. L. Carney, *J. Am. Chem. Soc.* **71**, 2476 (1949).
(130) H. Fraenkel-Conrat, R. S. Bean, and H. Lineweaver, *J. Biol. Chem.* **177**, 385 (1949).
(131) C. A. Janeway, personal communication.
(132) I. M. Klotz and J. M. Urquhart, *J. Am. Chem. Soc.* **71**, 1603 (1949).
(133) J. L. Crammer and A. Neuberger, *Biochem. J.* **37**, 302 (1943).
(134) W. L. Hughes and R. Straessle, *J. Am. Chem. Soc.* **72**, 452 (1950).
(135) C. H. Li, *J. Am. Chem. Soc.* **67**, 1065 (1945).
(136) A. Bonot, *Bull. soc. chim. biol.* **21**, 1422 (1939).
(137) J. Muus, A. H. Coons, and W. T. Salter, *J. Biol. Chem.* **139**, 135 (1941).
(138) B. K. Shakrokh, *J. Biol. Chem.* **151**, 659 (1943).

Considerable use has been made recently of iodinated serum albumin as a tracer in physiological studies.[24,25] The assumption is usually made that iodinated albumin behaves identically with normal albumin, and to a large degree this appears to be true. Occasionally, however, differences in performance have been used to elaborate some hypothetical physiological mechanism without careful consideration of protein differences. Some of the more obvious of these include:

(*1*) *Lability of the Iodo-Protein Bonds.* The experiments of Pitt-Rivers on the conversion of diiodotyrosyl residues to thyroxine are of interest here.[139,140]

(*2*) *The Instability of the Iodinated Protein.* This may have been increased by the iodination process, if considerable iodine has been introduced, or it may result from adverse handling (or storage) of the albumin at any time prior to injection. Such albumin might disappear more rapidly due to phagocytosis or catabolism, or more slowly due to its smaller diffusibility across the capillary membrane.

(*3*) *The Heterogeneity of the Iodinated Albumin.* Usually very small degrees of iodination have been used in the expectation that such material would most closely resemble normal albumin and also would be least heterogeneous in terms of iodine content (i.e., every molecule would either have one iodine atom or none). However, there is no reason to believe that the tyrosines are all equivalent so that isomers would certainly exist and some of the iodine would also be on the imidazole groups. Furthermore, the first two equivalents of iodine to react with albumin are not bound at all but rather oxidize the sulfhydryl.[134]

The carboxylate groupings of serum albumin, in spite of their large number, are relatively unreactive toward coordinating metals, including protons, as pointed out above. This low reactivity may be a result of salt linkages, since, as seen above in the titration curve, their reactivity increases as the charge on the protein increases.

Successful chemical modification of the carboxyl groups is also limited. This is partially due to the fact that simple esterification requires the carboxylic acid in its un-ionized form. Thus the protein must first be converted to its acid salt—a particularly harsh treatment involving rupture of all the salt linkages. This procedure has been used by Olcott and Fraenkel-Conrat, who treated serum albumin with hydrogen chloride in dry methanol. They reported quantitative and specific methylation of the carboxyl groups, although the physicochemical

(139) R. Pitt-Rivers, *Biochem. J.* **43**, 223 (1948).

(140) It has recently been shown that 3,5,3′-triiodothyronine is the active hormone, and is produced from thyroxine. J. Gross and R. Pitt-Rivers, *Biochem. J.* **53**, 645, 652 (1953).

properties of the product indicated marked structural changes to have occurred.[141]

Milder procedures of esterification involving reaction with the carboxylate ion can be carried out with ethylene oxide and aliphatic diazo compounds. The ethylene oxide reaction at neutrality readily covers a fraction of the carboxylate groups. However, the reaction is neither quantitative nor specific. Approximately equivalent amounts of reagent react with other groupings including amino groups.[142] Wilcox has investigated the reactivity of a series of diazo compounds of the formula $RCHN_2$.[143] He found $N_2CH—COOC_2H_5$ and $N_2CH—CONH_2$ to be the most useful, reacting at pH 5–6 and 0°C. Both of these proved much less reactive than diazomethane but specific for carboxyl groupings. (Diazomethane has been shown to react with practically all groupings containing active hydrogen.[144]) Both ethylene oxide-treated and "diazo-esterified" albumin showed the expected alkaline shifts in the isoelectric point.

The reactivities of the remaining groupings in serum albumin have been but little investigated, partly because mild chemical reactions are unknown. The aliphatic hydroxyls have been sulfated by treatment with concentrated sulfuric acid at low temperatures.[145] The amides have been hydrolyzed by sulfonic acids, although not without splitting some peptide bonds.[146]

The disulfide links have been found relatively resistant to reduction in the native protein. However, some reversible attack by thioglycolate ions has been observed at pH 8–10.[121] Not more than two disulfides, on the average, were attacked at thioglycolate concentrations which would completely reduce the denatured protein. Whether this indicated disulfides of varying reactivity or merely denoted a marked shift of equilibrium in favor of protein disulfide in the native protein was not decided.

No method of selectively attacking the peptide bonds by enzymes to produce large fragments has been found. The hydrogen bonds of the peptide chain appear particularly reactive toward urea (i.e., denatura-

(141) H. Fraenkel-Conrat and H. S. Olcott, *J. Biol. Chem.* **161,** 259 (1945).
(142) H. Fraenkel-Conrat, *J. Biol. Chem.* **154,** 227 (1944).
(143) P. E. Wilcox, Abstracts 12th Intern. Congr. Pure and Appl. Chem. New York, 1951, p. 60.
(144) H. Fraenkel-Conrat *in* D. Greenberg, Amino Acids and Proteins, 2nd ed., C. C Thomas Publ. Co., Springfield, Ill., 1951, p. 532.
(145) H. C. Reitz, R. E. Ferrell, and H. Fraenkel-Conrat, *J. Am. Chem. Soc.* **68,** 1024 (1946).
(146) J. Steinhardt, C. H. Fugitt, and M. Harris, *J. Research Natl. Bur. Standards* **26,** 293 (1941); **28,** 201 (1942).

tion), even 4 M urea producing marked effects.[103] However, as already mentioned, most of the changes are reversible.

Some reactions of serum albumin are at present uninterpretable in terms of protein structure. The most studied of these, anion binding, have been comprehensively covered by Klotz (Chap. 8) and will be merely summarized here as to their implications for protein structure. Perhaps the most surprising feature is the magnitude of the interaction constants observed—several orders of magnitude greater than those of models of any of the proposed reactive sites in aqueous solution. Thus Scatchard *et al.* found an intrinsic association constant of 1000 for the combination of SCN^- and albumin.[147] This suggests that the reactions may take place in a region of very different effective dielectric constant from water.[148]

A second feature of the albumin interactions is the marked preference for anions, a preference which appears to be mirrored by an actual dislike for cations—at least relative to simple model substances. Thus any proposed explanation should be in harmony with both effects.

A third feature is the apparent marked structural alteration accompanying binding as evidenced by the large entropy term (Chap. 8).

While not enough of the topography of the protein molecule is known to explain these phenomena in detail, it would appear to the author that the general features are interpretable in terms of competitive interactions with charged groups of different hydration energies in regions of widely varying dielectric constant.

It seems important to stress competitive interactions since in aqueous solution all of the valencies of atoms or radicals may be considered saturated with solvent, if not with another grouping, and the driving force for some reactions may be the relative affinity for solvent molecules of the groups involved. Thus, for example, a salt linkage[149] (e.g., an amino group plus a carboxyl group of the albumin molecule) existing in an

(147) G. Scatchard, I. H. Scheinberg, and S. H. Armstrong, Jr., *J. Am. Chem. Soc.* **72,** 535, 540 (1950).

(148) J. Schellman, *J. Phys. Chem.* **57,** 472 (1953); D. J. R. Laurence, *Biochem. J.* **51,** 168 (1952).

(149) K. Linderstrøm-Lang and C. F. Jacobsen [*Nature* **164,** 411 (1949)] feel that they have eliminated the possibility of a high frequency of "salt-linkages" by their studies of volume changes upon titrating a protein with acid. The volume change for the reaction: $RCOO^- + H_3O^+ \rightarrow RCOOH + H_2O$ was found to be +11 ml. per mole for acetate and for β-lactoglobulin. On the other hand solid glycine, which they took as a possible model of protein salt-linkages, showed a volume increase of only 4 ml. per mole upon solution in aqueous acid. However, it seems doubtful that solid glycine would adequately represent a salt linkage in solution where considerable electrostriction of solvent about the interacting groups might still occur.

anhydrous region[150] may be split by a competing anion (e.g., SCN^-) which is less strongly hydrated than the carboxylate group. The carboxylate grouping is thereby freed to move into a more hydrated region, and the energy of thiocyanate binding becomes the difference in the hydration energies of the carboxylate and thiocyanate groups.

Conversely, metal cations will be bound more weakly than in model substances, since they are presumably more strongly hydrated than the cationic nitrogen group which they would replace. Such a theory of course further complicates binding studies, since the interpretation must consider pairs of groups instead of isolated groupings. However, the probable importance of such mechanisms is emphasized when one remembers that the average charge density of a protein molecule is approximately that of a molar salt solution so that even a random arrangement would result in many closely adjacent charges.

h. Physiological Functions

In addition to its roles in osmotic regulation and as a nitrogen reserve, serum albumin may well assume a transport function by virtue of its ion-binding ability.[151,152] In this way albumin is linked with *many anionic substances in blood*, serving to "buffer" their activity and thus making them equally available at all points in the circulation. In the same sense it may be functionally involved in detoxication as, for example, by the binding of glycine conjugates and thus transporting them to the kidney.

In a similar way, the thiol group may act as a scavenger for heavy metals. It is interesting that mercurials are carried in the blood mainly by albumin[153] and deposited in the kidney prior to elimination.[154] This may be an unloading phenomenon caused by the greater acidity of the kidney.

The slight permeability of the capillary membranes to albumin may be of significance in transport phenomena. As an example, the excretion of mercurials may involve filtration through the glomerulus of the

(150) "Anhydrous" regions will also be regions of low dielectric constant as hypothesized by Schellman.[148]

(151) H. Bennhold, E. Kylin, and St. Rusznyak, Die Eiweisskörper des Blutplasmas, T. Steinkopf, Dresden, 1938.

(152) B. Davis, *in* C. M. McLeod's Evaluation of Chemotherapeutic Agents, Columbia Univ. Press, New York, 1949, p. 44.

(153) The serum albumin contains 80 per cent of the thiols of plasma. Also, when fractionating plasma containing merthiolate as a preservative, most of the mercury was found in Fraction V, the albumin fraction.

(154) K. K. Mustakallio and A. Telkkä, *Science* **118**, 320 (1953).

albumin conjugate followed by selective reabsorption of the albumin in the tubule at a lower pH where binding of mercurials is weaker.

To discuss a transport function for albumin in the extravascular fluid would simply compound our ignorance at the present time since extravascular albumin levels are unknown. The importance of such transport will depend upon the relative rates of diffusion of free and albumin-bound metabolite across the capillary membrane. Considering the much greater diffusibility of the small free metabolite, albumin transport would appear unimportant unless more than 99 per cent of the metabolite were bound to albumin.

III. Proteins Functioning as Transporters of Metabolites

Transport phenomena, as suggested by albumin, become the *sina qua non* for the respiratory proteins involved in oxygen transport (see Chap. 14). The limited oxygen capacity of these pigments has required special modifications in their physical states to avoid too great osmotic effects. Thus the hemocyanins are largely associated in solution with molecular weights of one million or more. In the case of hemoglobins with molecular weights around 68,000, the osmotic effects have been circumvented by transporting the pigment within specialized cells, the erythrocytes.

The oxygen affinity of these pigments varies from species to species in the interest of providing the greatest possible "buffering" power of O_2 tension at the environmental levels which the species inhabits.[6] The shape of the oxygen association curve and the pH dependence of oxygen affinity are further tailored in the interest of maximum efficiency.[155] Extravascular functions for hemoglobins or hemocyanins are obviously lacking as judged by the complete retention of pigment within the circulation. Owing to the much greater permeability of the lipoidal capillary membrane for oxygen than for more polar molecules, an extravascular transport function would indeed appear unnecessary.

1. Iron-Binding Globulin, Transferrin

One plasma protein which has been quite convincingly linked with specific transport is the iron-binding globulin, transferrin (siderophilin). Chemically, it is closely related to the iron-binding protein of egg white, conalbumin. This has been discussed in detail by Warner in Chapter 17. Immunologically, conalbumin from chicken egg white and transferrin from chicken plasma were found to be identical.[43] The chemistry of the metal interaction of conalbumin has also been covered in Chapter 17 and the similarity to transferrin pointed out. (Compare Chapter 8.)

(155) L. J. Henderson, Blood—A Study in General Physiology, Yale Univ. Press, New Haven, 1928.

Present to the extent of 2.5 g./l.[156] (3.3 per cent of the plasma protein), it travels as a β_1-globulin during electrophoresis, and represents ⅓ of this envelope of mobilities. It appears in the globulin fraction by salt fractionation[157] and in fraction IV-4 of Cohn's ethanol procedure (method 6). However, the solubility is markedly dependent upon the presence or absence of iron (Fe^{+++}) in the protein, the iron-containing form being much more soluble.[158] The iron is readily split off by acid, the linkage being unstable even at pH 6.[158] Thus fractionation procedures employed below pH 6 are relatively independent of the iron concentration, since dissociation permits the less soluble iron-free protein to precipitate. However, fractionation near pH 7 will be complicated by the normal partial saturation of this component with respect to iron (normally ⅓ saturated).[156]

The iron-combining globulin has been crystallized from aqueous alcohol by Koechlin[158] following its concentration in fraction IV-7.[159] Crystallization is not very reproducible, being impeded by small amounts of impurities. Nevertheless, Koechlin was able to obtain a small yield of material homogeneous by *electrophoresis* in barbiturate buffer pH 8.6. A molecular weight of 90,000 was calculated on the basis of iron-binding capacity[160] and sedimentation constant. This material still contained 1.8 per cent carbohydrate[161] but no detectable sulfhydryl groups. Transferrin is very soluble in water even at its isoelectric point (pH 5.8).[158,162] It has the largest dielectric increment of any protein yet studied.[163] Attempts to crystallize the iron complex have thus far failed because of the much smaller solubility of the iron-free protein at pH 5.8. At higher pH values, where the iron complex would be stable, the solubility becomes so great that alcohol precipitation is impossible.[164]

(156) C. B. Laurell and B. Ingelman [*Acta Chem. Scand.* **1,** 770 (1947)] have reported an iron-binding capacity of serum of 310 μg. % (or 0.05_5 millimolar). Assuming 2 atoms of iron per protein molecule and a molecular weight of 90,000, one calculates a plasma level of 2.5 g./l.

(157) G. Barkan and O. Schales, *Z. physiol. Chem.* **248,** 96 (1937).

(158) B. A. Koechlin, *J. Am. Chem. Soc.* **74,** 2649 (1952).

(159) D. M. Surgenor, L. E. Strong, H. L. Taylor, R. S. Gordon, Jr., and D. M. Gibson, *J. Am. Chem. Soc.* **71,** 1223 (1949).

(160) Minimum molecular weight = 45,000.

(161) Bain and Deutsch report conalbumin to contain no carbohydrate. *J. Biol. Chem.* **172,** 547 (1948).

(162) An extrapolated isoelectric point for the iron-containing protein at pH 4.4 has been reported.[156]

(163) J. L. Oncley and N. Hollies, personal communication.

(164) R. C. Warner and I. Weber [*J. Biol. Chem.* **191,** 173 (1951)] have recently reported the crystallization of conalbumin both with and without iron. The more neutral isoelectric point (pH 6.8) of this protein may aid preparation of the iron-containing crystals.

2. Zinc and Copper Components of Plasma. Ceruloplasmin

Zinc and copper also circulate in plasma, each in amounts approximately equal to iron (*ca.* 1 part per million). Since they are known to combine with transferrin, and since this component is normally only one-third saturated with iron, their transport by the same protein has been postulated. However, in view of the truly infinitesimal nutritional requirements for these elements, the transport of such large amounts in the circulation might seem pointless.

Zinc, like iron, is quite firmly bound at neutral pH but is readily split off at pH 5. At neutral pH it may be readily differentiated from zinc ions added *in vitro* by its slow rate of exchange to a cation-exchange resin. It may be displaced *in vitro* by slightly greater than stoichiometric amounts of cupric ion.[165] Thus the presence of a specific zinc protein in plasma seems very likely.

Plasma copper has been identified with a specific blue component, ceruloplasmin, from which copper cannot be reversibly removed.[166] This protein appears to be the same as "plasma hemocuprein" noted by Mann and Keilin,[167] by Laki,[168] and by McMeekin *et al.*[169] However, it may differ fundamentally from erythrocyte hemocuprein in being reversibly reducible by hydrosulfite,[166] the blue color reappearing on shaking with air. Hemocuprein is said to be destroyed by the action of hydrosulfite.[167]

Ceruloplasmin appears to interact strongly with other plasma proteins, its solubility being markedly dependent on the purity of the preparation. Thus in purification it is first carried down with the globulins precipitated between 40 and 50 per cent saturated ammonium sulfate, but when purified it is soluble under these conditions. Following the lead of Mann and Keilin, Holmberg and Laurell have further purified this component by selectively denaturing the other globulins in this precipitate by alcohol and chloroform and then extracting the blue pigment with salt solution.[166,170] This protein has recently been isolated and crystallized from fraction IV-1 (Cohn's method 6) by Kominz.[171]

Holmberg and Laurell report a copper content, like hemocuprein, of

(165) W. L. Hughes, unpublished observations noted in studying zinc removal from plasma by ion-exchange resins.

(166) C. G. Holmberg and C. B. Laurell, *Acta Chem. Scand.* **2,** 550 (1948).

(167) T. Mann and D. Keilin, *Proc. Roy. Soc.* (*London*) **B126,** 303 (1948).

(168) K. Laki, *Abst. commun. 17th Intern. Physiol. Congr., Oxford,* p. 373 (1947).

(169) E. J. Cohn, T. L. McMeekin, J. L. Oncley, J. M. Newell, and W. L. Hughes, *J. Am. Chem. Soc.* **62,** 3386 (1940).

(170) C. G. Holmberg and C. B. Laurell, *Acta Chem. Scand.* **1,** 944 (1947).

(171) D. Kominz, personal communication.

0.34 per cent but a larger molecular weight (by ultracentrifuge) of 151,000 (4× that of hemocuprein), from which they calculate 8 copper atoms per molecule. Ceruloplasmin is isoelectric at about pH 4.4 in salt solution, although it is unstable at this pH. It has the mobility of an α-globulin.

Ceruloplasmin has the enzymatic activity of a laccase, although less active than plant laccases. It catalyzes the oxidation of *p*-phenylenediamine (also ascorbic acid, and other aromatic diamines and diphenols) by oxygen with a maximum activity at pH 5–6.[172]

3. Lipoproteins

Lipides are very excellent examples of substances requiring a protein carrier for plasma transport, and the lipides of plasma have long been known to be protein-bound (this of course refers to transparent plasma, the fat droplets, chylomicrons, of opaque plasma being a physical suspension). Studies of these lipoproteins have proceded along two lines: (*1*) isolation of the lipides themselves from plasma by the gentlest means possible and subsequent study of the separated lipides and proteins, and (*2*) isolation of specific lipide–protein complexes and their characterization. The first method developed from the view that the lipides are merely an impurity (like salt) which confuses the protein picture, since their removal, if properly executed,[173,174] markedly "stabilizes" the proteins for subsequent investigations. However, accumulating evidence points to a specific combination of the lipides with a rather small group of the plasma proteins which can be fractionated quite intact.

A large part of the lipide in human plasma can be concentrated by increasing the density of the medium until the lipoprotein specifically rises in a gravitational field. Thus one obtains x-protein of McFarlane[175] and Pedersen,[232] which has been shown not to be an artifact since variation of solvent by addition of different salts or of glycine gives a similar separation as long as density requirements are fulfilled.[176] One of Cohn's fractions (III-0) contains the bulk of this same component over 50 per cent pure.[177]

Similarly, electrophoretic separations have permitted the preparation of α- and β-lipoprotein components,[178] a separation which has also been

(172) C. G. Holmberg and C. B. Laurell, *Acta Chem. Scand.* **5,** 476, 921 (1951).
(173) W. B. Hardy and S. Gardiner, *J. Physiol.* **40,** lxviii (1910).
(174) A. S. McFarlane, *Nature* **149,** 439 (1942).
(175) A. S. McFarlane, *Biochem. J.* **29,** 407, 660, 1175, 1202 (1935).
(176) J. L. Oncley, personal communication.
(177) J. L. Oncley, M. Melin, D. A. Richert, J. W. Cameron, and P. M. Gross, Jr., *J. Am. Chem. Soc.* **71,** 541 (1949).
(178) G. Blix, A. Tiselius, and H. Svensson, *J. Biol. Chem.* **137,** 485 (1941).

TABLE III
COMPOSITION OF VARIOUS PURIFIED PLASMA LIPOPROTEINS

Lipoprotein	Per cent composition of lipoprotein					Mole ratio		
	Cholesterol					Cholesterol	Free cholesterol	Cholesterol
	Total	Free	Ester	Phospholipide	Peptide	Phospholipide	Total cholesterol	Nitrogen
Macheboeuf[183]								
Cenapseacide (horse)	10.4	0	17.9	22.7	59.3	0.91	0	0.040
Blix, Tiselius, and Svensson[178]								
β-Globulin	8.6	...	....	10.0	....	1.7	....	.....
α-Globulin	4.4	...	....	7.2	....	1.2	....	.....
Albumin	1.1	...	....	2.2	....	1.0	....	.....
γ-Globulin	0.4	...	....	1.0	....	0.8	....	.....
Adair and Adair[186]	16.4	...	....	8.5[a]	40	3.8[a]	....	0.093
Pedersen's X-protein[232]	....	...	....	23.4	....	....	....	.....
Oncley, Gurd, and Melin[185]								
β-Lipoprotein	30.9	8.3	39.1	29.3	23.0	2.1	0.27	0.28
Cohn, Strong, and Blanchard[b]								
α_1-Lipoprotein	12.0	3.3	15	21	57	1.36	0.27	0.053
Lewis, Green, and Page[191]								
α_2-Lipoprotein	....	...	....	....	....	2.6	....	0.24
Gofman *et al.*[197]								
S_f 4	30	8	40	25	25	2.4	0.25	0.27

[a] Adair and Adair report an organic phosphorus in their lipoprotein of 0.67 per cent, corresponding to 16.7 per cent phospholipide if all the phosphorus was from this source. Only about half this amount was found in lipide extracts, however. A fatty acid determination indicated that 20.4 per cent of the lipoprotein was fatty acid. This is higher than could be explained on the basis of cholesterol ester and phospholipide, which might indicate some free triglyceride in their fraction, unless their phospholipide figure is low, as discussed above.

[b] The composition reported here is that obtained in a recent analysis by Dr. Robert Rosenberg of Fraction IV-1,1 from normal human plasma.

(Courtesy of Oncley and Gurd in Blood Cells and Plasma Proteins Their State in Nature, edited by James L. Tullis, Memoirs of the University Laboratory of Physical Chemistry Related to Medicine and Public Health, Harvard University, No. 2, p. 338, Academic Press Inc., New York, New York, 1953.)

duplicated by the ethanol fractionation procedure at low temperatures.[77] The similar lipide–protein relationships observed in a variety of experimental procedures indicate that selective interaction between lipides and certain proteins must have existed under physiological conditions.

The earlier analytical studies have given us information as to the types of lipide (Table III) and their distribution in the plasma fractions. Considerable variation in the amounts of lipides found will be noted. However, a remarkable constancy in the ratio of esterified to "free" (non-esterified) cholesterol appears in human plasma and its fractions.

Very surprisingly, Macheboeuf's data on a lipoprotein [cenapse acide (CA)] from horse blood shows no free cholesterol.[179,180] This is true in spite of the fact that horse serum contains more than 20 per cent of its cholesterol in the free form.[181]

Several explanations for this suggest themselves: First, Macheboeuf states that dog serum produces no CA when treated by his procedure.[180] Therefore the difference may be a species peculiarity. However, he has also shown that addition of 7 per cent alcohol, or less, is sufficient to labilize the lipide–protein link in plasma since it markedly accelerates the rate of extraction by ether.[182] Thus in the pooled human plasma a redistribution of the lipides may have occurred prior to, or during, processing.

The most likely explanation of this phenomenon follows Macheboeuf's finding that the amount and composition of the lipoprotein are dependent on the age of the horse serum, even 3 days standing at 4°C. prior to fractionation producing profound changes in the direction of higher cholesterol and lower phospholipide contents of CA.[183] This is in accord with Vandendriessche's report that the free cholesterol and phospholipide content of horse serum *decrease* on sterile incubation at 37°C. while the inorganic phosphorus increases[184] (labile phosphoric ester?). Thus, Macheboeuf's lack of free cholesterol may result from the length of time between drawing the blood and analysis. It would be interesting to know whether Vandendriessche's observation explains the presence of

(179) M. Macheboeuf, *Bull. soc. chim. biol.* **11,** 268, 485 (1929).

(180) M. Macheboeuf and P. Rebeyrotte, "Lipoproteins," *Discussions Faraday Soc.* **6,** 92 (1949).

(181) O. Muhlbock, *Acta Brevia Neerland. Physiol. Pharmacol. Microbiol.* **7,** 76 (1937) [*C. A.* **31,** 6302 (1937)].

(182) M. Macheboeuf, *in* J. L. Tullis, Blood Cells and Plasma Proteins, Academic Press, New York, 1953, p. 358.

(183) M. Macheboeuf and P. Rebeyrotte, "Lipoproteins," *Discussions Faraday Soc.* No. **6,** 62 (1949).

(184) L. Vandendriessche, *Natuurw. Tijdschr.* (*Belg.*) **26,** 62 (1944); [*C. A.* **40,** 6510 (1946)].

both free and esterified cholesterol *in vivo*, representing an equilibrium between cholesterol ester hydrolysis and its formation from phospholipide.

a. Cenapse Acide

Macheboeuf's cenapse acide (CA) is prepared by adjusting to pH 3.9 horse serum which previously has been half saturated with ammonium sulfate and filtered. The small precipitate which forms at pH 3.9 is dissolved at pH 7 and reprecipitated at pH 3.9 repeatedly until constant composition is obtained.[179,183]

The resulting lipoprotein is soluble above pH 6.5, shows a single component upon electrophoresis in pH 7.7 phosphate buffer, and has a mobility almost identical with that of serum albumin.[183] It sediments more slowly in the ultracentrifuge (S = 2.5–4) than albumin, but this is due to its high lipide content and resulting low density; Macheboeuf has calculated a molecular weight of 85,000.[183] Cenapse acide contains approximately 18 per cent cholesterol esters and 23 per cent lecithin (CA from malnourished horses contained as little as 12 per cent lecithin and only 4 per cent cholesterol esters).[183] Taking Macheboeuf's estimate of the serum concentration of this component as 2.5 g./l.,[183] it follows that CA accounts for approximately ⅓ of the 77 mg. per cent cholesterol in horse serum.[181]

The possibility that CA is an artifact produced by the rather rigorous treatment of pH 3.9 in half-saturated ammonium sulfate solution has been considered in some detail by Macheboeuf,[179,183] since the rather peculiar solubility characteristics of CA (insoluble below pH 6.5) are similar to those of "altered" serum albumins. However, Macheboeuf has been able to produce CA by the much milder treatment of precipitation at pH 5.5.[183] Furthermore, he has shown that CA may be "denatured" by precipitation with concentrated phosphate solutions.[182] A precipitate forms which is rich in lipides and which is insoluble even at pH8 until the lipides have been removed. The proteins remaining in solution after phosphate treatment are heterogeneous by electrophoresis.[182] In contrast to the lability of plasma lipide toward oxidation, he reports striking stability for the isolated CA. This is in marked contrast to the experience of Gurd *et al.* with isolated β_1-lipoprotein who find that it undergoes rapid autoxidation.[185]

b. β_1-Lipoprotein (X-Protein)

β_1-Lipoprotein of human plasma has been obtained in a state of high purity by ultracentrifugation of its concentrate (Cohn's fraction III-0) in

(185) J. L. Oncley, F. R. N. Gurd, and M. Melin, *J. Am. Chem. Soc.* **72,** 458 (1950).

a high density medium (1 *M* NaCl)[185,186] in which it rises away from the other proteins. So purified, it appears homogeneous in the ultracentrifuge. It appears to be a spherical highly solvated molecule of molecular weight 1.3×10^6 (anhydrous). Its composition is shown in Table III and schematically in Fig. 8 which dramatizes the extremely small portion of water-solubilizing groups. In spite of this composition the molecule is remarkably like a protein in physicochemical character being soluble

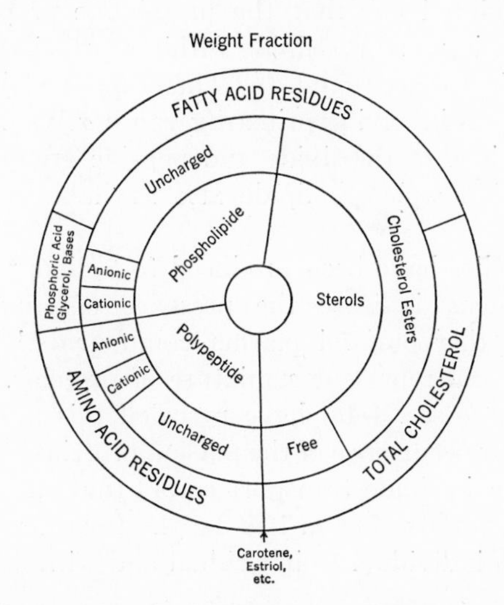

FIG. 8. Composition of β-lipoprotein. (Courtesy J. L. Oncley.)

in dilute salt solutions at the isoelectric point. Extraction of the lipide by cold procedures left a protein residue which was then soluble in distilled water.[185]

The instability of this lipoprotein places sharp restrictions on its adequate study: it may not be frozen or dried without denaturation, the lipide oxidizes readily, and no satisfactory antioxidants have been found.[186] This air-oxidation appears to affect first the traces of carotenoids present and eventually the unsaturated fatty acid esters. It is accompanied by marked changes in the absorption spectrum and by increased molecular density and electrical mobility (at pH 8.6).

Instability also appeared during attempts to measure the colloid osmotic pressure of this protein: high, unstable values indicating dissociation of the lipoprotein.[61] Similar difficulties were met with Mache-

(186) G. S. Adair and M. E. Adair [*J. Physiol* **102,** 17 (1943)], have prepared a crude lipoprotein fraction, probably B_1, by precipitation between 50 and 60 per cent saturated $(NH_4)_2SO_4$.

boeuf's CA.[187] In contrast, even lipemic serum presents no difficulties in osmotic-pressure measurements.[188] Whether instability is only a manifestation of oxidative changes or whether the previous alcohol fractionation has labilized the protein cannot be determined until anaerobic preparation of the β_1-lipoprotein has been carried out.

Some details of β-lipoprotein structure are evolving which appear important for interpretations of the physicochemical properties. Thus McFarlane has pointed out that the proteinlike properties cannot be explained by enclosing a ball of lipide within a protein sheath as there is not enough protein to go around.[189] Therefore Oncley and Gurd have proposed interactions involving primarily van der Waals' forces between the nonpolar portions of the lipide and peptide groupings, leaving the charged groups of the phospholipide also available on the lipoprotein surface.[185]

Specific antibodies have been produced to β_1-lipoprotein. However, precipitin phenomena indicate the presence of *several* antigens not identifiable with other purified plasma components.[190] Consequently, this fraction may still represent a mixture of proteins. Immuno-histochemical studies (see sec. I-4*b*) have revealed this protein to have the highest intracellular occurrence of the plasma proteins studied, suggesting it may be involved in lipide transport across the cell membrane. Normally present in amounts of from 1 to 2 g./l. of plasma, its concentration varies widely from individual to individual and with age and sex.[191,192]

c. α-Lipoproteins

The α-lipoproteins, first separated electrophoretically by Blix *et al.*,[178] have received less attention than the β-lipoproteins, partly because their solubility characteristics result in the coprecipitation of other proteins (including albumin) and partly because the low pH (5.1 in method 6) required for precipitation might have caused some alteration. In support of this, a marked tendency for the solubility of the α-lipoproteins to decrease continuously during repeated precipitation at pH 4.5, or below, was observed. The proteins purified at this pH bore solubility characteristics reminiscent of Macheboeuf's CA. With the aid of zinc ions these proteins together with the albumins can be precipitated at pH 5.8,

(187) M. Macheboeuf and G. Sandor, *Bull. soc. chim. biol.* **13,** 745 (1931).
(188) G. Popjak and E. F. McCarthy, "Lipoproteins," *Discussions Faraday Soc.* No. **6,** 97 (1949).
(189) A. S. McFarlane, *Discussions Faraday Soc.* **6,** 74 (1949).
(190) D. Gitlin, *Science* **117,** 591 (1953).
(191) L. A. Lewis, A. A. Green, and I. H. Page, *Am. J. Physiol.* **171,** 391 (1952).
(192) D. P. Barr, E. M. Russ, and H. A. Eder *in* J. Tullis, Blood Cells and Plasma Proteins, Academic Press, New York, 1953.

or higher, and the method has proved satisfactory for analytical purposes which make use of the lipide content of the precipitate.[193]

These proteins have a much lower cholesterol to phospholipide ratio (approximately equimolar amounts) than the β_1-lipoprotein (Table III). The cholesterol is remarkably labile being readily extracted from plasma by standing with ether. Surprisingly, "denaturation" by heating to 60°C. makes the cholesterol unextractable.[194]

According to Page *et al.*,[192] the principal component of α-lipoprotein is an α_1-component. Normally present in amounts of 1–2 g./l., its concentration varies quite as much as that of β_1-lipoprotein.

An additional lipoprotein component intermediate between β_1 and α_1 in flotation by Gofman's technic has been designated α_2-lipoprotein by Page *et al.*[191] It is present in the much smaller amounts of 0.1–0.2 g./l.[195]

d. Gofman's Components–Lipide Transport

The studies of Gofman *et al.*[196] (ultracentrifugal analysis in high density solutions) have revealed a variety of molecules rising both slower and faster than β_1-lipoproteins and point to the possibility of isolating the α-lipoproteins as well by upward sedimentation. Such methods should provide much better materials for further chemical studies, although the possible alterations caused by high ion concentrations must be kept in mind, particularly if ions are used which are known to interact strongly with proteins. This may be true of bromide ion as used by Page to produce the density of 1.21 for separation of all of the lipoproteins of plasma.[191]

The faster rising components of Gofman appear to form a series progressing up to visible fat droplets and contain appreciable amounts of neutral fats.[197] This may reflect the emulsifying power of lipoproteins for fats and thus demonstrates a transport function as postulated by Frazer.[198] Further evidence of their transport function may be found in their small but significant contents of fat-soluble vitamins and hormones, and in the rapid equilibration of their cholesterol with cellular

(193) W. F. Lever, F. R. N. Gurd, E. Uroma, R. K. Brown, B. A. Barnes, K. Schmid, and E. L. Schultz, *J. Chem. Invest.* **30,** 99 (1951).
(194) W. Batchelor, personal communication.
(195) L. A. Lewis and I. H. Page, *Circulation* **7,** 707 (1953).
(196) J. W. Gofman, F. T. Lindgren, and H. A. Elliott, *J. Biol. Chem.* **179,** 973 (1949).
(197) H. B. Jones, J. W. Gofman, F. T. Lindgren, T. P. Lyon, D. M. Graham, B. Strisower, and A. V. Nichols, *Am. J. Med.* **11,** 358 (1951).
(198) A. C. Frazer, *Physiol. Revs.* **26,** 103 (1946); *Discussions Faraday Soc.* **6,** 81 (1949).

cholesterol. This was demonstrated by equilibrating plasma with erythrocytes.[199]

IV. Protective (Anti-Infection) Functions—Antibodies

The protection of the vascular system from invasion is primarily the responsibility of external membranes. When these fail, as for example in a wound, blood pressure generally forces out fluid thus preventing contamination. However, in the case of invading organisms even small inocula may prove fatal unless an intravascular method of combat exists. Consequently, the organism has developed complex mechanisms for dealing with these, including specialized cells (leucocytes) and proteins (antibodies and complement). For a detailed discussion of this, the reader is referred to Chap. 22 (Boyd). However, since these components represent an appreciable fraction of the plasma components (γ-globulins alone account for 11 per cent of the plasma protein), their nonimmunological properties will be discussed here. (See also reviews by Campbell,[200] Marrack,[201] Smith and Jager,[202] and Haurowitz.[203])

1. Definition of Antibodies

Antibodies arise in response to stimulation by a suitable foreign substance (an antigen—usually a protein or polysaccharide) which has been introduced into the vertebrate's body. They are produced by certain cells of the reticulo-endothelium (of which the plasma cells at present appear most likely to be involved), and can be demonstrated in the circulation a week or more after the innoculation of the antigen. Antibodies are recognized by their ability to combine avidly with the original antigen and usually show extreme specificity, not reacting with even closely related molecules. As Landsteiner has shown,[203a] this specificity appears to be directed toward specific groupings which may be highly localized on the antigen's surface. The essential features necessary in such groupings in order to elicit the immunological response appear to be (*1*) the presence of one or more charged groups, (*2*) a relatively rigid spatial relationship between various parts of the antigenic group, and

(199) J. A. S. Hagerman and R. G. Gould, *Proc. Soc. Exptl. Biol. Med.* **78,** 329 (1951).
(200) D. H. Campbell and F. Lanni, *in* D. M. Greenberg, Amino Acids and Proteins, 2nd ed., C. C Thomas, Springfield, 1951, p. 649.
(201) J. R. Marrack, *Ann. Repts. Progr. Chem.* (Chem. Soc. London) **48** (**1951**), p. 2.
(202) E. L. Smith and B. V. Jäger, *Ann. Rev. Microbiol.* **6,** 207 (1952).
(203) F. Haurowitz, *Biol. Revs. Cambridge Phil. Soc.* **27,** 247 (1952).
(203*a*) K. Landsteiner, The Specificity of Serological Reactions, 2nd ed. Harvard University Press, Cambridge, 1945.

(*3*) an over-all configuration which is foreign to the cell producing the antibody.[204]

Combination of antibody and antigen can be recognized in a variety of ways (see Chap. 22) of which the most useful to the chemist include (*1*) change in reactivity of antigen (such as loss in enzyme function), (*2*) change in charge of antigen or antibody (i.e., by electrophoretic analysis),[205] and (*3*) change in degree of aggregation of immune components. This is usually recognized by an extreme decrease in solubility, i.e., the formation of a precipitate (precipitin reaction). However, at the high dilutions usually employed, such precipitation may require appreciable lengths of time and much earlier stages of aggregation may be followed by light-scattering measurements[206] or by ultracentrifugal analysis.[207]

The high insolubility of such precipitates and their ready solubility in the presence of excess antigen (and also in some instances, in excess antibody) has led Heidelberger,[208] Marrack,[209] Pauling[210] and others to postulate bivalent antibodies which form chains or lattices of alternating antigen and antibody molecules.[211] (Such chains would be terminated in small soluble units when antigen was in excess.) Antibodies can frequently be shown to neutralize the antigen's activity without the forma-

(204) Certain of the animal's own proteins which are highly isolated, such as the crystalline lens, may thus produce antibodies upon injection.[204a] Furthermore, immune responses of this type against the animal's own proteins have been proposed as an explanation for certain diseases such as rheumatic fever, lupus erythymatosis, and glomerular nephritis.[204b] Sensitization may here have been caused by a foreign antigen, such as a component of streptococcus, which produces an antibody cross-reacting with the patient's own tissues.

(204*a*) H. Gideon Wells, The Chemical Aspects of Immunity, The Chemical Catalog Co., New York, 1929, pp. 31, 71, 72; L. Hektoen, *J. Infectious Diseases* **31,** 72 (1922).

(204*b*) A. R. Rich, *Harvey Lectures* **42,** 106 (1947).

(205) J. R. Marrack, H. Hoch, and R. G. S. Johns, *Brit. J. Exptl. Pathol.* **32,** 212 (1951).

(206) D. Gitlin and H. Edelhoch, *J. Immunol.* **66,** 67 (1951).

(207) S. J. Singer and D. H. Campbell, *J. Am. Chem. Soc.* **73,** 3543 (1951). J. L. Oncley, E. Ellenbogen, D. Gitlin, and F. R. N. Gurd, *J. Phys. Chem.* **56,** 85 (1952).

(208) M. Heidelberger and F. E. Kendall, *J. Exptl. Med.* **61,** 563 (1935).

(209) J. R. Marrack, Chemistry of Antigens and Antibodies, 2nd ed., Great Britain Medical Research Council, Spec. Rept., Series No. 230, 1938.

(210) L. Pauling, *in* K. Landsteiner, The Specificity of Serological Reactions, 2nd ed., Harvard Univ. Press, Cambridge, 1945; L. Pauling, *J. Am. Chem. Soc.* **62,** 2643 (1940).

(211) A review of this subject has recently been given: A. M. Pappenheimer, Jr., The Nature and Significance of the Antibody Response, Columbia University Press, New York, 1953, p. 111.

tion of a visible precipitate. Such "incomplete" antibodies would then be described in the above theory as "univalent" since they cannot take part in chain formation. Whether in fact small aggregates of univalent antibody with antigen might be sufficiently insoluble to explain the precipitin reaction can not be predicted from our present knowledge of protein solubility, particularly when the structure of the complex in terms of the neutralization of solubilizing groups is considered. However, the presence of at least a few bivalent antibody molecules linking the primary aggregates would appear attractive.

While the multivalency of antibody is thus not firmly established, the multivalency of antigen appears more certain since specific precipitates may contain more than 20 moles of antibody per mole of antigen (see Chap. 22). Thus unless one assumes nonspecific adsorption of specific antibody, multiple sites on the antigen seem to be necessary.

Since these multiple sites on the antigen are probably not identical in structure, an antibody preparation even purified as a specific precipitate to a *pure protein* may still contain a *variety of antibody molecules* directed toward different groupings.[200]

2. Heterogeneity

Thus it is seen that antibodies may be heterogeneous in terms of the nature of their interaction (valency) and also in terms of the particular antigenic configuration against which they are directed. Further discussion of heterogeneity of antibodies might appear fruitless as it could readily be imagined that antibodies directed toward different antigenic groupings would vary widely in their physicochemical properties. However, sufficient similarities in molecular weight, electrophoretic mobility, solubility, and composition exist to make it profitable to discuss these components in terms of the homogeneity of the plasma fractions in which they are contained.

The possibility exists that all of the plasma components classed as γ-globulins by electrophoresis have antibody function. At least in the newborn, the plasma antibody and γ-globulin levels acquired from the mother decline together in the first weeks after birth and the subsequent appearance of *natural* immunity follows soon after the appearance of the individual's own γ-globulins.[212,213] Furthermore, in some pathological conditions lacking γ-globulins, antibodies are also lacking.[214] The ques-

(212) A. S. Wiener, *J. Exptl. Med.* **94,** 213 (1951).
(213) J. L. McGirr, *Vet. J.* **103,** 345 (1947).
(214) B. Schick and K. W. Greenbaum, *J. Pediat.* **27,** 241 (1945); E. G. Krebs, *J. Lab. Clin. Med.* **31,** 851 (1946).

tion remains: Is antibody formation an "altered" γ-globulin synthesis in the presence of antigen as postulated by Haurowitz or are *all* γ-globulins antibodies? While *known* antibody titers can account for only an infinitesimal fraction of the γ-globulins in normal sera,[215] the number of unknown components may be indefinitely large. Perhaps of greater significance is the inability of certain individuals with multiple myeloma to produce antibodies while producing enormous quantities of apparently normal γ-globulins.[215a]

Some difficulty arises in attempting a precise definition of γ-globulins since more refined measurements of electrophoretic mobility by reversible boundary-spreading technique (Chap. 6) reveal a continuous spectrum of mobilities of components whose relative amounts approximate a Gaussian probability curve with an isoelectric midpoint at pH 7.3 for human γ-globulin (Chap. 6). Partial resolution of these differently charged components has been achieved by fractional precipitation at the edge of the isoelectric zone[216] and by the Cann-Kirkwood electrophoresis convection apparatus.[217]

Sometimes two Gaussian envelopes of mobilities are observed in the region of the electrophoretic diagram usually assigned to the γ-globulins. These have then been called, γ_1- and γ_2-globulins. This phenomenon is particularly common in hyperimmune horse plasma where the faster component (the γ_1-globulin) is also designated as T-globulin. γ_1-Globulin may also correspond to fraction III-1 of human plasma,[177] containing the isohemagglutinins and certain antibodies such as typhoid 0 agglutinin, which has been designated as β_2-globulin by Oncley.[215]

The presence of the T-globulin in hyperimmune sera has been correlated with (*1*) the nature of the injected antigen (carbohydrate versus protein), (*2*) the stage of immunization, (*3*) the method of immunization —whether intravenous or interstitial injection of antigen and hence "cellular" versus "circulating" antibody production, and (*4*) the nature of the antibody formed—whether precipitating or "incomplete" univalent antibody.[218] However, these are not "all or none" effects, and antibody to a given antigen is usually found in both γ_1- and γ_2-compo-

(215) J. L. Oncley, *in* J. L. Tullis, ed., Blood Cells and Plasma Proteins, Academic Press, New York, 1953, p. 180.

(215*a*) D. L. Larson and L. J. Tomlinson, *J. Lab. Clin. Med.* **39,** 129 (1952).

(216) Such a fractionation has been observed in Cohn's method 9 between fractions II-1,2 and II-3 (see footnote 177).

(217) J. R. Cann, J. G. Kirkwood, R. A. Brown, and O. J. Plescia, *J. Am. Chem. Soc.* **71,** 1603 (1949).

(218) G. Edsall, *in* A. M. Pappenheimer, Jr., The Nature and Significance of the Antibody Response, Columbia Univ. Press, New York, 1953, p. 69.

nents so that the value of the above generalizations is limited.[219–221] It might seem attractive to assume that these two electrophoretic types of antibody have different sites of origin in the tissues and that their production will thus reflect the properties and consequent fate of the antigen. Nevertheless, both γ_1- and γ_2-globulins show great similarity in chemical properties and in amino acid composition,[222] (Table IV) suggesting that they may be more closely related than their electrophoretic differences indicate.[223]

TABLE IV

COMPOSITION OF IMMUNE PROTEINS[202]

The data are given as grams of constituent per 100 grams of anhydrous protein. The values for bovine γ_2-globulin are averages for two preparations.

Constituent	Human γ-globulin II-1 (110)	Human γ-globulin II-1,2 (222)	Human γ-globulin II-3 (222)	Bovine γ_2-globulin (222)	Bovine T-(γ_1)-globulin (222)	Equine γ_2-globulin (222)	Equine T(γ_1)-globulin (222)
Arginine	4.8	5.1	3.7	5.8	4.8	3.8	2.8
Aspartic acid	8.8						
Cystine (+ cysteine)	3.1	2.6	2.7	2.9	2.8	2.6	2.5
Glutamic acid	11.8						
Glycine	4.2						
Histidine	2.50	2.01	1.91	2.05	2.01	2.44	2.43
Isoleucine	2.7	2.8	2.0	3.2	3.0	4.4	3.3
Leucine	9.3	9.3	9.5	7.4	8.6	9.0	7.5
Lysine	8.1	7.2	6.3	6.7	6.4	8.6	6.7
Methionine	1.06	1.12	0.87	1.18	1.00	0.95	0.72
Phenylalanine	4.6	4.5	4.7	3.2	4.5	4.4	4.1
Proline	8.1						
Serine	11.4						
Threonine	8.4	8.8	7.4	10.0	9.5	11.1	8.7
Tryptophan	2.86	2.6	2.8	2.6	2.6	2.7	2.8
Tyrosine	6.75					6.8	6.8
Valine	9.7	9.7	9.7	10.0	9.5	10.1	10.4
Hexose		2.3	2.3	2.1	2.5	2.5	2.6
Hexosamine		1.27	1.23	1.31	1.50	1.18	1.53

(219) D. Gitlin, C. S. Davidson, and L. H. Wetterlow, *J. Immunol.* **63,** 415 (1949).
(220) M. Faure, R. Larny, and M. H. Coulon, *Ann. inst. Pasteur* **74,** 19 (1948).
(221) J. J. Perez and C. Mazureke, *Compt. rend. soc. biol.* **144,** 1639 (1950).
(222) E. L. Smith, R. D. Greene, and E. Bartner, *J. Biol. Chem.* **164,** 359 (1946); E. L. Smith and R. D. Greene, *J. Biol. Chem.* **171,** 355 (1947).
(223) J. R. Cann, D. H. Campbell, R. A. Brown, and J. G. Kirkwood [*J. Am. Chem. Soc.* **73,** 4611 (1951)] have reported a component moving *slower* than the normal γ-globulin in rabbit serum hyperimmune to *p*-azophenylarsonic acid protein.

While in mammalian plasma all antibodies appear to travel as γ-globulins (or as slow β-globulins), in chicken plasma antibodies of considerably higher mobilities are encountered (α-globulins).[224] The consequent high charge on these antibodies may explain the peculiar solubility properties of their specific precipitates[225] which at physiological pH are only formed at high salt concentrations such as 0.5 M NaCl. (That is, the formation of specific precipitate at pH 7.5 with a chicken antibody isoelectric at pH 5 might correspond to the formation of a specific precipitate with mammalian antibody at pH 9.)

The molecular weights of many mammalian antibodies appear to lie in the 150,000 range ($s_{20} = 7\ S$). However, heavier components ($s_{20} = 9\text{–}11\ S$ and $s_{20} = 18\text{–}20\ S$) have also been observed. Thus Kabat reported that horse antibody recovered from a specific precipitate with antipneumococcus had a molecular weight of one million ($s_{20} = 20\ S$).[226] Petermann and Pappenheimer[227,228] also obtained antibody in this molecular weight range and showed it could be "depolymerized" by peptic digestion to $s_{20} = 5.2\ S$ components, half of which material was inactive (i.e. nonprecipitable with antigen) and the other half contained *twice* the original specific reactivity. Similarly, Kabat showed that antibody recovered from specific precipitates by $Ba(OH)_2$ was "depolymerized" without loss of activity.[226]

Thus the concept arises that antibodies may consist of a 150,000 mol. wt. unit or polymers of this. (The small $s = 5.2$ unit has only been observed in peptic digests.) In fact, some of the polymers may be artifacts. Thus Heidelberger *et al.* have shown that horse antibodies to rabbit γ-globulin are so altered by chemical fractionation that several times as much nonspecific nitrogen is carried down in the specific precipitate when purified fractions are used instead of the original serum.[229] Similarly, human γ-globulin, purified as fraction II, is reported to contain up to 25 per cent of heavy components ($s_{20} = 9\text{–}20\ S$),[61] whereas Cann *et al.*[230] have shown that human γ-globulin separated by electrophoresis–convection is almost entirely free of such components. (For further discussion see *stability* under following sec. IV-4.)

(224) H. F. Deutsch, J. C. Nichol, and M. Cohn, *J. Immunol.* **63,** 195 (1949).
(225) H. R. Wolfe and E. Dilks, *J. Immunol.* **61,** 251 (1949); M. Goodman, H. R. Wolfe, and S. Norton, *J. Immunol.* **66,** 225 (1951).
(226) E. A. Kabat, *J. Exptl. Med.* **69,** 103 (1939).
(227) M. L. Petermann and A. M. Pappenheimer, *J. Phys. Chem.* **45,** 1 (1941).
(228) A. M. Pappenheimer, Jr., ed., The Nature and Significance of the Antibody Response, Columbia Univ. Press, New York, 1953, p. 111.
(229) M. Heidelberger, R. C. Krueger, and H. F. Deutsch, *Arch. Biochem. and Biophys.* **34,** 135, 146 (1951).
(230) J. R. Cann, R. A. Brown, S. J. Singer, J. B. Shumaker, Jr., and J. G. Kirkwood, *Science* **114,** 30 (1951).

With regard to artifacts, it should be borne in mind that antibody preparations have frequently been treated in ways not conducive to keeping the protein unaltered. Thus a heating step (56°C. for ½ hour) has frequently been employed for the purpose of destroying complement (for further discussion, see Chap. 22), and incubations of specific precipitates at 37°C. (or higher) are usually employed to aid flocculation. Furthermore, owing to the difficulty of obtaining potent antiserum, antisera are frequently stored for months (or even years) prior to study.

Nevertheless, some studies are hard to interpret except on the basis that native antibody globulins exist which are larger than the principal $s_{20} = 7\ S$ component. Thus Heidelberger and Pedersen[231,232] showed that horse antiserum to type I pneumococcus lost part of its heavy components ($s_{20} = 10$–$20\ S$) by treatment with specific polysaccharide, and Kabat[233] claims that all of the precipitating antibody in one horse serum could be removed in a preparative ultracentrifugal cell by sedimenting just long enough to remove the $s_{20} = 18\ S$ component. However, the age and past history of these sera were not indicated. Antibody recovered by salt dissociation of specific precipitates of horse and pig sera showed a preponderance of a "homogeneous" $s_{20} = 18\ S$ component, rather than a series of components of varying mobilities. This appears hard to explain as an artifact produced by storage or purification processes. Thus it would appear that ungulate antipneumococcus sera differ from other antipneumococcus sera and also differ from other antisera of the ungulate.[234] (A smaller partial specific volume (0.715) has also been reported for horse antipneumococcus antibody.[233] See also Table VI.)

A particularly significant piece of evidence establishing the interrelation of different antibodies and the total γ-globulin fraction comes from the studies of Porter who has shown by the dinitrophenol (DNP) technique of Sanger (see Chap. 2) that rabbit antiovalbumin isolated as the specific precipitate has the same terminal amino acid sequence as rabbit γ-globulin; namely, alanyl-leucyl-valyl-aspartyl.[235] Such a result is in keeping with Pauling's postulate of specific antibodies as arising from a single "unspecialized" globulin by specific folding of the polypeptide chain complementary to the antigen's surface.[210] However, Porter's report may oversimplify the situation since human γ-globulin has been reported to contain at least 7 different terminal amino groups in *fractional*

(231) M. Heidelberger and K. O. Pedersen, *J. Exptl. Med.* **65**, 393 (1937).
(232) K. O. Pedersen, Ultracentrifugal Studies on Serum and Serum Fractions, Almquist and Wikesells Boktryckeri, Uppsala, 1945.
(233) E. A. Kabat, *J. Exptl. Med.* **69**, 103 (1939).
(234) R. A. Kekwick and B. R. Record, *Brit. J. Exptl. Pathol.* **22**, 29 (1941).
(235) R. R. Porter, *Biochem. J.* **46**, 473 (1950).

amounts, suggesting that human antibodies may be polydisperse in regard to end groups.[236]

3. Purification of Antibodies

The most elegant methods of antibody purification make use of its specific precipitation by its antigen. For this purpose the proper amount of antigen to give complete precipitation is usually determined in advance by suitable pilot experiments. Following the mixing of antiserum (or plasma) with antigen, an incubation period of some minutes (or hours) at 37°C. is frequently employed to hasten aggregation of the specific precipitate, and this may be followed by one or more days standing at 0°C. to complete flocculation. The specific precipitate is generally so insoluble that it may be repeatedly washed without loss of antibody. Such washed precipitates appear to contain *only* antibody protein and antigen (perhaps also complement). However, the separation of antigen from antibody poses a major problem, which has only been solved in certain special instances. For a detailed discussion of these see Campbell and Lanni.[200]

Since any antiserum contains a spectrum of antibodies with varying reactivities toward a given antigen, a portion of the antibody may be recovered by adsorbing under conditions where union is most avid and then eluting under conditions where union is looser. Thus adsorption at 0°C. followed by elution at 40–60°C. has been frequently used to recover a small portion of the activity. More efficient methods, recovering up to 25 per cent of the activity of antipneumococcus globulin have been developed by Heidelberger and Kendall using 15 per cent NaCl as the eluting agent.[237] Obviously, these methods may not isolate truly "representative" antibody but rather that fraction with the weakest affinity for the antigen. However, they possess the advantage of simplicity in that the antigen, remaining insoluble with the rest of the antibody, may be readily removed.

If conditions can be found under which a dissociated specific precipitate can be chemically fractionated, more quantitative recovery of antibody may be obtained. However, dissociating conditions usually require extremes of pH, and the antibody may then be separated from the denatured antigen as in Campbell's method for purifying antiovalbumin by incubating the specific precipitate at room temperature and

(236) H. Van Vunakis. Ph.D. Dissertation, Columbia Univ., New York, 1951; Microfilm Abstracts (University Microfilms,) Ann Arbor, Mich., 1951, Vol. 11, p. 822.

(237) M. Heidelberger and F. E. Kendall, *J. Exptl. Med.* **64,** 161 (1936); see also: W. F. Goebel, P. J. Olitsky, and A. C. Saenz, *J. Exptl. Med.* **87,** 445 (1948).

pH 3 and then filtering off the denatured antigen. He reported yields and purities of up to 90 per cent by this method.[200] However, it is obviously a rather drastic way to treat a protein.

When the specificity of the antibody is directed against known haptenic groupings as in antibodies to artificial antigens, the specific precipitate may be dissociated under much milder conditions by a high concentration of small molecules containing the haptenic group. The hapten–antibody complex may then be fractionated from the antigen by suitable precipitants and the hapten–antibody complex may be dissociated by dialyzing away the small hapten molecule. This technique has been used successfully by Campbell *et al.*[238] for the purification of *p*-azophenylarsonic acid antibody from rabbit serum. They obtained over 90 per cent yields of antibody which was up to 98 per cent precipitable by antigen.[239]

A more promising general method involves fixing the antigen irreversibly to a suitable solid surface, adsorbing the antibody on this, and then eluting it from its specific complex. This procedure possesses the advantages of more rapid reaction (since aggregation is no longer necessary) and a permanent solid phase so that dissociation automatically permits separation. Furthermore, it may be used to purify nonprecipitating antibodies. Coupling of bovine serum albumin to cellulose has been accomplished by reacting the cellulose with *p*-nitrobenzyl chloride followed by reduction and diazotization of the aromatic amine and then coupling this to the serum albumin.[240] Human serum albumin has been coupled to a synthetic resin containing acid chloride groups.[241] The resulting antigen powder was then used to adsorb antibody either by suspension in the antiserum or chromatographically on a column.

In these cases partial elution of the antibody from the specific adsorbent was achieved with acid. However, a somewhat similar instance involving adsorption of isohemagglutinins on red cell surfaces has been reversed under much milder conditions by some of the sugar components of the blood group substance. Thus Isliker has recovered up to 60 per cent of the isohemagglutinins adsorbed on the stroma of cells by elution with galactose or lactose at 37°C.[241,242]

(238) D. Campbell, R. H. Blaker, and A. B. Pardee, *J. Am. Chem. Soc.* **70,** 2496 (1948).

(239) F. Karush [*Federation Proc.* **12,** 448 (1953)] has described a more elegant procedure using a more insoluble antigen (diazotized fibrinogen) permitting complete separation at neutral pH by ammonium sulfate precipitation.

(240) D. Campbell, E. Luescher, and S. G. Lermann, *Proc. Natl. Acad. Sci. U. S.* **37,** 575 (1951).

(241) H. Isliker, *Ann. N. Y. Acad. Sci.* **57,** 225 (1953).

(242) H. Isliker, *Intern. Congr. Biochem. Abstr. of Communs. 2nd Congr.*, p. 393.

Separation of the antibodies as a class (i.e., the γ-globulins) from other components of plasma has been accomplished by various methods. Since γ-globulins represent only 11 per cent of the total proteins of human plasma, this permits at once a five- to tenfold increase in the purity of the antibodies and also removes substances which may lower the preparation's stability (i.e., lipoproteins) or have other adverse properties. In the case of hyperimmune sera, where a specific antibody may represent most of the immune protein, relatively high concentrations of precipitating antibodies have been achieved.[243]

Electrophoretic separation of the γ-globulins, first achieved by Blix *et al.*, has been carried out on a larger scale by Cann, Kirkwood, *et al.*[217] in the electrophoresis–convection apparatus (see Chap. 6). In this way an 85 per cent yield of γ_2-globulins of 50 per cent purity was achieved from bovine serum on the first passage through the apparatus.[244] The impurities (mostly γ_1-globulin) were removed by a second separation in the apparatus thereby achieving a 75 per cent yield of γ_2-globulins of 88 per cent purity.

If the component to be purified consists of a mixture of proteins of closely related mobilities these may be further resolved in the apparatus by a longer treatment. Thus, starting with fraction II from human antipertussis plasma (a γ_2-globulin of over 95 per cent purity and with a mean isoelectric point of 7.5) a series of fractions was obtained with mean isoelectric points ranging from 7.0 to 7.9.[245] Some separation of immune activities was also observed, a greater protective action being found in the fraction of most alkaline isoelectric point.

Since the γ-globulins have the most alkaline isoelectric points of the plasma proteins and since protein solubility increases rapidly away from the isoelectric point, separation of these components from the remaining plasma proteins on the basis of solubility is relatively easy, and a variety of satisfactory methods have been devised. Of historic interest are the methods of (*1*) dialysis against water (or dilution with water) giving a euglobulin precipitate and (*2*) salting-out with ammonium sulfate.[246] Both methods are far from ideal. The former operates in a region where protein–protein interaction is enhanced due to the very low ionic strength. Furthermore, only a portion of the γ-globulins are sufficiently insoluble to be precipitated, leaving considerable γ-pseudoglobulin in solution.

(243) B. Chow and W. F. Goebel, *J. Exptl. Med.* **62,** 179 (1935).

(244) J. R. Cann, R. A. Brown, and J. G. Kirkwood, *J. Am. Chem. Soc.* **71,** 1609 (1949).

(245) J. R. Cann, R. A. Brown, J. G. Kirkwood, and J. H. Hink, Jr., *J. Biol. Chem.* **185,** 663 (1950).

(246) L. D. Felton, *J. Infectious Diseases* **43,** 543 (1928).

The second method operates in a region where the pH dependence of solubility is at a minimum so that separation on the basis of protein charge is minimized. Furthermore, the large amounts of salt present can only be removed by a tedious dialysis. Nevertheless, γ-globulin of high purity and in good yield has been so produced.[247]

The separation of γ-globulin of high purity in good yield has been extensively investigated by Cohn's group as part of his inclusive fractionation system. For details the reader is referred to the paper of Oncley *et al.*[177] A description of method 9 also appears at the end of this chapter. In general it was found that a good separation of γ-globulin from the β-lipoprotein could not be obtained in the primary precipitation from plasma. Therefore a fraction II + III, designed to contain all of the γ-globulin (excepting a few per cent lost in the preliminary fibrinogen precipitate—fraction I, and traces remaining in fraction IV) and most of the β-lipoprotein, was precipitated at pH 6.9 $0.09\Gamma/2$, and 25% ethanol at $-5°C$. The resulting simplified protein system was then further fractionated (Fig. 15) by going to a very low ionic strength at pH 7.2 to extract the β-lipoprotein followed by an extraction at pH 5.2, $0.015\Gamma/2$, and 17 per cent alcohol to extract the γ-globulin from remaining proteins of more acid isoelectric points such as the isohemagglutinins. The γ-globulin was then reprecipitated at pH 7.4 and 25 per cent ethanol. Such procedures have resulted in γ-globulin preparations of over 95 per cent purity (mobility of γ_2-globulin $= 0.8\text{–}1.5 \times 10^{-5}$ in pH 8.6, $0.1\Gamma/2$ veronal buffer) in yields of up to 80 per cent of that present in plasma. Additional antibodies, particularly of certain agglutinating types, are found in the residue from the γ-globulin extract. These have been called γ_1-globulins by Deutsch *et al.*, and a method of fractionating them from β-globulins and remaining γ_2-globulins has been described.[248] Such preparations have the immunological properties of fraction III-1 (method 9) although appearing much more homogeneous in electrophoresis (up to 95 per cent γ_1-globulin).

The application of methods 6 and 9 of Cohn's alcohol fractionation scheme to other animal species, both normal and hyperimmune has also been studied. Thus Smith fractionated bovine plasma, obtaining quite pure γ_1- and γ_2-globulins.[249] As already mentioned, these showed strikingly similar chemical composition, differing only in their contents of phenylalanine.[222,250] In the case of bovine fetal plasma, the γ-globu-

(247) H. Svensson, *J. Biol. Chem.* **139,** 805 (1941).
(248) H. F. Deutsch, R. A. Alberty, and L. J. Gosting, *J. Biol. Chem.* **165,** 21 (1946).
(249) E. L. Smith, *J. Dairy Sci.* **31,** 127 (1948); *J. Biol. Chem.* **164,** 345 (1946).
(250) The striking difference between γ_1 and γ_2 bovine globulin with regard to

lins were missing and appeared only following birth and the ingestion of colostrum.

The plasma of horses hyperimmune to tetanus toxin studied by method 6 showed an even wider spread of antitoxic activity. Large amounts of activity appeared in fraction IV, which contained only traces of γ-globulin by electrophoretic analysis. The slowest mobility in this fraction of any component present in appreciable amount (21 per cent) was -3.1×10^{-5}.[251]

The use of zinc ions as a precipitant for γ-globulins has been reported by Kunkel, who developed an analytical method on this basis.[252] However, his criteria for efficacy were based primarily on a comparison of yields of precipitate with the yields obtained in other analytical methods. More extensive studies of this system have shown that zinc ions show little promise of being more specific than the alcohol process already described, and that a complete isolation will only be achieved by a multiple-step process.[193]

More promising leads for efficient fractionating reagents would appear to be certain organic anions.[253,193] Thus acid phthalate has been used in a simple method of isolating horse antibody to type I pneumococcus polysaccharides.[254] In this process the horse antiserum was first diluted with water according to Felton.[246] The precipitate (60 per cent precipitable by specific polysaccharide) was dissolved in 0.9 per cent NaCl to give a 3 per cent protein solution, and remaining inert proteins were precipitated by the addition of 0.2 *M* pH 3.6 potassium acid phthalate at 20°C. The amount of acid phthalate required varied from 0.4 to 0.7 volumes with different lots of serum. In this way up to 90 per cent yields of over 98 per cent precipitable antibody were achieved. The most soluble portion of this antibody (that precipitating above 0.17 *M* ammonium sulfate) could be crystallized although the crystals always appeared imperfect, and usually became worse upon attempts at recrystallization. Although completely precipitable with type I pneumococcus polysaccharide, it showed a smaller *in vivo* protective action than antibody dissociated from the specific precipitate. Also, it proved unstable,

---

ultraviolet absorption [E. L. Smith and N. H. Coy, *J. Biol. Chem.* **164,** 367 (1946)] appears worthy of further study. Beaven and Holiday [*Advances in Protein Chem.* **7,** 382 (1952)] believe that the difference is due to tyrosine rather than phenylalanine as postulated by Smith.

(251) E. L. Smith and T. D. Gerlough, *J. Biol. Chem.* **167,** 679 (1947).

(252) H. G. Kunkel, *Proc. Soc. Exptl. Biol. Med.* **66,** 217 (1947).

(253) J. L. Oncley, F. H. Gordon, F. R. N. Gurd, and R. A. Lontie, Abstract of paper presented at the Division of Biological Chemistry, 119th Meeting of the American Chemical Society, Boston, Mass., April 1–5, 1951, p. 250.

(254) J. H. Northrop and W. F. Goebel, *J. Gen. Physiol.* **32,** 705 (1948).

denatured material forming continuously during the crystallization procedure. This may have been due to the low pH of 3.6 used in the phthalate isolation procedure.

4. Physical and Chemical Properties

The *physicochemical properties* of the antibody fractions have been most thoroughly investigated[61] in the form of the γ-globulins (fraction II) obtained from pooled normal human plasma. As already indicated, these contain a principal component with $S_{20} = 7.2$ from which a molecular weight of approximately 150,000 could be calculated by assuming an intrinsic viscosity of 0.06 and a partial specific volume of 0.739. (Osmotic-pressure measurements gave a molecular weight of 156,000. However, this value should be too large by the amount of $S_{20} = 10$ component present.) These data also permitted calculating a length of 230 A. and an axial ratio of 6:1.[61] Human γ-globulins have an average isoelectric point of 7.3 in $0.1\Gamma/2$ cacodylate–NaCl buffer.[255]

Fraction II is extremely heterogenous in solubility tests, and subfractions separated by repeated partial precipitation have varied a thousandfold in solubility.[253] This polydispersity is also demonstrated upon dialysis against distilled water, whereupon a portion precipitates (euglobulin fraction) leaving the pseudoglobulin in solution. However, the fractionation is not sharp since further dialysis of the pseudoglobulin fraction usually produces more euglobulin. More euglobulin may also be obtained by first subjecting the pseudoglobulin fraction to some other fractionating procedure, such as alcohol or salt precipitation, and then redialyzing the fractions so obtained. Thus it is extremely difficult to define these components in terms of solubility. In general, they show the usual solubility properties of proteins, being salted-in at low ionic strengths (in the range 0–0.1) and salted-out at high ionic strengths (i.e., greater than 2). They show solubility minima near to and related to the isoelectric point. Since all precipitates contain molecules with a range of isoelectric points, they contain presumably both isoelectric molecules and salt complexes of oppositely charged proteins.

The stability of immune proteins depends to a considerable degree on the method of assay. Some specific reactivity frequently remains in preparations which would otherwise be considered quite denatured. However, subtle changes in immunological properties are frequently the first changes observed in such preparations. Thus the "avidity" of the antibody (the rate of the secondary flocculation reaction) appears extremely sensitive to unfavorable treatment. Similarly, the precipitat-

(255) R. A. Alberty, E. A. Anderson, and J. W. Williams, *J. Phys. & Colloid Chem.* **52,** 217 (1948).

ing action of an antibody may be destroyed before its *in vivo* protective action has been affected.

The stability of antibodies is markedly affected by the presence of other proteins. Thus heating whole serum produces a much greater effect on the precipitin reaction of horse antiserum to type I pneumococcus than does a similar heat treatment of the euglobulin fraction (containing the antibody) separated from it.[256] Heat treatment of mixtures of casein and antipneumococcal euglobulin similarly affected the precipitin reaction more than equivalent heat treatment of the separated fractions, and loss of precipitability ran parallel to the appearance of a new electrophoretic component of mobility intermediate between that of casein and γ-globulin. Furthermore, antibody (γ-globulin) isolated electrophoretically from partially inactivated mixtures showed the original reactivity.[257]

Urea denaturation (8 *M*) of rabbit antiserum to ovalbumin similarly produced a preparation which gave larger precipitates with ovalbumin than did the original serum, whereas urea treatment of the purified antibody showed no such effect.[258] Rabbit antiserum to pneumococcus exposed to pH 4 or lower also produced a larger amount of specific precipitate,[259] as did horse antiserum to rabbit γ-globulin under conditions frequently used in antibody purification. One wonders whether the high "purity" reported for some antibody preparations may not also be an artifact resulting from the coupling of impurities to the antibody so that they are determined with it.

Thus antibodies appear to show considerable tendency to polymerize with other proteins without loss of specific reactivity, although the ability to precipitate may be affected. The explanation for this may lie in the resultant change in net charge on the antibody molecule thus affecting the solubility of the specific precipitate at physiological pH. From this point of view, complexes might also be formed between antibody molecules themselves, but would not be detected since the net charge and consequent solubility and electrophoretic characteristics would not be appreciably altered. Many high-molecular-weight antibodies could thus be artifacts as suggested above. It would be interesting to know whether such polymerizations involve exchanges of —SH with S—S as found for albumin,[105] since γ-globulin also has small amounts of sulfhydryl.

The stability of concentrated solutions of γ-globulins (fraction II) has been studied by following changes in their turbidity or viscosity since these seem to precede measurable changes in immunological titers. (This may merely mean that the percentage of denatured material measured physicochemically is too small to affect the immune titers.) In this way it has been found that γ-globulins have a stability optimum in isotonic salt solutions near pH 7. They are stable for some hours at 50°C., and the stability is markedly increased by the addition of various substances of which sugars (e.g., lactose) and amino acids (e.g., glycine)

(256) R. K. Jennings and L. D. Smith, *J. Immunol.* **45,** 108 (1942).
(257) L. E. Krejci, R. K. Jennings, and L. D. Smith, *J. Immunol.* **45,** 111 (1942).
(258) G. G. Wright, *J. Exptl. Med.* **81,** 647 (1945).
(259) A. J. Weil, A. M. Moos, and F. L. Clapp, *J. Immunol.* **37,** 413 (1939).

have proved most effective.[260] Sugars used for preservation should not be reducing sugars since these give rise to the "browning reaction." Large organic anions, which proved so effective in albumin stabilization, have little effect on the γ-globulins.

Antibodies are remarkably stable (judged by specific reactivity) to extremes of pH as indicated by various procedures used in their purification. Thus rabbit antiovalbumin and horse antipneumococcus globulins have withstood pH values of 2.5 and 3.7, respectively.[200,243] Above pH 12 rabbit γ-globulin changes its physical properties (marked increase in viscosity) while retaining most of its specific (antibody) activity.[261] Similarly, horse antipneumococcus antibody remains active after barium hydroxide treatment.[226] These data suggest that salt linkages are unimportant in maintaining the structure of the specific combining site on the antibody molecule.

The stability of horse antibody to high concentrations of guanidine chloride[262] and of rabbit antibody to 8 *M* urea[258] also suggests that the peptide chain may be "uncoiled" reversibly in these proteins as in the serum albumins.

γ-Globulins are especially sensitive to surface denaturation. Complete denaturation has been achieved by repeatedly raising and lowering a glass rod in a γ-globulin solution.[263] This may explain why antibody films have usually been reported devoid of specific activity (in contrast to antigen films).[264]

The chemical composition[265] of these γ-globulins is given in Table IV (p. 712). They appear to be composed exclusively of amino acids and carbohydrates. Sugars and amino sugars are present in small but constant amounts with the usual ratio of two moles of hexose per mole of amino sugar (see sec. VIII). γ-Globulins contain considerably more hydroxy amino acids than most proteins, but otherwise their composition is not particularly distinctive (Fig. 9).

The chemical reactivities of these components, other than their specific reactivities, have been but little investigated. This is in striking contrast to the enormous amount of study on the chemical modification of antigens[203a] (see Chap. 22).

While, as already mentioned in discussing stability, salt linkages do not appear important for the maintenance of the immunologically

(260) P. M. Gross, Jr., personal communication.
(261) L. A. Sternberger and M. L. Petermann, *J. Immunol.* **67,** 207 (1951).
(262) J. O. Erickson and H. Neurath, *J. Gen. Physiol.* **28,** 421 (1945).
(263) R. Lontie, personal communication.
(264) A. Rothen, *Advances in Protein Chem.* **3,** 123 (1947).
(265) Chemical composition (Table IV) is taken from Smith and Jäger.[202]

specific configuration, they are important in the specific reaction. Thus, acid and base have been used to dissociate specific precipitates for preparative purposes (see above). Complete dissociation should be distinguished from the solubilization of the specific precipitates which occurs at more moderate pH values due to the introduction of sufficient charge on the antigen–antibody complex. However both effects may represent the same type of reaction in varying degrees. The specific reaction has been similarly weakened or destroyed by removing the

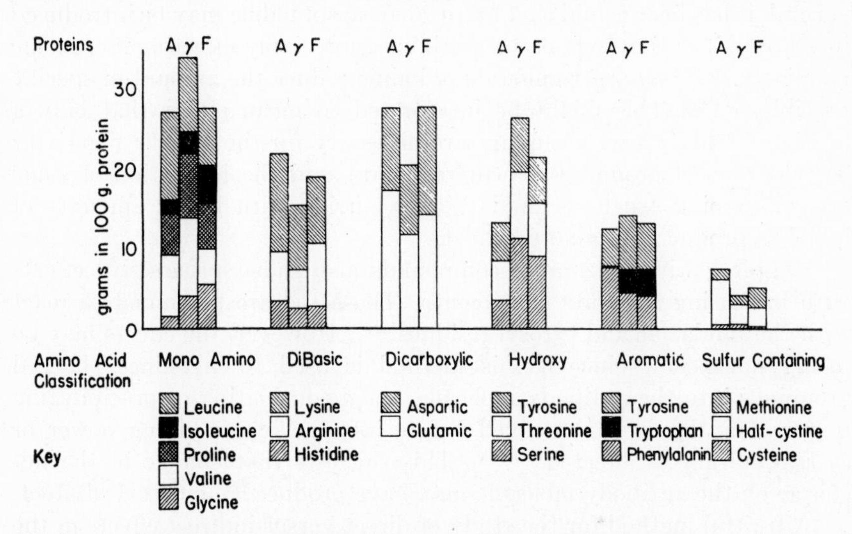

FIG. 9. Amino acid composition of some proteins (serum albumin, γ-globulin and fibrinogen) from human plasma. (Data from E. Brand.)

charge from the amino groups of the type I antipneumococcus by treatment with ketene.[266]

One or two molecules of certain isocyanates, which presumably react with amino groups,[267,268] may be introduced per antibody molecule without affecting the specific reaction.[269] This is the basis of Coon's method of preparing fluorescent antibodies for use as histological stains in following antigen distribution in tissues.

Ninhydrin, which is commonly used in estimating α-amino groups,[270]

(266) B. F. Chow and W. F. Goebel, *J. Exptl. Med.* **62,** 179 (1935).
(267) S. J. Hopkins and A. Wormall, *Biochem. J.* **27,** 740, 1706 (1933).
(268) H. Fraenkel-Conrat, *J. Biol. Chem.* **152,** 385 (1944).
(269) These studies[32] were carried out on a crude rabbit antibody to type III pneumococcus. However, unless the fluorescein was bound preferentially to the antibody, this fact should not affect the calculations.
(270) S. Moore and W. H. Stein, *J. Biol. Chem.* **176,** 367 (1948).

markedly increases the rate of flocculation of rabbit antisera.[271] However, in view of the small number of α-amino groups in rabbit antibody (1 per mole, $M = 150{,}000$),[235] it would seem probable that other groupings must also be attacked by this reagent.

Iodination has a marked effect on the specific reactivity of antibodies. It reacts presumably with the tyrosyl groupings, although, in view of the variation in relative reactivities of the tyrosyl residues in various proteins,[272] direct proof of this course of the reaction would be desirable. In general, it has been found that up to 20 atoms of iodine may be introduced into a rabbit antibody molecule without appreciably affecting its specific reactivity.[273,274] Larger amounts of iodine reduce the amount of specific precipitate.[275] This might be interpreted to mean either that only a portion of the tyrosyl groupings are necessary for the specific reactivity or that tyrosyl groupings are unimportant and the iodination of some other grouping (such as imidazole) occurring with larger amounts of iodine is producing the noted effect.

Coupling with diazonium compounds also shows progressive effects with increasing amounts of reagent. These too are presumed to react with the imidazole and tyrosyl residues.[276] However, the effects may be rather nonspecific since if this method is used to introduce charged groupings into the antibody molecule, the agglutinating or precipitating power of antibody is destroyed before its specific combining power or *in vivo* efficacy is affected.[277,278] This suggests that change in the net charge on the antibody molecule may have produced the observed effect.

A fruitful method for the study of direct versus indirect effects in the modification of specific groupings of antibodies has been suggested in a report by Pressman and Sternberger on the iodination of antibody in the presence and absence of its antigen.[275] They found that rabbit antibodies to ovalbumin *p*-azobenzene arsonate and to ovalbumin *p*-azobenzoate when iodinated in the presence of their haptenes (sodium benzene arsonate and sodium benzoate, respectively) required much larger amounts of iodine for inactivation.

The function of antibody[278a] may not be completed by its union with

(271) F. Tayeau, R. Pautrizel, and F. Leglise, *Abstr. 2nd Intern. Congr. Biochem. Abstr. of Communs.*, 1952, p. 108.
(272) C. H. Li, *J. Am. Chem. Soc.* **67,** 1065 (1945).
(273) D. Pressman and L. A. Sternberger, *J. Am. Chem. Soc.* **72,** 2226 (1950).
(274) S. Cohen, *J. Immunol.* **67,** 339 (1951).
(275) D. Pressman and L. A. Sternberger, *J. Immunol.* **66,** 609 (1951).
(276) H. Pauly, *Z. physiol. Chem.* **42,** 508 (1904).
(277) H. Eagle, D. E. Smith, and P. Vichess, *J. Exptl. Med.* **63,** 617 (1936).
(278) R. Coombs, L. S. Mynors, and G. Weber, *Brit. J. Exptl. Path.* **31,** 640 (1950).
(278*a*) The presence in serum of a "non-immune" bactericidal substance has been shown by: L. Pillemer, L. Blum, I. H. Lepow, O. A. Ross, E. W. Todd, and A. C. Wardlaw, *Science* **120,** 279 (1954).

antigen since further changes frequently follow in plasma through the action of complement (e.g., if the antigen is bacterial, lysis occurs of the bacterial cells added to the plasma). However, since very little is known of the chemical properties of complement,[278b] except that it consists of several protein components with markedly different chemical stabilities and solubilities, no further discussion will be given in this chapter. (However, see Chap. 22 and reviews by Heidelberger and Mayer[279] and by Pillemer.[280]) Heidelberger has shown that at least some of the components of complement combine in "stoichiometric" amounts with antigen–antibody precipitates. Perhaps these data should be reviewed in the light of the nonspecific coupling reactions of antibody mentioned above.

V. Clotting Components

Blood is of necessity a self-sealing fluid. The simplest mechanism for accomplishing this might be considered the use of blood cells to plug a leak, and such mechanisms predominate in the lower forms of life. However, beginning with the crustacea, more complicated systems appear, involving soluble plasma proteins.[281] Finally, in man the cellular mediation is relegated to the initiating action of the blood platelets, and the structure of the clot depends on one of the major protein constituents of plasma, fibrinogen. (See also Chap. 23.)

1. Fibrinogen

Fibrinogen, present to the extent of 3–4 g./l. of human plasma, is one of the least soluble of the plasma proteins and thus is readily concentrated by selectively precipitating it with small concentrations of salt or alcohol. In the classical method of Hammarsten, half-saturated sodium chloride has been used as a precipitant.[282,283] More recently, 8 per cent ethanol at pH 7 and 0°C. has been preferred (fraction I).[77] Lower ethanol concentrations can be used at more acid reactions, but the fibrinogen precipitate is no purer and is in fact more liable to spontaneous clotting on further fractionation. In the ethanol system, subsequent purification involved re-solution in citrate buffer (to prevent clotting due to the action of calcium or any prothrombin present), followed by fractional precipitation of the fibrinogen at two different pH values, separating it both from less soluble and more soluble impurities. The

(278*b*) The reader is referred to recent studies on the metal activation of complement by: L. Levine, K. M. Cowan, A. G. Osler, and M. M. Mayer, *J. Immunol.* **71,** 359 (1953).
(279) M. Heidelberger and M. M. Mayer, *Advances in Enzymol.* **8,** 71 (1948).
(280) L. Pillemer, *Chem. Revs.* **33,** 1 (1943).
(281) W. B. Hardy, *J. Physiol.* **13,** 165 (1892).
(282) O. Hammarsten, *Pflügers Arch. ges. Physiol.* **19,** 563 (1879).
(283) One molar ammonium sulfate is equally effective [L. Nanninga, *Arch. Neerland physiol.* **28,** 241 (1946)].

rationale for this method can be seen from the solubility diagrams (Fig. 10). In this way preparations containing over 98 per cent clottable nitrogen[284,285] have been obtained.[286]

Fraction I from bovine plasma, which is best precipitated at somewhat lower ethanol concentrations, has been further purified by ammonium sulfate fractionation for recent physicochemical studies.[287] This was accomplished[288] by dissolving dried bovine fraction I (Armour) in 0.1 *M* pH 6.4 phosphate buffer, then diluting with an equal volume of water, and letting stand overnight at 0°C. The precipitated impurities were filtered off, and the fibrinogen was precipitated with ⅓ volume of saturated ammonium sulfate. It was then dialyzed against 0.3 *M* KCl to remove the ammonium sulfate for physicochemical studies.

The molecular weight of fibrinogen has recently undergone major downward revision to 330,000 as the result of extensive studies at Wisconsin.[289,290] Proper evaluation of fibrinogen has been hampered by its instability. Aggregation in variable degrees (incipient clotting) usually appears to take place during purification, with consequent overestimation of the molecular weight. Therefore, these recent studies, giving the lowest values for the molecular weight and complete consistency between the entirely different technics of light scattering and sedimentation plus diffusion, would appear most reliable. (The low intrinsic viscosity of 0.25, which they obtained, also indicates their preparations to be as free of aggregates as the best previously described.[61])

The physical constants are given in Table V for bovine fibrinogen. The "best" value is followed by other values obtained for both human and bovine since these two proteins appear so similar that reported differences are presumably due to technical difficulties.

Physicochemical studies thus indicate fibrinogen to be a rod approximately 700 × 40 A. with a molecular weight of 330,000. Electron

(284) Owing to the splitting of small peptides (3–4 per cent of fibrinogen nitrogen) during clotting, as discussed in the text, previously published figures of "per cent clottable" have been increased by 3 per cent to allow for this as suggested by Shulman.[290]

(285) Assay of fibrinogen contents by determining the amount of clottable nitrogen may err considerably owing to occlusion of impurities unless adequate precautions are taken [P. R. Morrison, *J. Am. Chem. Soc.* **69,** 2723 (1947)].

(286) P. R. Morrison, J. T. Edsall, and S. G. Miller, *J. Am. Chem. Soc.* **70,** 3103 (1948).

(287) P. R. Morrison, S. Shulman, and W. F. Blatt, *Proc. Soc. Exptl. Biol. Med.* **78,** 653 (1951).

(288) K. Laki, *Arch. Biochem. Biophys.* **32,** 317 (1951).

(289) S. Katz, K. Gutfreund, S. Shulman, and J. D. Ferry, *J. Am. Chem. Soc.* **74,** 5706 (1952).

(290) S. Shulman, *J. Am. Chem. Soc.* **75,** 5846 (1953).

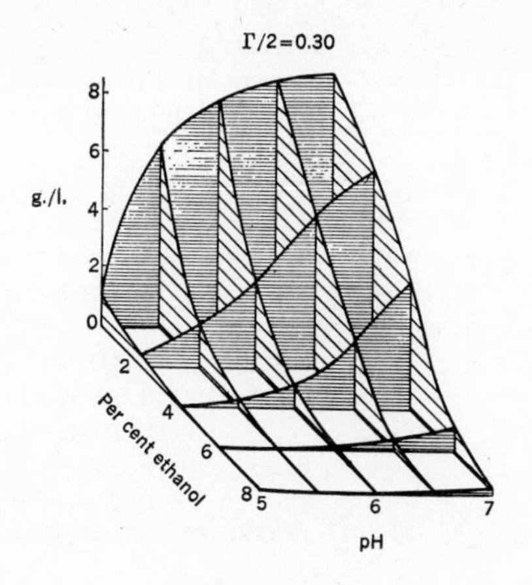

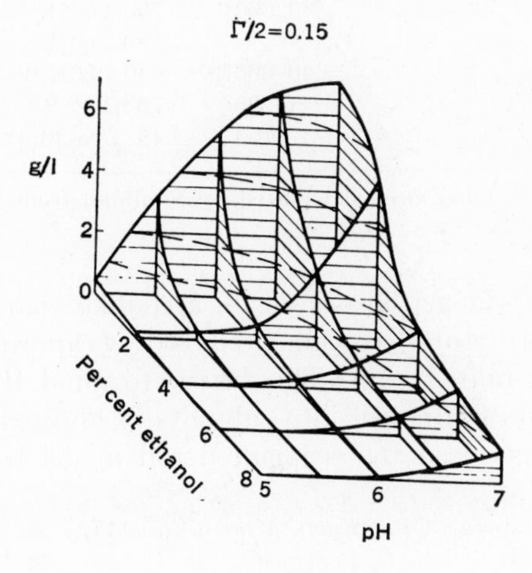

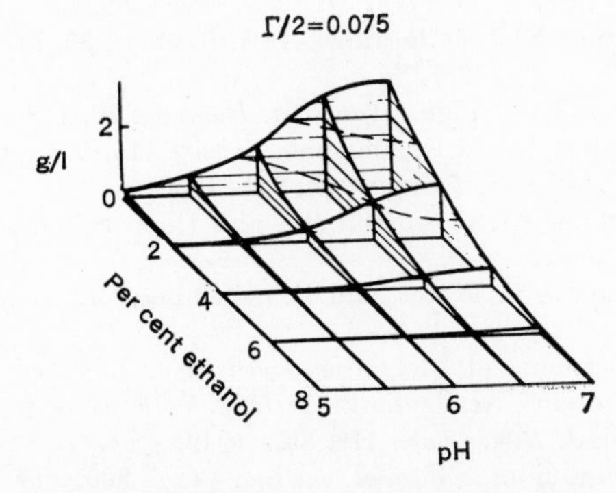

FIG. 10. Solubility of human fibrinogen as a function of ethanol concentration and pH at 0°C. (From Morrison *et al.*)[286]

TABLE V
PHYSICAL CONSTANTS OF FIBRINOGEN
(B = bovine; H = human)

	"Best" value	Other values
Sedimentation constant ($S_{20,w}$)	7.9B[290]	8.5–9H,[291] 9H[61]
		8.4–8.6B,[292] 8.4B,[293] 7.3–7.9B[294]
Diffusion constant, sq. cm./sec. ($\times 10^7$)	2.02B[290]	1.1H[291]
Partial specific volume	0.72	0.725H,[22] 0.723H[296]
		0.706B,[297] 0.717B[295]
Intrinsic viscosity, H_0	0.25	0.25H[61]
		0.25B,[290] 0.34B,[293] 0.25B[295]
Rotatory diffusion const., θ_{20} sec.$^{-1}$	39,000B[298]	35,000H[299]
Molecular weight		
Sedimentation and diffusion	330,000B[290]	700,000H[291]
Osmotic pressure		580,000H[61]
Light scattering	340,000B[289]	440,000H,[300] 407,000[298]
Molecular length, A.	700	700H,[299,61] 670B[298]
Axial ratio	20:1	19:1 to 30:1[a] B

[a] The value of 30:1 has been calculated by Shulman from his measurements of S_{20} and D_{20}.[290]

micrographs of fibrinogen also indicate a rodlike character and further suggest a fine structure, a segmented rod of approximately 100,000 molecular weight units.[301–303] This brings to mind Holmberg's finding that plasmin splits fibrinogen into subunits of approximately this same size. Shulman has used the segmented rod model to explain the dis-

(291) C. G. Holmberg, *Arkiv. Kemi, Mineral Geol.* **17A,** No. 28 (1944).
(292) V. L. Koenig and K. O. Pedersen, *Arch. Biochem.* **25,** 97 (1950).
(293) S. Shulman and J. D. Ferry, *J. Phys. Chem.* **55,** 135 (1951).
(294) V. L. Koenig and J. D. Perrings, *Arch. Biochem.* **36,** 147 (1952); *ibid.* **40,** 218 (1952).
(295) K. Bailey and F. Sanger, *Ann. Rev. Biochem.* **20,** 118 (1951).
(296) T. L. McMeekin and K. Marshall, *Science* **116,** 142 (1952).
(297) V. L. Koenig, *Arch. Biochem.* **25,** 241 (1950).
(298) C. S. Hocking, M. Laskowski, Jr., and H. A. Scheraga, *J. Am. Chem. Soc.* **74,** 775 (1952).
(299) J. T. Edsall, J. F. Foster, and H. Scheinberg, *J. Am. Chem. Soc.* **69,** 2731 (1947).
(300) P. R. Morrison and R. Lontie *quoted by* J. T. Edsall *in* Blood Cells and Plasma Proteins, Academic Press, New York, 1953, p. 123.
(301) C. E. Hall, *J. Biol. Chem.* **179,** 857 (1949).
(302) P. Kaesberg and S. Shulman, *J. Biol. Chem.* **200,** 293 (1953).
(303) For a very different interpretation of electron microscope data see: K. R. Porter and C. v. Z. Hawn, *J. Exptl. Med.* **90,** 225 (1949).

crepancies in frictional ratios calculated from his sedimentation plus diffusion and viscosity data.[290]

Fibrinogen is one of the least stable of the plasma proteins, being readily coagulated from plasma by heating at 56°C. (as in complement inactivation). However, the instability usually observed is due to its conversion to fibrin under the action of thrombin. Other substances have from time to time also been reported as clotting agents.[304] However, it is difficult to rule out the possibility that their action rather potentiates the action of traces of thrombin still present.

While the enzymatic nature of thrombin has long been assumed, difficulty is encountered in recovering the enzyme from clots presumably due to entrapment. Certainly it operates in minute amounts, being capable of converting at least 10^6 times its weight of fibrinogen to fibrin.

The reaction it catalyzes is now known to be the splitting from fibrinogen of peptides[305] representing 3–4 per cent of the total nitrogen.[306] This may explain the clotting action of certain proteolytic enzymes such as papain.[304] Sherry and Troll have discovered synthetic substrates for thrombin in the form of certain arginyl esters.[307] Tosyl arginine methyl ester is rapidly attacked with the liberation of methanol and this activity so closely paralleled the clotting activity in various preparations that they now propose it for the assay of thrombin activity. The proteolytic action of thrombin changes the *N*-terminal amino acids from glutamyl and tyrosyl (one of each in bovine fibrinogen) to glycyl and tyrosyl (in fibrin) in amounts suggesting that two glycyl end groups have replaced the terminal glutamyl residue.[308,309] However, the picture may be more complicated than this since at least two different peptides have been obtained from the clot liquor.[309]

The "clotting" of fibrinogen under the action of thrombin has been prevented by carrying out the conversion in 0.4 *M* hexamethylene glycol. In this medium aggregation is limited to the formation of soluble rods containing approximately 15 fibrinogen units.[310,311] Further, aggrega-

(304) J. H. Ferguson and P. H. Ralph, *Am. J. Physiol.* **138,** 648 (1948).
(305) K. Bailey, F. R. Bettelheim, L. Lorand, and W. R. Middlebrook, *Nature* **167,** 233 (1951).
(306) L. Lorand, *Nature* **167,** 992 (1951); K. Laki, *Science* **114,** 435 (1951).
(307) S. Sherry and W. Troll, *J. Biol. Chem.* **208,** 95 (1954).
(308) L. Lorand and W. R. Middlebrook, *Biochem. J.* **52,** 196 (1952).
(309) F. R. Bettelheim and K. Bailey, *Biochim. et Biophys. Acta* **9,** 578 (1952).
(310) S. Shulman and J. D. Ferry, *J. Phys. Chem.* **55,** 135 (1951).
(311) S. Shulman, *Discussions Faraday Soc.* **13,** 109 (1953), describes a variety of molecules which inhibit the clotting of fibrinogen in concentrations less than 0.5 *M*. These include, besides glycols, amides, amine hydrochlorides, and some large anions.

tion of *fibrin* appears to take place in two stages of which the first is reversible, since clots of purified fibrinogen can be dissolved in 3.5 M urea to give a protein of molecular weight similar to fibrinogen as judged by viscosity,[312] light-scattering[313] and ultracentrifugal measurements.[314] Fibrin may be reclotted from such urea solutions by dilution.[314,315]

However, clots as normally obtained from plasma are insoluble in urea.[315] This is possibly due to the formation of "irreversible" disulfide bonds between molecules (similar to the gelation of albumin in urea), since it can be prevented by the addition of a mercurial or iodoacetamide to plasma, whereupon clots remain soluble in urea.[316]

2. Prothrombin

Thrombin is obtained by the activation of a plasma precursor, prothrombin. This component (present to the extent of 70 mg. per liter of plasma) has been purified manyfold by an alcohol fractionation scheme. It was separated in fraction III-2, indicating it to be a relatively insoluble component of acid isoelectric point.[177] However, more convenient adsorption methods resulting in much greater purification are now available.

Thus by a procedure combining isoelectric precipitation with adsorption on magnesium hydroxide and salt fractionation, Seegers *et al.* have purified bovine prothrombin 700-fold over plasma. Alexander found barium sulfate to be a more specific adsorbent for prothrombin[317,318] permitting a high degree of purification of human prothrombin by adsorption and elution alone. A study of various reagents showed that only citrate was effective for the elution of prothrombin, whereas a variety of chelating agents for calcium removed inert proteins. Consequently, oxalated but not citrated plasma may be used for the adsorption step (also, plasma decalcified by resin treatment).

Seegers *et al.* found their best prothrombin to be a single component by the phase rule (the solubility being independent of excess saturating body).[319] A preparation containing $\frac{4}{5}$ of this activity was 80–90 per cent homogeneous by electrophoresis with an isoelectric point below

(312) E. Mihalyi, *Acta Chem. Scand.* **4,** 344 (1950).
(313) R. F. Steiner and K. Laki, *Arch. Biochem. and Biophys.* **34,** 24 (1951).
(314) S. Shulman, P. Ehrlich, and J. D. Ferry, *J. Am. Chem. Soc.* **73,** 1388 (1951).
(315) L. Lorand, *Nature* **166,** 694 (1951); K. Laki and L. Lorand, *Science* **108,** 280 (1948).
(316) A. G. Loewy, *quoted by* Edsall *in* Blood Cells and Plasma Proteins, Academic Press, New York, 1953, p. 133.
(317) B. Alexander and G. Landwehr, *Am. J. Physiol.* **159,** 322 (1949).
(318) J. Bordet and L. Delange [*Ann. inst. Pasteur* **26,** 657 (1912)] had used barium sulfate to render blood incoagulable.
(319) W. H. Seegers, E. C. Loomis, and J. M. Vandenbelt, *Arch. Biochem.* **6,** 85 (1945).

pH 5.0 (estimated 4.8).[320] It contained 4.6 per cent tyrosine, 3.3 per cent tryptophan, 0.96 per cent sulfur, and 4.3 per cent carbohydrate (orcinol test).[319] Laki *et al.* have just reported a complete amino acid analysis of bovine prothrombin.[321]

Bovine prothrombin, purified by Seegers, has been physico-chemically characterized by Lamy and Waugh,[322] who report a sedimentation constant of 4.84s, a diffusion constant of 6.24×10^{-7} an intrinsic viscosity of 0.041 and a partial specific volume of 0.70. From these data they calculated a molecular weight of 62,700 and axial ratios of 3.7 and 3.4 from sedimentation plus diffusion and viscosity data respectively. This molecule thus closely resembles albumin in regard to size, shape, and iso-electric point.

The relatively specific adsorption of prothrombin on alkaline earth salts may be related to the catalytic action of calcium in the conversion of prothombin to thrombin. However, the eluting action of citrate, which is out of proportion to its chelating ability, suggests that citrate may form a complex directly with prothrombin. Seegers also holds this view for the quite different reason that high citrate concentrations actually converted his best prothrombin preparations to thrombin.[323,324]

The mechanism of conversion of prothrombin to thrombin is still but poorly understood, although various elaborate theories involving more than a dozen additional components have been developed. A concise outline of some of these has been given recently.[325]

3. Plasminogen

Fibrin is readily digested by a variety of proteolytic enzymes. This has prompted the attempts to use trypsin *in vivo* for the dissolution of clots.[326] Plasma after treatment in appropriate ways contains an active fibrinolytic enzyme (fibrinolysin), which has also been called plasmin since it possesses proteolytic activity toward other proteins, as for example, casein.[327] While the simplest method of activation involves

(320) W. H. Seegers, E. C. Loomis, and J. M. Vandenbelt, *Proc. Soc. Exptl. Biol. Med.* **56**, 70 (1947).

(321) K. Laki, D. R. Kominz, P. Symonds, L. Lorand, and W. H. Seegers, *Arch. Biochem. and Biophys.* **49**, 276 (1954).

(322) F. Lamy and D. F. Waugh, *J. Biol. Chem.* **203**, 491 (1953).

(323) W. H. Seegers, R. T. McClaughry, and J. L. Fahey, *Blood* **5**, 421 (1950).

(324) The autocatalytic conversion to thrombin in solutions of 25% sodium citrate suggests a proteolytic mechanism, since Lamey and Waugh have observed the formation from homogeneous prothrombin of small inactive fragments, and large active fragments.

(325) E. C. Albritton, Standard Values in Blood, W. B. Saunders Co., Phila., 1952.

(326) I. Innerfeld, A. W. Schwartz, and A. A. Angrist, *Bull. N. Y. Acad. Med.* **28**, 537 (1952); H. G. Reiser, L. C. Roettig, and G. M. Curtis, *Surg. Forum, Proc. 36th Clin. Congr. Am. Coll. Surgeons* **1950**, 17 (Publ. 1951).

(327) L. R. Christensen and C. M. McLeod, *J. Gen. Physiol.* **28**, 363, 559 (1945).

merely shaking plasma with chloroform, more active preparations are obtained by treatment with a bacterial enzyme, streptokinase.[327]

Plasmin shows optimal activity near pH 7 like trypsin. However, it appears to be more specific in action since a substrate exhaustively attacked by this enzyme is further split by trypsin but not vice versa.[327]

The plasmin precursor, plasminogen, is one of the least soluble of the plasma proteins, being found in the residues after extracting the other components of Cohn's fractions I and II + III. It appears to have an acid isoelectric point, being precipitated best near pH 5. This has permitted some purification, following precipitation with fibrinogen, by selective extraction of the fibrin with acetic acid. The low solubility of this component would also seem to explain its ready adsorbability on bentonite or kaolin,[328] which unfortunately is not very specific.

The extreme stability of plasminogen toward acid and alkali (a property not found in the activated enzyme)[327] has been used for the preparation of fractions of the highest reported activity (425$\times$ serum) by selective denaturation of interacting contaminants.[329] Such preparations, which could be crystallized, were obtained first by acid extraction (0.05 M H_2SO_4) of Cohn's fraction III followed by adjustment to pH 11 for 3 min. Upon neutralization to pH 5.3, followed by readjustment to pH 2, most of the protein could be removed as a gel, and the now partially purified material was readily further purified by fractional precipitation at pH 6. Such material possessed the activity quoted above, which was not further enhanced by crystallization which took place in salt-free solution at pH 8.[329]

The function of plasminogen as a lytic agent for fibrin is far from proven. In fact no physiological mechanism of activation is known. However, the fact that fibrin clots slowly dissolve spontaneously is certainly suggestive.

VI. Enzymes

A large number of other enzymes have now been reported in plasma[330] for which no function (in plasma) is known.[331] The amounts present are so infinitesimal that the author believes they may largely represent leakage from cells which enter the blood either directly or via the lymphatics. Indeed, if one takes into account the continuous death of cells, the perme-

(328) L. F. Remmert and P. P. Cohen, *J. Biol. Chem.* **181,** 431 (1949).
(329) D. L. Kline, *J. Biol. Chem.* **204,** 949 (1953).
(330) E. J. Cohn, *Experientia* **3,** 125 (1947).
(331) Ceruloplasmin, discussed above is presumably a plasma enzyme, as are some of the components of complement. Hence enzymatic functions may be found in the protective components and elsewhere as well as in the clotting components.

ability of cells toward plasma protein (sec. I), and the relatively small size of many enzymes, it is surprising that more enzymes are not found. It is reasonable to suppose that with a suitable analytical procedure any cellular component can be identified in plasma and that the unusually sensitive analyses available for many enzymes have made a beginning of this possible.

The importance of enzyme investigations must not be underestimated since abnormal plasma levels of properly chosen constituents should be referable to pathological changes in the cells from which they originate and hence may have great diagnostic value. The relation between high serum acid phosphatase and carcinoma of the prostate is a case in point.[332]

VII. Hormones

The protein hormones (Chap. 20) are by definition transient components of plasma, molecules on their way from the site of formation (or storage) to the site of utilization. A classical way of demonstrating hormone action is by coupling the blood supplies of two individuals and showing that a stimulus in the first animal causes the characteristic response in the second.[333] Nevertheless, blood levels of hormones are so low that not only does blood make a very poor source of supply but, in fact, even assay may be impossible due to nonspecific effects produced by the very large doses which must be administered.

Hormones must enter and leave the circulation readily. Hence a study of their molecular sizes may be useful in interpreting capillary porosities. In line with such reasoning, the molecular weight of many purified hormones fall considerably below those of the plasma proteins (see Chap. 20).

Even in the case of insulin, where the molecular weight was first reported as 48,000, further study showed dissociation to a 12,000 molecular weight unit (actually a dimer of 6000) to occur rapidly. Consequently, in plasma, where the level is so low as to defy assay, aggregation can hardly occur. Another complication in using hormones to interpret capillary porosities arises from the methods used for their isolation, which frequently involve conditions known to be strongly hydrolytic. As a result, some workers consider the isolated substance to be an "active fragment." Certainly, the circulating hormone might be much larger.

VIII. Isolated Components of Unknown Function

Besides the components already discussed, several proteins have been isolated from plasma whose function is still obscure. The most

(332) T. J. Sullivan, E. B. Gutman, and A. B. Gutman, *J. Urol.* **48,** 426 (1942).

(333) This technique has recently been exploited by C. M. Williams to demonstrate the hormonal control of insect metamorphosis [*Federation Proc.* **10,** 546–52 (1951)].

abundant of these are the glycoproteins or mucoproteins.[334] The differentiation of plasma proteins on the basis of carbohydrate content is not very helpful since all the purified proteins except albumin contain hexoses, and most of those examined also contain hexosamines.[335] However, a few proteins of plasma contain much larger amounts of carbohydrate than the others so that fractionation may frequently be followed by analyzing for carbohydrate.

1. Carbohydrate-Containing Albumin (McMeekin)

In this way, using ammonium sulfate, McMeekin purified an "albumin" containing carbohydrate from horse serum.[339] (He called it albumin because of its occurrence with the albumins in ammonium sulfate fractionation, i.e., soluble in 2 M $(NH_4)_2SO_4$. This also served to relate it to the carbohydrate-containing "albumins" of Hewitt[340] and Kekwick[341] and to differentiate it from seromucoid—a truly globulin-like constituent.) This component was isolated from the mother liquors of albumin crystallizations and finally obtained in a crystalline form (hexagonal plates) which was quite distinct from the horse albumins (long, thin prisms). It appeared homogeneous by electrophoretic analysis in pH 7.7 and $\Gamma/2$0.2 phosphate buffer giving a mobility of 5.0. This is typical of an α-globulin. This component contained 15.1 per cent nitrogen and 5.5 per cent carbohydrate and must represent an appreciable fraction of the plasma protein—perhaps several times his reported yield of 1 per cent of the total protein because of the inherent losses involved in the laborious isolation procedure. Like seromucoid, this protein was not coagulated by heating at pH 4.8, but differed in being precipitated by 2 per cent trichloroacetic acid.

(334) K. Meyer [*Advances in Protein. Chem.* **2,** 249 (1945)] has defined the glycoproteins as those containing mucopolysaccharide held by a firm chemical bond but containing less than 4 per cent hexosamine. Whereas the mucoproteins are similarly constituted but with more than 4 per cent of hexosamine. He pointed out that this was an arbitrary division but reflected the appearance of marked solubility changes (more soluble in alcohol) with increasing carbohydrate content.

(335) The chemical bases for this statement are tests for (1) substances giving furfural when heated with acid (pentoses and hexoses give this reaction as judged by color formation in the presence of orcinol or carbazole)[336,337] and (2) substances giving a color with acetylacetone.[338]

(336) M. Sørensen and G. Haugaard, *Biochem. Z.* **360,** 247 (1933).

(337) F. B. Seibert and J. Atno, *J. Biol. Chem.* **163,** 511 (1946).

(338) J. W. Palmer, E. M. Smyth, and K. Meyer, *J. Biol. Chem.* **119,** 491 (1937).

(339) T. L. McMeekin, *J. Am. Chem. Soc.* **62,** 3393 (1940).

(340) L. F. Hewitt, *Biochem. J.* **31,** 360, 1047, 1534 (1937).

(341) R. A. Kekwick, *Biochem. J.* **32,** 552 (1938).

2. Seromucoid (Rimington)

Many of the mucoproteins are remarkable for their resistance to coagulation upon heating in aqueous solution. This may reflect unusual thermal stability or merely that the denatured form is also soluble in water. The latter phenomenon may be a result of the high carbohydrate content and a concomitantly large number of acidic groups. These may be either sulfuric or phosphoric acid esters of the sugar moiety. (Dische states that hexuronic acids are not present in plasma.)[342]

Heat coagulation of the bulk of the plasma proteins served as one of the earliest methods of isolating the serum mucoproteins, and was investigated in some detail by Rimington[343] who showed that at least two types of proteins were present: the more soluble (in aqueous alcohol) containing a large amount of carbohydrate and the less soluble (seromucoid) containing 11 per cent carbohydrate and one-half this amount of glucosamine.[344] Seromucoid (optical rotation, $(\alpha) = -54°$) appears to be the seroglycoid isolated by Hewitt.[345]

3. Acid Glycoprotein (M1 of Winzler and Mehl)

Proteins which remain soluble following heat treatment show several components. The most soluble of these may be obtained following removal of the bulk of the serum proteins with less drastic protein precipitants: i.e., 2.7 *M* ammonium sulfate at 4°C.[346] or 0.02 *M* Zn acetate in 19 per cent ethanol at $-5°C$.[347] to remove the bulk of the protein.

In the ammonium sulfate procedure, purification involves stepwise acidification at 2.7 *M* $(NH_4)_2SO_4$ with the removal of successive precipitates until pH 3.7 is reached. The mucoprotein remaining soluble at this point is then precipitated by saturated ammonium sulfate. Further purification involves refractionation at pH 3.7 with ammonium sulfate.[346]

In the zinc salt–ethanol procedure, after precipitation of the major components by 0.02 *M* Zn acetate in 19 per cent ethanol at pH 5.8 and $-5°C$., the remaining proteins are absorbed on zinc hydroxide. Further purification is accomplished by dissolving the zinc hydroxide with acetic

(342) Z. Dische, *J. Biol. Chem.* **167,** 189 (1947).
(343) C. Rimington, *Biochem. J.* **34,** 931 (1940).
(344) This ratio of 1 glucosamine to 2 hexose molecules (2 glucosamine + 4 mannose in egg albumin) [A. Neuberger, *Biochem. J.* **32,** 1435 (1938)] appears to be a fundamental building block in many mucoproteins.[334]
(345) L. F. Hewitt, *Biochem. J.* **32,** 26 (1938).
(346) H. E. Weimar, J. W. Mehl, and R. J. Winzler, *J. Biol. Chem.* **185,** 561 (1950).
(347) K. Schmid, *J. Am. Chem. Soc.* **72,** 2816 (1950); **75,** 60 (1953).

acid in aqueous ethanol (to pH 5.8) and then precipitating the α_2- and β_1-globulin impurities with small amounts of barium acetate. Following this procedure the protein has been crystallized as a lead salt.[347]

The proteins isolated by both methods in yields of 0.5 g./l. plasma appear to be identical in most respects, samples having been exchanged between the two groups of workers. A compilation of these properties from Refs. 346, 347, and 348 appears in Table VI. Differences observed in the sedimentation constant appeared to be due to the different ultracentrifuges used, and differences in the isoelectric point appeared to be due to the buffers used, very large specific salt effects being observed.[347]

This protein contains approximately 30 per cent carbohydrate, measured as hexose and hexosamine[346,347] and a complete amino acid analysis has been reported.[346] However, a fundamental and unresolved difference exists as to whether there are acid groups of phosphorus or of sulfur present.[349] And yet if they are present, whether the number of these is sufficient to explain the extremely acid isoelectric point seems doubtful in view of the large number of basic groups (30/mole). An alternative explanation can be sought in the apparently strong anion affinity of this molecule as observed in the effects of salts on the isoelectric point.[347]

This mucoprotein, which normally has the mobility of an α_1-globulin,[347,352] may be isolated electrophoretically at pH 4.5, where it carries a charge opposite to the remaining plasma proteins.[352] The amounts so observed in plasma seem to be in rough agreement with those obtained by chemical isolation procedures.[353,354]

(348) E. L. Smith, D. M. Brown, H. E. Weimar, and R. J. Winzler, *J. Biol. Chem.* **185,** 569 (1950).

(349) A recent report states that the acid groups contain neither phosphorus nor sulfur but that sialic acid,[350] identified by color reaction,[351] is present in large amount: L. Odin and I. Werner, *Acta Soc. Med. Upsaliensis* **57,** 227, 230 (1952).

(350) Sialic acid, $C_{14}H_{24}NO_{11}$, is a crystalline acid carbohydrate of equivalent weight 382 obtained from submaxillary mucin: G. Blix, *Z. physiol. Chem.* **240,** 43 (1936).

(351) G. Blix, L. Svennerholm, I. Werner, *Acta Chem. Scand.* **6,** 358 (1952).

(352) J. W. Mehl, F. Golden, and R. J. Winzler, *Proc. Soc. Exptl. Biol. Med.* **72,** 110 (1949).

(353) The electrophoretic pattern of mucoproteins obtained by acid precipitation of the bulk of the plasma proteins showed two additional components of higher isoelectric point than the isolated α_1-mucoprotein. This should explain the difficulty of Rimington and Hewitt in obtaining homogeneous fractions following heat coagulation of the bulk of the protein.

(354) H. Hoch and A. Chanutin have reported a component in *undialyzed* plasma moving faster than albumin in 0.15 *M* NaCl at pH 8 (or higher) in amounts of 0.2–0.5 per cent of the plasma proteins [*J. Biol. Chem.* **200,** 241 (1953)].

4. Fetuin

The use of electrophoresis as a method of separating proteins of varying carbohydrate content was explored by Blix *et al.*[178] Their values (carbohydrate contents of 7 per cent for α- and β-globulins vs. 3 per cent for γ-globulin and 1 per cent for albumin) indicate the presence of much more glycoprotein in plasma than we have thus far accounted for (in fact, the α_1 acid mucoprotein described above could easily hide in the albumin peak without accounting for all of the carbohydrate found there). Perhaps the more interesting observation is the wide distribution of carbohydrate, which has been found to be present in all of the fractions thus far studied, excepting albumin.

In fetal bovine serum, where proteins with the mobility of α-globulins become the major constituent,[251] a ready separation of such components is possible. Pedersen[355] accomplished this by precipitation with 0.4–0.45 saturated ammonium sulfate. This fraction proved relatively homogeneous in the ultracentrifuge but showed a large concentration dependence indicating considerable asymmetry. Pedersen calculated a molecular weight of 50,000, though Deutsch has recently given a molecular weight of 35,000.[356] This material was isoelectric at pH 3.5 and contained 12.5 per cent nitrogen and 0.6 per cent phosphate.[355] A recent report states it contains 9 per cent carbohydrate and 9 per cent glucosamine.[356] It does not appear to be the acid mucoprotein (M1) but might be the M2 component of Mehl *et al.* (an anionic shoulder on serum albumin in electrophoresis at pH 4.5). It might also be the seromucoid of Rimington[343] and the seroglycoid of Hewitt.[345]

The presence of fetuin in the fetal blood implies a considerable barrier in the placenta between the maternal and fetal blood stream. Further evidence for this is the very low concentration of albumin[251] and the absence of antibodies[357–359] in fetal bovine blood. Considerable differences exist in the comparative anatomy of the bovine and human pla-

(355) K. O. Pedersen, *J. Phys. Chem.* **51,** 164 (1948).
(356) H. F. Deutsch, *Federation Proc.* **12,** 196 (1953).
(357) M. L. Orcutt and P. E. Howe, *J. Exptl. Med.* **36,** 291 (1922).
(358) R. Aschaffenburg, S. Bartlett, S. K. Kon, P. Terry, S. Y. Thompson, D. M. Walker, C. Briggs, E. Cotchin, and R. Lovell, *J. Dairy Sci.* **3,** 187 (1949).
(359) The newborn calf obtains its antibodies from colostrum,[360] which contains anti-enzymes to permit absorption of antibodies prior to digestion.[361]
(360) The absorption of γ-globulin from colostrum was shown by R. G. Hansen and P. H. Phillips, *J. Biol. Chem.* **171,** 223 (1947).
(361) M. Laskowski, Jr. and M. Laskowski, *J. Biol. Chem.* **190,** 563 (1951).

centa.[362,363] In contrast to the multicellular barrier of the bovine placenta, the human placenta interposes but a single layer of cells between the maternal and fetal circulations so that permeability relations similar to those in the general capillary bed might be expected to exist. In agreement with this, human maternal and fetal blood are similar in composition not only in regard to major components, as judged electrophoretically,[364,365] but even with regard to certain antibody titers.[366] Such permeability also explains Rh sensitization of the human fetus (erythroblastosis foetalis). It seems unnecessary at present to invoke an active transport mechanism across the placental membrane when the relative permeability of proteins is so poorly understood.

Some differences in protein composition between maternal and fetal plasma have been noted in the human species. Thus the α- and β-globulin concentrations normally are elevated in the mother at the end of pregnancy and are higher than in the fetus. The explanation for this may lie in the presence of large molecules within the α- and β-envelopes. Thus the *β_1-lipoprotein* has a molecular weight of 1.3 million (see sec. III-3*b*), and the S_{20} component of plasma appears to move as an *α-globulin.* Thus neither of these should pass the placental membrane.

A similar picture seems to exist in antibody distribution, the fetal levels at term of the smaller antibodies (mol. wt. = 150,000) being generally similar to the maternal levels, whereas larger antibodies such as isohemagglutinins (mol. wt. = 500,000) have much lower fetal titers. Vahlquist[366] points out that the rate of transfer to the fetal blood stream may vary widely during the course of pregnancy, rising late in pregnancy and perhaps then falling again at term. This may explain some of the contradictions[366] to the above general picture. However, antibodies synthesized in the placental reticulo-endothethial system might also be shed directly into the fetal circulation.

An interesting study of the rabbit showed rapid transfer of homologous antibodies to the fetus.[367] The same workers found much slower transfer of heterologous γ-globulin to the fetus. However, these equilibria were studied by precipitin reaction and the authors did not prove specificity for the γ-globulin in the heterologous protein used. The heterologous protein disappeared extremely rapidly from the maternal blood as though

(362) A. Kutner and B. Ratner, *Am. J. Diseases Children* **25,** 413 (1923).
(363) J. H. Mason, T. Dalling, and W. S. Gordon, *J. Path. Bact.* **33,** 783 (1930).
(364) D. H. Moore, R. Martin Du Pan, and C. L. Buxton, *Am. J. Obstet. Gynecol.* **57,** 312 (1949).
(365) L. G. Longsworth, R. M. Curtis, R. H. Pembroke, *J. Clin. Invest.* **24,** 46 (1945).
(366) B. Vahlquist, *Etudes Neo-natales* **1,** No. 2, 31 (May 1952).
(367) S. G. Cohen, *J. Infectious Diseases* **87,** 291 (1950).

phagocytized. This implies a large molecular weight; therefore the antibody used for following distribution may have been directed against some impurity in the γ-globulin fraction injected.[368]

5. Electrophoretic Components—a Tally Sheet

Since Tiselius' fine development of electrophoretic analysis as a quantitative analytical tool, many workers have applied the method to plasma with generally concordant results.[67] (See also Chap. 6.) Most workers now rely on a 0.1 ionic strength sodium veronal (diethyl barbiturate) buffer at pH 8.6 since this has been found to give adequate resolution of α_1-globulin from albumin (Fig. 11). (See also Fig. 12 of

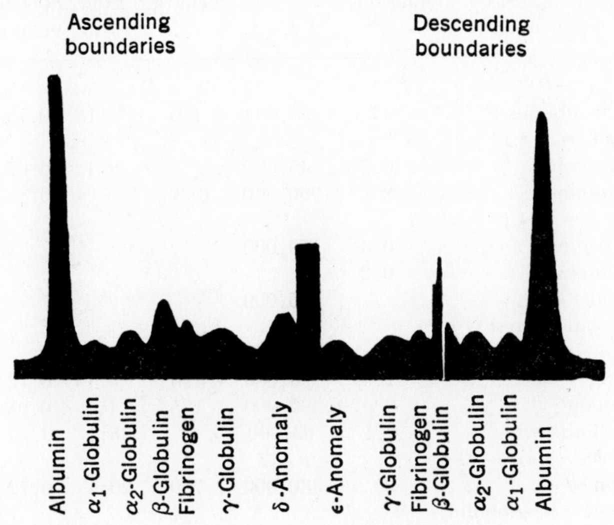

Fig. 11. Electrophoretic analysis of normal human plasma in 0.1 ionic strength veronal buffer at pH 8.6.

Chap. 6.) However, Hoch has shown that pH 8.8 phosphate buffer of 0.15 ionic strength is also satisfactory. It will be noted that fibrinogen (ϕ) falls between the β- and γ-globulins thus obscuring the β_2- or γ_1-component. In some mammalian species additional components appear of which the T-globulin (γ_1) of hyperimmune horse serum and a slower γ in hyperimmune rabbit serum have already been mentioned.[223] Other components so observed in electrophoresis include α_3-globulins in monkey, cat, and guinea pig,[369] and in calf plasma,[249] and a component in small

(368) In the earlier stages of gestation other routes may be involved: F. W. R. Brambell, W. A. Hemming, W. T. Rowlands, *Proc. Roy. Soc.* (*London*) **B135,** 390 (1948); **136,** 131 (1949); **137,** 239 (1950); **138,** 195, (1951).

(369) H. F. Deutsch and M. B. Goodloe, *J. Biol. Chem.* **161,** 1 (1945).

amount (less than ½ per cent of total protein) moving *faster* than albumin in human plasma.[354]

In this section it might be pertinent to sum up the protein components of plasma, recognized by chemical tests, in terms of these several mobility envelopes with the intention of pointing out which electrophoretic com-

TABLE VI
PROPERTIES OF PROTEIN COMPONENTS OF PLASMA
(Species is human unless otherwise noted)

Component	Concn. in plasma, g./l.	Mol. wt.	Length, A.	Axial ratio	Specific volume	$S_{20,w}$	Iso-electric pt.
Albumins μ = −5.9[a]	43						
Crystalline albumins	43	65,000	150	4:1	0.73	4.3	5.2[b]
$_1$-*Globulins* μ = −5.1	3						
α_1-Mucoprotein	0.5	45,000		8:1	0.67	3.5	3
α_1-Lipoprotein	2	(200,000)	(300)		(0.84)	(5)	(5.2)
α_2-*Globulins* μ = −4.1	7						
Ceruloplasmin	0.3	150,000					4.4
α_2-Lipoprotein	0.2						
Fetuin (calf)		35,000					
(Complement C[1] and C[2])							
β-*Globulins* μ = −2.8 to −3.5	11						
Transferrin	2.5	90,000	190	5:1	0.72		5.8
β-Lipoprotein	4	1,300,000	185	1:1	0.95	7	5.4
Prothrombin (bovine)	0.1	63,000		35:1	0.70		4.2
ϕ *and* γ_1-*Globulins* μ = −2.1	5						
Fibrinogen	3	330,000	700	20:1	0.72	8	(5)
(Antihemophylic globulin)							
Isohemagglutinins		(500,000)				9	(6)
γ_2-*Globulins* μ = −1.0	9	150,000	240	6:1	0.74[c]	7	6.5–8
Diphtheria antitoxin	10^{-3}						
Total plasma proteins	78						

[a] Electrophoretic mobilities[67] and electrophoretic concentrations[370] are those found in veronal buffer of pH 8.6 and Γ/2 0.1.

[b] Isoelectric point following "desalting" and fatty acid removal by ion-exchange resins or by dialysis and ether extraction.[68] Owing to ion-binding, isoelectric point is markedly affected by salts (=pH 4.7 in 0.15 Γ/2 acetate buffer).

[c] Bovine γ-globulin has a partial specific volume of 0.720 [M. O. Dayhoff, G. E. Perlmann, and D. A. MacInnes, *J. Am. Chem. Soc.* **74,** 2515 (1952)].

(370) G. E. Perlmann and P. Kaufman, *J. Am. Chem. Soc.* **67,** 638 (1945). The estimates of these workers have been chosen for this table since they investigated the effect of protein concentration (degree of dilution of the plasma) on the apparent amounts of components.

ponents are now largely accounted for, and which major components are still unidentified chemically. Examination of Table VI then shows:

1. Apparently all the albumin component (43 g./l.) tallies with the crystalline albumins (or with substances so closely related as to cross-react immunologically. However, no such cross-reaction to antialbumin serum has been observed with other components.[219] Trace components, of course, such as Hoch's fast component could readily be hidden in the experimental error.

2. Most of the α_1-component (3 g./l.) may be similarly related to the α_1-lipoprotein plus the α_1-mucoprotein. The figure of 2 g./l. for the α_1-lipoprotein was estimated by the author from the data of Lewis and Page[191] and may be in error by 50 per cent, owing to the wide individual variation they reported.

3. The α_2-proteins (7 g./l.), however, are largely unidentified, known components constituting less than 1 g./l. The α_2-globulins generally appear quite rich in carbohydrate and further glyco- or mucoproteins will probably emerge from their midst.

4. The β_1-globulins (11 g./l.) are made up in large measure of β-lipoprotein and transferrin, though approximately 4 g./l. is still unaccounted for.

5. The fibrinogen cannot be separated (electrophoretically) from the γ_1-(or β_2-)globulins. However, by clotting the fibrinogen, the γ_1-globulin can be made visible.[248] This component, like γ_2-globulin, contains antibodies. However, it differ from γ_2 in containing a preponderance of agglutinating antibodies such as the isohemagglutinins and typhoid O agglutinin.[177]

6. The γ_2-globulins (γ-globulins) (9 g./l.) contain most of the plasma antibodies, and it is conceivable that immune proteins make up the bulk of this envelope. However, components so identified, for example, diphtheria antitoxin (10^{-3} g./l.) make up an infinitesimal fraction of this component.

IX. Appendix—Inclusive Systems of Plasma Fractionation

While most procedures for isolating proteins are carried out with but a single component in mind, there is obvious merit in inclusive systems which will simultaneously isolate several components. Besides the economy in material effected, such procedures, stressing yields (or conservation of specific activity) and degrees of separation achieved, are especially useful for the development of the best conditions for fractionation. The study of protein–protein interactions, inherent in such an approach, may also shed light on the nature of the interaction *in vivo*.

However, even if one's interest is the isolation of but a single component, some study of the accompanying components is necessary since purity can only be defined in terms of *lack of specified impurities.* Moreover, in view of the complexity and lability of proteins one must always be on guard for subtle structural changes during an isolation procedure. Because of our limited knowledge of protein structure, evidence for lack of change must be largely inferred and is best based on the quantitative conservation of all of the properties of the component for which tests can be devised. This may be strengthened by proof that other labile components have also retained their activities through the procedures of fractionation.

In spite of the foregoing advantages, inclusive fractionation of so complex a system as plasma is so great an undertaking that it has but seldom been attempted until stimulated by the economic considerations involved in the large-scale fractionation of human plasma. However, it is now possible to obtain from the same plasma (*1*) albumin for the treatment of shock and albumin deficiencies, (*2*) γ-globulin for prevention and prophylaxis of certain contagious diseases, (*3*) isohemagglutinins for blood typing, (*4*) thrombin to promote blood clotting in open wounds, and (*5*) fibrinogen for the treatment of hemophilia (presumably its efficacy is due to an impurity "anti-hemophilic globulin") and for the preparation of fibrin clots, which have useful plastic properties in surgery —and still other clinically useful components[371] (Fig. 12).

Inclusive fractionations of plasma have been undertaken in the salting-out region of protein solubility using concentrated salt solutions and in the salting-in region using organic solvents or certain ionic precipitants such as heavy metals or large anions. In the interests of brevity it has seemed best to describe here the most detailed method (methods 6 and 9 of Cohn's fractionation scheme) which is given schematically in Figs. 13, 14, and 15, and then to discuss briefly the advantages and disadvantages of other methods with reference to this framework. This task is simplified by the fact that most of the plasma components show similar relative solubilities in the different schemes so that the discussion can center mainly around yields and the preservation of biological activities.

The original plan of the Cohn ethanol scheme was to separate a series of protein fractions of successively lower isoelectric points by adjustment of the pH to increasing acidities from neutrality (Fig. 14). The ethanol concentration and ionic strength would also be suitably varied to achieve the best fractionation at each step. Fractionation would be followed by electrophoretic analysis (which describes the protein system in terms of net charge on the different components, and consequently tends to

(371) C. A. Janeway, *Advances in Internal Med.* **3,** 295 (1949).

identify components in terms of their isoelectric point). This scheme of fractionation should minimize protein–protein interaction, which is strongest between proteins bearing opposite electric charges, since all the proteins would be negatively charged except the protein to be precipitated, which would have zero charge.

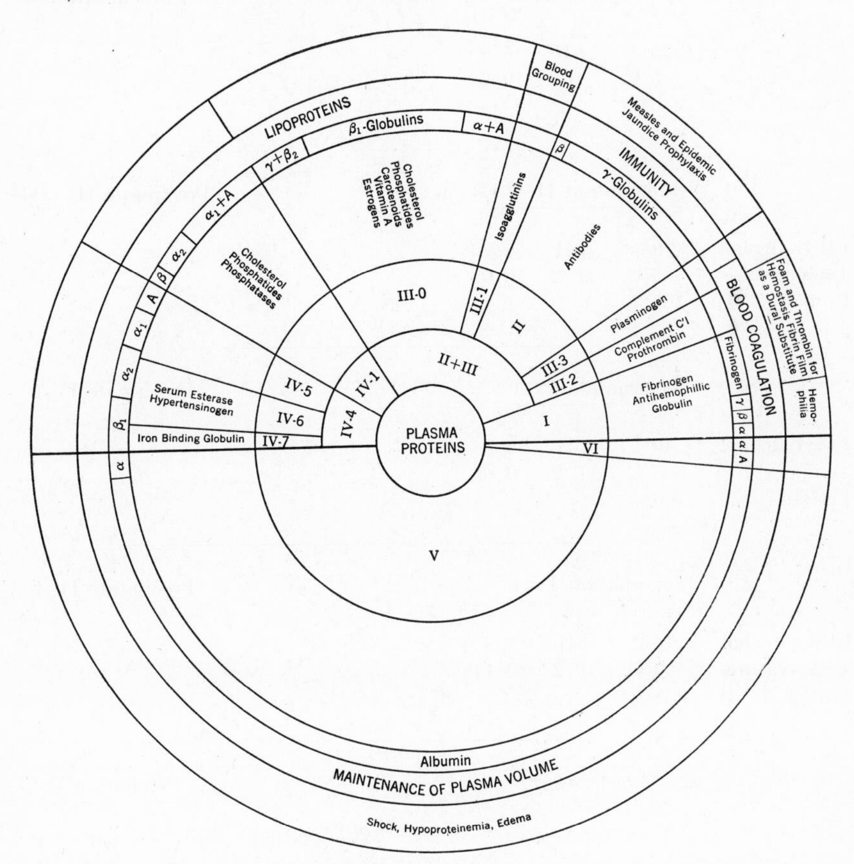

FIG. 12. Plasma proteins—their natural functions and clinical uses and separation into fractions.

The effectiveness of this scheme will, of course, depend on the sharpness of the solubility minima as a function of pH for the various components present, and, as will be seen in Figure 13, it failed in the first fraction obtained, where fibrinogen with a relatively acid isoelectric point was precipitated first because of its very low solubility even when charged. The remaining fractions precipitated in the order predicted although complete separation of electrophoretic components could not

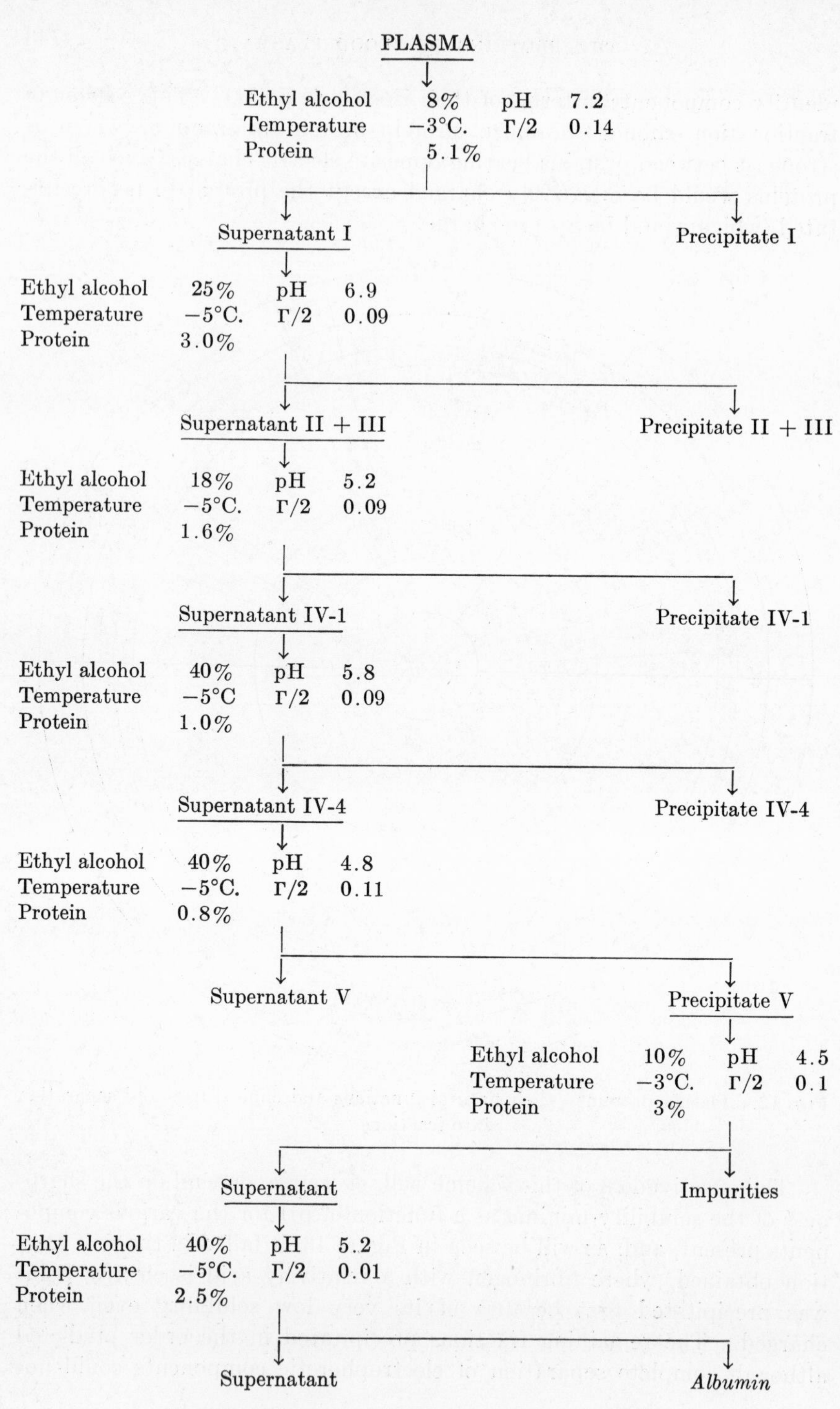

FIG. 13. Diagrammatic representation of method 6.

be achieved. Considerable amounts of β-globulins precipitated with the γ-globulin in fractions II + III, and the remaining β-globulins precipitated with the α-globulins in fractions IV-1 and IV-4.

Therefore the scheme as finally evolved in method 6 was adjusted so as to precipitate: *in fraction I*, as much fibrinogen and as little of the antibodies as possible; in *fractions II + III*, essentially all the antibodies (including isohemagglutinins) (this fraction also contains the β_1-lipoprotein); *in fraction IV-1*, the remaining lipoproteins (mostly α-globulins). (It will be noted that here the rule of increasing stepwise acidity is

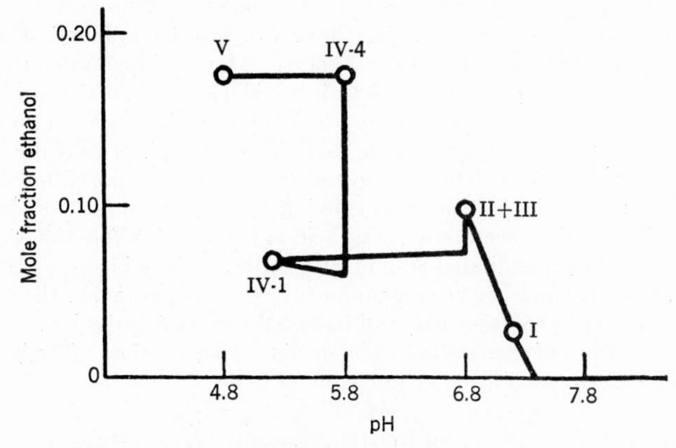

FIG. 14. Method 6.

broken, the succeeding fraction IV-4 being precipitated at a *higher pH*. This was done because the α-lipoproteins could not be precipitated at a more alkaline reaction except by increasing the ethanol concentration to the point where it grossly denatured this lipoprotein.) Fraction IV-4 was then designed to sweep out the remaining globulins with the least precipitation of albumin. Fraction V then contained most of the albumin; fraction VI contained the remaining protein: traces of albumin and a little mucoprotein (α_1- and α_2-mucoproteins).

1. METHOD 6

The plasma for which this method has been developed is obtained from bleedings in which 500 ml. of human blood is collected in 50 ml. of 4 per cent sodium citrate. Other diluents may be used. The plasma is stirred gently but thoroughly and cooled as quickly as possible to 0°C. without permitting formation of ice. The stirring is continued while sufficient sodium acetate–acetic acid buffer in a 53.3 per cent (measured at 25°C.) ethyl alcohol–water mixture is added through capillary jets to bring the pH to 7.2 ± 0.2 and the final ethyl alcohol concentration of the system to 8 per cent. The addition rate is 80–100 ml. per jet per minute, and the over-all time for the addition should be about 1½ hours. During the addition, the temperature is allowed to fall so that the system is maintained close to its freezing point and so that the final temperature is between −2.5 and −3°C. This first step requires 0.177 liter (measured at −5°C.) of 53.3 per cent ethyl alcohol for each liter of plasma (measured at 0°C.). About 1 ml. of 0.8 molar sodium acetate, buffered at pH 4.0 with acetic acid, for each liter of plasma should suffice for the pH adjustment. This buffer has a mole ratio of sodium acetate to acetic acid of 0.2 and is conveniently made up by taking: 200 ml. of 4 molar sodium acetate; 400 ml. of 10 molar acetic acid; and water to make 1 liter. When diluted with water eighty times, it should have a pH of 4.00 ± 0.02 when measured in a glass electrode at 25°C.

Precipitate I consists principally of fibrinogen. The precipitate is removed by centrifugation at a temperature between −2 and −3°C.

Supernatant I is next brought to 25 per cent ethyl alcohol and a pH of about 6.9 by the addition of cold 53.3 per cent alcohol containing a sodium acetate–acetic acid buffer. Capillary jets are used as before; the rate of addition is about 100 ml. per jet per minute, and the over-all time is about 5 hours. During this addition the temperature is held at the freezing point until −5°C. is reached and then maintained at −5°C. throughout the remainder of the addition. This step requires for each liter of supernatant I an alcohol buffer mixture made as follows: 601 ml. of 53.3 per cent ethyl alcohol at −5°C.; 0.88 ml. of 10 molar acetic acid at 25°C.; 0.44 ml. of 4 molar sodium acetate at 25°C.; 2.30 ml. of 95 per cent ethyl alcohol. The buffer used in this step has a mole ratio of sodium acetate to acetic acid of 0.2 and if, before the 53.3 per cent ethyl alcohol is added, it is diluted 80-fold with water, it should have a pH of 4.00 ± 0.02 at 25°C. No attempt has been made to vary the composition of the buffer so as to adjust the system to an exact pH, and indeed the system varies by several tenths of a pH unit as a result of changes in carbon dioxide concentration. The buffer added in the two additions contributes 14 milliequivalents of acetic acid for each liter of plasma.

Precipitate II + III is removed from supernatant I by centrifugation at −5°C.; by electrophoretic measurements it consists principally of β- and γ-globulins. It contains nearly all of the immune globulins and the isoagglutinins. Nearly all the prothrombin and plasminogen are precipitated in this fraction. It also contains large amounts of cholesterol and other lipide substances. The further subdivisions of precipitate II + III are described in the following section (Method 9).

Supernatant II + III is brought to a pH of 5.2 ± 0.1, and 18 per cent ethyl alcohol by the addition of water and a sodium acetate–acetic acid buffer. This addition is carried out in two steps: (*1*) the addition of 311 ml. of water at 0°C. for each liter of supernatant II + III at the rate of 400 ml. per jet per minute, holding this temperature constant at −5°C.; and (*2*) the addition of 78 ml. of water at 0°C. per liter of supernatant II + III, and enough of sodium acetate–acetic acid buffer to lower the pH to 5.1 ± 0.2; the mole ratio of this sodium acetate–acetic acid buffer is 0.2. The amount of the buffer necessary is determined by preliminary titration. The rate of additions is 100 ml. per jet per minute. The temperature is held at −5°C.

After the addition is complete, this system is stirred for 1 hour and allowed to stand 6–8 hours at −5°C. in order to complete the formation of the precipitate.

Precipitate IV-1 is removed from supernatant II + III by centrifugation at a temperature of −5°C.; by electrophoretic measurements it consists primarily of α-globulin and a considerable quantity of lipide material. This lipide-rich α-globulin is readily denatured by standing at low pH in the presence of high alcohol concentration even at low temperatures. If the fraction is to be preserved in soluble form, it should be dissolved in about four volumes of water at a pH near 6, and then frozen and dried immediately from the frozen state. Although rich in lipoproteins, this fraction can be dissolved in water at a pH near 6.0 and dried from the frozen state to give a material that is readily soluble in water. However, if the fraction is held for any length of time at low temperature, either as a paste or as a frozen solution, a part of the protein becomes insoluble.

Supernatant IV-1 is next brought to a pH of 5.80 ± 0.05, an ionic strength of 0.09 molar, and an ethyl alcohol concentration of 40 per cent at −5°C. The buffer is added first over a period of about ½ hour or more and the temperature is held at −5°C. throughout. (For each liter of supernatant IV-1, a mixture was made up containing: 1.14 grams of sodium bicarbonate; 7.90 ml. of 4 molar sodium acetate; enough water to make 77 ml. of solution. If a preliminary titration indicates that this mixture is not adequate to give the proper pH, it should be adjusted by altering the ratio of bicarbonate-ion to acetate-ion concentration, while the total sodium concentration is held constant. If the fraction is precipitated at pH 6.2, the albumin yield will be somewhat increased in precipitate V, but the purity and stability will be reduced.) Following this addition, the system is brought to 40 per cent ethyl alcohol by the addition of 456 ml. of cold 95 per cent alcohol for each liter of supernatant IV-1, while the temperature is maintained constant at −5°C.

Precipitate IV-4 is removed from supernatant IV-1 by centrifugation at −5°C.; by electrophoretic measurements it consists principally of α- and β-globulins with some albumin. The esterase activity in plasma is largely concentrated in this fraction, as well as some of the hypertensinogen. The hypertensinogen activity found in precipitate IV-4 represents the part that is not destroyed by 40 per cent ethyl alcohol at −5°C. The precipitate can be dissolved in cold water and dried from the frozen state.

Preliminary experiments indicate that quantitative precipitation of the hypertensinogen from supernatant IV-1 can be effected at 25 per cent ethyl alcohol and pH 4.8. Precipitate IV-4 formed under these conditions contains all the albumin ordinarily found in precipitate V, and experiments are in progress to find conditions for the separation of the albumin and globulin fractions.

Supernatant IV-4, after centrifugation, contains a small amount of suspended material and is therefore clarified by filtration. This has been done by suspending ½ per cent of washed standard Super-Cel in the supernatant and filtering at −5°C. through Republic S-1 sheets precoated with standard Super-Cel and washed. It is essential to the stability of the final albumin that the filtrate be highly clarified so that it possesses at most only a slight Tyndall effect.

After clarification is complete, all filtrates and washings are combined and the pH is lowered to 4.8 by the addition of a sodium acetate–acetic acid buffer, while the temperature is held at −5°C. and the ethyl alcohol concentration is held at 40 per cent. The addition takes about 2 hours and when completed the system is allowed to stand for at least 3 hours without stirring. During the addition, the temperature is held between −5 and −6°C. For this addition the buffer is made up by taking for each liter of clarified supernatant: 5.0 ml. of 10 N acetic acid; 2.5 ml. of 4 molar sodium acetate; 10.5 ml. of 95% ethyl alcohol; water to make 25 ml.

Precipitate V contains the bulk of the albumin present in human plasma. It generally contains less than 3 per cent of α-globulin as measured by electrophoresis at pH 8.6 in a barbiturate buffer of ionic strength 0.1. By the same measurement it should contain less than 0.5 per cent of β-globulin. The precipitated protein is removed by centrifugation or filtration at a temperature between −5 and −6°C. Supernatant V should be almost completely clear and should not contain more than 1 per cent of the plasma protein as albumin.

For certain clinical applications it is advantageous to reduce the electrolyte content of the albumin to a minimum. Impurities with a stability inferior to that of albumin must also be removed. If this is not done, the final 25 per cent solution will not remain clear when heated at 50°C., and visible particles will form.

Precipitate V, as obtained from the centrifuge, is in the form of a paste, each liter of which contains about 250 grams of protein, 0.045 mole of sodium acetate, 0.075 mole of acetic acid, and 0.035 mole of salt from the plasma, of which about 85 per cent is sodium chloride.

Purification is carried out by the removal of substances insoluble in 10 per cent ethyl alcohol at 0.01 molar salt concentration, a temperature between −2 and −3°C., a protein concentration of 3 per cent, and a pH between 4.5 and 4.7. Precipitate V is dissolved in six volumes of water at 0°C., and to the solution is added one volume of 53.3 per cent alcohol over a period of about 2 hours. During this addition, the temperature is reduced to between −2 and −3°C. The resulting turbid solution is stirred gently but thoroughly for about 2 hours and then clarified by filtration.

(When salvaging albumin from certain contaminated plasmas or in other cases where the level of unstable impurities may be high, still further purification may be desirable. If this is the case, the ethyl alcohol concentration should be raised from 10 to 15 per cent and the temperature lowered to −5°C. The precipitate is then removed, the solution clarified, and the albumin precipitated in the regular way. While this procedure will result in loss of albumin, a substantial purification is effected. It may be possible to recover some of the loss by recycling the precipitate through a second rework.)

Albumin is precipitated from the filtrate by raising the ethyl alcohol concentration to 40 per cent, lowering the temperature to −5°C., and, in order to minimize the absorption of acetic acid by the precipitate, raising the pH to 5.2. This is done by adding sufficient sodium bicarbonate to bring the pH to between 5.0 and 5.2, followed by the addition of 545 ml. of cold 95 per cent ethyl alcohol to each liter of filtrate over a period of about 2 hours at a temperature of −5°C.

An adverse situation also arises if albumin is precipitated at a salt concentration below 0.001 molar. At pH 4.8 the addition of ethyl alcohol results in a colloidally dispersed precipitate, which does not separate well in the centrifuge. Addition of electrolyte to produce aggregation at this point is not nearly as effective as it is when added before the ethyl alcohol concentration is raised.

Albumin is removed by centrifugation or filtration at −5 to −6°C. and dried from the frozen state at as low a temperature as is practical. By electrophoretic analysis at pH 8.6, the albumin generally contains less than 3 per cent of globulin impurities. When made up to a 25 per cent solution in 0.04 molar acetyltryptophan solution with the pH adjusted to 6.8 with sodium bicarbonate, application of heat of 50°C. for several months should not result in the formation of particles nor in materially increased cloudiness. This test is very sensitive to the presence of unstable impurities and therefore reflects the quality of the separations. At a temperature of 57°C. the albumin solution should not change in clarity markedly until after 100 hours of heating. This test is most sensitive to the stability of the albumin itself and therefore reflects the care with which the albumin was handled throughout the processing. When the solution is heated for 10 hours at 60°C., there should not be a visible change. The sodium content of the 25 per cent albumin solution prepared by this procedure should be less than 0.33 gram of sodium per 100 ml.

Fractional precipitation by adjusting the pH toward neutrality should also be possible starting with all the proteins positively charged. However, the low pH (below 4.5) initially necessary for fractionation plasma is probably harmful to many proteins, but when applied to plasma fractions (such as II and III, which contain a much more restricted range of isoelectric points) lesser acidities are required and a completely different scheme of fractionation is thus possible. Separations are further aided by the low ionic strengths which may be achieved in the absence of the plasma salts.

As finally worked out in method 9 (Fig. 15), fraction II + III from method 6 was first extracted at very low ionic strength at neutrality to remove the β_1-lipoprotein (fraction III-0); the remaining β-globulins were then precipitated at pH 5.2 from 17 per cent ethanol at $0.015\Gamma/2$, and purified γ-globulins precipitated at neutral pH with ethanol.

2. Method 9

Precipitate II + III is used as the starting material for this preparation. Each kilogram of this precipitate is suspended in 2 kg. of water containing ice (about one-quarter of the water should be frozen, in the form of very fine ice crystals). When this suspension is fairly uniform, add 3 kg. of 0° water to which 112 ml. of 0.5 molar disodium phosphate (pH 9.2) has been added. This suspension should be stirred slowly and kept at a temperature of 0°C. until all lumps are dissolved and a nearly complete solution is obtained. After the suspension is complete, add it to 20 kg. of 0° water, and stir slowly at 0° for 30–60 minutes. The pH of this suspension should read 7.2 ± 0.2. Then bring to 20 per cent ethyl alcohol by adding 15 liters of 53.3 per cent ethyl alcohol, keeping the temperature as low as possible until −5°C. is reached. This suspension should stand at −5°C. with slow stirring for several hours before centrifuging. Centrifuge at a rate of about 30 liters per hour.

Precipitate III-0 contains most of the lipoid and carotenoid pigment originally in precipitate II + III. It is precipitated from supernatant II + III-W by bringing the pH to about 5.7, with pH 4.0 calcium acetate buffer, and the ethyl alcohol concentration to 25 per cent. The buffer (80-fold) is that used in the fractionation of plasma, except that the sodium acetate is replaced by calcium acetate. It consists of 4.0 molar acetic acid and 0.4 molar calcium acetate. This suspension should be allowed to stand for about 48 hours at −5°C. before centrifugation, and then centrifuged at a rate of about 20 liters per hour. Since little is known of the conditions for the storage of this fraction, or its possible uses, supernatant II + III-W can be discarded until further studies indicate its usefulness.

Precipitate II + III-W consists of practically all of the γ-globulins, isoagglutinins, plasminogen, and prothrombin originally present in precipitate II + III. It contains very little cholesterol and carotenoid pigment, and a large part of the depressor activity has been removed. From a pool of normal plasma, it should represent about 60 per cent of the original II + III paste, and will contain 50–55 per cent of γ-globulin by electrophoresis at pH 8.6.

Precipitate II + III-W is resuspended in acetate buffer, and taken to pH 5.2, 0.015 ionic strength, and 17 per cent ethyl alcohol. This can be done by suspending each kilogram of precipitate II + III-W in 2 kg. of water and ice, as before, and then adding 2 kg. of cold water, to which 0.35 mole of sodium acetate has been added. When this suspension is complete, add sufficient pH 4.0 acetate buffer, diluted with 1 liter of cold water per kilogram of precipitate II + III-W, to lower the pH of the suspension to 5.2 ± 0.1. Stir for several hours. Then add 13.5 liters of cold water

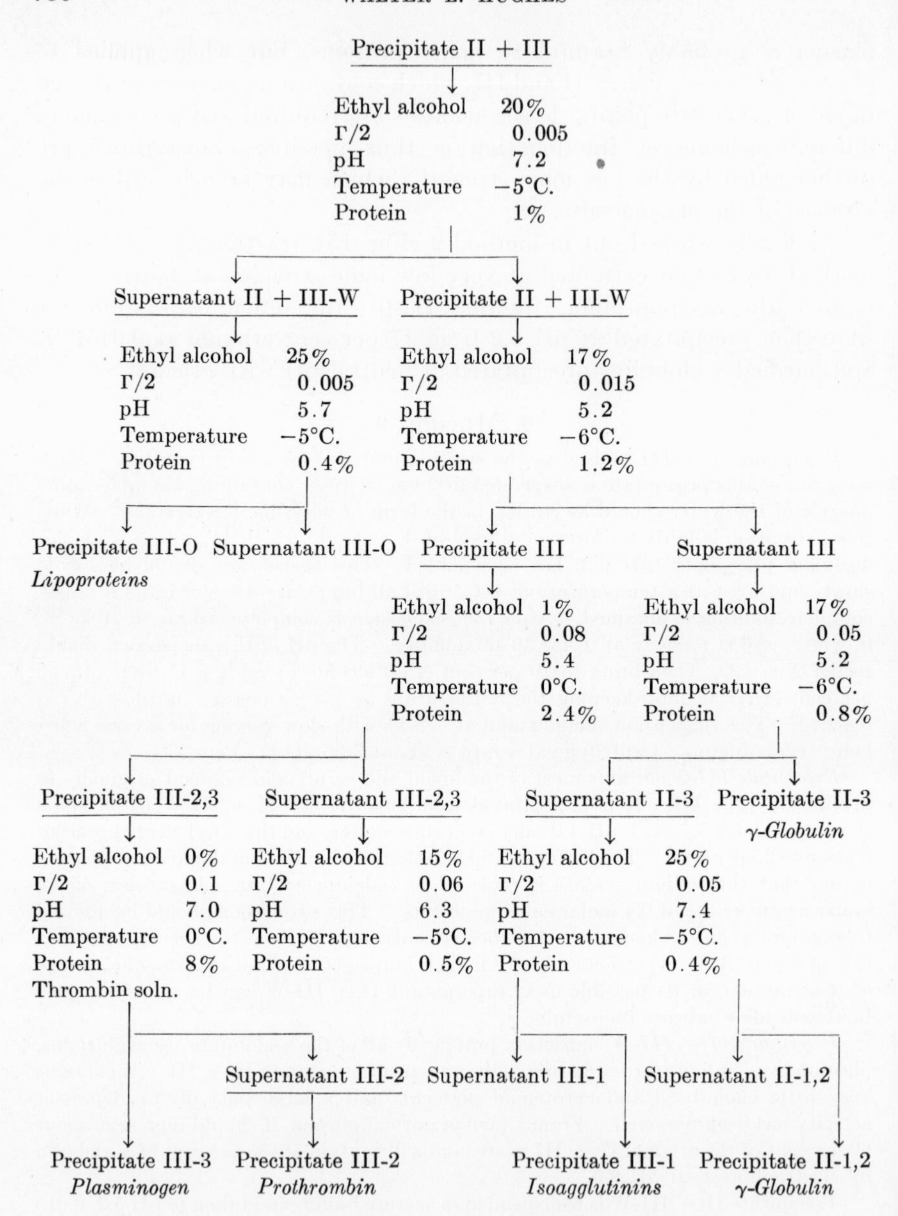

FIG. 15. Diagrammatic representation of method 9.

and then 8.66 liters of 53.3 per cent ethyl alcohol per kilogram of precipitate II + III-W, raising the ethyl alcohol concentration to 17 per cent. The temperature should be lowered during the addition of the alcohol, keeping the suspension near the freezing point until it is cooled to $-6°C$. The precipitate is removed by centrifugation at a rate of about 30 liters per hour and at a temperature of $-6.0 \pm 0.5°C$. Supernatant III from this centrifugation should then be clarified by filtration at a temperature of $-6.0 \pm 0.5°C$.

Precipitate III contains the isoagglutinins, plasminogen, and prothrombin. It has been stored at $-5°C$. for 2 weeks without appreciable loss of prothrombin, but should be processed and the prothrombin converted to thrombin as soon as possible to prevent loss of this activity.

Each kilogram of precipitate III is suspended in 2 kg. of water and ice, and then diluted with 2 kg. of 0.5 ionic strength pH 5.4 acetate buffer cooled to 0°C. This suspension is stirred until uniform, and then diluted with 7.5 kg. of 0° water. This suspension should be at pH 5.4, ionic strength 0.08, and 0 to $+1°C$., and should be stirred for several hours. The precipitate is removed by centrifugation at a relatively slow rate of speed and at 0 to $+1°C$.

Precipitate III-2,3 contains prothrombin, plasminogen, and a small amount of fibrinogen, which was not removed in precipitate I. Each kilogram of this precipitate should be suspended as soon as possible in 1 liter of 0° water and the pH adjusted to 6.8–7.0 with sodium glycinate buffer. (Sodium glycinate buffer is prepared by half-neutralizing a glycine solution with sodium hydroxide. The solution that may be used contains 1 mole of glycine and 0.5 mole of sodium hydroxide per liter. It has a pH of about 9.5 and has the advantage of causing less of a drift in pH upon aeration than is observed when sodium bicarbonate is used for increasing the alkalinity of the solutions.) Then add 1 liter of 0° water containing enough sodium chloride to give a total ionic strength of 0.1. After 1 to 2 hours of stirring, test the solution for preformed thrombin and remove a sample to see if the fibrinogen has clotted. To do this, centrifuge a small sample and then add thrombin to the supernatant solution to see whether more fibrin is formed. If all the fibrinogen has clotted, do not add additional thrombin. If there is still fibrinogen left in solution, then add enough thrombin to give a total amount (including what is already there) of 10,000 units per kilogram of precipitate III-2,3 in solution. Stir this suspension for an additional hour and centrifuge at 0°C. either for about an hour using a batch bowl, or at a very slow rate if continuous centrifugation is required.

Supernatant III-2 can either be dried directly if prothrombin is desired, or immediately converted to thrombin before drying. To accomplish the conversion of prothrombin to thrombin, calcium chloride and thromboplastin are added, and the temperature is raised to 20–25°C. This solution is then clarified, sterilized, and dried.

Precipitate III-3, obtained from the centrifugation of the solution of precipitate III-2,3 after the addition of thrombin, is a fibrous material that contains a high concentration of plasminogen. It should be suspended in sodium chloride and bicarbonate solution to give an ionic strength of 0.15 and a pH of 7.1 ± 0.1, and allowed to stand at 0°C. until all of the fibrin has been lysed. After clarification and sterilization, this material can be dried from the frozen state.

Each kilogram of supernatant III-2,3 is adjusted to pH 6.3 by the addition of sodium bicarbonate (approximately 20 ml. of 1 molar sodium bicarbonate should be required), and then the ethyl alcohol is adjusted to 15 per cent by the addition of an equal volume of 30 per cent ethyl alcohol, keeping the temperature at or near the freezing point until lowered to $-5°C$., where it should be maintained. Precipitate

III-1 can then be removed by centrifugation at −5°C., and at a speed of about 80 liters per hour. If the plasma used for this process is not group-specific, the isoagglutinin concentration is too low to make useful blood-grouping globulin. Since at present no other useful components in this fraction are known, supernatant III-1 can be discarded until further studies indicate its usefulness.

Precipitate III-1 contains the isoagglutinins. It can be dried from the frozen state, and reconstituted to about 5 per cent protein in isotonic saline at pH 7 for a blood-grouping solution, if the pool was group-specific and of sufficiently high titer.

Each kilogram of supernatant III is taken to about 0.05 ionic strength by the addition of 50 millimoles of sodium chloride, holding the pH at 5.2, the ethyl alcohol concentration at 17 per cent, and the temperature at −6°C.

Precipitate II-3 consists largely of γ-globulin. (The immunological titers of the antibodies studied are almost identical with precipitate II-1,2 except the typhoid "O" agglutinin.)

Precipitate II-3 can be removed by centrifugation of supernatant III at a rate of about 50 liters per hour and at a temperature of −6°C. It may be dried from the frozen state, preferably without the addition of sodium chloride or glycine.

Each kilogram of supernatant II-3 is taken to pH 7.4 ± 0.2 by the addition of about 15 millimoles of sodium bicarbonate, and to 20–25 per cent ethyl alcohol by the addition of 40–114 ml. of 95 per cent ethyl alcohol. The temperature is maintained at −5°C.

Precipitate II-1,2 consists largely of γ-globulin. This represents the more soluble part of the γ-globulin. Such material has been used clinically in both measles and epidemic jaundice. This precipitate should be dissolved to make solutions containing 16.5 grams of γ-globulin per 100 ml., and sterilized. Precipitate II-1,2 can be removed by centrifugation of supernatant II-3 at a rate of about 50 liters per hour and at a temperature of −5°C. It may be dried from the frozen state, preferably without the addition of sodium chloride or glycine.

A similar but abbreviated method of fractionation has been described by Pillemer *et al.*[372] in which methanol replaces ethanol as the precipitating reagent. About the same separations seem to be obtained as with ethanol at similar volume fractions of reagent. It is claimed that less denaturation is produced so that higher temperatures (0°C. instead of −5°C.) may be used in processing. It would be interesting to see objective data on the relative denaturing effects of these two reagents. The ethanol process described above is limited to some extent by the freezing point of the system, and the author feels that the real advantage in the use of methanol may lie in the lower temperatures that can be attained because of the greater mole fractions of methanol required for equivalent precipitation.

A partial fractionation of human plasma using ethyl ether as a precipitant has been reported by Kekwick.[373] Owing to the limited solubility of ether, only fibrinogen and γ-globulin could be precipitated.

A further study of the ethanol system with the intent of finding milder

(372) L. Pillemer and M. C. Hutchinson, *J. Biol. Chem.* **158,** 299 (1945).
(373) R. A. Kekwick, B. R. Record, and M. E. Mackay, *Nature* **157,** 629 (1946).

conditions has been reported. Precipitation by metallic ions, such as Zn^{++}, has been studied, and use made of protein–protein interactions to facilitate precipitation under mild conditions.[193] However, no better separations than in the older studies have been reported, and evidence that the conditions are really milder is unfortunately lacking. Zinc ions (at pH 6–6.5) precipitate a fraction which corresponds roughly to fractions I + II + III. (At higher temperatures greater precipitation, possibly with denaturation, is observed.) Maximal precipitation is obtained at about 0.02 M Zn^{++}, above which the proteins become soluble again (salting-in). After removing the precipitate, the remaining protein [designated SPP (stable plasma proteins)] can be precipitated by adding ethanol (15–20 per cent by volume is sufficient), or if the solution is made sufficiently alkaline to precipitate $Zn(OH)_2$, most of the remaining protein is carried down with it. The zinc ions may be readily removed by ion-exchange resins so that lyophilization may be unnecessary in therapeutic preparations.

Salting-out, one of the oldest methods of protein fractionation, has frequently been used in the purification of plasma components. Attempts at inclusive systems have been made by Cohn *et al.*[169] and by Svensson.[374] These studies, carried out respectively at room temperature and pH 6 and at "cold room temperature" and blood pH, showed similar separations as a function of ammonium sulfate concentration. One-third saturation with ammonium sulfate precipitated most of the γ-globulin. The remaining γ-globulin could be precipitated by increasing the ammonium sulfate to 0.4 of saturation, although not without coprecipitation of considerable amounts of α- and β-globulins. Most of the latter could then be precipitated by half saturation with ammonium sulfate, leaving essentially all of the albumin in the supernatant.

X. General References

Besides the references cited in the text, the following general references are included for further reading:

1. J. L. Tullis, ed., Blood Cells and Plasma Proteins, Academic Press, New York, 1953. This book covers, *inter alia:*
   a. Historical (18th and 19th Century) concepts of blood (E. J. Cohn).
   b. Recent innovations in protein fractionation by solubility methods (E. J. Cohn *et al.*).
   c. Some aspects of the clotting process (B. Alexander, J. H. Ferguson, and J. T. Edsall).
   d. Human antibodies (C. A. Janeway, J. F. Enders, J. L. Oncley, and W. E. Ehrich).
   e. Lipoproteins (J. L. Oncley, F. R. N. Gurd, M. Macheboeuf, and J. Folch-Pi).

(374) H. Svensson, *J. Biol. Chem.* **139,** 805 (1941).

2. E. C. Albritton, ed., Standard Values in Blood, W. B. Saunders Co., Phila., Pa., 1952.
3. Lipoproteins, *Discussions Faraday Soc.* **6** (1949).
4. Blood Clotting and Allied Problems (Trans. Josiah Macy, Jr. Conf., N.Y.), Vol. 1–4, 1948–51.
5. T. Astrup, Biochemistry of Blood Coagulation, Store Nordiske Videnskabsboghandel, Copenhagen, 1944.
6. P. A. Owren, The Coagulation of Blood, *Acta Med. Scand. Suppl.* **194** (1947).
7. The Physiology and Pathology of Hemostasis, Lea & Febiger, Phila., 1951.
8. "Blood Coagulation," Chap. 35 *in* The Enzymes, Vol. I, part 1, ed. by J. B. Sumner and K. Myrbäck, Academic Press, New York, 1950.
9. A. M. Pappenheimer, Jr., ed., The Nature and Significance of the Antibody Response, Columbia University Press, New York, 1953.
10. Immunochemistry, *Biochem. Soc. Symposia* No. **10**, Cambridge University Press, London, 1953.

CHAPTER 22

The Proteins of Immune Reactions

BY WILLIAM C. BOYD

Page

I. Immunity 756
II. Antigens 757
1. Conditions of Antigenicity 759
2. Nature of Antigenic Determinants 763
a. Altered Proteins 765
b. Conjugated Proteins 770
c. Importance of Polar Groups 773
d. Effect of Isomers 773
e. Specificity of Peptides 774
f. Influence of Optical Isomers 774
3. Number of Antigenic Determinants per Molecule 777
4. Specificity of Native Proteins 779
a. Functional Specificity 779
b. Species Specificity 779
c. Toxins 781
III. Antibodies 782
1. Multiple Functions of Antibodies 782
a. Precipitating Antibodies (Precipitins) 783
b. Agglutinating Antibodies (Agglutinins) 783
c. Neutralizing Antibodies 783
d. Complement-Fixing Antibodies 784
e. Blocking Antibodies 785
f. Allergic Antibodies (Reagins) 787
g. Opsonins 788
h. Natural Antibodies or Lectins 788
2. Chemical Properties of Antibodies 790
a. Antibodies as Proteins 790
b. Analyses of Antibodies 791
c. Molecular Weight of Antibodies 791
d. Shape of Antibody Molecules 793
e. Isoelectric Points of Antibodies 796
f. Electrophoretic Mobility 796
g. Susceptibility of Antibodies to Enzymes, etc 798
h. Specific Differences between Antibodies 799
i. Valence of Antibody Molecules 800
j. Heterogeneity of Antibodies 805
3. Nature of Antibody Determinants 806
4. Purification of Antibodies 807
5. Quantitative Determination of Amount of Antibody 809
6. Formation of Antibodies 809

Page
IV. Antibody–Antigen Combination 814
1. First Stage: Combination 815
a. Forces Involved 815
b. Rate of Reaction 817
c. Heat of Antibody–Antigen Combination 817
d. Effect of Electrolytes 820
2. Second Stage: Precipitation, Agglutination, Neutralization, etc 820
a. Neutralization of Toxins and Viruses 820
b. Agglutination 821
c. Precipitation 822
V. Complement and Complement Fixation 839
1. Complement Fixation 841
2. Quantitative Estimation of Complement 841
3. Theories of Complement Fixation 842

I. Immunity

Proteins are involved in the mechanism of immunity in several ways. Protein substances from the blood of the immune animal may combine with invading microorganisms and thereby render them susceptible of being engulfed by the phagocytes of the reticulo-endothelial system. Protein substances from the blood may combine with the foreign cells and cause them to agglutinate. Other plasma proteins may then combine with these coated cells and cause their death or lysis. Soluble toxins and other products of the metabolism of the invaders may be combined with, and neutralized, by certain blood proteins of the host. Under suitable conditions the soluble products of the invading microorganism are rendered insoluble and precipitated. Proteins may be produced by the host which have the property of rendering certain of his tissues (or the tissues of other individuals injected with his plasma or serum) peculiarly susceptible (hypersensitive) to substances produced by the invader. The proteins produced by the host which produce these effects are divided into two classes, antibodies and complement.

In addition to antibodies and complement, serum and plasma contain protein substances that enhance cellular agglutination. This was observed by Muir and Browning[1] and by Bordet and Gay.[2] Bordet and Streng[3] proposed the name conglutinin for the heat-stable, colloidal substance responsible. This conglutinin does not act without complement (for review see Hole and Coombs).[4] Wiener[5] extended the meaning of the term to cover the factor in serum and plasma which brings about the

(1) R. Muir and C. Browning, *J. Hyg.* **6**, 20 (1906).
(2) J. Bordet and F. Gay, *Ann. inst. Pasteur* **20**, 467 (1906).
(3) J. Bordet and O. Streng, *Zentr. Bakt. Parasitenk.* (*Abt. Orig.*) **49**, 260 (1909).
(4) N. H. Hole and R. R. A. Coombs, *J. Hyg.* **45**, 480 (1947).
(5) A. S. Wiener, *Am. J. Clin. Path.* **15**, 106 (1945).

agglutination of cells sensitized with blocking antibody (see p. 785). The British workers (see Hole and Coombs)[4] have not accepted Wiener's[5] extension of the term conglutinin to cover the factor in serum and plasma which brings about the agglutination of cells sensitized with blocking antibody (see p. 785). Dean[6] found such activity in the serum fraction containing the midpiece of complement, and Diamond and Denton[6a] found concentrated serum albumin to have the activity. Other fractions such as γ-globulin are also active as are substances such as gum acacia and gelatin.[7]

In addition, it is found that the soluble products of microorganisms and other infectious agents which cause the production of the antibodies by the host are themselves often proteins, although carbohydrates and possibly a few other compounds can act in this way. The substances that can stimulate the production of antibodies are called antigens.

II. Antigens

Antigenicity is not restricted to the proteins and carbohydrates produced by microorganisms and parasites. We use the term antigen to mean any substance capable of stimulating the production of antibodies in a host. Protein poisons such as snake venom may cause the production of neutralizing antibodies; harmless foreign cells such as erythrocytes, and innocuous proteins such as ovalbumin will also act as antigens. It is true that the result may not be to make the animal producing the antibodies more resistant; for Portier and Richet[8] discovered that dogs treated with the poisonous extracts of the tentacles of certain sea anemones became more instead of less susceptible, and a guinea pig injected once with ovalbumin, which is in the first instance perfectly harmless, may react fatally to a later injection, but there is reason to include these happenings under the general classification of immune phenomena.

Immunity to the poisonous protein of the Castor bean (*Ricinus communis*) can be produced by feeding,[9] and hypersensitivity to food allergens, which are usually proteins, results from ingestion of the antigen. Proteins degraded by digestive enzymes (or in other ways) to their constituent amino acids or to simple polypeptides are no longer antigenic, so immunization in the above examples evidently results from the absorption of sufficient protein, intact or relatively so, from the gastrointestinal tract.

(6) H. R. Dean, *J. Hyg.* **12,** 259 (1912).
(6a) L. Diamond and R. Denton, *J. Lab. Clin. Med.* **30,** 821 (1945).
(7) J. A. Flick and A. L. Villafañe, *J. Immunol.* **68,** 41 (1952).
(8) P. Portier and C. Richet, *Compt. rend. soc. biol.* **54,** 170 (1902).
(9) P. Ehrlich, *Deut. med. Wochschr.* **17,** 976 (1891).

Since hydrolyzed proteins are not antigenic, we ordinarily restrict the term antigen to substances that stimulate the production of antibodies when they find their way relatively unaltered into the circulation or tissues of an animal, or are injected parenterally. Here we shall be concerned with protein antigens only.

Not all proteins are antigenic. At least one derived protein, gelatin, is nonantigenic,[10] and the protamines are nonantigenic.[11] In addition to these proteins, which are nonantigenic no matter what the test animal, the proteins that are normally found in an animal's own circulation, or the same proteins from the blood of another member of the same species, are nonantigenic to it. This principle was given the name *horror autotoxicus* by Ehrlich.

Although it is true that no antibodies for the proteins of the circulation of an animal are ever found, certain immunologists have argued that they are nevertheless produced, but there is reason for doubt. If an animal did produce antibodies to its own hemoglobin or plasma proteins, these antibodies would be combined with by the great excess of the antigen constantly present, and the antibody–antigen compound would be promptly removed from the circulation, so that no trace of the antibody would ever be detected. It is thus conceivable that each animal is constantly producing antibodies to its own hemoglobin, serum albumin, serum globulins, and so on, and that these antibodies are removed as fast as they are formed. But it may be argued on broad biological grounds that this is unlikely, for it supposes that each animal is constantly producing a supply of precisely engineered molecules which are promptly eliminated without ever having served any useful purpose. The amino acids that went into the composition of such molecules might be recovered and used again, but the thermodynamic work involved in elaborating them would be permanently lost. It seems likely that natural selection would have eliminated such a wasteful mechanism ages ago.

A more experimental basis for not believing that the normal proteins of his own blood are antigenic to an individual is the fact that some of the other body proteins are antigenic, but antibodies for them do not normally occur. Casein, which is not found in the blood stream, will act as an antigen if injected into another individual of the same species, or even into the individual that produced it.

Lewis[12] was able to cause lactating goats to form antibodies to their own casein. Similarly, it seems that the proteins of the lens of the eye

(10) S. B. Hooker and W. C. Boyd, *J. Immunol.* **24,** 141 (1933).
(11) H. G. Wells, The Chemical Aspects of Immunity, Chemical Catalog Co., New York, 1925.
(12) J. H. Lewis, *J. Infectious Diseases* **55,** 168 (1934).

are autoantigenic[13] and it has been claimed that a guinea pig can be sensitized anaphylactically to the proteins contained in one of its eyes and, after an appropriate interval, shocked by injection of the proteins of the other eye.[12]

In line with the nonantigenicity of an animal's own blood proteins, it is found that the proteins from closely related animals, which are chemically very similar, are nonantigenic, or at any rate poorly antigenic. In general, proteins are better antigens if they come from organisms taxonomically remote from the experimental animal. Most plant proteins seem to be good antigens. If we consider especially rabbits, which have served so extensively as immunological test animals, we find that hemocyanin is a powerful antigen, while horse hemoglobin, for example, is a poor antigen. In fact hemoglobin is such a poor antigen that it has several times been erroneously reported to be nonantigenic. More careful work has established its antigenicity.[14,15] Egg albumin from the hen is a good antigen for rabbits, but Hooker and Boyd[16] failed to obtain a response to it in ducks. The proteins of horse serum, notably the albumins, are good antigens for the rabbit, and unfortunately also for man. The very small amount of horse serum (about 0.01 ml.) which was contained in the toxin–antitoxin mixtures once used for immunization to diphtheria regularly sensitized more than 25 per cent of the individuals receiving the injections.[17] Crystalline bovine albumin was discarded as a blood substitute because of the incidence of allergic reactions ("serum disease") it produces. In contrast, in only one case has it been possible to produce antibodies in a chimpanzee by the injection of human blood.

1. Conditions of Antigenicity

It is to be hoped that eventually we shall be able to answer the obvious question: What makes a given protein antigenic to a given animal? At present we are far from having a complete answer. If we attempt to generalize from the scanty available data, we may make the following statements:[14]

(*a*) Antigens must be foreign to the circulation of the experimental animal, and the more foreign (i.e., the more remote the source taxonomically) the more antigenic the protein will be.

(13) L. Markin and P. Kyes, *J. Infectious Diseases* **65**, 156 (1939).
(14) W. C. Boyd and S. Malkiel, *J. Infectious Diseases* **75**, 262 (1944).
(15) F. Haurowitz, Chemistry and Biology of Proteins, Academic Press, New York, 1950.
(16) S. B. Hooker and W. C. Boyd, *J. Immunol.* **26**, 469 (1934).
(17) L. Tuft, Clinical Allergy, W. B. Saunders, Philadelphia, 1938.

Proteins fulfilling very similar functions in different species are similar chemically, and this doubtless accounts for the low antigenicity of such proteins. Hemoglobin from the horse is a poor antigen for rabbits, although the plasma proteins are good antigens. It is tempting to speculate that the plasma proteins, especially the globulins, have a more specialized role than does hemoglobin, whose function is much the same in all mammalian species. Similarly, insulin, even from such taxonomically remote animals as the hog and the cow, is seldom an antigen in man.[18] Insulin fulfils the same function in all the higher animals, and seemingly does not vary much chemically from one species to another (see Chap. 20).

(*b*) Antigens must have more than a certain minimal degree of complexity and a certain molecular size. This is suggested by the nonantigenicity of the protamines and of the relatively simple molecule of gelatin, and by the high antigenicity of the large molecules of the hemocyanins.

According to Haurowitz,[19] another prerequisite for antigenicity is rigid structure of the determinant groups. He believes that the highly specific action of the aromatic diazo compounds is due to the rigidity of their benzene rings, while the inability of the long-chain fatty acids to act as determinant groups is due to the fact that the paraffin chains are easily distorted, and their shape alters constantly.

Various workers have been led to speculate further as to the causes of antigenicity, and some speculations have been the result of comparing gelatin with other proteins. The first suggestion of this sort is due to Obermayer and Pick.[20] In a passage which has often been misquoted, they said: "Therefore it seems probable to us that the species specific groupings in the protein molecule are mainly influenced by groups which are connected with the aromatic nuclei of the protein. It hardly has to be mentioned that the aromatic groups in themselves naturally would not suffice to explain the enormous number of possible variations which nature calls for, and that our conception of the role of the aromatic groups is that they are so to speak the center around which the species specific side chains group themselves; the entrance of substituents smooths out these species differences." It is apparent that these authors were more concerned with specificity than with antigenicity, but they have often been cited[21] as proposing that the possession of aromatic amino acids is requisite for antigenicity.

(18) F. C. Lowell, *J. Clin. Invest.* **23**, 225, 233 (1944).
(19) F. Haurowitz, *Biol. Revs. Cambridge Phil. Soc.* **27**, 247 (1952).
(20) F. Obermayer and E. P. Pick, *Wien. klin. Wochschr.* **17**, 265 (1904).
(21) H. G. Wells, The Chemical Aspects of Immunity, 2nd ed., Chemical Catalog Co., New York, 1929.

It is quite certain that aromatic amino acids in an antigen do not suffice to make it antigenic, and the absence of aromatic acids does not mean that a substance cannot be antigenic. Proteins that have been "racemized" (which probably involves denaturation or hydrolysis) by treatment with alkali[22] are no longer antigenic, although any aromatic amino acids they may have possessed are still present. The pneumococcus polysaccharides, which contain no aromatic groups, are antigenic for men and mice.[23] Sulfonated polystyrene (I), although full of aromatic groups, is not antigenic.[24]

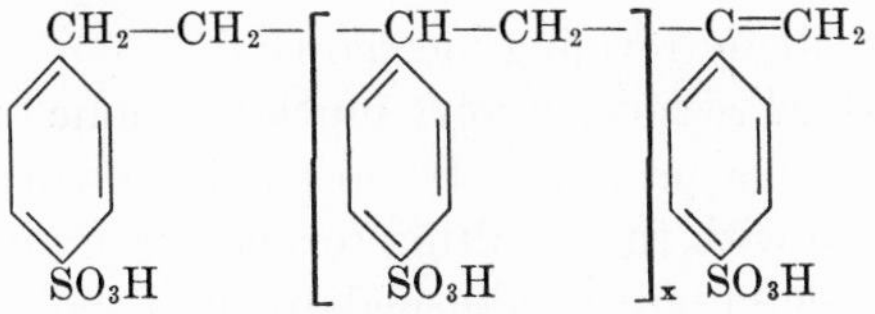

$CH_2—CH_2—[—CH—CH_2—]_x—C{=}CH_2$

SO_3H SO_3H SO_3H

I. Sulfonated polystyrene

Proteins that have been altered to the point that they no longer show the characteristic absorption bands of aromatic rings are still antigenic.[25,26]

Campbell and Bulman[27] suggest that a substance must be at least partly susceptible to the action of hydrolytic enzymes of the injected animal to be antigenic. The general validity of this rule seems doubtful, but the nonantigenicity of sulfonated polystyrene is in accord with it.

There is no evidence for the existence of the "species-specific side chains" postulated by Obermayer and Pick.[20] Species specificity is undoubtedly a matter of general chemical differences in the antigen molecules.

Furthermore, nonantigenic proteins can be rendered antigenic by treatments that do not add to their content of aromatic residues. Alkali-treated proteins can be made antigenic again by nitration[28] and iodination.[29] Gelatin can be rendered antigenic by coupling it with a variety of chemicals such as arsanilic acid.[10] It is true that arsanilic acid is an aromatic compound and couples with aromatic groups (tyrosine and

(22) C. Ten Broeck, *J. Biol. Chem.* **17**, 369 (1914).
(23) R. J. Dubos, The Bacterial Cell in Its Relation to Problems of Virulence, Immunity and Chemotherapy, Harvard University Press, Cambridge, 1945.
(24) W. C. Boyd, unpublished experiments, 1952.
(25) L. R. Wetter and H. F. Deutsch, *Arch. Biochem.* **28**, 399 (1950).
(26) R. O. Prudhomme and P. Grabar, *Bull. soc. chim. biol.* **29**, 122 (1947).
(27) D. H. Campbell and N. Bulman, *Fortschr. Chem. org. Naturstoffe* **9**, 443 (1952).
(28) K. Landsteiner and C. Barron, *Z. Immunitätsforsch.* **26**, 142 (1917).
(29) L. R. Johnson and A. Wormall, *Biochem. J.* **26**, 1202 (1932).

histidine) in the protein molecule, but it is doubtful if the aromatic character of the introduced group is as important as its acidic character.[30]

Haurowitz[15] attributes the nonantigenicity of gelatin to three causes: (*1*) It is a heat-treated denatured protein and thus has no definite fixed internal structure. (*2*) It is not deposited in the organism into which it is introduced, but is rapidly excreted. (*3*) It contains large amounts of glycine, the one amino acid having no side chain in the alpha position. Such linkages allow free rotation around the longitudinal axis, and as a result the molecule has no fixed configuration.[31] It seems reasonable that an antigen must have a definite configuration for it to be copied, in reverse, by the antibody-forming mechanism.

The first and third reasons involve much the same argument. It does seem reasonable that a protein with no fixed configuration could never be copied in reverse with the resulting formation of an antibody. However, it has been definitely established by two independent groups of workers[10,32] that coupling of certain chemical groups to gelatin will make it antigenic. It is doubtful if the introduction of these groups confers on the gelatin molecule a fixed configuration of the sort which it previously lacked.

It is a fact that gelatin is rapidly excreted from the body when it is injected. For this reason it is unsatisfactory as a blood substitute. However, there is no reason to think that gelatin coupled with arsanilic acid is retained any longer by the injected animal. Actually, most antigens soon disappear from the circulation, usually more rapidly if they are good antigens; the amount that is retained to serve as a mold for the construction of antibody is too small to be detected. It would be literally impossible to prove that a similarly small amount of gelatin is not retained for a similar time. Furthermore, the argument does not work in reverse. Gum acacia, used as a blood substitute in World War I, is retained in the body forever, but does not seem to cause the production of antibodies. Although mice can be immunized by the right dose of pneumococcus polysaccharide, too large a dose not only does not immunize, but renders the mice permanently incapable of being immunized. A small amount, undetectable by present methods, of the polysaccharide is evidently retained by the organism, but no antibody production results.

No theory of antigenicity yet offered seems entirely adequate. Antigens must be foreign to the circulation, and must not be too simple in structure; that is about all we can say. We may speculate on the problem

(30) K. Landsteiner, The Specificity of Serological Reactions, Harvard University Press, Cambridge, 1945.

(31) H. Neurath, *J. Am. Chem. Soc.* **65,** 2039 (1943).

(32) R. F. Clutton, C. R. Harington, and M. E. Yuill, *Biochem. J.* **32,** 1111 (1938).

of antigenicity of coupled derivatives of gelatin. The only glimmer of an idea to date is that the chemical treatment *accentuates* the structure of the gelatin, calling it to the attention, as it were, of the reticulo-endothelial cells of the rabbit. Nitration and iodination apparently do not accentuate it enough.

None of the special theories of antigenicity thus far put forward seem to explain the low antigenicity of insulin and hemoglobin; this is probably accounted for by the chemical similarity of these proteins, whether they are of equine, bovine, or porcine origin, to the corresponding protein of the experimental animal.

2. Nature of Antigenic Determinants

The antigenic specificity of a protein molecule is not a function of the whole molecule, but evidently resides in smaller portions which we give the name of antigenic determinants. This is suggested by the fact that protein antigens combine with several molecules of antibody (see below), showing that there exist in the antigen molecule numerous points of specific attachment for the antibody molecule. It is also made likely by the discovery from studies on haptens (see below, p. 773) that antibodies can be specific for, and react with, molecules much smaller than intact protein molecules.

Landsteiner[33] studied the reaction of antibodies for silk with hydrolysis products of silk. He found that silk-derived peptides of molecular weight of the order of 600 to 1000 would still react specifically with anti-silk antibody. The antigenic determinants in the intact silk molecule are evidently of this order of size. It is even possible that they may be smaller, for it is not impossible that smaller portions of the molecule, even if they contained the same amino acids, in the same order, as the parts of the intact molecule which function as the antigenic determinants, might assume a somewhat different spatial configuration. For when separate they would be freed from some of the original intermolecular forces such as hydrogen bonds, etc., and thus present a different pattern of charges to the antibody from the one presented by the intact molecule, and might therefore no longer be immunologically the same.

There is no reason to think that the antigenic determinants in proteins are prosthetic groups of carbohydrate or of other nonprotein nature. The surfaces of globular proteins possess negatively charged groups ($RCOO^-$) of aspartic and glutamic acid residues, and positively charged groups (RNH_3^+) of lysine and arginine residues. In addition, the phenolic hydroxyls of the tyrosyl and the imidazolyl groups of the histidine residues are polar; the work of Landsteiner[30] on "synthetic"

(33) K. Landsteiner, *J. Exptl. Med.* **75**, 269 (1942).

antigens has shown that the polar groups are of decisive significance for specificity.

Marrack[34] supposed that several of these amino acid residues acting together constituted an "active patch" on the surface of the antigen, with sufficient polarity to act as a strong combining site, and with a characteristic combination of, and spacing between, the polar groups to make the patch specifically characteristic of the protein in question. He supposed further that there would be a number of these "active patches," all with the same or very similar specificity.

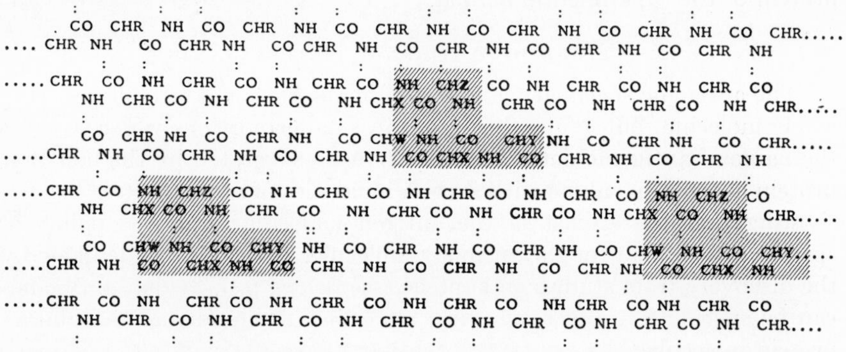

```
      CO  CHR  NH  CO  CHR  NH  CO  CHR  NH  CO  CHR  NH  CO  CHR  NH  CO  CHR.....
.... CHR  NH   CO  CHR NH   CO CHR  NH  CO  CHR  NH  CO  CHR  NH  CO  CHR  NH
.....CHR  CO  NH  CHR  CO  NH  CHR CO  NH  CHZ  CO  NH  CHR  CO  NH  CHR  CO
     NH  CHR CO  NH  CHR CO   NH CHX CO  NH   CHR CO  NH  CHR CO  NH  CHR.....
     CO  CHR NH  CO  CHR NH   CO CHW NH  CO  CHY NH  CO  CHR NH  CO  CHR.....
.....CHR  NH  CO  CHR  NH  CO  CHR  NH  CO CHX NH  CO  CHR  NH  CO  CHR  NH
.....CHR  CO  NH  CHZ  CO  NH  CHR  CO  NH CHR CO  NH  CHR  CO  NH  CHZ  CO
     NH  CHX CO  NH  CHR  CO  NH  CHR CO  NH  CHR CO  NH CHX  CO  NH  CHR.....
     CO  CHW NH  CO  CHY  NH  CO  CHR  NH  CO  CHR NH  CO CHW  NH  CO  CHY.....
.....CHR  NH  CO  CHX NH  CO  CHR  NH  CO  CHR  NH  CO  CHR NH  CO  CHX  NH
.....CHR  CO  NH  CHR CO  NH  CHR  CO  NH  CHR  CO  NH CHR  CO NH  CHR  CO
     NH  CHR  CO  NH  CHR  CO  NH  CHR  CO  NH  CHR  CO NH  CHR CO  NH  CHR....
```

FIG. 1. Protein polypeptide chains folded together into a layer held together by iminocarbonyl hydrogen bonds. *R* represents amino acid residues. Boxes indicate areas (Marrack's active patches) where amino acid residues *W*, *X*, *Y* and *Z*, which for some reason—perhaps being more polar—are of greater significance for antigenic determinants, recur in the same pattern at intervals.

Haurowitz[15] suggests that the antigenic determinants in proteins consist of definite arrangements of tyrosyl groups, free amino groups, and perhaps other groups, on the surface of the protein molecule. In such determinants polar groups play a prominent role.

The role of individual amino acid residues in the antigenic determinants of native proteins has not been established. Possibly any portion of a protein molecule of sufficient size can function as such a determinant, the specificity being determined by the number and arrangement of the various amino acids. Kleczkowski[35] found that when iodine atoms were introduced into all of the tyrosine of horse serum globulin, the ability to combine with the antibody for native horse serum globulin was lost. This would seem to demonstrate that the tyrosyl residue is an essential part of the determinant groups of this particular protein (assuming

(34) J. R. Marrack, The Chemistry of Antigens and Antibodies, Med. Res. Council, Special Report Series No. 230, London, 1938.

(35) A. Kleczkowski, *Brit. J. Exptl. Path.* **26**, 41 (1945).

halogenation produced no other changes). It does not, apparently, play such a key role in certain other proteins.

Pressman and Sternberger[36] found that a hapten combined specifically with an antibody protected it against the effects of iodination. Since the introduction of small amounts of iodine affects only the tyrosyl, histidyl and cysteinyl residues, it seems that some or all these residues may be essential parts of the specific combining groups. Porter[37] found that, although all the histidyl residues of rabbit γ-globulin and of ovalbumin were available for combination with fluorodinitrobenzene, one histidyl residue of the ovalbumin rabbit antiovalbumin was not available. This seems to indicate that histidine forms part of the combining group of the antibody or of the antigen.

a. Altered Proteins

The immunological reactions of chemically altered proteins have been studied in attempts to elucidate the chemical basis of antigenic specificity. We may summarize the results of such studies briefly.

Denaturation, even by drastic methods such as treatment with strong acids, acetone, alcohol, urea, or heating, does not always completely abolish the antigenic specificity of a protein (see Vol. I, Chap. 9). Depending on the extent of denaturation, denatured proteins may still react more or less with antibodies to the native protein. In some cases the denatured product will no longer react in one way (precipitate), but will in another (inhibit specifically the reaction of antibody and native antigen).[20,38–40] Injection of a denatured protein may on occasion produce antibodies reacting with native as well as denatured protein.[41] Heating most proteins in solution soon destroys their power to react visibly with their antibodies, although casein resists heating to 100°C. In such cases, however, antigenicity is not always lost, for such heated proteins will often produce antibodies if injected. In some cases the antibodies to a heated protein will react with heated proteins of other sorts.[30]

Heidelberger *et al.*[42–44] found immunological differences paralleling physical and chemical differences (viscosity, electrophoretic mobility,

(36) D. Pressman and L. Sternberger, *J. Immunol.* **66**, 609 (1951).
(37) R. R. Porter, *Biochem. J.* **46**, 473 (1950).
(38) M. Spiegel-Adolph, *Biochem. Z.* **170**, 126 (1926).
(39) F. Haurowitz and F. Bursa, *Rev. faculté sci. univ. Istanbul* **B10**, 283 (1945).
(40) J. O. Erickson and H. Neurath, *J. Exptl. Med.* **78**, 1 (1943).
(41) B. F. Miller, *J. Exptl. Med.* **58**, 625 (1933).
(42) P. H. Maurer and M. Heidelberger, *J. Am. Chem. Soc.* **73**, 2070 (1951).
(43) P. H. Maurer, M. Heidelberger, and D. H. Moore, *J. Am. Chem. Soc.* **73**, 2072 (1951).
(44) P. H. Maurer and M. Heidelberger, *J. Am. Chem. Soc.* **73**, 2076 (1951).

sedimentation in the ultracentrifuge) between egg albumin denatured by deamination and acid-denatured egg albumin. Neither product had reached the final step in the denatured state characterized by Astbury *et al.*[45] as a completely random one devoid of any specific structure. The studies of Heidelberger *et al.* support the suggestions that various degrees of denaturation exist.

The suggestion is that the sort of changes involved in denaturation do not invariably affect the fundamental nature of the antigenic determinants of a protein, but in some cases may produce changes of the same or similar nature in different proteins, giving them an immunological similarity which they previously lacked. We have little idea of the exact nature of such changes. Most of the chemical procedures listed below may denature the proteins more or less as well as making specific chemical changes in them.

Oxidation of proteins with permanganate has yielded products which would immunize. Antibodies were obtained which would precipitate the antigen injected, but not other proteins so treated, and not the untreated protein.[30] Thus species specificity seemed to be preserved, although the immunological nature of the protein was completely altered. We have no idea which of the changes resulting from permanganate treatment was responsible for the change.

Reduction of serum albumin with thioglycolic acid diminished the serological reactivity somewhat, but did not seem to affect that of ovalbumin.[46]

Digestion of proteins by enzymes, if allowed to go far, destroys their ability to precipitate with antibodies for the untreated protein, but injection of partially degraded protein resulting from treatment with pepsin may produce antibodies which react with the metaprotein and also with untreated protein.[30]

Acid and alkali treatment of proteins decreases their antigenic activity, alkalies being more active in this respect. Acid-treated proteins sometimes acquire the power of reacting with antibodies to unrelated proteins so treated.[30]

Deamination of casein did not change the more obvious antigenic characteristics.[47] This might suggest that free amino groups are not an important feature of the antigenic determinants of this protein. Maurer and Heidelberger[44] found that removal of about one third of the free amino groups from egg albumin did not influence its serological specificity.

Plakalbumin is formed from ovalbumin by the action of an enzyme of

(45) W. T. Astbury, S. Dickenson, and K. Bailey, *Biochem. J.* **29**, 2351 (1935).
(46) D. Blumenthal, *J. Biol. Chem.* **113**, 433 (1936).
(47) J. H. Lewis, *J. Infectious Diseases* **55**, 203 (1934).

Bacillus subtilis[48] with the loss of a peptide.[49,49a] It does not precipitate all the antibody from an antiserum against ovalbumin[50] which might suggest that the peptide plays some role in the antigenic determinants of native ovalbumin.

Formaldehyde-treated proteins may react with antibodies to formolized proteins of other species. Jacobs and Sommers[51] thought the changes due to formaldehyde resembled those due to heating more than they did the more marked changes following halogenation or nitration. Toxoids prepared by treating toxins with formaldehyde lose their toxic properties but retain their antigenicity and precipitability by antibodies virtually unimpaired. This suggests that the groups altered by formaldehyde are essential for the toxic groups of the protein, but not for the groups responsible for its immunological specificity. The two sorts of groups are clearly different.

It may be that the free amino groups are essential to the toxicity and not to the antigenicity, although Hewitt[52] believes that the reaction of formaldehyde with proteins in such cases is different from that with the free amino groups, being slow and irreversible, and involving smaller amounts of formaldehyde. In any case, other treatments which involve reaction with free amino groups, such as treatment with ketene gas, abolish the toxicity of toxin.[53,54]

It is of interest that the optical rotation of toxin is not altered by formaldehyde treatment, suggesting that the optically active carbon atoms adjacent to the peptide linkages have not been racemized.[55]

Esterification of proteins by treatment with alcohol in the presence of acid, or treatment with diazomethane, produced insoluble products which were still antigenic, but had lost their ability to react with antibodies for the unchanged protein.[30]

Acylation by treatment with acetic anhydride gave products behaving somewhat like methylated proteins. Landsteiner[30] concluded that the amino and hydroxyl groups are the ones most affected. Some cross reaction was found between treated proteins containing different acyl groups.

(48) K. Linderstrøm-Lang and M. Ottesen, *Nature* **159,** 807 (1947).
(49) N. Eeg-Larsen, K. Linderstrøm-Lang, and M. Ottesen, *Arch. Biochem.* **19,** 340 (1948).
(49*a*) M. Ottesen and A. Wollenberger, *Nature* **170,** 801 (1952).
(50) M. Kaminski and P. Grabar, *Bull. soc. chim. biol.* **31,** 684 (1949).
(51) S. L. Jacobs and S. C. Sommers, *J. Immunol.* **36,** 531 (1939).
(52) L. F. Hewitt, *Biochem. J.* **24,** 983 (1930).
(53) H. Goldie, *Compt. rend. soc. biol.* **126,** 974, 977 (1937).
(54) A. M. Pappenheimer, *J. Biol. Chem.* **125,** 201 (1938).
(55) M. D. Eaton, *Bact. Revs.* **2,** 3 (1938).

Introduction of benzoyl groups by treatment with benzoyl chloride gives a new specificity common to the benzoylated proteins.[56]

By treatment with carbobenzoxy chloride, Gaunt and Wormall[57] produced carbobenzoxy proteins. They considered that the reaction was concerned mainly with the free amino groups. The original species specificity was almost completely destroyed, and the antibodies produced reacted with other proteins similarly treated. The reactions were completely inhibited by carbobenzoxyamino acids, indicating that the most essential part of the new determinant is $—NHCOOCH_2C_6H_5$. This is supported by the observation that phenylcarbamido-amino acids also inhibited somewhat.

Halogenation of proteins has been widely studied.[58,59] Iodinated proteins acquire a new specificity, and antisera to them react with other iodoproteins. The brominated proteins behave only slightly differently from the iodinated proteins, and cross-react with them. Wormall[59] found that the reaction of iodoprotein with its antibody was specifically inhibited by 3,5-diiodo- or 3,5-dibromotyrosine. Snapper and Grunbaum[60] found that any substance containing the 3,5-diiodo-4-hydroxy group would inhibit the reaction. Thus the essential grouping in the antigenic determinants of halogenated proteins turns out to be the halogenated tyrosine (and possibly halogenated histidine, which would doubtless behave very similarly). It is possible, however, that the antibodies produced are strictly specific, not for halogenated tyrosine as such, but for halogenated tyrosine in the framework of the immediately adjacent groups of the protein.

Proteins into which small amounts of iodine have been introduced react serologically like untreated protein, and are stable.[61]

Nitration of proteins can be accomplished by treatment with nitric acid or with tetranitromethane.[30,59] The xanthoproteins produced cross-react serologically with each other. This is probably due to the nitrotyrosyl and nitrotryptophyl groups, for Mutsaars[58] demonstrated specific inhibition with compounds containing nitro and hydroxyl groups in a benzene ring, plus a carboxyl group, free or esterified. Nitration of gelatin, which, because of its relative deficiency in aromatic amino acids,

(56) A. Medveczky and A. Uhrovits, *Z. Immunitätsforsch.* **72,** 256 (1931).
(57) W. E. Gaunt and A. Wormall, *Biochem. J.* **33,** 908 (1938).
(58) W. Mutsaars, *Ann. inst. Pasteur* **62,** 81 (1939).
(59) A. Wormall, *J. Exptl. Med.* **51,** 295 (1930).
(60) I. Snapper and A. Grunbaum, *Brit. J. Exptl. Path.* **17,** 361 (1936).
(61) F. D. S. Butement, *Nature* **161,** 731 (1948); *idem, J. Chem. Soc.* **1949,** 408; H. M. Eisen and A. G. Keston, *J. Immunol.* **63,** 71 (1949); W. C. Knox and F. C. Endicott, *J. Immunol.* **65,** 532 (1950).

takes up few nitro groups, does not render it antigenic,[62,58] but nitrated gelatin will inhibit the reaction of antibodies to other nitrated proteins with their antigens.

Residual Native Specificity. In the case of coupled proteins some antibodies specific for the untreated protein are generally formed. Even the antibodies directed toward the introduced groups react better with that group when it is coupled to tyrosine. Hooker and Boyd,[63] using antibodies for horse serum proteins coupled with arsanilic acid and casein–arsanilic acid as the test antigen, found that the specific inhibitory activity per mole of compound tested increased rapidly in the order: arsanilic acid, phenol–azoarsanilic acid (or imidazole–azoarsanilic acid), gelatin–azoarsanilic acid. Since the inhibitory power of a hapten is greater, the greater its resemblance to the actual determinant in the antigen responsible for the production of the antibody, this indicated that the "antiarsanilic" antibodies were directed not solely toward the introduced arsonic acid group, but toward the tyrosyl (and histidyl) azoarsanilic residues, and showed some influence of other (presumably adjacent) parts of the intact protein molecule. Landsteiner,[30] using antibodies to methylated, acylated, halogenated, and nitrated proteins, where antibodies to the native protein are not always produced, observed that if the tests were carried out with diminishing quantities of antibodies, the reactions were stronger with the exactly homologous antigen, showing that the specificity depends to some extent upon the specific protein background.

Diazotization. Treatment of proteins with nitrous acid results in deamination and introduction of diazo groups into the tyrosine and tryptophan (and possibly histidine) nuclei[64] and produces antigens that are almost indistinguishable from the nitroproteins.[30,59,64a,65]

Mustard gas (**II**) couples with proteins, as does the corresponding sulfone (**III**). In each case a new specificity was conferred, although

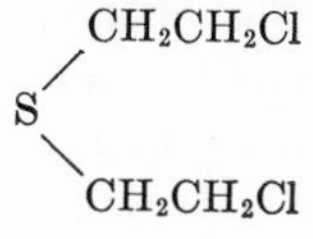

$$S(CH_2CH_2Cl)_2 \quad \textbf{II} \qquad O_2S(CH_2CH_2Cl)_2 \quad \textbf{III}$$

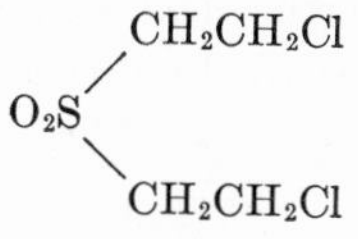

(62) W. C. Boyd, Fundamentals of Immunology, 2nd ed., Interscience Publishers, New York, 1947.

(63) S. B. Hooker and W. C. Boyd, *J. Immunol.* **25**, 61 (1933).

(64) J. St. L. Philpot and P. A. Small, *Biochem. J.* **32**, 542 (1938); A. Morel and P. Siley, *Bull. Soc. chim.* (**4**), **43**, 881 (1928); R. M. Herriott, *Advances in Protein Chem.* **4**, 169 (1947).

(64*a*) K. Landsteiner and E. Prášek, *Z. Immunitätsforsch.* **20**, 211 (1913).

(65) W. Mutsaars, *Compt. rend. soc. biol.* **129**, 510, 511 (1938).

it was not as striking as in some of the cases discussed above. The mustard gas and sulfone derivatives did not cross-react, indicating an important influence of the oxygen group and/or the state of oxidation of the sulfur.[66] Berenblum and Wormall[66] considered the reaction to involve the free amino groups of the proteins.

Phenyl isocyanate combines with the free amino groups of lysine.[67] Although not all the free amino groups are substituted, cross reactions with untreated protein are much reduced (suggesting the importance of amino groups in specificity) and a new specificity conferred. The reactions can be specifically inhibited by compounds of phenylisocyanate with lysine or ε-aminohexanoic acid, while compounds of other amino acids are less effective. The antisera react weakly with altered proteins prepared by treating proteins (including gelatin) with diazotized aniline, presumably because of the similarity of the groups $—CONHC_6H_5$ and $—N{=}NC_6H_5$.

β-Naphthoquinone sulfonate coupled with proteins yields antigens which produce antibodies which react with unrelated proteins similarly treated.[68] It is thought that the reaction involves the amino groups.

Wetter and Deutsch[69] studied the effects of acetylation, esterification, iodination, and coupling with diazo compounds of an antigen (ovomucoid), and Grabar and Kaminski[70] studied the effects of conversion of ovalbumin to plakalbumin, diazotization, acetylation, denaturation by heat, shaking, and ultrasonic vibration.

Malkiel[71] studied the effects of ultrasonic vibration on tobacco mosaic virus, and concluded that the treatment uncovered other antigenic groups in the molecule.

b. Conjugated Proteins

Conjugated proteins may be distinguished (rather arbitrarily) from altered proteins by the fact that the reactions of antibodies produced can be studied with the introduced group and its homologs independently of its protein matrix. This enables information to be obtained on the precise limit to the complexities of chemical structure of the antigen which antibodies may reflect in their structure, and is the chief source of knowledge of this important subject. The influence of antibodies

(66) I. Berenblum and A. Wormall, *Biochem. J.* **33**, 75 (1939).
(67) S. J. Hopkins and A. Wormall, *Biochem. J.* **27**, 740, 1706 (1933).
(68) O. Fujio, *J. Biochem.* (Japan) **33**, 241 (1941).
(69) L. R. Wetter and H. F. Deutsch, *Arch. Biochem.* **28**, 122 (1950).
(70) P. Grabar and M. Kaminski, *Bull. soc. chim. biol.* **32**, 620 (1950).
(71) S. Malkiel, *J. Immunol.* **57**, 51 (1947).

specific for the protein carrier (which are usually also produced) is avoided by studying the reactions of the antibodies with an unrelated protein similarly coupled. Three general methods of attaching the introduced group have been utilized: (*a*) coupling, through the sulfhydryl groups, with organic halogen compounds; (*b*) coupling with azides; (*c*) coupling with diazotized amines. The last method has been used far more than the other two.

(*1*) *Coupling through Sulfhydryl Groups.* Pillemer, Ecker, and Martiensen[72] introduced chemical groupings into proteins by reducing the disulfide sulfur to sulfhydryl groups and allowing these to react with organic halogen compounds according to the formula:

$$ASH + RX \rightarrow ASR + HX$$

This method is especially applicable to keratins because of the high percentage of disulfide sulfur that they contain.

(*2*) *Treatment with Azides.* Clutton, Harington, and Mead[73] and Clutton, Harington, and Yuill[32] introduced the structure and its *N*-carbobenzoxy derivative into proteins by use of the azide. This enabled

$$C_6H_{11}O_5—OC_6H_4CH_2CHNH_2CO—$$

them to study the effect of the introduced *O*-β-glucosido-*N*-carbobenzoxytyrosyl group. The original specificity was entirely masked and a new specificity produced. Such treatment rendered gelatin antigenic. Thyroxyl groups introduced in a similar way into proteins produced antigens which produced antibodies reacting slightly with thyroglobulin, suggesting that thyroxine is an important part of the antigenic determinants of this protein.

(*3*) *Treatment with Diazotized Amines.* Any compound that can be obtained coupled with a primary aromatic amine can be introduced by this technique, and the method has enabled the specificity of antibodies to peptides, carbohydrates, alkaloids, and simple compounds such as dicarboxylic acids to be studied. We may take up first the conclusions drawn by Landsteiner[30] from a study of antisera to proteins coupled with a variety of substituted aromatic amines. A general idea of the sort of result obtained can be had from Table I.

In some cases the specificity seemed to be complete; that is, the antibodies reacted only with the homologous antigen, but more frequently cross reactions were obtained, although the reaction with the homologous antigen was nearly always more intense. The antibody-forming mechanism seems able to distinguish rather fine differences in the structure of

(72) L. Pillemer, E. E. Ecker, and E. W. Martiensen, *J. Exptl. Med.* **69,** 191 (1939).
(73) R. F. Clutton, C. R. Harington, and T. H. Mead, *Biochem. J.* **31,** 764 (1937).

TABLE I

EXAMPLE OF THE SPECIFICITY OBSERVED IN ANTISERA TO ARTIFICIAL ANTIGENS OBTAINED BY COUPLING PROTEINS WITH SIMPLE AROMATIC AMINES[a]

Antisera made with	Tested against antigens made with					
	NH_2	NH_2, COOH	NH_2, COOH	NH_2, Cl, COOH	NH_2, CH_3, COOH	NH_2, SO_3H
NH_2, COOH	0	+++	0	++++	+++	+
NH_2, SO_3H	0	0	0		0	++++

0, no reaction; +, positive reaction; ++++, very strong reaction.

[a] K. Landsteiner and H. Lampl, *Biochem. Z.* **86**, 343 (1918).

these introduced groupings (haptens), but it is not absolutely specific. Evidently the antibodies, which reflect in some way the electronic configuration of the hapten, do not reflect every detail perfectly—"There is some blurring of outlines."

c. *Importance of Polar Groups*

Landsteiner concluded that the nature of the acid substituents in a hapten was of decisive influence. Antibodies to one sulfanilic acid-treated protein reacted with several sulfonic acids but showed little reaction with carboxylic acids and vice versa. The determining influence of the arsonic acid group was even more striking.

Introduction of "neutral" groups such as methyl, halogen, methoxy and nitro groups had less influence. A great many cross reactions were observed among such compounds, except those containing carbonyl groups or acetylamino groups. Strongly basic groups are also of decisive influence on specificity.[74]

The importance of charged groups in haptens was further shown by Landsteiner and van der Scheer[75] when they demonstrated that antibodies to the "neutral" haptens reacted also with proteins coupled with the methyl ester of *p*-aminobenzoic acid, whereas this coupled protein was hardly affected by antibodies for *p*-aminobenzoic acid. Hydrolysis of the coupled protein, however, to remove the methyl groups, gave an antigen no longer reacting with antibodies for the "neutral" haptens, and reacting strongly with antibodies to the acidic *p*-aminobenzoic acid.

From the pronounced effect of esterification, Landsteiner[30] concluded that the terminal group of an introduced hapten has a particularly significant influence on the specificity. The mere presence of acid groups also seems to define more sharply the specificity of the nucleus in which they occur, enabling other substituents in this nucleus to exert more influence than they otherwise would.

d. *Effect of Isomers*

Position in the benzene nucleus of substituents was found to have an important effect. This was shown by the specificity of the three different aminobenzoic acids and aminocinnamic acids. The position of the acidic group relative to the azo group (the point of attachment to the protein) determined the specificity and whether or not cross reactions would occur.[30]

Landsteiner investigated aliphatic chain haptens by attaching them,

(74) F. Haurowitz, *J. Immunol.* **43,** 331 (1942).
(75) K. Landsteiner and J. van der Scheer, *J. Exptl. Med.* **45,** 1045 (1927).

usually by a —CONH— linkage, to a benzene ring bearing another amino group. The aliphatic part could be varied at will.[30]

Antibodies for oxalic and succinic acids were quite specific, but with the higher acids (adipic and suberic) overlapping cross reactions were observed, and lengthening the chain by one methylene group did not make so much difference as it had in the case of oxalic acid.[30]

Pauling and Pressman[76] studied the inhibition by haptens of the precipitation with antisera homologous to *o*-, *m*-, and *p*-azophenylarsonic groups, and concluded from the degrees of cross reaction observed that the combining site on the antibody was adapted to conform spatially to the homologous hapten, although it could, presumably by slight stretching (about 1 A.), accommodate also the isomeric haptens. They also found good agreement between calculated van der Waals' attraction energies and free energy changes, $RT \ln K_0'$.

e. Specificity of Peptides

The specificity of amino acids and peptides was studied[30] by similarly condensing them with aromatic amines. The antibodies produced were quite specific, and cross reactions were obtained only with closely related amino acids, as between glycine and alanine, valine and leucine, and aspartic and glutamic acids. Even these related amino acids could be differentiated without difficulty.

Antibodies to dipeptides precipitated most strongly the homologous antigen, but gave overlapping reactions with other peptides in which the terminal amino acid was the same as in the immunizing antigen. Again, the influence of the terminal group of a hapten (and, by inference, of any antigenic determinant) was in evidence. With longer peptides, cross reactions traceable to some of the nonterminal amino acids were observed.

The experiments with peptide haptens support the idea that the antigenic determinants in native proteins are patches composed of characteristic arrangements of amino acids. The variety observed in the relatively simple peptides studied (up to pentapeptides) suggests that there is ample room for the wide variety of specificities actually observed with protein antigens.

f. Influence of Optical Isomers

The above observations on isomers indicated that the spatial arrangement of groups, as well as their chemical nature, was of importance. This was directly in line with the behavior of enzymes and optical isomers of drugs and dyestuffs. It was to be expected therefore that the antibody-forming mechanism could distinguish optical isomers.

(76) L. Pauling and D. Pressman, *J. Am. Chem. Soc.* **67**, 1003 (1945).

Landsteiner and van der Scheer[76a] were able to differentiate serologically between the D- and L-isomers of *p*-aminobenzoylphenylaminoacetic acids:

$$H_2N-C_6H_4-CONH-C(H)(COOH)-C_6H_5 \quad \text{and} \quad H_2N-C_6H_4-CONH-C(COOH)(H)-C_6H_5$$

It was also possible to distinguish the three forms of tartaric acid, dextro, levo, and meso, one from another. The antisera produced reacted in the predicted way with the isomers of malic acid. It was pointed out that unknown optical configurations might even be established by serological means, where other easier methods are not available.

Goebel and Avery[77] carried the serological differentiation of steric isomers still further. They prepared antibodies that clearly distinguished glucose and galactose, and others that differentiated α- and β-glucose. The distinction in the latter case was less sharp, and one gains the impression that this is about the limit of the powers of the immune mechanism to distinguish stereoisomeric differences.

Substitution of an acetyl group for the hydrogen of the sixth carbon atom of a β-glucoside produced a new hapten which did not react with antibodies for the α-glucoside, and which reacted less than did the homologous hapten with the antibodies to the unacetylated β-glucoside. In studies of the disaccharides as haptens,[78] it was found that the determining features were the molecular pattern of the saccharide as a whole, the spatial and chemical configuration of the terminal hexose, and the position of the linkage between the two sugars. Antibodies to glucose and glucuronic acid were entirely distinct, showing no crossing.[79] This again illustrates the strong influence of strongly polar groups.

Antibodies specific for various organic compounds have been obtained, including heterocyclic compounds,[80,81] thyroxine,[32] pyrazolone derivatives,[82,83] carcinogenic hydrocarbons,[84] strychnine and brucine,[85] and sulfonamides,[86,87] but the results have not added very much to the

(76*a*) K. Landsteiner and J. van der Scheer, *J. Exptl. Med.* **48**, 315 (1928).
(77) W. F. Goebel and O. T. Avery, *J. Exptl. Med.* **50**, 521, 533 (1929).
(78) W. F. Goebel, O. T. Avery, and F. H. Babers, *J. Exptl. Med.* **60**, 598 (1934).
(79) W. F. Goebel, *J. Exptl. Med.* **64**, 29 (1936).
(80) H. Erlenmeyer and E. Berger, *Arch. exptl. Path. Pharmakol.* **177**, 116 (1935).
(81) K. Landsteiner and N. W. Pirie, *J. Immunol.* **33**, 265 (1937).
(82) E. Berger, *Schweiz. med. Wochschr.* **17**, 1309 (1936).
(83) R. A. Harte, *J. Immunol.* **34**, 433 (1938).
(84) H. J. Creech and W. R. Franks, *Am. J. Cancer* **30**, 555 (1937).
(85) S. B. Hooker and W. C. Boyd, *J. Immunol.* **38**, 479 (1940).
(86) I. E. Gerber and M. Gross, *J. Immunol.* **48**, 103 (1944).
(87) A. G. Wedum, *J. Infectious Diseases* **70**, 173 (1942).

existing knowledge of the nature and size of the antigenic determinants in native protein antigens.

Pauling and collaborators[88] carried out extensive studies on antibodies to artificial antigens made by coupling *o*-, *m*-, and *p*-arsanilic acid and *p*-(*p*-aminophenylazo)phenylarsonic acid to various proteins and their reaction with the corresponding haptens. They made quantitative observations which they considered to establish the multivalency (or bivalency at any rate) of antibody, and to throw light on the bond energies involved, both in the reaction of hapten with homologous antibodies, and with antibodies to isomeric haptens. The regular precipitability of bivalent haptens made from diazotized arsanilic acid was considered by Pauling to be strong evidence for the multivalency of antibody (a subject then in dispute), but it was later shown[89,90] that many, perhaps most, of the bi- and trivalent haptens used were not molecularly dispersed in the solutions used. In fact Pauling *et al.*[91] themselves (except with one hapten) failed to obtain such precipitation with antibodies homologous to the *p*-azobenzoic acid group, thus confirming the earlier work of Hooker and Boyd.[92]

Karush[93] has pointed out that large proportions of such organic compounds when added to serum are absorbed by the albumin and that this affects the quantitative aspects of the reactions between antibody and the inhibiting or precipitating haptens. This has been confirmed by Pardee and Pauling.[94] Marrack states[95] "the calculations by Pauling and his colleagues, based on the results of experiments in which these dyes were added to whole serum, are invalid. Those concerned with inhibition are invalid for one reason—the adsorption of the inhibitor on the albumin—and those concerned with the amount of precipitate formed

(88) L. Pauling, D. Pressman, D. H. Campbell, C. Ikeda, and M. Ikawa, *J. Am. Chem. Soc.* **64,** 2994 (1942); L. Pauling, D. Pressman, D. H. Campbell, and C. Ikeda, *J. Am. Chem. Soc.* **64,** 3003 (1942); L. Pauling, D. Pressman, and C. Ikeda, *J. Am. Chem. Soc.* **64,** 3010 (1942); D. Pressman, D. H. Brown, and L. Pauling, *J. Am. Chem. Soc.* **64,** 3015 (1942); D. Pressman, J. T. Maynard, A. L. Grossberg, and L. Pauling, *J. Am. Chem. Soc.* **65,** 728 (1943); L. Pauling, D. Pressman, and A. L. Grossberg, *J. Am. Chem. Soc.* **66,** 784 (1944); L. Pauling and D. Pressman, *J. Am. Chem. Soc.* **67,** 1003 (1945).

(89) W. C. Boyd and J. Behnke, *Science* **100,** 13 (1944).

(90) A. B. Pardee and S. M. Swingle, *J. Am. Chem. Soc.* **71,** 148 (1949).

(91) D. Pressman, S. M. Swingle, A. L. Grossberg, and L. Pauling, *J. Am. Chem. Soc.* **66,** 1731 (1944).

(92) S. B. Hooker and W. C. Boyd, *J. Immunol.* **42,** 419 (1941).

(93) F. Karush, *J. Am. Chem. Soc.* **72,** 2705 (1950).

(94) A. B. Pardee and L. Pauling, *J. Am. Chem. Soc.* **71,** 143 (1949).

(95) J. R. Marrack, *Ann. Repts. on Progress Chem.* (Chem. Soc. London) **48,** 249 (1951).

by dye with antiserum for three reasons—the polymerization of the dye, the adsorption of the dye on albumin, and the co-precipitation of inert protein." Eisen and Karush[96] comment on the large amount of antibody recovered by Campbell *et al.*,[97] and Marrack[95] suggests that some of it was inert protein precipitated and reprecipitable by the dye used.

Pardee and Pauling[94] believe that in serum the equilibrium between monodisperse and aggregated dye is shifted in the direction of the monomeric form, and Pressman and Siegal[98] concur that most of the conclusions based on the experiments with dye haptens are still justified. In favor of this, Campbell and Bulman[27] point out that a dye containing two different haptenic groups, R and X, did not precipitate with either anti-R or anti-X serum alone, which it should have done if it were aggregated, but did precipitate with a mixture of the antisera.

3. Number of Antigenic Determinants per Molecule

It is not possible at present to calculate how many antigenic determinants, or "active patches," there are on the surface of a given protein molecule. The only experimental approach is the analysis of specific precipitates, which makes possible an estimation of the number of antibody molecules that can unite simultaneously with one molecule of antigen. Clearly, there must be at least as many antigenic determinants per molecule as the largest value so obtained for any given antigen. Typical results that have been obtained are shown in Table II.

It will be seen that there is a rough sort of correlation between molecular weight and "valence" of an antigen. In fact, by making the

TABLE II
Calculated "Valence" of Certain Protein Antigens

Antigen	Approx. molecular weight,[a] $\times 10^{-3}$	Valence
Ovalbumin	43–44	5
Serum albumin	70	6
Diphtheria toxin	70	8
Thyroglobulin	650	40
Busycon hemocyanin	6500	74
Viviparus hemocyanin	6500	231

[a] See Chap. 7.

(96) W. Eisen and F. Karush, *J. Am. Chem. Soc.* **71,** 363 (1949).
(97) D. H. Campbell, R. H. Blaker, and A. B. Pardee, *J. Am. Chem. Soc.* **70,** 2496 (1948).
(98) D. Pressman and M. Siegal, *J. Am. Chem. Soc.* **75,** 686 (1953).

assumption that antibody has a molecular weight of 160,000 and attaches itself to the surface of the antigen by "lying on its side," when antibody and antigen combine in equivalent proportions (see below), it is possible to predict that the relation between molecular weight (M) of the antigen and the ratio by weight (R) of antibody to antigen in such "equivalence point" compounds should be[98a]

$$R = 37{,}800\ M^{-0.8} + 179\ M^{-0.35}$$

This equation predicts the results of actual analyses moderately well (see Fig. 2).

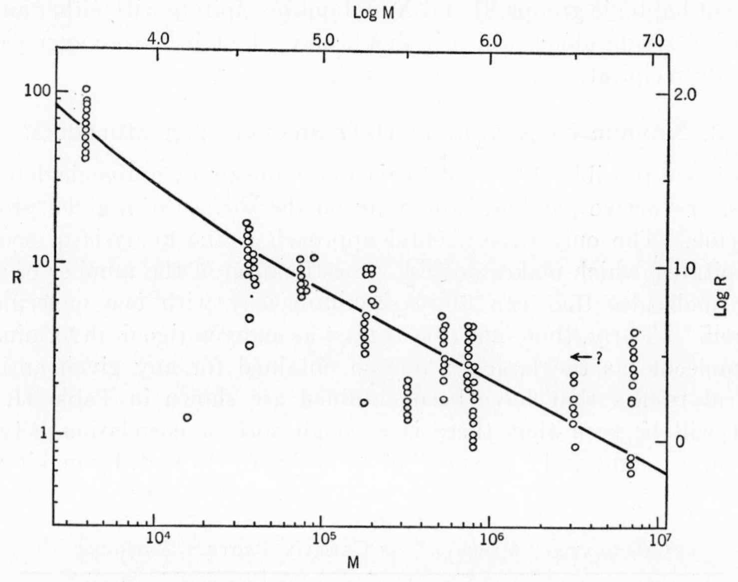

FIG. 2. Expected relation between molecular weight of antigen (M) and ratio by weight of antibody to antigen in equivalence-point precipitates, calculated from geometrical considerations. Circles show experimental data.

It is apparent that values such as those in Table II can only be a minimum estimate of the "valence" of antigens; for at the point at which the surface of the antigen is just covered, for which the above equation is approximately valid, combining sites exist which are as yet unused, as shown by the fact that if more antibody is added, more will combine specifically, and compounds richer in antibody will be obtained. Even when the maximum possible number of antibody molecules have combined, some unused combining sites may remain, but no antibody molecules reach them because of the steric hindrance by the antibody

(98*a*) W. C. Boyd and S. B. Hooker, *J. Gen. Physiol.* **22**, 281 (1939).

molecules already in position. In fact if this were not the case, it would indicate a rather remarkable spacing of the antigenic determinants on the surface of the antigen molecule. They would have to be arranged at just such distances that each could be reached by a molecule of antibody, and that when this was accomplished the surface would be completely covered by antibody molecules "standing on their heads," for this is what the analytical figures indicate. It is more likely that some unused sites exist, and that the estimates of the number of antigenic determinants based on the analyses of precipitates are minimum values.

4. Specificity of Native Proteins

Our ideas of the chemical nature of the antigenic determinants of native proteins are the result of deduction and extrapolation from the above experiments. On the whole they agree well with the facts as we know them. Before discussing the specificity of native proteins we should define two terms which are often employed in this connection: "functional specificity" and "species specificity."

a. Functional Specificity

Various proteins of the body which are adapted to different physiological and chemical functions have, naturally, different chemical structures. The percentage of the various amino acids may vary, or the serial arrangement of these amino acids in the polypeptide chain may differ. It is thus not surprising that they also differ immunologically. Antibodies that are produced by injecting these proteins into an animal of a different species readily distinguish hemoglobin and serum albumin. On the other hand, serum albumins of related species, all evidently fulfilling roughly the same function, are chemically similar, and at one time during World War II it was hoped that bovine serum albumin could replace human serum albumin as a blood substitute. More than 100 volunteers had been injected before the first untoward reaction occurred. Later two rhesus monkeys were injected with bovine serum albumin, and showed no reaction, even though kept under observation for a long period of time.[99] Serum albumins of different mammalian species are evidently antigenically similar, but not absolutely identical.

b. Species Specificity

Even when proteins are fulfilling, so far as we can tell, the same function in different species, they are still likely to be somewhat different antigenically. More than one chemical structure can be efficient in a given process, and evidently large numbers, similar but different, exist.

(99) W. C. Boyd, Unpublished experiments (1942).

Each species has its own pattern for protein synthesis, and these seldom or never are duplicated exactly. As species diverge in the course of evolution, chemical as well as morphological differences develop.

The classical investigation of species specificity is that of Nuttall,[100] who studied the reactions of rabbit antisera to the serum proteins of various animals. He found that the degree of cross reaction paralleled, on the whole, the taxonomic similarity between the species. If the amount of specific precipitate produced when antihuman serum was mixed with a test portion of human serum were rated as 100 per cent, the amount of precipitate produced with gorilla serum as test antigen was 64 per cent, and only 3 per cent with the serum of the dog (Table III).

TABLE III

RELATIVE AMOUNTS OF PRECIPITATE WITH THREE DIFFERENT ANTI-HUMAN RABBIT SERA AND EQUAL AMOUNTS OF BLOOD SERUM OF VARIOUS SPECIES[a]

Species	Immune serum number[b]		
	1	2	3
Man	100	100	100
Chimpanzee	—	—	130[c]
Gorilla	—	—	64
Orangutan	47	80	42
Cynocephalus mormon	30	50	42
Cercopithecus petaurista	30	50	—
Ateles vellerosus	22	25	—
Cat (*Felix domesticus*)	11	—	3
Dog (*Canis familiaris*)	11	—	3

[a] G. H. Nuttall, Blood Immunity and Blood Relationship, Cambridge Univ. Press, London, 1904.

[b] Amount of precipitate obtained with human blood taken arbitrarily as 100.

[c] Loose precipitate.

The results obtained depend somewhat upon the individual serum employed (practically all from rabbits), but are more or less consistent.

Reichert and Brown,[101] in a study on the crystallography of hemoglobins, found differences in the crystal forms in all the hemoglobins they studied, although the hemoglobin of the mule was, not surprisingly, crystallographically similar to that of the horse, and also that of the ass. Heidelberger and Landsteiner[102] found that antibodies to hemoglobins of the horse, ass, and mule indicated just such differences and similarities.

(100) G. H. Nuttall, Blood Immunity and Blood Relationship, Cambridge, 1904.
(101) E. T. Reichert and A. P. Brown, *Carnegie Inst. Wash. Pub.* No. **116** (1909).
(102) M. Heidelberger and K. Landsteiner, *J. Exptl. Med.* **38**, 561 (1923).

Crystallography and serology evidently reflect the chemical differences in these molecules about equally well.

It is now known that at least five kinds of human hemoglobin[102a] with differing electrophoretic mobilities, and sometimes different solubilities, exist. Fetal hemoglobin (hemoglobin F) is different from the hemoglobin of adults.[103] In sickle-cell anemia, another hemoglobin, called hemoglobin S, occurs in sufferers of this rare disease, which is almost entirely restricted to Africa and parts of India. Another abnormal hemoglobin, first called hemoglobin III and now hemoglobin C, seems to be due to the action of a dominant gene.[104] Another type, not producing disease by itself, has been reported.[105]

c. Toxins

The antigenic poisons which we call toxins are discussed elsewhere in this book (Chap. 15). In addition to the toxic substances produced by bacteria, animal poisons, such as snake venom, scorpion venom, and spider poisons; and plant toxins, such as ricin from the castor bean (*Ricinus communis*) and abrin from the seeds of the Indian licorice (*Abrus precatorius*) have been studied immunologically.

The animal toxin of the rattlesnake (*Crotalus terrificus*) has been crystallized by Slotta and Fraenkel-Conrat.[106] The protein so obtained seemed to be homogeneous and to possess a molecular weight of about 30,000.[106a] Both the toxicity and the hemolytic effect *in vitro* seemed to depend on the lecithinase activity of the molecule. Cobra venom has been similarly purified.

The immunological behavior of toxins is not different from that of other protein antigens in any significant respect. In injecting them to produce antibodies, small doses must at first be used to avoid killing the experimental animals, but otherwise there is nothing distinctive about the immunology of toxins. The neutralization of toxins by antitoxins is discussed below (p. 820).

No toxins have been studied in which the toxic property resided in a separable prosthetic group. The toxicity seems to be an integral part of the molecule, evidently determined by some particular grouping of amino acids. It is interesting that it can be destroyed, in the case of diphtheria and certain other toxins, by the action of formaldehyde or

(102*a*) J. V. Neel *et al.*, *Science* **118,** 116 (1953).
(103) G. H. Beaven, H. Hoch, and E. R. Holiday, *Biochem. J.* **49,** 374 (1951).
(104) J. V. Neel and H. A. Itano, *Trans. Assoc. Am. Physicians* **45,** 203 (1952).
(105) H. Ranney, in press.
(106) K. H. Slotter and H. L. Fraenkel-Conrat, *Ber.* **71B,** 1076 (1938).
(106*a*) N. Grälen and T. Svedberg, *Biochem. J.* **32,** 1375 (1938).

ketene, leaving the antigenic and antibody-combining powers of the molecule intact.

III. Antibodies

Mammals (and possibly other organisms) produce modified γ-globulins in response to bacterial invasion, to other infections, and often to artificial immunization. These γ-globulins have the property of combining firmly and specifically with the antigens that caused their production.

Antibodies are recognized by their specific reaction with the antigen. The nature of the reaction may be much the same in all cases, but the effects, as visible in the test tube, are often rather different. Antibodies, when they react with their antigens, may (*a*) precipitate them (in the case of soluble protein, or carbohydrate, antigens); (*b*) agglutinate them (in the case of microorganisms or other particulate antigens); (*c*) cause complement to be "fixed" (in the case of practically any antibody–antigen combination); (*d*) sensitize them to the lytic action of complement; (*e*) occupy the combining sites of the antigen so as to prevent the expected action of another antibody of similar specificity (blocking antibodies); (*f*) produce a system that can be agglutinated or precipitated by an antibody to the antibody.

1. Multiple Functions of Antibodies

The different names for antibodies: precipitins, agglutinins, lysins, amboceptors, etc., which appear in the immunological literature, should not be taken to imply that these antibodies are necessarily different. Originally a separate name was given to the then hypothetical cause of each serological phenomenon, but it was later recognized that the same antibody molecule might perform all or most of these various functions. For example, Heidelberger and Kabat[107] found that the actual amount of antibody in a type I antipneumococcus serum, whether determined as precipitin or as agglutinin, was exactly the same. Doerr and Russ[108] found the anaphylactic sensitizing effect of antisera to be quantitatively proportional to the precipitin content, and Chow, Lee and Wu[109] found that immunologically pure precipitin would protect mice from type I pneumococcus, produce anaphylactic sensitization, or fix complement.

It is true that a single homogeneous antigen may give rise to more than one antibody, and in fact this is probably usually the case.[110-116]

(107) M. Heidelberger and E. A. Kabat, *J. Exptl. Med.* **63,** 737 (1936).
(108) R. Doerr and V. K. Russ, *Z. Immunitätsforsch.* **3,** 181 (1909).
(109) B. F. Chow, K. H. Lee, and H. Wu, *Chinese J. Physiol.* **11,** 139 (1937).
(110) A. G. Cole, *Arch. Path.* **26,** 96 (1938).
(111) K. Goodner and F. L. Horsfall, *J. Exptl. Med.* **66,** 413, 425, 437 (1937).

However, each of these antibodies can usually perform all the functions listed below. There is some evidence that suggests that certain kinds of antibodies, e.g., agglutins and lysins, may under some circumstances be at least partially distinct.

a. Precipitating Antibodies (Precipitins)

Certain antibodies, when mixed with the antigen, cause the formation of a precipitate. Analysis of such precipitates reveals that they consist of definite molecular compounds of antibody and antigen. Such antibodies are called *precipitins*. Both carbohydrate and protein antigens may be precipitated by appropriate antibodies. The nature of the chemical reactions involved will be discussed below (p. 815).

b. Agglutinating Antibodies (Agglutinins)

Antibodies to pathological microorganisms (and also to harmless foreign cells such as erythrocytes) very often cause the particles with which they react to stick together in clumps. This phenomenon is called agglutination. There is no particular reason to think that agglutination is essentially different from precipitation; it is merely that one of the reagents, the bacteria or red cells, is composed of relatively large particles, and the sticking together of these particles is a conspicuous phenomenon.

c. Neutralizing Antibodies

Antibodies to toxins (antitoxins) will neutralize the toxic effect when mixed with the toxin in the proper proportions. This can be demonstrated by injecting the mixtures into susceptible animals. The neutralization of lytic antibodies can be demonstrated by testing the lysin–antilysin mixtures against susceptible cells in the presence of complement.

Toxins are not altered chemically by the combination with antitoxins, as can be demonstrated by destroying the antitoxin and recovering the toxin.[117–119] The function of the antitoxin is probably merely to cover the surface of the toxin molecule and thereby prevent its toxic

(112) M. Heidelberger and F. E. Kendall, *J. Exptl. Med.* **61**, 559 (1935).
(113) S. B. Hooker and W. C. Boyd, *J. Immunol.* **30**, 41 (1936).
(114) K. Landsteiner and J. van der Scheer, *J. Exptl. Med.* **63**, 325 (1936).
(115) H. G. Wells, J. H. Lewis, and D. B. Jones, *J. Infectious Diseases* **40**, 326 (1927).
(116) H. G. Wells and T. S. Osborne, *J. Infectious Diseases* **12**, 341 (1913).
(117) A. Calmette, *Ann. inst. Pasteur* **9**, 225 (1895).
(118) J. Morgenroth, *Berlin klin. Wochschr.* **42**, 1550 (1905).
(119) A. Wassermann, *Z. Hyg. Infektionskrankh.* **22**, 263 (1896).

groups from combining with the susceptible tissue. It has been suggested[120] that virus-neutralizing antibodies act by preventing the entry of the virus into the susceptible cell.

Some enzymes are strongly inhibited by antibodies specifically directed against them,[121,122] and Krebs and Najjar[122] believe that in some such cases one of the points of attachment of the antibody to the antigen is at the "active center" of the enzyme. Other enzymes are still 70–95 per cent active when combined with antibody, even when precipitation of the enzyme by the antibody occurs.[123]

In the case of enzymes like urease, this is doubtless because the molecules of substrate are small enough to enter the interstices between the molecules of enzyme and antienzyme, and thereby reach the enzymatically active groups.

d. Complement-Fixing Antibodies

When antibodies combine with their antigens, a group of proteins collectively called complement (see p. 839) usually combines with the complex. In the case of precipitation reactions this produces no visible effect, and it was not until 1940 that Heidelberger[124] was able to measure, by nitrogen determinations, the relatively small amount of protein involved. In the case of lytic antibodies (i.e., antibodies that cause the distintegration and partial solution of the cells against which they are directed), the presence of complement is essential for the action. The lysis of an erythrocyte by a so-called hemolysin is actually brought about not by the hemolytic antibody but by the complement, which combines with the cell as a consequence of the antibody having combined. Most antibody–antigen systems will combine with ("fix" in the jargon of the serologist) complement, but a few exceptions have been observed.[125] When cells susceptible of being lysed are involved, the combination of complement usually completes the system, and lysis occurs.

(120) M. H. Salaman, *Proc. Intern. Congr. Microbiol., 3rd Congr., New York,* 356, 1940.

(121) F. Kubowitz and P. Ott, *Biochem. Z.* **314,** 94 (1943); P. C. Zamecnik and F. Lipman, *J. Exptl. Med.* **85,** 395 (1947); M. McCarty, *J. Gen. Physiol.* **29,** 123 (1946); M. G. Macfarlane and B. C. Knight, *Biochem. J.* **35,** 884 (1941); E. G. Krebs and R. R. Wright, *J. Biol. Chem.* **192,** 555 (1951).

(122) E. G. Krebs and V. A. Najjar, *J. Exptl. Med.* **88,** 569 (1948).

(123) J. S. Kirk and J. B. Sumner, *J. Immunol.* **25,** 495 (1934); D. H. Campbell and L. Fourt, *J. Biol. Chem.* 385 (1939); J. Smolens and M. G. Sevag, *J. Gen. Physiol.* **26,** 11 (1942); M. Adams, *J. Exptl. Med.* **76,** 175 (1942).

(124) M. Heidelberger, H. P. Treffers, and M. Mayer, *J. Exptl. Med.* **71,** 271 (1940).

(125) F. L. Horsfall and K. Goodner, *J. Immunol.* **31,** 135 (1936).

e. Blocking Antibodies

Antibodies are known that combine specifically with their antigen, but produce no visible effect.[126] Many of the known examples are antibodies directed against human or other mammalian erythrocytes.[127–129] In some cases the addition of a fairly concentrated (15–20 per cent) serum albumin (both human and bovine have been successfully used and no doubt others would serve) will cause the agglutination of the cells that have combined with such antibody. In other cases, the fact that the antibody has combined can be detected only by observing that the action of an agglutinating antibody of the same specificity is inhibited. It is then said that the first antibody "blocks" the action of the second. The picture is that the blocking antibody combines with the receptors ("active patches") on the surface of the cell, and thus prevents the combination of the second antibody which, if it did combine, would cause agglutination. It would look as if the blocking antibody, when it combines with a receptor, produces a greater decrease in free energy than would the combination of the agglutinating antibody. Such antibodies are quite important in the study of the Rh blood groups in man.

It is not known how blocking antibodies differ from agglutinating antibodies. It is possible that their chemical properties (solubility, for example) are such that their union with the cell fails to produce agglutination. Another possibility is that they have only a single combining group for the antigen (are "univalent")[130] and therefore, though able to combine, are not able to link the antigen particles together in an agglutinate. This will be discussed under the mechanism of serological reactions (p. 804), but at present it seems unlikely that this suggestion is correct.

(*1*) *Coombs Test.* It occurred to various workers, of whom Coombs *et al.* are best known,[131] that if antibody combined with a cell but did not cause agglutination it might nevertheless be possible to cause agglutination by adding an antibody for the antibody. In this way the antibody molecules might be linked together, and the cells would thereby be linked together and thus agglutinate. Or possibly the mere addition of the extra molecules of proteins to the surface of the cell would "load

(126) A. F. Coca and M. F. Kelly, *J. Immunol.* **6**, 87 (1921).
(127) L. K. Diamond, *cited in* R. R. Race and R. Sanger, Blood Groups in Man, Oxford Univ. Press, London, 1950.
(128) R. R. Race, *Nature* **153**, 771 (1944).
(129) A. S. Wiener, *Proc. Soc. Exptl. Biol. Med.* **56**, 173 (1944).
(130) A. M. Pappenheimer, *J. Exptl. Med.* **71**, 263 (1940).
(131) R. R. A. Coombs, A. E. Mourant, and R. R. Race, *Brit. J. Exptl. Path.* **26**, 255 (1945).

it down," or alter its surface charge, or something of the sort, and permit agglutination (p. 821). In the Coombs test, human antibody, not of the agglutinating type, is added to human red cells. These cells are then washed free of nonspecific human proteins, and a rabbit antibody for human γ-globulin is added. If the reaction is positive, the cells then agglutinate.

It is perhaps possible that "Coombs antibodies" and blocking antibodies are not essentially different, and the important thing is the range of concentrations in which the various tests will work. It seems certain, however, that the blocking antibodies are different from the agglutinating antibodies, for the blocking antibodies have been found to be more stable to heat[132] and high pressures[133] than the agglutinating antibodies.

Hill, Haberman, and Guy[134] have proposed the name "cryptagglutinoids" or "third-order antibodies" for antibodies that can combine with antigen, but exhibit neither agglutinating nor blocking action.

Other types of antibody have been proposed, but the evidence is too controversial to discuss in detail here.

It may be mentioned that although an antibody is in a certain sense the opposite of its antigen, an antibody for an antibody is not therefore identical with the original antigen. The way in which an antibody differs from "normal" γ-globulin is minor (p. 792) and apparently is not recognized by the antibody-forming mechanism of another species. So when an antibody is injected into a different species, the antibody-forming mechanism recognizes only the major features of the injected protein, and forms in response simply anti-γ-globulin, which is itself a γ-globulin characteristic of the second species.

The number of specific combining groups in an antibody molecule is not large (not more than two, in fact; see p. 800) and these groups may not differ from the remainder of the globulin molecule in a way that is easily recognized by the antibody-forming mechanism of another animal. It is not surprising, therefore, that the evidence so far available, though it at times suggests that antibodies to an antibody are to some extent directed toward the specific combining group,[135] is far from conclusive.

Serological differences between antibody (antipneumococcus I and II) and normal globulin were observed by Erickson and Neurath,[136] and these authors concluded that antibody and normal globulin were chemically different. They stated, however, that "those groups of the protein

(132) A. Kleczkowsski, *Brit. J. Exptl. Path.* **22,** 192 (1941).
(133) W. C. Boyd, *J. Exptl. Med.* **83,** 401 (1946).
(134) J. M. Hill, S. Haberman, and R. Guy, *J. Clin. Path.* **19,** 134 (1949).
(135) M. Krüpe and E. Powilleit, *Z. Hyg. Infektionskrankh.* **134,** 198 (1952).
(136) J. O. Erickson and H. Neurath, *J. Gen. Physiol.* **28,** 415 (1945).

which are responsible for antibody activity do not contribute to the antigenic structure."

f. Allergic Antibodies (Reagins)

In allergic conditions, antibodies can often be demonstrated in the patient's blood. The behavior of these antibodies differs in a number of respects from that of ordinary antibodies, and some authors use the noncommittal term "reagin" for them.[137] Reagins do not precipitate or neutralize the antigen (although neutralizing antibodies may also be found). They may be demonstrated by taking advantage of their power of passively sensitizing the human skin.

It is not known how the physical and chemical properties of reagins differ from those of ordinary antibodies. It is possible that the only difference is that the reagins are present in very minute quantities, too small to be detected by the usual precipitin or neutralization techniques. Kabat and Landow[138] obtained evidence that about 0.01 μg. of antibody nitrogen might be sufficient to sensitize the area of skin used in the usual test. The next most sensitive test, complement fixation, could detect not less than twenty times this much, and agglutination and precipitin tests are even less sensitive. This suggestion has not been accepted by all allergists, however.

Kuhns and Pappenheimer[139] found that potent human antitoxic sera fell into two main types. In persons with a personal or a family history of allergy, or of both, antibody of a skin-sensitizing type, which remained in the injected area for many weeks, was found. This antibody was nonprecipitating with its antigen, and fixed complement poorly. Heating it to 56°C. for 4 hours destroyed its skin-sensitizing properties. In other persons a nonsensitizing, precipitating type of antibody was found.

Vaughan and Kabat[140] are not convinced that the skin-sensitizing antibody in such experiments is necessarily of the "incomplete" or nonprecipitating type. In their studies of the capacity of rabbit antiovalbumin to sensitize human skin they found that the capacity of the antisera to sensitize was completely unrelated to the antiovalbumin present, and was probably due to the presence of one, or possibly two, antibodies (of the precipitating type) directed against impurities in the crystalline ovalbumin. These impurities were different from conalbumin, ovomucoid, and lysozyme.

(137) A. F. Coca, Essentials of Immunology for Medical Students, Williams & Wilkins, Baltimore, 1925.
(138) E. A. Kabat and H. Landow, *J. Immunol.* **44,** 69 (1942).
(139) W. J. Kuhns and A. M. Pappenheimer, *J. Exptl. Med.* **95,** 375 (1952).
(140) J. H. Vaughan and E. A. Kabat, *J. Exptl. Med.* **97,** 821 (1953).

In some allergic conditions, such as drug allergy, antibodies of any sort have never been convincingly demonstrated.

g. Opsonins

It was found[141] that the destruction of bacteria by phagocytes was greatly increased when immune serum was added to the leucocytes. The word "opsonins" was coined for the hypothetical substances that made the bacteria more attractive to the leucocytes. It has been suggested that opsonins consist of two parts, one thermostable and one thermolabile at 56°C. The thermostable portion is probably ordinary antibody, and the thermolabile portion may be some fraction of complement, although certain differences have been reported.[142–144]

h. Natural Antibodies and Lectins

Substances reacting with a given infectious agent, or with chemical products derived from it, are sometimes encountered in the blood of an animal that has not been immunized nor, so far as is known, been infected with this particular agent. Also, substances acting on the erythrocytes of other species, or even on different individuals of the same species, are of frequent occurrence. The isoagglutins of the human blood groups are the best known example of the latter. All these substances act so much like antibodies that they are collectively called normal or natural antibodies.

There can be little doubt that these natural antibodies fall into two classes. The first class includes the antibodies that have resulted from a mild ("subclinical") infection which escaped notice, but which caused the production of some antibody. An invasion that produced no detectable illness might have been sufficient. Another cause for the presence of such antibodies is invasion by microorganisms that are antigenically related to the pathogens we are interested in (or by their normal presence in parts of the body such as the digestive tract). Apparently, certain pathogenic organisms may be present in the gastrointestinal tract without causing illness.[145]

It has been maintained[146] that the isohemagglutinins in man are the result of immunization by antigens related to the blood group A and B

(141) J. Denys and J. Leclef, *La Cellule* **11,** 177 (1895).
(142) J. Gordon, H. R. Whitehead, and A. Wormall, *J. Path. Bact.* **32,** 57 (1929).
(143) J. Gordon and F. C. Thompson, *Brit. J. Exptl. Path.* **16,** 101 (1935).
(144) H. Zinnser and E. G. Cary, *J. Exptl. Med.* **19,** 345 (1914).
(145) G. Ramon and E. Lemétayer, *Rev. immunol.* **1,** 209 (1935).
(146) A. S. Wiener, *J. Immunol.* **66,** 287 (1951).

substances. Other workers[146a,146b] have suggested that the agglutinins are formed as a direct consequence of the action of the blood group genes possessed by the individual.

In addition to the natural antibodies which are actually of immune origin, there undoubtedly exists a second class of such substances which develop without any antigenic stimulus. Landsteiner[30] considered that most of the normal hemagglutinins and hemolysins acting on the blood of foreign species were in this class. This is supported by certain regularities in the zoological distribution of these substances. Also, some plants,[147–150] especially certain beans, contain proteins that act more or less specifically on the erythrocytes of certain species or individuals. Boyd[149,150] has reported the study of certain Lima beans that yield a protein almost completely specific for the blood group antigen A.

It is possible that the specificity of the second class of natural antibodies is accidental. Their surface may have one or more areas where the amino acids are arranged in such a way as to constitute an antideterminant which corresponds to the molecular configuration of some antigen. It might be that no element of design is involved; the phenomenon may reflect the fact that the number of different configurations of charges on a given area of protein surface, though large, is not infinite, and some configurations will be duplicated by chance. At any rate, we have at present no inkling of any evolutionary trend which could account for it. At least we cannot at present affirm that the anti-A of Lima beans has been formed in response to the stimulus of the presence of blood group A substance.

It would appear to be a matter of semantics as to whether a substance not produced in response to an antigen should be called an antibody, even though it is a protein and combines specifically with certain antigens only. It might be better to have a different word for these substances, and the present writer would like to propose the word *lectin*, from the Latin *lectus*, the past participle of *legere*, meaning to pick, choose, or select.

(146*a*) T. Furuhata, *Japan Med. World* **7**, 197 (1924).
(146*b*) S. Filitti-Wurmser, Y. Jacquot-Armand, and R. Wurmser, *J. Chimie Physique* **47**, 419 (1950).
(147) G. W. G. Bird, *Indian J. Med. Research* **40**, 289 (1952).
(147*a*) P. Cazal and M. Lalaurie, *Acta Haematol.* **8**, 73 (1952).
(147*b*) R. Koulumies, *Ann. Med. Exptl. et Biol. Fenniae* (Helsinki) **27**, 185 (1949).
(147*c*) M. Krüpe, *Z. Immunitätsforsch.* **107**, 450 (1950).
(147*d*) M. Krüpe, and C. Braun, *Naturwissenschaften* **39**, 284 (1952).
(148) K. O. Renkonen, *Ann. Med. Exptl. Biol. Fenn.* **26**, 66 (1948).
(149) W. C. Boyd, *J. Immunol.* **65**, 281 (1950).
(150) W. C. Boyd and R. M. Reguera, *J. Immunol.* **62**, 333 (1949).

2. Chemical Properties of Antibodies

a. Antibodies as Proteins

The most conspicuous feature of the chemistry of immune antibodies is that they are proteins, and, with a few doubtful exceptions, serum globulins. Most seem to be γ-globulins (see below). The modifications, whatever they are (see p. 814), which give antibodies their power of uniting specifically with the antigen, do not seem to produce effects on the chemical properties. If, as Pauling believes,[151] these modifications consist simply of a refolding of the polypeptide chain, leaving the order of amino residues in that chain unaltered, no large effect would be expected. If, as supposed by Alexander,[152] Mudd,[153] and Breinl and Haurowitz,[154] the antibody molecule consists in part of a "patch" where the amino acids are arranged somewhat differently, this would still not be expected to affect the chemical properties noticeably, for the size of these active patches would be small in comparison with the whole globulin molecule (p. 791), and it is not likely (p. 800) that there are more than two or three of them in each molecule.[34]

In their susceptibility to most denaturing and destructive agents, antibodies do not differ from other proteins.[34] They are destroyed by heat, and the process has a very high heat of activation. Heating causes aggregation of the proteins of immune serum[155] as does treatment with high pressures (3000–5000 atmospheres).[133] Treatment with alcohol at room temperature destroys antibodies rapidly, but at low temperatures they may be precipitated and fractionated from alcohol–water mixtures.[156] Treatment with diazonium compounds, iodine, bromine, formaldehyde, or ketene usually reduces or destroys their specific activity.

When only 3–5 halogen atoms are introduced into each antibody molecule, its reactivity is practically unchanged; this enables antibodies to be labeled with radioactive iodine for quantitative studies.[157]

Antibodies seem in general to be more resistant to the action of proteolytic enzymes.[158,159,159a] This property has been taken advantage

(151) L. Pauling, *J. Am. Chem. Soc.* **62**, 2643 (1940).
(152) J. Alexander, *Protoplasma* **14**, 296 (1932).
(153) S. Mudd, *J. Immunol.* **23**, 423 (1932).
(154) F. Breinl and F. Haurowitz, *Z. physiol. Chem.* **192**, 45 (1930).
(155) F. C. Bawden and A. Kleczkowski, *Brit. J. Exptl. Path.* **23**, 178 (1942).
(156) E. J. Cohn, T. L. McMeekin, J. L. Oncley, J. M. Newell, and W. L. Hughes, *J. Am. Chem. Soc.* **62**, 3386 (1940).
(157) F. D. S. Butement, *J. Chem. Soc., Suppl.* **2**, 408 (1949).
(158) I. Parfentjev, U. S. Patent 2,065,196 (1936).
(159) J. W. Williams, *Fortschr. Chem. org. Naturstoffe* **7**, 270 (1950).
(159a) C. G. Pope, *Brit. J. Exptl. Path.* **20**, 201 (1939).

of in the purification of antibodies (see p. 807). Haurowitz[19] considers that the higher resistance of antibodies to the action of proteolytic enzymes indicate that their peptide chains are folded in a more stable manner than those of other serum globulins.

b. Analyses of Antibodies

Various antibodies have been analyzed for total nitrogen, amide nitrogen, mono- and diamino nitrogen, and amino acids.

A table summarizing the results up to 1938 will be found in Marrack's book.[34] No differences were found which could not be accounted for by the fact that the various workers used different preparations of different degrees of purity. Smith and Green[160] found antibodies to contain the same amino acids as normal serum globulins.

A summary of studies on the amino acid composition of antibodies is given in Table IV, taken from Smith and Jager.[161]

c. Molecular Weight of Antibodies

Antibodies produced by the rabbit (and probably human antibody also) seem to have a molecular weight of about 160,000, or in some cases somewhat higher.[162–165] For human antipneumococcus antibody a sedimentation constant of 7.4 and a calculated molecular weight of 195,000 have been reported.[162] Davis *et al.*[166] found that Wassermann antibody (syphilitic reagin) of human origin, disassociated from the antibody–antigen precipitate, showed two components with sedimentation constants of 7 and 19, respectively.

Horse antidiphtheria toxin has a sedimentation constant of about 7 and a molecular weight of about 160,000.[167] Horse antipneumococcus antibody, however, shows a sedimentation constant of 18, and its molecular weight is estimated to be nearly 1,000,000.[168] This is one striking feature of horse antipneumococcus sera, although a small amount of globulin with this high molecular weight occurs in normal horse, bovine,

(160) E. Smith and B. Greene, *J. Biol. Chem.* **171,** 355 (1947).
(161) E. L. Smith and B. V. Jager, *Ann. Rev. Microbiol.* **6,** 207 (1952).
(162) E. A. Kabat and A. E. Bezer, *J. Exptl. Med.* **82,** 207 (1945).
(163) J. L. Oncley, G. Scatchard, and A. Brown, *J. Phys. & Colloid Chem.* **51,** 184 (1947).
(164) M. L. Petermann and A. M. Pappenheimer, *J. Phys. Chem.* **45,** 1 (1941).
(165) T. Svedberg and K. O. Pedersen, The Ultracentrifuge, Oxford Univ. Press, London, 1940.
(166) B. D. Davis, D. H. Moore, E. A. Kabat, and A. Harris, *J. Immunol.* **50,** 1 (1945).
(167) A. M. Pappenheimer, H. P. Lundgren, and J. W. Williams, *J. Exptl. Med.* **71,** 247 (1940).
(168) N. Fell, K. G. Stern, and R. D. Coghill, *J. Immunol.* **39,** 223 (1940).

TABLE IV

COMPOSITION OF IMMUNE PROTEINS (SMITH AND JAGER)[161]

The data are given as grams of constituent per 100 g. of anhydrous protein. The values for bovine γ-2-globulin are averages for two preparations. Determinations for the bovine immune lactoglobulins are averages for two preparations of each, one from milk and one from colostrum.

Constituent	Human γ-globulin II-1	Human γ-globulin II-1, 2	Human γ-globulin II-3	Bovine γ-2-globulin	Bovine T-(γ-1)-globulin	Bovine immune eulactoglobulin	Bovine immune pseudolactoglobulin	Equine γ-2-globulin	Equine T(γ-1)-globulin
Arginine	4.8	5.1	3.7	5.8	4.8	4.9	3.5; 5.6;[a] 5.2[b]	3.8	2.8
Aspartic acid	8.8						9.4;[a] 9.3[b]		
Cystine (+ cysteine)	3.1	2.6	2.7	2.9	2.8	3.2	3.0	2.6	2.5
Glutamic acid	11.8						12.3;[a] 10.7[b]		
Glycine	4.2								
Histidine	2.50	2.01	1.91	2.05	2.01	1.89	2.1; 2.3;[a] 2.2[b]	2.44	2.43
Isoleucine	2.7	2.8	2.0	3.2	3.0	3.1	3.1	4.4	3.3
Leucine	9.3	9.3	9.5	7.4	8.6	10.4	9.1; 8.5;[a] 8.3[b]	9.0	7.5
Lysine	8.1	7.2	6.3	6.7	6.4	6.3	7.2; 6.1;[a] 6.1[b]	8.6	6.7
Methionine	1.06	1.12	0.87	1.18	1.00	0.98	1.08; 1.3;[a] 1.1[b]	0.95	0.72
Phenylalanine	4.6	4.5	4.7	3.2	4.5	3.6	3.8; 3.9;[a] 4.3[b]	4.4	4.1
Proline	8.1						10.0[a]		
Serine	11.4								
Threonine	8.4	8.8	7.4	10.0	9.5	10.5	10.1; 9.0;[a] 10.2[b]	11.1	8.7
Tryptophan	2.86	2.6	2.8	2.6	2.6	2.4	2.7; 3.2[a]	2.7	2.8
Tyrosine	6.75						6.7[a]	6.8[c]	6.8[c]
Valine	9.7	9.7	9.7	10.0	9.5	10.4	9.4; 8.7;[a] 9.3[b]	10.1	10.4
Hexose		2.3	2.3	2.1	2.5	2.9	2.8	2.5	2.6
Hexosamine		1.27	1.23	1.31	1.50	1.45	1.32	1.18	1.53

[a] R. G. Hansen, R. L. Potter, and P. H. Phillips, *J. Biol. Chem.* **171,** 229–32 (1947).

[b] L. M. Henderson and E. E. Snell, *J. Biol. Chem.* **172,** 15–29 (1948).

[c] E. L. Smith and B. V. Jager, *Ann. Rev. Microbiol.* **6,** 207 (1952).

and probably even human serum. The difference in the two sorts of horse antibodies may possibly be due to the different routes of injection employed in immunization.[169] Protein antigens such as toxins are usually injected subcutaneously; bacteria such as pneumococci are usually administered intravenously.

The rabbit, irrespective of the route of injection, seldom seems to produce antibodies of the larger size.[170] However, Païc[171] found that rabbit anti-sheep hemolysin had a sedimentation velocity in the ultracentrifuge which indicated that the molecule, if spherical, has a molecular weight of 420,000. Since it is likely the molecule is not spherical, the true molecular weight is probably greater than this. Païc thinks this lysin is different in other ways from the other proteins of rabbit serum.

Results of ultracentrifugal molecular-weight determinations of certain antibodies are shown[172,27] in Table V.

Filitti-Wurmser, Aubel-Lesure and Wurmser[172a] determined the sedimentation constants of the anti-B (β) agglutinin which occurs in the serum of human individuals of group O and group A. There was a marked difference in the anti-B from persons of genotype A_1O, A_1A_1, and OO. For these three types of serum the anti-B was found to have sedimentation constants s_{20} of 15.5, 11, and 6.5 respectively. From these values the authors estimated the molecular weights to be 500,000 for the anti-B from genotype A_1O, 300,000 from A_1A_1, and 170,000 from OO.

d. Shape of Antibody Molecules

The antibody molecules that have been measured are asymmetric. Rabbit antibody molecules have been calculated to have a short axis of 37 A. and a long axis of 274 A., giving an axial ratio of 7.5;[173] the corresponding values for human antipneumococcus antibody are 37 A. and 338 A. with an axial ratio of 9.2. However, these calculations neglect hydration, assume shapes of prolate ellipsoids of revolution, and yield maximum asymmetries.[174] For horse antitoxin, an axial ratio of 7.0 was calculated.[164] For horse antipneumococcus antibody, Neurath[173] calculated axes of 47 A. and 950 A. and an axial ratio of 20. The horse

(169) M. Heidelberger, H. P. Treffers, and J. Freund, *Federation Proc.* **1,** 178 (1942).

(170) E. A. Kabat, *J. Exptl. Med.* **69,** 103 (1939).

(171) M. M. Païc, *Bull. soc. chim. biol.* **21,** 412 (1939).

(172) E. A. Kabat and M. M. Mayer, Experimental Immunochemistry, C. C. Thomas, Springfield, 1948.

(172*a*) S. Filitti-Wurmser, G. Aubel-Lesure, and R. Wurmser, *J. Chim. phys.* **50,** 236 (1953).

(173) H. Neurath, *J. Am. Chem. Soc.* **61,** 1841 (1939).

(174) H. Neurath, personal communication.

TABLE V

MOLECULAR WEIGHTS AND ASYMMETRY VALUES FOR ANTIBODY MOLECULES[a]

The computations of molecular weights from these data involve also the partial specific volume of the proteins. The partial specific volumes of many proteins are near 0.75 cc./g. For horse antipneumococcus antibody the value is 0.715, and for human serum γ-globulin 0.739. (H. P. Lundgren and W. H. Ward *in* D. M. Greenberg, Amino Acids and Proteins, C. C Thomas, Springfield, 1951.)

Source of antibody	Antigen	Per cent protein in solution examined in the ultra-centrifuge	Sedimentation constant, S_{20}, Svedbergs	Diffusion constant, $D_{20} \times 10^7$, sq. cm./sec.	Molecular weight	Frictional ratio f/f_0	Dimensions, A.[b]
Horse	Pneumococcus[f] (1, 2, 3)	0.22	19.3	1.80	920,000	2.0	946 × 47
Horse	Diphtheria antitoxin[c] (4)		7.2	3.90	184,000	1.4	286 × 39
Horse	Antitoxin[d] (5)		5.7	5.0	113,000	1.3	204 × 36
Horse	Antitoxin[e] (6, 7)		5.5	5.7	90,000	1.23	166 × 35
Cow	Pneumococcus (2, 3)	0.64	18.1	1.69	910,000	2.0	
Pig	Pneumococcus (2, 3)	0.58	18.0	1.64	930,000	2.0	
Rabbit	Pneumococcus[f] (1, 2, 3)	0.19	7.0	4.23	157,000	1.4	272 × 37
Rabbit	Egg albumin (1, 2, 3)	0.53	6.5	3.75	165,000	1.6	244 × 24
Rabbit	Sheep red cells (8)		18.9				
Monkey	Pneumococcus[f] (2, 3)	0.31	6.7	4.08	157,000	1.5	312 × 34
Man	Pneumococcus[f] (2)	0.39	7.4	3.60	195,000	1.5	338 × 37
Man	Isoagglutinin (10)		19.8				
Man	Syphilitic Wassermann (9)	0.2	{18.5–19.1 6.5– 7.0				

[a] D. H. Campbell and N. Bulman, Forschritte der Chemie Organischer Naturstoffe **9,** 443 (1952); E. A. Kabat and M. M. Mayer, Experimental Immunochemistry, C. C Thomas, Springfield, 1948.

[b] Calculated for a model of an anhydrous prolate ellipsoid of revolution.

[c] The water-soluble fraction.

[d] The water-soluble fraction after treatment with pepsin.

[e] Crystallized preparation obtained by trypsin treatment of toxin–antitoxin precipitates.

[f] Antibody dissociated from specific precipitates by salt.

References:

1. M. Heidelberger and K. O. Pedersen, *J. Exptl. Med.* **65,** 393 (1937).
2. E. A. Kabat and K. O. Pedersen, *Science* **87,** 372 (1938); E. A. Kabat, *J. Exptl. Med.* **69,** 103 (1939).
3. A. Tiselius and E. A. Kabat, *Science* **87,** 416 (1938); *idem, J. Exptl. Med.* **69,** 119 (1939).
4. A. M. Pappenheimer, Jr., H. P. Lundgren, and J. W. Williams, *J. Exptl. Med.* **71,** 247 (1940).
5. M. L. Petermann and A. M. Pappenheimer, Jr., *J. Phys. Chem.* **45,** 1 (1941).
6. J. H. Northrop, *J. Gen. Physiol.* **25,** 465 (1942).
7. A. Rothen, *J. Gen. Physiol.* **25,** 487 (1942).
8. M. Païc, *Bull. soc. chim. biol.* **21,** 412 (1939).
9. B. D. Davis, D. H. Moore, E. A. Kabat, and A. Harris, *J. Immunol.* **50,** 1 (1945).
10. K. O. Pedersen, Ultracentrifugal Studies on Serum and Serum Fractions, Almquist and Wiksell, Upsala, Sweden, 1945.

antipneumococcus antibody thus seems to be about three times as long as the antitoxin.

Malkiel[71] photographed the precipitate that resulted when tobacco mosaic virus was reacted with its specific (rabbit) antibody, and found that the molecules of the virus, which was clearly resolved by the microscope, were separated by a definite, constant space, about 270 A., presumably the length of the antibody molecule.

Campbell and Bulman[27] calculated, from the Perrin equation, the axial ratios of various antibodies. The results are shown in Table V.

e. Isoelectric Points of Antibodies

Rabbit antibodies seem to have an isoelectric point not far from that of normal rabbit γ-globulin, about 6.6.[175,176] The antipneumococcus antibody of the horse again differs from normal horse globulin, having an isoelectric point of about 4.8[176–178] as opposed to a normal value of about 5.7.[176]

f. Electrophoretic Mobility

The electrophoretic mobilities of antibody molecules reflect their isoelectric points. In most cases the measurements center around -1.0×10^{-5} sq. cm./sec./v. at a pH of about 8.0 and an ionic strength of 0.15. After prolonged immunization, especially of horses, a faster-moving component (T) often appears which has a mobility between -1.0 and -2.0×10^{-5}. A detailed discussion of the electrophoretic properties of normal and immune sera has been given by Williams.[159] Antibodies usually migrate with the other γ-globulins, but in horse antiprotein sera (such as antitoxin) a part of the antibody migrates with the β-globulins,[178–181] and Seibert and Nelson[182] reported found in rabbit serum an antibody to tuberculin which migrated with the α-globulins. The evidence presented fails to stand up under critical examination, however.[172] Davis[166] found Wassermann antibody (syphilitic reagin) to have a mobility between that of the β- and γ-globulins.

When horses are immunized by subcutaneous injections of certain

(175) P. Girard and M. Louran, *Compt. rend. soc. biol.* **116**, 1010 (1934).
(176) M. Heidelberger, K. O. Pedersen, and A. Tiselius, *Nature* **138**, 165 (1936).
(177) A. Tiselius, *Biochem. J.* **31**, 313 (1937).
(178) A. Tiselius and E. A. Kabat, *J. Exptl. Med.* **69**, 119 (1939).
(179) J. van der Scheer, J. B. Lagsdin, and R. W. G. Wyckoff, *J. Immunol.* **41**, 209 (1941).
(180) R. A. Kekwick, *Chemistry & Industry* **60**, 486 (1941).
(181) R. A. Kekwick and B. R. Record, *Brit. J. Exptl. Path.* **22**, 29 (1941).
(182) F. B. Seibert and J. W. Nelson, *Proc. Soc. Exptl. Biol. Med.* **49**, 77 (1942).

toxoids, or rabbit serum albumin, egg albumin, or hemocyanin,[183] the resultant antibodies appear mostly in the fraction of globulins precipitated between 33 and 50 per cent saturation with ammonium sulfate. Electrophoretic studies of such immune sera frequently show an increase in the T- or γ_1-globulin fraction. Such antibodies differ from rabbit antibodies and from antibodies prepared in the horse by the repeated intravenous injection of bacteria, pneumococcal nucleoprotein, bacterial polysaccharides, and rabbit globulin[183–186] in that when mixed with their antigen they give a very narrow zone of precipitation. Precipitation in such cases is inhibited by excess of antibody as well as antigen. Boyd[186] suggested calling such antisera the H type, as opposed to the more usual R type.

Measurements of the electrophoretic mobilities of various antibodies are given in Table VI.

TABLE VI

ELECTROPHORETIC MOBILITIES OF THE COMPONENTS OF SERA FROM VARIOUS ANIMAL SPECIES[a]

Species	Antibody for	$AN/\Sigma N^f$ %	$\mu \times 10^5$ sq. cm./sec./v.				
			Al	α	β	T	γ
Horse	Pneumococcus (types I, II, V, VII, VIII, XXIII)	20–30	5.1	3.6	2.9	—	0.9
Horse[b]	[e]	0	5.7	3.8	3.1	—	1.0
Pig[b]	Pneumococcus (type I)	1.6	5.7	3.5	2.7	—	1.1
Rabbit[b]	Ovalbumin	36.4	6.0	3.6	2.9	—	1.1
Monkey[b]	Pneumococcus (type III)	6.6	5.2	4.3	3.0	—	0.7
Horse[c]	Diphtheria toxin	18.9	—	4.9	—	2.6	—
Horse[d]	Diphtheria toxin	—	—	4	—	—	—

[a] *References:* A. Tiselius and E. A. Kabat, *J. Exptl. Med.* **69**, 119 (1939); J. van der Scheer, J. B. Lagsdin and R. W. G. Wyckoff, *J. Immunol.* **41**, 209 (1941).

[b] Determined at pH 7.72 ± 0.02 in buffer containing 0.15 *M* NaCl and 0.02 *M* PO_4^{---}. Temperature, +0.5°C.

[c] Determined at pH 7.35, ionic strength 0.1.

[d] At pH 7.3.

[e] Normal horse serum: α, β, γ = globulin fractions; A = antibody, Al = albumin, T = (special) antibody component in horse antiprotein sera.

[f] Ratio of antibody nitrogen to total nitrogen in per cent.

(183) H. P. Treffers, M. Heidelberger, and J. Freund, *J. Exptl. Med.* **86**, 95 (1947).
(184) H. P. Treffers, M. Heidelberger, and J. Freund, *J. Exptl. Med.* **86**, 83 (1947).
(185) H. P. Treffers, M. Heidelberger, and J. Freund, *J. Exptl. Med.* **86**, 77 (1947).
(186) W. C. Boyd, *J. Exptl. Med.* **74**, 369 (1941).

It seems that horse antidiphtheria toxin can have the characteristics either of a $\beta(\beta_2)$- or of a γ-globulin. The two kinds of antitoxin display definite differences in serological behavior. Pure β-antitoxin flocculates (precipitates) more slowly with toxin than does the γ-antibody, and has a lower L/L_f ratio (this ratio is a rough measure of the *in vivo* effectiveness of the toxin). The precipitate formed by the most rapidly flocculating mixture of antibody and toxin contains, when β_2-antitoxin is used, only half as much antibody (measured biologically) as when γ-antitoxin is used.[180,181]

g. Susceptibility of Antibodies to Enzymes, etc.

Some antibodies have been found more resistant to the action of certain enzymes, such as pepsin, than are normal serum proteins.[94] Treatment of diphtheria antitoxin serum with pepsin until 70–80 per cent of the protein was rendered noncoagulable by heat gave a product with a considerably higher proportion of immunologically active protein.[158] However, similar treatment of horse antipneumococcus serum destroys practically all the mouse protective power, although considerable precipitating power remains,[179] and digested antipneumococcus antibody (types I and II) combines with twice as much specific polysaccharide per milligram of antibody nitrogen as does the untreated antibody.[187] Peterman and Pappenheimer[187] found that no matter what the molecular weight was of the antibody in the starting material for this product, they obtained a sedimentation constant of 5.2, suggesting a final molecular weight of less than 100,000. In a more recent study Williams, Baldwin, Saunders, and Squire[188] have found pepsin-digested γ-globulins actually to be much more heterogeneous than this. They cautiously do not give an estimate of the mean molecular weight of the product. Practically none of the original antibody activity (*in vitro*) was lost, suggesting that the enzyme treatment preferentially destroys the immunologically inactive portions of the molecule. Similar conclusions emerge from the work of Northrop,[189] who treated a precipitate of diphtheria toxin–antitoxin with pepsin and obtained a fully active antitoxin which had only half the molecular weight of the original antitoxin. This material could be crystallized.

This preparation of Northrop's is the only antibody which has been crystallized, and its crystallizability may result partly from the enzyme

(187) M. L. Petermann and A. M. Pappenheimer, *Science* **93,** 458 (1941).
(188) J. W. Williams, R. L. Baldwin, W. M. Saunders, and P. G. Squire, *J. Am. Chem. Soc.* **74,** 1542 (1952).
(189) J. H. Northrop, *J. Gen. Physiol.* **25,** 465 (1942).

treatment. Most antibodies are probably too heterogeneous to lend themselves to ready crystallization (p. 805).

By precipitation with acid potassium phthalate and subsequent fractionation with ammonium sulfate, Northrop and Goebel[190] obtained a preparation of horse antipneumococcus antibody completely precipitable by the homologous type I polysaccharide.

One fraction of this antibody preparation could be crystallized in poorly formed rounded rosettes, which, however, were not any purer than the remainder and were accompanied by the formation of an insoluble protein.

Comparison of these various fractions by salt solubility or immunological reactions indicated that immune horse antipneumococcus serum contains a large number of different proteins capable of precipitating with the specific pneumococcus polysaccharide but having very different protective values in the animal, different reactions with antihorse serum, and different solubility in salt solutions.

Another procedure for determining the heterogeneity of antigens or antibodies is the method of agar diffusion developed by Oudin and by Onchterlony.[191] In this procedure, antiserum is added to a solution of agar and allowed to gel in a small tube. The agar gel is then overlaid with a solution of the antigen. Antigen diffusing into the antibody-containing agar produces a visible precipitate, the boundary extending with time. When the antigen is inhomogeneous and the antiserum in the agar contains several antibodies, two or more boundary zones will be produced. Smith and Jager[161] point out, however, that this method will not detect low concentrations of impurities, and has in addition certain other limitations.

Differences between different antibodies in their resistance to heat, acid, and alkali have been reported,[192,193] and Boyd[133] found that anti-Rh human antibody of the "incomplete" or "blocking" type withstood a pressure 1000 atmospheres higher than did the agglutinating antibody. It is not known to what feature of the molecule these differences can be attributed.

h. Specific Differences between Antibodies

It is to be expected that antibodies would exhibit the species differences which characterize the normal plasma proteins of the various species

(190) J. H. Northrop and W. F. Goebel, *J. Gen. Physiol.* **32,** 705 (1949).
(191) J. Oudin, *Ann. inst. Pasteur* **75,** 30, 109 (1948); *idem, Compt. rend.* **228,** 1890 (1949); Ö. Ouchterlony, *Arkiv Kemi* **1,** 43, 55 (1949); S. D. Elek, *Brit. J. Exptl. Path.* **30,** 484 (1949); J. Munoz and E. L. Becker, *J. Immunol.* **65,** 47 (1950).
(192) K. Goodner, F. L. Horsfall, and R. J. Dubos, *J. Immunol.* **33,** 279 (1937).
(193) W. D. Harkins, L. Fourt, and P. C. Fourt, *J. Biol. Chem.* **132,** 111 (1940).

involved, and this is found to be so. Antibodies to antibodies react with the normal globulins of the animal from which the original antibody preparation was obtained, and antibodies to normal γ-globulin precipitate antibody of the species that provided the γ-globulin. In addition, cross reactions of the expected degree are observed.

Antibodies from different species may sometimes show characteristic differences in behavior. Horse antitoxin is a more soluble protein ("pseudoglobulin") than is the antitoxin from rabbits. Antipneumococcus antibodies in the horse, cow, and pig tend to be large molecules of molecular weight about 930,000, while those of man, rabbit, and monkey seem to be of the same molecular weight as the normal globulin, about 160,000.[170]

Goodner and Horsfall[194] observed a partition of antipneumococcal antibodies from various species into two groups, as regards complement-fixing ability and lipide composition. They found that hemolytic complement of the guinea pig was fixed by antibodies from rabbit, rat, guinea pig, and sheep, while it was not fixed by antibody from horse, man, dog, mouse, cat, and goat. Dingle *et al.*[195] observed similar differences. Horsfall and Goodner[125] found that lecithin was necessary for the combining and precipitating activity of horse antipneumococcus serum, while cephalin was necessary for the activity of rabbit antipneumococcus serum.

Other species differences have been reported. The guinea pig is a poor producer of precipitating antibodies,[196] but sometimes produces antisera that have good bactericidal power *in vivo,* although not able to precipitate or agglutinate.[195]

The difference in solubility between horse and rabbit antiprotein antibodies is reflected in their behavior in flocculation.[186,169] This will be discussed later (p. 826).

i. Valence of Antibody Molecules

By valence in this connection we mean the number of (complex) specific combining groups for antigen ("antideterminants") per molecule.

It was at first supposed that antibody would naturally be multivalent, and theories based upon the postulate of a "multivalence" of unspecified value were sometimes proposed.[34,197] Such theories (of antibody forma-

(194) K. Goodner and F. L. Horsfall, *J. Exptl. Med.* **64,** 201 (1936).
(195) J. H. Dingle, L. D. Fothergill, and C. A. Chandler, *J. Immunol.* **34,** 357 (1938).
(196) C. A. Colwell and G. P. Youmans, *J. Infectious Diseases* **68,** 226 (1941).
(197) M. Heidelberger and F. E. Kendall, *J. Exptl. Med.* **61,** 563 (1935).

tion which have been proposed), however, do not always make it easy to see how *multivalent* molecules of specific antibody could be built up.

When speculation about this subject began, there were two principal points of view. Heidelberger and Kendall[197] and Marrack[34] suggested that antibody was "multivalent," and Haurowitz[198] and Hooker and Boyd[199] (although in 1934 they had made calculations assuming quadrivalent antibody)[200] suggested that the assumption of multivalent antibody was not required to explain any experiments thus far reported. By the principle of *entia non multiplicanda,* Hooker and Boyd[199] preferred univalence as the simplest hypothesis.

In the nearly twenty years that have elapsed since the controversy about antibody valence began, the question has been answered by experiments in a way that seems pretty conclusive, and the "multivalent" view has been proven to be correct, if by "multivalence" the earlier writers mean bivalence. There is now adequate experimental evidence, in the opinion of the present writer, that most antibodies are bivalent.

The first such evidence to carry much conviction was the ultracentrifugal study of Pappenheimer, Lundgren, and Williams[167] of the products of the reaction of horse antidiphtheria toxin and diphtheria toxin. Their results were compatible with the notion that the antitoxin molecule was capable of combining with two molecules of toxin, and was therefore bivalent.

Pauling and collaborators[88] calculated the valence of antibodies from analytical studies of precipitates made with various precipitable haptens, and concluded that the antibody was bivalent, with the possible admixture of some trivalent molecules (required by Pauling's theory of antibody–antigen reactions); but Boyd and Behnke[89] pointed out that some of these haptens were aggregated in solution and therefore consisted effectively of multivalent aggregates, which might be conceived to precipitate in the same way as other multivalent molecules such as protein antigens. Kabat[201] also pointed out that Pauling and co-workers had not always allowed sufficient time for the attainment of equilibrium in their reactions.

Eisen and Karush,[96] by plotting the reciprocal of the number of moles of univalent hapten (*p*-(*p*-hydroxyphenylazo)phenylarsonic acid) per mole of antibody ($1/r$) versus the reciprocal of the free hapten concentration ($1/c$) and extrapolating $1/c$ to zero, obtained values of $1/r$ of 0.5 ± 10 per cent, indicating a valence of two for the antibody.

(198) F. Haurowitz, *Z. physiol. Chem.* **245,** 23 (1936).
(199) S. B. Hooker and W. C. Boyd, *J. Immunol.* **33,** 337 (1937).
(200) W. C. Boyd and S. B. Hooker, *J. Gen. Physiol.* **17,** 341 (1934).
(201) E. A. Kabat, *Ann. Rev. Biochem.* **15,** 505 (1946).

From an electrophoretic study of the compounds formed with excess antigen (crystalline horse serum albumin and crystalline bovine serum albumin) and antibody, Marrack, Hoch, and Johns[202] obtained evidence that the antibody was bivalent. They suggested that the second valency of the antibody molecule is weak, and is reinforced, especially when antibody is in excess, by a nonspecific attraction due to the fact that the polar groups of the antibody are brought into opposition and attract each other instead of water molecules.

Plescia, Becker, and Williams[203] determined the ratio of antigen to antibody in preparations made by dissolving the precipitate of human serum albumin and its (rabbit) antibody in excess antigen by electrophoretic deterinination of the amount of free antigen and calculation

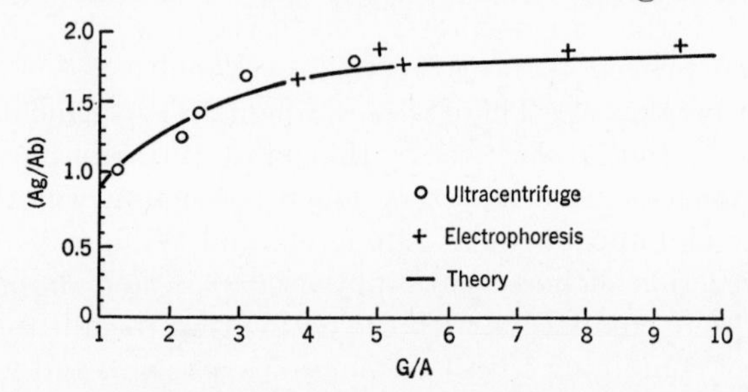

FIG. 3. Observed molecular ratios of antigen to antibody in the region of antigen excess, compared with values predicted by the equation of Goldberg and Williams.[205]

of the antigen–antibody ratio from the amount remaining. The limiting value, as the amount of antigen in excess increased, was effectively two (2.04); the bovine albumin system yielded a higher value (2.81) and did not positively suggest that an upper limit had been reached, but the authors point out that the difficulties of analysis in the region of extreme antigen excess are such that this value should be accepted with some reserve.

Singer and Campbell[204] concluded from an electrophoretic and ultracentrifugal study of the soluble antigen–antibody complex in the system lightly iodinated bovine serum albumin and rabbit antibovine serum antibody that the principal complex was mostly AG_2.

Figure 3, taken from Goldberg and Williams,[205] illustrates the trend of various studies toward the maximum valence of two, which may be

(202) J. R. Marrack, H. Hoch, and R. G. S. Johns, *Brit. J. Exptl. Path.* **32,** 212 (1951).

(203) O. J. Plescia, E. L. Becker, and J. W. Williams, *J. Am. Chem. Soc.* **74,** 1362 (1952).

(204) S. J. Singer and D. H. Campbell, *J. Am. Chem. Soc.* **74,** 1794 (1952).

(205) R. J. Goldberg and J. W. Williams, *Trans. Faraday Soc.* **13,** 226 (1953).

regarded as firmly established, unless study of exceptional kinds of antibody (such as the horse antipneumococcus antibody of higher molecular weight) demonstrates the existence of antibodies with higher valences.

In a number of cases[206,124,130] antibodies are found that have the power of combining with antigen but do not form a precipitate. These antibodies are similar in behavior to the blocking antibodies observed with agglutinating systems. Such antibody has been variously called low-grade, incomplete, imperfect, and univalent. Some authors[124] who have used the term "univalent" have stated that they did not mean it in a literal sense, however. Other factors, such as differences in solubility, may be involved.

In the study of the Rh blood groups,[129] antibodies were encountered that could combine with erythrocytes but did not cause agglutination.

In 1944, Diamond[127] reported that a preparation of concentrated globulin from an anti-Rh serum showed prezone effect and was incapable of inactivating a serum containing an anti-Rh agglutinin of high titer. Later, Race[128] and Wiener[129] independently recognized the phenomenon and elucidated its nature. The serum which could inactivate other anti-Rh sera contained an antibody which could combine with Rh-positive erythrocytes and prevent their agglutination by anti-Rh sera, thus blocking the reaction of the ordinary agglutinating sera. Wiener therefore termed such antibodies "blocking" antibodies. Race termed them "incomplete" antibodies.

Pappenheimer in 1940[130] had found a nonprecipitating antibody in the serum of a horse immunized against crystalline egg albumin. Believing that this antibody must be univalent, he termed it "incomplete" antibody, implying that it was antibody that had for some reason not developed its proper complement of two combining groups. Heidelberger, Treffers, and Mayer[124] also used the term "incomplete" antibody for similar reasons. Wiener[5] published diagrams in which the "incomplete" antibody was represented simply as one-half of a "complete" bivalent molecule.

Adequate investigations of the molecular weight of "incomplete" antibody have not yet been published. Preliminary studies by the present author suggest that the molecular weight, instead of being less than that of normal antibody, is, if anything, rather more.

If the incomplete antibody owes its peculiar properties to being univalent, any process that polymerizes protein molecules might be expected to convert incomplete into "complete" agglutinating antibody. It is known that high pressure (4000–5000 atm.) causes extensive poly-

(206) M. Heidelberger and F. E. Kendall, *J. Exptl. Med.* **62**, 697 (1935).

merization of the molecules of serum. Nevertheless, Boyd[133] found that no pressure up to that which completely inactivated the incomplete antibody converted any of it into agglutinating antibody. The incomplete antibody required a pressure of about 1000 atm. more to inactivate it than did the agglutinating antibody.

It was found that suspension of the red cells in albumin[6] or treatment with trypsin[207] made the cells agglutinable by "incomplete" antibody. These observations made the hypothesis much less likely that the failure of the "incomplete" antibody to agglutinate was due to the presence of only one combining group on each molecule.

In 1950, Race and Sanger[208] observed that the incomplete anti-D Rh antibody agglutinated promptly and powerfully the red cells of a human blood which appeared to possess the Rh formula —D—/—D—. Evidently, with the proper sort of antigen the "incomplete" antibody is quite capable of causing agglutination. Pappenheimer[209] says "the failure of 'blocking' antibody to agglutinate untreated cells cannot be attributed to its univalence." Pappenheimer goes on to suggest that it remains to be proved that the behavior of nonprecipitating and nonagglutinating antibodies can be correctly ascribed to univalence.

Goldberg and Williams[205] state that the ultracentrifugal study of antibody–antigen systems has the advantage that it gives direct evidence of the existence of several distinct antigen–antibody complexes in solution, which electrophoresis does not resolve. They consider that, in the present stage of development and practice, electrophoretic study gives higher precision, especially in the region of high antigen excess. From a study of the available evidence they conclude that the maximum valence of precipitating antibody is two (see Fig. 3, p. 802).

Gitlin and Edelhoch[210] studied the reaction between bovine serum albumin and horse antibody by means of light-scattering measurements. They found that an equilibrium was reached when the reagents were mixed, more rapidly in the zone of antigen excess than in the zone of antibody excess. Large complexes consisting of soluble aggregates of antigen and antibody appeared to exist over all possible ratios of antigen to antibody.

Goldberg and Campbell[211] studied the precipitation of bovine serum

(207) J. A. Morton and M. M. Pickles, *Nature* **159,** 779 (1947).
(208) R. R. Race and R. Sanger, The Human Blood Groups, Blackwell, Oxford, 1950.
(209) A. M. Pappenheimer, The Nature and Significance of the Antibody Response, Columbia Univ. Press, New York, 1953.
(210) D. Gitlin and H. Edelhoch, *J. Immunol.* **66,** 67 (1951).
(211) R. J. Goldberg and D. H. Campbell, *J. Immunol.* **66,** 79 (1951).

albumin by rabbit antibody by light-scattering methods. They found that the rate of aggregation of antibody and antigen molecules into large complexes was dependent on the ratio in which antibody and antigen were mixed. They concluded that, at least to a first approximation, the reactions occurring were bimolecular.

Light-scattering methods are discussed elsewhere in this book by Edsall (Chap. 7).

j. Heterogeneity of Antibodies

It is not known just what determines whether a particular configuration of amino acids will be recognized as an antigenic determinant by the antibody-forming mechanism, but presumably more than one sort of determinant could exist in the same molecule. This would allow one and the same antigen to give rise to more than one antibody. There is considerable evidence that this actually happens.

Antibodies to proteins artificially conjugated with haptens (p. 773) may be of three sorts: directed toward the native protein, directed toward the hapten, or (sometimes) directed toward both simultaneously. Antisera to pure native proteins often behave as if they contained a variety of antibodies of different specificity.[16,212]

Boyd has suggested[62] that we should think of these different antibodies for a single antigen as not necessarily constituting distinct species, but as members of a rather large family, with varying degrees of variation from a mean. Some may be directed toward one antigenic determinant, some toward another, and even those directed toward the same determinant may vary in the sharpness with which they reproduce (in reverse) the full electronic configuration. It is likely that the weakly reactive antibody molecules reflect the structure of the determinant imperfectly, while the more "avid" molecules have reflected it more faithfully. Hooker and Boyd[213] obtained evidence that the sharpness of the specificity of the antibodies directed toward a hapten decreased as immunization progressed.

The work of Northrop and Goebel[190] on the fractionation of horse antipneumococcus antibody seemed to demonstrate the presence of a variety of antibody molecules, for none of the fractions obtained were uniform. The work of Cann, Brown, and Kirkwood[214] on fractionation of antibodies by electrophoretic convection further demonstrated their heterogeneity.

(212) K. Landsteiner and J. van der Scheer, *J. Exptl. Med.* **71,** 445 (1940).
(213) S. B. Hooker and W. C. Boyd, *Proc. Soc. Exptl. Biol. Med.* **29,** 298 (1931).
(214) J. R. Cann, R. A. Brown, and J. G. Kirkwood, *J. Biol. Chem.* **185,** 663 (1950).

3. Nature of Antibody Determinants

There exists no experimental evidence as to the chemical nature of the specific reactive groups (determinants or, perhaps more accurately, "antideterminants"). Tests of the specificity of portions of the antibody molecule prepared by enzymatic digestion and other methods of molecular fragmentation, coupled with amino acid analysis, may be expected to yield valuable information in the future. In this connection, the natural antibodies ("lectins") from plants may find some application (p. 788).

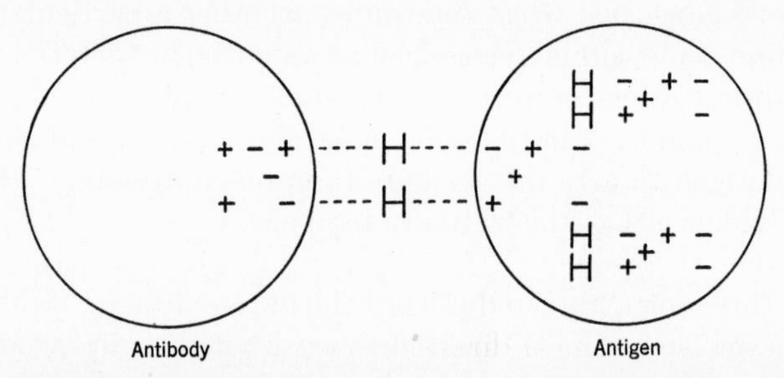

Fig. 4. Schematic example of how charges of opposite sign and groups capable of entering into hydrogen bond formation might determine the specific reaction between antigen and antibody.

At present our ideas of the structure of the antideterminants of antibody molecules derive from our hypothetical pictures of the way in which antibodies are formed (p. 809) and the forces that cause antibodies and antigens to unite (p. 815).

It is supposed that the specific groups of the antibody, like those of antigens, consist of characteristic arrangements of amino acids at certain points—or rather in certain areas—of the surface of the molecule. Several changes, probably both positive and negative, and hydrogen-bonding sites, must be involved, in order to account for the high degree of specificity that antibodies exhibit, especially their ability to distinguish stereoisomers. The specificity conferred on an antigenic determinant by the introduction of a small chemical group is considered to result partly from its own structure and partly from alteration in the local electric field.[34]

A purely schematic representation of one of the simplest cases of correspondence between antibody and antigen is shown in Fig. 4. It is based on the concept that the specific combining group of the antibody will possess charged groups corresponding (in reverse) to the charges of the antigenic determinant, plus hydrogen-bonding sites, all arranged in the fashion of a mirror image.

There is no evidence that antibodies contain prosthetic specific groups, and it is generally supposed that the reactive patch is an integral part of the protein structure. Haurowitz[215] supposes that the surface of the antibody corresponds to the surface of the antigen as does an electrotype to an electrode of complicated shape. Hooker and Boyd[200] and later Pauling[88] have proposed diagrammatic pictures in which the hapten or antigenic determinant fits into a pocket in the antibody, and the fit is supposed to be a close one (Fig. 5).

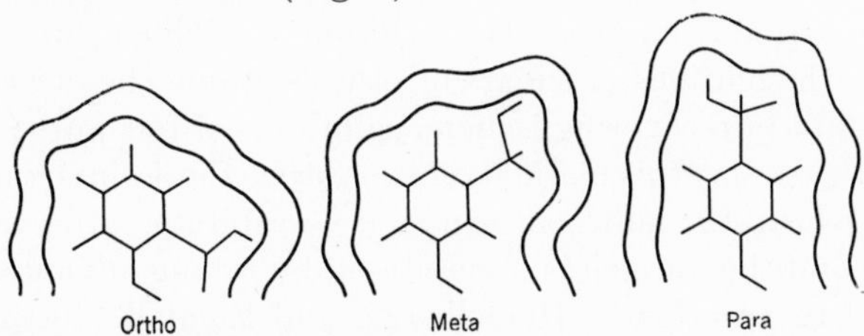

FIG. 5. Scale drawings of hypothetical antibody cavities specific for the *o*-, *m*-, and *p*-azophenylarsonic acid haptenic groups, with circumferential contours corresponding to a close fit and to a radial dilatation of 1.0 A. (Modified from Pauling.)[88]

It has been pointed out by Neurath[174] that this closeness of fit would allow the van der Waals' forces, which are short-range forces, to come into play.

4. PURIFICATION OF ANTIBODIES

Purely chemical methods of purification of antibodies have never yielded a 100 per cent pure antibody, e.g., one completely precipitable by its antigen. Methods that have been used include salting-out procedures,[216] precipitation with alcohol in the cold,[217,218] and nonspecific adsorption.

The only methods that have produced antibody preparations of high purity have depended on reacting the antibody with antigen, washing the resulting precipitate (or agglutinate) free from nonspecific proteins, and then employing chemical means to destroy or remove the antigen. In some cases, warming the suspended complex will split off considerable amounts of the antibody; in other cases, an alteration in the salt concentration is effective.[219] Sometimes, the antigen can be destroyed without

(215) F. Haurowitz, *Fortschr. Allergielehre,* Kalløs, Ed., Basel, 1939, p. 19.
(216) L. D. Felton, *Bull. Johns Hopkins Hosp.* **38,** 33 (1936).
(217) L. D. Felton, *J. Immunol.* **21,** 357 (1931).
(218) E. J. Cohn, J. L. Oncley, L. E. Strong, W. L. Hughes, Jr., and S. H. Armstrong, Jr., *J. Clin. Invest.* **23,** 417 (1944).
(219) M. Heidelberger, F. E. Kendall, and T. Teorell, *J. Exptl. Med.* **63,** 819 (1936).

damaging the antibody too much, and in this way Ramon[220] and Locke, Main, and Hirsch[221] have recovered antitoxin free from toxin–antitoxin precipitates. Sumner and Kirk[222] obtained antibody from urease–anti-urease precipitates by making the urease insoluble by treatment with acid. The only methods for concentrating antibody on a large scale are salting-out and fractionation with ethanol–water or methanol–water mixtures in the cold.

Adsorption of agglutinating antibody for red cells on boiled red cells or stroma prepared from red cells, in chromatographic columns, followed by elution with solutions of sugars or other suitable chemicals, has been fairly successful in recovering isohemagglutinins of fair purity.[223]

The most successful methods have been those involving soluble antigen in which the antibody–antigen precipitate, after washing, is redissolved, and the antigen is precipitated by taking advantage of some pecularity of its reactions. Heidelberger and Kendall[224] employed high salt concentrations (1.71–3.42 M NaCl) to dissociate antibody from pneumococcus–horse antipneumococcus precipitates. Felton[224a] was able to precipitate the antigenic polysaccharide out of dissolved pneumococcus polysaccharide–antipneumococcus precipitates by the addition of calcium chloride or strontium chloride and by adjusting the pH. Sternberger and Pressman[225] have devised a more general method, applicable to protein–antiprotein systems. In this procedure, the antigen is coupled with a suitable amount of a chemical reagent which introduces large numbers of highly polar groups (for instance it can be treated with diazotized arsanilic acid). These foreign groups do not prevent the native antigenic determinants, whatever their chemical nature may be, from reacting with the antibody to the native antigen. The resulting precipitate is then washed and dissolved at about pH 12, and the modified antigen is precipitated by the addition of calcium aluminate. The pH of the supernatant is then adjusted back to the physiological pH. This seems to be a general method capable of yielding antibody preparations of high purity. About 30 per cent of the antibody present can be recovered by this technique.

A new technique developed by Kirkwood and his group,[226] which

(220) G. Ramon, *Compt. rend. soc. biol.* **88,** 167 (1923).
(221) A. Locke, E. R. Main, and E. F. Hirsh, *J. Infectious Diseases* **39,** 126, 484 (1926).
(222) J. Sumner and J. S. Kirk, *Z. physiol. Chem.* **205,** 219 (1932).
(223) H. Isliker, personal communication, 1951.
(224) M. Heidelberger and F. E. Kendall, *J. Exptl. Med.* **64,** 161 (1936).
(224*a*) L. D. Felton, *J. Immunol.* **22,** 453 (1932).
(225) L. A. Sternberger and D. Pressman, *J. Immunol.* **65,** 219 (1950).
(226) J. R. Cann and J. G. Kirkwood, *Cold Spring Harbor Symposia Quant. Biol.* **14,** 9 (1950).

combines electrophoresis and convection, has in some cases separated antibodies that otherwise could not be separated by physical means. Some separation of the blocking antibody and the "cryptagglutinoids" of a human anti-Rh serum was achieved. Treatment of a rabbit anti-arsanilic serum gave some separation of antibodies directed toward the arsanilic hapten and antibodies directed toward the native protein carrier (bovine serum globulin). Like all electrophoretic methods, this method is limited in the amount of material that can be processed, and does not seem suitable as a method of large-scale purification of antibodies.

The observation[147,147e,148,149] that natural antibodies that are completely specific for the blood group A antigen exist in certain seeds (e.g., Lima beans) and can be extracted in practically unlimited amounts in high purity (in some cases apparently crystalline) makes available a source of purified antibody on a scale not known heretofore. This may facilitate the quantitative study of the antibody–antigen reaction and of the chemical nature of specific antibody combining groups.

5. Quantitative Determination of Amount of Antibody

The first use of the micro-Kjeldahl nitrogen method to analyze specific antibody–antigen precipitates of the antigen that did not contain nitrogen, or else contained a chemical marker such that the antigen nitrogen in the precipitate could be calculated, seems to have been due to Wu *et al.*[227] and was followed immediately by the work of Heidelberger and Kendall.[228] The analysis by such methods of specific precipitates (or agglutinates) has become standard procedure.[172]

The amount of antibody in rabbit serum can vary from amounts too small to be detected to amounts of antibody protein of the order of 12 mg./ml.,[229] 14 mg./ml., or an average value of 18 mg. in hyperimmune animals.[230] At least one rabbit has been found to contain 42 mg./ml. of antibody protein in his serum.[230]

6. Formation of Antibodies

Such evidence as there is suggests that antibodies are formed in the same part of the body as the serum proteins in general. It is the conviction of Madden and Whipple,[231] who have reviewed this subject, that the liver is the most important organ in the formation of the plasma proteins.

(227) H. Wu, L. H. Cheng, and C. P. Li, *Proc. Soc. Exptl. Biol. Med.* **25**, 853 (1928).
(228) M. Heidelberger and F. E. Kendall, *J. Exptl. Med.* **50**, 809 (1929).
(229) W. C. Boyd and H. Bernard, *J. Immunol.* **33**, 111 (1937).
(230) J. van der Scheer, E. Bahnel, F. A. Clarke, and R. W. G. Wyckoff, *J. Immunol.* **44**, 165 (1942).
(231) S. C. Madden and G. H. Whipple, *Physiol. Revs.* **20**, 194 (1940).

There is considerable evidence that antibodies are formed in the reticulo-endothelial system: that system of macrophagic cells found in the spleen, lymphatic glands, thymus, liver, lymph sinuses, adrenal capillaries, pituitary sinuses, and also wandering free in the blood. Attempts to prove this have often relied on experiments intended to "block" the animal's reticulo-endothelial system, as by the injection of nonantigenic particulate material, or the removal (and sometimes the transplanting) of lymph nodes. Chemical examinations for injected antigen have been made, especially with the use of tagged antigens, starting from the hypothesis that where the deposited antigen is found, there antibody synthesis is probably taking place. In this connection, the work of Haurowitz and Breinl,[232] using arsenic-containing antigens; of Rous and Beard,[233] using paramagnetic iron-containing antigens; and of Sabin,[234] using highly colored antigens; were the most significant before the technique of radioactive labeling was developed.

Haurowitz and Crampton[235] studied the deposition in the organs of rabbits of injected ovalbumin which had been iodinated with I^{131}-containing iodine. About 2000 molecules were found per liver cell (spleen yielded similar results), and study of the rate of elimination suggested that 300 days would have to elapse for the cell content to be reduced to 200 molecules, and about 3000 days for it to fall to 20 molecules per cell. The antigen was first bound by the microsomes, then more and more was found in the mitochondria. The high antigen concentrations in the cytoplasmic granules was considered significant.

The demonstration of γ-globulin in lymphocytes led to investigations of the antibody content of the lymphocytes of immunized animals,[236,237] and it was shown that extracts of lymphocytes of immunized animals contained specific antibodies to the injected antigen, sometimes in greater concentration than did the serum. Extracts of the lymphocytes from normal animals did not contain the antibodies.

It was further shown that the release of antibody from lymphocytes was controlled by the adrenal cortical hormones, as was the release of normal γ-globulins.[238]

(232) F. Haurowitz and F. Breinl, *Z. physiol. Chem.* **205**, 259 (1932).
(233) P. Rous and J. W. Beard, *J. Exptl. Med.* **59**, 577 (1934).
(234) F. R. Sabin, *J. Exptl. Med.* **70**, 67 (1939).
(235) F. Haurowitz and C. F. Crampton, *J. Immunol.* **68**, 73 (1952).
(236) T. F. Dougherty, J. H. Chase, and A. White, *Proc. Soc. Exptl. Biol. Med.* **58**, 135 (1944).
(237) T. N. Harris, E. Grimm, E. Mertens, and W. E. Ehrich, *J. Exptl. Med.* **81**, 73 (1945).
(238) T. F. Dougherty, A. White, and J. H. Chase, *Proc. Soc. Exptl. Biol. Med.* **59**, 172 (1945).

Nevertheless, it would seem that the weight of evidence at the present time suggests that the plasma cell, rather than the lymphocyte, is the main source of antibody production. This evidence has been summarized by Fagraeus[239] somewhat as follows: (*1*) The injection of an antigen, such as ovalbumin, into rabbits produces a sharp rise in the number of plasma cells in the spleen, whereas the injection of similar but nonantigenic substances such as gelatin, peptone, and dextran[239a] produces no such effect. (*2*) In the course of a secondary antigenic response, cytological studies show first a proliferation of large reticulum cells with an increase in cytoplasmic basophils (transitional cells) followed by their development in the direction of typical plasma cells. At the time antibodies are being most actively produced, immature plasma cells predominate. (*3*) In tissue-culture experiments with spleen fragments from immunized animals, the red pulp produced more antibody than did the white pulp, and the capacity of the red pulp to produce antibodies was directly correlated with its content of plasma cells, particularly the immature variety.

Bjørneboe, Gormsen, and Lindquist[242] found that in hyperimmunized animals the antibody content of tissues is correlated with the number of plasma cells visible in histological sections.

Burnet and Fenner[243] summarize ("with some diffidence") our present notions of antibody production somewhat as follows: (*1*) Antibody production against particulate antigens introduced into local tissues is predominantly a function of the plasma cells in the lymph nodes draining the region. (*2*) The spleen plays an essentially similar part when the antigenic material circulates in the blood. (*3*) Antibody produced in the lymph node passes to the circulation largely in the cytoplasm of the lymphocytes and plasma cells in the efferent lymph stream. (*4*) Liberation of antibody from these cells may be due in part or wholly to cytoplasmic dissolution induced or controlled by hormones of the adrenal cortex.

Burnet and Fenner emphasize that these conclusions are meant to apply only to serum antibodies developed in response to the inoculation of particulate antigens. They believe there is no certain evidence as to

(239) A. Fagraeus, *Acta Med. Scand.* **130,** *Suppl.* 204 (1948).

(239*a*) Dextrans are doubtless poorer antigens than proteins such as ovalbumin and may actually be nonantigenic in rabbits.[240] In man, however, the injection of dextrans results in cutaneous sensitization and the production of precipitins, showing that dextrans are antigenic in man.[241]

(240) A. Gronwall and B. Ingelman, *Acta Physiol. Scand.* **7,** 97 (1944).

(241) E. A. Kabat and D. Berg, *Ann. N. Y. Acad. Sci.* **55,** 471 (1952).

(242) M. Bjørneboe, H. Gormsen, and F. Lindquist, *J. Immunol.* **55,** 121 (1947).

(243) F. M. Burnet and F. Fenner, The Production of Antibodies, 2nd ed., Macmillan, Melbourne, 1949.

where the antibodies concerned with sensitization reactions or the response to tissue grafts are produced and virtually no evidence as to how they are carried to the general tissues of the body.

Most modern theories of antibody formation presuppose that the antibody molecule is formed in contact with a molecule (or by a cell modified by such a molecule) of antigen. The theory that seems most suggestive, and most in line with what we know about protein synthesis in general, is that of Haurowitz.[15] He points out that protein synthesis seems to take place in or around the nucleolus and in small granular particles of the cytoplasm. (It takes place also in virus particles, of course, but with this we are not at present concerned.) These sites of protein synthesis are rich in desoxyribonucleic acid.

Haurowitz points out that the high degree of specificity of the synthesized protein can be accounted for, within the boundaries of our present knowledge, only by assuming that the cell contains a repeatedly functioning pattern which acts as a *template*. Furthermore, the idea that the reproduction of proteins can take place only in the expanded state, in a unimolecular film,[244,245] is the only concept reconcilable with our present knowledge of intermolecular forces. Therefore, one of the first questions we must ask is: what keeps the template (and the molecule of antigen) in the expanded unimolecular form? Haurowitz believes that this is done by molecules of nucleic acid, which are polar molecules, and are known to be able to combine with proteins.

The first step of antibody formation, according to Haurowitz, is the deposition of amino acids upon the surface of the template, each deposited amino acid being the same as the one beneath it in the template. This process is similar to the process of crystallization, which is itself often quite specific. Next, these deposited amino acids are linked together into a polypeptide chain by enzymes of the cathepsin type. Then the new molecule is cast off into the circulation and, freed from the polar forces of the nucleic acid, is free to assume its typical globular structure.

In the formation of antibodies, if the template consisted of a molecule of antigen, the new molecule, being a positive replica, would be not a molecule of antibody but another molecule of antigen. This, of course, does not happen. Haurowitz supposes that the template is in reality some innate pattern of the body cell adapted for the purpose of directing the synthesis of other molecules like itself, but modified in configuration by the presence of some of the antigen. It is this modification which is important, and which when duplicated in a new molecule constitutes the "active patch" of the antibody molecule.

(244) L. T. Truland, *Am. Naturalist* **51**, 321 (1917).
(245) H. J. Muller, *Sci. Monthly* **44**, 210 (1937).

There have been other theories of antibody formation, but only two of them have more than historical interest. Burnet and Fenner[243] suggested that certain enzymes of the body, in destroying the introduced foreign antigen, become altered in a specific manner, and that these "trained enzymes" then have the power of synthesizing specific antibodies.

Burnet and Fenner[243] assume that antibodies can be formed in the absence of antigen. In their opinion (*a*) enzymes involved in the destruction of normal body constituents (called "self-markers") become adapted to acting on similar molecules of foreign substances, and (*b*) these adapted enzymes are self-reproducing and continue to multiply after the elimination of the antigen, and (*c*) the antibodies are (enzymatically inactive) partial replicas of these adapted enzymes.

Haurowitz points out a number of objections to this view. Of these the most serious is that antibodies can be manufactured against artificial heptens, which do not resemble any constituent of the body, and for whose destruction no enzymes exist in the body. Also, Burnet's objections to current theories of antibody formation are based partly on his disinclination to believe that molecules of antigen, or fragments of such molecules, can persist in the animal body for as long as antibody production can sometimes go on (over fifty years in the case of yellow fever virus). Nevertheless, it has been well established that pneumococcus antigen can persist in some animals (mice) for life, and carbon particles and pigments may sometimes be stored in the organisms for extended periods of time.

Pauling[151] proposed the hypothesis that in the process of antibody formation the previously formed globulin molecule was denatured and thus its characteristic folded configuration was partly lost. If it were then "renatured" when in contact with a molecule of antigen, it would fold up again in a way such that some of its polar groups would conform in reverse to the configuration of the antigen, and would thus become a molecule of antibody. The order of the amino acids in the chain would not be changed, but the change would affect the way in which the chain folded up to give a globular molecule. Pauling suggested that there were a very large number of such ways, all having about the same internal energy, and thus all about equally probable; the presence of the antigen in the body simply determined which of these would be adopted.[245a]

(245*a*) Pauling[151] assumed that it is only the ends of the polypeptide chain which have thus accessible to them a large number of configurations with nearly the same energy, and that this results from a difference in amino acid composition between the ends and the main body of the chain. He postulated that "the end parts of the globulin molecule contain a very large proportion (perhaps one-third or one-half) of proline and hydroxyproline and other residues which prevent the assumption of a stable layer con-

The specific arrangement of amino acids in any given antibody has not yet been determined. Porter,[37] however, found that the γ-globulin from rabbit serum containing the antiovalbumin had the same terminal arrangement of amino acids (alanyl-leucyl-valyl-aspartyl, with glutamyl probably occupying the fifth position) as did the immunologically inactive γ-globulin from the same animal. This at least does not contradict Pauling's hypothesis that the order of the amino acids in the polypeptide chain of antibody is not different from that of normal γ-globulin.

Pauling insists (just why is not clear to the present writer) that his theory of antibody formation requires that the two ends of the globulin molecule be modified simultaneously, and that if two different antigens or determinant groups, A and B, are present, antibody with one end modified by A and the other by B (anti-AB) ought to be formed along with anti-A and anti-B. This has never been confirmed experimentally.[246]

Pauling and Campbell[247] attempted to prove this theory experimentally, and claimed to have synthesized antibodies *in vitro* by bringing denatured globulin into contact with simple antigens, renaturing the globulin, and then removing the antigen. In the ten years that have elapsed since this claim was published there has been no confirmation of it,[248,249] even from Pauling's own laboratory. In view of the immense practical importance of any method of producing antibodies *in vitro*, this lack of confirmation suggests that the experiment does not invariably work.

Pauling's picture of a cavity in the antibody, specifically adapted in shape to the antigenic determinant, is shown in Fig. 5.

The existence of natural plant antibodies of high specificity,[147,147a–d,148–150] although not fatal to the above concepts of antibody formation, raises some interesting questions. In particular, one wonders what the similarity between these plant lectins and immune antibodies can be. Perhaps the arrangement of amino acids on certain parts of the surface of these plant proteins accidentally enables them to unite specifically with the antigen.

IV. Antibody–Antigen Combination

There are several features of the reaction between antibody and antigen which make it convenient to divide the reaction into two stages.

---

figuration." It would be possible to test this assumption about the internal composition of γ-globulins, for example by analyses of partially digested antibodies which still retain full activity (p. 791). There does not seem to be any analytical evidence bearing on it available at present.

(246) F. Haurowitz and P. Schwerin, *J. Immunol.* **47**, 111 (1943).
(247) L. Pauling and D. H. Campbell, *J. Exptl. Med.* **76**, 211 (1942).
(248) F. Haurowitz and P. Schwerin, *Arch. Biochem.* **11**, 515 (1946).
(249) A. M. Kuzin and N. A. Nevraeva, *Biokhimiya* **12**, 49 (1947).

This does not necessarily imply that the two stages are entirely separate, or that the mechanisms involved are different.

Among the reasons for this division are: (*a*) The first stage proceeds without visible alteration, and can be detected only indirectly, by some secondary effect, while the second stage is conspicuous and easily detected. (*b*) In the case of most small haptens reacting alone with antibody, no second stage follows the first. (*c*) Under some conditions, as at low ionic strength, combination may occur, but the later steps leading to aggregation may be suppressed.[249a] (*d*) The speeds of the reactions of the two stages may be very unequal, the first stage being always very rapid, and the second often quite slow. (*e*) The energy change takes place almost entirely during the first stage. (*f*) In some cases, the specificity of the second stage seems to be less. (*g*) The forces leading to aggregation are weaker than those involved in the primary interactions between antibody and antigen.[250]

1. First Stage: Combination

a. Forces Involved

The forces that cause antibody molecules to unite with their antigens are probably complex. The electrostatic attraction of charged groups (e.g., $—COO^-$ for $—NH_3^+$, and vice versa), electronic van der Waals' attraction, steric hindrance depending on molecular sizes and shapes, and hydrogen-bond formation may all play a role. The forces uniting the artificially introduced group $—N{=}N—C_6H_4AsO_3H^-$ to its specific antibody would consist of the attraction of the negative charge for a complementary charge in the antibody, the hydrogen bond formed with the antibody by the undissociated hydrogen of the group, the van der Waals' attraction between the group and the adjacent parts of the antibody, and possibly a hydrogen bond formed by one of the nitrogens of the —N=N— group with a hydrogen-donating group of the antibody. Steric factors would ensure suitable distances of contact between the atoms of the haptenic group and the atoms of antibody.

It seems likely that the main forces in point of strength are the coulomb forces between negatively and positively charged groups. There may be, in the case of natural antigens, several charged groups, doubtless not all of the same sign, in each determinant or "active patch." This is supported by several lines of evidence, such as the striking influence of polar groups in specificity (see above), and the high heat of reaction between antibody and antigen (p. 817).

(249*a*) J. T. Duncan, *Brit. J. Exptl. Path.* **18,** 108 (1937).
(250) W. J. Kleinschmidt and P. D. Boyer, *J. Immunol.* **68,** 257 (1952).

Goebel and Hotchkiss[251] found that antipneumococcus horse sera of types I, III, and VIII gave strong precipitation with artificial antigens containing benzenecarboxylic and sulfonic acid radicals, and considered that this was due to the attraction between the ionized carboxyl groups of the uronic acids and the ionized amino groups of the antibody. This idea was supported by the observations that treatment of the antibody with formaldehyde, which probably reacts with the amino groups, destroyed the specific precipitating power of these antibodies, and acetylation was almost equally effective.

Kleinschmidt and Boyer[250] concluded from a study of the pH effects on specific precipitate formation that the charged groups of the glutamic acid, aspartic acid, and lysine residues are of importance for the reaction between ovalbumin and its antibody, and that histidine residues are not of prominent importance. However, 25–30 per cent of the amino groups of ovalbumin can be removed without decreasing its specific reactivity.[42–44]

Experiments with thin films of antigen and antibody separated by inert "barriers" led Rothen[252] to postulate that long-range forces may be involved in the reaction between antibody and antigen. These experiments have been criticized by Singer,[252a] and Campbell and Bulman[27] state that: "As most of the experimental evidence in immunochemistry is in accord with the well established principles involving short-range forces, at present there is not sufficient evidence to warrant the acceptance of specific long-range forces." With this opinion the present writer is in agreement.

It is pointed out by Filitti-Wurmser *et al.*[253] that the heat of reaction calculated by them for the combination of anti-B agglutinin from serum of an individual of genotype A_1O ($\Delta H = -16$ Kcal.) would correspond to the formation of 3 or 4 hydrogen bonds, or 16 van der Waals attractions.

(251) W. F. Goebel and R. D. Hotchkiss, *J. Exptl. Med.* **66,** 191 (1937).

(252) A. Rothen, *J. Biol. Chem.* **168,** 75 (1947); *idem, J. Am. Chem. Soc.* **70,** 2732 (1948).

(252*a*) S. J. Singer, *J. Biol. Chem.* **182,** 189 (1950).

(253) S. Filitti-Wurmser and Y. Jacquot-Armand, *Arch. Sci. physiol.* **1,** 151 (1947). R. Wurmser and S. Filitti-Wurmser, *Biochem. et Biophys. Acta* **4,** 238 (1950). S. Filitti-Wurmser, Y. Jacquot-Armand and R. Wurmser, *J. Chim. phys.* **47,** 419 (1950); **49,** 550 (1952).
S. Filitti-Wurmser, G. Aubel-Lesure, and R. Wurmser, *J. Chim. phys.* **50,** 236 (1953); **50,** 317 (1953).
S. Filitti-Wurmser, Y. Jacquot-Armand, and R. Wurmser, *J. Chim. phys.* **50,** 240 (1953).
Y. Jacquot-Armand and S. Filitti-Wurmser, *Arch. Sci. physiol.* **7,** 233 (1953).
R. Wurmser and S. Filitti-Wurmser, *Biochim. et Biophys. Acta* **12,** 92 (1953).

b. Rate of Reaction

If agglutinin is added to cells and the tubes are immediately centrifuged, the agglutination is as strong as would be obtained if the mixtures had been allowed to stand before centrifuging, except in the case of certain weak antibodies such as are met with in the Rh system. This suggests that the greater part of the antibody combines in a few seconds. When precipitin and antigen are mixed, the precipitate often seems to appear instantaneously. Heidelberger, Treffers, and Mayer[124] found that the combination between antibody for ovalbumin and its antigen was complete in 20 sec., even at 0°C. Mayer and Heidelberger[254] considered that the combination between pneumococcus polysaccharides and their antibodies was at least 90 per cent complete in less than 3 sec. at 0°C. The rate at which heat is evolved when antibody and antigen react (see below) also suggests that the reaction, even when it involves mixing throughout a volume as large as 800 ml., is complete in 3 min. or less.

c. Heat of Antibody–Antigen Combination

Boyd *et al.*[254a] measured the heat evolved at 31°C. when horse antihemocyanin reacted with hemocyanin (from *Busycon caniliculatum*). In the region of antibody excess where, in the case of horse antiprotein antibodies, no precipitate is formed, a value of about 3.0 cal. per gram of antigen nitrogen was found. Since the molecular weight of this hemocyanin was reported by Svedberg[255] as 6,800,000, this amount of heat corresponds to about 3300 kcal. per mole of antigen. By extrapolation from the results of analyses of specific precipitates made with this system, it was calculated that the result corresponded to about 40 kcal. per mole of antibody. The large value obtained suggests that a considerable number of electric charges are involved in the union of each molecule of antigen with one of antibody, as suggested above.

Since the work of Boyd *et al.*,[254a] various authors have calculated, from equilibrium measurements, the energy changes involved in other antibody–antigen reactions, making various assumptions. The results are shown in Table VII, modified from Marrack.[95] The binding energy of bovine serum albumin for the dye, Orange I, is shown for comparison. It will be seen that considerable variation in the values exists, although the order of magnitude is the same.

(254) M. Mayer and M. Heidelberger, *J. Biol. Chem.* **143,** 567 (1942).
(254*a*) W. C. Boyd, J. C. Conn, D. C. Gregg, G. B. Kistiakowsky, and R. M. Roberts, *J. Biol. Chem.* **139,** 787 (1941).
(255) I. Eriksson-Quensel and T. Svedberg, *Biol. Bull.* **71,** 498 (1936).

The positive entropy changes are stated by Haurowitz[19] to show that water molecules are liberated as a result of antibody–antigen combination. The negative changes would be expected to result from the aggregation, and might also indicate restricted rotational degrees of freedom of the protein molecules in the resulting aggregate, and possibly some binding of water molecules as a result of the reaction.

TABLE VII

THERMODYNAMIC VALUES MEASURED OR CALCULATED FOR ANTIBODY–ANTIGEN REACTIONS

(After Marrack)[95]

Antigen	Antibody	ΔF, kcal./mole	ΔH, kcal./mole	ΔS, e.u.	K
Hemocyanin[a]	Horse anti-hemocyanin	-10^{c}	-40	-100	
Arsanil-azo-bovine serum globulin[b]	Rabbit antiserum	*ca.* -8.5	-2	21	10^{6}–10^{7}
Sulfanil-azo-ovalbumin	Rabbit antiserum	-8.5	-2.8	21	10^{6}–10^{7}
Anthranil-azo-bovine serum[d] globulin	Rabbit antiserum		-70		
Hapten[e]	Rabbit antibody	-7.7			3.5×10^{5}
Orange I[f]	Bovine serum albumin	-6.36	-3.91	8.75	9.93×10^{4}
Divalent arsanilic hapten[g]	Rabbit antibody	-7.6			

[a] W. C. Boyd, J. B. Conn, D. C. Gregg, G. B. Kistiakowsky, and R. M. Roberts, *J. Biol. Chem.* **139,** 787 (1941).

[b] F. Haurowitz, C. F. Crampton, and R. Sowinski, *Federation Proc.* **10,** 560 (1951).

[c] Assumed value.

[d] F. Haurowitz and L. Etili, *Federation Proc.* **8,** 404 (1949).

[e] M. J. S. Dewar, *Discussions Faraday Soc.* **2,** 261 (1947).

[f] F. Karush, *J. Am. Chem. Soc.* **72,** 2705 (1949).

[g] S. Epstein, *Science* **118,** 756 (1953).

Campbell and Bulman[27] have suggested that the calculations of Haurowitz *et al.*[256] should be received with some reservation until it is known that the equation assumed by the authors actually represents the reaction taking place and that a temperature variation in the solubility of the complexes themselves was not an influencing factor.

(256) F. Haurowitz, C. F. Crampton, and R. Sowinkski, *Federation Proc.* **10,** 560 (1951).

The skillful and patient investigations of Mme. Filitti-Wurmser[253] and co-workers deserve a paragraph to themselves. These workers made use of a modification of the Ashby technique of counting the number of free and agglutinated erythrocytes in mixtures of erythrocytes and serum to measure the degree of agglutination in mixtures kept at different temperatures. From measurements of the degree of agglutination as a function of the concentration of free agglutinin at different temperatures the change in enthalpy ΔH was calculated.

The surprising result emerges from these calculations that the anti-B (β) agglutinin, although homogeneous in any particular serum, varies considerably in its heat of reaction with the antigenic sites on group B erythrocytes, depending upon the blood group or subgroup of the individual producing the agglutinin, and even varying within group A in a way depending on the genotype of the individual. A summary of these results is shown in Table VIIA.

TABLE VIIA

THERMODYNAMIC VALUES OF ANTI-B (β) AGGLUTININS FROM INDIVIDUALS OF DIFFERENT BLOOD GROUPS AND SUBGROUPS[253]

Group of subgroup	ΔF, kcal./mole	ΔH, kcal./mole	ΔS, e.u.	$K37°$
A_1O	−9.8	−16.0	−20	0.8×10^7
OO	−9.2	− 1.7	+24	3×10^6
A_1A_1		− 8.0	+ 8	
A_2		− 9.0		
A_3		− 3.0		

Filitti-Wurmser *et al.* point out that the homogeneity of the agglutinins in any one person is in contrast to the heterogeneity of antibodies produced in response to immunization with an antigen, and suggest that this is because the isoagglutinins are produced as a direct consequence of gene action.

It is concluded that the differences in order of magnitude in the heat of reaction of the different agglutinins can hardly be attributed to differences in the specific reactive groups. Filitti-Wurmser *et al.* suggest instead that the increase in entropy which results when the anti-B of group O serum combines with B may be connected with a perturbation of the entire protein molecule. Perhaps the change produced at a specific group touches off a sort of reversible denaturation, resulting in a production of disorder and consequent absorption of heat. These effects mask the evolution of heat and decrease in entropy which result from the local reaction of the specific group. In the case of a molecule of anti-B from an individual of genotype A_1O, which is a larger molecule (p. 791), the perturbation produced in the specific group is not sufficient to modify particularly the whole molecule.

d. *Effect of Electrolytes*

Bordet[257] claimed that the presence of salts was necessary for the agglutination of bacteria, but that combination of antibody with bacteria took place in the absence of salts. Landsteiner and Welecki[258] found that the combination of hemolysin with red cells was considerably reduced by hypertonic (1 *M*) sodium chloride solution, while Heidelberger and Kabat[259] found that considerable antibody could be dissociated from specific precipitates and from agglutinated pneumococci by the action of 2.5 *M* sodium chloride.

Marrack[34] suggested that the reduced combination of antigen and antibody in high salt concentrations might be due to the establishment of an atmosphere of electrolyte ions around the oppositely charged polar ions of the combining groups of antibody and antigen, reducing the attraction of these for one another.

2. Second Stage: Precipitation, Agglutination, Neutralization, etc.

a. *Neutralization of Toxins and Viruses*

Antitoxins combine with toxins in varying proportions, depending on the ratio in which the reagents are mixed. By determinations of the toxicity of the various mixtures, the ratio in which the two reagents combine can be calculated. The results of a typical experiment are shown in Fig. 6.

It will be observed that the amount of antitoxin bound to a unit of toxin increases as the ratio of antitoxin to toxin in the original mixture is increased, but not in linear relation. The results are typical of those observed when substances with multiple combining groups are involved. For this reason, equations of the adsorption type, e.g., $R = kx^n$, where R is the ratio of antitoxin to toxin bound, x is the final concentration of free antitoxin in the mixture, and k and n are constants, are found to fit the quantitative results fairly well.

(*1*) *Avidity.* It has been observed that the protective effect of antitoxin sera, as determined by animal experiments, does not always parallel the antitoxin content as determined by *in vitro* titrations. Kraus[260]

(257) J. Bordet, Traite de L'Immunité, Masson, Paris, 1939.
(258) K. Landsteiner and S. Welecki, *Z. Immunitätsforsch.* **8**, 397 (1910).
(259) M. Heidelberger and E. A. Kabat, *J. Exptl. Med.* **65**, 885 (1937).
(260) R. Kraus, *Zentr. Bakt. Parasitenk.* **34**, 408 (1903).

found in 1903 that normal goat or horse serum took about an hour to neutralize the toxin of the Nasik vibrio, whereas immune antitoxin neutralized it at once. Kraus seems to have suggested that the immune sera had a higher "avidity" for the toxin. Glenny and Barr[261] considered that the important feature of avidity is firmness of union between antibody and antigen.

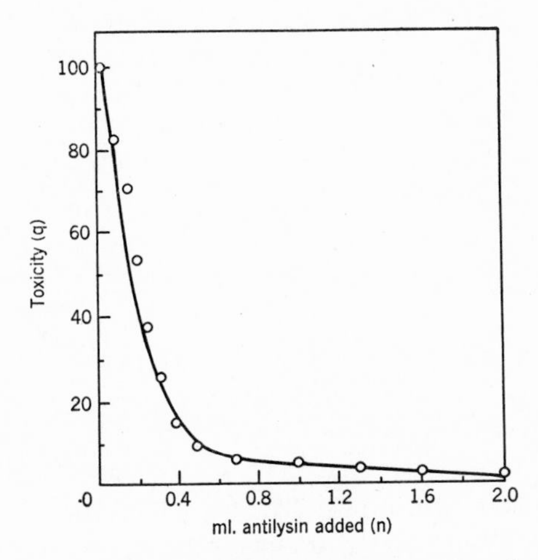

FIG. 6. Relative toxicity (q) of a constant amount of tetanolysin after addition of varying amounts of antibody (n).[286]

It is of obvious practical importance that antisera should contain not only an adequate amount of antibody, but antibody of satisfactory avidity. This requirement comes up in the evaluation of blood-grouping reagents, among other cases.[262]

b. Agglutination

The amount of antibody involved in agglutination reactions was determined by Heidelberger and Kabat[263] by nitrogen determinations of agglutinated and untreated bacteria. Pneumococci, in amounts that removed most of the agglutinin from the serum, took up from 0.2 to 1.9 g. of antibody nitrogen per gram of bacterial nitrogen. Even smaller amounts, of the order of 1/100 of this, were sufficient to bring about agglutination.

(*1*) *Zones in Agglutination.* If an agglutinating serum is sufficiently diluted, it will, of course, no longer agglutinate the cell suspension. The

(261) A. T. Glenny and M. Barr, *J. Path. Bact.* **35,** 91 (1932).
(262) W. C. Boyd, *Ann. N. Y. Acad. Sci.* **46,** 927 (1946).
(263) M. Heidelberger and E. A. Kabat, *J. Exptl. Med.* **60,** 643 (1934).

"zone" of antibody concentrations, in which no agglutination is observed because of insufficient antiserum, is called the "postzone."

With some sera a "prezone" is observed, in which too much antiserum is present. In some cases this is due to the presence of an inhibiting substance which interferes with agglutination by reducing the cohesiveness of the bacteria after they have been combined with antibody. This interference is specific. Jones and Orcutt[264] found by direct measurement of the cohesive force between smears of bacteria on slides that the presence of this inhibiting substance reduced the cohesive force between smears of bacteria treated with agglutinin to the values observed with untreated bacteria. Such treated films could be restored to normal cohesiveness by washing, and treated bacteria from the prezone would agglutinate after washing. An interesting feature of the observations of Jones and Orcutt is that the presence of this inhibiting substance did not increase the volume of the antibody-coated bacteria, though the increase due to the presence of the agglutinin was readily measurable. It would thus seem that the film of inhibitor is either very thin or, more likely, consists of a few molecules in scattered loci on the bacterial surface.

In blood-grouping sera, especially anti-Rh sera,[127–129] specific inhibiting substances are often found. It has been suggested that these consist of univalent antibody which is able to combine with the specific combining sites of the red cells, but, possessing only one combining group per molecule, cannot link the cells together and so bring about agglutination. If this is so, the mechanism would seem to be different from that studied by Jones and Orcutt. In addition, Race and Sanger[208] have reported observations that do not seem consistent with the idea that this blocking antibody is univalent (p. 804).

c. Precipitation

When a soluble antigen and its antibody are mixed, an insoluble combination of the two substances often results. Most of the theoretical studies of serological reactions have made use of this phenomenon.

(*1*) *Amount of Precipitate; Zones.* The amount of precipitate produced depends upon the absolute amounts of the two reagents and upon the ratio in which they are mixed. The precipitate is usually determined by the micro-Kjeldahl method,[265,172] but Heidelberger and MacPherson,[266] using the Folin-Ciocalteau phenol reagent, have proposed a method that gives determinations with even smaller amounts (as little

(264) F. S. Jones and M. Orcutt, *J. Immunol.* **27,** 215 (1934).
(265) J. K. Parnas and R. Wagner, *Biochem. Z.* **125,** 253 (1921).
(266) M. Heidelberger and C. F. C. MacPherson, *Science* **97,** 405 (1943).

as 10 μg.) of protein. When antibody and antigen are mixed in "equivalent" proportions, both are usually quantitatively precipitated.

When the proportions of the two reagents are not "equivalent," precipitation may be incomplete or lacking. If antigen is in excess, i.e., insufficient antibody to precipitate the antigen is present, no precipitate results, and a postzone occurs. Unfortunately, this is usually called a "prezone" because in the study of precipitating systems it used to be customary to use a constant amount of antibody and to test different dilutions of the antigen, starting with the strongest concentration. Strong antigen was said to "inhibit" the precipitation.

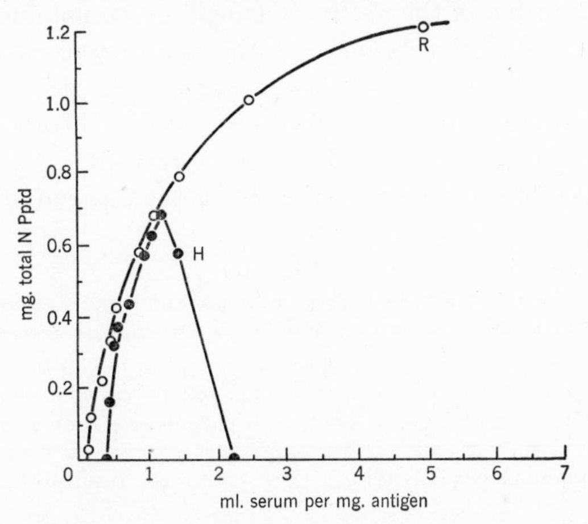

FIG. 7. Amount of precipitate formed by the addition to a constant amount of antigen of increasing amounts of antibody. R = rabbit antibody (antiovalbumin), H = horse antibody (antitoxin).

Excess of antibody does not affect the completeness of precipitation of antigen very much in the case of rabbit antibodies, nor does it in the case of horse antipolysaccharide antibodies. In the case of horse antiprotein antibodies, such as antitoxins, it is observed that excess antibody interferes with, or prevents, precipitation, and soluble compounds of antibody and antigen are formed. This is referred to as inhibition with antibody excess.

Because of the difference in behavior of the two sorts of antibodies, quantitative study of the precipitin reaction yields different results when such antibodies are involved. With antibodies of the "R" type (rabbit antibodies and horse antipolysaccharide antibodies), addition of increasing amounts of antibody to a given amount of antigen yields larger and larger amounts of precipitate, because compounds richer in antibody are formed and precipitated. With antibodies of the "H" type (horse antiprotein), the addition of increasing amounts yields increasing amounts of precipitate up to a maximum, after which the amount begins to decrease sharply because the compounds with the larger amounts of antibody are soluble. This is illustrated in Fig. 7.

(*2*) *Equivalence Zone.* There is usually (probably always in the case of pure antigens and pure antibodies) a proportion or proportions in which the two reagents can be mixed, so that the supernatant liquid, after separation of the precipitate, will contain no detectable amount of antibody or antigen (or small traces of both). This is called the equivalence zone.[267] This may coincide with the proportion that gives the maximal precipitation of the antibody from a given volume of antiserum.

(*3*) *Composition of Precipitate.* In the majority of cases, the specific precipitate consists mostly of antibody (p. 778). In addition to the antibody and antigen, minor constituents such as lipide material and components of complement (p. 839) may be present.

The influence of the molecular weight of the antigen on the composi-

TABLE VIII

RATIO OF ANTIBODY TO ANTIGEN IN PRECIPITATES MADE WITH VARYING PROPORTIONS OF ANTIBODY TO ANTIGEN FOR TWO DIFFERENT SYSTEMS[a]

One cubic centimeter of antiserum used for all experiments

A = antibody, T = toxin, S = specific capsular polysaccharide of type III pneumococcus, N = nitrogen. Figures in parentheses somewhat uncertain

Horse antipneumococcus S III				Horse antidiphtheria toxin			
S added mg.	ΣN pptd., mg.	R = (N/S) × 6.25	Reagent present in excess	Toxin N added, mg.	ΣN pptd., mg.	R = antibody N/ toxin N	Reagent present in excess
0.02	0.45	140	A	0.023	0		A
0.04	0.79	124	A	0.046	0		A
0.05	0.97	121	A	0.069	0.386	(6.9)	(A)
0.06	1.08	112	A	0.081	0.554	5.8	(A)
0.08	1.29	101	A	0.092	0.564	5.1	
0.10	1.54	96	A	0.103	0.579	4.6	
0.12	1.68	87	A	0.138	0.612	3.4	
0.15	1.71	71	A	0.184	0.661	2.6	(T)
0.20	1.75	?	S	(0.196)		(2.4)	(T)
				0.207	0.652		(T)
				0.230	0.359		T
				0.276	0		T

[a] *References:* M. Heidelberger and F. E. Kendall, *J. Exptl. Med.* **61,** 559, 563 (1935); A. M. Pappenheimer and E. S. Robinson, *J. Immunol.* **32,** 291 (1937).

(267) M. Heidelberger, *Bact. Revs.* **3,** 49 (1939).

tion of the precipitate has already been discussed (p. 778). With large antigen molecules the ratio by weight of antibody to antigen in the precipitate tends to be small (less than 1); with small antigen molecules it tends to be large (10–100).

The ratio of antibody to antigen in the precipitate increases in proportion to the amount of antibody added to a fixed amount of antigen. The data obtained fit the absorption equations of Freundlich:[268] $R = a(A)^b$, and of Langmuir:[269] $R = \frac{ab(A)}{1 + a(A)}$, fairly well. R is the ratio of antigen in the precipitate, (A) is the concentration of antibody in the supernatant, and a and b are constants. Typical results are shown in Table VIII.

TABLE IX

MOLECULAR COMPOSITION OF ANTIBODY–ANTIGEN COMPOUNDS[a]

R = rabbit, H = horse, A = antibody, G = antigen. Formulas in parentheses are somewhat uncertain

Antigen	Antibody	Empirical composition of precipitate at				Composition of soluble compound in zone of partial inhibition
		Extreme antibody excess	Equivalence zone		Zone of partial inhibition	
			Antibody excess	Antigen excess		
Ovalbumin	R	A_5G	A_3G	A_5G_2	A_2G	(AG)
Dye–ovalbumin	R	(A_5G)	(A_3G)	A_5G_2	A_3G_4	(AG_2)
Serum albumin	R	A_6G	A_4G	A_3G	A_2G	(AG)
Thyroglobulin	R	$A_{40}G$	$A_{14}G$	$A_{10}G$	A_2G	(AG)
Viviparus hemocyanin	R	—	$A_{120}G$	$A_{83}G$	$A_{36}G$	—
Tobacco mosaic virus	R	$A_{900}G$	$A_{450}G$	—	—	—
Diphtheria toxin	H	A_8G	A_4G	A_3G_2	AG	(AG_2)
Ovalbumin	H	(A_4G)	A_2G	—	AG	(AG_2)

[a] *References:* M. Heidelberger, *Bact. Revs.* **3**, 49 (1939); S. Malkiel and W. C. Boyd, *J. Exptl. Med.* **66**, 383 (1937); A. M. Pappenheimer, *J. Exptl. Med.* **71**, 263 (1940); M. Heidelberger, H. P. Treffers, and M. Mayer, *J. Exptl. Med.* **71**, 271 (1940); A. M. Pappenheimer, H. P. Lundgren, and J. W. Williams, *J. Exptl. Med.* **71**, 263 (1940); E. A. Kabat, *Ann. Rev. Biochem.* **15**, 505 (1946).

(4) Molecular Composition of Precipitates. Analyses of specific precipitates, plus a knowledge of the molecular weights of antibody and antigen, enable the molecular composition of specific precipitates to be calculated. Typical results for several systems are shown in Table IX.

(268) H. Freundlich, Kapillarchemie, Akadem. Verlag, Leipzig, 1923.

(269) I. Langmuir, *J. Am. Chem. Soc.* **40**, 1361 (1918).

It is obvious that the precipitate obtained at any particular point of the range of antibody–antigen proportions, is a mixture of two or more of these compounds.

(*5*) *Solubility of Precipitate.* Specific precipitates made with antibody and antigen in the usual proportions are very insoluble (of the order of 0.05–0.005 mg. protein nitrogen/ml. salt solution). Nevertheless, when small amounts of antibody are being measured, it is necessary to apply corrections for the solubility of the precipitate. When precipitates are made with antigen excess, or (in the case of horse antiprotein antibodies) with antibody excess, the precipitates may be so soluble that collection, washing, and analysis become difficult. The more soluble portions of such precipitates tend to wash out first, leaving compounds not strictly representative of the whole.

(*6*) *Velocity of Precipitation: Optima.* There is a certain ratio in which antibody and antigen may be mixed which gives a mixture which precipitates (flocculates) most rapidly. Determinations of the time of flocculation of antibody–antigen mixtures have practical and theoretical interest. The first method is due to Ramon,[270] who applied it to the titration of toxins and antitoxins. Varying amounts of the antitoxin were added to a fixed amount of toxin, and the particular mixture that flocculated before the others was observed. This has been called the β-procedure by British writers.[34]

Dean and Webb[271] added varying concentrations of antigen to a fixed amount of antibody, and observed the mixture that flocculated most rapidly. This has been called the α-procedure by British writers.[34] The two methods do not give exactly the same antibody–antigen ratio for the α- and β-optima, and it is easy to see mathematically why they cannot be expected to agree exactly.[272]

The α-optimum (amount of serum constant; serial dilutions of antigen used) is generally used for antibody–antigen titrations, but for titrations of diphtheria toxin (or toxoid) against horse antitoxin the historically earlier β-system (antigen constant, with decreasing amounts of antiserum) is still customary.

It is only antisera of the H-type (antibodies that inhibit the precipitation when present in excess) that give both a well-defined α- and β-optimum. Sera of the R-type (rabbit antisera and horse anticarbohydrate antisera) give a good α-type optimum, but the β-type does not exist or is poorly defined. A typical experiment is shown in Fig. 8, where the dotted line indicates the α-optimum and the dashed line the β-optimum.

(270) G. Ramon, *Compt. rend. soc. biol.* **86**, 661, 711, 813 (1922).
(271) H. R. Dean and R. A. Webb, *J. Path. Bact.* **29**, 473 (1926).
(272) W. C. Boyd and M. A. Purnell, *J. Exptl. Med.* **80**, 289 (1944).

The values given by the α-procedure are found by examining the times of flocculation found in any column (antibody constant); and those for the β-procedure, in any column (antigen constant). It may be seen that they are far from agreeing.

With sera of the extreme H-type such as (horse) diphtheria antitoxin, the contours shown in Fig. 8 appear as extremely eccentric parabolas,

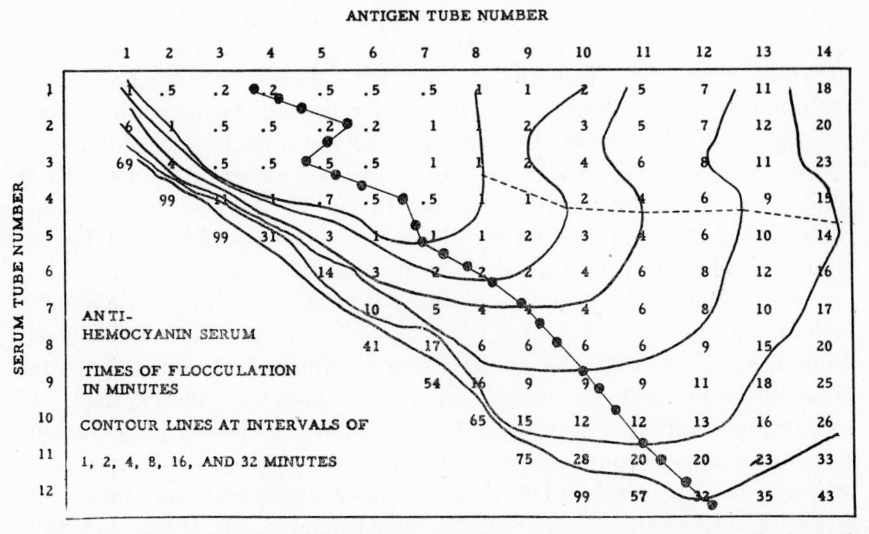

FIG. 8. Times of flocculation of mixtures of various amounts of antibody and antigen in various concentrations. Line of solid circles connects points of constant antibody (α) optima, dashed line connects points of constant antigen (β) optima. Solid lines connect points of equal flocculation times. Note ratio of antibody concentration to antigen concentration is sensibly constant for α optimum, shifting for β optimum.[62]

and the difference between the optima determined by the two methods is very small, and difficult to measure experimentally. This has led certain writers,[273] whose work has been mainly with toxins, to deny that the two optima are different, but careful measurement reveals even in this case a slight difference. The extreme H-type of serum represents but a special case of the more general phenomenon.

Teorell[274] has proposed a mathematical theory of serological flocculation rates, which is based on the reasonable assumption that the velocity of flocculation depends upon the amount of antigen precipitated.[275] He was able to predict rather well the shape of the isochrones observed, both with sera of the R-type and of the H-type.

(273) S. D. Elek and E. Levy, *Nature* **164**, 353 (1949).
(274) T. Teorell, *J. Hyg.* **44,** 227, 237 (1946); *idem, Nature* **151,** 696 (1943).
(275) S. B. Hooker and W. C. Boyd, *J. Gen. Physiol.* **19**, 373 (1935).

(*7*) *Relation of Flocculation Optima to Neutralization.* The practical application of the β-optimum lay in the fact that the mixture that gave the β-optimum was usually found to be neutral (the toxin was neutralized but not "overneutralized"). This is not always the case, however. Glenny and O'Kell[276] found some of the most rapidly flocculating mixtures to be neutral, some toxic, and some overneutralized. Other workers[277–279] obtained similar results.

Dean and Webb[271] found neither antibody nor antigen in the supernatants of the (α) optimum precipitates. Similar results have been obtained by other workers.[34,186] It is apparent that the optimum proportions point is occasionally, at least, a guide to the neutral mixture.

(*8*) *Mechanism of the Precipitation Reaction.* There have been three main theories of the precipitation reaction. The earliest that has any interest today was proposed by Bordet,[257] who was impressed by the similarities between serological reactions and colloidal reactions.

Since a sufficient lowering of the ζ-potential on the particles of a colloid, as by the addition of electrolytes, will cause it to flocculate, Bordet supposed that the combination of antibody with antigen discharged the particles of antigen and allowed them to come together. In explanation of Bordet's notion, we may recall that many antigenic suspensions are none too stable even by themselves, and especially certain suspensions of bacteria and erythrocytes often agglutinate spontaneously, sometimes to the annoyance of the experimenter. It would not take much alteration of the surface properties of particles which are already on the verge of agglutination to cause them to stick together into clumps which would be visible.

Against Bordet's idea may be mentioned that this clumping of discharged particles of antigen would be, as he states, nonspecific. That is, antigen A plus antibody–anti-A, mixed with antigen B plus anti-B, ought to yield heterogeneous clumps containing both A and B. This experiment has been tried by numerous workers, and, although the predicted phenomenon is sometimes observed, in other cases the clumps are homogeneous, containing A, or B, but rarely both.[280] This casts doubt on the validity of Bordet's hypothesis as the only mechanism of the second stage.

If Bordet's idea of the mechanism of precipitation were correct, mixed independent serological systems, say antigen A plus anti-A, mixed with

(276) A. T. Glenny and C. C. O'Kell, *J. Path. Bact.* **27,** 187 (1924).
(277) J. T. Duncan, *Brit. J. Exptl. Path.* **13,** 489 (1932).
(278) G. L. Taylor, G. S. Adair, and M. Adair, *J. Hyg.* **34,** 118 (1934).
(279) F. M. Burnet, *J. Path. Bact.* **34,** 471 (1931).
(280) H. A. Abramson, W. C. Boyd, S. B. Hooker, P. M. Porter, and M. A. Purnell, *J. Bact.* **50,** 15 (1945).

B and anti-B, ought to flocculate faster than either system alone, owing to the formation of mixed aggregates; this was in fact observed by Hooker and Boyd[199] and by Duncan.[281] Lanni,[282] however, by studying mixed precipitation in the dark-field microscope was able to observe that mixed aggregates did not form until the particles were already nearly large enough to precipitate. Aggregation was thus found to be specific during most of the process.

Also, Pressman, Campbell, and Pauling,[283] experimenting with erythrocytes to which they had coupled a small number of chemical groups (of the order of 60 per cell), found that an antibody to the introduced group would agglutinate the cell. It could be computed that the antibody attached to the cells could have affected, at the most, 1 per cent of the surface. This would hardly seem sufficient to bring the Bordet mechanism into play.

Another concept of the mechanism of the second stage was offered independently by Marrack[34] and by Heidelberger and Kendall.[197] According to these workers, the antibody, having two or more combining groups per molecule, linked the molecules (or cells) of antigen together *specifically*, to form a "lattice." Since the hypothesis involved the supposition of an alternate arrangement, in three dimensions, of antibody–antigen–antibody–antigen . . . , Hooker and Boyd[92] proposed calling it the "alternation" theory.

Marrack conceived that the attraction between antibody–antigen compounds which led to their aggregation was not a loss of solubility but a specific attraction between them: "This mutual attraction is due to the link provided by further antigen molecules . . ."[34] His picture of the process is shown in Fig. 9.

Heidelberger's[267] concept of the precipitin reaction depended on the idea that antigen (G) and antibody (A) are chemically and immunologically multivalent with respect to each other. The reaction is considered as a series of successive bimolecular reactions which take place before precipitation occurs. The mass law is supposed to apply, in the sense that the rates of formation of the products of these reactions are proportional to the concentrations of the reacting substances.

The first reaction postulated is

$$G + A \rightleftharpoons GA$$

followed, in the region of antibody excess, by competing bimolecular

(281) J. T. Duncan, *Brit. J. Exptl. Path.* **19,** 328 (1938).
(282) F. Lanni, *J. Exptl. Med.* **84,** 167 (1946).
(283) D. Pressman, D. H. Campbell, and L. Pauling, *J. Immunol.* **44,** 101 (1942).

reactions due to the mutual multivalence of the components:

$$\text{GA} + \text{A} \rightleftharpoons \text{GA.A}$$
$$\text{GA} + \text{GA} \rightleftharpoons \text{GA.GA}$$

This is followed by other steps such as

$$\begin{aligned} &\text{GA.A} + \text{A} \rightleftharpoons \text{GA.A.A} \\ &\text{GA.GA} + \text{A} \rightleftharpoons \text{GA.GA.A} \\ &\text{GA.A} + \text{GA.A} \rightleftharpoons \text{GAA.GAA} \\ &\text{GA.A} + \text{GA.GA} \rightleftharpoons \text{GAA.GA.GA} \\ &\text{GA.GA} + \text{GA.GA} \rightleftharpoons \text{GAGA.GAGA} \end{aligned}$$

The product formed is supposed to be a three-dimensional structure, called by Marrack a "lattice." "The process of aggregation as well as the initial hapten–antibody combination is considered to be a chemical reaction between definite molecular groupings."[197]

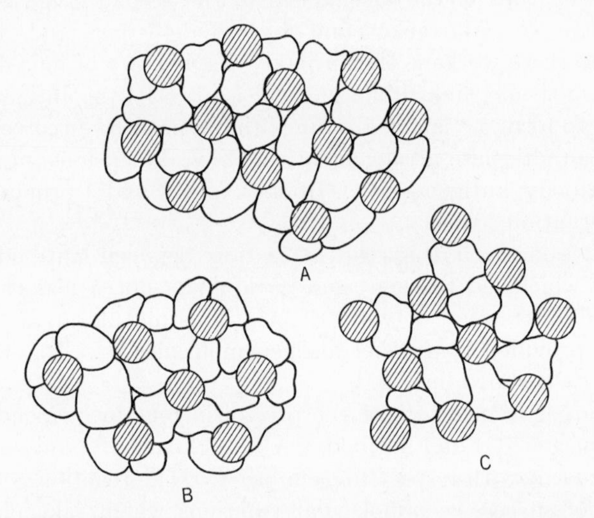

FIG. 9. Marrack's "lattice theory" of the antibody–antigen compound. *A* = typical precipitate, B = antibody excess, *C* = antigen excess. (Shaded areas = antigen, white = antibody.) (Modified from Marrack.)[34]

It is apparent that the Heidelberger theory is equivalent to the postulation of irreversible reactions. Otherwise the result would be independent of the order in which the hypothetical bimolecular reactions (whether competing or not) took place, and the mass law would apply to the equilibrium between the composition of the precipitate and the concentration of the reagents left in the supernatant. This objection was pointed out by Pauling *et al.*[283a]

(283*a*) L. Pauling, D. H. Campbell, and D. Pressman, *Physiol. Revs.* **23**, 203 (1943).

Teorell[274] proposed a theory that is not open to this objection (see p. 836). In his theory, the antibody, although thought to be bivalent, behaves as if it were univalent, and the following simultaneous equilibria are established:

$$\begin{array}{lll} A + G & \rightleftharpoons AG & \text{with dissociation constant } K_n \\ A + AG & \rightleftharpoons A_2G & \text{`` `` `` } K_{n-1} \\ A + A_2G & \rightleftharpoons A_3G & \text{`` `` `` } K_{n-2} \\ A + A_{n-1}G & \rightleftharpoons A_nG & \text{`` `` `` } K_1 \end{array}$$

Precipitation is ascribed simply to insolubility of compounds such as AG or A_2G, and not to a "lattice" formation.

There are certain ways in which the predictions of these three theories are different, and it is of interest to compare their predictions with the experimental findings.

Certain objections to Bordet's nonspecific mechanism have already been mentioned (p. 828). Contrary to Bordet,[257] the alternation theory would predict that an antigen (or hapten) having two or more combining groups ought to be able to form a precipitate, since it could take part in the formation of a lattice with multivalent antibody. This has been tested experimentally,[88,99,284,285] and although some bivalent haptens failed to precipitate with antibody, others did so. In the case of those that did precipitate, however, the possibility remained that the substances might not be molecularly dispersed in solution, but aggregated, so that the effect was that of a multivalent hapten. One such hapten was tested by diffusion measurements[89] and found to be aggregated to the extent of about 55 molecules per particle. Landsteiner and van der Scheer[75] had previously commented on the colloidal character of one of their haptens which proved to be precipitable. Nevertheless, the fact remains that some bivalent haptens have been found to precipitate specifically, and no univalent hapten has done so. The preponderance of the evidence seems to be in favor of the alternation theory.

Another experiment that favors the alternation theory is Heidelberger's[267] observation that type I pneumococci, agglutinated with an excess of antibody and washed until detectable free antibody does not appear in the washings, then resuspended, are reagglutinated by the addition of fresh type I pneumococci or of type I-specific polysaccharide. Such agglutination did not result from the addition of heterologous pneumococci or carbohydrates of other types. This was taken to show the role of intermediate links, due to the presence of more than one combining group on the antibody molecule, in building up specific antibody–antigen aggregates.

(284) M. Heidelberger and K. O. Pedersen, *J. Exptl. Med.* **65**, 393 (1937).
(285) S. B. Hooker and W. C. Boyd, *J. Immunol.* **23**, 465 (1932).

(*9*) *Mathematical Treatments.* The first attempt at a quantitative theory of antibody–antigen reactions was proposed by Arrhenius[286] before anything was known about the molecular weight of the reagents or the molecular ratios in which they combine. His theory involves assumptions that are now known to be incorrect.

The credit for the first treatment consistent with modern knowledge belongs to Heidelberger and Kendall.[206,287] On the basis of their theory of "the mutual multivalence of antibody and antigen," they developed a theory of the precipitin reaction which led to the equation:

$$R = a - bG$$

where R is the ratio of antibody to antigen in the precipitate, G is the number of milligrams of antigen nitrogen *precipitated*, and a and b are constants determined from the data.

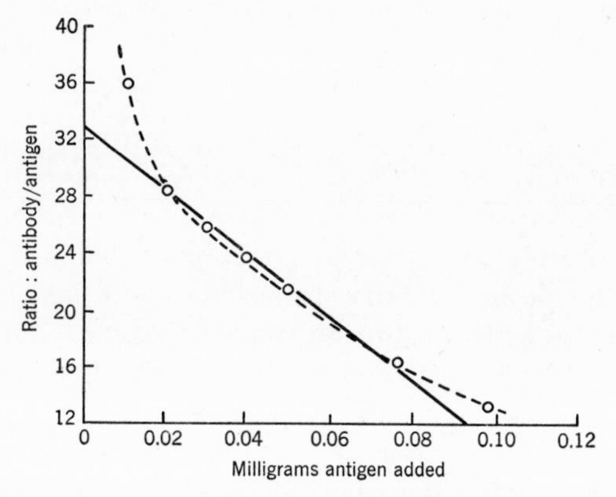

FIG. 10. Relation between composition of precipitate (R) and amount of antigen added to unit amount of antibody. Circles, experimental points; straight line, Heidelberger and Kendall's theoretical linear relation; curve (dashes), fitting by curvilinear equation.[62]

The derivation of this equation assumes that the reactions leading to precipitate formation are irreversible, and is thus open to objection, since precipitates are easily dissolved if an amount of antigen that would have prevented the formation of the precipitate in the first place is added.[288] In addition, the equation is proposed as valid only for the zone of antibody excess, and has to be "reversed" to apply to the range where antigen is in excess. Neither as given nor as reversed will it apply to systems where the antibody is of the H-type.

(286) S. Arrhenius, Immunochemistry, Macmillan, New York, 1907.
(287) M. Heidelberger and F. E. Kendall, *J. Exptl. Med.* **62,** 467 (1935).
(288) W. C. Boyd, *J. Immunol.* **38,** 143 (1940).

The equation of Heidelberger and Kendall fitted the experimental results in a limited range fairly well, although the equation predicts a linear relation between R and G, and the relation is in fact not linear. This is shown in Fig. 10.

Equations of the type

$$R = a + bG + cG^2$$

fit the experimental data in the range proposed better[62] than the equation of Heidelberger and Kendall (see Fig. 9), and these authors[206,287] themselves found better agreement by using the equation

$$R = a - bG_a^{1/2}$$

where G_a represents the milligrams of antigen nitrogen added.

Equations of the Freundlich[268] and of the Langmuir[269] types (p. 825) give even better agreement with the experimental data in some cases, as shown in Table X.

Hershey[289] has given the quantitative precipitin reaction a much more thorough mathematical treatment. He shows that if we assume the series of reactions is essentially

$$\begin{aligned} G + A &\rightleftharpoons GA \\ GA + A &\rightleftharpoons GA_2 \\ &\cdots\cdots\cdots \\ GA_{n+1} + A &\rightleftharpoons GA_n \end{aligned}$$

and that these reactions reach an equilibrium not affected by the aggregation, an equation is obtained:

$$R = abA/(1 + bA)$$

which is identical with the equation of Langmuir. It has been pointed out[62,289] that merely the assumption that the dissociation constants are equal for the reaction between antibody molecules and the various combining sites of antigen leads directly to the Langmuir equation. This is thoroughly discussed in this book by Klotz (Chap. 8).

Hershey concludes that even after the compounds first formed have aggregated to form the precipitate, this initial equilibrium is not greatly disturbed, and a similar equation continues to apply. For the region of antigen excess, he obtains an analogous equation.

Equations of this type account best for the striking linear relation between antibody–antigen ratio in the precipitate, and the concentration

(289) A. D. Hershey, *J. Immunol.* **42,** 455, 485, 515 (1941); *ibid.* **43,** 39, 249, 381 (1942).

TABLE X

QUANTITATIVE RELATION BETWEEN AMOUNT OF (BRUCELLA) ANTIGEN ADDED, AMOUNT OF ANTIGEN AND ANTIBODY PRECIPITATED, AND COMPOSITION OF PRECIPITATE,[a] WITH PREDICTED RESULTS FROM VARIOUS EQUATIONS

Amounts are given in milligrams.

The equations used for the calculations of the predicted values of antibody precipitated were:

(1) $R = 1.325 - 0.714G_a$ [b,c]

(2) $R = 1.58 - 0.968G_a^{1/2}$ [b]

(3) $R = 14.84p/(1 + 957p)$ [d,e,f]

(4) $R = 1.57p^{0.31}$ [g]

In these equations R = the ratio by weight of antibody to antigen in the precipitate, G_a = antigen added, p = the concentration of unprecipitated antibody, Σx^2 = the sum of the squares of the deviations from observed values.

Antigen added	Antigen pptd.	Antibody pptd. (observed)	"Free antibody" (p)	Antibody pptd. (Eq. 1)	Antibody pptd. (Eq. 2)	Antibody pptd. (Eq. 3)	Antibody pptd. (Eq. 4)	Antibody/antigen in ppt. (R)
0.05	All	0.066	0.557	0.064	0.068	0.066	0.066	1.32
0.10	All	0.128	0.495	0.125	0.127	0.128	0.127	1.28
0.15	All	0.175	0.448	0.182	0.180	0.189	0.184	1.16
0.20	All	0.229	0.394	0.236	0.229	0.244	0.236	1.14
0.30	All	0.324	0.299	0.333	0.318	0.324	0.324	1.08
0.50	0.338	0.362	0.261	0.367	0.349	0.375	0.352	1.06
0.70	0.616	0.506	0.117	0.545	0.505	0.505	0.498	0.82
1.00	0.946	0.575	0.048	0.614	0.605	0.464	0.576	0.61
1.20	0.998	0.580	0.043	0.611	0.612	0.449	0.579	0.58
Σx^2	—	—	—	4.428	2.156	29.991	0.297	—

References:

[a] R. B. Pennell and I. F. Huddleson, *J. Exptl. Med.* **68**, 73 (1938).
[b] M. Heidelberger and F. E. Kendall, *J. Exptl. Med.* **62**, 467 (1935).
[c] M. Heidelberger and F. E. Kendall, *J. Exptl. Med.* **62**, 697 (1935).
[d] B. N. Ghosh, *Indian J. Med. Research* **23**, 285 (1935).
[e] B. N. Ghosh, *Indian J. Med. Research* **23**, 287 (1935).
[f] I. Langmuir, *J. Am. Chem. Soc.* **40**, 1361 (1918).
[g] H. Freundlich, Kapillarchemie, Akadem. Verlagsgesellschaft, Leipzig, 1923.

of antibody (or, in the region of antigen excess, of antigen) present in the supernatant, which is shown in Figs. 11 and 12.

Ghosh[290] has derived an equation for antibody–antigen reactions which also is algebraically identical with the Langmuir[269] equation.

The derivations of Hershey help to explain the applicability of "adsorption" equations, especially that of Langmuir, to the results of quantitative analysis of precipitates.

(290) B. N. Ghosh, *Indian J. Med. Research* **23**, 285 (1935); *ibid.* **23**, 837 (1936).

Pauling *et al.*[88] have proposed a treatment of a very much simplified case. Considering the two reactions

$$\begin{aligned} A + G &= AG, \text{ equilibrium constant } 4k \\ AG + G &= AG_2 \qquad \text{''} \qquad \text{''} \qquad k \end{aligned}$$

where the compound AG is supposed to be insoluble and AG_2 soluble, and postulating that the bond strength in each of the two bonds in

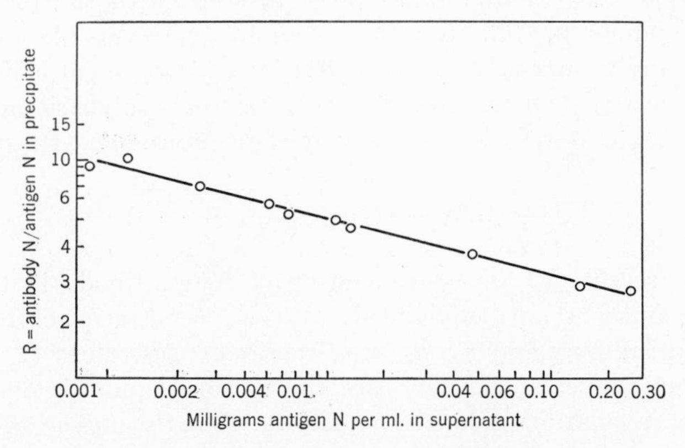

FIG. 11. Linear relation between log R (ratio of antibody to antigen in compound) and logarithm of free antibody in supernatant. Region of antibody excess.[62]

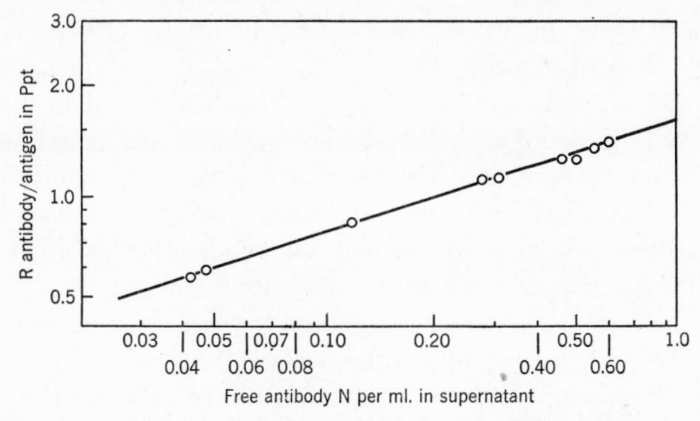

FIG. 12. Linear relation between log R (ratio of antibody to antigen in compound) and logarithm of free antigen in supernatant. Region of antigen excess.

G—A—G is the same as the one in A—G, and assuming that the equilibrium constants differ only by the entropy factor 4, they obtain the equation:

$$AG(\text{ppt}) = G - s - \frac{1 + 2ks}{2 + 2ks}\{G - A + [s(1 + ks)/k + (G - A)^2]^{1/2}\}$$

where AG is the amount of precipitate, A and G are the total amounts (per unit volume) of the antibody and antigen involved, and s is the solubility of the precipitate. Pauling *et al.* found that this equation fitted fairly well the results they obtained with the simple haptens they were studying, though it does not the data for antibodies and protein antigens.

Teorell[274] derived equations for the antibody–antigen reaction, considering that the reaction takes place in steps with the formation of definite compounds, and that the reactions are reversible so that the mass-law equilibrium applies (see p. 831). The treatment is in principle identical with that of the dissociation of a weak polybasic acid. It is assumed that antibody *behaves* as if it were univalent. He obtains an equation:

$$(A) = A - (G)(p_1 + p_2 + \cdots)$$

where (A) represents the concentration of free antibody in the supernatant, A, and total antibody added; (G) the concentration of free antigen in the supernatant; and p_1, p_2, etc., represent the values $(A)/k_n$, $(A)/k_nk_{n-1}$, etc., where k_n, k_{n-1} etc. are the various equilibrium constants. Numerical calculations from this equation, when reasonable assumptions are made concerning solubilities and antigen valence, lead to very good agreement with experimental results. Teorell's treatment is the only one that allows the prediction with fair accuracy of the entire course of the precipitin reaction, even when antibody of the H-type is involved. An example is shown in Fig. 13.

Morales *et al.*,[291] assuming multivalent antigen and univalent antibody, derived, by conventional statistical methods, relations between measurable concentrations of the reactants and the dissociation constant characterizing the equilibrium. It is suggested that certain effects usually neglected, such as interaction between adjacent antibody molecules on the same antigen molecule, may not be negligible.

In all these mathematical treatments it was found convenient to ignore the known heterogeneity of the antibodies present in immune sera.

Goldberg[292] has applied to antigen–antibody systems the method of calculating the most probable distribution of species used by Flory[293] and by Stockmayer[294] to study the distribution of molecular size in branched-chain polymers. This procedure is so well founded mathematically,

(291) M. F. Morales, J. Botts, and T. L. Hill, *J. Am. Chem. Soc.* **70,** 2339 (1948).
(292) R. J. Goldberg, *J. Am. Chem. Soc.* **74,** 5715 (1952).
(293) P. J. Flory, *J. Am. Chem. Soc.* **58,** 1879 (1931); *ibid.* **63,** 3083, 3091, 3096 (1941).
(294) W. H. Stockmayer, *J. Chem. Phys.* **11,** 45 (1943).

and the predictions of Goldberg's theory agree so well with experiment, that it alone will be discussed in any detail.

It is assumed that antibody is bivalent and antigen multivalent, although the existence of some univalent antibodies is not asserted to be impossible.

It is assumed that antibody and antigen unit to produce chains and branched structures until the system reaches a "critical point" at which

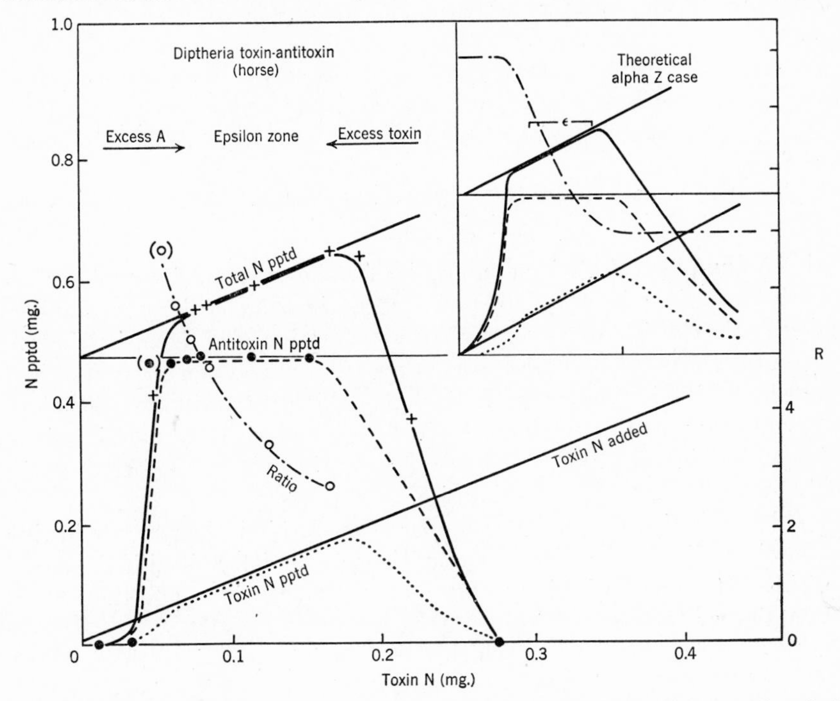

FIG. 13. Showing the success obtained in fitting Teorell's equation to the whole course of the precipitation reaction, including equivalence zone and zones of antibody excess and antigen excess. The example chosen (horse antitoxin) is not easily fitted by other equations.[274]

it changes from one composed chiefly of small aggregates to one composed chiefly of relatively few exceedingly large aggregates, and precipitation begins.

Goldberg also recognizes the influence of the differences in solubility between horse antibody and rabbit antibody, which was pointed out by Boyd[186] but ignored in most other quantitative treatments of the precipitin reaction.

It is assumed that intra-aggregate reactions yielding cyclical structures cannot occur, which fixes the number of bonds in an aggregate of a given size. It is also assumed that any unreacted site is as reactive as any other site, regardless of the size or shape of the aggregate to which it is attached. These assumptions are found necessary to prevent the mathematical treatment from becoming too unwieldy, but it is not considered that they introduce any large errors.

From these assumptions Goldberg finds m_{ijk}, the number of aggregates each of which is composed of i bivalent antibody molecules, j univalent antibody molecules, and k antigen molecules, to be

$$m_{ijk} = fG \frac{(fk - k)!}{(fk - 2k + 2 - q - j)!k!q!j!} r^{k-1}\rho^{k+i-1} \times p^{k+i+j-1}(1 - p)^{fk-k-i-j+1}(1 - \rho pr)^{i-k+1}(1 - \rho)$$

where

$$i = k - 1 + q$$
$$0 \leq q + j \leq fk - 2k + 2$$

and G = number of antigen molecules in the system, f = number of effective reaction sites (valence) of each antigen molecule, q = number of free antibody sites on an aggregate, p = fraction of antigen sites in the system that have reacted (also called the extent of reaction), ρ = fraction of antibody sites in the system that belong to bivalent antibody molecules, A = number of antibody molecules in the system with two reactive sites (bivalent antibody), $r = fG/2A$, $s = fG/D$ where D = number of antibody molecules with one reactive site.

This expression enables the number of every kind of aggregate in the system including free antibody and antigen to be determined if the composition of the system initially, the valence of the antigen, and the extent of reaction are known.

The extent of reaction at which the material passes into the form of very large aggregates is dependent on the valence of the antigen and the initial composition of the system.

Goldberg's theory has many advantages over previous treatments. It contains no arbitrary parameters for curve-fitting purposes. The calculated antibody–antigen ratios agree with experimental data for all compositions of the system. They are valid for systems in which the composition is varied by varying the antibody content. The theory offers a method of determining the limits of inhibition in antigen excess and antibody excess.

Two example of the agreement of the theory with experiment are shown in Tables XI and XII. The antibody–antigen ratio (by moles) of the precipitate (R) for systems starting with various ratios of antibody to antigen, is compared with the ratio i/k for the maximum extent of the reaction and for the critical point. Goldberg points out that the experimentally observed R should lie between the corresponding values of i/k. On the whole, agreement is good.

TABLE XI[a]

THEORETICAL AND EXPERIMENTAL VALUES OF PRECIPITATE RATIOS FOR HORSE SERUM ALBUMIN–RABBIT ANTISERUM ALBUMIN

i/k represents the molecular ratio of antibody to antigen, max refers to the maximum ratio, and c to the critical ratio at which (and beyond) the smaller aggregates are disappearing from the system.

A/G	R	$(i/k)_{max}$	$(i/k)c$
17	6.0	6.0	5.3
11	5.6	6.0	4.3
8.2	5.1	6.0	3.8
5.1	4.0	4.7	3.0
4.4	3.6	3.8	2.8
4.1	3.3	3.5	2.7
3.6	3.1	3.1	2.5
3.3	2.9	2.8	2.4

[a] R. J. Goldberg, *J. Am. Chem. Soc.* **74**, 5717 (1952).

TABLE XII[a]

THEORETICAL AND EXPERIMENTAL VALUES OF PRECIPITATE RATIOS FOR DIPHTHERIA TOXIN–HORSE ANTITOXIN

i/k represents the molecular ratio of antibody to antigen, max refers to the maximum ratio, and c to the critical ratio at which (and beyond) the smaller aggregates are disappearing from the system.

A/G	R	$(i/k)_{max}$	$(i/k)c$
2.9	2.7	2.5	2.2
2.6	2.4	2.2	2.1
2.3	2.1	2.0	2.0
1.7	1.6	1.6	1.7
1.3	1.2	1.2	1.5

[a] R. J. Goldberg, *J. Am. Chem. Soc.* **74**, 5715 (1952).

The Goldberg equation predicts well the observed composition of the antibody–antigen complex in the region of antigen excess (see Fig. 3).

V. Complement and Complement Fixation

The substances collectively called complement, which are necessary for the lysis observed in some serological reactions, can be separated into four different components, depending on their sensitivity to destruction by various reagents. A component that we designate as C_1' (once called "midpiece"),[294a] is insoluble in electrolyte-free water, destroyed by heating in solution to 56°C. for 30 min., and adsorbable by lead phosphate or titanium dioxide.[297] The portion called C_2' (formerly "endpiece") is soluble in electrolyte-free water, also inactivated by heating to 56°C. for 30 min., and adsorbable by kaolin or magnesium hydroxide (as is also C_1'). Yeast cells absorb out still a third component of complement, called C_3'. This component is stable to heat up to about 63°C. Another relatively heat-stable component, C_4', is not removed by yeast but is inactivated by treatment with ammonia.[298]

All these components of complement are ordinarily found in the serum of the guinea pig, which is the customary source, partly because guinea pigs are usually at hand in serological laboratories, but still more because guinea pig serum contains relatively more complement activity than do certain other sera. For instance, the serum of the horse is relatively poor as a source of (hemolytic) complement for most antibody–antigen systems,[299] as is that of the elephant.[300] Human serum is somewhat intermediate. One strain of guinea pigs is hereditarily deficient in C_3'.

Pillemer *et al.*[301] were able to obtain C_1', C_2', and C_4' in pure form, as well-defined proteins, by fractional precipitation of guinea pig serum with ammonium sulfate and dialysis. C_1' proved to be a globulin, with a sedimentation constant S_{20} of 6.4 and an isoelectric point of pH 5.2–5.4. It was destroyed by heating to 50°C. for 30 min. C_2' and C_4' were obtained together as an apparently pure mucoprotein containing 10.3 per cent carbohydrate, with an isoelectric point of pH 6.3–6.4. Both of these types of complement activity may therefore possibly reside in one and the same molecule. The C_2' activity was destroyed by heating to 50°C. for 30 min., and the C_4' activity by heating to 66°C. for 30 min. C_3' activity was found in all the globulin fractions and even to some extent in the albumin. It was thought that it might be a phospholipide of phosphoprotein.

C_1' was found to comprise 0.72 per cent of the total serum proteins of the guinea pig, and the mucoprotein containing the C_2' and C_4' activity comprised 0.17 per cent of the total serum proteins.

(294*a*) The present nomenclature for the components of complement was devised jointly by Heidelberger[295] and by Pillemer and Ecker.[296]

(295) M. Heidelberger, *J. Exptl. Med.* **73,** 681 (1941).
(296) L. Pillemer and E. E. Ecker, *Science* **94,** 437 (1941).
(297) E. E. Ecker and L. Pillemer, *Ann. N. Y. Acad. Sci.* **43,** 63 (1942).
(298) J. Gordon, H. R. Whitehead, and A. Wormall, *Biochem. J.* **20,** 1028, 1036 (1926).
(299) R. Muir, *J. Path. Bact.* **16,** 523 (1911/1912).
(300) M. Heidelberger, personal communication, 1951.
(301) L. Pillemer, F. Chu, S. Seifter, and E. E. Ecker, *J. Immunol.* **45,** 51 (1942).

1. Complement Fixation

By adding to a mixture in which an antibody–antigen reaction has occurred a second system (such as erythrocytes combined with hemolytic antibody) which requires complement for its completion, it can be shown that complement was taken up by the first antibody–antigen reaction as the antibody–antigen compound formed. Neither antigen nor antibody, under ordinary conditions, possesses this power of combining with the components of complement.

2. Quantitative Estimation of Complement

Routine complement-fixation tests, such as the Wassermann test for syphilis, are quite intricate, and will not be described here. They involve quantitative titrations of the second antibody used (hemolysis), of the complement in the presence of hemolysin and susceptible erythrocytes, titration of the Wassermann "antigen," and preliminary estimation of the best proportions in which to mix syphilitic serum and "antigen" to obtain best complement fixation. Although quantitative in a sense, they give no information about the actual amounts of complement protein involved or of the amounts of the various components used.

Heidelberger[295,302] devised quantitative techniques by which the actual amounts of complement taking part in serological reactions could be measured. By accurate nitrogen determinations, he was able to demonstrate that when complement was taken up by an antibody–antigen compound such as a precipitate, extra nitrogen appeared. He estimated that 1000 "hemolytic units" corresponded to 0.10–0.14 mg. of complement nitrogen. A "hemolytic unit" was defined as the amount of complement required to hemolyze 0.2 ml. of a minimally sensitized (i.e., combined with the minimal amount of hemolysin that would bring about hemolysis) 2–2.5 per cent sheep erythrocyte suspension. It was concluded that ordinary guinea pig serum containing 200–250 units contains about 0.04–0.06 mg. of complement nitrogen per cubic centimeter. Haurowitz and Yenson[303] computed that 1.5×10^{-14} g. of complement is required to hemolyze a single erythrocyte.

Interpretation of hemolytic titers in connection with complement fixation tests is, however, rendered difficult by the fact that each of the four components of complement must take part in the reaction for hemolysis to occur, although the amounts of the various components are not the same. The hemolytic titer of a complement source is limited

(302) M. Heidelberger, *Science* **92,** 534 (1940).
(303) F. Haurowitz and M. M. Yenson, *J. Immunol.* **47,** 309 (1943).

by the component present in lowest titer. The titer of guinea pig complement is limited by its content of C_3',[195] although this component is only partially removed from the guinea pig serum during the process of fixation. Hog serum, on the other hand, contains as much as ten times the amount of C_3' as do guinea pig and human sera. By its use Heidelberger *et al.*[304] attempted to titrate quantitatively the C_2' in mixtures in which it was the least abundant component. Ordinarily, guinea pig and human sera contain more than enough C_1' and C_2' in proportion to the C_3'. When mixtures in which C_2' was the least abundant component were used, results were obtained showing that the increase in hemolytic effect was not due solely to the increasing amounts of C_2', but depended also to some extent on the amount of C_3' present. The two components therefore seem to some extent to have a reciprocal action.

3. Theories of Complement Fixation

Since complement can be adsorbed by various particulate materials, it was early proposed that complement fixation is primarily an adsorption reaction, and depends largely upon the size of the antibody–antigen aggregates. Goldsworthy[305] found that the maximum complement-fixing of an antibody–antigen mixture did not appear at once when the two reagents were mixed, but seemed to increase until, so Goldsworthy supposed, the aggregates had had time to grow to a certain maximum size. Heidelberger *et al.*[306] observed that resuspended, finely divided specific precipitates took up complement from guinea pig sera almost as well as did precipitates allowed to form in the presence of complement. Hambleton[307] found that tubercle bacilli, by the addition of calcium chloride and magnesium chloride, could be made to fix complement avidly, although no antibody was present. One can hardly escape the conclusion that adsorption on the surface of aggregates plays an important role in complement fixation.

Heidelberger, Weil, and Treffers[308] have, however, proposed a picture of complement fixation in which surface adsorption plays no role. According to these authors, complement molecules possess combining

(304) M. Heidelberger, J. Jonsen, B. H. Waksman, and W. Manski, *J. Immunol.* **67,** 449 (1951).

(305) N. E. Goldsworthy, *J. Path. Bact.* **31,** 220 (1928).

(306) M. Heidelberger, M. Rocha e Silva, and M. Mayer, *J. Exptl. Med.* **74,** 359 (1941).

(307) A. Hambleton, *Can. J. Research* **7,** 583, 596 (1952).

(308) M. Heidelberger, A. J. Weil, and H. P. Treffers, *J. Exptl. Med.* **73,** 695 (1941).

groups capable of loose combination with antibody—too loose to enable antibody alone to fix complement. Possibly they are capable of forming loose, readily dissociable unions with antigen molecules too. These unions are supposed to become progressively firmer as antibody–antigen aggregates are formed, and the complement molecules are gradually surrounded by antibody and antigen molecules. The picture is not, however, merely one of complement molecules being embedded inside antibody–antigen aggregates; but Heidelberger *et al.* suggest that stabilization of the originally weak bonds with complement "might result either through the attraction of approaching ionized groupings of opposite sign, through hydrogen bonding, through spatial accommodation of large groupings of C_1' (combining component of complement) and A (antibody), or through the presence, on C_1' as on antigen and antibody, of more than one grouping capable of reacting with A molecules brought into opposition." They make the interesting suggestion that such union could possibly be demonstrated without the presence of an antibody–antigen complex, or even without the presence of normal globulin, if there were any way of bringing a sufficient number of C_1' molecules into suitable apposition and holding them there. Thus far the demonstration of the anticomplementary (i.e., complement-combining) power of purified normal γ-globulin[309] seems to be the nearest thing to a test of this suggestion that has been carried out.

Bowman, Mayer, and Rapp[310] have demonstrated that the combination between red cells and antibody is reversible, and that after complement has caused lysis of antibody-coated red cells, antibody may be transferred from the antibody-stromata complex to fresh cells, and antibody molecules can thus perform hemolytic work over and over again (provided sufficient complement is available). This means that the kinetic behavior of the antibody has a catalytic character. Bowman, Mayer, and Rapp are inclined to believe that immune hemolysis is actually a catalytic process, in which the antibody acts like an enzyme while the complement is a reactant that is used up. Sevag,[311] in his interesting book *Immuno-catalysis*, has for a number of years been calling attention to the analogies between enzyme reactions and antibody–antigen reactions, although Sevag emphasized chiefly the similarities he saw between

(309) B. D. Davis, E. A. Kabat, A. D. Harris, and D. H. Moore, *J. Immunol.* **49,** 223 (1944).

(310) W. M. Bowman, M. M. Mayer, and H. J. Rapp., *J. Exptl. Med.* **94,** 87 (1951).

(311) M. G. Sevag, Immunocatalysis, C. C Thomas, 2nd ed., Springfield, Illinois, 1951.

enzymes and antigens. Haurowitz[19] states that if we disregard the instances of immunization with enzymes to obtain antibodies which may or may not neutralize the enzyme (see p. 784), we have no reason to ascribe enzyme properties to antigens. The determinant groups of the conjugated antigens (p. 770) are certainly devoid of enzyme activity. The most rigorous criterion of a catalyst is that it reduces the activation energy of the catalyzed reaction, and we do not know of any such action of antigens.

Chapter 23

Structure Proteins. I

By J. C. KENDREW

Page

I. General Introduction 846
II. Silk Fibroin 849
1. Introduction 849
2. Chemical Composition 850
3. Configuration of the Polypeptide Chain in Silk: X-Ray and Spectroscopic Investigations 851
4. The Biosynthesis of Silk 855
5. Chemical Properties of Silk: Evidence for Large Units of Structure 856
III. The Keratin Group 859
1. Introduction 859
2. Wool Keratin and Related Materials 860
a. Occurrence: Histology 860
b. Mechanical Properties 861
c. X-Ray Pattern 864
d. Spectroscopy 870
e. Chemical Properties 872
f. Soluble and Regenerated Keratin 875
3. Fibrinogen and Fibrin 876
a. Fibrinogen 876
b. The Conversion of Fibrinogen to Fibrin 878
c. Fibrin 882
4. Other Members of the Keratin Group 885
a. Introduction 885
b. Epidermin 885
c. Bacterial Flagella 887
5. Synthetic Polypeptides 889
6. The Structure of the Keratins 892
a. The Configuration of the Polypeptide Chain in β-Keratin 892
b. The Configuration of the Polypeptide Chain in α-Keratin 896
c. Evidence for Larger Units of Structure in the k-m-e-f Proteins 905
IV. The Collagen Group 909
1. Introduction 909
2. Properties of Mammalian Collagen 912
a. Morphology and Fine Structure 912
b. Physical Properties 917
c. Chemical Properties 920
d. Gelatin 923
e. Soluble and Regenerated Collagen 925
f. The Formation of Collagen in Living Organisms 930

Page
3. Theories of the Structure of Collagen.................................. 931
a. Introduction.. 931
b. General Features of the Structure of Collagen....................... 932
c. Elaborations of the Unidimensional Theory of Structure............. 934
d. Configuration of the Polypeptide Chain in Collagen................ 936
4. Other Members of the Collagen Group.................................. 942
a. In Mammalian Tissue.. 943
b. In the Tissues of Other Vertebrates.................................. 944
c. In Other Phyla... 945
d. The Secreted Collagens... 945
V. Miscellaneous Structure Proteins.. 946
1. Elastin... 946
2. Bacterial Polyglutamic Acid.. 949

I. General Introduction

The skeletal structures of animals consist in large part of protein; on the other hand in plants (and generally in microorganisms) such structures are principally made up of polysaccharide. Structure protein is often macroscopically fibrous; hence the term "fibrous protein" is more or less synonymous with structure protein, though some proteins which are evidently fibrous—e.g., muscle—have functions other than mere support. It has, in fact, become conventional to divide the whole range of proteins into two main classes: the fibrous and the globular. This classification is clearly realistic, though, as we shall see, it is still empirical, since the structural relations between the two classes remain obscure. In this chapter we shall be concerned with fibrous proteins other than muscle, i.e., with those whose function is, so far as is known, purely structural.

Animal structural materials often contain major constituents which are not proteins: e.g., bone is made up of the protein collagen together with deposits of the mineral substance hydroxyapatite (containing calcium, carbonate, and phosphate) and some carbohydrate. The exoskeleton of many invertebrates is made of chitin, which contains protein together with the carbohydrate which forms its major ingredient.

Owing to their industrial importance, the chemical behavior of fibrous proteins has been studied more intensively than that of any other class of protein. This emphasis somewhat obscures the fact that fibrous proteins are actually extremely inert by comparison with all other types of protein; for example, they are often entirely unaffected by water, dilute acid, alkali, or ordinary proteolytic enzymes, at least in the native condition. Such unreactivity is evidently to be correlated with their passive functions. From the point of view of the biochemist, their inactivity, and, in particular, their insolubility, place serious obstacles in the way of their study by conventional techniques; the first of these obstacles is the almost complete impossibility of purifying them, or of extracting from them a chemically homogeneous product. This difficulty has in all probability a more fundamental origin—that many of the structure proteins are genuinely heterogeneous *in vivo*. It is doubtful, in fact, whether the term "keratin," for example, would have any unambiguous meaning even if ideal methods of extraction were available. These difficulties are exemplified by the fact that it has not been possible to assign a molecular weight to any member of the class; it is still not quite clear whether this failure is due to the imperfection of the available techniques or whether the term molecular weight is intrinsically meaningless in connection with structure proteins.

Apart from the basic chemical techniques of amino acid analysis, end-group analysis, and partial hydrolysis, and apart from the more technological studies, the methods applied to a study of the structure proteins have, for the reasons just outlined, been mainly physical in nature. We shall consider in some detail the results of applying such techniques as x-ray diffraction, ultraviolet and infrared spectroscopy, and electron microscopy; the techniques themselves have been discussed elsewhere in this work. For obvious reasons our attention will be focused more on structure than on function. Historically, the emphasis on the structure of fibrous proteins derives from the belief that in them might be found the simplest exemplification, the archetype, of structural principles applicable throughout the whole range of proteins. This belief has been largely justified. We shall unfortunately be able to say very little in clarification of another most fundamental question—the biogenesis of the structure proteins. So far very little is known about that very important topic.

Historically, the first clearly recognized members of the class were keratin, in hair, horn, nails, and wool; collagen, in tendon; and silk. A classification of fibrous proteins was slow in developing. It was early realized that there were specific differences in amino acid composition; thus keratin contained much cystine; collagen much proline, hydroxyproline, and glycine; and silk much glycine and alanine:

	Wool keratin	Collagen	Silk
cystine and cysteine	**11.9**	2.3	none
pyroline + hydroxyproline	9.5	**29.1**	0.74
glycine	6.5	**27.1**	**43.6**
alanine	4.1	9.5	**29.7**

(grams amino acid/100 g. protein)

Again, the staining properties and the elastic behaviors were found to be characteristic and specific. However, when other tissues, such as skin, bone, and muscle, were taken into consideration, the classification was found to be by no means clear-cut. In this field the outstanding contribution was made by Astbury and his collaborators, in their use of x-ray diffraction patterns for purposes of diagnosis and classification.

They discovered that wool keratin yielded two highly characteristic x-ray patterns, one in its normal unstretched state, and another after mechanical extension when damp and warm. They concluded that two distinct molecular configurations of the polypeptide chain were involved; they called these α-keratin and β-keratin. It then emerged that a great range of structural materials yielded one or other of these patterns, and accordingly they proposed that these should be grouped together as the keratin–myosin–epidermin–fibrinogen (k-m-e-f) group. Some members of it (e.g., porcupine quill) were found only in the α-configuration: others only in the β-configuration (e.g., myosin, epidermin). Wide variations in amino acid composition, in chemical and mechanical behavior, and in function, are found among members of the group; nevertheless the classification based on x-ray pattern has proved more useful than any other, and would appear, as originally postulated by Astbury, to be justified in a very fundamental way as the expression of a common underlying scheme of molecular structure. It has since become probable that an analogous classification might equally well be based on regularities in infrared spectra. Similarly, collagen was found, also by Astbury and his coworkers, to yield a characteristic diffraction pattern which could be used diagnostically: connective tissues, bone, and many structural materials from many kinds of animal have accordingly been grouped as collagens, in spite of wide variations in composition. Indeed, the two main groups of the Astbury classification appear to comprehend virtually all known structure proteins whose x-ray pattern has been examined, with the exception of elastin (a component of some types of elastic tissue) and actin in muscle. We shall discuss the k-m-e-f and the collagen groups in turn (secs. III and IV), leaving a few oddments (which include those whose x-ray patterns have not been examined) until last (sec. V). Silk can in many respects be classified as a β-protein in the k-m-e-f group, but for reasons mainly of a historical nature it is more convenient to consider it separately (sec. II).

Broadly speaking, the structure proteins have received most attention from those concerned with their industrial applications. Here we shall be concerned with industrial problems only incidentally, and the main focus of our attention will be on structure and matters relevant thereto. This is, nevertheless, a difficult moment at which to discuss structural questions; in this field great advances have been made during the last eighteen months, but there has not yet been time to make a new synthesis of the whole field in terms of them. We shall thus be in a position to record many recent discoveries of great interest, but not to reach many general conclusions.

II. Silk Fibroin

1. Introduction

Silk fibroin is, as it were, the fibrous protein par excellence. It was the subject of some of the earliest structural investigations of proteins—on the chemical side, Fischer obtained dipeptides from it by hydrolysis as early as 1902[1] and thus gave the first direct demonstration of the presence of the peptide bond in a protein; while on the physical side the first x-ray diffraction pattern to be obtained from a protein was that of silk, discovered by Herzog and Jancke in 1920.[2] These early studies and many which followed them were undertaken partly because of the importance of silk as an industrial product, partly because on account of its secretory origin, silk is readily obtained free from contaminants, and thence the fibroin itself can be isolated in pure form by easy methods.

Silk, then is an extracellular fiber extruded by the silkworm (*Bombyx mori* is the common cultivated species) from a pair of glands. The silk thread is in the form of a double filament of the protein fibroin, enveloped in an outer coating of silk glue or sericin, which acts as an adhesive. The sericin may be removed by boiling in water to which detergent has been added, leaving very pure fibers of fibroin.

Sericin has not been studied in detail. Its preparation and properties were first adequately described by Kodama;[3] the amino acid composition has been qualitatively examined by Shaw and Smith.[4] The latter authors divided the sericin into three fractions on the basis of solubility: in each fraction the amino acids principally present were serine, threonine, glycine, aspartic acid; with moderate amounts of glutamic acid, arginine, alanine, leucines, valine, and tyrosine. The fractions differ in minor constituents, but in any case the general picture is of a composition quite different from that of fibroin (see below).

Silk fibroin has several times been examined in the electron microscope, with generally disappointing results. In the earlier work[5–7] fine fibrils of varying thickness down to 40–80 A. were observed (with an unconfirmed suggestion of cross-striations of period 240 A.). In a recent

(1) E. Fischer, *Chem.-Ztg.* **26,** 939 (1902).
(2) R. O. Herzog and W. Jancke, *Ber.* **53,** 2162 (1920).
(3) K. Kodama, *Biochem. J.* **20,** 1208 (1926).
(4) J. T. B. Shaw and S. G. Smith, *Nature* **168,** 745 (1951).
(5) E. Franz, F. H. Müller, and L. Wallner, *Die Chemie* **45,** 75 (1942).
(6) H. Zahn, *Kolloid-Z.* **112,** 91 (1949).
(7) R. Hegetschweiler, *Makromol. Chem.* **4,** 156 (1949).

study by Mercer,[8] in which the fibers were partly digested by trypsin, rather distinct fibrils about 100 A. across were reported.

2. Chemical Composition

The amino acid composition of silk fibroin has been studied in detail by many workers. The most up-to-date figures have been collected by Sanger[9] (see also the review of Tristram[10] and Table XXII, p. 220, Vol. I part A), incorporating recent redeterminations by Levy and Slobodiansky.[11] The approximate numbers of residues per 10^5 gm. of protein are as follows:

Gly	563	Tyr	63	Leu	20	Phe	10	Lys	3
Ala	343	Asp	20	Val	30	Thr	13	His	3
Ser	133	Glu	13	Pro	7	Arg	7		
				Total	1230				

It will be noted that this composition can be roughly represented as $(Gly_3Ala_2X_2)_n$, where X is any residue other than glycine or alanine. This simple relation, and earlier less accurate versions (which generally assumed that the Ala to Gly ratio was 1:2) have formed the subject of many speculations and theories. For example, fibroin was one of the classical proteins to which the Bergmann-Niemann theory of residue periodicities was applied: these authors postulated[12] a regular sequence

$$(\text{—Gly—Ala—Gly—X—Gly—Ala—Gly—X—})_n$$

in which for example every sixteenth X was tyrosine and every two-hundredth was arginine. In fact this theory was more or less untenable even before the recent more accurate amino acid data were collected, both for reasons connected with the x-ray diagram (see p. 854) and in the light of the results of many partial hydrolysis studies carried out specifically to test the theory.

The most important evidence of the latter kind came from the work of Abderhalden and Bahn,[13] who isolated a tetrapeptide Tyr.Ser.Pro.Tyr from the partial hydrolyzate of fibroin; the existence of such a sequence, containing neither glycine nor alanine, is clearly incompatible with any scheme like the above. Again, Stein, Moore, and Bergmann[14] isolated

(8) E. H. Mercer, *Australian J. Sci. Research Ser. B* **5**, 366 (1952).
(9) F. Sanger, *Advances in Protein Chem.* **7**, 1 (1952).
(10) G. R. Tristram, *Advances in Protein Chem.* **5**, 83 (1949).
(11) M. Levy and E. Slobodiansky, *Cold Spring Harbor Symposia Quant. Biol.* **14**, 113 (1949).
(12) M. Bergmann and C. Niemann, *J. Biol. Chem.* **122**, 577 (1938).
(13) E. Abderhalden and A. Bahn, *Z. physiol. Chem.* **210**, 246 (1932); *ibid.* **219**, 72 (1933).
(14) W. H. Stein, S. Moore, and M. Bergmann, *J. Biol. Chem.* **154**, 191 (1944).

Gly.Ala and Ala.Gly; Levy and Slobodiansky,[11,15] repeating this work in more detail, obtained Ala.Gly in much larger amounts than Gly.Ala, and found hardly any Gly.Gly; they also obtained the tripeptide Gly.Ala.Gly in quantity.[16,17] They were thus led to propose a different scheme

$$(\text{—X—Ala—Gly—Ala—Gly—X—Gly—})_n$$

in which one-third of the X's are serine, one in 12 is aspartic acid, and one in 18 is glutamic acid.

It will be convenient here to anticipate a description of more recent work leading to different conclusions by mentioning some results of Drucker and Smith[18] who were able, by treatments described below (see p. 857) to obtain two fractions from fibroin, one a peptide of molecular weight about 7000 containing *only* the amino acids glycine, alanine, and serine; and the other containing only tyrosine, valine, leucine, arginine, phenylalanine, aspartic acid, glutamic acid, and proline. Whatever detailed scheme is proposed, it seems justifiable to conclude that the fibroin "molecule" is not homogeneous, but consists of two (or more) regions, in one of which are concentrated the alanine, glycine, and serine, and in the other the rest of the amino acids.

3. Configuration of the Polypeptide Chain in Silk: X-Ray and Spectroscopic Investigations

The early discovery by Herzog and Jancke[2] of an x-ray diffraction pattern from silk fiber was soon followed by a detailed study carried out by Brill,[19] who examined silks from several species of insect and found that they all gave very similar (though not identical) patterns (see also Trogus and Hess).[20] The typical silk x-ray pattern is shown in Fig. 1; it will be seen to contain a number of spots of varying sharpness, lying on definite layer lines. Brill indexed these reflections and arrived at a unit cell whose identity period along the fiber axis was 7.00 A.; the other dimensions were 9.27 and 10.4 A. He calculated that each cell must contain 4 glycyl and 4 alanyl residues; and concluded further that his unit cell could not accommodate any of the bulkier amino acid residues present in smaller proportions in silk, and hence that these residues must form a second, amorphous phase, not contributing to the pattern. Somewhat later Kratky and Kuriyama[21] found that they could induce double orien-

(15) M. Levy and E. Slobodian, *J. Biol. Chem.* **199,** 563 (1952).
(16) E. Slobodian and M. Levy, *Federation Proc.* **11,** 288 (1952).
(17) E. Slobodian and M. Levy, *J. Biol. Chem.* **201,** 371 (1953).
(18) B. Drucker and S. G. Smith, *Nature* **165,** 196 (1950).
(19) R. Brill, *Ann.* **434,** 204 (1923).
(20) C. Trogus and K. Hess, *Biochem. Z.* **260,** 376 (1933).
(21) O. Kratky and S. Kuriyama, *Z. physik. Chem.* **B11,** 363 (1931).

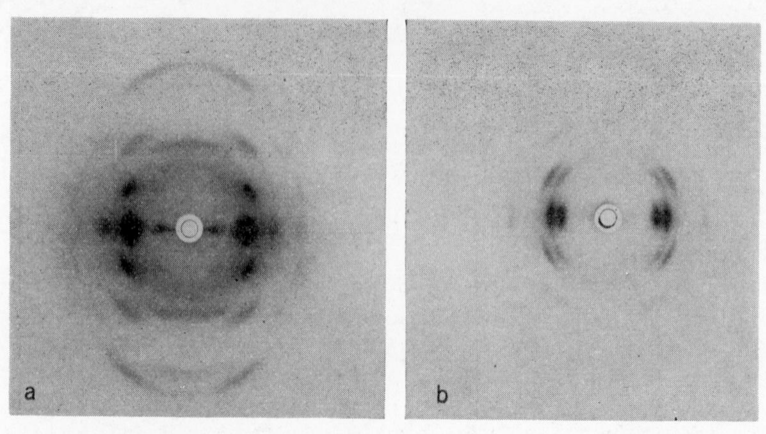

FIG. 1. X-ray pattern of silk: (*a*) *Bombyx mori*, (*b*) Tussah silk (fiber axis vertical). By courtesy of Bamford, Brown, Elliott, Hanby, and Trotter.

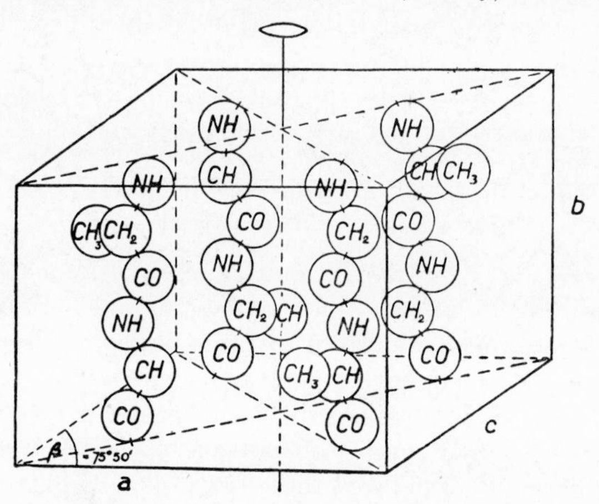

FIG. 2. Unit cell II of silk fibroin (diagrammatic); see Table I. After Meyer.[22]

tation in silk fibers by rolling out the contents of the gland into ribbons. Hence they were able to show that the spots could be indexed equally well in terms of several different unit cells (see Table I). Recently, Bamford *et al.*[23] have reported a very strong axial reflection of spacing 1.156 A.; this is the sixth order of the main axial repeat, the most accurate value for the spacing of which is therefore $6 \times 1.156 = 6.94$ A.

Meyer and Mark[24] concluded that a consistent interpretation of the

(22) K. H. Meyer, Natural and Synthetic High Polymers, Interscience Publishers, New York, 1942.

(23) C. H. Bamford, L. Brown, A. Elliott, W. E. Hanby, and I. F. Trotter, *Nature* **171,** 1149 (1953).

(24) K. H. Meyer and H. Mark, *Ber.* **61,** 1932 (1928).

TABLE I

UNIT CELLS AND INDEXING OF REFLECTIONS OF SILK OF BOMBYX MORI
(After Meyer,[22] Table 53)

	Unit cells				
	a	*b*	*c*	*β*	Residues/cell
I	10.1 A.	6.95 A.	4.72 A.	66°2′	2
II	9.90	6.95	8.92	76°40′	4
III	9.60	6.95	9.02	73°52′	4

Reflection	Spacing and indexing of reflections		Indices		
	Spacing, A.	Intensity	I	II	III
Equatorial A_1	9.2	Strong	100	100	100
A_2	4.6	Strong	200	200	201
A_3	4.3	Very strong	001	002	002
A_4	3.1	Moderate	300	300	300
A_5	2.4	Weak	202	303	401
1st. layer I_1	5.5	Weak	110	110	110
I_2	3.6	Moderate	210	210	211
I_3	2.2	Strong	011	012	012
2nd. layer II_0 (merid.)	3.5	Moderate	020	020	020
II_1	3.3	Moderate	120	120	120
II_2	2.7	Weak	220	220	221
II_3	2.2	Weak	320	320	320
3rd. layer III_1	2.3	Strong	130	130	130
III_2	2.1	Moderate	230	230	231

x-ray and other results could be secured only by assuming that the crystalline regions of the fiber consisted of long alanyl-glycyl chains running parallel to the fiber axis, the axial repeat of 6.95 A. corresponding to two residues. The general arrangement proposed is shown in Fig. 2. The known dimensions of the peptide chain are such that a repeat distance of 6.95 A. corresponds to two residues almost fully stretched out

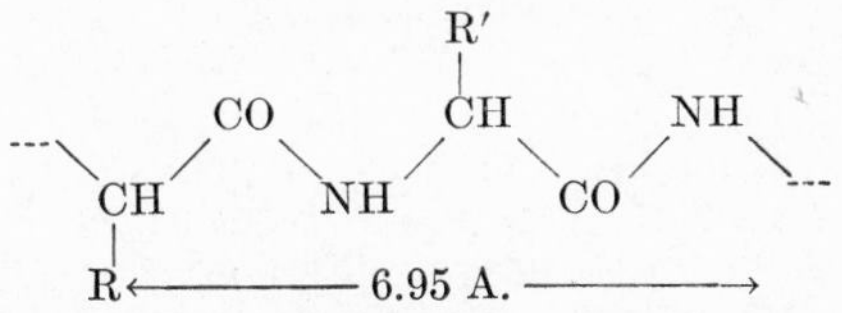

```
                          R'
                          |
            CO            CH            NH
   ---\    /   \        /    \        /   \
       CH         NH             CO         ---
       |
       R<———————— 6.95 A. ————————>
```

The latest figures[25] for these dimensions would give a *fully* stretched distance of 7.27 A.

(25) R. B. Corey and J. Donohue, *J. Am. Chem. Soc.* **72,** 2899 (1950).

Meyer and Mark confirmed the earlier conclusion that the unit cell would not accommodate the larger amino acid residues (though alanine could perhaps be replaced by serine). Direct evidence for this was obtained by Goldschmidt and Strauss,[26] who found that on treating silk with hypobromite part of the fiber dissolved, leaving a residue of highly birefringent flakes which gave a normal fibroin x-ray pattern, but more intense than that of the original fiber, and which contained only alanine and glycine, in a ratio which they found to be 3:1. It would appear that in these experiments they effectively isolated the crystalline component of silk. It was further shown by Meyer, Fuld, and Klemm[27] that if the tyrosine groups of silk are coupled with diazonium compounds, their bulk being thus greatly increased, the diffraction pattern of the silk does not alter; similar results have been obtained by the action of 2,4-dinitrofluorobenzene on the tyrosine residues.[28] Thus the x-ray results lead to the same general conclusion about the inhomogeneity of the fiber as had been reached on the basis of chemical studies.[29] However, until lately no detailed structure had been proposed, in the same sense that the crystal structure of a low-molecular-weight organic molecule is described in detail, though tentative suggestions toward such a structure were made (see e.g., Brill.)[30] Very recently, however, models have been proposed by Pauling and Corey which we shall discuss in connection with β-keratin (see p. 895).

In conclusion we may consider the results of spectroscopic studies of silk in the infrared and ultraviolet. In the infrared, observations with polarized radiation have shown that the >N—H and >C=O stretching frequencies exhibit high perpendicular dichroism (see Table IV, and Vol. I, part A, p. 427, Fig. 6) and that these bonds are therefore oriented perpendicular to the chain direction:[31,32] this, of course, is what would be expected in a structure consisting of stretched polypeptide chain (see p. 893). There is, further, a close resemblance between the infrared spectrum of silk and those of synthetic polyglycine and poly-L-alanine,[32,33]

(26) S. Goldschmidt and K. Strauss, *Ann.* **480,** 263 (1930).
(27) K. H. Meyer, M. Fuld, and O. Klemm, *Helv. Chim. Acta* **23,** 1441 (1940).
(28) H. Zahn and A. Würz, *Biochem. Z.* **322,** 327 (1952).
(29) There is, however, evidence that the tyrosine residues cannot be distributed merely statistically in amorphous regions; see H. Zahn, O. Kratky, and A. Sekora, *Z. Naturforsch.* **6b,** 9 (1951); also, H. F. Friedrich-Freksa, O. Kratky, and A. Sekora, *Naturwissenschaften* **32,** 78 (1944).
(30) R. Brill, *Z. physik. Chem.* **B53,** 61 (1943).
(31) J. D. Bath and J. W. Ellis, *J. Phys. Chem.* **45,** 204 (1941).
(32) E. J. Ambrose and A. Elliott, *Proc. Roy. Soc.* (London) **A206,** 206 (1951).
(33) W. T. Astbury, C. E. Dalgliesh, S. E. Darmon, and G. B. B. M. Sutherland, *Nature* **162,** 596 (1948).

which have the β- or stretched configuration. Ambrose and Elliott[32] consider, however, that a component in the α- or folded configuration may be present in addition to stretched chains.

In the ultraviolet interesting effects are observed when silk is treated with alkali; Fixl, Schauenstein, and others have recently suggested that changes in the ultraviolet absorption in the 2500–3000 A. region are largely due to enolization of the peptide bond itself,[34–36a] but these views have been strongly criticized by Beaven and Holiday[37] (see also pp. 439 ff., Vol. I, part A).

4. The Biosynthesis of Silk

The silk gland contains a highly viscous solution whose protein content is up to 30 per cent of its wet weight. This solution contains both sericin and a soluble precursor of fibroin (which has been called fibroinogen).[38] By mechanical action such as stirring or extrusion, or even on standing, the fibroinogen is removed from solution: indeed this component is highly surface-active so that the liquid contents of the gland immediately form a tough surface skin. The sericin also comes out of solution on prolonged standing.

The process of silk formation has been studied by x-rays with interesting results:[38]

(*a*) If the contents of the gland are stretched they behave like rubber, showing reversible long-range elasticity: but if the strip of material is held under tension for a few seconds, or if the stretching is repeated a number of times, it loses its elasticity, and then—but only then—gives the characteristic x-ray pattern of silk. The process is evidently one of crystallization, like that which takes place with rubber at low temperatures.

(*b*) If the gland contents are carefully dried, without stretching, they give an x-ray powder diagram.[39] This has been studied in detail by Kratky *et al.*[40,41] who showed that the powder rings, which are very sharp ones, do not correspond to those of normal silk (which they call Silk II), but are characteristic of a new crystalline phase which they call Silk I. These spacings are as follows:

(34) J. O. Fixl, O. Kratky, and E. Schauenstein, *Monatsh.* **80,** 153 (1949).
(35) E. Schauenstein, *Monatsh.* **80,** 843 (1949).
(36) O. Kratky, E. Schauenstein, and A. Sekora, *Nature* **166,** 1031 (1950).
(36*a*) E. Schauenstein, *Melliand Textilber.* **33,** 591 (1952).
(37) G. H. Beaven and E. R. Holiday, *Advances in Protein Chem.* **7,** 319 (1952).
(38) K. H. Meyer and J. Jeannerat, *Helv. Chim. Acta* **22,** 22 (1939).
(39) R. Brill, *Naturwissenschaften* **18,** 622 (1930).
(40) O. Kratky, E. Schauenstein, and A. Sekora, *Nature* **165,** 319 (1950).
(41) O. Kratky and E. Schauenstein, *Discussions Faraday Soc.* No. **11,** 171 (1951).

7.5 A.	mod.	3.6 A.	str.
4.9	weak	3.1	str.
4.5	v. str.	2.6	weak
4.0	v. weak	2.4	rather weak

The authors observe that these reflections have a distinct resemblance to those of actin. Even slight stretching of Silk I causes an immediate and irreversible transformation to Silk II.

This transformation has been further studied by Mercer,[42,8] who showed that the original contents of the gland (Silk I) dried on to collodion films gave no recognizable structure under the electron microscope, but that fibrous objects become visible on standing.

The suggestion has been made that fibroinogen (Silk I) is a globular precursor of the truly fibrous fibroin (Silk II), and that the conversion from one to the other is a process of end-to-end aggregation like that proposed in many other instances (see p. 907); the existence in silk fibers of discrete fibrils with uniform diameter (see p. 850 above) and the possibility of obtaining long x-ray spacings from them (see next section) have been taken to favor this hypothesis, but at present it cannot be said that there is evidence definitely excluding an unrolling process akin to denaturation.

5. Chemical Properties of Silk: Evidence for Large Units of Structure

Silk is in general an unreactive and insoluble material, and in its natural state it is not attacked by trypsin. In water it swells to some extent, but less than any other fibrous protein:[43] the amount of water absorbed per residue is almost identical with that of polyglycine and is almost independent of the physical state of the fiber.[44] The tensile strength of silk fiber is high (about 35 kg./sq. mm.) but there is appreciable reversible elasticity, presumably due to the large amount of amorphous material present;[45] unlike keratin, silk fibroin does not give a new x-ray pattern on stretching.

In spite of its general inertness, silk does show reactivity towards a number of special reagents: thus in concentrated formic acid the fiber swells and shortens, exhibiting rubberlike elasticity and losing its crystalline x-ray pattern.[46]

(42) E. H. Mercer, *Nature* **168,** 792 (1951).
(43) D. J. Lloyd and R. H. Marriott, *Trans. Faraday Soc.* **32,** 932 (1936).
(44) E. F. Mellon, A. H. Korn, and S. R. Hoover, *J. Am. Chem. Soc.* **71,** 2761 (1949).
(45) H. Mark, *Chem. Eng. News* **27,** 138 (1949).
(46) K. H. Meyer and H. Mark, Der Aufbau der hochpolymeren organischen Naturstoffe, Akad. Verlagsges., Leipzig, 1930.

An important reaction is with a solution of lithium bromide, in which fibroin completely dissolves in high concentration (25–50 g. fibroin/100 ml.).[47] Nothing is known of the mechanism of this reaction, but recently an examination of the solution by infrared spectroscopy[48] has strongly suggested that in it the fibroin molecules are in the folded or α-configuration. Similar reactions occur with certain other inorganic salts, e.g., calcium chloride or nitrate, and thiocyanate, and also with dichloroacetic acid.[49]

Another important reaction was discovered by Coleman and Howitt, who showed,[50,51] following Takamatsu,[52] that silk fibroin forms a soluble complex with cupric ethylenediamine: on dialysis the copper is entirely removed, leaving an opalescent solution whose concentration may be as high as 10 per cent. This solution remains clear for about 14 days and then suddenly sets to a gel. It is highly surface-active, being invariably covered by a tough skin; vigorous stirring rapidly removes all the protein from it. This solution contains *all* the original fibroin, which can be precipitated from it by acetone, alcohol, etc. On acidifying the solution, part of the protein is precipitated: but both the precipitable part and the part remaining in solution consist of particles of molecular weight of the order 33,000.

Coleman and Howitt explain these very interesting findings by supposing that when the fibrous silk fibroin comes into solution, the individual chains first become disentangled, and can then roll up to form globular particles apparently similar to the original fibroinogen in the silk gland; for this reason they call the process "renaturation," a term which perhaps implies more knowledge than we actually possess of the nature of the materials involved. The part of the solution which is precipitated by acid is still dispersed or denatured, and there is an equilibrium between this and the renatured form, an equilibrium displaced by the copper ethylenediamine. This conception of the mechanism is supported by the work of Drucker and Smith,[18] who have shown that the original fibroin, the dispersed material, and the renatured as well, all have the same amino acid composition. They also partially hydrolyzed renatured fibroin (by trypsin) and obtained an insoluble peptide of molecular weight 7000, containing *only* glycine, alanine, and serine; the remainder in

(47) P. P. von Weimarn, *Ind. Eng. Chem.* **19,** 109 (1927).
(48) B. A. Toms and A. Elliott, *Nature* **169,** 877 (1952).
(49) E. J. Ambrose, C. H. Bamford, A. Elliott, and W. E. Hanby, *Nature* **167,** 264 (1951).
(50) D. Coleman and F. O. Howitt, *Nature* **155,** 78 (1945).
(51) D. Coleman and F. O. Howitt, *Proc. Roy. Soc.* (London) **A190,** 45 (1947).
(52) Y. Takamatsu, *J. Soc. Chem. Ind. Japan* **36,** 566 (1933).

solution contained the other amino acids present in fibroin. They accordingly imagine the molecule of renatured fibroin (mol. wt. 33,000) to have a structure ABABA, where A represents regions containing only the three smallest residues.

Physical investigations of the Coleman-Howitt reaction have led to contradictory results. The original authors claim from x-ray studies that both denatured and renatured fibroin have the extended β-configuration; on the other hand, Ambrose *et al.*[49] consider on the basis of their infrared work that renatured fibroin has the α- or folded configuration. Kratky, Schauenstein, and Sekora[53] have found that by careful drying of a solution of renatured fibroin they can obtain a powder giving a weak x-ray pattern similar to that of Silk I—i.e., to that of the original fibroinogen.

Space does not permit a discussion of the theory of Coleman and Howitt[51] explaining the action of copper in renaturing fibroin, but in essence they imagine a process of folding the chain round proline "hinges," induced by chelation onto the copper atoms. The whole interpretation of these very interesting phenomena is still very confused, and all that can be said is that there do seem to be some grounds for supposing that renaturation is the reversal of the fibroinogen–fibroin transformation. The really significant question, whether this transformation, in either direction, involves linear aggregation of globular fibroinogen monomers, or whether it consists of a side-to-side aggregation of unrolled chains, is still quite unsettled. There is nevertheless a little evidence to suggest that ordinary silk fibroin in its natural state contains large units, from the work of Kratky and Schauenstein[41] and of Zahn and Würz[28] who have been able to treat silk fibers, either by alkali or by nitration, in such a way that they exhibit long x-ray spacings; in alkali-treated silk (described as "Silk III") are observed meridional spacings of 21 A. and the second and third orders thereof,[54] while nitrated silk has meridional long spacings of 10.42 and 20.84 A. Incidentally, alkaline treatment simultaneously produces an orientation of the tyrosine group (the fiber axis lying in the plane of the benzene rings), as shown by the pleochroism of the ultraviolet absorption;[41] this effect suggests that there is an oversimplification in the older theories according to which large residues such as tyrosine

(53) O. Kratky, E. Schauenstein, and A. Sekora, *Nature* **166,** 1031 (1950).

(54) Silk treated in this way also exhibits equatorial spacings of 8.9, 11, 15, 23, 31, and 45 A. [O. Kratky, E. Schauenstein, and A. Sekora, *Nature* **165,** 527 (1950)], but these are now[41] attributed to the presence of waxy impurities [see also O. Kratky, E. Schauenstein, and A. Sekora, *Nature* **170,** 796 (1952)]. Similar equatorial reflections, but at a spacing of 36 A., are found in silk from Italian silkworms without any treatment.

are confined to amorphous regions of the fiber. Indeed it is clear that there is urgent need to repeat much of the earlier work, using up-to-date techniques of amino acid analysis etc., in order to see whether the theories based on it can stand up to more detailed scrutiny.

III. The Keratin Group

1. Introduction

As we have indicated, wool keratin exists in an unstretched and a stretched form, the distinct x-ray patterns of which, known respectively as the α-keratin and the β-keratin patterns, show that different molecular configurations are present in the two states. Similar patterns are given by very many fibrous protein materials; and we may, with Astbury, define the keratin group as consisting of all materials exhibiting either.

This basis for classification was first proposed and elaborated in the early 1940's: Astbury now refers to the group of proteins thus defined as the keratin–myosin–epidermin–fibrinogen (k-m-e-f) group. It may perhaps be further subdivided into those substances which normally exist in the α-form and those normally found in the β-form. In the first class may be included keratin itself, epidermin, fibrin, and myosin, as well as a number of less important materials; in the second, feather keratin and silk. We have discussed silk fibroin separately: this was a matter of convenience and to some extent of historical accident, but from a different point of view silk should perhaps be regarded as the prototype of the keratin class. Myosin, one of the main components of muscle, will be dealt with in the following chapter.

There is reason to suppose that at least some of the globular proteins are built up of chains in the same configuration as α-keratin; the evidence for this has been discussed in Chap. 4, Vol. I, part A.

Members of the keratin group occur both intracellularly (e.g., keratin) and extracellularly (e.g., fibrin); the group is extremely widespread in animal tissues. Those members of it which have been studied most have a primarily architectural function, but there is no evidence that this is true of all.

We shall describe in most detail the properties of wool keratin, before discussing more briefly the other members of the group. A vast amount of work has been done on the behavior of wool, and in the space available here it is impossible to do more than touch on the more important aspects of it. A valuable review, containing many useful references, has recently been published by Zahn.[55]

(55) H. Zahn, *Das Leder* **1,** 222, 265 (1950); *ibid.* **2,** 8 (1951).

2. Wool Keratin and Related Materials

a. Occurrence: Histology

Wool, hair, horn, and nails form a closely related group of materials occurring in the mammals: the feathers of birds and the scales of fish are very similar. Here we shall mainly be concerned with the keratin of wool and hair. There is a large literature on the histology of wool: recent summaries have been given by Zahn[56] and by Lindberg *et al.*[57] Broadly, mammalian hair is made up of three concentric zones which, from outside inwards, are known as the cuticle, the cortex, and the medulla. The latter is missing in thinner hairs: the cuticle, itself consisting of several layers, surrounds the cells of the cortex. Keratin as commonly understood is contained in the cortex: little is known of the composition of the cuticle, though it is certainly complex. Among other components it appears to contain a continuous and highly resistant tubular membrane which has been called the "subcutis" because it was thought to lie between cuticle and cortex.[58–61] Mercer and his colleagues[57,62] have given evidence that in fact this membrane is external and not internal. If wool is treated with peroxyacetic (peracetic) acid and then dissolved in dilute alkali (see p. 875) there remains a residue of about 7–10 per cent of the original material: electron microscope examination shows that this consists chiefly of these tubular membranes.[63] The residue gives a disoriented β-keratin x-ray pattern. Alexander and Earland[63] suggest that the membrane consists of a network of chains with the β-configuration, but this conclusion has been criticized[62,64] by those who consider the β-pattern to be due to fragments of cortex remaining attached to the tubes. (After the chemical treatment to which the fibers have been subjected such fragments would probably be in the β-form.) Whatever final conclusions may be reached on this and related matters, it is clear that the heterogeneity of the wool fiber as a whole must be taken seriously to account in any discussion of keratin structure (see Zahn).[55]

(56) H. Zahn, *Textil-Praxis* **1,** 3 (1948).
(57) J. Lindberg, E. H. Mercer, B. Philip, and N. Gralén, *Textile Research J.* **19,** 673, 678 (1949).
(58) H. Reumuth, Dissertation, Aachen, Forschungsheft 3d. Deutsch. Forsch.-Inst.f.Text.-Ind., M.-Gladbach, 1938.
(59) E. Lehmann, *Melliand Textilber.* **25,** 1 (1944).
(60) E. Lehmann, *Kolloid-Z.* **108,** 6 (1944).
(61) E. Elöd and H. Zahn, *Naturwissenschaften* **33,** 158 (1946).
(62) E. H. Mercer, *Nature* **168,** 824 (1951).
(63) P. Alexander and C. Earland, *Nature* **166,** 396 (1950).
(64) N. Peacock, J. Sikorski, and H. J. Woods, *Nature* **167,** 408 (1951).

b. Mechanical Properties

Wool and hair are highly extensible materials: under suitable conditions they can be stretched 100 per cent or more, and on removal of tension there may be shortening right back to the original length. It was the first great contribution of Astbury and his colleagues to show[65–67] that this reversible change is accompanied by a reversible change in the x-ray pattern. Hair in its natural state is known as α-keratin, and exhibits the α-type x-ray pattern; stretched hair is β-keratin and exhibits the β-pattern. We shall be concerned later with the nature (p. 864) and structural origin (p. 892) of the patterns; in the meantime we need only note their existence and their value in characterizing the two distinct molecular states of the protein.

The phenomena attendant upon stretching of hair under various conditions of temperature and humidity are complicated, and only a brief outline of them will be given here (for a detailed account see Astbury).[68]

At room temperature dry wool fibers can be stretched by about 20 per cent: as the temperature and humidity are increased the maximum extension tolerated increases too, until in steam 100 per cent extension can be achieved. When completely dry, stretched hair does not contract at all if the tension is removed; on the other hand, if it is wet, the hair can under ideal conditions revert completely to its original length. This ideal behavior is exhibited only if the original stretching was carried out under cool, not too wet conditions, and was limited in extent; as extension, humidity, or temperature are increased the phenomenon of "setting" progressively develops—i.e., a loss of tension and a slowing down of the rate of contraction when the force is removed.

Another type of behavior is observed if the hair, after stretching in the cold, is kept extended in steam for a *very short* time (about 2 min.); on releasing the tension (the fiber still being kept in steam) it contracts to about two-thirds of its *original* length. This phenomenon is known as *supercontraction*. Very little is known of its nature: the x-ray diagram of supercontracted hair is similar to (though much less well oriented than) that of ordinary β-keratin,[69] so it would appear at first sight that no new type of molecular folding has taken place. (It should be noted that

(65) W. T. Astbury and H. J. Woods, *Nature* **126,** 913 (1930).
(66) W. T. Astbury and A. Street, *Phil. Trans.* **A230,** 75 (1931).
(67) W. T. Astbury and H. J. Woods, *Phil. Trans.* **A232,** 333 (1933).
(68) W. T. Astbury, Fundamentals of Fibre Structure, Oxford Univ. Press, 1933, p. 134 ff.
(69) E. Elöd and H. Zahn, *Melliand Textilber.* **28,** 2 (1947).

supercontraction can be induced in a number of different proteins and by a number of different reagents. There is no reason to suppose that the mechanism is always the same.)

In order to explain these phenomena, extensive use has been made of the cross-linking hypothesis, according to which the insolubility and inertness of keratin, and the phenomena of set, may be related to the presence of many cystine —S—S— bridges between adjacent chains. This hypothesis was proposed and developed by Stary,[70] Astbury,[66,67] Speakman,[71,72] etc. and others. In the present connection it postulates that the first action of steam or hot water is a chemical one—a rupture of the cross links, allowing the individual chains to unfold into the stretched or β-configuration. But if the action of steam is continued, new cross-links are formed—at first ones which are labile at high temperatures ("temporary set"), then ones which are stable indefinitely in the presence of steam ("permanent set"). Supercontraction occurs only when the tension is released under conditions such that the individual chains are free from mutual restraint—i.e., when the fiber is relaxed in steam after the original cross links have been destroyed and before the new ones have been formed.

It must be admitted that the correctness of this scheme has never been rigidly proved, although it has been widely accepted as giving a plausible picture of the processes undergone by stretched wool fibers. In detail it should therefore be treated with some reserve although, of course, there is no doubt that sulfur bridges are present and play some part in determining the mechanical behavior of wool.

In stretching at room temperature it would appear from the reversibility of the length–tension diagram that no covalent bonds are broken: or, in other words, that α-keratin can be extended into the β-configuration without the rupture of S—S bonds. This important question has never been finally settled by conclusive experiments, however. The suggestion which has been made is that the —S—S— bridge is hydrolyzed to a mercapto derivative and a sulfenic acid, and that the latter may then break down to an aldehyde

$$\text{—C—S—S—C—} \rightarrow \text{—C—SH} + \text{HO—S—C—}$$
$$\downarrow$$
$$\text{OHC—}$$

It has been further suggested that the new cross links formed by the action of steam connect these aldehyde groups with the $—NH_2$ groups of

(70) Z. Stary, *Z. physiol. Chem.* **175,** 178 (1928).
(71) J. B. Speakman, *J. Soc. Dyers Colourists* **52,** 335, 423 (1936).
(72) J. B. Speakman and C. S. Whewell, *J. Soc. Dyers Colourists* **52,** 380 (1936).

lysine and arginine, a hypothesis supported by the finding that setting power is much reduced by preliminary deamination of the fibers (or treatment with fluorodinitrobenzene).[73] Generally speaking, satisfactory explanations of the observed phenomena can be made on these lines. On the other hand, it is fairly clear that it would be an oversimplification to suppose that sulfur bridges or other covalent links are alone responsible for cross-linking and "set." Rudall[74] showed that set could be partly reversed by urea and concluded that hydrogen bonds were involved; further experiments of the same kind have been carried out by Alexander,[75,76] Sikorski,[77] and others; they used various agents known to attack hydrogen bonds (LiBr, urea, formic acid), and also cuprammonium hydroxide which alone induced complete reversal of set in keratin. The phenomenon is evidently complex, involving the participation of more than one type of bond, and we are as ignorant of its details as we are of cross-linking in proteins generally.

The long-range elasticity of keratin has formed the subject of a number of thermodynamic studies, particularly so far as concerns its resemblance or otherwise to the elasticity of rubber. There have been two extreme views: (*a*) that in the interconversion of α- and β-keratin, involving a specific molecular rearrangement, entropy plays a subsidiary part—essentially the system is an "internal energy" machine and contraction is due to specific forces acting between definite groups in the peptide chain (Astbury school); (*b*) that the thermoelastic behavior is closely analogous to that of rubber and other elastomers, the elastic forces being exerted by the thermal agitation of segments of the chain in random motion. The latter view carries the implication that the reversible changes in the x-ray pattern are in a sense irrelevant to the issue (Meyer school).[78,79] Some recent exchanges in this controversy may be referred to.[80] In spite of the formal analogies between the thermodynamics of the stretching of wool and that of rubber, it is hard to see how the two specific x-ray diagrams of keratin—which we shall now proceed to consider—can be reconciled with the extreme views of the out-and-out entropy school. On the other hand, it has recently been concluded from

(73) A. J. Farnworth and J. B. Speakman, *Nature* **161,** 890 (1951).
(74) K. M. Rudall, *Symposium Soc. Dyers Colourists* 15 (1946).
(75) P. Alexander, *Research* (London) **2,** 246 (1949).
(76) P. Alexander, *Ann. N. Y. Acad. Sci.* **53,** 653 (1953).
(77) J. Sikorski, *Nature* **170,** 275 (1952).
(78) K. H. Meyer, A. J. A. van der Wyk, W. Gonon and C. Haselbach, *Trans. Faraday Soc.* **48,** 669 (1952).
(79) K. H. Meyer, G. von Susich, and E. Valkó, *Kolloid-Z.* **59,** 208 (1932).
(80) K. H. Meyer, and C. Haselbach; H. J. Woods; K. H. Meyer; and W. T. Astbury, *Nature* **164,** 33ff. (1949).

a study of x-ray diffraction and infrared spectra[81] that in both unstretched and stretched hair the largest part of the keratin is in an amorphous, disoriented condition; the well-known x-ray diagram is produced by the relatively small fraction of the material which is oriented and crystalline. The elastic properties of the wool, like its infrared spectrum, would be mainly conditioned by the major, amorphous component. The contribution of this amorphous component has not been sufficiently considered by the adherents of the internal energy school.

c. X-Ray Pattern

Like those of so many other biologically important materials, the x-ray diffraction pattern of keratin would appear to have been first recorded

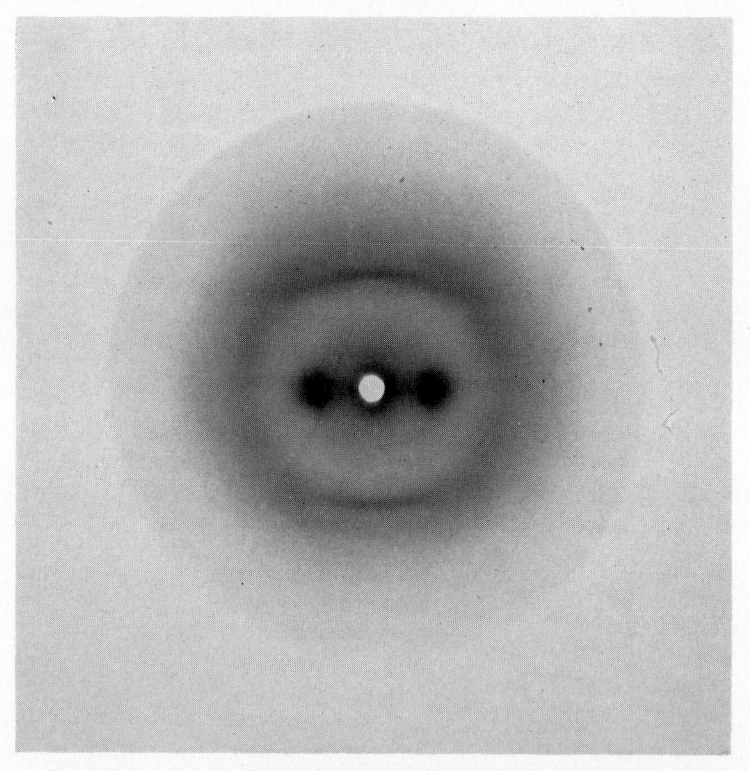

FIG. 3. Wide-angle x-ray pattern of α-keratin (Cotswold wool): fiber axis vertical. By courtesy of W. T. Astbury.

by Herzog and Jancke in 1921.[82] The earliest detailed study, however, was made by Astbury and his collaborators in the early 1930's.[65–67]

(81) A. Elliott, *Textile Research J.* **22,** 783 (1952).

(82) R. O. Herzog and W. Jancke, *Festschrift Kaiser-Wilhelm Gesellschaft,* 1921.

Characteristic examples of the typical α- and β-patterns are shown in Figs. 3, 4*a*, and 4*b*, and Tables II and III list the principal spacings.

TABLE II
PRINCIPAL WIDE-ANGLE X-RAY REFLECTIONS OF α-KERATIN

	Spacing, A.	Intensity[a]
Equator	27	s.
	9.8	v.s.
	3.5	v.w.
Meridian (1st layer)	10.3	v.w.
(2nd layer)	5.1_1	v.v.s.
	3.9_5	s.
	3.4_0	w.–m.
	3.0_1	m.–s.
	2.5_6	w.–m.
	1.48_6	m.–s.
1st. layer (non-merid.)	4.5	w.
2nd. layer (non-merid.)	4.1	m.
3rd. layer (non-merid.)	3.0	w.

[a] v = very; w = weak; m = medium; s = strong.

TABLE III
PRINCIPAL WIDE-ANGLE X-RAY REFLECTIONS OF β-KERATIN

	Spacing, A.	Intensity[a]
Equator	9.7	s.
	4.65	v.s.
	2.4	w.
Meridian (2nd. layer)	3.33	s.
(3rd. layer)	2.2	w.
1st. layer, non-merid.	4.7	m.
	3.75	s.
	2.2	v.w.
2nd. layer, non-merid.	2.7	w.
3rd. layer, non-merid.	2.0	w.

[a] v = very; w = weak; m = medium; s = strong.

The most obvious features are, in α-keratin, the prominent equatorial reflections at 9.8 A. and the meridional reflection at 5.1 A.; and in β-keratin, the two equatorial reflections at 9.7 and 4.65 A. and the meridional reflection at 3.33 A.

Besides these reflections of medium spacing, we must notice two other groups of reflections which have recently assumed great importance and

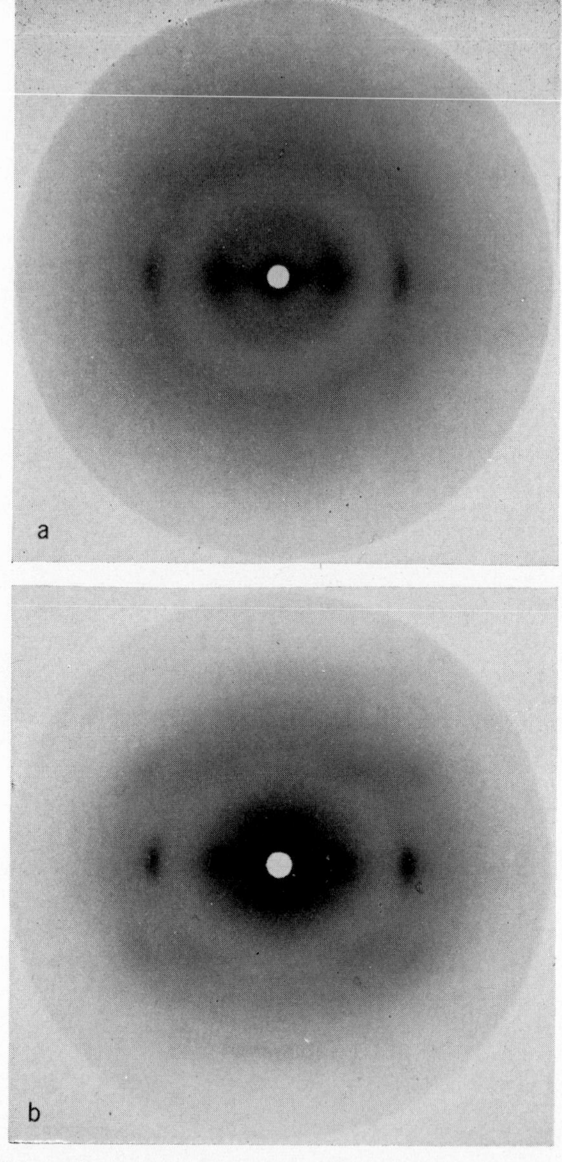

FIG. 4. Wide-angle x-ray patterns of β-keratin: fiber axis vertical. (*a*) stretched horn (by courtesy of W. T. Astbury); (*b*) stretched epidermin; and (*c*) cross β-pattern of epidermin (by courtesy of K. M. Rudall).

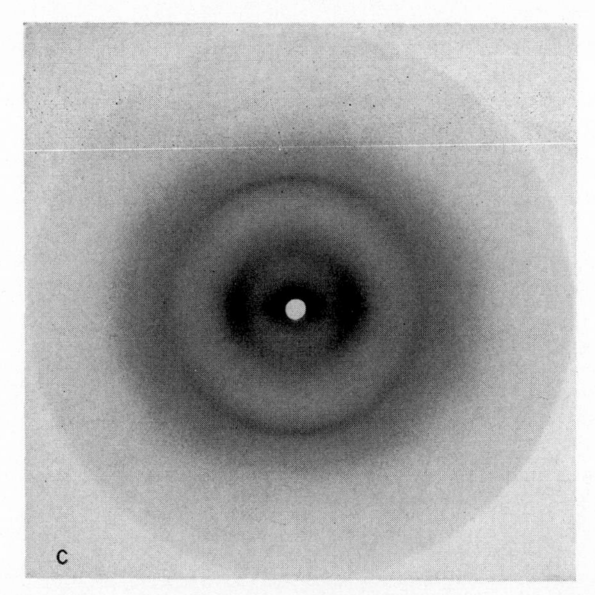

FIG. 4. *Continued.*

which are usually not observed in the diffraction pattern unless special techniques are used. The first group, the reflections of short spacing, has as its principal member a strong meridional reflection of 1.5 A. This was listed by MacArthur in 1943[83] as a reflection of medium strength from porcupine quill, a material giving the α-keratin pattern, but no special significance was attached to it at that time.

Now by conventional methods of fiber photography (in which x-ray beam and fiber axis are perpendicular), spacings if truly meridional can only be observed if the fiber is so badly oriented that some of its strands are distorted from parallelism by an angle equal to the Bragg angle—in the case of the 1.5-A. reflection and Cu Kα x-radiation this angle is as much as 31°; thus the more perfect the specimen the less well the spacing would be observed.

In order that reflection from planes of spacing d may take place, the angle between the incident radiation and the planes must be the Bragg angle θ, where $n\lambda = 2d \sin \theta$. Meridional spacings in a fiber photograph are derived from planes perpendicular to the fiber axis: hence to record them the incident beam must make an angle $(90 - \theta)$ with the fiber axis, assuming the fiber to be perfectly oriented.

Perutz,[84] realizing that x-ray photographs of keratin had never been taken with the fiber tilted into the reflecting position, found that when the calculated angle of tilt was used the 1.5-A. reflection became very

(83) I. MacArthur, *Nature* **152,** 38 (1943).
(84) M. F. Perutz, *Nature* **167,** 1053 (1951).

strong, in fact one of the most important reflections in the pattern. Others less strong have been reported at 2.56, 3.0, 3.4, and 4.0 A.[85,86] The 1.5-A. reflection has been observed in most materials known to have the α-keratin structure (synthetic polypeptides, epidermin, myosin, tropomyosin, fibrin, bacterial flagella) but not in any β-keratins; indeed on stretching wool or hair the reflection disappears. The structural significance of this reflection will be discussed below (p. 901). In β-keratin an analogous reflection of 1.104 A. has recently been reported.[23]

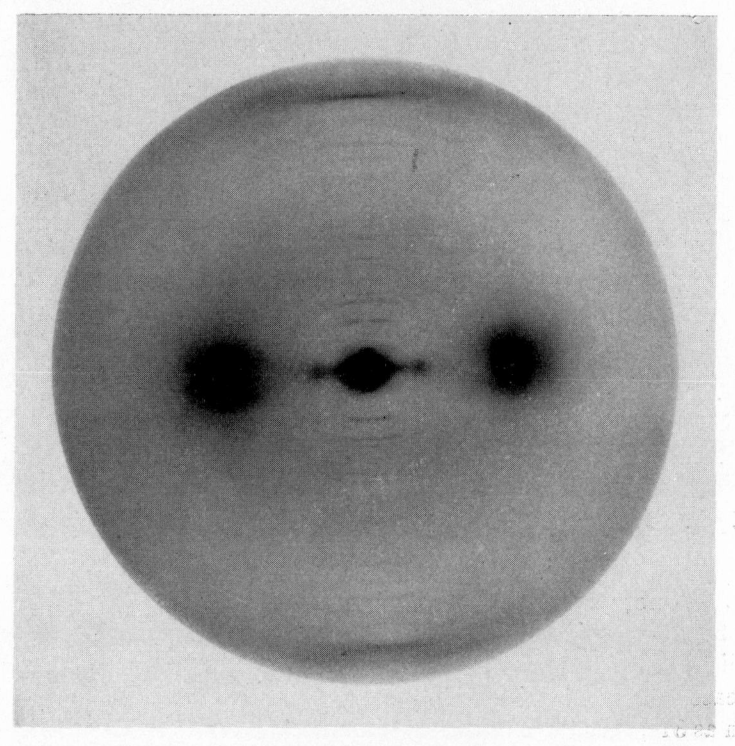

FIG. 5. Low-angle x-ray pattern of porcupine quill tip: fiber axis vertical. By courtesy of I. MacArthur.

At the other end of the scale are reflections of long spacing. Wool keratin and also the closely related porcupine quill and feather keratin all yield such reflections, which (apart from a 24-A. reflection observed by Astbury and Marwick in feather keratin in 1932)[87] were first described by Corey and Wyckoff in 1936.[88]

(85) W. T. Astbury, *Proc. Roy. Soc.* (London) **B141,** 1 (1953).
(86) W. T. Astbury and J. W. Haggith, *Biochim. et Biophys. Acta* **10,** 483 (1953).
(87) W. T. Astbury and T. C. Marwick, *Nature* **130,** 309 (1932).
(88) R. B. Corey and R. W. G. Wyckoff, *J. Biol. Chem.* **114,** 407 (1936).

Porcupine quill has been used by many workers because it gives a diffraction pattern similar to that of α-keratin of hair, but more highly developed. The diffraction patterns of both wool and quill at low angles have been described in great detail by Bear and his co-workers[89–91] and

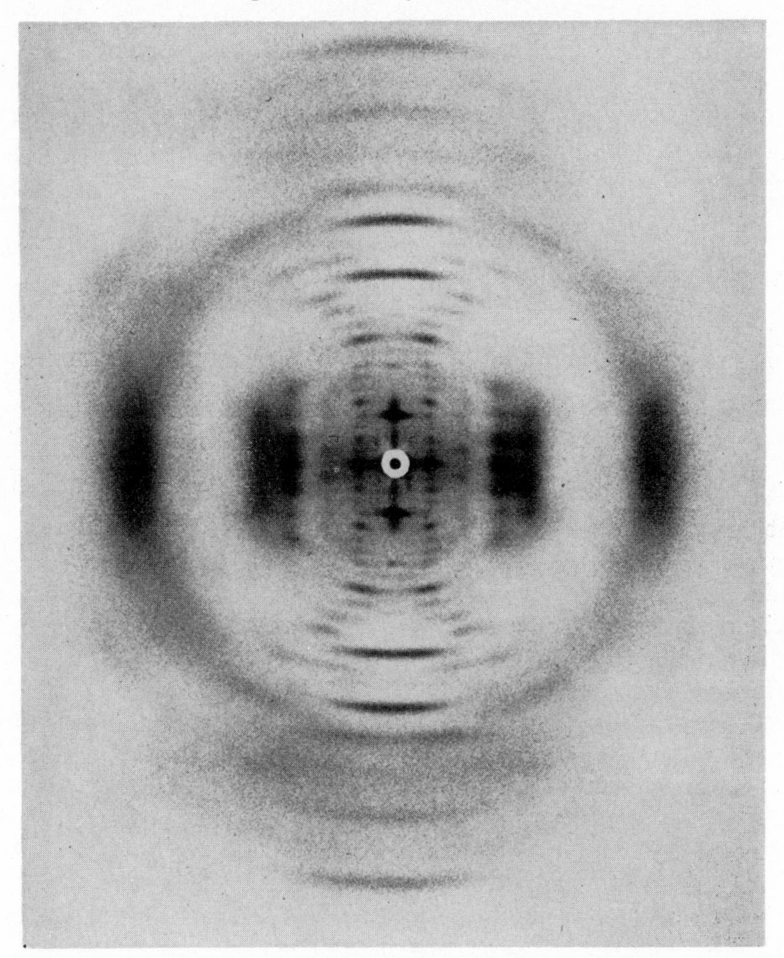

FIG. 6. X-ray pattern of sea-gull feather keratin: fiber axis vertical. After Bear and Rugo.[91]

by MacArthur.[83] They can be indexed in terms of an axial repeat of 198 A. and a lateral one of 83 A. (MacArthur originally preferred the longer meridional repeat of 658 A.); Fig. 5 shows an example of the quill

(89) R. S. Bear, *J. Am. Chem. Soc.* **65,** 1784 (1943).
(90) R. S. Bear, *J. Am. Chem. Soc.* **66,** 2043 (1943).
(91) R. S. Bear and H. J. Rugo, *Ann. N. Y. Acad. Sci.* **53,** 627 (1951).

pattern. Up to 19 meridional and 3 lateral orders of reflections are observed. On stretching the specimen to the β-form (not easily done in the case of quill) the long spacings disappear; on reversion to the α-form they are to some extent regenerated.

Feather keratin gives an entirely different pattern (Fig. 6). This material was first studied by Astbury and Marwick,[87] who showed that, like silk, it gives a pattern apparently of the β-type in the natural condition, with a principal meridional reflection at 3.1 A.—i.e., slightly less than that of silk fibroin. No contracted form of feather keratin is known. Surprisingly, swan quill does not give the 1.1-A. axial reflection observed in silk and β-keratin;[23] nor, apparently, does it give an α-type 1.5-A. reflection (but see p. 896). Feather keratin gives a rich low-angle x-ray pattern, which can be indexed in terms of an axial repeat of 95 A. and one of 34 A. transverse to the fiber direction. Up to 21 meridional orders of reflection are observed.[90,91]

The x-ray diffraction pattern of the keratins will be discussed further in connection with theories of structure (see p. 892).

d. Spectroscopy

The infrared spectra of keratins have been studied principally by Ambrose, Hanby, Elliott, and Temple[92,93,32,81] (early results were also reported by Goldstein and Halford).[94] As always, the most illuminating results have come from studies with polarized radiation: these are summarized in Table IV, which shows for comparison some observations made on other materials.

The interpretation of infrared spectra has been discussed in detail in Chap. 5, Vol. I, part A. It will be seen that all the results in the Table agree with the idea that in α-keratin, in porcupine quill, and in the α-form of synthetic polypeptides, both C=O and N—H bonds lie more or less parallel to the fiber axis, whereas in feather keratin, silk, nylon, and the β-form of synthetic polypeptides they lie perpendicular to it (see Vol. I, part A, pp. 424–8, Figs. 5–7). These conclusions will be relevant to our discussion of the configurations of the polypeptide chain present in these materials.

Ambrose and Elliott[32,81] find that in almost all cases there is some admixture of configurations: e.g., they observed a weak perpendicular component of the N—H deformation frequency at 1695 cm.$^{-1}$ in silk, and in feather both N—H deformation and C=O stretching frequencies are double, with a lower dichroism on the high-frequency side. Both these

(92) E. J. Ambrose and W. E. Hanby, *Nature* **163**, 483 (1949).
(93) E. J. Ambrose, A. Elliott, and R. B. Temple, *Nature* **163**, 859 (1949).
(94) M. Goldstein and R. S. Halford, *J. Am. Chem. Soc.* **71**, 3854 (1949).

TABLE IV

INFRARED DICHROISM OF SOME KERATINS AND RELATED MATERIALS[a]

Approximate			Keratins			Synthetic polypeptides, etc.		
Frequency, cm.$^{-1}$	Wavelength, μ	Significance of band	Hair	Porcupine quill	Feather	α-	β-, nylon	Silk
1530–1570	6.4	N—H deformation	$\perp$	$\perp$	$\parallel$	$\perp$	$\parallel$	$\parallel$
1640–1660	6.1	C=O stretching	$\parallel$	$\parallel$	$\perp$	$\parallel$	$\perp$	$\perp$
3050–3070 3300–3320	3.3 3.0	N—H stretching	$\parallel$	$\parallel$	$\perp$	$\parallel$	$\perp$	$\perp$
4600 approx	2.2	C=O deformation (overtone)		-				$\parallel$
4830	2.1	N—H combination (overtone)		$\perp$				$\parallel$

[a] $\parallel$ = maximum absorption for electric vector parallel to fiber axis.
$\perp$ = maximum absorption for electric vector perpendicular to fiber axis.

results suggest the presence of a component in the α-configuration. Similarly, the N—H deformation frequency of elephant hair and wool[81] shows signs of the presence of a β-component; these results might, however, be connected with the discovery by Alexander and Earland[63] of membranous material in the β-configuration after oxidation of wool fibers (see p. 875).

The ultraviolet spectrum of keratin has been studied by Beaven, Holiday, and Jope[95] and, using polarized radiation, by Perutz, Jope, and Barer.[96] As would be anticipated, the absorption bands can be satisfactorily accounted for in terms of the content of aromatic amino acids, but in no case was any dichroic effect observed.

e. Chemical Properties

It has already been pointed out that at the histological level hair is a very complex material. Furthermore, it is very difficult to fractionate it into even its microscopically distinguishable components. It is probable that there is complexity at the chemical level too;[55] to quote only one piece of evidence for this, Lindley[97] succeeded in fractionating wool into several components having different amino acid compositions. Hence, although very much work has been done on the amino acid composition of keratin, it is highly doubtful whether the results have much direct bearing on problems of structure. The published figures should be used only as a general guide to the character of the protein.

A summary has been given by Tristram in Chap. 3 of this work (see Vol. I, part A, p. 220, Table XXII): most of the available figures are those collected by Astbury,[98] with some new values determined by Graham, Waitkoff, and Hier[99] and by Lang and Lucas.[100] The approximate numbers of residues per 10^5 gm. of protein are as follows:

(Cys-)	99	Arg	60	Ala	46	His	7
Glu	96	Leu }	86	Val	40	Try	9
Ser	95	Ileu }		Tyr	26	Meth	5
Gly	87	Thr	54	Phe	22		
Pro	83	Asp	54	Lys	19		
					Total		888

(95) G. H. Beaven, E. R. Holiday, and E. M. Jope, *Discussions Faraday Soc.* No. **9**, 406 (1950).
(96) M. F. Perutz, M. Jope, and R. Barer, *Discussions Faraday Soc.* No. **9**, 423 (1950).
(97) H. Lindley, *Nature* **160,** 190 (1947).
(98) W. T. Astbury, *J. Chem. Soc.* **1942,** 337.
(99) C. E. Graham, H. K. Waitkoff, and S. W. Hier, *J. Biol. Chem.* **177,** 529 (1949).
(100) J. M. Lang and C. C. Lucas, *Biochem. J.* **52,** 84 (1952).

The most noteworthy feature is, of course, the large percentage of cystine residues. It is to be observed, however, that considerable amounts of glycine and proline, generally thought of as characteristic especially of collagen, are also present in keratin.

Using the fluorodinitrobenzene method, Middlebrook[101] obtained a complex mixture of amino end-groups (Val, Ala, Gly, Thr, Ser, Glu, Asp, in approximate ratios 4:2:8:8:2:2:1). This result is not unexpected in view of the chemical heterogeneity of keratin; likewise, partial hydrolysis[102–104] results in the production of very complex mixtures of small peptides (for summary, see the review of Sanger).[9] The latter studies were often carried out with the object of testing Astbury's α-configuration chain model, according to which polar and nonpolar residues might be expected to alternate along the chain. The evidence obtained was against such alternation, and also against the validity of the Bergmann-Niemann hypothesis (p. 906).

We may now turn to the chemical reactions of wool keratin.[105] Owing to the industrial importance of wool, these have been studied in great detail: but the mass of experimental results which have thus been accumulated should not blind us to the fact that the most salient characteristic of keratin is its *lack* of chemical reactivity compared with proteins in general. This inertness is exemplified by its partial indifference to boiling water and to proteolytic enzymes—an indifference which, however, appears to be at least to some extent morphological in origin; thus, wool is readily degraded by enzymes if first finely ground,[106] and powdered horn is partly soluble in water;[107] in fact in all circumstances this inertness to enzymes and water is relative rather than absolute.[108,55]

Much research has been devoted to a study of the reactions of the —S—S— bridges of cystine which, as has already been indicated, appear to play a large part in stabilizing keratin. As a general rule, destruction of these links renders keratin soluble, a process which is of great importance in wool technology, the manufacture of depilatories, etc. The —S—S— bridges can, for example, be reduced: by hydrogen sulfide, sulfides, or thioglycolic acid to S—H groups: by bisulfite to one —SH and one —S—SO_3H group: by cyanide to one S—H and one SCN group,

(101) W. R. Middlebrook, *Biochim. et Biophys. Acta* **7,** 547 (1951).
(102) A. J. P. Martin, *Symposium Soc. Dyers Colourists* 1 (1946).
(103) R. Consden, A. H. Gordon, and A. J. P. Martin, *Biochem. J.* **44,** 548 (1949).
(104) R. Consden and A. H. Gordon, *Biochem. J.* **46,** 8 (1950).
(105) For a more detailed account see the review already referred to.[55]
(106) J. I. Routh, *J. Biol. Chem.* **123,** civ (1938).
(107) H. Cohen, *Arch. Biochem.* **4,** 145 (1944).
(108) E. Elöd and H. Zahn, *Melliand Textilber.* **25,** 361 (1944).

etc.[109] The reduced material exhibits lower mechanical strength, is readily attacked by proteolytic enzymes, and is soluble in acid or alkali. Its reactivity persists if the S—H groups are reoxidized to S—S, but after reoxidation the fiber largely recovers its original mechanical properties; this reversal is used in permanent waving for "cold setting." Again, the original —S—S— bridges of native keratin can be oxidized, by a variety of ordinary oxidizing agents (including hydrogen peroxide, hypochlorite, and peracetic acid) to give reactive and soluble products. These oxidizing reactions are complex and involve a variety of groups in the protein in addition to the —S—S— bridges: for a detailed account, including references to earlier work, see the series of papers by Alexander *et al.*[110–114] (see also Blackburn and Lowther).[115] Among other things, these workers showed[110] that the cystine could be differentiated into two fractions differing in their susceptibility to oxidation. Cuthbertson, Lindley, and Phillips[116,117] were also able to subdivide the cystines into fractions behaving differently on alkaline hydrolysis. The reaction with alkali is interesting because it may result in the abstraction of one sulfur atom from the —S—S— bridge with the production of a residue of lanthionine.[116,117,117a]

$$\begin{array}{c} | \\ NH \\ | \\ CH-CH_2-S-S-CH_2-CH \\ | \\ CO \\ | \end{array} \text{(NH, CO on each CH)} \rightarrow \begin{array}{c} | \\ NH \\ | \\ CH-CH_2-S-CH_2-CH \\ | \\ CO \\ | \end{array} + S$$

Free lanthionine is released if the fiber is subsequently hydrolyzed.

Finally, a good deal of work has been done to study reactions of keratin whereby cross links additional to those originally present are formed: in particular by the reaction of formaldehyde, well-known as a cross-linking reagent for proteins in general. Reference should be made to the review of French and Edsall,[118] and to the work of Alexander

(109) D. R. Goddard and L. Michaelis, *J. Biol. Chem.* **106,** 605 (1934); *ibid.* **112,** 361 (1935).
(110) P. Alexander, R. F. Hudson, and M. Fox, *Biochem. J.* **46,** 27 (1950).
(111) P. Alexander, D. Carter, and C. Earland, *Biochem. J.* **47,** 251 (1950).
(112) P. Alexander, D. Gough, and R. F. Hudson, *Biochem. J.* **48,** 20 (1951).
(113) P. Alexander and D. Gough, *Biochem. J.* **48,** 504 (1951).
(114) P. Alexander, M. Fox, and R. F. Hudson, *Biochem. J.* **49,** 129 (1951).
(115) S. Blackburn and A. G. Lowther, *Biochem. J.* **49,** 554 (1951).
(116) W. R. Cuthbertson and H. Phillips, *Biochem. J.* **39,** 7 (1945).
(117) H. Lindley and H. Phillips, *Biochem. J.* **39,** 17 (1945).
(117*a*) M. H. Horn, D. B. Jones, and S. J. Ringel, *J. Biol. Chem.* **138,** 141 (1941).
(118) D. French and J. T. Edsall, *Advances in Protein Chem.* **2,** 277 (1945).

et al.,[119] who concluded that the cross link in the formaldehyde reaction is possibly formed between ε-amino groups of lysine and ≥CH groups of tyrosine.

f. Soluble and Regenerated Keratin

In conclusion we shall briefly consider the solubilization of keratin. This can be achieved by the action of reducing and oxidizing agents, as discussed above, and the resulting solution can be studied by conventional techniques. It is unfortunate that no simple picture emerges: the solutions are invariably polydisperse. One of the most clear-cut experiments reported is that of Alexander, Earland, and Happey;[63,120] they oxidized the —S—S— bridges with peracetic acid and dissolved the resultant material in dilute alkali. From seven to 10 per cent of the original protein was left as an insoluble tubular membrane giving a disoriented β-type x-ray pattern (see p. 860); the solution was in the main electrophoretically homogeneous, containing particles of molecular weight about 70,000. The soluble material could be precipitated from very dilute solution in the α-form; by dissolving this in concentrated formic acid and distilling the latter off *in vacuo*, a β-protein was obtained. This effect of formic acid in inducing a β-configuration is paralleled in the behavior of synthetic polypeptides (see p. 890).

Much work has been done on the solubilization of keratin by means of mixtures of reducing agent and detergent (see Jones and Mecham,[121] Lundgren and O'Connell,[122] and Ward, High, and Lundgren[123]); solutions are obtained which can be extruded into fibers in the β-configuration. The solutions are found electrophoretically to contain two main species, of molecular weights 55,000–75,000 and 35,000–40,000.

Again, solutions have been obtained using strong urea containing a reducer.[122] Mercer[124] has found that these can be salted out to give an orientable fibrous material, having the α-configuration, and generally resembling epidermin. Mercer and Olofsson[125] found that the solutions from which the fibers are salted out are polydisperse, but have as their main constituent highly asymmetric particles of molecular weight 84,000; on further reduction at high pH, extensive disaggregation takes place to a molecular weight of 8000, and the molecule becomes less asymmetric.

(119) P. Alexander, D. Carter, and K. G. Johnson, *Biochem. J.* **48,** 435 (1951).
(120) F. Happey, *Nature* **166,** 397 (1950).
(121) C. B. Jones and D. K. Mecham, *Arch. Biochem.* **3,** 193 (1943).
(122) H. P. Lundgren and R. A. O'Connell, *Ind. Eng. Chem.* **36,** 370 (1944).
(123) W. H. Ward, L. M. High, and H. P. Lundgren, *J. Polymer Sci.* **1,** 22 (1946).
(124) E. H. Mercer, *Nature* **163,** 18 (1949).
(125) E. H. Mercer and B. Olofsson, *J. Polymer Sci.* **6,** 671 (1951).

The power to form fibers is simultaneously lost. Woodin has found[125a] that, when dissolved in urea and bisulfite, feather keratin disaggregates to particles whose molecular weight may be as low as 10,000; the solution is apparently monodisperse. Again, Wormell and Happey[126] have regenerated keratin from a cuprammonium solution, but the product, which was poorly oriented, was in the β-configuration.

Thus no very precise conclusions can be drawn from these experiments. In general it would appear that a substantial proportion of wool keratin can be brought into solution by any treatment combining destruction of the —S—S— bridges with solubilization (achieved either by moving away from the isoelectric point or by destroying hydrogen bonds with urea). The resulting solution is polydisperse, presumably owing to the complex nature of the original keratin; however it apparently contains substantial components with molecular weights in the range 30,000–60,000. Material can be precipitated from the solutions in either the α- or the β-forms, according to circumstances.

3. Fibrinogen and Fibrin

The conversion of the soluble protein fibrinogen to an insoluble gel of fibrin is the basis of the clotting of blood, and as such has been discussed under the heading of plasma proteins in Chap. 21 of this work. From another point of view it is appropriate to mention these substances in the present context, for they are members of the k-m-e-f group, yielding diffraction patterns of either the α- or β-type according to the conditions. Early x-ray work by Katz and de Rooy in 1933[127] indicated that fibrin as they prepared it was mainly in the β-configuration: 10 years later, Bailey, Astbury, and Rudall[128] showed that an oriented fibrinogen preparation precipitated from plasma was initially in the α-configuration (with some admixture of β-), but could be squeezed into threads of β-protein. Fibrin was also obtained in α- and β-forms.

a. Fibrinogen

Fibrinogen comprises 0.25 per cent of human plasma, making up 4 per cent of its protein content. Its separation and purification from other plasma proteins have already been discussed (see p. 725). It has formed the subject of very many physicochemical studies, including

(125a) A. M. Woodin, *Biochem. J.* **57,** 99 (1954).
(126) R. L. Wormell and F. Happey, *Nature* **163,** 18 (1949).
(127) J. R. Katz and A. de Rooy, *Naturwissenschaften* **21,** 559 (1933).
(128) K. Bailey, W. T. Astbury, and K. M. Rudall, *Nature* **151,** 716 (1943).

measurements of flow birefringence,[129] viscosity,[130,131] sedimentation,[130] osmotic pressure, and light scattering.[131,132] One of the most recent determinations is that of Shulman,[133] who deduces from measurements of sedimentation and diffusion that the molecular weight is 330,000 ± 10,000 in good agreement with the light-scattering value. There is now general agreement that the molecular weight is in the region of 330,000, and that the molecule is a highly asymmetric rod-shaped object about 600 A. long and having axial ratio about 18:1. A recent investigation by Caspary and Kekwick[133a] suggests that in dilute solution (less than 0.15 gm./100 ml.) the human fibrinogen molecule may dissociate into smaller particles, of molecular weight 130,000 ± 20,000 and axial ratio 6:1.

In its amino acid composition, fibrinogen bears considerable resemblance to myosin, a fact which originally led to the suggestion that its x-ray pattern might be re-examined with a view to its classification in the k-m-e-f group.[134] Detailed figures have been given in Chap. 3 (Table XVII, p. 215, Vol. I, part A); these may be conveniently summarized as follows:

Glu	99	Leu	54	Ileu	37	His	17
Asp	98	Thr	52	Val	35	Met	17
Gly	75	Pro	49	Tyr	30	Try	16
Ser	67	Arg	45	Phe	28	CySH	3
Lys	63	Ala	41	(CyS—)	19		

(numbers of residues/10^5 g. protein)

The amino acid composition of fibrin is extremely similar to that of fibrinogen: indeed until recently they were assumed to be identical. It appears that fibrin (and presumably fibrinogen as well) contains polysaccharide as an essential component: about 2 per cent reducing sugars (mannose and galactose) and 0.6 per cent hexosamine have been reported.[135,136]

Little is known of the structure of the fibrinogen molecule. In the electron microscope it is near the present limit of effective resolution:

(129) J. T. Edsall, J. F. Foster, and H. Scheinberg, *J. Am. Chem. Soc.* **69,** 2731 (1947).

(130) J. L. Oncley, G. Scatchard, and A. Brown, *J. Phys. & Colloid Chem.* **51,** 184 (1947).

(131) C. S. Hocking, M. Laskowski, and H. A. Scheraga, *J. Am. Chem. Soc.* **74,** 775 (1952).

(132) S. Katz, K. Gutfreund, S. Shulman, and J. D. Ferry, *J. Am. Chem. Soc.* **74,** 5706 (1952).

(133) S. Shulman, *J. Am. Chem. Soc.* **75,** 5846 (1953).

(133*a*) E. A. Caspary and R. A. Kekwick, *Biochem. J.* **56,** xxxv (1954).

(134) K. Bailey, *Advances in Protein Chem.* **1,** 289 (1944).

(135) R. Consden and W. M. Stanier, *Nature* **169,** 783 (1952).

(136) St. Szára and D. Bagdy, *Biochim. et Biophys. Acta* **11,** 313 (1953).

it has been reported by Hall[137] that fibrinogen molecules are 600 A. long and 30–40 A. wide, in conformity with the data obtained in solution. The particles "appear nodose, not unlike a string of beads." Recent electron-microscope studies by Siegel, Mernan, and Scheraga[138] lead to similar conclusions; but these workers find that the fibrinogen molecule is polydisperse, the number of "beads" varying from individual to individual. Some other recent results by Mitchell[139] appear to be not inconsistent with this picture.

b. *The Conversion of Fibrinogen to Fibrin*

This reaction is of unique interest not only on account of its physiological importance but because it is the only example of the conversion of a soluble protein into an insoluble structure protein which can be carried out *in vitro* with some assurance that the process followed is identical with that which occurs *in vivo*. It thus provides us with an invaluable opportunity of studying fibrogenesis under controlled conditions.

What is actually observed is that fibrinogen is converted to a clot of fibrin under the influence of another substance known as thrombin. We shall not concern ourselves here with the way in which thrombin appears in the blood stream at the required time and place (see p. 730). We shall rather consider its mode of action. All the evidence suggests that this is purely catalytic—thrombin can transform at least 10^5 times its own weight of fibrinogen—and since thrombin is a protein we must classify it as an enzyme. There have been several theories as to the type of change in fibrinogen which is catalyzed; thrombin has been thought of as a denaturase, as an enzyme liberating —SH groups (which would subsequently be oxidized by quinone), and as a protease. It now seems certain that it is in fact a protease, but of a highly specific nature—so specific that until very recently fibrinogen was its only known substrate[140] (in earlier days this high specificity was not fully appreciated since thrombin preparations were generally contaminated with plasmin, a protease of more normal type). The nature of the change which it catalyzes has been recently elucidated by the work of Bettelheim, Bailey, Laki, Lorand,

(137) C. E. Hall, *J. Biol. Chem.* **179,** 857 (1949).

(138) B. H. Siegel, J. P. Mernan, and H. A. Scheraga, *Biochim. et Biophys. Acta* **11,** 329 (1953).

(139) R. F. Mitchell, *Biochim. et Biophys. Acta* **9,** 430 (1952).

(140) It has now been shown by S. Sherry & W. Troll [*J. Clin. Invest.* **32,** 603 (1953)] that thrombin can catalyze the hydrolysis of the synthetic peptide tosyl-arginyl-methyl ester, splitting off methyl alcohol in the process.

and Middlebrook.[141–148] It was first shown that whereas (bovine) fibrinogen has 3–4 tyrosine amino end groups and 2 glutamic acid end groups per molecule, in fibrin the same weight of material has 3–4 tyrosine and 4 glycine end groups. It turned out that these changes were accompanied by the shedding into solution of peptide material which had earlier completely escaped notice. Concerning the nature of this peptide material, different workers have reached different conclusions. Lorand[145] reported that the main product was a single peptide, which he called fibrinopeptide and which contains the following amino acid residues: Gly, Asp, Glu, Ser, Arg, Lys, Ileu, Phe, Val, Ala, Thr.[147] On the other hand, Bettelheim and Bailey[142,143] have come to the conclusion that at least two types of peptide molecule are thrown off, two molecules of each type coming from one fibrinogen molecule. These both have a minimum molecular weight of about 3000, but their amino acid compositions are different from each other and from that of Lorand's fibrinopeptide. Both contain considerable amounts of Asp, Glu, Gly, and Pro and moderate amounts of Thr, Val, Leu, Phe, and Arg; but peptide A contains much Ser and no Ala, Tyr, or Lys, while peptide B has much Ala and some Tyr and Lys, but probably no Ser. More striking, peptide A has all the lost glutamic acid end groups of fibrinogen, while peptide B apparently possesses no *N*-terminal groups reactive to fluorodinitrobenzene. Bettelheim and Bailey suggest that the four peptide molecules, two of peptide A and two of peptide B, may be attached to four glycine α-amino groups in fibrinogen; this scheme would account for the disappearance of two glutamic acid end groups and the appearance of four glycine end groups when fibrinogen is transformed to fibrin.

Yet another result is reported by Laki,[148] who obtains a complex mixture of peptides (which he calls cofibrin) by the action of thrombin on fibrinogen; one of these may be identical with Lorand's fibrinopeptide. In view of the divergent results obtained by different workers it is clear that any definite scheme proposed for the changes involved must be accepted with reserve; it is quite possible, for example, that the specimens of thrombin used in the various experiments were not identical.

In human fibrinogen *N*-terminal alanine is found in place of glutamic

(141) K. Bailey, F. R. Bettelheim, L. Lorand, W. R. Middlebrook, *Nature* **167,** 233 (1951).
(142) F. R. Bettelheim & K. Bailey, *Biochim. et Biophys. Acta* **9,** 578 (1952).
(143) F. R. Bettelheim, Ph.D. Dissertation, University of Cambridge, 1953.
(144) L. Lorand, *Nature* **167,** 992 (1951).
(145) L. Lorand, *Biochem. J.* **52,** 200 (1952).
(146) L. Lorand and W. R. Middlebrook, *Biochem. J.* **52,** 196 (1952).
(147) L. Lorand and W. R. Middlebrook, *Biochim. et Biophys. Acta* **9,** 581 (1952).
(148) K. Laki, *Federation Proc.* **12,** 471 (1953).

acid; otherwise the general picture seems to be similar to that in the bovine material.[149,150]

Whatever the details may eventually turn out to be, we are clearly concerned in these reactions with an "unmasking" process of the type which has been made familiar in the chymotrypsin and pepsin systems and also in ovalbumin; thus

$$\text{pepsinogen} \xrightarrow{\text{pepsin}} \text{pepsin} + \text{peptide material}$$

$$\text{ovalbumin} \xrightarrow{\textit{Bacillus subtilis}} \text{plakalbumin} + \text{peptide material}$$

$$\text{fibrinogen} \xrightarrow{\text{thrombin}} \text{fibrin} + \text{peptide material}$$

It is still not clear whether the task of thrombin is completed when the peptides have been removed, i.e., whether the "unmasked" fibrinogen units then polymerize spontaneously. The results of experiments with urea (see below, p. 881) make this very probable, however.

The next question which arises concerns the nature of the links involved in binding together the units of fibrin. Here we are on much less certain ground. It was at one time suggested that —S—S— bridges were primarily responsible. It is now clear, however, that the oxidation of S—H to S—S is not essential: pure fibrinogen contains no free S—H groups, and normal clotting can take place in the presence of ferricyanide.[151–153] Nevertheless S—H groups do appear to be involved in some secondary way, as the following results show. The fibrin produced from purified fibrinogen is soluble in urea[154–158] and in dilute acids and alkalies;[157] but clots formed in the presence of calcium ions and a serum factor (of unknown nature) are insoluble in these reagents. If now an —SH reagent such as mercuric chloride or iodoacetamide is also added to this reaction mixture, the clot is again soluble.[158] We thus arrive at the notion of two types of clot; the urea-soluble type, in which —S—S— bridges are not involved, and the urea-insoluble type, in which —S—S— bridges reinforce the normal binding.

(149) L. Lorand and W. R. Middlebrook, *Science* **118,** 515 (1953).
(150) K. Bailey, personal communication.
(151) L. B. Jaques *in* Blood Clotting & Allied Problems, Transactions of the First Conference of the Josiah Macy, Jr. Foundation, New York, 1948, p. 58.
(152) D. Bagdy, F. Guba, L. Lorand, and E. Mihalyi, *Hung. Acta Physiol.* **1,** 197 (1948).
(153) J. T. Edsall and W. F. Lever, *J. Biol. Chem.* **191,** 735 (1951).
(154) L. Lorand, *Nature* **166,** 694 (1950).
(155) L. Lorand, *Hung. Acta Physiol.* **1,** 6 (1948).
(156) K. Laki and L. Lorand, *Science* **108,** 280 (1948).
(157) K. C. Robbins, *Am. J. Physiol.* **142,** 581 (1944).
(158) A. G. Loewy, *quoted in* Edsall & Lever.[153]

Little is known about the forces of association concerned in the formation of urea-soluble clots. The retarding effect of high salt concentration[159] shows that electrostatic forces must be involved: attention has mainly been focused on hydrogen-bonding as the prime agent because of the striking effects of urea, which is of course known to act mainly by saturating sites which might otherwise form hydrogen bonds. It is significant that the solutions of fibrin in urea contain particles which are indistinguishable in size and shape from those of the original fibrinogen.[160,161] Furthermore the material dissolved in urea can be reclotted merely by dialyzing out the urea:[162,163] this process is fully reversible and closely resembles the original clotting in that its course can be modified similarly by altering the pH and by adding various reagents (see below). Reclotted fibrin closely resembles the original material in electron micrographs.

Much effort has been devoted to studies of the effects of various reagents on the clotting process (for a recent summary, see Shulman),[164] as well as of the kinetics of the process (Waugh and Livingstone).[165] The simplest effects are those of pH and ionic strength, which appear to be related in a fairly simple way to the electrical charge of the particles—rate of clotting is diminished at high ionic strength, and is maximal at pH 7 but decreases on either side of this value until there is complete inhibition at pH's 5.3 and 10.[153,166] Substances which increase clotting time may be classified into retarders and inhibitors,[164] which apparently differ qualitatively as well as quantitatively. These materials include alcohols,[167] amides, cations, anions, and dipolar ions; particularly striking effects have been noted with urea, guanidine, cystine, and glycols. Under the influence of certain reagents there is evidence of definite intermediates in the clotting process.[160,168–170] In acid inhibition of clotting it appears[160,170,143,168] that there is an intermediate very similar to or identical with that obtained by complete depolymerization of fibrin with urea;

(159) W. F. H. M. Mommaerts, *J. Gen. Physiol.* **29,** 103, 113 (1945).
(160) P. Ehrlich, S. Shulman, and J. D. Ferry, *J. Am. Chem. Soc.* **74,** 2258 (1952).
(161) F. R. Steiner and K. Laki, *J. Am. Chem. Soc.* **73,** 882 (1951); *idem, Arch. Biochem. Biophys.* **34,** 24 (1951).
(162) F. R. Steiner, *Science* **114,** 460 (1951).
(163) E. Mihalyi, *Acta Chem. Scand.* **4,** 334, 351 (1950).
(164) S. Shulman, *Discussions Faraday Soc.* No. **13,** 109 (1953).
(165) D. F. Waugh and B. J. Livingstone, *Science* **113,** 121 (1951).
(166) S. Shulman and J. D. Ferry, *J. Phys. & Colloid Chem.* **54,** 66 (1950).
(167) J. D. Ferry and S. Shulman, *J. Am. Chem. Soc.* **71,** 3198 (1949).
(168) K. Laki and W. F. H. M. Mommaerts, *Nature* **156,** 664 (1945).
(169) K. Laki, *Arch. Biochem. Biophys.* **32,** 317 (1951).
(170) J. D. Ferry, S. Shulman, and J. F. Foster, *Arch. Biochem. Biophys.* **39,** 387 (1952).

this may provisionally be identified with "unmasked" fibrinogen. In systems containing moderate concentrations of urea or hexamethylene glycol, another type of intermediate has been detected in the ultracentrifuge: it has about twice the cross section and ten times the length of fibrinogen, but exists in a range of sizes although in the ultracentrifuge only a single peak is observed.[160] The truth is that we still know relatively little of the processes intermediate between the removal of the peptide material and the appearance of the clot: nor can the very plentiful experimental results yet be interpreted in structural terms. The following scheme has been proposed by Ehrlich *et al.*[160] as a preliminary flow sheet:

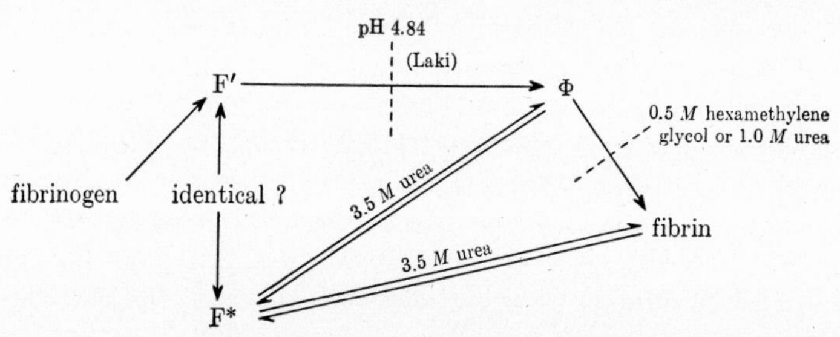

(F′, acid intermediate; Φ, intermediate found in urea or hexamethylene glycol; F*, product of action of urea on fibrin).

c. *Fibrin*

The fine structure of fibrin has been studied in the electron microscope by Wolpers and Ruska,[171,172] by Hawn and Porter,[173] and by Hall.[137,174] It consists of fibrils (apparently identical *in vitro* and *in vivo*) cross-striated with an average periodicity of 230 A. The periodicity is complex, consisting of an alternation of broad and narrow stain-absorbing bands separated by nonabsorbing regions: the bands themselves have a fine structure (see Fig. 7). Under high resolution the fibrils appear to be made up of particles of diameter 30–50 A. There is no definite evidence as to the manner in which fibrinogen filaments associate to form fibrin fibrils.

The structure of the fibrin gel, whose protein content may be as low as 0.004 per cent,[175] has been much discussed (for a review, see Ferry).[176]

(171) C. H. Wolpers and H. Ruska, *Klin. Wochschr.* **18,** 1077, 1111 (1939).
(172) C. H. Wolpers, *Klin. Wochschr.* **24–25,** 424 (1947).
(173) C. van Z. Hawn and K. Porter, *J. Exptl. Med.* **86,** 285 (1947); *ibid.* **90,** 225 (1949).
(174) C. E. Hall, *J. Am. Chem. Soc.* **71,** 1138 (1949).
(175) J. D. Ferry and P. R. Morrison, *J. Am. Chem. Soc.* **69,** 388 (1947).
(176) J. D. Ferry, *Advances in Protein Chem.* **4,** 1 (1948).

There are two main types of clot:[175,176] the fine, translucent variety, and a coarse, opaque type. The former is obtained under conditions such that molecular interaction is minimal (e.g., at high pH); it is elastic and friable, whereas coarse clots are nonfriable and plastic. It has been plausibly suggested that in coarse clots there is much side-to-side aggregation of fibrils to form bundles, while in fine clots the tangled fibrils

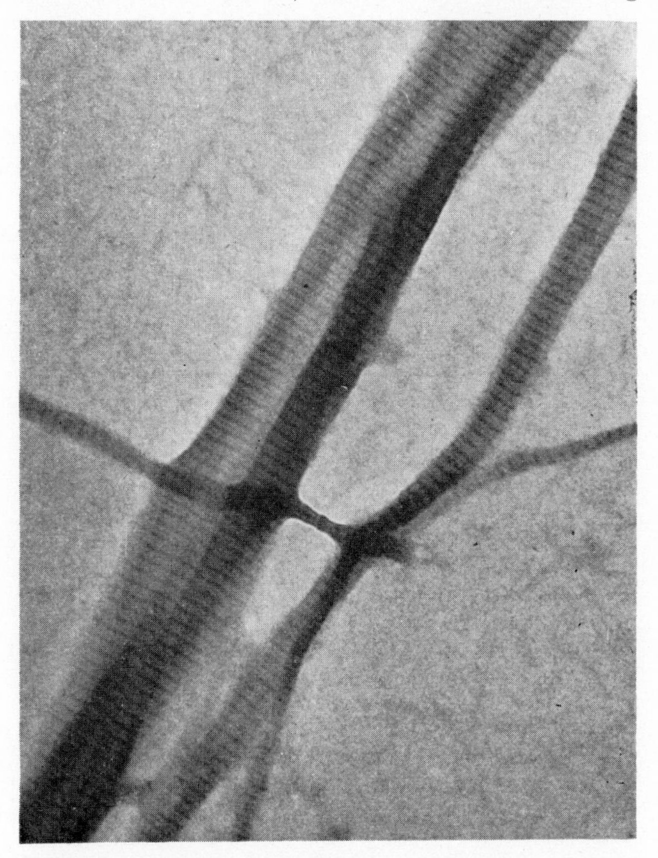

FIG. 7. Electron micrograph of bovine fibrin, stained with phosphotungstic acid: magnification 80,000×. After C. E. Hall.[137]

maintain their individuality (see Fig. 8). The striking feature of these systems is, of course, the very low concentrations of protein required to give a coherent gel. One is driven to assume the presence of a tenuous three-dimensional framework of rather rigid filamentous objects, connected together at rare intervals: this is the essence of the scheme proposed for fine clots (Fig. 8*a*). In the coarse clots the model of Fig. 8*b* contains larger inhomogeneities which would explain their higher opacity;

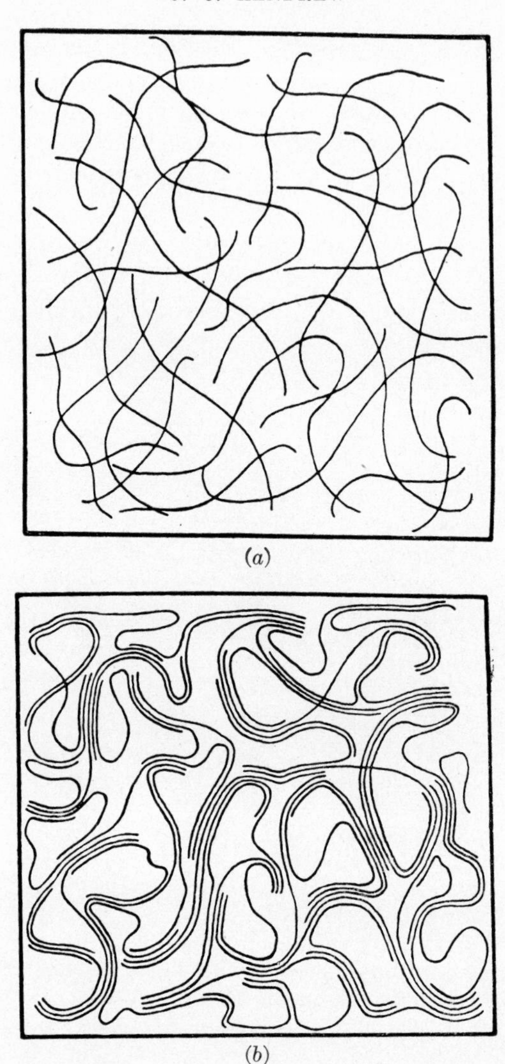

FIG. 8. Schematic illustrations of structures proposed for fibrin clots: (a) fine clot, (b) coarse clot. After J. D. Ferry and P. R. Morrison.[175]

thick strands can actually be observed with the electron microscope in coarse clots,[171] but not in fine clots.

Fibrin has found important uses[177,178] in the form of fibrin foam—used in hemostasis—and as fibrin film, used as a dural substitute in surgery and for other purposes.

(177) J. D. Ferry and P. R. Morrison, *J. Clin. Invest.* **23,** 566 (1944).
(178) E. A. Bering, *J. Clin. Invest.* **23,** 586 (1944).

4. Other Members of the Keratin Group

a. Introduction

In this section we shall briefly consider less well-known proteins which give the α-pattern. Some of these, customarily included under the name "keratin," would probably be found on close study to differ sufficiently from wool keratin to deserve specific names. As it is, the name keratin is generally taken to include substances as widely different as hair, wool, horn, toe- and finger-nails, feathers, fish scales, etc. The distribution of such materials has been discussed by Rudall.[179]

An early classification of the keratins into *eukeratins* and *pseudokeratins* was based on the greater resistance to enzymes of the former and on differences between their amino acid compositions.[180,181] Eukeratins included hair, horn, feather, etc.; in their amino acid make-up the ratios His:Lys:Arg have the characteristic values 1:4:12. Pseudokeratins, including epidermin, nerve proteins, ovokeratin of hen's egg, and whalebone, have widely different values of these ratios.[182] The His:Lys:Arg ratios of eukeratins are empirical constants: no significance has been proposed for them.

Two members of the group deserve separate treatment.

b. Epidermin

The deeper region of mammalian skin, known as the cutis, contains connective tissue elements made up of collagen (see p. 909); above it lies the epidermis whose main protein constituent is epidermin, a member of the keratin group. Space prevents us from discussing this important material in detail: for a more extended treatment and bibliography, reference should be made to the recent review by Rudall.[183]

Like other keratins, epidermin is not a chemical individual. In transverse sections of skin, several zones of the epidermis can be distinguished: below is the *stratum mucosum*, containing actively dividing cells, while above is the *stratum corneum*, consisting mainly of dead and keratinized cells. The lower zone has been shown to be rich in free —SH groups and poor in —S—S— bridges; in the upper zone the reverse is the case.[183–186] The suggestion is that "keratinization," a process occurring subsequent to the initial deposition of the protein, is essentially one of

(179) K. M. Rudall, *Biochim. et Biophys. Acta* **1,** 549 (1947).
(180) R. J. Block and H. B. Vickery, *J. Biol. Chem.* **93,** 113 (1931).
(181) R. J. Block, *J. Biol. Chem.* **121,** 761 (1937).
(182) R. J. Block, *Ann. N. Y. Acad. Sci.* **53,** 608 (1951).
(183) K. M. Rudall, *Advances in Protein Chem.* **7,** 253 (1952).
(184) A. Giroud, H. Bulliard, and C. P. Leblond, *Bull. histol. appl. physiol. et path. et tech. microscop.* **11,** 129 (1934).
(185) A. Giroud and H. Bulliard, *Arch. Anat. microscop.* **31,** 271 (1935).
(186) M. Chèvremont and J. Frederic, *Arch. biol.* (Liége) **54,** 589 (1943).

cross-linking. Insofar as the amino acid figures for epidermin have any meaning at all, it would appear that, compared with wool keratin, its main features are much lower arginine and cystine contents, and a much higher methionine content.

Mechanically, epidermin is much less stable than wool. Thus, on placing it in water at 70–85°C., a 20 per cent contraction occurs; it is significant that the exact transformation temperature depends on the origin of the protein, rising by almost 20°C. in passing from mucosal to corneal protein. The elastic behavior has been studied in detail by Rudall.[183,187]

Epidermin is insoluble in neutral salt solutions, but almost completely soluble in 6 *M* urea (cf. fibrin). Extraction by urea has been much used as a method of isolating epidermin; but it is clear that the product is by no means homogeneous. Measurements of sedimentation and diffusion constants show[188] that the main component has a molecular weight of 60,000 and frictional ratio 3.5, which indicates that the particles are very asymmetric. Since they are isolated by solution in urea, the subunits are presumably linked together in epidermin by hydrogen bonds rather than by —S—S— bridges; but it is interesting to note that on adding bisulfite to reduce —S—S— bridges, further disaggregation occurs and a more homogeneous solution is obtained, containing predominantly particles of lower molecular weight.

The x-ray pattern normally given by epidermin, either in its native state or as a dried-down extract, is the typical α-pattern. Rather strangely the transformation to the β-form is difficult to achieve; simple stretching, even in great degree, has little effect on the pattern, presumably because for some reason the chains very readily slip past one another. The β-pattern can, however, be produced by stretching in ethanol or ammonium sulfate,[183] though why these reagents should inhibit the slipping which normally takes place is obscure (see Fig. 4*b*).

We meet a new phenomenon in following the x-ray pattern of thermally contracting epidermin, either native or extracted. A β-pattern is developed on contraction, with the unusual feature that the normally equatorial strong reflection at 4.65 A. is found on the meridian: as if, in other words, the polypeptide chain were oriented perpendicular to the fiber axis instead of parallel to it (see Fig. 4*c*). This pattern, known as the cross β-pattern,[183,74] can also be observed after isometric thermal treatment: the essential conditions for its appearance appear to be high temperature and humidity. The change can be reversed: fibers showing the cross β-pattern are reconverted to α- by saturated urea solutions.[183]

(187) K. M. Rudall, *Proc. Roy. Soc.* (London) **B141,** 39 (1953).
(188) E. H. Mercer and B. Olofsson, *J. Polymer Sci.* **6,** 261 (1951).

Cross β-patterns, though studied in most detail for epidermin, have been encountered elsewhere; the first to be noted was obtained by Astbury, Dickinson, and Bailey,[189] by stretching films of heat-denatured ovalbumin. More recently Mercer[124] obtained a cross β-pattern from wool after treatment with urea and reducer followed by extension of the fibers. We shall touch briefly on the structure of the cross β-form in a later section (see p. 904).

The infrared spectrum of epidermin has been studied[183] with results closely analogous to those obtained with wool keratin. In the cross β-form the C=O stretching frequency is, in contrast to normal β-structures, oriented parallel to the fiber axis.

c. Bacterial Flagella

Following the discovery by Gard[190] that the flagella of certain bacteria could be detached from their parent cells and separated from them by centrifugation, a study has been made by Weibull[191–194] of isolated flagella of *Proteus vulgaris* and of *Bacillus subtilis*. He has shown that suspensions of flagella can be treated just as if each flagellum were a very asymmetric protein molecule of high molecular weight; they can be reversibly precipitated by ammonium sulfate, behave normally in electrophoresis, and have an invarient chemical composition typical of a protein. In acid solution they decompose to smaller particles of molecular weight of the order 41,000 (this is a lower limit).

Electron micrographs of detached flagella by Astbury and Weibull[195] showed that their diameters were not much greater than 100 A. More recently, very well-resolved pictures have been taken by Starr and Williams[196] of flagella from motile diphtheroid bacteria. These have a diameter of 190 A. and can clearly be seen to consist of a triple-thread helix with axial period about 500 A.; the helix is always left-handed (see Fig. 9). This is probably the first instance where well-defined internal structure has been directly observed in a molecule, or organ (it is still not clear which term is more appropriate) of such small dimensions; and it is particularly interesting in connection with recent suggestions (which we shall discuss below) that fibrous proteins have a helical structure. X-ray diffraction studies of flagella have been made by Astbury and

(189) W. T. Astbury, S. Dickinson, and K. Bailey, *Biochem. J.* **29,** 2351 (1935).
(190) S. Gard, *Arkiv Kemi, Mineral. Geol.* **19A,** No. 21 (1944).
(191) C. Weibull, *Biochem. et Biophys. Acta* **2,** 351 (1948).
(192) C. Weibull, *Acta Chem. Scand.* **4,** 260 (1950).
(193) C. Weibull, *Acta Chem. Scand.* **4,** 268 (1950).
(194) C. Weibull, *Biochim. et Biophys. Acta* **3,** 378 (1949).
(195) W. T. Astbury and C. Weibull, *Nature* **163,** 280 (1949).
(196) M. P. Starr and R. C. Williams, *J. Bact.* **63,** 701 (1952).

Weibull.[195,197] They found that films or fibers made of oriented flagella gave the α-keratin pattern, and on squeezing they were able to achieve a partial conversion to the β-form. Hence the material of the flagella—or the flagellum itself—must be classified in the k-m-e-f group. Ast-

FIG. 9. Electron micrograph of flagellum of Congo diphtheroid bacterium, shadowed with uranium: magnification 100,000×. After M. P. Starr and R. C. Williams.[196]

bury[197a] has found meridional long-spacings very probably corresponding to a period of 410 A., and thus resembling those of muscle and of F-actin (though the distributions of intensities between the orders are not identical).

In its amino acid composition, and particularly in the absence of cystine, flagellar protein resembles myosin. This raises the interesting

(197) C. Weibull, *Nature* **165**, 482 (1950).
(197*a*) W. T. Astbury, personal communication.

question whether a bacterial flagellum is to be regarded as a primitive muscle, or, in view of its small diameter, even (to quote Astbury)[198] as a unimolecular muscle. First it is necessary to be certain that the flagella are active, motile organs; and on this subject there is still active controversy.[199,200] If we provisionally assume that the flagella are really motile, we find ourselves faced with a whole series of formidable problems. What is the mechanism of the contraction, and how is it transmitted down the length of the flagellum?[201,198] What is the source of energy? How are the movements coordinated? About none of these topics have we any information at all.

In conclusion we may note that the term flagellum may be used to describe widely differing organs. Astbury and Saha[202] have recently examined algal flagella. These appear to be made of protein, but are much thicker than the bacterial flagella described above, and give a new type of x-ray diagram which, insofar as it is reminiscent of anything previously described, resembles that of β-protein. These flagella are still in a preliminary stage of investigation, but it is clear that they greatly differ from the bacterial variety.

5. Synthetic Polypeptides

The main concern of this work is the proteins; it will nevertheless be convenient to digress briefly at this point to mention some physical investigations which have been made of the synthetic polypeptides, because of the relevance of these investigations to a discussion of the molecular structures of the k-m-e-f group of proteins.

Many methods of synthesizing poly-α-amino acids are now known: the recent review by Katchalski[203] may be consulted for details of these and of the chemical properties of the resulting polymers. In many of their properties they resemble fibrous proteins, and as models of proteins they have the advantage of simplicity of composition—thus we can be certain that, if we so desire it, all the side chains are identical. In many instances it has been possible to prepare oriented films or fibers which can be studied by means of x-ray diffraction or infrared absorption with polarized radiation, just as can the structure proteins which we have been discussing.

Early x-ray studies were those of Meyer and Go in 1934,[204] of Miller,

(198) W. T. Astbury, *Sci. American* **184,** 21 (1951).
(199) A. Pijper, *Nature* **168,** 749 (1951).
(200) C. Weibull, *Nature* **168,** 750 (1951).
(201) C. Weibull, *Nature* **167,** 511 (1951).
(202) W. T. Astbury and N. N. Saha, *Nature* **171,** 280 (1953).
(203) E. Katchalski, *Advances in Protein Chem.* **6,** 123 (1951).
(204) K. H. Meyer and Y. Go, *Helv. Chim. Acta* **17,** 1488 (1934)

Fankuchen, and Mark,[205] of Astbury *et al.*,[33] and of Brown, Coleman and Farthing.[206] At that stage only disoriented or poorly oriented specimens were available, but it was already clear that there were significant resemblances to the k-m-e-f group. For example, the copolymer of DL-β-phenylalanine and L-leucine gave a recognizable α-pattern, with diffraction maxima at 5.2 and 11.7 A.[206] Most materials however gave a β-pattern or something similar to it: e.g., poly-DL-isoleucine.[33]

By far the largest contribution to our knowledge of the x-ray patterns of synthetic polypeptides comes from the work of Bamford, Hanby, Happey, Brown, Trotter, and their colleagues.[207–210a] This group, besides greatly improving methods of specimen orientation and hence the quality of the x-ray photographs, made the important discovery that in some instances the same polymer could be prepared in either α- or β-form: the first by deposition from *m*-cresol, the second from formic acid. Further, the β-form could be induced to fold at least partially to the α-form by subsequent treatment with *m*-cresol.[208] It is interesting that formic acid alone of all the solvents investigated is capable of inducing the formation of β-polypeptide; the reason for this specificity is not understood.[211] Fairly clearly, however, the effect of formic acid must be due to a preferential hydrogen-bonding onto the NH groups of the peptide link, which are thus inhibited from forming intrachain hydrogen-bonds with peptide CO groups, so that the chains crystallize out in the β- instead of the α-form.

Examples of the x-ray photographs obtainable are shown in Figs. 10 and 11. By far the best pictures so far obtained are those of poly-L-alanine and poly-γ-methyl-L-glutamate, in the α-form. A detailed table of reflections from the latter material is given by Bamford *et al.*;[209] further data are given by Yakel, Pauling, and Corey.[212] The resemblance of the pattern to that of α-keratin is obvious; the close relationship was emphasized by the discovery by Perutz that the synthetic fibers also

(205) E. Miller, I. Fankuchen, and H. Mark, *Research* (London) **1,** 720 (1948).
(206) C. J. Brown, D. Coleman, and A. C. Farthing, *Nature* **163,** 834 (1949).
(207) W. T. Astbury, *Nature* **163,** 722 (1949).
(208) C. H. Bamford, W. E. Hanby, and F. Happey, *Nature* **164,** 751 (1949).
(209) C. H. Bamford, L. Brown, A. Elliott, W. E. Hanby, and I. F. Trotter, *Nature* **169,** 357 (1952).
(210) C. H. Bamford, L. Brown, A. Elliott, W. E. Hanby, and I. F. Trotter, *Proc. Roy. Soc.* (London) **B141,** 49 (1953).
(210*a*) C. H. Bamford, L. Brown, A. Elliott, W. E. Hanby, and I. F. Trotter, *Nature* **173,** 27 (1954).
(211) C. H. Bamford, W. E. Hanby, and F. Happey, *Proc. Roy. Soc.* (London) **A205,** 30 (1951).
(212) H. L. Yakel, L. Pauling, and R. B. Corey, *Nature* **169,** 920 (1952).

give a strong 1.5-A. reflection[84] (on the other hand in β-polypeptides there is a strong wide-angle reflection at about 1.14 A.).[23]

In parallel with the x-ray work, a close study has been made of the infrared spectra of synthetic polypeptides. Here, too, the resemblance to protein fibers of the k-m-e-f group is striking.[33,213] Elliott, Ambrose, Hanby, and others have published detailed results.[210,92,214–217] These have been summarized above in Table IV (see also Fig. 5, Vol. I, part A, p. 424). As in the proteins, the C=O and N—H bonds are oriented

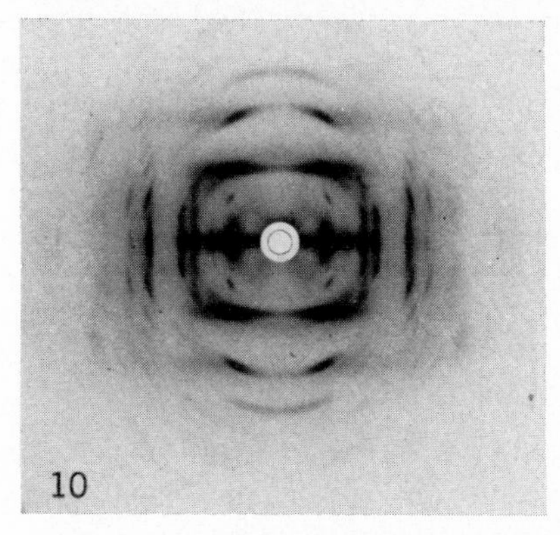

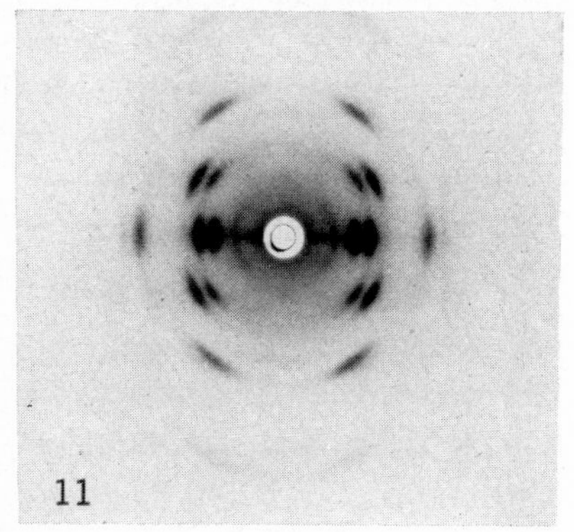

FIG. 10. X-ray pattern of poly-L-alanine, α-form; fiber axis vertical. After Bamford *et al.*[210a]

FIG. 11. X-ray pattern of poly-L-alanine, β-form; fiber axis vertical. After Bamford *et al.*[210a]

more or less parallel to the fiber axis in the α-form, and perpendicular to it in the β-form. The N—H stretching frequency at 3305 $cm.^{-1}$ may show a dichroic ratio as high as 14:1 in α-structures.[216] It has also been possible to establish certain changes in frequency consequent on a change of configuration: thus the C=O stretching frequency is about 1659 $cm.^{-1}$ in α- and 1629 $cm.^{-1}$ in β-structures, while the N—H deformation frequency which is 1527 $cm.^{-1}$ in α-forms shifts to 1550 $cm.^{-1}$ in β-forms. This frequency shift has been used as a criterion to decide whether an unknown structure has the α- or the β-configuration, or a mixture of both.[218] A recent paper by Ehrlich and Sutherland[219] suggests that

(213) S. E. Darmon and G. B. B. M. Sutherland, *J. Am. Chem. Soc.* **69,** 2074 (1947).
(214) A. Elliott and E. J. Ambrose, *Discussions Faraday Soc.* No. **9,** 246 (1950).
(215) A. Elliott and E. J. Ambrose, *Nature* **165,** 921 (1950).
(216) E. J. Ambrose and A. Elliott, *Proc. Roy. Soc.* (London) **205,** 47 (1951).
(217) C. H. Bamford, W. E. Hanby, and F. Happey, *Proc. Roy. Soc.* (London) **A206,** 407 (1951).
(218) E. J. Ambrose and A. Elliott, *Proc. Roy. Soc.* (London) **A208,** 75 (1951).
(219) G. Ehrlich and G. B. B. M. Sutherland, *Nature* **172,** 671 (1953).

caution should be exercised in applying it, at any rate to proteins, where a great variety of side chains is present; absorption in the 1500–1600 cm.$^{-1}$ region is sensitive to the state of ionization of carboxyl groups. But indeed we are still so ignorant of the origins and mechanisms of most of the absorptions which make up the highly complicated infrared spectra of proteins that caution must be a general rule in interpreting them.

It is very significant that in dilute solution (in nonpolar solvent) an α-polypeptide such as poly-DL-phenylalanine shows only those bands characteristic of hydrogen-bonded N—H (3305 cm.$^{-1}$); the band characteristic of unbonded N—H (at 3460 cm.$^{-1}$) is completely absent. On the other hand in polymers containing C=O and N—H groups in the side chain (e.g., polycarbobenzoxy-DL-lysine) the unbonded N—H frequency is readily visible under the same conditions. The conclusion would appear to be that in the α-form all the N—H groups of the peptide links are completely bound by hydrogen bonds.

6. The Structure of the Keratins

Controversies have raged for the last 20 years about the topics now to be discussed, and today, though the air shows some signs of clearing, the dust of battle is still rather thick and in some directions quite impenetrable. Here we shall not attempt to catalog all theories of structure which have at some time been seriously proposed, but merely those which either have made a substantial contribution to the development of ideas in the subject or are still to be regarded as being "in the running." The problems to be solved fall naturally into two classes: the configuration of the peptide chain itself, and the disposition of chains into larger units, if any can be shown to exist. Much more is known about the first topic than the second. In this account we shall first consider the chain configuration in β-proteins, then that in α-proteins, and finally we shall devote some space to theories of the larger-scale structure of members of the k-m-e-f group. By adopting this order we shall proceed by degrees from problems which are relatively better understood to fields where speculation has still not begun to harden into established fact. Our discussion will to some extent follow the same lines as that in Chap. 4, Vol. I, part A, where similar topics are considered in a wider context.

a. The Configuration of the Polypeptide Chain in β-Keratin

In an earlier section (p. 864) we have indicated that the principal features of the x-ray pattern of stretched hair, or β-keratin, are two equatorial reflections at 9.7 and 4.65 A., and a meridional reflection at 3.33 A. Astbury and Street[66] noted the close similarity between this pattern and that of silk, with its strong equatorial reflections at 9.2 and

4.6–4.3 A. and meridional reflection at 3.5 A. They adopted essentially the same interpretation as Meyer and Mark had done for silk, namely that the chain was in a more or less stretched configuration, with a basic axial repeat of 6.7 A. corresponding to two amino acid residues; to explain the fact that this axial repeat is slightly less than that of silk (6.95 A.) they supposed that in β-keratin the chains were less completely stretched. The equatorial reflection at 4.65 A. was interpreted as a "backbone" spacing corresponding to the distance between two neighboring chains, and the reflection at 9.7 A. as a "side-chain" spacing, also between chains, but in a direction at right angles to the backbone spacing, and such that side chains intervene between the two main chains. The arrangement is made clear by Fig. 12, p. 257, Vol. I, part A; the main chains lie in sheets with side chains projecting up and down from them; the backbone spacing is that between chains of a single sheet, and the side-chain spacing is the distance between neighboring sheets. This scheme implies that in a doubly oriented specimen the two equatorial spacings should be at right angles; this was shown to be the case by Astbury and Sisson[220] who achieved double orientation by squeezing keratin fibers laterally. Within sheets the chains are held together by hydrogen bonds between NH and CO groups of peptide bonds.

In their general outline these early proposals of Astbury and Woods have been accepted without serious criticism since they were first put forward, and today there seems no doubt of their essential correctness. At various times slight variants of them have been proposed, e.g., by Huggins[221] and others. Experimentally, confirmation has come from studies of infrared dichroism (see p. 870) which have shown that in β-keratins the peptide C=O and N—H bonds lie perpendicular to the fiber axis and that they are probably all hydrogen-bonded.

Recently the picture has been modified in detail, though not in its general nature, as a result of the work of Pauling, Corey, and their collaborators. A large-scale program of studies of the crystal structures of amino acids and simple peptides, extending over many years, has enabled this school of workers to formulate a definitive set of interatomic distances and angles within the peptide chain[25,222–224] (see p. 360 ff. and Fig. 56, p. 369, Vol. I, part A). These dimensions have been used to make a study of possible configurations for polypeptide chains, with the following requirements as limiting conditions: (*a*) that each amide group

(220) W. T. Astbury and W. A. Sisson, *Proc. Roy. Soc.* (London) **A150,** 533 (1935).
(221) M. L. Huggins, *Chem. Revs.* **32,** 195 (1943).
(222) R. B. Corey, *Chem. Revs.* **26,** 227 (1940).
(223) R. B. Corey, *Advances in Protein Chem.* **4,** 385 (1948).
(224) R. B. Corey and L. Pauling, *Proc. Roy. Soc.* (London) **B141,** 10 (1953).

in the peptide bond must be virtually planar, owing to the nearly 50 per cent double-bond character of the CO—NH bond; either a *cis*-configuration C=N (with O and H) or a *trans*-configuration C=N (with O and H) are considered to be acceptable; (*b*) that each CO and NH group is involved in the formation of a hydrogen bond with the N—H...O atoms nearly linear and the N—O distance about 2.8 A.; and, lastly but less strictly, (*c*) that the orientation of groups about the single bonds NH—CHR and CHR—CO

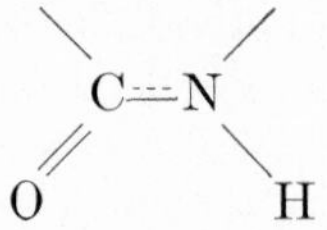

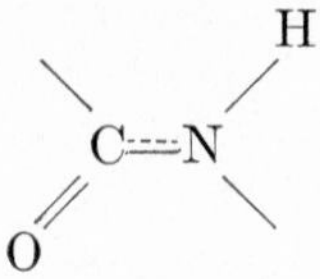

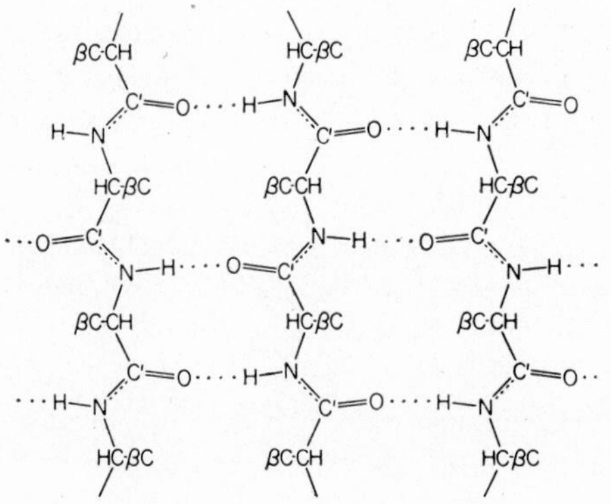

FIG. 12. Diagrammatic representation of the antiparallel-chain pleated sheet. After Pauling and Corey.[226]

should correspond to the positions of maximum stability found in studies of simple organic molecules with Raman spectra and by other means.[225–227] Pauling and Corey have built models of a large number of polypeptide chain configurations which satisfy their criteria, and have assigned some of them to various types of protein which we shall discuss later. Among them are three configurations which they call pleated sheets and which they discuss in connection with silk and the β-keratins.[226–228] They point out that the *fully* extended structure of plane sheets proposed by Astbury is unacceptable if the canonical interatomic distances and angles are to be preserved, owing to steric hindrance between neighboring side chains; the fully extended configuration would only be possible for poly-

(225) L. Pauling, R. B. Corey, and H. R. Branson, *Proc. Natl. Acad. Sci. U. S.* **37,** 205 (1951).
(226) L. Pauling and R. B. Corey, *Proc. Natl. Acad. Sci. U. S.* **37,** 729 (1951).
(227) L. Pauling and R. B. Corey, *Proc. Roy. Soc.* (London) **B141,** 21 (1953).
(228) L. Pauling and R. B. Corey, *Proc. Natl. Acad. Sci. U. S.* **37,** 251 (1951).

glycine. The first pleated sheet structure of Pauling and Corey, known as the polar pleated sheet,[228] has now been abandoned: but the other two, the antiparallel-chain and the parallel-chain pleated sheets,[226,227] are believed by Pauling and Corey to be the basis of the structures of β-keratins, or silk, and of synthetic polypeptides in the β-form (four other sheet configurations have also been described by the same authors,[229,230] but are not believed to occur in proteins). Figures 12 and 13 represent these structures schematically, while perspective drawings of models have been given on p. 388 of Vol. I, part A. The configuration of individual chains is identical in the two structures, and has the property that chains possessing it may be assembled either in antiparallel or in parallel array, acceptable interchain hydrogen bonds being formed in either case;

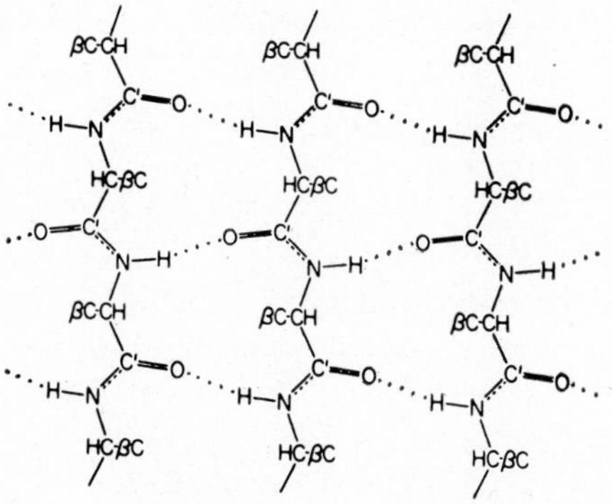

FIG. 13. Diagrammatic representation of the parallel-chain pleated sheet. After Pauling and Corey.[226]

indeed, mixed antiparallel and parallel structures would appear to be possible. In the parallel-chain version the axial repeat is 7.00 A., corresponding closely to the observed value for silk; in the antiparallel-chain sheet the repeat is 6.50 A.[230] Detailed proposals for the structures of the different β-proteins have not yet been published; the suggestion made is that the structures of silk and of polyglycine are based on the antiparallel sheet, while that of β-keratin is a form of parallel sheet. In the case of silk it has been proposed that alternate residues are glycines and that the structure is in the form of double sheets, all the larger side chains being inside the "sandwich" and the hydrogen atoms of glycine on the outside; adjacent "sandwiches" would fit together like the sheets in polyglycine.[227] Pending detailed specification of the proposed structures it is not possible to say that they are proved correct in detail, but it seems rather probable

(229) L. Pauling and R. B. Corey, *Proc. Natl. Acad. Sci. U. S.* **39,** 247 (1953).
(230) L. Pauling and R. B. Corey, *Proc. Natl. Acad. Sci. U. S.* **39,** 253 (1953).

that the actual structures are of the general type suggested. Some criticism of the proposals has been made in detail by Bamford *et al.*[23]

To conclude our discussion of β-structures we should refer to feather keratin, whose x-ray pattern in many ways resembles that of wool in the β-form, though the principal meridional reflection has a definitely shorter spacing (3.1 A.). Feather keratin has generally been classed as a member of the β-keratin family: nevertheless there are difficulties about such an interpretation. Thus it does not yield a meridional reflection at 1.1 A. as do silk and β-keratin;[23] again, its infrared spectrum suggests that a substantial α-component is present (see p. 870). Pauling and Corey[231] proposed a structure involving layers of pleated sheets interleaving double layers of α-helices, but this structure does not satisfy all the experimental data, and other solutions have been formulated by the same authors involving complex arrangements of α-helices alone.[227,232] Even the experimental facts are disputed; some have claimed,[227] but others have denied,[23,84] that feather keratin gives the 1.5-A. reflection characteristic of α-structures.

b. *The Configuration of the Polypeptide Chain in α-Keratin*

Discussion of the structure of proteins has centered about this topic, and a very large number of proposals has been made to account for the facts observed. We have indicated that in β-keratin the chains have a nearly stretched configuration: in α-keratin the problem is to find a more folded configuration which will satisfactorily account for, *inter alia:*

(*a*) *The x-ray diffraction pattern and infrared spectrum of α-keratin.*

(*b*) *The mechanics of the α-β transformation.* In the past there has been a good deal of discussion regarding the precise degree of extension required to complete the transformation, and at a very early stage Astbury and Street,[66] for example, claimed than an almost exact doubling of length was involved. As time has passed, however, it has become more and more difficult to draw precise conclusions: partly because it emerged that keratin was far from being a homogeneous substance, either at the histological level (p. 860) or at the chemical level (p. 872), partly because every specimen contains a high proportion of amorphous material, and partly because in both unstretched and stretched states hair yields an infrared spectrum showing mixtures of α- and β-components (for observations on this point, together with a general discussion, see Elliott).[81] New data on the extensibility of k-m-e-f proteins have recently been published by Rudall.[187]

(231) L. Pauling and R. B. Corey, *Proc. Natl. Acad. Sci. U. S.* **37**, 256 (1951).

(232) L. Pauling and R. B. Corey, *in* Les protéines, Institut International de Chimie Solvay, Brussels, 1953, p. 63.

(*c*) *The densities of α- and β-keratin.* α-Keratin and β-keratin both have the same density, viz., about 1.3; hence in broad terms the structures must be close-packed to the same degree in both configurations. There has been much discussion of the precise densities to be expected for various models; this discussion has been rather unprofitable for the same reasons as have just been mentioned in connection with the degree of extensibility.

It has almost universally been supposed that an explanation of the mechanical behavior must be sought in terms of a reversible change from one specific molecular configuration to another. However, as we have mentioned above (p. 863), one important school, that of K. H. Meyer, has maintained that the change in x-ray pattern on extension is irrelevant to the mechanism involved.

(*1*) *The Astbury Structure.* We shall first discuss the Astbury model which occupied the center of the stage for so long. This model, first proposed in 1941[233,234] in supersession of an earlier version,[66,67,220] consists in essence of a series of planar loops held together by hydrogen bonds. The use of hydrogen bonds to maintain the structure in a folded configuration is common to all models later proposed, and is supported by all the newer evidence from infrared spectra. A diagrammatic sketch of the structure is shown in Fig. 14. Its essential features are the following:

(*a*) The unit repeat, which is just greater than 10 A., contains 6 amino acid residues—thus the translation per residue in 10/6 or 1.7 A.

(*b*) Only one-third of the peptide bonds are hydrogen-bonded.

(*c*) Side chains stick out alternately above and below the plane of the loops, and the side chains on either side form close-packed triads (e.g., R, R_5, R_3 in the diagram).

(*d*) Stretching to the α-configuration involves that 6 residues occupying 10 A. in the α-form occupy 6 × 3.33 or 20 A. in the β-form, giving a 2:1 extension.

(*e*) Stretching should not involve rupture of —S—S— bridges between neighboring chains, since these would lie perpendicular to the plane of the fold.

There is a huge literature describing and criticizing this model. Chief among criticisms which have been made of it are the following. First, only one-third of the peptide CO and NH groups form hydrogen bonds; such a structure would be less stable than one in which all CO and NH groups were hydrogen-bonded.[221,225] Secondly, the correlation between degree of extension and the observed change in x-ray pattern cannot be

(233) W. T. Astbury and F. O. Bell, *Nature* **147,** 696 (1941).
(234) W. T. Astbury, *Chemistry & Industry* **60,** 491 (1941).

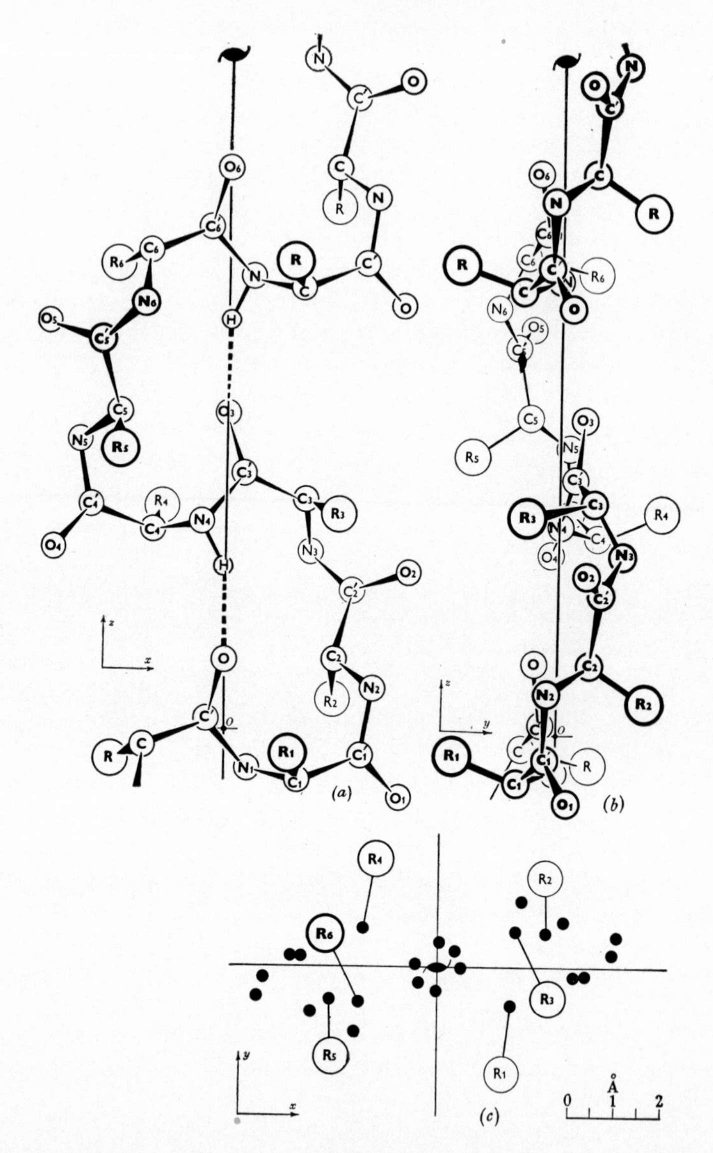

FIG. 14. Diagrammatic representation of Astbury's model for α-keratin; (*a*) and (*b*) side views; (*c*) end-on view. Side chains are marked "R." After Bragg, Kendrew, and Perutz.[241]

made quantitative. Thirdly, the structure does not explain the strong meridional reflection of 1.5 A. (p. 867). Fourthly, the observed infrared dichroism suggests that the C=O and N—H bonds lie predominantly in the axial direction; inspection of the diagram shows that in the model most of these bonds are not so oriented. On the other hand, the very fact that this model held largely undisputed sway for so many years indicates that it did account for very many of the experimental data.

(*2*) *Early Helical Structures.* The first helical model to be proposed was that of Huggins[221] and Taylor;[235] it contained three amino acid residues in a repeat of just over 5 A. Every peptide CO group is hydrogen-bonded to the NH group of a residue in the adjoining turn of the helix. This model is the prototype of a whole series of helical structures which have at various times been proposed, and like all the rest of the earlier ones it is characterized by having an integral number of amino acid residues in each turn of the helix. This is a restriction on the possible types of helical structure, which seemed natural at the time but which really involves the application of a crystallographic principle of symmetry in a context where it is not justified; and it was the abandonment of this principle which, in part, led Pauling and Corey to the discovery of the α-helix which we shall shortly describe. On the other hand, the earlier helical structures did overcome a number of difficulties inherent in the Astbury model, for example in having all possible hydrogen bonds completed and in qualitatively accounting for the infrared dichroism by causing the C=O and N—H bonds to be more or less aligned along the fiber axis.

(*3*) *Ribbon Structures.* The first structure of this kind was described by Huggins;[221] a similar one was proposed by Zahn[236] and independently by Mizushima *et al.*,[237] and discussed in great detail by Ambrose, Bamford, Elliott, Hanby, and their co-workers in a number of publications[92,93,211,216,217] with especial reference to the synthetic polypeptides. Their model has been illustrated in Fig. 27, p. 377, Vol. I, part A; each NH group is linked to a CO group of the next peptide bond along the chain to form a 7-membered ring (in a variant model described by Huggins[221] the NH group is linked to the nearest CO along the chain in the other direction, giving an 8-membered ring). In many ways the structure is a plausible one. All possible hydrogen bonds are completed; the repeating unit is close to the observed value. As in the Astbury structure, —S—S— bridges between chains would not interfere with stretching. The chain

(235) H. S. Taylor, *Proc. Am. Phil. Soc.* **85,** 1 (1941).
(236) H. Zahn, *Z. Naturforsch.* **2b,** 104 (1947).
(237) S. Mizushima, T. Simanouti, M. Tsuboi, T. Sugita, and E. Kato, *Nature* **164,** 918 (1949).

can fold back on itself to make a U-turn without breaking hydrogen bonds. The bond orientation agrees with the observed infrared dichroism. On the other hand, the structure has been criticized from many points of view, among them the following.

First, it has been stated by Pauling and Corey[238] that the structure cannot be built as originally described if the canonical bond dimensions are used, and in particular that the fundamental postulate of a planar amide bond is not satisfied; similarly, Donohue[239] has shown that a minor variant of the structure described by Robinson and Ambrose[240] is unacceptable in terms of these dimensions. On the other hand, Donohue[239] has shown that one version of the structure *can* be built if the requirement for a strict twofold axis is relaxed, and that it is not too improbable energetically.

Secondly, the maximum extensibility of the model is only 25 per cent, which seems rather little to account for the observed behavior of hair. Thirdly, a structure of this kind with screw dyad symmetry would be expected to give a strong second-order (2.6 A.) and a weak first-order (5.2 A.) meridional reflection; the reverse is observed, and to account for this it is necessary to invoke special arrangements of side chains. Fourthly, the structure would not be expected to give the wide-angle meridional reflection observed at 1.5 A., but rather one at 2.5–2.8 A.; special assumptions have to be made to overcome this difficulty. Thus the tide of opinion has been running rather strongly against this structure, in spite of its intrinsic elegance.

(4) General Surveys of Structures. In a systematic survey of possible chain configurations, Bragg, Kendrew, and Perutz[241] enumerated a large number of types of structure, and attempted, without reaching firm conclusions, to assess their relative merits; these configurations included most of those earlier proposed as well as a number of new ones, among them one topologically similar to the α-helix of Pauling and Corey. However, these authors retained the concept of an integral number of residues per repeating unit, and furthermore they adopted a less rigid set of standards of atomic dimensions than did Pauling and Corey in their subsequent work. Their paper might perhaps be regarded as the last full-scale discussion of structural possibilities within the framework of the older concepts, before the greater clarification of the whole situation which followed the work of Pauling and Corey. Similar surveys, includ-

(238) L. Pauling and R. B. Corey, *Proc. Natl. Acad. Sci. U. S.* **37,** 241 (1951).
(239) J. Donohue, *Proc. Natl. Acad. Sci. U. S.* **39,** 470 (1953).
(240) C. Robinson and E. J. Ambrose, *Trans. Faraday Soc.* **48,** 854 (1952).
(241) W. L. Bragg, J. C. Kendrew, and M. F. Perutz, *Proc. Roy. Soc.* (London) **A203,** 321 (1950).

ing more recently proposed structures, have been made by Robinson and Ambrose[240] and by Donohue.[239]

(*5*) *Structures Proposed by Pauling and Corey: the α-Helix.* We have already enumerated the main assumptions which form the foundation of these proposals. The most striking and successful of them is the α-helix, suggested in the first instance for the α-form of synthetic polypeptides, and hence for the α-form of the whole k-m-e-f family of proteins.[242–244,225,238] A drawing of it has been given in Figs. 58*a* and 58*b*, p. 379, Vol. I, part A. It is a nonintegral helix having about 3.6 residues per turn, and the length of a turn being about 5.4 A.; the structure accordingly repeats after 5 turns containing 18 residues and occupying 27 A., and the translation per residue is 5.4/3.6 or 1.5 A. Like helical structures proposed earlier, it is such that every CO group is hydrogen-bonded to an NH group, in this case in the third residue along the chain. The helices have approximately cylindrical symmetry and would naturally fit together in hexagonal close-packing.

Much evidence has come forward that the α-forms of synthetic polypeptides such as poly-γ-methyl-L-glutamate have this structure. The most striking single piece of experimental evidence in its favor is the 1.5-A. meridional reflection, originally observed by MacArthur in porcupine quill[83] and mentioned by Pauling and Corey in their discussion of reflections to be expected from keratin;[244] but first brought into prominence as an important diagnostic test for the α-helix by Perutz,[84] who observed it in many materials, both synthetic and naturally occurring, believed to have the α-structure (see p. 868), including α-forms of synthetic polypeptides and α-keratin. This strong reflection is believed to derive from planes normal to the fiber axis drawn through successive residues of the helix; its spacing is precisely that anticipated for the α-helix, and no structure previously proposed would be expected to give such a reflection. Further x-ray evidence for the correctness of the α-helix in synthetic polypeptides has been put forward by Cochran, Crick, and Vand,[245,246] who developed an elegant theoretical treatment of general validity whereby the helical nature of a structure may be inferred from qualitative examination of its x-ray pattern, and the unit repeat and number of residues per turn determined. They found that the x-ray pattern of poly-γ-methyl-L-glutamate was in excellent agreement with theory for a helix of the type proposed by Pauling and Corey; the pattern

(242) L. Pauling and R. B. Corey, *J. Am. Chem. Soc.* **71,** 5349 (1950).
(243) L. Pauling and R. B. Corey, *Proc. Natl. Acad. Sci. U. S.* **37,** 235 (1951).
(244) L. Pauling and R. B. Corey, *Proc. Natl. Acad. Sci. U. S.* **37,** 261 (1951).
(245) W. Cochran and F. H. C. Crick, *Nature* **169,** 234 (1952).
(246) W. Cochran, F. H. C. Crick, and V. Vand, *Acta Cryst.* **5,** 581 (1952).

was found to correspond to a helix of 18 residues in 5 turns occupying 27 A. A further comparison with new x-ray data has been made by Bamford *et al.*[210] Again, the infrared dichroism is in qualitative agreement with the model, in which the intrachain hydrogen bonds lie more or less parallel to the fiber axis.[247] The latest figures for the extensibility of k-m-e-f proteins[187] are in good agreement with the value calculated for the α-helix (122 per cent extensibility).

All in all, the majority of workers in the field now agree that the α-helix forms the basis of the structure of α-forms of synthetic polypeptides, even though up to the present no one has made a *quantitative* comparison between observed and calculated x-ray intensities taking into account the side chains as well as the main chains.

When we turn to α-keratin the position is rather less satisfactory, and here there is less general agreement as to the part played by the α-helix. On the one hand the x-ray pattern of α-keratin closely resembles that of less well-ordered specimens of synthetic α-polypeptides, to such an extent that it is hard to believe that they do not possess fundamentally the same structure. On the other hand, difficulties are encountered as soon as one tries to explain all the properties of α-keratin in terms of the α-helix. Thus it has generally been considered by chemists that S—S bonds, which are present in great numbers, need not be ruptured during extension to the β-form; a very little consideration suffices to show that extension of the α-helices would be prevented if these bridges are disposed more or less at random between neighboring chains.[248] This objection can perhaps be avoided by invoking the inhomogeneity of keratin and, for example, relegating the —S—S— bridges to amorphous regions of the structure; but this is an inelegant evasion of the difficulty. Again, there are incompletely resolved discrepancies between the observed and calculated densities.[248] More serious is the fact that α-keratin gives a strong x-ray reflection on the meridian at 5.15 A., unlike the synthetic polypeptides in which the reflections in this region are definitely off the meridian and have spacings of 5.4 A. as predicted for the α-helix. All attempts to split the meridional 5.15-A. arc of α-keratin into two off-meridional components have failed. One might avoid this difficulty by supposing that the α-helices lie at some small angle to the fiber axis; but

(247) Quantitative agreement cannot be expected at the present time; the transition moments involved in the infrared absorptions observed almost certainly diverge from the geometrical bond directions by an amount which cannot at present be estimated with accuracy. See W. C. Price and R. D. B. Fraser, *Proc. Roy. Soc.* (London) **B141,** 66 (1953) and *Nature* **170,** 490 (1952); also A. Elliott, *Nature* **172,** 359 (1953).

(248) W. T. Astbury, *Proc. Roy. Soc.* (London) **B141,** 1 (1953).

in this case the 1.5-A. reflection of keratin should be off-meridional, which it is not. The essential problem is to find an arrangement for which both 5.15-A. and 1.5-A. reflections are truly meridional.[249]

Two proposals have been made independently which go a long way to avoid this difficulty, at least qualitatively. Crick[250] has proposed that the whole helix might be distorted into a "super-helix" or coiled coil with pitch angle about 18°; this angle could be obtained, for example, from a super-helix of radius 10½ A. and axial spacing 198 A. (the repeat of porcupine quill). This scheme derives from model-building experiments which show that if the helices are distorted systematically in the manner proposed, the side chains of one helix can be made to interlock with the holes between side chains of its neighbor, a very plausible arrangement which is not possible if the helices are parallel. At the same time Pauling and Corey[251] made very similar suggestions; Fig. 15 shows the type of structure they proposed.

More recently Crick has developed these ideas in two papers, the first of which[252] gives a theory for the Fourier transform of a coiled coil, while the second[253] shows in detail how packing of α-helices such that side chains interlock would lead to an angle of 20° between neighboring helices and perhaps, for long chains, a coiled coil; he also shows by means of the theory how a coiled-coil arrangement would give a diffuse α-type x-ray pattern with meridional 5.1- and 1.5-A. reflections.

Before leaving the subject of the α-helix it should be mentioned that a number of other nonintegral helical structures satisfying the Pauling-Corey canonical dimensions has recently been proposed, without, however, any of them being definitely ascribed to any known material. First among these was the γ-helix of Pauling and Corey,[225,242] illustrated in Figs. 58*c* and 58*d*, p. 379, Vol. I, part A; this is a much more open helix than the α-helix, with about 5 residues per turn of about 5 A. Next is a helix described by Low and Baybutt,[254] which they have called the π-helix and which is intermediate between α- and γ-helices (see also

(249) In their original treatment Pauling and Corey discussed these difficulties by reference to the so-called Lotmar-Picken x-ray diagram of muscle, which they regarded as a manifestation of α-myosin in a highly crystalline form.[244] There is now evidence that the Lotmar-Picken is irrelevant to this issue, being due to the presence of an unknown water-soluble component, very probably not a protein and almost certainly not myosin; see H. E. Huxley and J. C. Kendrew, *Nature* **170,** 882 (1952).

(250) F. H. C. Crick, *Nature* **170,** 882 (1952).

(251) L. Pauling and R. B. Corey, *Nature* **171,** 59 (1953).

(252) F. H. C. Crick, *Acta Cryst.* **6,** 685 (1953).

(253) F. H. C. Crick, *Acta Cryst.* **6,** 689 (1953).

(254) B. W. Low and R. B. Baybutt, *J. Am. Chem. Soc.* **74,** 5806 (1952).

p. 381, Vol. I, part A). Finally several new helical structures have been discussed by Donohue, in a paper[239] in which a systematic exploration of possible helices is made, and those which can be constructed with acceptable bond dimensions are compared from the point of view of their energy contents. It is interesting that he finds a completely strain-free configuration only for the α-helix; for the ribbon structure and the π-helix he estimates an instability of at least 0.5 kcal./mole/residue, and for the γ-helix 2.0 kcal./mole/residue.

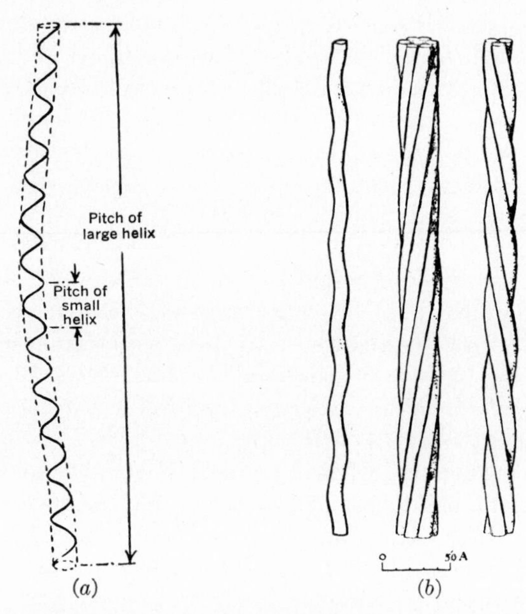

FIG. 15. (*a*) "Coiled coil" as proposed by Pauling and Corey; the pitch of the large helix is 12.5× that of the small helix, i.e., 68 A. and 5.44 A. in the case of the α-helix. (*b*) Two ways in which "coiled coils" might be fitted together to form compound helices. After Pauling and Corey.[251]

(*6*) *Supercontracted Keratin. The Cross β-Pattern.* The phenomenon of supercontraction has been described above (p. 861). It is not accompanied by the appearance of any new type of x-ray pattern: supercontracted fibers give only a very disoriented β-pattern. Supercontraction is nevertheless so striking a phenomenon macroscopically that in spite of this observation several attempts have been made to explain it in terms of a new molecular fold. However, recent electron-microscope studies by Mercer[255] show that during supercontraction of wool the fibrils of which it is composed appear to twist into spirals; it would thus appear very probable that we are concerned here with a phenomenon taking

(255) E. H. Mercer, *Textile Research J.* **22,** 476 (1952).

place at a higher level of organization than the individual chain. Its cause remains quite obscure, however. Some attempts have been made to relate the contraction of muscle to supercontraction, but there would appear to be no evidence in favor of this hypothesis.

Mention has been made of the cross β-pattern in the heat contraction of epidermin (see p. 886). The only hypothesis which has been proposed to account for this is one within the context of the Astbury α-fold; Fig. 17, p. 270, Vol. I, part A indicates the nature of this hypothesis (for discussion of it, see Rudall).[74,183] It is difficult to see how it could be reconciled with the α-helix model.

c. Evidence for Larger Units of Structure in the k-m-e-f Proteins

There are many reasons for thinking that the individual chains in their specific configuration are not the largest submicroscopic organized structure in the k-m-e-f proteins; indeed it is plausible to suppose that differences between larger units of some kind are at least partly responsible for the specific characteristics of different members of the family, since all of them appear to have one or other of the two chain configurations which we have described. The most important evidence for the existence of larger units is the following:

(*1*) *Electron Microscope Evidence.* Electron-microscope studies of fibrin have already been described (see p. 882); the fibrils exhibit cross-striations with period 230 A., and under high resolution they appear to consist of particles of diameter 30–50 A. Similarly, particles have been observed in fibrinogen filaments. Particulate structures in myosin will be described in the next chapter (p. 972). In the case of wool there have been technical criticisms of the electron-microscope studies; nevertheless it is worth mentioning, without too much conviction at this stage, that Farrant, Rees, and Mercer[256–258] have examined in the electron microscope the cortical cells resulting from enzymic disintegration of wool fibers, and have observed twisted protofibrils of fairly uniform size which appear to be made up of corpuscular units rather over 100 A. in diameter; the fibrils are embedded in an amorphous matrix composed of similar particles.

(*2*) *X-Ray Evidence.* Wool keratin, feather keratin, and porcupine quill all yield low-angle diffraction patterns containing reflections whose spacings are submultiples of a basic spacing of the order 100 A. or more

(256) E. H. Mercer and A. L. G. Rees, *Australian J. Exptl. Biol. Med. Sci.* **24,** 175 (1946).
(257) J. L. Farrant, A. L. G. Rees, and E. H. Mercer, *Nature* **159,** 535 (1947).
(258) E. H. Mercer, *Proc. Intern. Congr. Exptl. Cytol. 6th Congr., Stöckholm,* 1947, 60 (1947).

(see p. 868). Similar spacings in the muscle proteins will be described in the next chapter (p. 1034). Such spacings prove unambiguously that all these proteins contain regularly repeated structures whose dimensions correspond to the basic period and are thus large compared to the folds in individual chains.

(*3*) *Chemical Evidence.* Many members of the k-m-e-f group can be degraded by chemical means into small molecules which behave in solution like globular proteins. It is true that in no case (except perhaps fibrin) have strictly homogeneous products been obtained, nor is there more than a general resemblance between products derived from the same protein by different procedures. Nevertheless, the evidence is fairly convincing that these products are not artifacts, but are related to genuine subunits in the original structure. We may recapitulate some of these instances:

(*a*) In *keratin,* soluble particles of molecular weight 70,000 are obtained after oxidizing the —S—S— bridges; these particles contain chains in the α-configuration and are therefore not denatured in the ordinary sense of the term. Reduction and detergent action give particles of molecular weight 55,000–75,000 and 35,000–40,000. Urea gives particles of molecular weight 84,000, which can be further degraded to smaller ones of molecular weight 8000. Similar treatment of feather keratin yields particles of molecular weight about 10,000.

(*b*) *Myosin* can be degraded to small units whose preparation and properties are described in the next chapter (see p. 972).

(*c*) *Epidermin* dissolves in urea giving particles of molecular weight 60,000.

(*d*) *Fibrin* of the urea-soluble variety dissolves in urea to give particles indistinguishable from those of fibrinogen. This example is particularly interesting because, of course, we can observe the reverse process of aggregation of fibrinogen to fibrin either *in vivo* or *in vitro;* this is the only case at present available to us of the formation of a typical k-m-e-f protein from smaller units. We shall encounter a similar aggregation process in collagen (p. 925).

Two main types of theory have been proposed to account for these observations. In theories of the first type it is supposed that the polypeptide chains run indefinitely through the structure in the direction of the fiber axis, and that there are long periodicities in the structure of individual chains. Two kinds of periodicity have been postulated:

(*1*) Periodicities due to regularly repeating arrangements of amino acid residues along the chain. This is the essence of the Bergmann-Niemann hypothesis, and, as we have pointed out earlier (p. 873), chemical degradation studies have lent no support to it. The most detailed discussions of this theory in its application to keratin are those of Astbury[98] and MacArthur;[83] it has also been mentioned by Pauling and Corey.[244] We shall meet it again in connection with collagen.

(*2*) Periodicities due to long-period waves or wobbles in the coiled chains. Such periodicities are given by the coiled coil or super-helix arrangements of Crick[250] and of Pauling and Corey;[251] it has been mentioned above that a super-helix of pitch angle 18° would explain the 198-A. axial repeat of porcupine quill.

Theories of the second type postulate that fibrous proteins are made up by the linear (or helical) aggregation of large numbers of similar or identical subunits, each of which might resemble the molecule of a globular protein.[259] Within each subunit the polypeptide chains would be arranged so as to be more or less parallel to the fiber axis; thus the over-all picture at low resolution would be of continuous chains running along the fiber. Such an arrangement does not of course exclude the possibility of short segments of "coiled coils" within each subunit. Although at the present time there is nothing approaching proof that the subunit hypothesis is correct, the evidence we have just quoted appears to support it more strongly than any other. The chemical degradation results in particular are most easily explained in these terms, and a striking analogy has been found in the work of Waugh on the aggregation of globular proteins to fibers without irreversible denaturation taking place.

Waugh[260,261] found, using the electron microscope, that the thixotropic birefringent gel obtained by heating insulin solutions for a short time to 100° at pH 2–2.5 contained uniform fibrils of diameter about 150 A. and several microns long.[262] At high pH the fibrils redissolve on standing in the cold, and the resulting solution exhibits the original physiological activity of the insulin; the regenerated insulin can be crystallized and appears to be identical in every way to the original material.[263] This suggests that, unlike those cases where a globular protein is irreversibly unfolded to a β-configuration (e.g., at an interface) and subsequently drawn into fibers, we are here dealing with a reversible aggregation of globular particles. Waugh has given evidence that monomer units are not held together in the fibrils by covalent bonds. Similar results

(259) A very early suggestion that fibrous proteins might be built up in this way was made by W. T. Astbury and R. Lomax, *Nature* **133,** 795 (1934), on the basis of a comparison between the x-ray patterns of feather keratin and of pepsin crystals. More recently the idea of helical aggregation of subunits has been developed by L. Pauling, *Discussions Faraday Soc.* No. **13,** 170 (1953).

(260) D. F. Waugh, *J. Am. Chem. Soc.* **66,** 663 (1944).

(261) D. F. Waugh, *J. Am. Chem. Soc.* **68,** 247 (1946).

(262) In a more recent electron-microscope study, J. L. Farrant and E. H. Mercer [*Biochim. et Biophys. Acta* **8,** 355 (1952)] found fibrils whose diameters were as small as 50–80 A., but obtained no evidence of axial periodicity.

(263) D. F. Waugh, *J. Am. Chem. Soc.* **70,** 1850 (1948).

have been obtained with other globular proteins, e.g., serum albumin and ovalbumin.[264] A variety of techniques has been applied to studies of similar systems by Barbu and Joly, whose recent paper[265] should be consulted for details of the methods used by these workers, and for a bibliography.

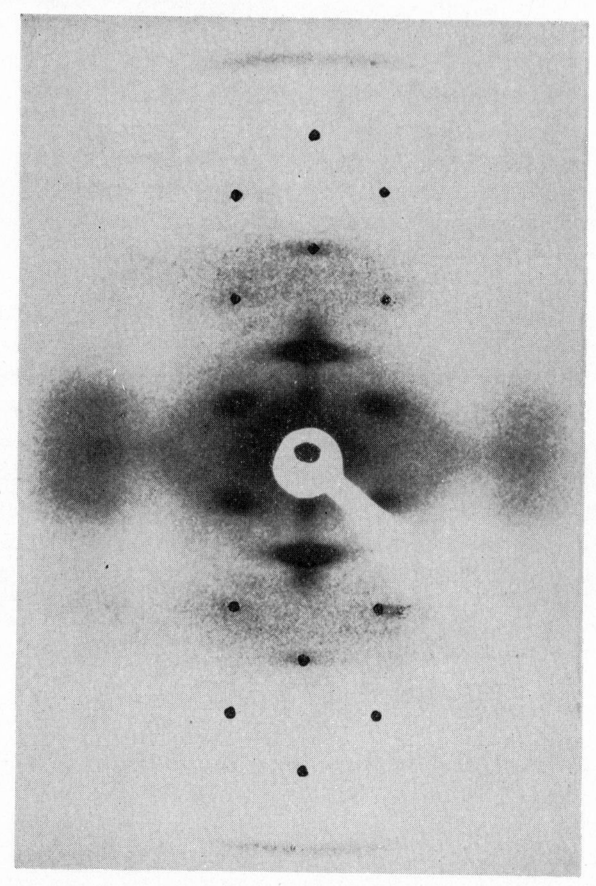

FIG. 16. Low-angle "net diagram" of sea-gull feather keratin: fiber axis vertical. Several faint spots are difficult to reproduce and have been indicated by dots. After Bear and Rugo.[91]

It is also relevant to note that Watson[266] has shown that the x-ray pattern of tobacco mosaic virus may be convincingly explained in terms of a helical arrangement of subunits.

Bear and Rugo[91] have interpreted the x-ray pattern of feather keratin in terms of these concepts. They find that by the action of various

(264) M. P. Jaggi and D. F. Waugh, *Federation Proc.* **9**, 66 (1950).
(265) E. Barbu and M. Joly, *Discussions Faraday Soc.* No. **13**, 77 (1953).
(266) J. D. Watson, *Biochim. et Biophys. Acta* **13**, 10 (1954).

agencies—especially heat, moisture, and lower alcohols—the low-angle pattern can be caused to alter with the production of what they term a "net diagram." This consists of meridional reflections on the 4th, 8th, and 12th layer lines, and pairs of off-meridional reflections on the 2nd, 6th, 10th, and 14th lines, displaced from the meridian by a constant amount corresponding to a lateral spacing of 34 A. They point out that this is a type of pseudo-face-centered pattern which would be expected from a two-dimensional network of globular particles. The net diagram is illustrated in Fig. 16. These proposals have not been elaborated in detail, but they are at least suggestive of the type of structural plan which may be present. It should be noted that the net diagram can probably be interpreted equally well in terms of subunits arranged helically.

We may thus rather tentatively conclude that the most useful working hypothesis to adopt at the present time is that k-m-e-f proteins result from a linear (or helical) aggregation of globular particles. Such a hypothesis has a wider application; we have seen in a previous section (p. 856) that there is considerable evidence that silk fibroin is fabricated in the silk gland by aggregation of "fibroinogen" monomers. On the other hand, in the collagen group we shall draw the conclusion that although subunits are very probably present, they are more likely to be linear than globular in shape.

IV. The Collagen Group

1. Introduction

The name collagen (meaning glue-forming; from κόλλα = glue) was given in the first instance to the main fibrous protein constituent of mammalian connective tissues, tendons, and bones.

Its physiological function is to act as an inert, inextensible material, either, as in tendons, to transmit the tensions exerted by muscles, or, as in cuticle, to form part of a protective covering, or, as in bone, to form a constituent of the skeletal framework. So far as is known it plays no part in the metabolism of the organism.

Classically, the principal diagnostic characteristics of collagen were its mechanical properties including its resistance to changes in length under physiological conditions, its chemical inertness (especially its indifference in the native state to common proteolytic enzymes), its unusual amino acid composition, and its conversion to gelatin on prolonged heating with water or alkali. More recently it has been found that some or all of these properties are shared by other proteins found in mammalian tissues, and by still others occurring in phyla only distantly related to the mammals; these materials have therefore also been classified as collagens. But since they exhibit considerable differences among themselves (e.g., in amino acid composition; see Tables XXIII and XXIV, Vol. I, part A, pp. 221, 222), and since in any case many of them are impossible to prepare in anything like pure form, there has been need for a more adequate definition of collagen. Meanwhile, different types of collagen have been studied by modern physical techniques, particularly the electron microscope and x-ray diffraction methods. As seen in the electron microscope, most collagen fibers exhibit a very characteristic cross-banded structure which will be illustrated and described below; but this has not yet been demonstrated in some materials which would on other grounds be regarded as collagens. On the other hand, the wide-angle x-ray diffraction pattern of collagen fibers does appear to be highly characteristic. This was shown by many workers in the 1920's to have as its principal features a spacing of about 2.86 A. along the fiber axis, and one of 10–17 A. (depending on the state of hydration) perpendicular to the fiber axis; the former spacing is quite unlike any reflection observed in other fibrous proteins, e.g., in the keratin group. While the interpretation of this pattern is still doubtful, its value as a diagnostic character has become more and more clear as more materials have been studied. It is, in fact, now permissible to *define* collagen as any fibrous protein material which gives these x-ray spacings: this basis of classification was first clearly enunciated by Astbury[267] in 1938 following earlier suggestions of Gonell and Kratky;[268] it has been placed on a firmer footing by all subsequent work, especially in the schools of Astbury and of Schmitt and Bear (see, e.g., Bear[269]).

Table V, which is adapted from Bear,[269] lists the principal materials which must be regarded as collagens if this definition is accepted. It contains examples from many phyla in spite of the fact that the search for collagens has not been a systematic one; collagens must clearly be very widespread constituents of living organisms. It will be observed that the nomenclature is not systematic; in some cases, e.g., the so-called ovokeratins and byssokeratins, it is misleading.

So far as is known, all collagens are extracellular; of their mode of formation next to nothing is known (but see p. 930). A few have been called "secreted collagens" in recognition of their more obviously secretory origin, usually from epithelial cells; these were in some cases originally classified as keratins, and in some respects they seem to be the least typical collagens, in their failure to give any low-angle x-ray pattern (see p. 916) and in the absence of cross-banding in electron micrographs.

(267) W. T. Astbury, *Trans. Faraday Soc.* **34,** 378 (1948).
(268) H. W. Gonell and O. Kratky, *Handbuch der physikalischen und technischen Mechanik* **4,** 317 (1931).
(269) R. S. Bear, *Advances in Protein Chem.* **7,** 69 (1952).

Most studies of collagen have been directed either toward elucidating the mechanism of two very important industrial processes—the production of leather by tanning of skins and the manufacture of gelatin and

TABLE V
DISTRIBUTION OF COLLAGENS
(Adapted from Bear[269] with additional data from Randall *et al.*[270])

Phylum	Class	Name	Source
Vertebrata	Mammalia	Collagen	Connective tissues, tendons, bones, skin corium
		Reticulin	Skin, spleen, lymph nodes, adipose tissue
		Vitrosin	Vitreous humor of eye
	Aves	Collagen	Tendon, bone
	Amphibia		Body wall of toad and frog
	Pisces		
	Teleostomi	Ichthyocol	Skin, tendon, swim bladder
		Ichthylepidin	Scales
			Body wall of eel
	Elasmobranchii	Elastoidin	Shark fins
Mollusca	Cephalopoda		Squid connective tissue
Annelida			Body wall of *Sabella*
Echinodermata			
	Holothuroida		*Thyone* body wall
	Echinoida		*Arbacia* peristome
	Asteroida		*Asterias* body wall
Coelenterata			
	Anthozoa	Cornein	Axial stalks, acontia
Porifera			
	Demospongiae	Spongin	Skeletal fibers
		Secreted collagens	
Vertebrata		"Ovokeratin"	Skate egg capsule
Mollusca		"Byssokeratin"	*Mytilus* (mussel) byssus threads
Annelida			Earthworm cuticle
Nematoda			*Ascaris* cuticle
Echinodermata			"Ejected filaments" of sea cucumber

glue—or toward explaining its very unusual properties in structural terms. In the latter connection it may be useful at this stage to summarize some of the peculiar features of collagen which differentiate it from other pro-

(270) J. T. Randall, R. D. B. Fraser, S. Jackson, A. V. W. Martin, and A. C. T. North, *Nature* **169**, 1029 (1952).

teins. First, its amino acid composition: ordinary mammalian collagen contains a very high proportion of proline and of hydroxyproline (which is not known to occur elsewhere except in elastin), which together make up over 20 per cent of the residues; the proportion of nonpolar residues (especially glycine and alanine) is also very high, over 50 per cent; aromatic amino acids are practically absent, as are those containing sulfur. Then, it exhibits unusual mechanical properties and a striking chemical inertness—features often explained in the keratins in terms of S—S cross links, which are however absent from collagen. Again, collagen can be dissolved in dilute acid buffers to form a solution from which it can be reprecipitated in a fibrous form which exhibits, in the electron microscope, all the fine-structure bands characteristic of the original material; this remarkable phenomenon will be described in more detail below (p. 925). Finally, there seems to be more than a suspicion that some of the so-called impurities invariably associated with collagen are in fact essential components; evidence is accumulating to show that in some way polysaccharide material forms an integral part of collagen fibrils (see p. 921).

The following sections will describe first the morphology and properties of mammalian collagen, secondly attempts which have been made to interpret these in terms of structure, and finally some of the less well-known members of the collagen class. Many of the matters treated are discussed in a recent review by Bear[269] in more detail than is possible in this chapter; and a very comprehensive bibliography on collagen has been prepared by Borasky.[271] Many recent studies were reported on at a conference held under the auspices of the Faraday Society in March, 1953.[272]

2. Properties of Mammalian Collagen

a. Morphology and Fine Structure

(*1*) Collagen has a fibrous appearance at all resolutions at present available; in the light microscope the fibers have diameters of the order of 100–200 μ in tendon, and four or five times less in skin; in the electron microscope, fibrils are seen whose diameters may be anything down to the resolving power of the instrument—say 50 A. The electron microscope has, of course, given the most complete picture of the morphology of col-

(271) R. Borasky, Guide to the Literature on Collagen, Bureau of Agricultural and Industrial Chemistry, Agricultural Research Administration, U.S.D.A., Eastern Regional Research Laboratories, Philadelphia, 1950.

(272) J. T. Randall (ed.), The Nature & Structure of Collagen, Faraday Society Discussion, Butterworth, London, 1953.

lagen fibrils; results have come mainly from the schools of Schmitt,[272a,273] Wolpers,[274,275] and Wyckoff.[276] Electron micrographs may be taken of material either shadowed or stained, e.g., with phosphotungstic acid; one example of each is given in Fig. 17. The principal feature is the pronounced cross-banding; its period averages about 640 A., though indi-

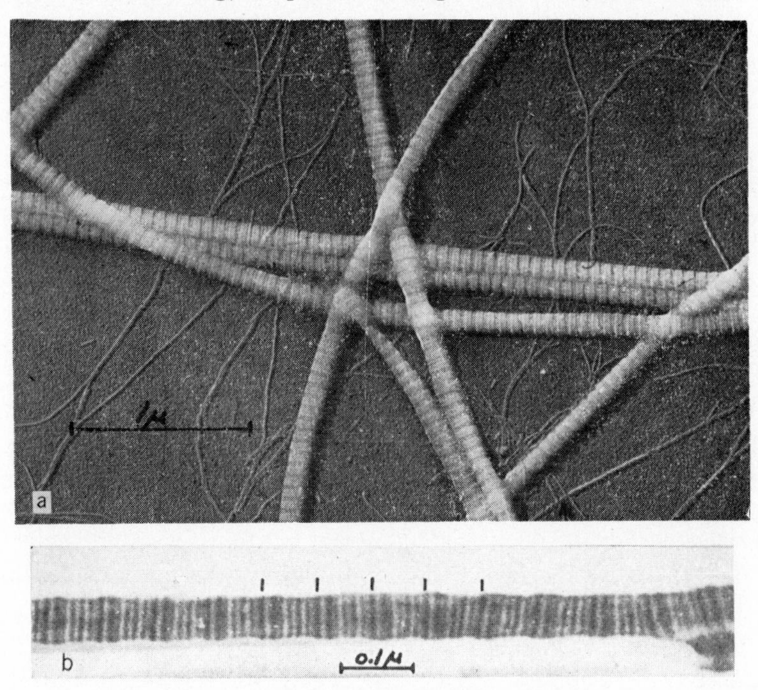

FIG. 17. Electron micrographs of collagen.

(*a*) Ratskin collagen, treated with trypsin, washed and shadowed with chromium; magnification 25,000×. The filaments in the background are derived from the trypsin. By courtesy of J. Gross.

(*b*) Chrome-tanned calfskin, stained with phosphotungstic acid; magnification 100,000. After Schmitt and Gross.[273]

vidual fibrils even in the same preparation may differ considerably (probably owing to irregularities in drying). In stained preparations it can be seen that the main period of 640 A. is broken up into a number of subbands; the number of these varies with the resolving power of the micro-

(272*a*) C. E. Hall, M. A. Jakus, and F. O. Schmitt, *J. Am. Chem. Soc.* **64,** 1234 (1942).

(273) F. O. Schmitt and J. Gross, *J. Am. Leather Chemists' Assoc.* **43,** 658 (1948).

(274) C. Wolpers, *Klin. Wochschr.* **22,** 624 (1943); *ibid.* **23,** 169 (1944).

(275) C. Wolpers, *Arch. path. Anat. Physiol.* (*Virchow's*) **312,** 292 (1944).

(276) A. W. Pratt and R. W. G. Wyckoff, *Biochim. et Biophys. Acta* **5,** 166 (1950).

scope and with the source of the material, but is generally 5, 6, or 7. There is little doubt that further improvements in technique will reveal still finer elements of structure; thus a recent report mentions a repeating unit of 10 or 11 subbands.[277] The subbands are not equally spaced, and the structure is a polar one; that is to say the subbands are disposed unsymmetrically within each main period so that the two directions along any fiber can be distinguished. The dark bands have slightly greater diameters than the light interband regions; their greater electron-absorptive power might perhaps be due to this, but is more probably caused by a preferential uptake of stain.

So much for the longitudinal fine structure. In the lateral direction there appears to be no regularity at all—the finest fibrils may be of any width, and in most materials (especially tendon) they tend to fray to filaments of diameter 50–100 A. or less; this would appear to set an upper limit to the diameter of the collagen molecule. It should however be noted that some other types of collagen (e.g., from skin) tend to fracture across the fiber axis rather than longitudinally.

Cartilage contains another type of fibril in addition to those of typical collagen. These have diameters of 300–500 A. and form a coarse woven network; they show traces of a longitudinal periodicity of 180–200 A. (see e.g., Porter and Vanamee,[278] Randall *et al.*,[270] and Jackson[279]). Their relationship to collagen is unknown; but fibrils with similar periodicities have been observed in embryonic and in reprecipitated collagens (see pp. 928, 931).

TABLE VI

WIDE-ANGLE X-RAY SPACINGS OF COLLAGEN

(Kangaroo tail tendon)

(Indexed in terms of a 20-A. meridional repeat; the figures given are the spacing components; in the column headed "meridian," parallel to the fiber axis; and in the other columns, transverse to the fiber axis. After Bear.[269])

Index of axial repeat	Meridian, A.	First row line, A.	Second row line, A.	Other reflections, A.
0 (Equator)	—	**11.6**	**5.7**	**4.6**, 2.15
2	**9.55**	**11.6**		
4	(5.0)	12.	6.	
5	**3.97**		6.	4.
7	**2.86**		6.	

(*2*) We now turn to the x-ray analysis of collagen. The wide-angle pattern, as has already been remarked, is characteristically different

(277) W. Grassmann, U. Hofmann, Th. Nemetschek, *Naturwissenschaften* **39,** 215 (1952).

(278) K. R. Porter and P. Vanamee, *Proc. Soc. Exptl. Biol. Med.* **71,** 513 (1949).

(279) S. Fitton Jackson *in op. cit.*,[272] p. 140.

from that of α- or β-keratin. A typical photograph is shown in Fig. 18*a*, and the principal reflections are listed in Table VI. We may note in particular the meridional spacing of 2.86 A. and the equatorial spacing of 10–11 A. The former has generally been identified as the main repeat of pattern in the axial direction—it is considerably shorter than that characteristic of a fully extended chain (cf. β-keratin and silk fibroin). The equatorial 10–11-A. reflection is humidity-sensitive (see p. 919).

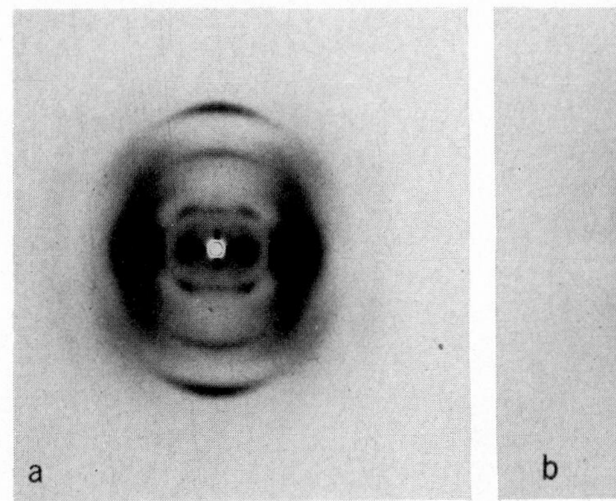

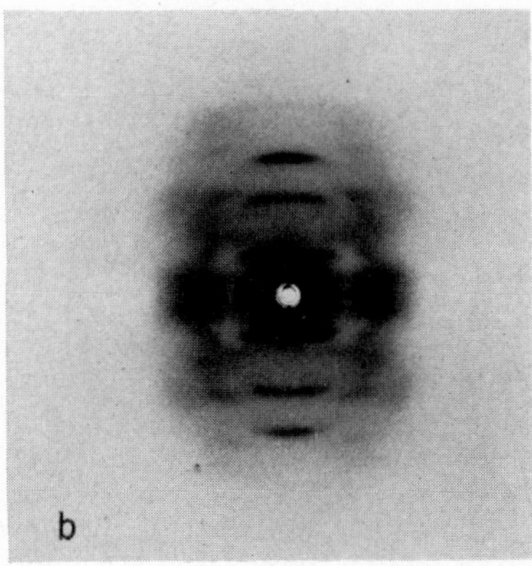

FIG. 18. Wide-angle x-ray patterns of collagen (rattail tendon): fiber axis vertical. (*a*) Normal, (*b*) stretched. After P. Cowan, A. J. North, and J. T. Randall.[282]

Indexing the reflections in terms of a simple unit cell is by no means a straightforward matter; the only clearly apparent layer line corresponds to an axial spacing of 9.55 A., but neither of the prominent meridional arcs (at 2.86 and 3.97 A.) will fit a unit cell based on this repeat. Some authors (e.g., Meyer)[280] have regarded the diagram as a complex arising from two distinct crystalline phases: others have assigned an axial period of about 20 A.,[269,281] in terms of which most of the reflections can be indexed fairly satisfactorily. The agreement is not perfect, however, and the truth is that in the absence of any convincing theory of structure the difficulty remains unexplained.

It has recently been shown by Cowan, North, and Randall[282] that if a collagen fiber is kept stretched while its diffraction pattern is being recorded, it yields an x-ray diagram in which the 2.86-A. meridional reflection may be increased by as much as 14 per cent. This pattern (see Fig. 18b) contains many more discrete reflections than that of a normal fiber.

(280) K. H. Meyer, Natural & Synthetic High Polymers, Interscience, New York, 1942, p. 456.
(281) R. O. Herzog & W. Jancke, *Z. physik. Chem.* **B12**, 228 (1931).
(282) P. M. Cowan, A. C. T. North, and J. T. Randall *in op. cit.*,[272] p. 241.

(3) The low-angle pattern of collagen was discovered more recently[283,284] than the wide-angle, but, as we shall see below, it has proved much more amenable to interpretation. It consists (Fig. 19*a*) of a large number of meridional reflexions; when it was first discovered, x-ray techniques were primitive and the fundamental spacing was variously

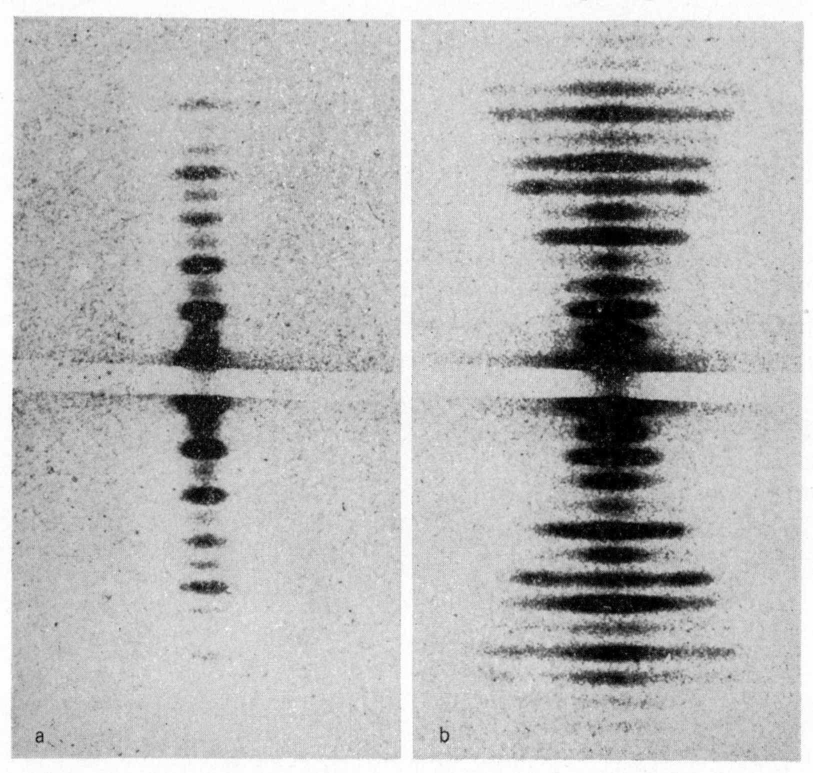

FIG. 19. Low-angle x-ray pattern of kangaroo-tail tendon (fiber axis vertical); (*a*) with excess moisture present; (*b*) after brief exposure to water and drying under tension. After Bear, Bolduan, and Salo.[301]

reported, but the matter was put onto a firm footing by Bear[285] (and nearly simultaneously by Kratky and Sekora)[286] who showed that all the lines were orders of a fundamental spacing of about 640 A. This figure compares well with the main banding observed in the electron microscope; a detailed correlation between the low-angle x-ray pattern and the band

(283) G. L. Clark, E. A. Parker, J. A. Schaad, and W. J. Warren, *J. Am. Chem. Soc.* **57,** 1509 (1935).
(284) R. W. G. Wyckoff, R. B. Corey, and J. Biscoe, *Science* **82,** 175 (1935).
(285) R. S. Bear, *J. Am. Chem. Soc.* **64,** 727 (1942); *ibid.* **66,** 1297 (1944).
(286) O. Kratky and A. Sekora, *J. Makromol. Chem.* **1,** 113 (1943).

structure has been described by Kaesberg and Shurman.[287] Up to 30 orders may be observed, indicating a high degree of axial order, and a high degree of reproducibility of period between one fibril and another—in contrast to the considerable scatter of the results obtained in the electron microscope. There is no obvious correlation between the narrow-angle pattern and the wide-angle pattern.

There have been reports of long equatorial spacings in collagen: e.g., Gerngross, Hermann, and Abitz[288] found a reflection of 23 A., and Corey and Wyckoff[88] reported 19.9, 30.0, and 47.6 A. However these reflections were weak and were not invariably observed: it seems probable that, as has been suggested more recently,[269] they can be ascribed to a lipide impurity. It may in fact be concluded that the low-angle diffraction pattern (of the wet fiber) is essentially one-dimensional, containing no true non-meridional reflections.

b. Physical Properties

(*1*) Under normal conditions, collagen fibers are only slightly extensible (the percentage extension at the elastic limit being about 7 per cent for wet fibers, and about half this for dry); in this respect collagen is sharply distinguished from keratin. The distinction may not be absolute, however; for under one set of special conditions very large extensions have been reported. In taking electron-microscope pictures of collagen, the supporting membrane sometimes fractures and occasionally a very thin fibril is left suspended across the gap; the fiber period is then seen to be very much—up to nine times—greater than the normal 640 A.[289,290] At such extreme extensions the fine structure disappears, but it can still be seen to be present up to an extension of about 100 per cent. Schmitt and Bear argue from the latter observation that the extensibility must be due to a concerted lengthening of each individual polypeptide chain: if it were due to slipping of one chain past another the fine structure bands would disappear. Hence, they conclude, "it seems inescapable that the protofibrils are capable of concerted extension." Unfortunately, no one has yet succeeded in producing appreciable extensions in collagen fibers except in this very special way. The question of the extensibility of the collagen chain must therefore be regarded as still unsettled, which is unfortunate since it is obviously of crucial importance as a test of theories of the molecular structure of this protein.

(287) P. Kaesberg and M. M. Shurman, *Biochim. et Biophys. Acta* **11,** 1 (1953).
(288) O. Gerngross, K. Hermann, and W. Abitz, *Biochem. Z.* **228,** 409 (1930).
(289) F. O. Schmitt, C. E. Hall, and M. A. Jakus, *J. Cellular Comp. Physiol.* **20,** 11 (1942).
(290) P. O. Mustacchi, *Science* **113,** 405 (1951).

(2) On heating to about 60°C., moist collagen suddenly contracts to one third or one fourth of its normal length. The contracted material has quite new properties: it is rubbery in behavior, and the low-angle x-ray pattern disappears completely, the wide-angle in part. Analogies have been drawn[291] between this change and the denaturation of globular proteins; the thermodynamics of the processes appear to be similar, large values of ΔH and ΔS being involved. The effect seems to be due to a thermal rupture of interchain bonds; the transition temperature is lowered by treatment with alkali and raised by the action of substances, such as formaldehyde, which produce covalent cross links. In certain members of the collagen group, especially elastoidin,[292] the effect is in part reversible, the fibers lengthening nearly to their original length on lowering the temperature. The thermodynamics of thermal contraction are discussed in detail by Bear.[269]

(3) Under the influence of acids, alkalies, or salts, collagen takes up a considerable amount of water. This swelling may be of one or other of two types, or a combination of the two.[43] The first is known in industry as "plumping"; it results from the action of acids or alkalies, is nonspecific, and is suppressed by salts; it has been interpreted as a Donnan effect due to charged groups in the fibers. It is minimal at the isoelectric point (which is ill-defined, but lies in the range pH 6.5–8.5);[293] as one leaves the isoelectric point in either direction it increases to a maximum and then (on the acid side) decreases.[294,295] This type of swelling makes the fibers shorter, thicker, and more translucent, and is reversed by the application of tension to the fiber. Alkaline swelling phenomena have been studied in great detail by Bowes and Kenten.[296]

The other type of swelling is induced by neutral salt solutions; it produces a thickening but no shortening of the fibers, and under the microscope the fibers are seen to have split into finer filaments; it cannot be reversed by tension, but may be by prolonged washing in water. This effect has been called lyophilic or Hofmeister swelling. It is undoubtedly due to specific ion effects, producing fracture of salt links and hydrogen-bond bridges between chains, but the details of its mechanism are unknown: the names given to it fulfil their usual role as cloaks for ignorance.

(4) The effect of hydration on the x-ray pattern is important enough

(291) L. Hermann, *Arch. ges. Physiol. (Pfluger's)* **7,** 417 (1873); *ibid.* **8,** 275 (1874).
(292) E. Fauré-Fremiet, *J. chim. phys.* **33,** 681 (1936); *ibid.* **34,** 125 (1937).
(293) J. H. Bowes and R. H. Kenten, *Biochem. J.* **43,** 358 (1948).
(294) J. H. Bowes and R. H. Kenten, *Nature* **160,** 827 (1947).
(295) D. M. G. Armstrong *in op. cit.,*[272] p. 91.
(296) J. H. Bowes and R. H. Kenten, *Biochem. J.* **46,** 1, 524, 530 (1950).

to deserve separate discussion. The *wide-angle pattern* alters very strikingly in one feature, the prominent equatorial reflection; this has a spacing of 10.4 A. in dry collagen, but on increasing the humidity it moves progressively inwards to a spacing of 15–16 A.[297] In gelatin the corresponding reflection behaves similarly, reaching a spacing of 17 A. under favorable circumstances.[288,298] The equatorial pattern of collagen of course includes other reflections—a weak spot at about 6–6.5 A., and diffuse reflections at 4–4.5 A. and at 2 A. What happens to these on hydration is obscure: it is sometimes stated that *only* the 10–11 A.-reflection is humidity-sensitive, but a search of the literature fails to reveal any experiment leading to this conclusion which has been carried out by modern techniques. However, very recently Randall *et al.*[270] have stated that the 2-A. reflection at any rate does not move. The wide-angle meridional reflections change little on hydration.

The *narrow-angle pattern* does change with varying humidity, the main axial period increasing with hydration,[285] from 628 A. at 2 per cent relative humidity to 672 A. at 100 per cent.[299] The relative intensities of the reflections also change greatly. If the water content of the fiber is greatly reduced by drastic drying, the character of the pattern changes: instead of consisting of a series of short meridional lines of more or less equal length, the pattern becomes "fanned out," the length of layer lines increasing with index and in some cases developing noncentral maxima,[300,301] (see Fig. 19*b*).

We shall return to the discussion of all these effects in considering the models which have been proposed for the structure of collagen.

(*5*) Collagen fibers exhibit both form and intrinsic birefringence of visible light, in each case positive with respect to the fiber axis.[302] This accords well with the conception of fibrillar elements parallel to the fiber axis, made up of chains also running in the axial direction.

The ultraviolet spectrum of collagen has received less attention than that of most other proteins, owing to the fact that collagen contains very few side chains with aromatic rings, which normally contribute the main features of interest in this spectrum. A study of it has, however, been made by Fixl, Kratky, and Schauenstein;[303] their work has been discussed in Chap. 5 (see p. 438 ff., Vol. I, part A).

(297) A. Kuntzel and F. Prakke, *Biochem. Z.* **267,** 243 (1943).
(298) J. R. Katz and J. C. Derksen, *Rec. trav. chim.* **51,** 513 (1932).
(299) B. A. Wright, *Nature* **162,** 23 (1948).
(300) O. E. A. Bolduan and R. S. Bear, *J. Polymer Sci.* **5,** 159 (1950).
(301) R. S. Bear, O. E. A. Bolduan, and T. P. Salo, *J. Am. Leather Chemists' Assoc.* **46,** 107 (1951).
(302) H. H. Pfeiffer, *Arch. exptl. Zellforsch. Gewebezücht.* **25,** 92 (1943).
(303) J. O. Fixl, O. Kratky, and E. Schauenstein, *Monatsh.* **80,** 439 (1949).

Dichroism of the infrared spectrum of collagen was first reported by Fraser,[304] and later by Ambrose and Elliott in gelatin as well as in collagen.[32] There is perpendicular dichroism in the C=O and N—H stretching modes, and parallel dichroism in the N—H deformation mode, suggesting that these bonds are oriented perpendicular to the fiber axis. These results are at first sight similar to those obtained with β-keratin; however the N—H stretching frequency of 3330 cm.$^{-1}$ is significantly higher than in β-keratin (*ca.* 3300 cm.$^{-1}$). More recently it has been reported by Randall *et al.*[270] that in heat-contracted collagen the band reverts to the normal β-keratin frequency. These authors conclude that the higher frequency is characteristic of interchain hydrogen bonds, and the lower of intrachain bonds; they suggest that a change of this nature is involved in heat contraction.

c. *Chemical Properties*

(*1*) The amino acid composition of collagen has not proved easy to determine accurately, owing to the difficulty of effecting complete separation from the other constituents of skin. The most complete and accurate figures are those determined by Chibnall and his collaborators and by Bowes and Kenten (see Table XXIII, Vol. I, part A, p. 221, also the summary by Bowes and Kenten);[293] some new values have been determined by Graham, Waitkoff, and Hier.[99] Together, the figures account for over 99 per cent of the nitrogen present and agree with the values obtained from titration curves. These figures may be summarized as follows:

Gly	363	Arg	49	Lys	31	HOLys	7
Pro	131	Asp	47	Val	29	Met	5
HOPro	107	Leu } Ileu }	43	Phe	15	His	5
Ala	107			Thr	19	Try	—
Glu	77	Ser	32	Tyr	5	Cys	—

Total: 1072 residues per 10^5 gm. of protein.
Amide nitrogen, 47.

They exhibit several features which distinguish collagen sharply from other proteins. For example:

(*a*) The presence of hydroxyproline, which is not known to occur in any other protein (except elastin); and the high proportion (22 per cent of all residues) of proline and hydroxyproline together. Many theories of collagen structure have been based on a generalization derived from earlier measurements, but shown by these figures not to be even approximately correct, that one in three of all residues is proline or hydroxyproline.

(304) R. D. B. Fraser, *Discussions Faraday Soc.* No. **9,** 378 (1950).

(*b*) The occurrence of hydroxylysine, which has not been shown to be present in any other protein.

(*c*) The high proportion—51 per cent—of nonpolar residues, and especially the high proportion (34 per cent) of glycine. Some theories of collagen structure depend on the proportion of glycine being 33.3 per cent.

(*d*) The unusually large amounts of hydroxy acids (hydroxyproline, 10.0%; serine, 3.0%; threonine, 1.8%; hydroxylysine, 0.7%); the complete absence of cysteine/cystine; and the small numbers of aromatic acids.

Bowes and Moss[305] have applied Sanger's amino end-group technique to collagen, and find, in the untreated material, less than one end-group per 1,500,000 mol. wt. units; they conclude that *either* the true molecular weight of collagen is greater than 1,500,000, *or* the molecule contains cyclic chains, *or* the end groups are masked, e.g., by carbohydrate (see below). A few end groups are liberated in collagen on treating it with various reagents, e.g., alkali, formic acid, heat. Surprisingly, only about half the ϵ-NH_2 groups of lysine react with Sanger's fluorodinitrobenzene reagent (see Vol. I, part A, p. 196) although all the lysines titrate normally: even more surprisingly this remains true after heat shrinkage; treatment with urea, alkali, or acid; and conversion to gelatin.

Very little is known of the sequence of amino acids in collagen: but a study of partial hydrolyzates has shown[306] that the Bergmann-Niemann theory is untenable for this protein, at any rate in the simple form earlier proposed, according to which the sequence was

$$(\text{— glycine — proline or hydroxyproline — some other residue —})_n$$

It has for long been known that cartilage, tendon, etc. contain substantial amounts of polysaccharide material, which persists in purified collagen (see Bowes and Moss).[305] The nature of the polysaccharide is not at all clear, though it is known to contain hexoses (glucose, galactose, mannose) and hexosamine;[307] nor is there much indication of the type of binding between polysaccharide and protein.[308] The names chondroitin sulfate and hyaluronic acid have often been used in this connection, but in fact one cannot yet say definitely which polysaccharide is actually present in *purified* collagen fibrils (as opposed to whole connective tissue).

(305) J. H. Bowes and J. A. Moss, *Nature* **168,** 514 (1951).
(306) A. H. Gordon, A. J. P. Martin, and R. L. M. Synge, *Biochem. J.* **35,** 1369 (1941).
(307) J. Gross, J. H. Highberger, and F. O. Schmitt, *Proc. Soc. Exptl. Biol. Med.* **80,** 462 (1952).
(308) K. Meyer, *Discussions Faraday Soc.* No. **13,** 271 (1953).

Thus the fact that tendon becomes much less stable after treatment with hyaluronidase[309] (e.g., swells to a greater extent, is more soluble in acid, and shrinks at a lower temperature), though very interesting from the point of view of tissue structure, does not necessarily give much indication of the type of polysaccharide intimately concerned in the structure of collagen itself. That polysaccharide is thus intimately concerned has been strongly suggested by the work of Highberger, Gross, and Schmitt[310] on reconstituted collagen (see p. 928).

(*2*) Pepsin reacts readily with collagen: other mammalian proteases (trypsin, cathepsin, etc.) react only after the material has been mechanically disintegrated.[311] Specific collagenases are however produced by many microorganisms (for references, see Smith;[312] see also the discussion by Robb-Smith[313]).

The action of some other reagents is dealt with elsewhere, e.g., of alkali yielding gelatin (see p. 923) and of dilute acids yielding soluble collagen (see p. 925).

Collagen is apparently immunologically inactive.[314]

(*3*) One class of chemical change undergone by collagen, namely the tanning reactions, is so very important industrially that it must be given separate mention. Space does not permit a full discussion of tanning; a detailed review from the biochemical point of view has been given by Gustavson,[315] while physicochemical aspects of tanning were discussed in a recent symposium.[315a] The common feature of all tanning agents is that they are polyfunctional, their action being to introduce numerous stable cross links between the chains of the collagen. Indeed the tanning of collagen is the longest known and most intensively studied of all protein cross-linking reactions. Chemically speaking, tanning agents are of very diverse kinds: the most common are basic chromic salts, vegetable tannins (complex polyphenols), aldehydes (e.g., formaldehyde), condensates of phenols with aldehydes (made soluble by introducing $-SO_3H$ groups), unsaturated oils, and quinones. The reactions involved have been studied intensively but are complicated and are still only partly understood.

Basic chromium salts act through coordination of the carboxyl groups

(309) D. S. Jackson *in op. cit.*,[272] p. 177.
(310) J. H. Highberger, J. Gross, and F. O. Schmitt, *Proc. Natl. Acad. Sci. U. S.* **37,** 286 (1951).
(311) I. W. Sizer, *Enzymologia* **13,** 288, 293 (1949).
(312) E. L. Smith, *in* J. B. Sumner and K. Myrbäck (eds.), The Enzymes, Academic Press, New York, 1951, Vol. I pt. 2, p. 862.
(313) A. H. T. Robb-Smith *in op. cit.*,[272] p. 14.
(314) B. H. Waksman and H. L. Mason, *J. Immunol.* **63,** 427 (1949).
(315) K. H. Gustavson, *Advances in Protein Chem.* **5,** 353 (1949).
(315*a*) *Discussions Faraday Soc.* No. 16 (1954).

on adjacent chains with a polynuclear chromium complex;[316] other groups, including the peptide links, are probably not involved. It is found, for example, that basic chromium salts are not fixed by collagen in which the carboxyl groups have been masked by methylation. The cross links formed by chromium are remarkably stable, chromed collagen being completely resistant to boiling water and proteolytic enzymes.[315]

It would appear that the vegetable tannins react both with basic side chains and with the peptide bonds, in the latter case presumably by hydrogen-bonding. The evidence for the former type of reaction is that deamination of collagen decreases its capacity for irreversible fixation of tannins;[317] for the latter, that after this treatment some binding capacity nevertheless still remains, and that the tannin-binding capacity of polyamides containing no ionizable side chains is very great.

The principal action of formaldehyde as a tanning agent is to combine with the ϵ-NH_2 groups of lysine and the guanidino groups of arginine.[315,318]

d. Gelatin

The manufacture of gelatin and glue (which is merely an impure and concentrated form of gelatin) is the second major industry in which collagen is involved, and therefore the conversion of collagen to gelatin has been the subject of much research (for a detailed review see Ferry).[176] The starting material is skin or bone; the material is treated with lime or sometimes with acid (pH 3.5),[319] and is extracted with water at 60°C. The extract is a solution of gelatin.

The main action of the alkali[320,321] is to hydrolyze amide groups to carboxyl, shifting the isoelectric point to the region of pH 4.8–5.0; there are also minor reactions—some arginine residues lose urea to form ornithine, and a few are further degraded to citrulline; some amino end groups (mainly glycine) are freed. Little is known of the effects of acid treatment; the isoelectric point is shifted to about pH 9, but the mechanism is obscure.

The property of gelatin which gives it industrial importance is that if a solution of concentration greater than about 1 per cent is cooled below about 40°C. it sets to a stiff gel: this transformation is reversible.

We may first of all consider the properties of gelatin in solution above the gelling temperature. Its amino acid composition (apart from a lower

(316) K. H. Gustavson, *J. Intern. Soc. Leather Trades' Chemists* **36,** 182 (1952).
(317) A. W. Thomas and S. B. Foster, *J. Am. Chem. Soc.* **48,** 489 (1926).
(318) J. H. Bowes and R. H. Kenten, *Biochem. J.* **44,** 142 (1949).
(319) Bone is given a preliminary treatment with acid to remove calcareous matter.
(320) J. H. Bowes and R. H. Kenten, *Biochem. J.* **43,** 365 (1948).
(321) J. H. Bowes, *Research* (London) **4,** 155 (1951).

figure for amide N) is extremely close to that of collagen. There have been various views about its molecular weight.[322] Scatchard *et al.*[323] examined gelatin solutions with the ultracentrifuge, by viscosity, and by flow birefringence, and concluded that although the solutions were far from being monodisperse, they could be regarded as more or less degraded versions of a "parent gelatin" of molecular weight about 110,000, containing 1170 residues, and having dimensions about 800 × 17 A.; these results accord well with more recent studies of various kinds of collagen solution, from which it has become probable that a definite collagen "molecule" does have real identity. On the other hand, Pouradier *et al.*[324] developed methods of fractionating gelatin solutions, and succeeded in preparing more or less homogeneous fractions whose molecular weights covered a continuous range from 15,000 to 250,000. They showed that gelatin solutions may be regarded as true, i.e., molecularly dispersed, solutions, and that the melting point increased with molecular weight to a limiting value;[325] on the other hand, the isoelectric point and chemical composition were quite, and the acid- and base-binding capacities almost, independent of molecular weight.[326] The axial ratio increases as one moves away from the isoelectric point, presumably owing to the mutual repulsion of like charges;[324] this effect, as would be expected, is suppressed by salts.

Next we may consider the mechanism of gelling. The most striking characteristic of gelatin gels is that they may contain as little as 1 per cent protein: this would appear to exclude a "solvation" mechanism of gelation. Another mechanism which has been proposed, involving long-range forces, does not appear to have experimental evidence in its favor. We are left with the "network" theories; and in fact the evidence agrees well with the idea of a network containing regions of local order[327] and stabilized by weak bonds—van der Waals attractions or hydrogen bonds. The fact that the melting point of gelatin gels is lowered by urea favors the latter type of bond. However, the exact nature of the linkage is still obscure, though it has been shown that the free guanidino groups of arginine are essential for gelation, whereas the ϵ-NH_2 groups of lysine are not involved.[328] Recently, light has been thrown on the role of hydrogen

(322) A. G. Ward, *Research* (London) **4,** 119 (1951).
(323) G. Scatchard, J. L. Oncley, J. W. Williams, and A. Brown, *J. Am. Chem. Soc.* **66,** 1980 (1944).
(324) J. Pouradier and A. M. Venet, *J. chim. phys.* **47,** 11 (1951).
(325) J. Pouradier and A. M. Venet, *J. chim. phys.* **47,** 391 (1951).
(326) J. Pouradier, J. Roman, and A. M. Venet, *J. chim. phys.* **47,** 887 (1951).
(327) K. Hermann and O. Gerngross, *Kautschuk* **8,** 181 (1932).
(328) P. Grabar and J. Morel, *Bull. soc. chim. biol.* **32,** 643 (1950).

bonding by the work of Robinson and Bott[329,330] who have studied the optical rotatory power of hot and cold gelatin solutions, which is known to double (approximately) on gelling. They were able to "freeze in" to a film prepared from hot solution the optical rotatory power characteristic of hot solutions: and they found that such films gave infrared N—H stretching frequencies (3310 $cm.^{-1}$) similar to those of α-protein, while films prepared in the cold exhibited a frequency of 3330 $cm.^{-1}$, characteristic of collagen. They suggest that gelling consists essentially of a change from a solution of independent chains in the α-configuration stabilized by internal, intra-chain, hydrogen bonds, to a network held together by inter-chain hydrogen bonds in the configuration we describe as the collagen structure.

The mechanical and other properties of gelatin gels have been studied in great detail; an account of them is to be found in the review previously referred to.[176] The x-ray diffraction results are of interest (for references, this review should be consulted, p. 34). Gelatin gels, and films dried from them, give wide-angle x-ray patterns closely resembling that of collagen itself (no low-angle pattern is observed even in oriented specimens), and behaving similarly on swelling and shrinkage; on the other hand, gelatin solutions, and films dried from them, give no typical collagen reflections. The progress of gelation can be followed by means of the x-ray pattern, and the process of "crystallization" is found to be a slow one: similarly on melting a gel the collagen reflections disappear gradually. Study of the mechanical properties confirms this picture of gelation.

Thus, although the details are not understood, it does seem that the sol–gel transition in gelatin must be regarded as involving a real structural change in the gelatin "molecule" itself, followed by cross linkage to a three-dimensional network structure.

e. Soluble and Regenerated Collagen

It was first shown in 1900 by Zachariadès[331] that rattail collagen can be dissolved in certain dilute acids—e.g., 1:25,000 acetic acid—to a clear solution; Nageotte[332] showed that on adding dilute salt (e.g., NaCl) this solution throws down a fibrous precipitate apparently closely related to the original collagen; alternatively, on dialyzing out the acid, a gel is formed.

(329) C. Robinson and M. J. Bott, *Nature* **168,** 325 (1951).
(330) C. Robinson *in op. cit.,*[272] p. 96.
(331) M. P.-A. Zachariadès, *Compt. rend. soc. biol.* **52,** 182, 251, 1127 (1900).
(332) J. Nageotte, *Compt. rend. soc. biol.* **96,** 172 (1927).

More recently a group of Russian workers has shown[333–335] that by a related procedure a soluble form of collagen, which they have named procollagen, can be extracted from the skin of many animals (birds, fish, and reptiles, as well as mammals). The method is to extract with organic acid buffers (pH 3.0–4.5), preferably citrate: the clear extract on dialysis against water yields a precipitate which is described as "crystalline" and which photographs show to consist of long needles (Randall *et al.*[336] conclude, however, that these objects are not true crystals). This type of extract is evidently very similar to the Nageotte material although, to judge by their different behavior on dialysis against water, the two cannot be identical.

The Nageotte material was early studied by Wyckoff and Corey[337] by x-ray methods: they found that when oriented by stretching it gave a diffraction pattern, including long spacings, which resembled that of native collagen although it was imperfectly ordered. Electron-microscope examination[289,338] showed that the precipitated fibrils exhibited periodic cross-striation apparently identical with that of native collagen. Recently, Noda and Wyckoff[339] have examined the effects of ionic strength, pH, and concentration on fibril formation and have found that 0.02–0.1 *M* citrate buffers of pH 4.6–5.0 give the best results.

The Russian procollagen has also been studied in some detail. Bresler *et al.*,[340] using physicochemical techniques, have concluded that in solution procollagen has a molecular weight of about 70,000 and dimensions of about 380 × 17 A. Randall *et al.*[336] found that procollagen resembles collagen in wide-angle x-ray diffraction and infrared spectrum. On the other hand, some differences in composition have been noted:[336,341] the conversion of collagen to procollagen involves the loss of some component rich in tyrosine and hexosamine and containing little hydroxyproline—perhaps a mucopolysaccharide, it is suggested.[336]

(333) A. A. Tustanovsky, *Biokhimiya* **12,** 285 (1947).
(334) V. N. Orekhovich, A. A. Tustanovsky, K. D. Orekhovich, and N. E. Plotnikova, *Biokhimiya* **13,** 55 (1948).
(335) V. N. Orekhovich, Communications au deuxième Congres International de Biochimie. Editions de L'Académie des Sciences de l'U.R.S.S., Moscow, 1952, p. 106.
(336) J. T. Randall, G. L. Brown, S. Fitton Jackson, F. C. Kelly, A. C. T. North, W. E. Seeds, and G. R. Wilkinson *in op. cit.*,[272] p. 213.
(337) R. W. G. Wyckoff and R. B. Corey, *Proc. Soc. Exptl. Biol. Med.* **34,** 285 (1936).
(338) G. Bahr, *Exptl. Cell Research* **1,** 603 (1950).
(339) H. Noda and R. W. G. Wyckoff, *Biochim. et Biophys. Acta* **7,** 494 (1951).
(340) S. E. Bresler, P. A. Finogenov, and S. Ya. Frenkel, *Compt. rend. acad. sci. U.R.S.S.* **72,** 555 (1950).
(341) J. H. Bowes, R. G. Elliott, and J. A. Moss *in op. cit.*,[272] p. 199.

FIG. 20. Electron micrograph of a mixture of FLS and collagen-type fibrils produced by dialysis of an acetic acid solution of ichthyocol and an intermediate concentration of human α_1 acid glycoprotein against water: magnification 20,925×. By courtesy of Highberger, Gross, and Schmitt.

Highberger, Gross, and Schmitt[310,342] have found in electron-microscope studies that the reprecipitated procollagen contains some fibrils whose cross-striations correspond closely with those of native collagen. But these authors made the discovery that, mixed with the "normal" fibrils, were some whose cross-striations were quite unlike any observed

(342) J. H. Highberger, J. Gross, and F. O. Schmitt, *J. Am. Chem. Soc.* **72**, 3321 (1950).

in native collagen, having a period ranging from 2000 to 3000 A. (see Fig. 20); these they call fibrous long-spacing or FLS fibrils. Unlike normal collagen, FLS fibrils have a band system which is nonpolarized. This remarkable observation has led to the discovery by the same and other authors of a series of different types of cross-striated fibrils obtained from collagen solutions under various circumstances. It is still not at all clear which factor in a particular set of conditions is responsible for a given type of striation; all one can do at the present time is to enumerate the results without much interpretation.

(*1*) In their experiments with the Nageotte type of extract, Noda and Wyckoff[339] found that occasionally fibrils were produced having a period of about 220 A. reminiscent of the fibrils with similar period present in connective tissue; they obtained these by extracting tendon with HCl (pH 3.8) and precipitating with 0.03 *M* potassium citrate (pH 5). Similar results have been reported by Highberger *et al.*[310]

(*2*) Highberger *et al.*[310] observed that where the original collagen was purified by the method of Bergmann and Stein,[343] involving treatment with 10% NaCl, dibasic phosphate, and ether, before acetate extraction, few or no FLS fibrils were obtained. Now procollagen can be precipitated from extracts of skin made with phosphate alone; and these fibrils are found to be of the FLS type. The conclusion drawn was that skin contains some factor soluble in phosphate which promotes the formation of FLS fibrils; procollagen extracted from skin which has previously been treated with phosphate does not contain this factor and therefore will not yield FLS fibrils. Bearing in mind the known association of mucopolysaccharide with collagen and the fact that mucoprotein is very soluble in phosphates, they proposed and tested by experiment the hypothesis that mucoprotein was the factor concerned. Using material from bovine plasma, they found that where mucoprotein was added in the ratio of about 1:10 to the amount of procollagen, the precipitate consisted almost entirely of FLS fibrils; with a ratio about 1:100 a mixture of FLS and normal fibrils, and at 1:1000 normal fibrils only, were obtained; and at much lower ratios an apparently structureless gel was formed. Similar results were obtained with other mucoproteins, including the highly purified preparation of Schmid.[344]

More recently the same authors[345] (and also Randall *et al.*[270,346]) have studied the effect of numerous test substances on the type of fibril

(343) M. Bergmann and W. H. Stein, *J. Biol. Chem.* **128,** 217 (1939).
(344) K. Schmid, *J. Am. Chem. Soc.* **72,** 2816 (1950).
(345) J. Gross, J. H. Highberger, and F. O. Schmitt, *Proc. Soc. Exptl. Biol. Med.* **80,** 462 (1952).
(346) S. Fitton Jackson and J. T. Randall *in op. cit.*,[272] p. 181.

obtained from acid extracts of ichthyocol. They have found a whole range of materials of widely different chemical types which can induce the formation of FLS fibrils, e.g., mucoproteins, thrombin, proteinases, adrenocorticotropic hormone (ACTH), heparin, hyaluronates, chondroitin sulfates; other materials, e.g., serum albumin, gliadin, and protamine, gave a negative result. In some cases normal and FLS fibrils were obtained, in others unstriated or FLS fibrils. The whole picture is com-

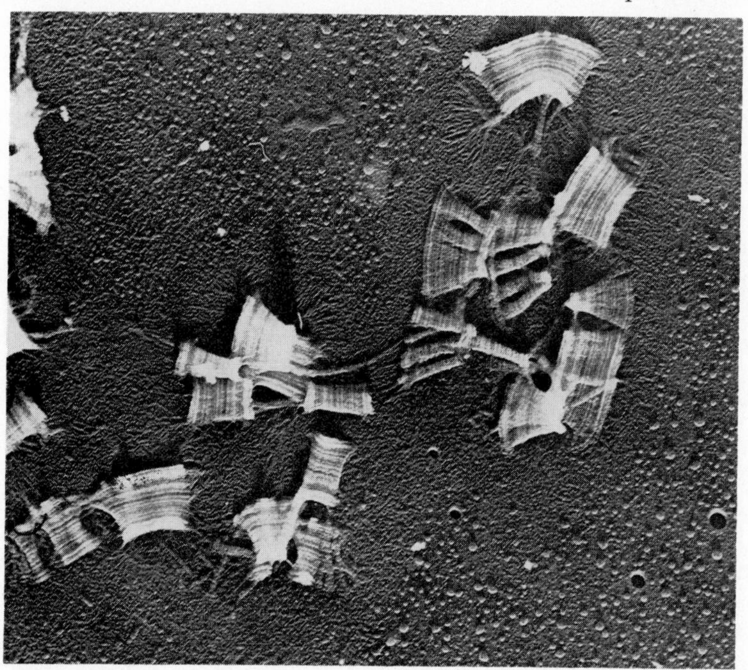

Fig. 21. Electron micrograph of SLS particles formed by adding acid ATP to an acetic acid solution of ichthyocol: fraying produced by washing in dilute acetic acid in absence of ATP. Magnification 30,150×. After Schmitt, Gross, and Highberger.[347]

plicated: the authors conclude that the acid extract itself must contain all factors necessary for the production of both normal and FLS fibrils, and that the action of the various added reagents is relatively nonspecific under the conditions of the experiment. It still appears probable, though not yet proved, that polysaccharide material—which is present in the original acid extract—is in some way involved.

(*3*) Schmitt, Gross, and Highberger[347] have now discovered yet another type of cross-striated particles by a slight variation in procedure, involving the dialysis of a phosphate extract (pH 8) against citrate buffer

(347) F. O. Schmitt, J. Gross, and J. H. Highberger, *Proc. Natl. Acad. Sci. U. S.* **39,** 459 (1953).

(pH 4). These are segments about 2000 A. long having a characteristic cross-banded pattern (containing at least 20 bands) differing both from normal collagen and from FLS material, in particular in that the banding is polarized but has a period the full length of the segment, i.e., of the order of 2000 A. (see Fig. 21). Electron micrographs clearly show that the segments are made up by the parallel aggregation of very thin filaments, as are normal collagen and FLS fibrils. These objects have been named SLS or segment long-spacing fibrils.

Since phosphate extracts of skin give absorption spectra similar to those of nucleic acid derivatives, the influence of such derivatives in promoting SLS formation was investigated. Many gave negative results, but adenosinetriphosphoric acid (ATP) was found to be extremely effective in producing SLS precipitates. The significance of this discovery is not clear, but it is a most interesting one in view of the intimate role played by ATP in biological systems. It should however be added that there is no evidence for the presence of SLS (or FLS) material in tissues.

By various manipulations normal collagen fibrils, FLS fibrils, and SLS material can be reversibly interconverted.

The interpretation and implication of all these results are far from clear, though they are evidently very relevant to problems of tissue genesis. In the meantime one cannot overstress the remarkable character of the collagen regeneration reactions, in which typical fibrils, exhibiting all the features of long-range order characteristic of those found in the living organism, are produced from a true solution by a test-tube, enzyme-free reaction. An understanding of these and analogous reactions would undoubtedly do much to dissipate our at present almost complete ignorance of a fundamental biological problem, namely the mechanism of production of highly organized structure in living organisms.

f. The Formation of Collagen in Living Organisms

Very little is known of the way in which collagen fibers are laid down *in vivo*. The precipitation of typical collagen fibrils from solution, coupled with the fact that all collagen fibers appear to be extracellular, might lead to the conclusion that a collagen "solution" is secreted by the cell, and that this is precipitated to form collagen fibers. However, tissue-culture studies[279] suggest that intracellular as well as extracellular processes are involved, though at least part of the growth of the fibers takes place outside the cells. But how they are oriented and organized remains unknown. There is, however, evidence that alkali-soluble collagen is a true precursor of collagen proper.[347a]

(347*a*) R. D. Harkness, A. M. Marko, H. M. Muir and A. Neuberger, *Biochem. J.* **56,** 558 (1954).

Some studies have been made of the connective tissue of young and embryonic organisms and of tissue cultures, and the progressive appearance of collagen fibers has been noted: they are often preceded by thin fibrils (less than 500 A. across) exhibiting a simple cross-striation of period between 150 and 300 A.[270,278]

This is the third context in which we have encountered such a periodicity; it will be recalled that similar fibrils are found associated with normal collagen in connective tissue, and in reprecipitated collagen under certain conditions. Fibrin has a similar axial period but there are no reasons to suggest a connection.

Investigations by Gross[348] and others have shown a progressive decrease with age in the proportion of fibers which are argyrophilic, i.e., which would classically be diagnosed as reticulin. In this connection it would be most interesting to have more information about the relation between reticulin and collagen proper (see p. 943). Differences have been reported between the ultraviolet and x-ray spectra of immature and mature collagens;[349,350] but the latter were not confirmed by Bear (quoted by Gross).[348] For a general discussion and bibliography see Fitton Jackson.[279]

3. Theories of the Structure of Collagen

a. Introduction

A number of structures has been proposed for collagen, but at the present time it seems rather improbable that any of them is correct, though some are probably on the right lines. Certain crucial pieces of information, in principle probably capable of being settled by direct appeal to experiment, have not in fact been definitely so settled—for example, whether the individual chains of collagen are or are not greatly extensible, and the exact nature of the changes undergone by the wide-angle equatorial x-ray pattern during swelling and shrinkage. All that can be done at present is to review the principal theories which have been proposed and to point out the respects in which they do or do not agree with experimental findings.

It is convenient to preface a discussion of the configuration of the chain itself by some consideration of relations between neighboring chains in the structure.

(348) J. Gross, *J. Gerontol.* **5,** 343 (1950).
(349) G. C. Heringa and A. Weidinger *Acta Neerl. Morphol.* **4,** 291 (1946).
(350) O. Kratky, M. Ratzenhofer, and E. Schauenstein, *Protoplasma* **39,** 684 (1950).

b. General Features of the Structure of Collagen

Regarding the individual polypeptide chain as the basic unit of structure, we may imagine various ways in which an assembly of them might be built up to give a "fibrous" structure:

(*1*) Chains running indefinitely along the fiber axis.

(*a*) Chains strongly bound in all directions, i.e., three-dimensional order.
(*b*) Neighboring chains in sheets, between which the bonding is weak: i.e., two-dimensional order.
(*c*) Weak bonding in all directions between chains, i.e., one-dimensional order.

(*2*) Chains coiled or folded into "globular" particles, which would then be joined together in some way to form fibrils (cf. the k-m-e-f group).

The balance of evidence is against structures of type 2. These would exhibit large transverse periodicities; but no equatorial x-ray reflections of spacing greater than 11–17 A. have been observed, and in the electron microscope the fibril width appears to be capable of any variation right down to the limit of resolution of the best instruments. This behavior may be contrasted with that of, e.g., myofibrils.

Structures of type 1*a* may also be excluded by virtue of the swelling properties of collagen: it is difficult to see how relatively large amounts of water could penetrate a structure bound three dimensionally by covalent bonds in such a way that this structure maintained its integrity (as demonstrated by x-ray diffraction) and mechanical strength, and reverted to its original condition on drying. We are left with what may be called sheet structures and protofibrillar structures. A choice might immediately be made between these alternatives if the configuration of the chain itself were established, since the inter-chain bindings open to it would probably be obvious: but in the absence of any proved theory of chain configuration it is necessary to consider other evidence.

Bear and Schmitt and their collaborators have been the chief protagonists of the protofibrillar theory, and they have assembled an impressive mass of evidence[269] in favor of a one-dimensional structure, i.e., a structure in which the largest chemically bound (i.e., covalently bonded) units are very long in the axial direction and very narrow in all directions normal to the axis. The evidence may be summarized as follows:

(*1*) *The Low-Angle Diffraction Pattern.* In spite of an intensive search, no equatorial long spacings have been found in the collagen low-angle pattern; it consists *only* of a set of meridional lines which, even under high resolution, exhibit none of the "row line" structure to be expected if there were transverse order. A detailed examination of this diffraction pattern has shown[300] that it can indeed best be explained on the basis of one-dimensional order. Further, a model consisting essentially of thin, rodlike molecules of indefinite length can be used to interpret the changes suffered by the low-angle diffraction pattern on staining, drying, etc. (see sec. *c* below).

(*2*) *The Equatorial Wide-Angle Pattern.* It has been indicated that, on swelling, the 11-A. reflection on the equator moves inwards to a spacing eventually of 17 A. This can readily be interpreted on the basis of cylindrical molecules, of the order 11–12 A. thick, which move apart when swelling takes place. The other equatorial reflections remain more or less unchanged on swelling (but see below): they are said to be due to short intramolecular spacings. Another feature of the wide-angle x-ray pattern of collagen which argues in favor of a fibrillar rather than a sheet structure is that no one has ever succeeded in preparing specimens of collagen exhibiting double orientation, as should be possible in the latter case. In this respect collagen may be contrasted with β-keratin.

(*3*) *The Electron Microscope Results.* Collagen fibrils generally tend to cleave longitudinally, and frayed filaments are a common feature of electron micrographs: the characteristic banded appearance persists in such filaments. Their diameters vary continuously over the whole resolvable range down to 50–100 A. This behavior is consistent with the "molecule" being very thin, of the order of a very few polypeptide chains across, in fact.

(*4*) *Degraded Collagens.* It is clear that both gelatin and "soluble collagen" are structurally very closely related to collagen itself—they are not denatured products, as witness the x-ray pattern of gelatin and the electron micrographs of reprecipitated "soluble collagen." Scatchard *et al.*[323] concluded that the molecule of "parent gelatin" has dimensions 800 $\times$ 17 A.: while Bresler *et al.*[340] found dimensions of 380 $\times$ 17 A. for procollagen. These figures are in remarkably good agreement with the model.

(*5*) *Light Scattering.* The results of measurements of light scattering suggest that in acid solution collagen particles have diameters of only 10–20 A.[351]

Such is the case for an essentially unidimensional structure for collagen. It has not been universally accepted. By implication some of the chain models to be discussed below could hardly be made to agree with such a conception; and more explicitly a contrary view has been proposed by Randall *et al.*[270] who have pointed out that if collagen really consists of cylinders in hexagonal close-packing one would expect that on swelling *all* the equatorial reflections would move in sympathy. They prefer to interpret the data in terms of two-dimensional sheets which maintain their integrity on swelling, while water penetrates between them: the 11-A. reflection would then be intersheet, and the other equatorials intrasheet reflections. But of course the observed reflections correspond to intersections in reciprocal space of the Fourier transform of the molecule with the reciprocal lattice. If that lattice is hexagonal and contracts and expands as a consequence of the swelling and shrinkage of the fiber, the equatorial reflections should all move in sympathy and their intensities change according to the intercepts of the corresponding reciprocal lattice

(351) M. B. McEwen and M. I. Pratt *in op. cit.*,[272] p. 158.

points with the Fourier transform of the molecule. However, at wide angles the reflections are closely crowded together and it may be that what we observe is not individual reciprocal lattice points, but as it were a smear of reflected intensity consisting of several reflections blurred together and essentially giving an outline of a region of high intensity in the Fourier transform. When the molecules move further apart this transform stays the same, and the blur corresponding to an intense region in it also remains apparently the same, though in fact it now originates as the summation of a new set of reflections.

In this way a situation could occur where a structure of the kind postulated by Bear and Schmitt would give rise to an equatorial pattern the outer parts of which would apparently remain unchanged on swelling: only in the inner parts, where reflections are more widely separated and can be clearly resolved, would their motion be apparent.

Whether this is what actually happens, or whether the interpretation of Randall *et al.* is correct, cannot yet be settled, since the experimental data are inadequate: more and better x-ray pictures should be decisive. Until these have been taken it can only be repeated that the body of data collected by Bear and Schmitt in favor of a one-dimensional structure is very impressive.

c. *Elaborations of the Unidimensional Theory of Structure*

Attention has been drawn to the changes in the low-angle meridional reflections of collagen on drying: the lines lengthen, the high orders most and the low orders least, and develop noncentral maxima of intensity. Bear and his collaborators have developed a detailed theory to account for this;[352,353,301] they explain it by assuming that the cylindrical molecule on drying develops regions of disorder regularly spaced along its length. The disorder may be resolved into axial and radial components, the former of which is responsible for the general fanning-out of the pattern, and the latter for the noncentral maxima. These ideas are correlated with the electron microscope results (see Fig. 22) by assuming that the visible dark bands, whose diameters are greater than that of the rest of the fibril, are regions where the long and polar side chains are concentrated: on swelling these preferentially take up water (they also preferentially stain), and on shrinkage they are left more disordered than the rest of the fiber because of the more-than-average size of the side chains in those regions (polar side chains are mostly larger than nonpolar ones).

Different theories have been proposed by other workers to account for the large periodicities: Kratky thought that the low-angle reflections

(352) R. S. Bear and O. E. A. Bolduan, *Acta Cryst.* **3,** 230, 236 (1950).
(353) R. S. Bear and O. E. A. Bolduan, *J. Applied Phys.* **22,** 191 (1951).

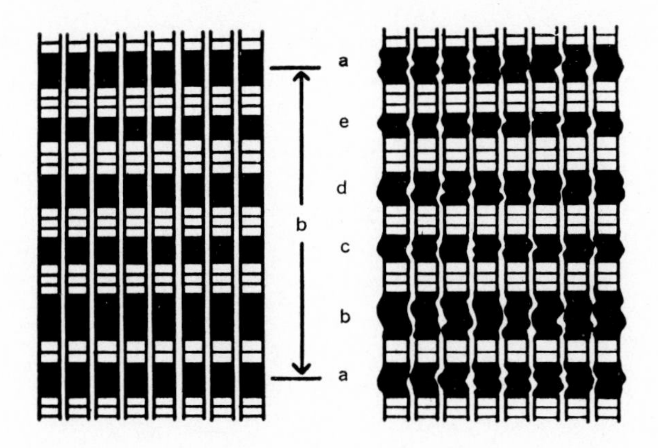

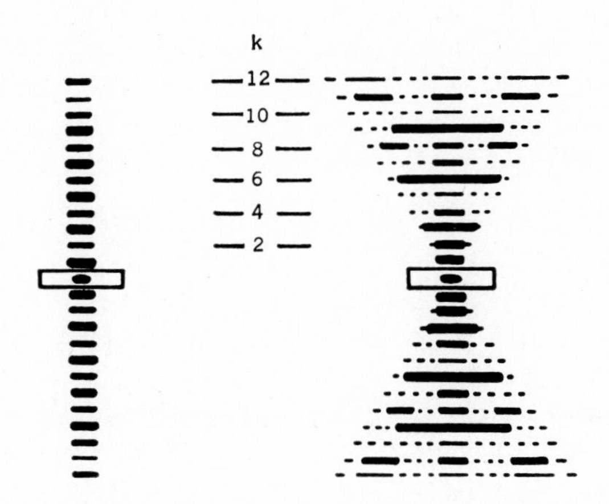

FIG. 22. Scheme suggested to account for the changes observed in the low-angle diffraction pattern of collagen on drying. Above, longitudinal sections through perfect (left) and mixed perfect and imperfect (right) models for the fibril structure; below, the corresponding x-ray patterns. Solid areas represent the electron-optically visible bands *a–e* which become distorted on drying; the other regions do not change on drying. "Perfect" regions give short meridional diffraction lines, on all layers in the wet fibril, and on some only in the dry. The latter also shows components of longitudinal imperfection, e.g., the long line at $K = 9$, and components of radial imperfection, e.g., the pair of noncentral maxima at $K = 8$. After Bear, Bolduan, and Salo.[301]

were due to regular sequences of amino acids,[354] while Randall *et al.*[270] suggested that the banding and low-angle reflections were due to the chain being systematically thrown into large folds in regions where a high concentration of polar side chains occurs.

There has naturally been a good deal of speculation about the relationship between collagen molecules in solution and the large periodicities of collagen fibrils. Bresler *et al.*[340] concluded that procollagen was a tightly coiled chain only 380 A. long; if fully stretched the polypeptide chain contained in it would have a length of the order of 2000 A. The obvious suggestion has been made[269] that FLS fibrils, with their period of about the same length, are built of stretched or "β-" procollagen molecules in antiparallel array (giving a nonpolarized structure). The implication would be that SLS segments, with their polarized banding, contain similar arrays of "β-" collagen, but in parallel array; while in collagen proper there would be parallel arrays of *folded* chains. This scheme is attractive but undoubtedly too simple as it stands to explain the very complicated phenomena of reconstitution, while it suffers from the serious defect that FLS fibers do not give a "β-" type wide-angle x-ray photograph, but instead one virtually identical with that of normal collagen,[347] suggesting that the two forms have the same chain configuration.

d. Configuration of the Polypeptide Chain in Collagen

(1) Introduction. We have now to pass in review a series of proposals for the configuration of the chain itself. Nearly all of these are based on the occurrence of a prominent meridional reflection at 2.86 A., quite unlike anything observed in the keratins: and also most of those who have proposed structures have been impressed by the high percentage of proline (which will be taken to include hydroxyproline for the purposes of this section) in collagen. There can be no reasonable doubt of the structural significance of the prolines. It must, however, be remembered that there exist materials giving the characteristic collagen diffraction pattern which have been reported to contain little of these amino acids, though it now seems rather probable that at any rate some of the analyses were at fault or that the materials were very impure (these "abnormal" collagens are generally derived from primitive and little-studied organisms); thus recent analyses by Gross[355] of spongin show, in contrast to earlier studies of this protein, that it contains as much proline as does ordinary skin collagen. Again, Neuman and Logan[356,357] have found

(354) O. Kratky, *Monatsh.* **77,** 224 (1947); *idem, J. Polymer Sci.* **3,** 195 (1948).
(355) J. Gross, personal communication.
(356) R. E. Neuman *Arch. Biochem.* **24,** 289 (1949).
(357) R. E. Neuman and M. A. Logan, *J. Biol. Chem.* **184,** 299 (1950).

only minor differences between the proline and hydroxyproline contents of collagens derived from a variety of sources, including a number of mammalian species as well as fish and turtle. There is an urgent need for more redeterminations using modern methods, particularly in the invertebrate collagens. In the meantime generalization is impossible, but even if it should prove that high proline content is a universal feature of collagens, the exact quantitative relationships have still to be settled; many structures which have been proposed are based on an assumption that one residue in three is proline, while in fact no modern analyses have indicated a higher proportion than about one in four.

The main special feature of proline from the structural point of view is that unlike all the other naturally occurring acids in proteins it is an imino and not an amino acid. This is to say that the hydrogen-bonded peptide link

```
          \
           CH—R
          /
     O=C
          \             /
           NH---O=C
          /             \
   R—CH
          \
```

which forms the basis of so many configurations proposed for polypeptide chains, is not possible for this acid, since its peptide link contains no free hydrogen atom

```
          \
           CH—R
          /
 ---O=C
          \
           N——CH2
           |       \
           |        CH2
           |       /
           CH—CH2
          /
```

Also, the ring of proline is sterically clumsy: if, for example, one attempts to insert a proline residue in the Pauling α-helix the difficulty is not merely that the sequence of hydrogen bonds is broken, but that the proline residue positively will not fit at all and imposes a twist or kink in the chain as a whole.

(*2*) *The Astbury Structure.*[358,359] The essence of this structure was to find a way of arranging the residues in a chain which should be inextensible and yet have a repeat (2.86 A.) less than that of the fully extended chains of β-keratin (3.34 A.); and this was done by postulating a *cis-*

(358) W. T. Astbury, *J. Intern. Soc. Leather Trades' Chemists* **24,** 69 (1940).
(359) W. T. Astbury & F. O. Bell, *Nature* **145,** 421 (1940).

configuration for every third peptide bond, the abnormal steric properties of the proline and glycine residues being held responsible for this arrangement

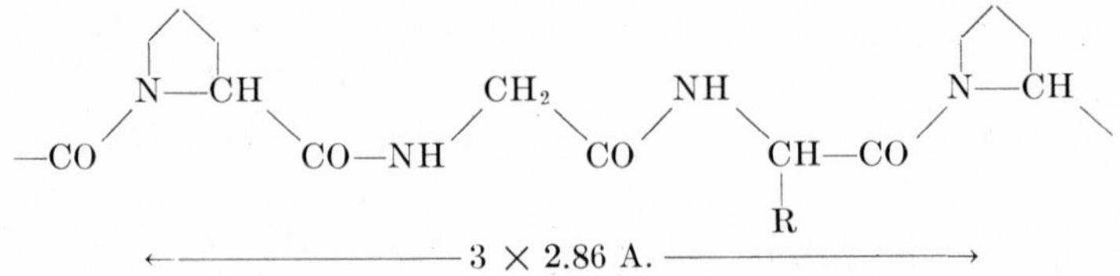

```
        N—CH               CH2        NH              N—CH
       /    \             /   \      /  \            /    \
  —CO        CO—NH              CO       CH—CO
                                         |
                                         R
  ←——————————————— 3 × 2.86 A. ———————————————→
```

The most serious difficulties about this model are that (*a*) its repeat only *averages* 2.86 A. per residue: the true repeat is 3 × 2.86 = 8.58 A., not a strong reflection; (*b*) the chain has no screw axis of symmetry and would probably tend to curl up;[221,241] and (*c*) the structure is based on a literal interpretation of Bergmann-Niemann periodicity principles since shown to be untenable.

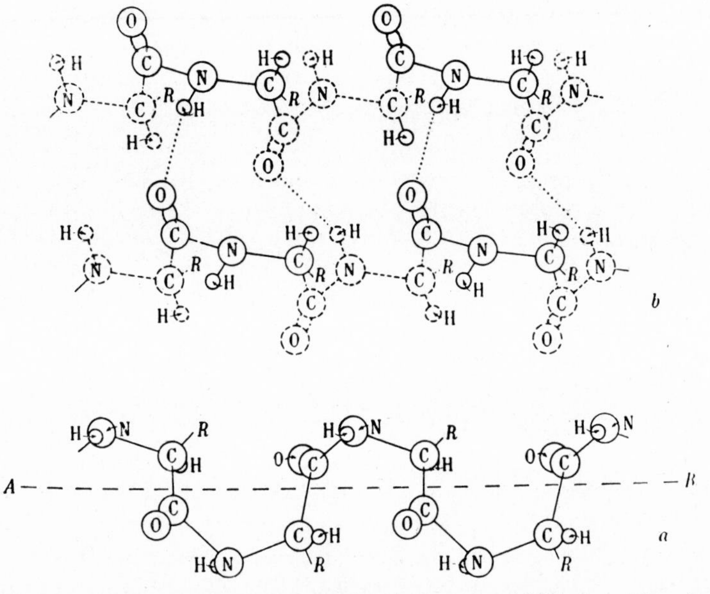

FIG. 23. Structure for collagen proposed by Ambrose and Elliott. After Ambrose and Elliott.[32]

(*3*) *The Huggins,*[221] *Ambrose-Elliott*[32] *and Related Structures.* Huggins thought in terms of two-dimensional sheets of chains held together by hydrogen bonds, the side chains sticking up or down from the sheet which could thus accommodate bulky residues (e.g., proline). A modification of this structure has been proposed by Ambrose and Elliott; the details are shown in Fig. 23. The purpose of this modification is to take account of the observed infrared dichroism, according to which both C═O and N—H bonds are perpendicular to the chain direction.

Other structures have been proposed, e.g., those of Zahn,[360] which are in essence similar to those so far discussed, in that (*a*) the shortening of the main repeat from the 3.34 A. of β-keratin to the 2.86 A. of collagen is achieved by the introduction of *cis*-peptide bonds, and (*b*) the whole structure is two-dimensional since inter-chain hydrogen bonds bind the structure into sheets. The model next to be discussed resembles these in the first particular, but differs from them in the second.

(*4*) *The Pauling-Corey Structure.*[361] Pauling and Corey accepted one of the main conclusions of the Bear school in regarding the collagen molecule as unidimensional, with all hydrogen bonds internally satisfied. Interpreting the equatorial wide-angle reflections as due to a hexagonal net, they deduced a unit cell of basal area 125 sq. A. and height 2.86 A.; from the observed density they calculate that this cell must contain three residues, and hence arrive at the notion of a triple helix, in which an essential feature is again the occurrence of *cis*-peptide links, two in every three (Fig. 24). The model naturally satisfies the dimensional requirements of Pauling, Corey, and Branson,[225] in particular the condition that the peptide bond should be planar; in this it is unlike models proposed earlier.

This structure qualitatively fits the infrared dichroism (but not quantitatively),[362] and it also provides a convincing explanation of the swelling phenomena which have been discussed above; furthermore, it is unlike structures previously proposed in having a true 2.86-A. repeat. A difficulty about it is that it does assume the same simple 1:1:1 ratio between, and regular sequence of, glycine, proline, and the other amino acids, which was a fundamental feature of most of the earlier structures and which we have seen cannot be sustained by the results of recent chemical analyses. This structure is now not believed by its authors to be correct,[232] partly because of difficulties in reconciling it in detail with the x-ray pattern and partly because it does not allow the chain to have any appreciable extensibility. It has been mentioned here because it is the prototype of helical collagen models on which more recent discussions have centered. This type of model seems to account very plausibly for the general nature of the wide-angle x-ray diagram, which has the appearance of being due to a helical structure.[363]

(360) H. Zahn, *Kolloid-Z.* **111,** 96 (1948).
(361) L. Pauling and R. B. Corey, *Proc. Natl. Acad. Sci. U. S.* **37,** 272 (1951).
(362) J. T. Randall, R. D. B. Fraser, and A. C. T. North, *Proc. Roy. Soc.* (London) **B141,** 62 (1953).
(363) C. Cohen and R. S. Bear, *J. Am. Chem. Soc.* **75,** 2783 (1953).

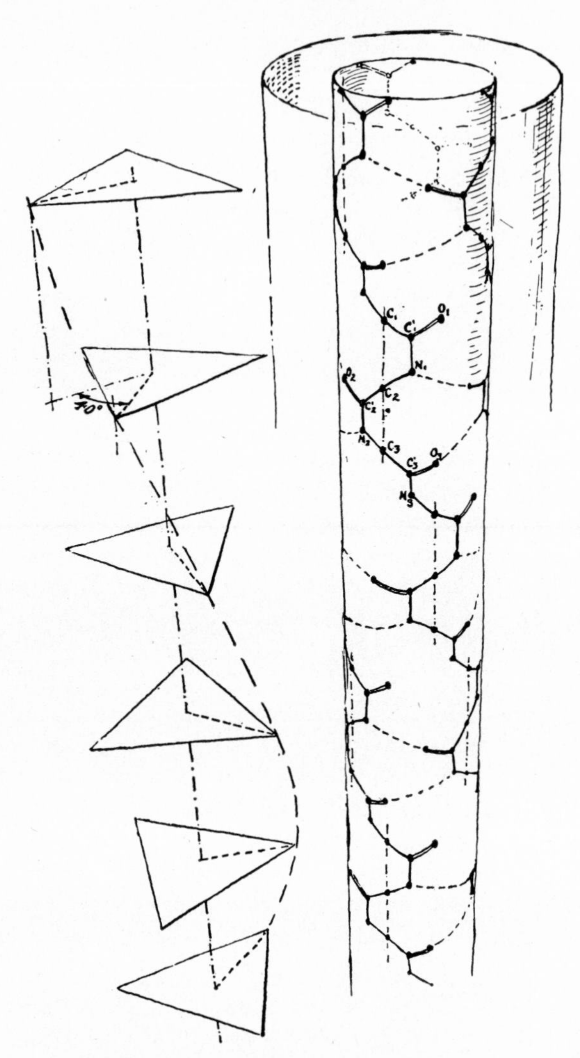

FIG. 24. Structure for collagen proposed by Pauling and Corey. After Pauling and Corey.[361]

(5) *Discussion of Helical Structures by Bear.*[269] Bear rejects all structures previously proposed on the grounds that collagen, so far from being an inextensible protein, is in fact the most extensible of all. At first sight there is an element of the paradoxical in this repudiation of the evidence of common observation: and indeed many attempts have been made to stretch collagen under various conditions, with complete lack of success except in the one situation discovered by Schmitt *et al.*[289] on the

electron-microscope grid, where the extensibility is great. But since one well-authenticated instance should be sufficient to establish the point, we must give at least provisional credence to it.

Bear has not made any firm proposal for a chain model. Cohen and Bear[363] have indicated a number of conditions which a model should satisfy, however. From an analysis of the wide-angle x-ray pattern of kangaroo tendon they conclude that the data would best fit a helix of two turns in 20 A. containing seven equivalent groups of residues; the density requires that there should be 21 amino acid residues in 20 A., so each equivalent group would contain three residues. Such an analysis does not determine uniquely the chemical connections of the residues; one can imagine the same spatial pattern of equivalent groups being given by multiple helices of steep pitch like the model of Pauling and Corey, or alternatively by a single helix of shallow pitch. The latter would have the merit of allowing several-fold extensibility. An example of such a single helix of shallow pitch would be the γ-helix of Pauling and Corey (see Figs. 58*c* and 58*d*, Vol. I, part A, p. 379); in this model one residue occupies 0.95 A. of axial length, so the possible extensibility (to a β-form) would be $3.34/0.95 = 3\frac{1}{2}$ times. However in other respects the γ-helix is not satisfactory as a model for collagen, since (*a*) the γ-helix would give parallel dichroism of the C=O and N—H stretching frequencies—the contrary is observed;[304,32] (*b*) proline, as has been mentioned, will not fit into the α-helix for steric reasons: *a fortiori* it would appear that it cannot pack into the γ-helix whose coils are closer; (*c*) no explanation is given of the great strength of the 2.86-A. reflection: it is necessary to assume special sequences of amino acids to explain it; (*d*) a very strong wide-angle reflection of spacing $2.86/3 = 0.95$ A. should be observed: in fact this is not a strong reflection;[364] and (*e*) in general there is the difficulty that the structure appears to be quite indifferent to, and take no account of, the abnormal amino acid composition of collagen.

Crick has recently proposed[364a] a new helical structure for collagen. This consists of two intertwined polypeptide chains, and although it accounts very satisfactorily for many features of the diffraction pattern it apparently cannot be reconciled with all the X-ray data; moreover some of its interatomic distances are distinctly short.

(*6*) *The Randall Structure.*[270,362] This model is based on the idea of a two-dimensional or sheet structure but it differs from all such structures previously proposed in that the abnormally short 2.86-A. periodicity is accounted for, not by the presence of *cis*-links, but by tilting an ordinary

(364) H. E. Huxley and M. F. Perutz, *quoted by* M. F. Perutz, *Ann. Repts. on Progr. Chem.* (Chem. Soc. London) **48,** 362 (1951).
(364*a*) F. H. C. Crick, *J. Chem. Phys.*, **22,** 347 (1954).

trans-link away from the fiber axis, the tilt being imposed presumably by the bulky proline groups. The model is shown in Fig. 25.

So much for theories of the structure of collagen. It should be clear that although much experimental work has been carried out and much thought has been expended it is impossible to conclude on any note other than one of doubt: it seems very probable that no model yet proposed is correct. This scepticism will perhaps be accepted the more readily

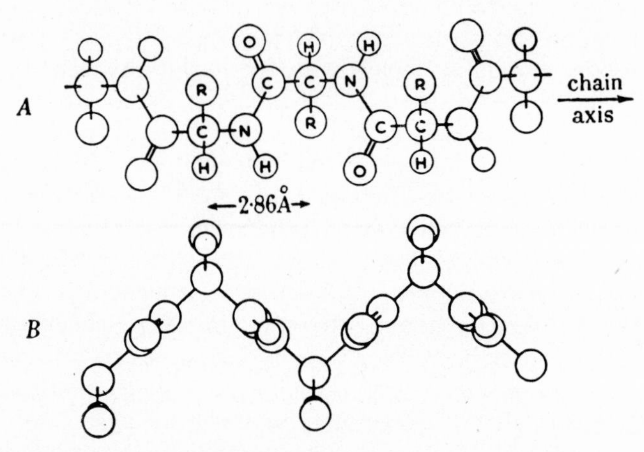

FIG. 25. Structure for collagen proposed by Randall *et al.* After Randall *et al.*[270]

when it is remembered that quite apart from other difficulties which have been mentioned, none of the theories has anything whatever to say about the heat contraction of collagen, one of its most striking properties; nor do any of them take account of the connections which have now been clearly demonstrated between collagen and mucopolysaccharides.

4. OTHER MEMBERS OF THE COLLAGEN GROUP

It has already been indicated that collagens, defined as materials yielding a wide-angle x-ray diagram similar to that of mammalian tendon, are widely distributed throughout the animal kingdom. No detailed description of these substances will be given here: in any case most of them have been little studied. Reference may be made to papers by Marks, Bear, and Blake,[365] by Randall *et al.*,[270] and by Bear,[269] for reviews and discussions. Present knowledge of the distribution of collagens has been summarized in Table V (p. 911); we may refer briefly to the more important of them.

(365) M. H. Marks, R. S. Bear, and C. H. Blake, *J. Exptl. Zoöl.* **111,** 55 (1949).

a. In Mammalian Tissue

The connective tissue of mammalian skin contains three types of fibrous material which can be distinguished histologically:[366–368] these are collagen, reticulin, and elastin. Separation of these materials is not easy. For a long time it was believed by many that elastin was a form of collagen, but it is now clear that this is not so, and that elastin should be placed in a separate class of proteins: we shall discuss it further below (p. 946). Reticulin, however, is much more closely related to collagen. Unlike collagen fibers, those of reticulin anastomose with one another to form a net (or reticulum—hence the name). This network is closely associated with typical collagen fibers in connective tissue, and it was therefore very difficult to get any indication of its properties: most of the earlier work on reticulin was probably vitiated by the presence of collagen in the samples. Recently, Kramer and Little[369] have found that the connective tissue of the renal cortex consists almost entirely of typical reticulin, and have used this material as a basis for a study of the properties of reticulin. According to their results, reticulin differs from collagen in the following respects: (*1*) It consists of a large number of minute anastomosing fibrils lying in an abundant amorphous matrix, (*2*) the fibrils are arranged at random rather than in parallel bundles, (*3*) reticulin forms membranous or lamellar structures rather than fibers, and (*4*) it apparently does not form cold-setting gelatin on boiling with water. On the other hand, reticulin fibrils exhibit typical 650-A. banding in the electron microscope and a normal collagen-type wide-angle x-ray pattern. Quantitative amino acid analyses of reticulin are not available: qualitatively it would appear that its composition is not unlike that of collagen though the proline and hydroxyproline contents may be somewhat lower.[368] However, in a study just published,[369a] Irving and Tomlin conclude that reticular fibers are identical with those of collagen.

Another collagenous material occurring in mammalian tissue is vitrosin, a fibrous component of the vitreous humor of the eye. As long ago as 1894 this material was thought to be collagenous because it dissolved in hot water to a solution which gelled on cooling;[370,371] in 1948 Pirie, Schmidt, and Waters[372] showed that it gave the collagen x-ray

(366) J. Nageotte and L. Guyon, *Am. J. Path.* **6,** 631 (1930).
(367) W. Jacobson *in op. cit.,*[272] p. 6.
(368) J. H. Bowes and R. H. Kenten, *Biochem. J.* **45,** 281 (1949).
(369) H. Kramer and K. Little *in op. cit.,*[272] p. 33.
(369*a*) E. A. Irving and S. G. Tomlin, *Proc. Roy. Soc.* (London) **B142,** 113 (1954).
(370) R. A. Young, *J. Physiol.* (London) **16,** 325 (1894).
(371) C. T. Mörner, *Z. physiol. Chem.* **18,** 233 (1894).
(372) A. Pirie, G. Schmidt, and J. W. Waters, *Brit. J. Ophthalmol.* **32,** 321 (1948).

pattern. More recently electron micrographs have been made by Matoltsy, Gross, and Grignolo,[373] who report the presence of three distinct types of fibril, two of which at least exhibit the normal collagen periodicity of about 600 A.

Finally, it is interesting to note that collagen from the remains of mammoths still exhibits typical fine structure in spite of prolonged cold storage.[374,270]

b. In the Tissues of Other Vertebrates

Collagen is widely distributed in the skins of amphibia[270,334] and birds.[270] More important are various substances obtained from fish, especially the ichthyocol of swim bladders, known commercially as isinglass (used in glues, jellies, etc.), and the elastoidin of scales and fins. These have x-ray patterns similar to that of collagen.

Ichthyocol is reported to contain appreciably less hydroxyproline than does mammalian collagen; hence although its serine and threonine contents are somewhat greater, the total number of hydroxyl groups is less.[375,356] It also has the property of being brought more readily and completely into solution by the usual methods than mammalian collagen; it has therefore been widely used in studies of soluble collagen.[310,347] Its low-angle x-ray pattern is similar to, but not identical with, that of mammalian collagen.[269]

Elastoidin exhibits a number of interesting points of difference from mammalian collagen, although it has a similar wide-angle x-ray pattern.[376,377] Its low-angle pattern has a somewhat shorter axial period (600 A.) and indicates that the structure must be considerably disordered.[269] Chemically elastoidin resembles ichthyocol in containing less hydroxyproline and slightly more serine and threonine than mammalian collagen—in all, fewer hydroxyl groups.[357] Its heat contraction differs sharply from that of mammalian collagen in that it takes place at a lower temperature,[378] and in that it is partly reversible;[292] on cooling, heat-contracted elastoidin partly regains its original length and its wide-angle x-ray pattern, while its resistance to trypsin digestion, lost during heat contraction, is to some extent recovered. Gustavson[379–381] has found

(373) A. G. Matoltsy, J. Gross, and A. Grignolo, *Proc. Soc. Exptl. Biol. Med.* **76,** 857 (1951).
(374) R. S. Bear, *J. Am. Chem. Soc.* **66,** 1300 (1944).
(375) J. M. R. Beveridge and C. C. Lucas, *J. Biol. Chem.* **155,** 547 (1944).
(376) W. T. Astbury and R. Lomax, *J. Chem. Soc.* **1935,** 846.
(377) G. Champétier and E. Fauré-Fremiet, *J. chim. phys.* **34,** 197 (1937).
(378) E. Fauré-Fremiet and R. Woelfflin, *J. chim. phys.* **33,** 666, 695, 801 (1936).
(379) K. H. Gustavson, *J. Intern. Soc. Leather Trades' Chemists* **33,** 332 (1949).
(380) K. H. Gustavson, *J. Am. Leather Chemists' Assoc.* **45,** 789 (1950).

that teleost fishes may be classified on the basis of the temperature at which their elastoidin suffers heat contraction, into the *bathybic* fish (including most marine teleosts), where the temperature is 40°C., and the *pelagic* fish (including most fresh-water species) where it is about 55°C. He correlates the lower contraction temperature of fish collagens with their lower content of hydroxy acids; in his view the most important interchain links in mammalian collagens are hydrogen bonds involving the side chains of hydroxy acids, while in fish collagens there are far fewer of these and the main binding agents are salt bridges, so that the whole structure is far less stable. Thus he finds that the strength of fish collagens is adversely affected by multivalent anions of sulfonic acids (which would be attached to groups otherwise capable of forming salt bridges), while mammalian collagens are hardly affected.

c. In Other Phyla

Fibers with the collagen periodicity have been found in many organisms in other phyla: the search for them has not been systematic, but they are evidently very widely distributed.

Of special interest are the proteins gorgonin and spongin, which form the residues left on decalcifying the skeletons of certain corals and sponges (for a detailed bibliography see the review by Roche and Michel).[382] The x-ray patterns of these materials show that they should undoubtedly be classed as collagens. Earlier work indicated that they were widely different in amino acid composition from normal collagens, particularly in containing very little proline; however the recent studies of Gross[355] to which we have referred (p. 936), suggest that revision is necessary. One special feature they do possess, however, is that they contain much tyrosine, largely in the form of the rare diiodotyrosine. The organisms containing these proteins may thus act as concentrators of the iodine of sea water.

d. The Secreted Collagens

Certain invertebrate collagens have been classed separately as "secreted collagens" since they are secretory products in a more obvious sense of the term than are the collagens proper. Among these are the misnamed ovokeratin (from the capsule of the skate egg) and byssokeratin (from byssus threads of bivalves, e.g., *Mytilus edulis*), which have been studied especially by Champétier and Fauré-Fremiet[383] and by

(381) K. H. Gustavson, *Svensk Kem. Tid.* **65,** 70 (1953).
(382) J. Roche and R. Michel, *Advances in Protein Chem.* **6,** 253 (1951).
(383) G. Champétier and E. Fauré-Fremiet, *Compt. rend.* **207,** 1133 (1938).

Mercer.[384] However, as an illustration of the danger of drawing general conclusions from an examination of only one or two members of a group of animals, it may be mentioned that Fitton Jackson *et al.*[385] have found that whereas the byssus threads of *Mytilus edulis* show many of the characteristics of collagen, those of *Pinna nobilis* appear to be quite unlike collagen and indeed give an x-ray pattern not resembling that of any previously examined fibrous proteins (having, e.g., meridional spacings of 12.3, 10.3, 4.5, and 3.9 A.).

The cuticles of the earthworms and *Ascaris*[386,387] should also be included in this group. Secreted collagens apparently differ from normal ones in that they originate in epithelial cells, and in showing no trace of long periodicities, either in x-ray patterns or in electron micrographs.

V. Miscellaneous Structure Proteins

1. Elastin

Many mammalian connective tissues contain, in addition to collagen and reticulin, a third fibrous component known as elastin. Elastin fibers are recognized by their yellow color, by certain selective staining reactions, and by their appearance under the microscope—unlike collagen fibers they may branch freely and anastomose with one another. Certain connective tissues contain mainly elastin and very little collagen; the *ligamentum nuchae* and the aortic media are particularly rich and have generally been used as sources of elastin. However, some collagen seems invariably to be present in such tissues, and this complication has made it very difficult to characterize elastin and to describe its properties. In particular, the earlier studies of amino acid composition were undoubtedly vitiated by collagen contamination. More recent work has shown with certainty that elastin differs considerably in composition from collagen;[99,356,357,368,388] a summary is given in Tables XXIII and XXIV, Vol. I, part A, pp. 221–2. Indeed the only resemblance between the two seems to be that they share a high proline content and are poor in histidine, cystine, tyrosine and tryptophan; unlike collagen, elastin contains very little hydroxyproline. The striking feature of the composition of elastin is that it contains only 7 per cent of polar side chains,

(384) E. H. Mercer, *Australian J. Marine and Fresh-water Research* **3,** 199 (1952).
(385) S. Fitton Jackson, F. C. Kelly, A. C. T. North, J. T. Randall, W. E. Seeds, M. Watson, and G. R. Wilkinson *in op. cit.*,[272] p. 106.
(386) L. E. R. Picken, M. G. M. Pryor, and M. M. Swann, *Nature* **159,** 434 (1947).
(387) R. Reed and K. M. Rudall, *Biochim. et Biophys. Acta* **2,** 7 (1948).
(388) W. H. Stein and E. G. Miller, *J. Biol. Chem.* **125,** 599 (1938).

whereas 34 per cent of the side chains of collagen are polar. In this respect it differs not only from collagen but also from nearly all other proteins.

FIG. 26. Electron micrograph of elastin fiber from rabbit aorta; magnification 30,000×. By courtesy of J. Gross.

Attempts to discover a definite fine structure in elastin have been uniformly unsuccessful. Electron-microscope studies have been made by Franchi and de Robertis,[389] by Wolpers,[390] and by Gross,[391] but in no

(389) C. M. Franchi and E. de Robertis, *Proc. Soc Exptl. Biol. Med.* **76,** 515 (1951).

(390) C. Wolpers, *Klin. Wochschr.* **23,** 169 (1944).

(391) J. Gross, *J. Exptl. Med.* **89,** 699 (1949). This paper should be read in

case has fine structure been observed (see Fig. 26). X-ray studies have been similarly unprofitable; earlier workers sometimes obtained a collagen-type diagram from elastin but, as was surmised by Astbury[358] and by Bear,[285] these results were due to the presence of small amounts of collagen in the specimens.

Elastin is more resistant than collagen to enzymes, to acids, and to bases; unlike all other proteins it cannot be heat-denatured, being apparently unaffected by boiling water. It exhibits long-range elasticity, and force–extension studies show that its mechanical behavior is closely analogous to that of rubber; hence it has been concluded by Meyer and Ferri[392] that its structure is of the same kind—i.e., that it consists of long-chain molecules in an "amorphous" tangle, and that the restoring force in contraction is due to thermal agitation of segments of the chain. In conformity with this view it is stated that elastin fibers only become birefringent on stretching;[393] further, a rise in temperature causes elastin fibers to contract. On the other hand, elastin fibers will remain stretched after tension has been removed if they are completely dry.[394] The almost perfectly analogous behavior of rubber and elastin finds a close parallel in the chemical composition of their "side chains," which determine the interaction between neighboring molecules: in rubber these are all nonpolar methyl groups, while in elastin, in sharp contrast to other proteins, over 90 per cent of them are nonpolar.

Elastin is readily attacked by a specific enzyme known as elastase, which was first obtained by Eijkman[395] from bacteria, but was later shown by Baló and Banga[396] to be present in pancreatic extracts. The latter authors found not only that this enzyme is distinct from the ordinary proteases (which do not attack elastin), but also that the elastin fibers disintegrate under its action without the production of amino acids. Banga and Schuber[397] have recently stated, however, that many amino and carboxyl groups are freed during the action of elastase, with the production of globular particles. Appreciable quantities of sulfuric acid are released during the process.

---

conjunction with a later publication by the same author [*Proc. Soc. Exptl. Biol. Med.* **78,** 241 (1951)] in which it is shown that the helical fibers observed in the earlier study are in fact derived from the trypsin used in specimen preparation, and have no connection with elastin.

(392) K. H. Meyer and C. Ferri, *Arch. ges. Physiol.* (*Pflüger's*) **238,** 78 (1936).

(393) W. J. Schmidt, Die Doppelbrechung von Karyoplasma Zytoplasma und Metaplasma, Gebrüder Borntraeger, Berlin, 1937.

(394) L. D. Jordan and M. Garrod *Trans. Faraday Soc.* **64,** 441 (1948).

(395) C. Eijkman, *Centr. Bakt. I. Orig.* **35,** 1 (1904).

(396) J. Baló and I. Banga, *Biochem. J.* **46,** 384 (1950).

(397) I. Banga and D. Schuber, *Acta Physiol. Acad. Sci. Hung.* 4, 13 (1953).

Purely chemical treatments too may lead to the production of smaller protein particles; thus Adair *et al.*[398] find that partial hydrolysis with oxalic acid gives components of molecular weight 84,000 and 6000, many free amino and carboxyl groups being produced in the process; the former of these components, unlike elastin itself, is denatured by heat. Hall[399] has shown that elastin is soluble in excess of boiling urea. The product is of low molecular weight and may be homogeneous; it is likewise denatured by heat. Polysaccharide material and acid are released into solution during this process.[400] Similar results have been obtained by Wood[401] using other reagents such as sodium hydroxide or peroxyformic (performic) acid.

To explain these and similar results both Banga[402] and Hall, Reed, and Tunbridge[400] have proposed that elastin is made up of globular subunits (named "G-elastin" by the former author and "pro-elastin" by the latter). The scheme put forward by Hall *et al.* involves mucopolysaccharide in intimate relation with the protein "pro-elastin" units, in fact acting as a binding agent between them; and the suggestion made is that elastase is actually a mucase. The presence of polysaccharide in elastic tissue has been known for a long time; and, as has been indicated, polysaccharide and acid are always released when elastin is broken down into smaller units.

For a general discussion of these problems reference should be made to the above-mentioned paper of Hall, Reed, and Tunbridge;[400] the properties of elastin have also been reviewed by Gross.[348]

2. Bacterial Polyglutamic Acid

It was found by Ivánovics and Bruckner in 1937[403] that certain species of bacillus produce a protein-like capsular substance of high molecular weight. This material has been isolated from *Bacillus anthracis*, *Bacillus subtilis*, etc., and on analysis was found to consist entirely of D(−)-glutamic acid residues; it will be noted that these are of the so-called unnatural configuration. It would further appear that these residues are linked together, either in part or wholly, by γ-peptide bonds instead of the normal α-links; some authors[404,405] have taken the former view,

(398) G. S. Adair, H. F. Davis, and S. M. Partridge, *Nature* **167,** 605 (1951).
(399) D. A. Hall, *Nature* **168,** 513 (1951).
(400) D. A. Hall, R. Reed, and R. E. Tunbridge, *Nature* **170,** 264 (1952).
(401) G. C. Wood, unpublished; quoted by Hall *et al.*[400]
(402) I. Banga, *Z. Vitamin-Hormon-u. Fermentforsch.* **4,** 49 (1951).
(403) G. Ivánovics and V. Bruckner, *Naturwissenschaften* **25,** 250 (1937).
(404) W. E. Hanby and H. N. Rydon, *Biochem. J.* **40,** 297 (1946).
(405) F. Haurowitz and F. Bursa, *Biochem. J.* **44,** 509 (1949).

others[406,407] the latter. It is not known how widely materials of this type are distributed. In respect both of optical configuration and of linkage, bacterial polyglutamic acid lies outside the normal definition of protein; it has been mentioned here to indicate that structural materials built from amino acids need not necessarily conform to this definition.[408]

In conclusion it should be stated that our classification of the main types of structure protein is probably far from complete. Knowledge of these materials depends on effective methods of characterizing and isolating them; studies of fibrous proteins have been biased in favor of those which bulk large in the few mammals forming the subject of almost all biochemical researches, and which, possibly by accident, have a striking appearance under the light microscope or electron microscope or yield well-defined x-ray patterns. Other materials, less easily characterized but often associated with biologically more significant structures than those which we have discussed, have been reported in the literature. Thus, to cite only a few examples, structure proteins have been described as components of chromosomes, of nerve axon and myelin sheath, and of mitochrondria and microsomes. At this time such materials do not properly come within the purview of a treatment such as this one, though in the future they will undoubtedly do so; today they lie more in the province of the cytologist. For a survey of such materials reference may be made to the review by Schmitt.[409] Still further widening of the field may be anticipated when biochemists devote serious attention to organisms more distantly related to their own than the common farmyard animals. In the meantime it should be remembered that the emphasis on a very few proteins which has necessarily been adopted here is dictated more by the convenience and availability of these materials for study than by their significance to biological function.

(406) J. Kovács, V. Bruckner, and K. Kovács, *J. Chem. Soc.* **1952,** 4255; *ibid.* **1953,** 145, 148.
(407) V. Bruckner, J. Kovács, & G. Dénes, *Nature* **172,** 508 (1953).
(408) A more detailed account of this material is to be found in a recent review by E. Bricas and Cl. Fromageot, *Advances in Protein Chem.* **8,** 1 (1953).
(409) F. O. Schmitt, *Advances in Protein Chem.* **1,** 25 (1944).

CHAPTER 24

Structure Proteins. II. Muscle

BY KENNETH BAILEY

	Page
I. Introduction	952
1. Historical	952
2. The Different Kinds of Muscle	954
a. Smooth Muscle	954
b. Striated (Skeletal) Muscle	955
c. Cardiac Muscle	955
3. Biochemical Investigations on Different Kinds of Muscle	956
4. Protein Components of Muscle (Introductory)	956
II. The Structure Proteins of Skeletal Muscle	957
1. Terminology	957
2. Myosin	960
a. General Physical and Chemical Properties of Myosin	960
b. The Sulfhydryl Character	963
c. The Adenosinetriphosphatase (ATPase) Activity	964
d. Characterization of Various Myosins by Physical Methods	969
e. Structural Features of Myosin	972
f. The Action of Trypsin on Myosin	976
3. Actin	977
a. General Properties of Actin	977
b. Preparation of Actin	979
c. Physical Studies on G- and F-Actin	983
d. The Nature of F-Actin	984
e. The Interaction of Actin and Myosin in Solution	986
f. The Action of ATP on Actomyosin in Solution	988
g. The Interaction of ATP and Actomyosin in Gel Form	990
h. The Nature of the Myosin–Actin–ATP Interaction by the G. Weber Polarization–Fluorescence Method	992
4. Tropomyosin	994
a. General Properties	994
b. Particle Size of Tropomyosin and Aggregation Phenomena	997
c. Viscosity Studies	999
d. Characterization of Tropomyosin by Other Techniques	1001
III. Particulate Components	1002
1. Granules	1002
2. The Isolated Myofibril	1004
3. The Granule ATPase	1004
IV. Extractability of Muscle Proteins	1005

	Page
V. Estimation of Muscle Proteins	1010
1. Adult Muscle	1010
2. The Changes During Development of Muscle	1013
3. Changes in the Muscle Proteins During Atrophy	1015
VI. Amino Acid Composition of the Structure Proteins and Its Significance	1018
VII. X-Ray and Electron Microscope Studies of Muscle and Muscle Proteins	1024
1. Wide-Angle Diffraction of Actomyosin	1024
2. Electron Microscope Studies of Skeletal Muscle	1027
3. Small-Angle Diffraction[1]	1034
4. Paramyosin	1038
VIII. Models of Muscle Contraction	1040
1. The Behavior of Muscle Fibers after Mincing	1040
2. The Thread and Fiber Models	1043
3. Mechanical Behavior of the Models	1044
4. Behavior of the Models as a Function of ATP Splitting and ATP Concentration	1045
5. Some Thermodynamical Considerations	1046
IX. Biological Activities Associated with the Sarcoplasm	1050
X. Appendix and Conclusions	1051
1. A and I Bands During Contraction and Stretch	1051
2. Relaxing Factors	1053

I. Introduction

1. Historical[1a]

"Let the waters bring forth abundantly the moving creature that hath life, and fowl that may fly above the earth in the firmament of heaven."

In the animal kingdom, life and motion are so inseparably linked that the form of an animal evokes first of all some concept of its mode and speed of movement, and this association of "being" and "moving" is nowhere more explicit than in the Biblical account of the Creation. It is such an ever-present quality, that from the time of Aristotle (384–322 B.C.) onward, both philosophers and scientists have sought to explain its nature. In the *De Incessu Animalium* Aristotle analyzed the nature of movement in creatures with limbs, and in the *De Motu Animalium* developed the idea that the moving parts are passive to an external force which initiates movement and exerts power. This ability, he believed "is precisely the characteristic of spirit. It contracts and expands naturally, and so is able to pull and to thrust from one and the same cause." The first true experimentalist, however, was Galen (A.D. 131–201), who at that early time understood the importance of muscle

(1) This section contributed by H. E. Huxley.

(1*a*) Drawn largely from J. F. Fulton, Muscular Contraction and the Reflex Control of Movement, Ballière, Tindall & Cox, 1926.

antagonists and of the necessity of "movement" to maintain posture. Galen's teaching had tremendous influence for many centuries, indeed, right up to the 19th century, and in one form or another, his view that contractility arose from the "animal spirits" which passed from the brain along the nerves to the muscles, causing them to swell and shorten, persisted in such outstanding investigators as Croone (1633–1684), Steno (1638–1686), and Mayow (1645–1679). But Mayow at the same time was cautious and saw great difficulties in Galen's theory and its later modifications, and he marks the decline of Galenian influence on the nature of contraction. The very subtle nature of the nervous influence was implicit from the demonstration of both Glisson (1597–1677) and Swammerdam (1637–1680) that contraction produced no change in muscle volume, and to von Haller (1708–1777) it was quite clear that the intrinsic property of muscle to contract lay in the muscle itself, and not instigated by the blowing up of elements which received some kind of humor or animal spirits from the nerve.

The provision made by Croone and implemented financially by his wife for the perpetuation of lectures devoted to the physiology of muscular motion has left us with a long series of names ever since 1738, which testify to the vast amount of thought which was given to the subject in England alone, and no doubt these Croonian lectures of the Royal Society have stimulated the interest of scientists whose fields of interest lay outside the domain of physiology proper.

In reality, current suggestions and hypotheses as to the nature of contraction are a long way removed from early physiological thinking, and have their basis in the "colloid-chemical" studies of early workers such as Kühne, Danilewsky, von Fürth, and Halliburton. Much of this work centered upon the clotting of muscle plasma and its possible relation to the hardening which is so evident in rigor mortis, and thus marked a stage in which the structural elements themselves were related to a physical change. Nevertheless, this early work was extraordinarily confused and confusing, for not only did observations differ from worker to worker, but terminology became so mixed up that students new to the muscle field must have felt very discouraged. Some of the confusion lasted until 1929, when Smith,[2] re-examining carefully the conditions of these older experiments, deduced that the requisite for a clottable plasma was the use of dilute salt as an extractant, and he related the increasing hardness of a muscle passing into rigor to progressive weakness in the plasma clot; that in fact as a muscle dies there is a change of the state of the protoplasm such that it does not yield to dilute salt solutions. A complete explanation of these phenomena is only possible now, however,

(2) E. C. Smith, *Proc. Roy. Soc.* (London) **B105**, 579 (1929).

in terms of the factors in muscle which affect extractability and equally the changes in the extracted portion which give rise to an actomyosin gel.

The colloid-chemical era led inevitably to studies aimed at an understanding of muscle and muscle proteins on a molecular basis, and the major advance in this direction was made when the properties of solutions made by extracting freshly minced muscle with strong salt were ascertained; particularly the study of flow birefringence by von Muralt and Edsall[3] who related the anisotropy of particles contained in them to the basic structural elements that play a part in the functional activity of muscle. It is with these studies that the present chapter is mainly concerned: it will not deal in any detail with those proteins and enzymes which are common also to other cells in which glycolysis and respiration occur.

2. The Different Kinds of Muscle

Histologically, muscular tissue is very diverse. It ranges from the relatively simple smooth muscle cell to the highly differentiated fibers of striated muscle which are especially rich in histological detail. But apart from tissues which for anatomical reasons can be classed as "muscle," the phenomenon of contractility is even more general: the movement of amebas, the contractility of bacterial flagella, even the streaming of protoplasm. It is quite possible that the mechanism underlying muscle contraction is applicable in these cases also. That it is the same in the very diverse types of muscle, no one seriously doubts, although the evidence is rather fragmentary.

a. Smooth Muscle

In mammals, smooth muscle cells are usually spindle-shaped, varying from 15 to 500 μ in length. They are found in the walls of the digestive and respiratory tracts, amongst the connective tissue fibers of the skin, in the urinary and genital ducts, in the uterus, and in arteries and veins. They do not possess any distinct enclosing sheath, and the protoplasm appears homogeneous except after acid treatment when longitudinal fibrillation is visible. Unlike the multinucleated fibers of skeletal muscle, the nucleus lies in the center of the cell and not at the periphery. In strong contractions, the cells appear to move relative to each other, and the greater uptake of stain suggests that water is squeezed out.[4]

Amongst the invertebrates, large accessible masses of smooth muscle can be found: the foot muscle of the snail; the foot and adductor muscle of mollusks such as *Pecten*, *Mya*, and *Anodonta;* and the retractor of the annelid *Phascolosoma*. Adductor muscles (except those of mussels like *Mytilus* and *Pinna*), contain two types of muscle which lie side by side; the one is hyaline and of a hard, brittle texture, yellow in color, while the other is white, soft, and readily teased into fibers. The former is said to have a primitive striation[5] and is responsible for rapid movement, either to close the shells or to enable their closure and opening to propel the animal; the second acts as a

(3) A. L. von Muralt and J. T. Edsall, *J. Biol. Chem.* **89**, 315, 351 (1930).

(4) See A. Bethe, Allgemeine Physiol., Springer-Verlag, Berlin, 1952, p. 250; and E. B. Meigs, *in* E. V. Cowdry, Special Cytology, Paul Hoeber, New York, 1932, Vol. II, p. 1113.

(5) F. Marceau, *Arch. zool. exptl. et gén.* **2**, 295 (1909)

"catch" muscle to keep the shell closed, and it contains the "paramyosin" component which will be discussed later on. In Crustaceae all muscle is striated.

b. Striated (Skeletal) Muscle

All the skeletal muscles of the body are striated, and together they make up a greater proportion of the body weight than any other tissue—almost half the weight of the human body. Striations are due to alternating bands of material of different refractive index, the anisotropic A band appearing bright under polarized light and the isotropic I band dark; though in actual fact the I band does possess a weak birefringence. Under ordinary light, the situation is reversed. The A bands stain especially well with iron–hematoxylin.

Skeletal muscle may be white or red, the color being due to myoglobin which in extreme cases—as in the seal and the whale—may amount to 5–7 per cent of the fresh flesh weight. Red muscle is also richer in particulate bodies or granules, originally called sarcosomes by Retzius, which seem to correspond to the mitochondria of other cells, since like them they contain the enzymes of the respiratory cycle. It may be recalled that much of the classical work on cell respiration was carried out on pigeon breast muscle.

A typical mammalian muscle is enclosed in a sheath, the perimysium, and the muscle itself may be permeated quite extensively by fat deposits and connective tissue (the endomysium), the latter determining the toughness of the meat. The contractile tissue proper consists of fibers of diameter 10–100 μ which eventually fuse with the tendon fibrils. Whether the fibers run parallel to the long axis of the muscle fiber, as in the psoas or sartorius, or are inclined, as in the gastrocnemius, depends largely on the role of the muscle in the body. Each fiber possesses an enclosing membrane, the sarcolemma, which is now considered to be a complex structure, one layer at least consisting of fine interlacing reticulin fibers. Besides the transverse striations, each fiber is split longitudinally into fibrils of about 1 μ diameter which are separated from each other by a sarcoplasmic gap of 0.5 μ. The fibrils are the ultimate morphological unit of muscle and contain the contractile material. In the electron microscope, the fibril itself is composed of thinner threads which are called the filaments. In the ensuing discussion, the distinction between fiber, fibril (or myofibril), and filament must be clearly understood.

The lengths of the A and I bands differ in different muscles; in vertebrates, they are each about 0.5 μ in length, but in arthropods the A band may be 8 μ and the I 2 μ. Both bands contain histological details which are more appropriately discussed in sec. VII-2. Likewise, the histogenesis of muscle will be discussed in relation to quantitative developmental aspects (sec. V-2).

The muscle fiber is a fusiform cell with many nuclei, and its contents will be referred to as the intracellular material. The extracellular components include the vascular and nervous tissue, connective tissue (elastin and collagen), and the materials of the interstitial space. All the insoluble components of the extracellular tissue, together with the insoluble components of the sarcolemma (probably mainly reticulin), will be called the stroma.

c. Cardiac Muscle

This too shows transverse striations which are less distinct than those of skeletal. The fibers are surrounded by a very thin sheath which is not easily isolated, and the fibers are divided by transverse septa or intercalated disks, through which the myofibrils pass without interruption. The fibers are connected by groups of fibrils which pass obliquely between adjacent fibers. Like smooth muscle, the nuclei are placed centrally. Cardiac muscle, in fact, in its histological structure and physiological properties is intermediate between smooth and skeletal and is especially adapted to rhythmical contraction. Skeletal muscle is voluntary muscle and is normally stimulated to contract by a nerve impulse; smooth and cardiac are largely under hormonal control and are therefore involuntary.

3. Biochemical Investigations on Different Kinds of Muscle

Most types of muscle have been studied from the standpoint of their metabolic activity, either their respiratory activity as a whole, or the enzyme systems associated with such activity or with glycolysis. The nonprotein components of muscle, the extractives, were once a favorite field of study, but less attention is paid to them now. They include carnosine, anserine, glutathione, carnitine, sarcosine, taurine, the common amino acids, and the purine bases. Between different muscles great differences exist with respect to the amounts of important metabolites such as adenosine triphosphate (ATP) or creatine phosphate. When the latter is absent it is replaced by arginine phosphate, just as carnosine is replaced by the related anserine.

Investigations on the proteins of muscle, or more specifically on the proteins of the myofibril, are not so wide. At first sight, it seems anomalous that most attention has been paid to mammalian skeletal muscle (rabbit), which is far more complex than smooth muscle. The reason is probably that the latter is not easy to secure quickly in large amounts from freshly killed animals, and investigators have found difficulty in extracting much of the representative protein; indeed, Mehl[6] concluded that the only globulin-like protein extracted with salt solutions probably arose from extracellular sources. More success has attended investigations of uterine muscle, but altogether little is known about the proteins of either vertebrate or invertebrate smooth muscle. Cardiac muscle, for no very obvious reason, has also received scant attention.

Except where the type and source of muscle are specifically stated, most of the present chapter will center round the proteins of the white skeletal muscle of the rabbit.

4. Protein Components of Muscle (Introductory)

For those unfamiliar with muscle and its proteins, it is necessary to outline the location, extraction, and character of the main components.

When a muscle is minced and diluted with an equal volume of water or of physiological saline, a viscous mass is obtained from which no juice can be obtained by simple pressure. In the course of half an hour the fibers undergo a syneresis, and a juice can be expressed by simple hand pressure which contains all the components of the glycolytic system. The nature of this change has only recently been analyzed (sec. VIII-1).

The press juice contains the sarcoplasmic proteins and some of the muscle granules, and to this fraction von Fürth gave the name myogen. Its albumin-like character was noted by von Fürth,[7] but Weber and Meyer,[8] following up these early investigations, showed that on dialysis

(6) J. W. Mehl, personal communication.

(7) O. von Fürth, *Arch. exptl. Path. Pharmakol.* **36**, 231 (1895).

(8) H. H. Weber and K. Meyer, *Biochem Z.* **266**, 137 (1933).

to a specific conductivity of 10^{-4} mho a precipitate could be obtained, most of which redissolved on addition of salt. To this globulin fraction they gave the name "globulin X." The enzymatic make-up of these fractions was rarely considered, and their behavior and properties were studied as though they were mono- or paucidisperse systems; nowadays, the biological and chemical heterogeneity of myogen is well established, and most probably all components are enzymatic in character.

After comminution and extraction of the muscle in this way, the residue consists of amorphous material and short pieces of contracted, shrunken fibers in which striations are still visible. This material is not suitable for the extraction of proteins which comprise the contractile material proper, since changes in solubility have been induced. These proteins are best obtained by following the classical methods of Danilewsky, Halliburton, and Edsall, whereby freshly minced muscle excised immediately after death is stirred up with strong salt solutions $\left(\frac{\Gamma}{2} = 0.5\right)$; on 20-fold dilution a globulin is precipitated which has until recently retained the name given by von Fürth[7,9]—myosin. Because of the large amount obtained by exhaustive extraction, and because of the asymmetric character of its particles, it was assumed that myosin formed the major component of the myofibril. On critical examination, such myosins were found to be heterodisperse, a characteristic readily explained when Szent-Györgyi and his school showed them to be a complex of two proteins, one of which, first isolated by Straub, is water-soluble and named actin, and the other, salt-soluble and fibrous, for which the name myosin has been retained. The interaction of actin and myosin, and of the actomyosin complex with adenosine triphosphate (ATP), appears to play the fundamental role in muscle contractility. A third protein component of the myofibril with interesting properties was discovered about the same time as actin, and because of its similarity to true myosin, was named tropomyosin. As far as we know, these three proteins constitute the myofibril, but we should not forget that other histological features, the material composing the Z band (bisecting the I) and the M band (bisecting the A) (see sec. VII-2) are probably also of protein nature and have not been isolated as specific proteins.

II. The Structure Proteins of Skeletal Muscle

1. Terminology

The name myosin was originally given by Kühne[10] to the substance in muscle press juice which on standing at room temperature set to a gel. Because it was later

(9) O. von Fürth, *Ergeb. Physiol.* **17**, 363 (1919).
(10) W. Kühne, *Arch. Anat. u. Physiol.* p. 748 (1859).

confused with protein having properties more closely resembling a true globulin,[7,11] the name myosin was still reserved for that fraction of muscle protein which appeared more abundantly in stronger salt solution than in simple press juice. Up to the discovery of actin, all preparations of "muscle globulin" were prepared by extracting with fairly strong salt solutions between pH 7 and 8.5, followed by dilution to a low ionic strength, and these preparations contained varying amounts of actin, according to the duration of the extraction and subsequent treatment. Edsall[12] recommended an extraction period of 1½–2 hours and filtration through a 1-cm. layer of paper pulp; the extracted solution was diluted to ionic strength 0.06–0.12 in the case of cow cheek, and 0.04–0.06 with rabbit muscle. Later,[13] the procedure was modified, and the former extracting medium of KCl–phosphate, ionic strength 1.36, was replaced by KCl–bicarbonate, ionic strength 0.53; again, the ionic strength on dilution was between 0.1 and 0.05, but for reprecipitation, dilution was carried only to ionic strength 0.2–0.25. Under these varying conditions, and particularly in the latter instance, the procedure was accidentally designed to recover actomyosin and to lose pure myosin in the supernatant, for the critical ionic strength in the separation of the two components is about 0.3.

The author's method[14] was designed in 1942 to obtain a water-clear sol, since it had been noticed that many extracts were rather turbid and that this could be removed by filtration through 2 inches of pulp; the extraction time was also reduced to 1 hour. It was noted that on progressive dilution at pH 7 "there appears a flocculent precipitate of myosin which exhibits a sheen on stirring." It is apparent now that the turbidity was due to extraction of actomyosin (and probably fat globules) which were largely removed on paper pulp,[15] and that the final product was very comparable with Szent-Györgyi's myosin A which was originally thought to be free of actin and to be a crystalline product. Nevertheless, the preparations have contained some actin since generally (though not always), they exhibited flow birefringence. Weber,[16] however, had used for extraction salt solutions of rather alkaline pH (between pH 8 and 9), and there is no doubt that this would favor the extraction of actin. He inclines to the view that some of the actin must have become inactive and remained in the supernatant fluid on dilution.

These comments serve to show that the evaluation of results obtained in the pre-actin era cannot be made without careful consideration of the methods employed, which differed quite widely even then.

In an effort to distinguish between preparations judged to contain actin, and those reported not to, various names have been suggested. Szent-Györgyi[17] has prepared a myosin A by a method which reduces the actin content, first by extracting with a somewhat acid salt solution, and then by subsequent removal of an actomyosin fraction. Weber[15] has shown, however, by ultracentrifugation, that this still contains actin, and Tsao[18] likewise has found that ammonium sulfate fractionation of myosin

(11) A. Danilewsky, *Z. physiol. Chem.* **5**, 158 (1881).
(12) J. T. Edsall, *J. Biol. Chem.* **89**, 289 (1930).
(13) J. P. Greenstein and J. T. Edsall, *J. Biol. Chem.* **133**, 397 (1940).
(14) K. Bailey, *Biochem. J.* **36**, 121 (1942).
(15) H. H. Weber and H. Portzehl, *Advances in Protein Chem.* **7**, 161 (1952).
(16) H. H. Weber, *Biochem. Z.* **158**, 443, 473 (1925).
(17) I. Banga and A. Szent-Györgyi, *Studies Inst. Med. Chem. Univ. Szeged* **1**, 5 (1941/2).
(18) T. C. Tsao, *Biochim. et Biophys. Acta* **11**, 368 (1953).

A according to Dubuisson's directions[19] gives two fractions with distinct properties. The first precipitates at 30 per cent saturation and behaves both osmotically and viscometrically quite differently from the main fraction which precipitates between 40 and 50 per cent saturation. This latter fraction is considered to be pure myosin.

Schramm and Weber,[20] making for the first time an ultracentrifugal study of myosin sols, observed a slow-sedimenting component $s_{20} = 6$ and a fast, $s_{20} = 20$. These were called S-myosin (schwer, schnell) and L-myosin (leicht, langsam). The S-myosin, of course, was actomyosin, and L-myosin was actin-free myosin, which later was prepared by Weber's group[21] in a very simple manner, making use of the salting-in thresholds of the two components. Dilution to ionic strength 0.3 causes the precipitation of actomyosin, leaving free myosin in solution, and this can be precipitated by further dilution.

To Weber's and Szent-Györgyi's terminology must be added that of Dubuisson,[22] which attempts to distinguish between myosin and actomyosin on an electrophoretic basis. Here, the convention α, β, and γ are employed to denote fractions of increasing mobility.

It is not possible at present to decide which myosin is most homogeneous and free from actin. Tsao[18] has observed that in a depolymerizing solvent such as guanidine-HCl, the average particle weight of the actomyosin portion of Szent-Györgyi's myosin A is much greater (294,000) than the fraction appearing between 40 and 50 per cent saturation with ammonium sulfate (110,000), and while Weber's L-myosin is homogeneous in the ultracentrifuge, it, like actomyosin, shows some turbidity if dissolved in concentrated urea, whereas Tsao's 40–50 per cent fraction is water-clear.

It may be necessary for a time to follow author's terminology wherever the properties of different preparations are to be compared. Ultimately, when there is some measure of agreement as to what constitutes a pure myosin, the term "myosin" without further qualification should be used, and actomyosins for preparations known, or assumed from their manner of preparation, to contain actin. "Myosin" preparations investigated in the pre-actin period could be denoted quite simply by (acto-)myosin.

The preparations to which reference will be made from time to time are as follows:

(Acto-)myosins of Weber, Edsall, Greenstein and Edsall, Bailey.

Myosin A, the actin-poor preparation of Szent-Györgyi.

Myosin B, the actin-rich myosin of Szent-Györgyi made by direct extraction from the muscle.

Artificial actomyosin, made by mixing actin and myosin.

Myosin T, the actin-free fraction of Szent-Györgyi's myosin A as prepared by Tsao.

L-Myosin, the actin-free myosin (or nearly so) of Weber.

Myosin α, myosin β, electrophoretic components corresponding to actomyosin and true myosin, respectively.

Myosin γ, nature unknown.

"Myosin" may also be used in a general sense to denote actin-free myosin when general properties are under discussion.

(19) M. Dubuisson, *Experientia* **2,** 413 (1946).
(20) G. Schramm and H. H. Weber, *Kolloid-Z.* **100,** 242 (1942).
(21) H. Portzehl, G. Schramm, and H. H. Weber, *Z. Naturforsch.* **5b,** 61 (1950).
(22) M. Dubuisson, *Experientia* **2,** 258 (1946).

2. Myosin

a. General Physical and Chemical Properties of Myosin

Myosin T is electrophoretically homogeneous at pH 9. It is readily soluble in 0.5 *M* KCl giving a water-clear solution which shows no birefringence at low shear rates. The relative viscosity at moderate shear rate is normal and shows neither rise nor fall in the presence of ATP. In Fig. 1 this behavior is contrasted with the 30–40 per cent fraction and with artificial actomyosin.

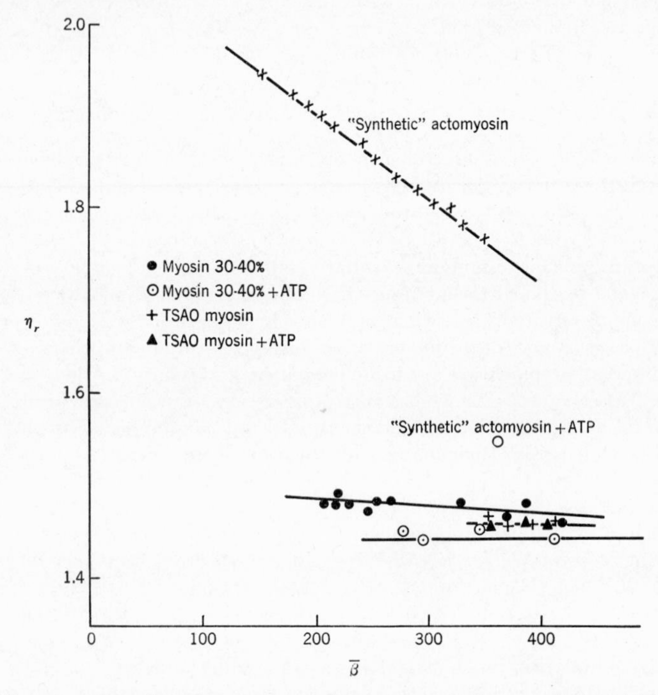

Fig. 1. The viscosity response of various myosins on adding ATP.

Actomyosin and myosin both possess the solubility properties of globulins, though the salting-in threshold of the latter at pH 7 is much lower $\left(\frac{\Gamma}{2} = 0.04 \text{ as against } 0.3\right)$. If all salt is washed away, the opalescent gel swells to a translucent form which gives a strong pseudophotoelastic effect. These gels, first described by Muralt and Edsall,[3] arise by virtue of a large Donnan effect, and have led to the mistaken belief that myosin is water soluble. Some of the seed globulins can likewise be dispersed in ion-free water or in isoelectric amino acid solutions, and are precipi-

TABLE I

SOME CONSTANTS FOR PROTEINS OF THE MYOFIBRIL (RABBIT)

	Tropomyosin	Actin (monomer)	Actin (dimer)	Actin (thixotropic)	Myosin (actin-free)	"Natural" actomyosin
1. Salting-in range $\left(\frac{\Gamma}{2} \text{ at pH } 7\right)^a$	Soluble in water	Soluble in water	Soluble in water	Soluble in water	0.2–0.3	0.3–0.5
2. Salting-out range $\left(\frac{\Gamma}{2} \text{ at pH } 7\right)^a$	5.1–7	?	?	?	5–6	?–5
3. Isoelectric point	5.1 (P.O.)	4.8 (P.O.)	4.8 (P.O.)	4.8 (P.O.)	5.4 (E)[79] 5.76[b]	5.6 (P.O.)
4. Isoelectric precipitation zone (pH)	4.6–6.5	4.5–6.5	4.5–6.5	4.5–6.5	4.8–7.5	4.5–8
5. Flow birefringence	+ (salt absent)	Nil	Nil	+ (spontaneous)	Nil	+ve
6. $S^0_{20} \times 10^{13}$	$2.6_{c=0.6}$[134]	2.7–4[100,107]		>750[81,97]	6.7,[80] 7.1,[21] 7.2[81] 7.2[81]	>90 ≫280[21]
7. $D^0_{20} \times 10^7$	$2.4_{c=0.6}$[134]	2.5[100,107]		—	0.9[82]	—
8. Particle weight: s and D	—	150,000[100,107]		—	850,000[82]	—
Light scattering	~55,000[c]	57,000,[d] 120,000,[101] 80,000[102]		—	—	—
Osmotic pressure	53,000[86]	74,000[103]	140,000[103]	Fairly low	840,000[82]	—
Polarization-fluorescence	—	70,000[103]	140,000[103]	140,000[103]	—	—
Electron microscope	—	—	—	1.5 × 10[6][99,e]	—	—
9. Partial specific volume (V)	0.71[134]	—	—	—	0.74[78]	—
10. Refractive index increment	0.190[47]	0.170[d]	—	—	0.185[47]	—
11. Intrinsic viscosity	0.523[86]	0.21[103]	[0.5][103]	—	—	—
12. Rotational relaxation time ($\rho_h \times 10^8$ sec. 25°)	8[86]	15[103]	30.5[103]	32[104]	—	—
13. Length, A.	~400[86] (v)	~290[103]	~580[103] (?)	1–5 μ[75,79,e]	$2{,}300 \frac{f}{f_0}$	—
14. Thickness, A.	~15	~24	~24 (?)	~100	$25 \frac{f}{f_0}$ 1500 (L.S.)[83]	—
15. Electrophoretic mobility[i] (ascending) (sq. cm./v./sec. × 10)[5]	−5.6[f] (NaCl-phosphate, pH 7.4 $\frac{\Gamma}{2} = 0.4$)	−4.6[g] (As col. 2)		—	−2.9[h] (As col. 2 pH 7.3)	−3.1[h] (As col. 6)

Symbols: P.O. = precipitation optimum; E = electrophoretic; v = viscosity; L.S. = light scattering.

[a] For more complete data, see Ref. 15; [b] Isoionic point (Ref. 25); [c] Doty and Sanders, communicated; [d] W. F. H. M. Mommaerts, *J. Biol. Chem.* **198**, 445 (1952); [e] Aggregates after drying (Ref. 99); [f] M. Dubuisson, *Experientia* **6**, 269 (150); [g] M. Dubuisson, *Biochim. et Biophys. Acta* **5**, 426 (1950); [h] M. Dubuisson, *Biol. Revs. Cambridge Phil. Soc.* **25**, 46 (1950); [i] See G. Hamoir, *Discussions Faraday Soc.* No. 13 (1953), and Ref. 15.

tated by a mere trace of salt. Solubility properties of fibrillar proteins are collected for reference in Table I.

From the cataphoresis of particles (KCl–buffer, $\frac{\Gamma}{2} = 0.05$), Weber[23] deduced an isoelectric point of 5.0, whereas the isoionic point deduced by prolonged dialysis is near pH 6.3.[12,24] Such dialyzed preparations were later shown by Hollwede and Weber to contain cation, which could be removed by adjusting the pH to 5.4. It is thus probable that the isoionic point of (acto-)myosin in the presence of alkali salts is near 5.5. Mihalyi,[25] in a very complete study of the hydrogen-dissociation curve of myosin A, has obtained the isoionic point by Sørensen's original method—the pH at which dialyzed protein on addition of salt shows a minimum of change; this occurs at pH 5.75.

The titration data of Mihalyi at different ionic strengths, different temperatures, and in the presence of formaldehyde reveal a not very exact correlation between total titration and analytical data (see p. 1022). It is unlikely that the differences reside in the amounts of actin present in different myosin preparations; nor is the disagreement in the acid-binding value due to terminal amino groups which appear to be absent in myosin.

On the acid and alkaline side of the isoelectric point the solubility of myosin increases in the absence of salt ions, as is usual with most globulins. In the presence of salt, the solubility is depressed on the acid side and elevated on the alkaline. This behavior again is typical of seed globulins such as edestin, and the reason is not altogether clear. When brought into solution by acid in absence of salt (pH 4.5 or lower), most globulins, when reneutralized, precipitate in the denatured form. It is curious, however, that neither myosin, actomyosin, actin, nor tropomyosin is denatured in the classical sense by this treatment, although 95 per cent of the sarcoplasmic proteins do become insoluble. It might be thought that this property could be used for the assay of fibrillar proteins; unfortunately, the interactions between insoluble and soluble proteins after acid treatment are complex and give rise to erratic values.

In other ways, myosin is easily denatured, whether by freeze-drying, dehydration with organic solvents, or by heat. In a sufficiently strong salt solution at pH 7, however, it cannot be heat coagulated, and in this respect again resembles globulins such as edestin. Depolymerizing solvents, e.g., urea and guanidine.HCl, induce a change to a salt-insoluble form.

(23) H. H. Weber, *Ergeb. Physiol. exptl. Pharmakol.* **36,** 109 (1934).
(24) W. T. Salter, *Proc. Soc. Exptl. Biol. Med.* **24,** 116 (1926).
(25) E. Mihalyi, *Enzymologia* **14,** 224 (1950).

b. The Sulfhydryl Character

Arnold[26] was the first to show that the proteins of animal tissues often gave a pronounced nitroprusside reaction which became stronger after heat coagulation; and when Mirsky and Anson[27] again resumed the study of sulfhydryl groups in proteins, the interest centered largely on conditions determining their activity in the native protein and the generation or revelation of new —SH groups on denaturation. This latter effect was explained in various ways until Neurath *et al.*,[28] reviewing the position in 1940, argued quite convincingly that some —SH groups of native proteins, whether for steric or other reasons, do not react with —SH reagents, but after denaturation will do so; they are not produced *de novo* during the denaturation process.

The profound effect of iodoacetate in suppressing lactic acid formation in living muscle, without at the same time abolishing the capacity to do work (the Lundsgaard effect),[29] stimulated an interest in the relation of —SH groups and enzyme activity. From the work of Green, Needham, and Dewan,[30] and of Rapkine,[31] it became clear that the enzyme most sensitive to iodoacetate in anaerobic glycolysis was triosephosphate dehydrogenase. The second world war greatly accelerated the study of enzymes whose activity was dependent upon the presence of —SH groups for the reason that lachrymators and other war gases were suspected of exerting their effect by combination with these groups.

Mirsky[32] first showed that (acto-)myosin in the native state contained —SH groups capable of reacting with iodoacetate, and representing some 75 per cent of the —SH groups found in the denatured protein. Subsequent authors have used different methods of estimation, and considering the ease of oxidation of thiol groups, and differences in the starting material, it is not surprising that the values are not very concordant (Table II). The total —SH groups found by titrating with porphyrindin in guanidine.HCl are equivalent to the difference of total and methionine-S, and suggest that no cystine is present.[13] The quantitative application to pure myosin of the nitroprusside test in guanidine-HCl suggests a lower value.[33]

(26) V. Arnold, *Z. physiol. Chem.* **70**, 314 (1911).
(27) A. E. Mirsky and M. L. Anson, *J. Gen. Physiol.* **18**, 307 (1935).
(28) H. Neurath, J. P. Greenstein, F. W. Putnam, and J. O. Erickson, *Chem. Revs.* **34**, 157 (1944).
(29) E. Lundsgaard, *Biochem. Z.* **120**, 1 (1930).
(30) D. E. Green, D. M. Needham, and J. G. Dewan, *Biochem. J.* **31**, 2327 (1937).
(31) L. Rapkine, *Biochem. J.* **32**, 1729 (1938).
(32) A. E. Mirsky, *J. Gen. Physiol.* **19**, 559 (1936).
(33) T.-C. Tsao and K. Bailey, *Biochim. et Biophys. Acta* **11**, 102 (1953).

While triosephosphate dehydrogenase is very sensitive to alkylating reagents, the —SH groups of myosin are not. Singer and Barron[34] first studied systematically the dependence of the ATPase activity of myosin in relation to the molarity of —SH reagent—alkylating, oxidizing, and mercaptide-forming. The inhibitory action increased in the order cited, and in the case of chloromercuribenzoate, was correlated chemically with the disappearance of —SH groups. Subsequent quantitative studies have confirmed this difference in reactivity toward different types of reagent. These will be discussed later in relation (*a*) to the enzyme activity of myosin, and (*b*) to its ability to form actomyosin (sec. II-2*c* and II-3).

TABLE II
THE REACTIVE SULFHYDRYL GROUPS OF NATIVE AND DENATURED MYOSINS

	—SH content, cysteine/100g.			
Type of myosin	Native	Denatured	Method	Author
(Acto-)myosin	0.31	0.67	Iodoacetate–ferricyanide	Mirsky[32]
(Acto-)myosin	0.27	—	Phenol 1:6-indophenol	Todrick and Walker[35]
(Acto-)myosin	0.42	0.65	Porphyrindin–urea	Greenstein and Edsall[13]
		1.15	Porphyrindin–guanidine	
Myosin A	0.36	0.66	Ferricyanide	Fredericq[36]
Actomyosin	0.5	0.8	Ferricyanide	Godeaux[37]
Myosin T	—	0.85	*N*-Ethylmaleimide–guanidine	Tsao and Bailey[33]

c. The Adenosinetriphosphatase (ATPase) Activity

The significance of the discovery of Liubimova and Engelhardt,[38] in 1939, that the ATPase activity of minced muscle was predominantly associated with the isolated (acto-)myosin and could not be separated from it, was generally appreciated for two reasons: first, studies relating glycolysis and mechanical activity had focused attention upon ATP as the probable source of free energy for the contractile process; and secondly, myosin had become accepted at that time as the contractile substance. The Russian work was confirmed in several laboratories, which,

(34) T. P. Singer and E. S. G. Barron, *Proc. Soc. Exptl. Biol. Med.* **56,** 120 (1944).
(35) A. Todrick and E. Walker, *Biochem. J.* **31,** 292 (1937).
(36) E. Fredericq, *Bull. soc. chim. Belges* **54,** 265 (1945).
(37) J. Godeaux, *Compt. rend. soc. biol.* **140,** 675 (1946).
(38) M. M. Liubimova and V. A. Engelhardt, *Biokhimiya* **4,** 716 (1939).

with the advent of war, worked in compulsory isolation. English and American work up to 1944 was summarized by Bailey,[39] the Russian by Engelhardt[40] in 1946, and the Hungarian by Szent-Györgyi[41] in 1947.

Much of the earlier work on ATPase was carried out with actomyosins of low actin content; but with the discovery of actin, Banga[42] showed that the enzyme character resided in the myosin rather than the actin. It was natural to suppose, with the example of enzymes which are "stuck" to particulate matter, that a protein occurring in such vast amounts in the animal body could not itself be the enzyme. On an absolute basis, the activity is 20–200 times less than other pure enzymes; the Q_p value of an average preparation (37°) is only 3000–6000, compared with that of 100,000 (30°) for a purified preparation of inorganic pyrophosphatase,[43] and that of 10^6 for a recently crystallized product.[44] By precipitating on basic lanthanum acetate and eluting, Polis and Meyerhof[45] were able to increase the activity 2–3-fold. This is not a decisive argument against the identity of myosin and ATPase until activity can be obtained with a myosin-free solution. The most recent claim[46] to having achieved a separation, using a method involving butanol which is highly successful for the dissolution of insoluble enzymes, has not been justified by further work. Morton and Tsao,[18,47] in fact, have made the most determined attempt so far to effect a separation. The butanol procedure consists in treating myosin gel under varying conditions of pH and salt concentration with butanol at 0–30°. The emulsions are broken up by centrifuging, the butanol is discarded, and enzyme activity is tested in gel and supernatant fluid. These authors drew the following conclusions:

(*1*) Activity in the gel is always associated with solubility in 0.5 *M* KCl; loss of solubility is always accompanied by loss of activity.

(*2*) When the gel is active, the supernatant fluid may or may not be; when the gel is inactive, the supernatant fluid is never active.

(*3*) ATPase is inactivated and myosin denatured by the butanol unless protected by ATP or actin. (The protective action of the latter may be due, as we shall see, to the fact that actin interacts with myosin at the ATPase centers, and that the substrate ATP increases the thermal stability, as was noted by Liubimova and Engelhardt.[40])

(39) K. Bailey, *Advances in Protein Chem.* **1,** 289 (1944).
(40) V. A. Engelhardt, *Advances in Enzymol.* **6,** 147 (1946).
(41) A. Szent-Györgyi, Chemistry of Muscular Contraction, Academic Press, New York, 1947.
(42) I. Banga, *Studies Inst. Med. Chem. Univ. Szeged* **1,** 27 (1941/2).
(43) K. Bailey and E. C. Webb, *Biochem. J.* **38,** 394 (1944).
(44) L. A. Heppel and R. J. Hilmoe, *J. Biol. Chem.* **192,** 87 (1951).
(45) B. D. Polis and O. Meyerhof, *J. Biol. Chem.* **163,** 339 (1946).
(46) R. K. Morton, *Nature* **166,** 1092 (1950).
(47) T.-C. Tsao, Ph. D. Dissertation, Cambridge, England, 1951.

(*4*) Other factors helping in the preservation of ATPase were: pH near 7, low ionic strength, and the presence of reducing agents such as glutathione.

Loss of solubility and inactivation do not always run parallel. Acidification inactivates, as also any treatment which blocks or oxidizes —SH groups, although no obvious change in solubility accompanies such treatment. Most surprising is the modification of myosin to a water-soluble form by the action of trypsin, without appreciable loss of enzyme activity.[48,49]

There seems no doubt that ATPase is either closely associated with myosin, or is part of the myosin molecule itself. The most impressive evidence that ATPase is a functional part of the myosin molecule is the correlation of the enzyme character and colloid properties (sec. II-3*f* *et seq.*).

(*i*) *Specificity.* Myosin ATPase acts only on ATP[39] and inosine triphosphate,[50] the latter more rapidly than the former, and it splits only the terminal phosphate group. In the presence of myokinase, which is only removed from myosin preparations by repeated precipitations, both phosphate groups are split according to the reaction

$$2 \text{ adenosine diphosphate (ADP)} \rightleftharpoons \text{adenylic acid (AMP)} + \text{ATP}$$

Since adenylic deaminase is also often present, the final products are inosinic acid and phosphate. For reasons not wholly elucidated, the terminal phosphate of apparently pure ATP is split only to the extent of 90 per cent.[51–53] ATP has now been prepared synthetically,[54] and it would be interesting to study the specificity requirements in greater detail: whether the purine or sugar or both can be exchanged for other purines and sugars, and whether the sugar can be substituted. In this way it may be possible to decide whether the enzyme has a two-point affinity or is specific only to the triphosphate portion. There is, at least, no action on inorganic polyphosphates, nor on substituted pyrophosphates, although these are able to dissociate actin and myosin in solution.

(48) G. Gergely, *Federation Proc.* **9,** 176 (1950).
(49) S. V. Perry, *Biochem. J.* **48,** 257 (1951).
(50) A. Kleinzeller, *Biochem. J.* **36,** 729 (1942).
(51) K. Bailey, *Biochem. J.* **45,** 479 (1949).
(52) S. L. Rowles and L. A. Stocken, *Biochem. J.* **47,** 489 (1950).
(53) However, no laboratory or commercial preparation of ATP analyzed by the resin method of W. E. Cohn and C. E. Carter [*J. Am. Chem. Soc.* **72,** 4273 (1950)] whether carried out in Cambridge or reported in the literature seems free of ADP.
(54) J. Baddiley, A. M. Michelson, and A. R. Todd, *J. Chem. Soc.* **1949,** 582.

The so-called "apyrases" such as occur in potato[55] act both on ATP and ADP and to some extent on adenylic acid. Other types of animal ATPase do not respond to Ca activation as does myosin ATPase, but without exception are Mg activated. This type of ATPase is usually associated with particulate bodies, mitochondria, microsomes and the like.

(2) *Effect of Ions.* Although there are two points of agreement, that certain bivalent metals are necessary for the action of ATPase, and that Ca is the most powerful activator at alkaline pH, there exists some confusion as to the effect of other metals. In the presence of Na and K ions, Bailey[14] found that 4× precipitated myosin, having no action on ADP, was activated by Ca and Mn and not by Mg. Indeed, Liubimova and Pevsner,[56] using the Ca salt of ATP, had noted that Mg ions antagonized the activating effect of Ca, an observation explored in greater detail by Greville and Lehmann.[57] Banga,[42] studying the action of many ions on actomyosin, noticed that univalent ions (K^+, Na^+, Li^+, NH_4^+) possessed an activating action about 40 per cent that of Ca; Mg (cf. above) was also found to activate.

Perhaps the most thorough study which emphasizes the *mutual* influence of ions is that of Mommaerts and Seraidarian.[58] Their results can be summarized as follows: (*a*) At pH 7 (glycine "buffer"), both myosin and actomyosin split ATP optimally when the potassium chloride concentration is about 0.3 *M*; in the presence of 0.001 *M* $CaCl_2$, the K optimum is shifted to lower concentrations, and the activity is three times as great. (*b*) In the presence of 0.1 *M* KCl, there is a very pronounced optimum at pH 9 when Ca is the activating ion, and a second less-marked optimum at about pH 6.5 (borate and glycine "buffers"). Without Ca, only this latter optimum was observed. (*c*) The effect of Mg is always to suppress the activation by other ions. In the case of yeast pyrophosphatase, it has been shown that the antagonism of Ca and Mg depends upon the ionic ratio.[43] (This enzyme is Mg activated.) The effect is observable to some extent with ATPase of actin-poor myosin, but with actomyosin, a 5:1 ratio of Mg and Ca does not altogether suppress the activity at pH 7. There is now sufficient evidence to accept that actin-rich myosins are in fact activated by Mg ions at neutral pH provided that K ions are absent.

To attempt an explanation of activation effects is certainly a difficult task. It might be thought that the activation of actomyosin by Mg at

(55) H. M. Kalckar, *J. Biol. Chem.* **153**, 355 (1944).
(56) M. M. Liubimova and D. Pevsner, *Biokhimiya* **6**, 178 (1941).
(57) G. D. Greville and H. Lehmann, *Nature* **152**, 81 (1943).
(58) W. F. H. M. Mommaerts and K. Seraidarian, *J. Gen. Physiol.* **30**, 201 (1947).

neutral pH is due to the presence of another enzyme, such as the mitochondrial ATPase studied by Kielley and Meyerhof.[59] This, however, is a labile enzyme and should contaminate equally well myosin or actomyosin. Weber[15] suggests that activation at pH 9 is that of L-myosin in solution; that at pH 6.5 of actomyosin in insoluble form.

The argument has been advanced that the influence of Mg in the concentration found in muscle would seriously diminish the activity of ATPase, and that the actual observed rate of phosphate formation in activity is 100 times as great.[58] It is not always sound to extrapolate from a disordered to a structurally organized system, and until the mechanisms of phosphate breakdown are discovered, one must be cautious in accepting the argument.[49] Perry,[49] in fact, has shown that the activation of intact myofibrils is quite different from that of the same myofibrils under otherwise identical conditions except that they have initially been dissolved in salt solution. With the intact myofibril, the following effects were observed: (*a*) As noted with myosin and actomyosin, there is an optimal K concentration for the activation by Ca. (*b*) The activation by Mg can be just as great as that by Ca, but occurs at a much lower concentration; above the optimum, Mg is strongly inhibitory, while for Ca the concentration is not so critical. (*c*) In the presence of 0.1 *M* KCl, fibrils which have first been dissolved before incubation with substrate do not respond to activation by Mg at all.

These latter results need no emphasis and their importance is very apparent. They provide a basis for the assumption that in the presence of the K ions of the sarcoplasm activation of ATPase could be brought about either by Ca or by Mg, according to the availability of these ions at the site of enzyme action.

(*3*) *Other Enzymes Associated with Myosin.* There seems no doubt that myosin preparations are often able to deaminate adenylic acid. Hermann and Josepovits,[60] indeed, have claimed that the original Schmidt deaminase preparation and Kalckar's preparation[61] are no more active than crystalline myosin A. Gergely,[62] however, has obtained a preparation 20 times more active than that of myosin, and suggests that the enzyme is closely associated but not identical.[63] Szent-Györgyi[41] has put forward the idea that tenaciously held impurities are really part of the

(59) W. W. Kielley and O. Meyerhof, *J. Biol. Chem.* **176,** 591 (1948).

(60) V. Sz. Hermann and G. Josepovits, *Nature* **164,** 846 (1949).

(61) H. M. Kalckar, *J. Biol. Chem.* **148,** 127 (1943).

(62) J. Gergely, *Federation Proc.* **10,** 188 (1951).

(63) V. A. Engelhardt (*Communs. 2nd Intern. Congr., Paris, 1952;* published by Editions de l'Academie des Sciences U.R.S.S., 1952) has successfully separated the deaminase activity from the ATPase (but not vice versa) by heat treatment. The deaminase fraction is a globulin.

myosin complex, and suggests the name protins. There is no justification for this special terminology nor for the definition that only the whole complex is capable of "contraction." It is quite obvious that myosin will act ideally as an adsorbent particularly at low ionic strength, just as calcium phosphate is employed under similar conditions. The purification of myosin by dilution at low ionic strength may not suffice to remove some enzymes quickly; once-precipitated myosin possesses quite a high myokinase activity, which requires about 4–5 precipitations to remove entirely. What was thought by Price and Cori[64] to be ATPase separated from myosin itself was found to be creatine phosphokinase which had been carried down on the myosin, despite the fact that this enzyme is extremely soluble. It is not unlikely that myosin could be used to carry down certain enzymes which might be eluted by stronger salt solutions at a pH where myosin itself is insoluble.

d. Characterization of Various Myosins by Physical Methods

It is appropriate at this stage to anticipate the discussion on actin and actomyosin, and to review some of the older work on what was thought to be myosin, and to compare it with recent studies on actin-free material.

(*1*) *Investigations on Myosins Contaminated with Actin.* In the pioneer work of Edsall and von Muralt, the asymmetry of the particles was inferred from the flow birefringence and high viscosity. It was noted that both acid and alkali destroyed flow birefringence. Much later, the effect of a whole series of substances on rabbit and lobster myosin was investigated.[65] Guanidine-HCl, arginine-HCl, Ca^{++}, Mg^{++}, I^-, and SCN^- in low concentration, or Li^+, NH_4Cl, Br_2, and urea in high concentration caused both the disappearance of flow birefringence at low shear rates, and a fall in viscosity. No deep-rooted change was apparent, and it was concluded that the very long molecules were broken up into less asymmetric chains. Extending these observations, Needham and co-workers[66] examined simultaneously the changes in flow birefringence and anomalous viscosity produced by electrolytes, change of pH, urea, and other substances. Of particular interest was the effect of ATP which in very low concentration ($\sim$0.004 M) caused a decrease in flow birefringence and in viscosity. All these studies would appear to relate to the dissociation of actin and myosin; the latter in particular anticipates the findings of Szent-Györgyi and his school.

(64) W. H. Price and C. F. Cori, *J. Biol. Chem.* **162,** 393 (1946).
(65) J. T. Edsall and J. W. Mehl, *J. Biol. Chem.* **133,** 409 (1940).
(66) M. Dainty, A. Kleinzeller, A. S. C. Lawrence, M. Miall, J. Needham, D. M. Needham and S.-C. Shen, *J. Gen. Physiol.* **27,** 355 (1944).

In some studies, flow birefringence and viscosity were treated quantitatively to obtain an estimate of particle size. For (acto-)myosin from rabbit, from smooth muscle of octopus, and from snail, an approximate length of 5000–10,000 A. was deduced.[67] Other estimates were derived from osmotic-pressure, sedimentation, and electron-optical studies. The osmotic measurements of Weber and Stöver[68] gave the first indication of the approximate particle weight, $\sim 10^6$ at 1 per cent protein concentration, falling to 10^5 in urea solutions. The asymmetry of the particles was confirmed by Ardenne and Weber[69] using the electron microscope; fibrils 5–10 μ thick and several microns long were thought to arise by aggregation of individual myosin molecules. Quite widely different species of animal (rabbit, frog, lobster, scallop, and clam) were found to give particles of fairly uniform width (50–250 A.) but of very variable length (usually $\sim$15,000 A.[70]).

The polydispersity of (acto-)myosin was first indicated by the ultracentrifuge and later confirmed by electrophoresis. The first serious ultracentrifugal study was carried out by Schramm and Weber,[20] who differentiated a slow component (L-myosin) and a fast, [(acto-)myosin] of sedimentation constants 6 and 20; the heavy component gave rise to others (29–36 S) as the solutions aged. Ziff and Moore,[71] however, observed a strong concentration dependence of the sedimentation, which had so far been neglected, though Bergold[72] was the first to derive the extrapolated value of s at infinite dilution (see Table I). The most recent work from several laboratories seems to agree that in natural actomyosins it is possible to observe the sedimentation of free myosins and several actomyosins side by side.[21]

Together with data on the over-all dimensions of (acto-)myosin particles, some information had been obtained on intramolecular structure. The x-ray diagram of stretched (acto-)myosin fibers permitted Boehm and Weber[73] to infer that "myosin" particles are quite elongated; birefringence and elasticity studies on the same material suggested that the particles contained chains in a folded configuration, which conferred a long-range elasticity. This was substantiated directly and unambiguously by the x-ray investigation of the α-β transformation and the supercontraction of actomyosin films. In native myosin, as in mam-

(67) See J. T. Edsall *in* Cohn and Edsall, Proteins, Amino Acids and Peptides, Reinhold Publishing Corp., New York, 1943, p. 540.
(68) H. H. Weber and R. Stöver, *Biochem. Z.* **259**, 269 (1953).
(69) M. V. Ardenne and H. H. Weber, *Kolloid-Z.* **97**, 322 (1941).
(70) C. E. Hall, M. A. Jakus, and F. O. Schmitt, *Biol. Bull.* **90**, 32 (1946).
(71) M. Ziff and D. H. Moore, *J. Biol. Chem.* **153**, 653 (1944).
(72) G. Bergold, *Z. Naturforsch.* **1**, 100 (1946).
(73) G. Boehm and H. H. Weber, *Kolloid-Z.* **61**, 269 (1932).

malian keratin and epidermis, fibrinogen, and tropomyosin, the wide-angle diffraction pattern is of α-type; it can be observed in living and dead muscle, even after drying.[74]

(*2*) *Investigations on "Crystalline" Myosin.* The electron microscope studies of Jakus and Hall[75] on rabbit myosin "crystals" have shown that they are composed of fibrils, oriented with their long axis roughly parallel to each other and to the long axis of the aggregate. The addition of 0.2 *M* KCl causes disaggregation into short rodlets 200–400 A. in width and 0.5–1 μ in length. These dimensions fall within the range of the particles found in actin-containing preparations. The shadows of the rodlets are very short, indicating that the particles are flattened on the supporting film. Similar results were obtained by Rosza and Staudinger[76] and by Snellman and Erdös,[77] the latter showing that the "crystals" can be broken up into threads of uniform thickness by supersonic waves. After dissolution in 0.3 *M* KCl, very little could be seen.

The sedimentation of "crystalline" myosin is predominantly that of pure myosin, and the addition of ATP does not produce any marked change. Although actomyosin is present, it is not visible, and the fast component which Snellman and Erdös observed was probably an aggregation product of myosin, since it was more abundant in twice-crystallized myosin[78] (cf. Weber).[15] These authors observed an interesting effect of urea on myosin which will be discussed later: that when first dissolved, the particles become less symmetrical, but after 7 days, components of lower sedimentation constant accumulate.

In confirmation of ultracentrifugal data, once "crystallized" myosin was found to have a second component after prolonged electrophoresis.[79] Above pH 5.4 in KCl solutions, myosin moves toward the anode; in the presence of Ca^{++} and Mg^{++} it becomes cathodic between pH 2 and 9. The further addition of KCl reduces the amount of bivalent ion bound, so that at pH 7 the protein can be made to move either way according to the ratio of bi- and univalent ions. It would be surprising if this effect were unique for myosin, although it may well be exaggerated because the net charge is not very high (sec. VI).

(*3*) *Investigations on Pure Myosin.* The sedimentation constant of the lightest component in myosin preparations, after extrapolation to

(74) W. T. Astbury and S. Dickinson, *Proc. Roy. Soc.* (London) **B129,** 307 (1940).
(75) M. A. Jakus and C. E. Hall, *J. Biol. Chem.* **167,** 705 (1947).
(76) G. Rozsa and M. Staudinger, *Makromol. Chem.* **2,** 66 (1948).
(77) O. Snellman and T. Erdös, *Biochim. et Biophys. Acta* **2,** 660 (1948).
(78) O. Snellman and T. Erdös, *Biochim. et Biophys. Acta* **2,** 650 (1948).
(79) T. Erdös and O. Snellmann, *Biochim. et Biophys. Acta* **2,** 642 (1948).

infinite dilution, has been reported to be 7.1,[21] 6.7[80] (L-myosin), and 7.2[81] ("crystalline" myosin). The observation that, on aging, well-defined aggregation products arise makes difficult the determination of diffusion constant.[21] How these products are formed is not known, and any process of gradual aggregation should not be evinced in such sharply defined stages. The possibility that polymerization occurs by oxidation of —SH groups to give —SS— bridges has not been explored. The higher sedimentation constant could arise by a side-to-side aggregation or by a decrease of asymmetry.

The diffusion constant has been determined by a rapid method[82] in 6–8 hours, and the homogeneity of the preparation ultimately checked by the symmetry of the diffusion diagram and by the ultracentrifuge; likewise, when allowance is made for the dependence of sedimentation on concentration, the sedimentation diagram also conforms to that for a monodisperse protein. Taking D_{20} as 0.87×10^{-7} and $S_{20} = 7.1$, the particle weight is $850{,}000 \pm 30{,}000$ and the axial ratio, 98, neglecting hydration. The particle weight has also been checked by the osmotic method[82] in the same laboratory ($840{,}000 \pm 33{,}000$). These values of Weber's group seem to be the most reliable that have been reported. The axial ratio derived from the frictional ratio represents a particle some 2300 A. in length and 23 A. wide. The latter dimension is so small that it seems relevant to question whether such a molecule really can be interpreted as a rigid structure. The angular dissymmetry of light scattering[83] suggests a shorter molecule, 1500 A. Although the viscosity of myosin varies little with viscosity gradient, it is not quite certain that the viscosity increment represents a random translation of particles.[84]

e. Structural Features of Myosin

Of the detailed molecular structure of myosin, little is known. Both *in situ* and after extraction, the diffraction pattern is of α-type,[74] and films made by drying down a solution on glass can be stretched to give an oriented α-pattern and then an oriented β. No small-angle diffractions have ever been reported, although the depolymerization of myosin in urea[68] and in alkali[18] shows that the native molecule is in reality built up from smaller units, which might be expected to give rise to such spacings.

Like tropomyosin, another protein of the α-keratin group, the only amino groups which react with Sanger's reagent, fluorodinitrobenzene,

(80) W. F. H. M. Mommaerts and R. G. Parrish, *J. Biol. Chem.* **88,** 545 (1951).
(81) P. Johnson and R. Landolt, *Nature* **165,** 430 (1950).
(82) H. Portzehl, *Z. Naturforsch.* **5,** 75 (1950).
(83) W. F. H. M. Mommaerts, *J. Biol. Chem.* **188,** 553 (1951).
(84) H. H. Weber, *Biochim. et Biophys. Acta* **4,** 12 (1950).

are those of the lysine side chains; α-amino groups have been detected in only small amount.[85] In the case of tropomyosin, there is strong supporting evidence for a cyclopeptide structure, and the striking similarities in the properties of the two proteins suggest some obscure relation which may well extend to the finer details of structure.[86] Of course, it is always possible that α-amino groups are present but are unreactive toward the reagent; nevertheless, the procedures used to uncover groups which are unreactive toward chemical agents—denaturation, treatment in urea solutions—do not reveal any extra groups in myosin and tropomyosin.

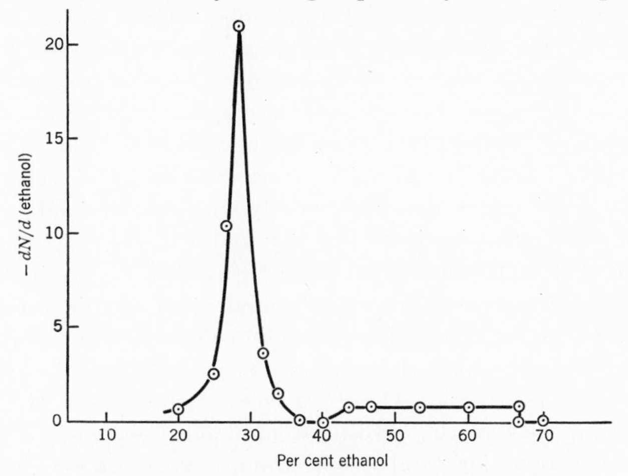

FIG. 2. The fractionation of urea-treated myosin by ethanol.

Tsao[18] has made considerable progress in the investigation of the subunits of myosin by following the conditions for their liberation, and determining their size from osmotic pressure and polarization–fluorescence. When myosin is dissolved in urea at room temperature, certain well-defined fractions can be obtained by addition of ethanol. Freshly dissolved, the whole of the protein precipitates between limits of 10–15 per cent (v/v) ethanol; after 1–2 months the major portion precipitates at 20–30 per cent, and at 6 months or later, at 65 per cent. As a function of time, no further change occurs, the intrinsic viscosity remaining at 0.6–0.7, indicative of particles of high asymmetry. The process, though so very slow, does not appear to involve the cleavage of peptide bonds, for where this was suspected to occur (at pH 13) N-terminal residues could readily be detected by Sanger's method.

In Fig. 2 the fractionation behavior at 2 months is shown; a main fraction (I) is followed by a small fraction (II) which continues to pre-

(85) K. Bailey, *Biochem. J.* **49,** 23 (1951).

(86) T.-C. Tsao, K. Bailey, and G. S. Adair, *Biochem. J.* **49,** 27 (1951).

cipitate up to 70 per cent ethanol. After removal of urea, fraction I is a synerizing gel, insoluble in water and salt; fraction II, a water-soluble protein which comprises some 4–5 per cent of the total protein. At 6 months and later, the fractionation follows a different course, for now the whole protein precipitates at 65 per cent ethanol. It can nevertheless be fractionated with ammonium sulfate to give two fractions, A and B, of which B was shown to be identical in all its properties with II; but it now amounts to nearly 10 per cent of the total protein. Fraction A, however, differs from I being now water-soluble, as though the continued action of urea has caused an intramolecular rearrangement of groups which impart a greater symmetry of charge.

(*1*) *The Size and Nature of the Subunits.* Osmotic-pressure measurement on urea-treated myosin gives the impression that extensive depolymerization has occurred. In reality, the low values are due to the small particle weight of fraction II, which by osmotic pressure is about 14,000 (Table III). This value agrees excellently with that derived for fraction B by fluorescence–polarization[86a] (16,000) and with the weight associated with one *N*-terminal group (also 16,000 g.). The *N*-terminal assay is commensurate with the "impurities" found in all myosin preparations, and includes two or three different amino acids (valine, alanine, and lysine); likewise, electrophoresis shows a large peak and a smaller associated peak. That fraction II cannot come within the ordinary definition of an impurity in the myosin seems likely for two reasons: first, its liberation is very slow; and second, the probable impurity, actin, does not after urea treatment pass through membranes which are permeable to fraction II.

Fraction I appears to be the myosin "framework," which except for changes in shape and solubility seems unaffected by urea. The water-soluble modification, B, by G. Weber's method (see later) has a particle weight much greater than 300,000 (the limit for that method), and its sedimentation constant in urea (8.7 S at c = 0.4) is still very high. At pH 10.7, however, the framework "explodes" and both osmotic pressure and fluoresence–polarization show that the particle weight is about 170,000 (Table III). Pure myosin itself gives the same value under the same conditions, and here it is important to note that fraction II cannot be separated from myosin treated in this way: it contributes little therefore to the average particle weight of the subunits.

The general picture that emerges from these studies is that the myosin framework is built up of about 5 units of particle weight 165,000, which do not possess an identifiable *N*-terminal residue and may thus be cyclic. Associated with them are as many small open-chain polypeptides of

(86*a*) G. Weber, *Biochem. J.* **51**, 145 (1952).

TABLE III

Average Particle Weight of Depolymerized Fragments of Myosin by Fluorescence–Polarization Measurements[18]

Fraction	Condition	Slope, $S \times 10^4$	$1/p_0$	$\rho_h{}^a \times 10^8$	a/b^b	ρ_0/ρ_h	$\rho_0 \times 10^8$	$V \times 10^5$	d^c	M
A	Water, pH 7	0	4.21	—	30	0.375	—	Very large		Very large
A	Borate, pH 10.7, cooling	0.166	4.62	37.9	30	0.375	14.2	1.31	1.35	177,000
									1.25	164,000
A	Borate, pH 10.7, heating	0.140	4.02	39.4	30	0.375	14.8	1.36	1.35	184,000
									1.25	170,000
B^d	Water, pH 7 (initial slope)	2.09	5.32	3.43	12	0.410	1.4	.129	1.35	17,400
									1.25	16,000

[a] Calculated for 25°C.

[b] From viscosity measurements.

[c] Assuming density of protein $d = 1.35$ (partial specific volume 0.74). Upper value calculated for anhydrous particles, lower value for particles with 30 per cent hydration.

[d] The corresponding fraction II (see text) has a particle weight of 14,000 by osmotic pressure.

average particle weight about 16,000. The forces which need to be overcome to liberate these two types of unit differ: the framework units are liberated at a pH where ϵ-amino groups become deionized, as though the binding is largely electrostatic; the small units remain attached and so far have only been liberated by concentrated urea and guanidine. Both are asymmetric, and the axial ratio of the framework subunit is between 30 and 40, no matter whether it has been subjected to the initial action of urea or not. Though tropomyosin can often be extracted from some myosin preparations, there was nothing to show that the subunits had any of the characteristic properties of tropomyosin.

f. The Action of Trypsin on Myosin

Gergely[48] and Perry[49] independently observed that a rapid treatment of myosin with trypsin produced a water-soluble product which showed no loss of enzyme activity. While the general properties of the enzyme are unchanged,[86b] the ultracentrifuge shows that an extremely rapid fission into two distinct components (2.5 and 5.3 *S*, respectively) occurs by an all-or-none reaction.[86c] This phase[86d] is followed by a slower one characterized by more extensive degradation and loss of enzyme activity.

The two proteins have been called meromyosins (*meros* (Gr.) meaning part or portion): the heavier, H-meromyosin; the lighter, L-meromyosin. The latter has the properties of a globulin and deposits on dialysis in the form of birefringent sheaths.[86e] The H-meromyosin contains all the

TABLE IV
Physical Data on the Meromyosins[86c,d,e]

	S^0_{20}	D^0_{20}	M	f/f_0	Suggested length and diam., A.
H-Meromyosin	6.96	2.91	232,000	1.78	430 × 30
L-Meromyosin	2.86	2.87	96,000	2.45	550 × 17

ATPase activity and combines with actin to give a more viscous, ATP-sensitive complex. The finding that the nitroprusside reacting groups are also associated supports the general thesis that sites containing these groups are responsible both for ATPase activity and for combination with actin. The yields of the two proteins, 43 per cent for L- and 57 per cent for H-meromyosin, combined with the particle weights from sedimenta-

(86*b*) J. Gergely, *J. Biol. Chem.* **200,** 543 (1953).
(86*c*) E. Mihalyi and A. G. Szent-Györgyi, *J. Biol. Chem.* **201,** 189 (1953).
(86*d*) E. Mihalyi, *J. Biol. Chem.* **201,** 197 (1953).
(86*e*) A. G. Szent-Györgyi, *Biochim. et Biophys. Acta* **42,** 305 (1953).

tion and diffusion data (Table IV) allow 4 molecules of L- and 2 of H-meromyosin per mole of myosin.

This important work, carried out in its later stages by Mihalyi and A. G. Szent-Györgyi, suggests that chemically different subunits occur in myosin, only one of which carries the biological properties which have been studied so widely. The action of trypsin and the reported presence of *N*-terminal groups in the meromyosins suggest that they are joined by peptide bonds, a finding of great significance in relation to the biogenesis of large molecules. However, it is difficult to correlate these findings as yet with those of Tsao. The average particle weight of H- and L-meromyosin in the proportions found in myosin would give a number average of 146,000, not greatly divergent from Tsao's experimental value of 170,000, and Tsao's data give no information as to the heterogeneity or otherwise of the urea-liberated subunits. Moreover, the "open" peptide component has not been reported by Mihalyi and A. G. Szent-Györgyi, and it is tempting to speculate that the major parts of the myosin molecule may be held together by such peptides and that these are most readily attacked by trypsin. A concept such as this would tend to bring together the two lines of approach, but it must be emphasized that the enzymatic approach is much the more promising since it preserves the biological activities of the original molecule.

3. Actin

a. General Properties of Actin

The discovery of actin as a major component of muscle followed from the observation that certain myosin-containing extracts of muscle (myosin B) lost their gel-like consistency on addition of ATP, while other extracts (myosin A) were much less affected by ATP.[17,41] Considering that actin can be prepared only by unconventional methods, the achievement of Straub[87,88] in isolating it so soon is notable. Before the methods of preparation can be fully appreciated, it is preferable to gain some knowledge of the properties of actin.

One striking property is that it can exist in two forms, one of which in solution is limpid and is converted by salt ions or acid into a highly viscous form which shows strong flow birefringence. Even more important is the ability of actin to interact with myosin to give a colloidal system whose state is profoundly affected by ATP.

The effect of ions on the transformation of G to F actin has been studied in some detail by viscometric techniques. Many ions are

(87) F. B. Straub, *Studies Inst. Med. Chem. Univ. Szeged* **2,** 3 (1942).
(88) F. B. Straub, *Studies Inst. Med. Chem. Univ. Szeged* **3,** 23 (1943).

capable of transforming to the same ultimate viscosity, although the rates vary widely.[88] Ca^{++} and Na^{+} (or K^{+}) show a distinct antagonism, while iodide (approx. 0.5 M) inhibits. The rate appears to be increased by small amounts of myosin. As normally prepared, G-actin preparations contain small amounts of Mg; if this is removed by sodium hexametaphosphate (Calgon), the subsequent addition of alkali salts does not increase the viscosity.[89] The addition of more Mg ion to that originally present appears to speed up some change such that the removal of Mg and addition of KCl now allows the transformation.

TABLE V
NUCLEOTIDE CONTENT OF ACTIN AND ISOLATED MYOFIBRILS
(Results expressed as μmoles/g.)

Material	Adenylic acid	ADP	ATP
Myofibrils	0.88	2.7	0.47[a]
Actin in myofibrils*	4.4	13.5	2.3[a]
F-Actin	—	—	8.5[b]
F-Actin	1.6	11.7	2.4[c]
G-Actin	—	—	3.4[b]
G-Actin	—	—	9.8[b]
G-Actin	—	—	23[d]
G-Actin	—	—	12–39[e]
G-Actin	2.1	1.9	7.8[c]
F-Actin	—	17.5	Trace[f]

* Calculation made on the basis that all the nucleotide is associated with the actin which makes up 20 per cent of the total myofibrillar proteins.

[a] Perry.[99]

[b] K. Laki, W. J. Bowen, and A. M. Clark, *J. Gen. Physiol.* **33**, 437 (1950).

[c] A. G. Szent-Györgyi, *Arch. Biochem. Biophys.* **31**, 97 (1951).

[d] Straub and Feuer.[90]

[e] M. Dubuisson and L. Mathieu, *Experientia* **6**, 103 (1950).

[f] W. F. H. M. Mommaerts, *J. Biol. Chem.* **198**, 469 (1952).

Actin contains small amounts of easily hydrolyzed P which in G-actin has been carefully identified as ATP by Straub and Feuer,[90] who have further suggested that the G → F transformation involves the reversible change of ATP to ADP in the actin molecules. Other workers have in fact confirmed that ADP is formed, but the wide variation in the nucleotide content of both G- and F-actin do not leave one entirely convinced. Table V, collected mainly by Perry,[91] shows the extent of variation.

(89) G. Feuer, F. Molnár, E. Pettkó, and F. B. Straub, *Hung. Acta Physiol.* **1**, 1 (1948).

(90) F. B. Straub and G. Feuer, *Biochim. et Biophys. Acta* **4**, 455 (1950).

(91) S. V. Perry, *Biochem. J.* **51**, 495 (1952).

More convincing perhaps is his demonstration that washed, isolated myofibrils contain fairly consistent amounts of ADP,[91] which from the probable actin content, would amount to 1 mole in 75,000. This latter figure is in good agreement with the molecular weight of the primary actin particle (see below). A. G. Szent-Györgyi's estimate[92] of ADP in isolated F-actin likewise gives a comparable figure (about 90,000).

By the quantitative nitroprusside reaction, actin contains 0.85 per cent of cysteine of which about 40 per cent reacts to give an adduct with *N*-ethylmaleimide.[33] This reagent, however, does not affect the characteristic properties of actin, while oxidants such as iodosobenzoate, cystine, and methylene blue have some effect.[89] The most effective reagents in preventing the transformation of G- to F-actin, or of reversing it, are heavy-metal compounds: chloromercuribenzoate, Salyrgan (a mercury-containing arsenical), and copper glycinate. Kuschinsky and Turba[93–95] have explained their effect as due primarily to combination with —SH groups which they consider necessary for what is generally believed to be a "polymerization" of G-actin particles. As we shall see, this direct interpretation is open to question, as indeed is the concept of polymerization.

b. Preparation of Actin

The main operations in the preparation of actin as developed by Straub,[87,88] by Guba, and by Szent-Györgyi[41] are: (*a*) the extraction of fresh muscle mince with salt solution at pH 6.5 to remove some of the myosin; (*b*) the dilution of the residue with strongly alkaline salt solution (pH 10); (*c*) drying the alkaline residue with acetone to denature residual myosin; and (*d*) extraction with water to obtain a solution of G-actin. The latter can be polymerized in two ways: either by addition of salt, or by isoelectric precipitation at pH 4.6 and reneutralizing to pH 7. Not all the protein extracted polymerizes, and F-actin can be separated from unpolymerized G-actin by ultracentrifugation.

By electrophoresis[33,96] or by the ultracentrifuge,[97] the purity of actin cannot be followed because of thixotropic effects, and it is necessary to employ a high or low pH where only the G-form exists.[33] Besides some enzymatic impurities (myokinase and creatine phosphokinase and phos-

(92) A. G. Szent-Györgyi, *Arch. Biochem. Biophys.* **31,** 97 (1951).
(93) G. Kuschinsky and F. Turba, *Naturwissenschaften* **37,** 425 (1950).
(94) F. Turba, G. Kuschinsky, and H. Thomann, *Naturwissenschaften* **37,** 453 (1950).
(95) F. Turba and G. Kuschinsky, *Biochim. et Biophys. Acta* **8,** 76 (1952).
(96) S. S. Spicer and J. Gergeley, *J. Biol. Chem.* **188,** 179 (1951).
(97) W. F. H. M. Mommaerts, *J. Biol. Chem.* **188,** 559 (1951).

pholipides largely removable by treatment of the acetone residue with chloroform), Straub's actin not only contains inactive actin, but a peak which is augmented when pure tropomyosin is added to the preparation; moreover, tropomyosin in crystalline form has been isolated from it.[33] The second impurity is some kind of flavoprotein, and both together may constitute some 10 per cent of the total protein.

Mommaerts[97] has purified actin by spinning down the gel form in the ultracentrifuge, and Szent-Györgyi[98] by extraction with KI (in the presence of ATP), followed by ethanol fractionation. Mommaert's actin, apart from requiring expensive apparatus, appears refractory to dissolve, and on subsequent centrifugation, not all of it appears to sediment. The KI–actin was found to contain at least 40 per cent of inactive form, and no electrophoretic data were applied to enable judgment as to the absence of the two above-mentioned impurities. At the present stage of development, it seems more important to ensure chemical purity rather than full activity, and in the preparation of Tsao and the author,[33] the purification was followed in the Tiselius apparatus at pH's 2 and 10.

In exploratory experiments, the factors affecting (*a*) the yield of actin and (*b*) the effect of pH (at the stage before the muscle residue is dried by organic solvents) upon electrophoretic purity were examined. While the yield from acetone-treated residue increases with rising pH, the specific viscosity after conversion to F-actin declines; if butanol followed by acetone treatment is employed, the yield again increases, but for residue treated at pH 9 large amounts (as much as 30 per cent) of tropomyosin are present as impurity. In the final method, four principles were combined: (*a*) Liberation of actin is favored by a fine state of subdivision in the fiber and also by (*b*) the liberation of actin from the complex with lipide and by (*c*) maintenance of a neutral pH during washing to preserve activity; and (*d*) the extracting medium must suppress the extraction of impurities. To remove lipide, the method incorporates the use of butanol, shown by Morton[46] to be effective for several particle-bound enzyme systems, and 30 per cent acetone as the final extracting medium, which leaves tropomyosin behind. Moreover, the flavoprotein is no longer found. Because by this method myosin, actin, and tropomyosin can all be prepared from the same muscle sample, it seems worthy of record here:

(*1*) *Method Allowing Preparation of Myosin, Actin and Tropomyosin from the Same Muscle Sample.*[33] Rabbit muscle is minced and stirred with 3 vol. of ice-cold acid phosphate–KCl solution (Guba and Szent-Györgyi).[41] After 10 min., the suspension is diluted with water, 4 vol. for every volume of salt used. The solution of myosin is then strained through a cloth and the residue washed first with 4 vol. of 0.4 per cent $NaHCO_3$ solution (pH of final suspension 7.0) and then with 10–15 vol. of distilled water. Each washing lasts for 20–30 min. under constant stirring at room temperature. The residue is pressed as dry as possible, and is macerated for 1 min. in the Waring blendor with 5 vol. of *n*-butanol which has been precooled so that the temperature does not rise above 20° during maceration. After standing for ½ hour, butanol is strained out through a cloth and the residue washed with 2–3 vol. of acetone and

(98) A. G. Szent-Györgyi, *J. Biol. Chem.* **192,** 361 (1951).

again strained out. Any lumps of fiber are disintegrated by blending with acetone for a few seconds, washing with acetone twice more, and finally drying in air overnight.

The following operations are carried out in the cold room with all reagents at 0°:

Every 10 g. of air-dried fiber are extracted with 200–300 ml. of 30 per cent (*v/v*) acetone containing 40 μg. each of Na ATP and neutralized ascorbic acid/ml. After extraction for 30 min. with constant slow stirring, the liquor is strained through a cloth onto a Büchner funnel containing a thin layer of paper pulp (washed with 30 per cent acetone) which removes fine debris. Molar acetate buffer (0.5 ml., pH 4.65) is added to the clear filtrate and the precipitate spun down at once. A few drops of 5 per cent $NaHCO_3$ are added to the precipitate which transforms into a transparent,

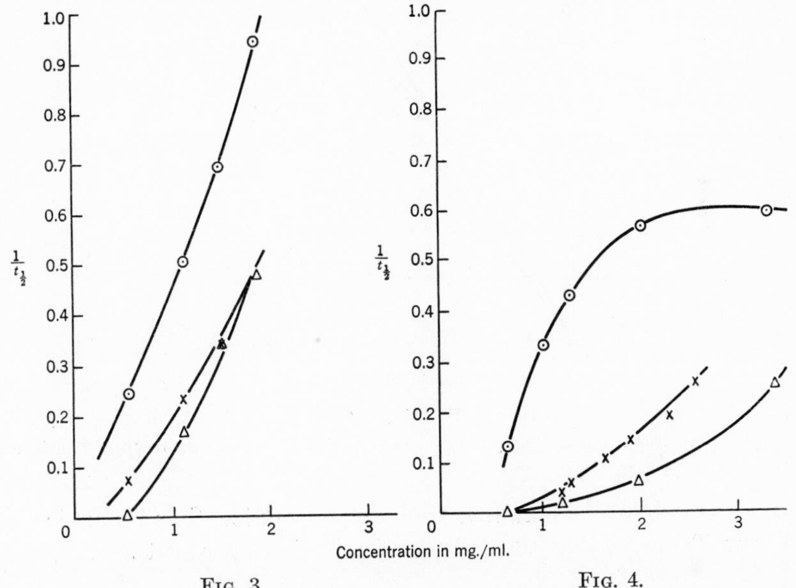

Fig. 3. Fig. 4.

Fig. 3. Influence of protein concentration and neutral salts on the rate of "polymerization" of butanol–actin at pH 7. Ordinate: $t_{1/2}$ signifies half the period taken to polymerize fully. Polymerization induced by 0.1 *M* KCl (×), 0.1 *M* KCl + 0.0001 *M* MgSO; (⊙), or 0.1 *M* KCl + 0.001 *M* $CaCl_2$ (△).

Fig. 4. As for Fig. 3, using Straub actin.[89]

highly thixotropic gel. This is dialyzed against several rapid changes of 0.1 *M* KCl containing ATP and ascorbic acid as above to remove acetone. If water instead of KCl is used, the F-form reverts to nonthixotropic G-actin.

The yield is rather lower than that for Straub actin, especially if the final acetone extraction is carried out without stirring. If impurities are not vital to the preparation, acetone extraction can be replaced by water. From the final fiber residue, tropomyosin can be extracted with *M* KCl.

(2) Properties of Butanol Actin. Some of the properties of this actin, compared with Straub actin, are given in Figs. 3–5. It will be seen from Fig. 3 that Straub actin does not give rise to F-actin in concentrations below 0.5 per cent, presumably because the concentration of active

molecules is then too low; from Fig. 4 it is evident that KCl + Mg^{++} are far more effective in encouraging viscosity increase than KCl alone or KCl + Ca^{++}. Figure 5 shows that for a given protein concentration,

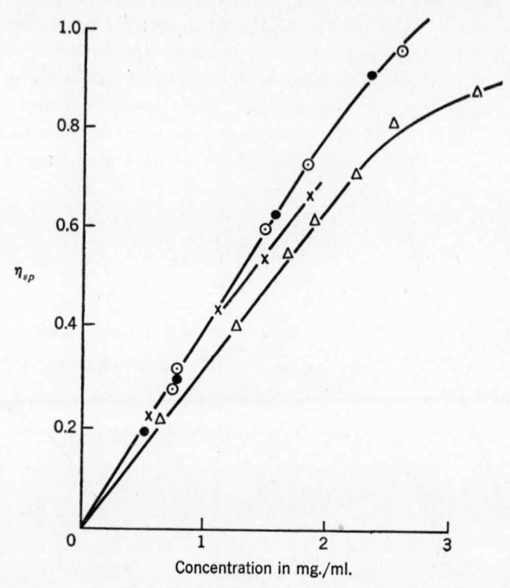

FIG. 5. "Specific viscosity" of F-actin. △ Straub actin; × fiber pretreated at pH 9 before acetone drying; ⊙ pretreated at pH 6.3 before acetone drying; • butanol–actin.

the specific viscosity of Straub actin is 20 per cent less than butanol actin, and from Table VI that the ATP sensitivity is 15 per cent less.

TABLE VI
ACTOMYOSIN-FORMING ABILITY OF ACTIN

Preparation	pH of pretreatment	(A + M) (η_{rel})	(A + M + ATP[a]) (η_{rel})	ATP[b] sensitivity
Feuer *et al.*[89]	10	1.85	1.32	1.21
Acetone	8	1.94	1.32	1.38
Acetone	6	1.98	1.32	1.45
Butanol–acetone	6	1.96	1.32	1.42
Butanol–acetone (homogeneous)	7	2.02	1.34	1.40

[a] The conditions for the test are as follows: myosin 1.50 mg./ml., actin 0.50 mg./ml., Na ATP 0.0008 *M*, phosphate buffer of pH 7 0.01 *M*, KCl 0.5 *M*; 22.0°C.; viscosity measured with Tsuda viscometer; capillary diameter 0.059 cm., length 29.4 cm., pressure head 67.2 cm. water.

[b] Defined (Weber) as $\frac{\log \eta_{rel} - \log \eta_{rel}\ \text{ATP}}{\log \eta_{rel}\ \text{ATP}} \times 100.$

c. *Physical Studies on G- and F-Actin*

Straub's interpretation that the G-F transformation is brought about by the linear aggregation of "globular" particles received strong support from electron microscope studies. Jakus and Hall[75] prepared F-actin by lowering the pH of the solution and observed that the filaments obtained on drying were between 80 and 140 A. in width, the length varying very widely but decreasing with increasing pH. Polymerization in 0.1 *M* KCl at pH 7 gave comparable results, but micrographs of G-actin showed only globules of indefinite size. An elegant study by Rosza *et al.*[99] revealed considerable detail in the various stages of polymerization. A single F-actin filament appeared to arise from a lengthwise association of ellipsoidal particles, 300 × 100 A., of probable particle weight 1.5×10^6, each inclined at about 20° to the axis of the filament. The lateral association of the filaments gave rise to a cross-banding sometimes perpendicular to the axis, but more often at an angle.

In the Tiselius apparatus as in the ultracentrifuge, F-actin gives a number of ill-defined peaks because of its thixotropy; indeed, in the ultracentrifuge, it behaves like a jelly, the sedimentation constant varying with speed. Johnson and Landolt[81] observed a main component of 50 *S* and another of 4.5–6 *S*, the latter due to the presence of inactive G-actin. From the gel character, the high sedimentation, the flow birefringence at very low shear rates, the viscosity, and the light scattering, it would indeed be difficult to avoid the conclusion that F-actin in solution is not some kind of large linear polymer. Nevertheless, another interpretation will be given below.

Active G-actin cannot be studied in the ultracentrifuge because the addition of salt to minimize charge effect converts it to the thixotropic form. Its "globular" nature has been generally accepted on the basis of a low relative viscosity, and a molecular weight of about 70,000 on a tryptophan content now known to be highly erroneous,[89] the correct value of 1.2 per cent giving a minimal molecular weight of 17,000. However, Feuer *et al.*[89] had obtained independent evidence with collodion membranes of graded porosity that this was the order of magnitude. In depolymerizing media (0.5 *M* KI, glycine pH 9) or in the presence of Calgon,[100] Straub actin was later found to be polydisperse in the ultracentrifuge, the sedimentation constants varying from 2.7 to 4. The value in hexametaphosphate (3.2 *S*) combined with the diffusion constant (2.5×10^{-7}) gave, however, a molecular weight of 150,000,[100] a value

(99) G. Rosza, A. Szent-Györgyi, and R. W. G. Wyckoff, *Biochim. et Biophys. Acta* **3,** 561 (1949).

(100) O. Snellman, T. Erdös, and M. Tenow, *Proc. Intern. Congr. Exptl. Cytol. 6th Congr., Stockholm,* 1947, p. 247.

of the order of magnitude found in light scattering (120,000).[101] The latter technique applied to actin in 0.5 *M* KI gave a lower value of 80,000.[102] Much of the confusion here existing has now disappeared by Tsao's demonstration[103] that G-actin can exist in two separate states, one a monomer of particle weight 70,000 and the other a dimer of 140,000. This newer evidence is so decisive, and so important also in the interpretation of the nature of F-actin, that it will be described separately in the next section.

d. The Nature of F-Actin[103,104]

The method of Weber[86a] gives information on the rotational relaxation time of molecules up to about 300,000 molecular weight. If fluorescent groups are attached to a protein, then on illumination, only those molecules whose absorption oscillator is parallel to the exciting light will absorb. If in the interval between excitation and emission (τ) the molecules become randomly oriented, then the emitted light will be completely depolarized; but if the disorientation is only partial, the emitted light too will be depolarized only partially. Perrin has shown that for spherical molecules and natural light,

$$\frac{1}{p} + \frac{1}{3} = \left(\frac{1}{p_0} + \frac{1}{3}\right)\left(1 + \frac{3\tau}{\rho}\right)$$

where p is the partial linear polarization emitted at right angles to the exciting light and p_0 the value of p when $3\tau/\rho \rightarrow 0$, i.e., in very viscous media, when ρ the rotational relaxation time is very large. For asymmetric molecules, an approximate equation can be formulated like that just given, if ρ is replaced by ρ_h, the mean harmonic relaxation time around the two axes, a long (a) and a short (b):

$$\frac{1}{\rho_h} = \frac{1}{2}\left(\frac{1}{\rho_a} + \frac{1}{\rho_b}\right)$$

For molecules of asymmetry >12, ρ_a can be neglected, and ρ_h then equals $2\rho_b$. But as the molecules get longer and longer, ρ_b never exceeds ρ_0 (the rotational relaxation time of a sphere of equal volume) by more than $4/3$, and in this case $\rho_0 = \dfrac{3\rho_n}{8}$.

ρ_h can be deduced from a plot of $1/p$ against T/η, since, if β is the

(101) P. Johnson and R. Landolt, *Discussions Faraday Soc.* No. 11 (1951).
(102) R. F. Steiner, K. Laki and S. Spicer, *J. Polymer. Sci.* **1**, 23 (1952).
(103) T.-C. Tsao, *Biochim. et Biophys. Acta* **11**, 227 (1953).
(104) T.-C. Tsao, *Biochim. et Biophys. Acta* **11**, 236 (1953).

slope of this curve, then

$$\frac{\beta}{\left(\frac{1}{p_0} + \frac{1}{3}\right)} \times \frac{T}{\eta} = \frac{3\tau}{\rho_h}$$

Tsao[103] found that the intrinsic viscosity of the monomeric form of G-actin is 0.24 and that of the dimer 0.52, indicative of asymmetries (about 12 and 24, respectively) to which the above simplifications can be applied. On the basis of viscosity, there seems little justification for what seems to have been a groundless belief that the nonviscous form of actin is globular.

The monomer–dimer transformation of actin was first suggested by osmotic pressure measurements, and confirmed by polarization–fluorescence. One factor which strongly influences the state of actin is a bivalent metal ion—either Ca or Mg—since chelating agents such as versenate above 21° cause a change in ρ_h from 31.5×10^{-8} to 13.8×10^{-8} sec. It may be that dimerization is effected through a metal–nucleotide link.

The dimer form of actin transforms to the thixotropic state when salt is added, and if, as previous evidence has indicated, long threads are formed in solution, then the conventional plot by Weber's method should give a horizontal line; but in actual fact, the slope does not differ from that of nonthixotropic dimeric actin. At its face value, this does not admit the existence of "ovoids," nor of polymeric fibrils made from them. An objection may legitimately be made, however, that such polymers are really present, but that the depolarization is due to actin dimers present in the interstices of the gel network. There are two reasons why this explanation seems invalid. In the ultracentrifuge, the fluorescence is associated predominantly with the fast-sedimenting gel and not with the small fraction of "unpolymerized" actin of sedimentation constant 3–4 S which stays in the supernatant. Actin, which is inactive in the sense that it does not transform to the thixotropic state, is not so trapped by the gel that it will *not* remain in the supernatant fluid, since this is one method of purifying the Straub preparation. The second objection anticipates a later section: the polarization of G-actin, whether active or inactive, is quite unaffected by addition of myosin sol, whereas that of F-actin changes instantaneously to the slope characteristic of the monomer.

Tsao[104] has inclined to the view that the actin dimers in the F-actin state are capable of building up "structure" in solution, and cites as examples the structural viscosity which he discovered in urea solutions: also the "swarms" which have a definable sedimentation when urea or

KI solutions are studied in the ultracentrifuge. The nature of forces operating to build up such structure in solution can only be guessed at, and it would seem desirable to study the simpler systems. There is another explanation, however, for the results are also explicable on the basis that polymeric molecules *are* formed and that the dimers have a certain freedom to rotate about their long axis. There appears to be some precedent for this view, since Oncley[104a] finds that in the serum mercaptalbumin dimer, there is a relaxation time associated with the molecule as a whole, and another characteristic of serum albumin itself.

e. The Interaction of Actin and Myosin in Solution

Three independent observations may be said to have foreshadowed the existence of actin: first, the finding by Banga and Szent-Györgyi[17] that myosin extracted quickly from muscle is less viscous than extracts made over long periods, and that the action of ATP is to reduce the viscosity of the latter type of extract to that of the former; second, the parallel observations of Needham and co-workers[66] that ATP diminishes both viscosity and flow birefringence of what were thought to be myosin sols; third, the observation of Ardenne and Weber[69] that myosin sols in the ultracentrifuge consisted of both light and heavy components. But the idea that ATP had its effect upon a *complex* of true myosin and another protein was the outstanding imaginative deduction of the Szeged school, leading Straub[87,88] to isolate an entirely new protein, actin.

In solution in water at pH 7, F-actin is not unlike a mucin in consistency, and is brought out of solution if myosin in gel form is added. The complex has a higher threshold of salting-in than the original myosin, and to study it in true solution an ionic strength of 0.4–0.5 must be used. The effect of ATP on actomyosin in solution is capable of a more direct interpretation than the effect on actomyosin gel, and this will be considered first.

It is difficult to infer directly the nature of the actomyosin complex. Of the various electron-microscope studies, Jakus and Hall[75] showed that actin-poor myosin (myosin A) in the process of dissolution in 0.2 *M* KCl gave rise to fairly short and discrete rodlets (0.5–1 μ). A preparation containing more actin contained particles of very variable length, suggesting that a random breaking of much longer particles had occurred during extraction. "Artificial" actomyosins were said to give a similar picture, and the conclusion was that actomyosin consists of actin filaments to which the myosin particles adhere. Other studies of artificial actomyosin reveal not so much the presence of discrete rods, but of anas-

(104*a*) J. L. Oncley, private communication.

tomosed networks of filaments,[77,105] and this interpretation is more in keeping with the gel-like qualities and the very high, anomalous viscosity as compared with myosin proper. But once more it is necessary to stress the fact that inferences as to structure apply directly only to the *dried* specimen and not necessarily to the state in solution.

The only visible form of myosin proper is the micelle of precipitated ("crystalline") myosin, in which the particles would seem to be aligned side by side. After complete solution, nothing can be seen; but after partial solution, a less robust type of aggregate is apparent.

Several authors have attempted to determine the proportion of actin and myosin which "saturate" each other. From viscosity measurements at low protein concentrations ($\sim$0.2 per cent) the maximum viscosity is found with 3 parts of myosin to 2 parts of actin.[106] The viscosity is here defined as that in which thixotropic effects are considered to be overcome, and is not, so far as can be judged, the limiting viscosity at high shear rates. By studying in the ultracentrifuge the proportions of the two proteins which give most of the "jelly" component and less of more slowly sedimenting material, Johnson and Landolt[101] give a myosin to actin ratio of about 3, though their parallel viscosity experiments indicate a value of 2.

In most studies, pure actin has not been employed. Not only is the Straub product contaminated with other proteins, but part, according to Mommaerts,[97] does not interact with myosin, though it could be trapped in the interacting portion of the system. Mommaerts, indeed, finds that as much as 60 per cent of Straub actin does not polymerize. It is very probable therefore that really active actin preparations which have recently been obtained will saturate more than 3 parts of myosin.

There exists some disagreement as to the nature of the components in the interaction of actin and myosin. Snellman and Gelotte[107] claim that with a myosin to actin ratio of 3, only the actomyosin peaks are discernible in the ultracentrifuge, and, in other proportions, the peaks of free actin or myosin appear. In Weber's laboratory also, it was possible to fractionate an actomyosin preparation such that no free myosin was observed. Johnson and Landolt,[101] on the other hand, whatever the relative proportions of actin and myosin, could observe peaks of one protein or the other. Moreover, the gel portion, on resuspending, gave rise to more of the slowly sedimenting peaks. In the light of this evidence, they have proposed that the formation of actomyosin is governed by the equilibrium reaction $A + M \rightleftharpoons AM$. In strong salt solutions,

(105) S. V. Perry and R. Reed, *Biochim. et Biophys. Acta* **1**, 379 (1947).
(106) F. Jaisle, *Biochem. Z.* **321**, 451 (1951).
(107) O. Snellman and B. Gelotte, *Exptl. Cell Research* **1**, 234 (1950).

there seems no doubt that actomyosin dissociates, and it would not be unexpected if some dissociation occurs at lower ionic strength; if the equilibrium is far to the left, it would be difficult nevertheless to demonstrate whether other components, inactive modifications of actin and myosin, are present, as they often can be. A more convincing experimental demonstration has been provided by Laki, Spicer, and Carroll[108] who show that a system giving a major actomyosin peak and a small myosin peak in the ultracentrifuge at 27° gives way to the reverse picture at 5°, suggesting that the combination is endothermic. The effect of ATP is thus equivalent to raising the ionic strength or to lowering the temperature.

f. The Action of ATP on Actomyosin in Solution

The visible effect of ATP in reducing the viscosity of actomyosin gel was thought by the Szeged school to indicate a dissociation of the complex, and most investigations since have supported this view. Parallel with a decrease in viscosity, flow birefringence and turbidity also diminish, and in the ultracentrifuge, the myosin component reappears. Even with minute amounts of ATP the dissociation can be shown to be complete since the relative viscosity of the mixture is an additive property of both components, i.e., $\log \eta_{AM} = \log \eta_A + \log \eta_M$.[84] Usually the viscosity increases again as the ATP is enzymatically split, and it is thus difficult to establish a stoichiometric relation. By addition of Mg ions, splitting is inhibited, and the viscosity drop becomes maximal at an ATP to myosin ratio of 1 mole to 300,000 g. protein.[109]

Of reagents so far tested, only ATP and inosine triphosphate (ITP) possess this apparent dissociating action which is self-reversing as the reagent is hydrolyzed.[109] Inorganic pyrophosphate in small amount will produce a permanent effect at 0° but not at 23°. Nonspecific effects[110] have been noted with urea and guanidine even when used in concentrations much lower than those required to denature and to dissolve the denatured protein, and presumably all the reagents noted by Edsall and Mehl[65] (p. 969) as having an effect on Edsall's myosin act by dissociation. Adenylic acid, orthophosphate, and metaphosphate have no effect, and adenosine diphosphate (ADP) only when myokinase is present.

Godeaux[111] was the first to show that the myosin–actin interaction and the extent of its modification by ATP was affected by treatment with —SH reagents. His experiments were carried out on actomyosin gel

(108) K. Laki, S. S. Spicer, and W. R. Carroll, *Nature* **169,** 328 (1952).
(109) F. W. H. M. Mommaerts, *J. Gen. Physiol.* **31,** 361 (1948).
(110) F. W. H. M. Mommaerts, *Arkiv. Kemi, Mineral, Geol.* **19A,** No. 18.
(111) J. Godeaux, *Bull. soc. roy sci. Liége* p. 216 (1944).

and appear to have been unknown to English and Hungarian workers up to 1947. Meanwhile, Bailey and Perry,[112] having failed to find any general physicochemical explanation, examined the effect of treating the myosin component with —SH reagents of various types, removing excess reagent, and determining in parallel the ability to combine with actin (a simple viscometric test) and the ability to split ATP. There appeared to be a remarkable parallelism between the two properties, particularly when oxidants or heavy-metal substituents were used; moreover, the —SH groups involved were those which react almost stoichiometrically with inhibitor and are best described as the reactive or accessible —SH groups. With alkylating reagents, however, the parallelism is not so close. It is possible here that secondary effects intervene; that in fact the reactivity of groups varies according to the reagent used, and that in this case, a much more general substitution occurs before inhibition is complete. According to Gilmour and Calaby,[113] it is possible by oxidation with ninhydrin to obtain a myosin almost free from adenosinetriphosphatase (ATPase) activity and still capable of producing a viscosity increase with actin. These findings are not altogether acceptable, (*1*) because the measured viscosity increases were rather small, and (*2*) where a large rise of viscosity was recorded on mixing myosin and actin, the actomyosin had become so modified that it precipitated from solution.

The results as a whole suggested what still appears to be the most attractive hypothesis of the actin–myosin–ATP interaction: that certain centers, of which —SH groups form part, possess ATPase activity and these are also the centers which interact with actin. The affinity for the true substrate ATP is so much greater than for actin that actomyosin is always virtually dissociated in the presence of ATP, and this explains why actin cannot be regarded as a competitive inhibitor in any practical sense. This interpretation was clearly stated at the time, but criticism of the hypothesis appears to have overlooked it.[114] It was also shown that actin does possess a detectable amount of acid-labile P, which was considered too small to be the interacting center of the actin. In view of Straub's unambiguous characterization of ATP in G-actin, this further hypothesis must seriously be considered.

It is not surprising that —SH reagents can react like ATP in dissociating actin and myosin after combination; but the two recorded by Turba and Kuschinsky,[95] copper glycinate and Salyrgan (salicyl-(γ-hydroxymercuri-β-methoxypropyl) amide-*O*-acetate) are metallic compounds

(112) K. Bailey and S. V. Perry, *Biochim. et Biophys. Acta* **1,** 506 (1947).
(113) D. Gilmour and J. H. Calaby, *Australian J. Sci. Research* **2,** 216 (1949).
(114) A. Szent-Györgyi, Chemistry of Muscular Contraction, 2nd ed., Academic Press, New York, 1951, p. 73.

which could influence also the actin component by combination with the prosthetic group.

g. The Interaction of ATP and Actomyosin in Gel Form

Actomyosin can either be precipitated as a loose flocculus by tenfold dilution of its solution in 0.5 *M* KCl, or can be obtained as threads by extrusion through an orifice into a very dilute salt solution. Weber[115,116] was the first to use such threads in an oriented form as objects whose mechanical properties, birefringence, and x-ray diffraction were very similar to those of the fibrillar substance. One of the earliest reports of the effect of ATP on the mechanical properties of extruded threads came from the Russian school,[117] that a thread under a slight load showed a greater extensibility in the presence of ATP. This observation was in keeping with the idea that the anastomosis of actin and myosin particles is diminished in the presence of ATP, though the nonspecific interaction between the two proteins remain. Engelhardt *et al.*[117] showed fairly clearly that there was a parallelism between the enzyme activity of the thread and the mechanical response; further that inactivation of ATPase with Ag ions completely abolished the response. This latter observation is not so readily explained since the destruction of the specific sites of interaction by poisons ought to result, as with ATP, in a more extensible structure. Some poisons, mainly Hg and Cu compounds, do dissociate actin and myosin,[94,95] though the effect may be slow; with metals, it is possible that the less specific interactions are strengthened. It is important to note at this stage that Engelhardt obtained an effect with other pyrophosphate compounds, such as DPN, thiamine pyrophosphate, and inorganic pyrophosphate, which are not split by the enzyme.

Using threads or actomyosin precipitates which were probably richer in actin, the Szeged school[118] observed that in solutions of ionic strength below 0.15 (usually 0.05 *M* KCl) and in the presence of 0.0001 *M* $MgCl_2$, there is an intense syneresis in the presence of ATP, so that the thread shrinks in all directions, or, in the case of a flocculus, packs to a dense precipitate (superprecipitation). The shrinkage is essentially a phenomenon which occurs in the absence of tension; if loaded, even so slightly that an increase in length of 0.1 per cent is observed, the addition of ATP produces elongation.[119] In this last work it was claimed that the

(115) H. H. Weber, *Ergeb. Physiol. u. exptl. Pharmakol.* **36,** 103 (1934).
(116) H. H. Weber, *Arch. ges. Physiol. (Pflügers)* **235,** 205 (1934).
(117) V. A. Engelhardt, M. N. Liubimova, and R. A. Meitina, *Compt. rend. acad. sci U.R.S.S.* **30,** 644 (1941).
(118) A. Szent-Györgyi, *Studies Inst. Med. Chem. Univ. Szeged* **1,** 17 (1941/2).
(119) F. Buchtal, A. Deutsch, G. G. Knappeis, and A. Munch-Petersen, *Acta Physiol. Scand.* **13,** 167 (1946).

effects could be produced with threads which were almost inactive enzymatically, though adequate enzymatic data were not presented. Such a finding runs contrary to experience in Cambridge, and to the fact, as both Godeaux[111] and Buchtal have noted, that —SH reagents abolish or diminish the shrinkage response after addition of ATP.

In fact, all recent work shows a parallelism in the loss of ability to split ATP, to shrink an actomyosin gel, or to cause contraction in a model fiber. Kuschinski and Turba[120] have shown this in the case of mercurials and benzaldehyde, and Korey,[121] using the glycerol fiber (p. 1043), in the case of iodosobenzoate and mapharsan. The precipitation of myosin in the isoelectric region will affect the enzyme activity to a greater or lesser degree, least, if no salt is present. Here again, a parallelism between activity and the ability to form a superprecipitate after addition of actin and then of ATP was noted.[122] Weber[15] has particularly stressed that the rates of ATP breakdown and "contraction" can be modified and correlated, simply by studying the effect of ATP concentration on the two processes, and without the use of inhibitors (sec. VIII-3):

"The rate of ATP breakdown and the extent of 'contraction' can, however, be modified without the use of inhibitors, for there are optimal concentrations of ATP for both processes; above these concentrations, the values for tension and for breakdown again decrease. This is the case for the (psoas) fiber and the (actomyosin) thread, and true also for the splitting and shrinkage which occurs in the superprecipitation of actomyosin. The optimal concentration of ATP for tension and breakdown, and also apparently, for superprecipitation all decrease as the temperature is lowered, and for an exact comparison of the optimal ATP concentrations for contraction and splitting, two conditions must be fulfilled when the measurements are made: (*a*) the rate of ATP breakdown must represent the initial velocity rate; (*b*) the contractile system must be in the same state of shortening.[123] At 2°, the optimum ATP concentrations are $10^{-2.25}$ *M* for the development of tension and $10^{-2.3}$ *M* for the splitting of ATP; at 20° the values are $10^{-1.65}$ and $10^{-1.7}$ *M*, respectively."

Two aspects of the problem clearly emerge. One concerns the attachment of ATP at specific sites, and the other, the splitting of ATP; more exactly, the *rates* of both these processes. Ions may affect each to a different extent: for a given ATP concentration, one ion may be more effective in the first stage than in the second, and the reverse may be true for actual breakdown.

(120) G. Kuschinsky and F. Turba, *Experientia* **6**, 103 (1950).
(121) S. Korey, *Biochim. et Biophys. Acta* **4**, 58 (1950).
(122) S. Spicer, *Arch. Biochem.* **25**, 369 (1949).
(123) E. Heinz and F. Holton, quoted by H. H. Weber, Ref. 15.

As long as an ion does not completely inhibit the enzyme, syneresis will ensue, even though the rate of shrinking is very much reduced; for the energy requirement of a syneresing gel of low protein concentration is so small that it may give the impression, as we have seen, that such gels are not enzymatically active.

Between the statement of these effects and their explanation, there is a wide gap. The simple extrapolation, that the syneresis of actomyosin gel is a true model of contraction, has been severely criticized, for the reason that such shrinkage may involve nothing more than a side-to-side aggregation of random particles, as in a fibrin clot. Perry *et al.* from electron microscope and x-ray studies could find no evidence that more than this was involved.[124] The true model for contraction should be a thread of oriented actomyosin particles, which contracts on addition of ATP, maintaining its original volume. Matoltsty and Varga[125] have shown that, according to the pretreatment, both the actomyosin thread and the glycerol-treated psoas fiber, will give any kind of volume response, increase, decrease, or no change. In the absence, therefore, of any pointer to changes on a molecular level, we shall need to examine very carefully whether, in spite of volume alteration, they reflect the fundamental act of contraction. If they do, it follows that syneresis *must* signify an act of contraction in all possible directions and that changes of configuration are involved. The volume increase need not be discussed, because to obtain it the threads were first dried. Further discussion will be postponed until various models are described in more detail.

h. The Nature of the Myosin–Actin–ATP Interaction by the G. Weber Polarization–Fluorescence Method[104]

There sometimes appears in the history of a subject a finding which seems quite contrary to views that are sound on all other evidence. This stage seems to have been reached in muscle. The cautionary light has been seen in the rejection of F-actin as a polymer in solution, and although in the preceding sections the use of words like "combination" have been circumvented lest they be taken too literally, the impression must have been given that actin and myosin "combine" and that ATP dissociates the complex. On all other published evidence, no other conclusion seems possible. But again, G. Weber's method applied to the system *in solution*, throws doubt upon this conventional interpretation.

If labeled F-actin forms a physical complex with myosin, then the $1/p$ vs. T/η curve should be horizontal, since the relaxation time of

(124) S. V. Perry, R. Reed, W. T. Astbury, and L. C. Spark, *Biochim. et Biophys. Acta* **2,** 674 (1948).

(125) G. Matoltsty and L. Varga, *Enzymologia* **14,** 264 (1950).

molecules of greater size than 300,000 is too large. This condition should hold for myosin itself, and although it is not possible to label myosin itself without denaturation, the water-soluble framework which we have discussed in sec. II-2*e* does fulfil the prediction. The addition of F-actin to myosin, however, gives a plot that is neither horizontal, nor that of F-actin, but that of the monomer; further, the addition of ATP translates the curve to that of the dimer. The process is quite reversible; as ATP is split the monomer is again formed (Fig. 6). Even more surprising was the finding that labeled F-actin, brought either reversibly or irreversibly to the G-form shows no change of polarization when myosin was added.

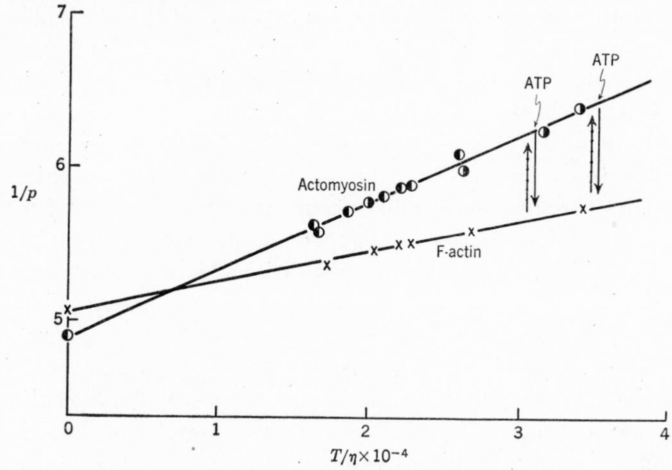

FIG. 6. Fluorescence–polarization of F-actin and F-actomyosin in 0.5 *M* KCl + 0.006 *M* phosphate pH 7. F-Actin in presence of 10^{-4} *M* ATP. Weight ratio of actin to myosin = 1:2 ◐ or 1:4 ◑. Effect of ATP shown for 26° and 22°C.

These facts permit of four tentative conclusions: (*1*) Only the thixotropic form of actin is "reactive"; (*2*) the interaction with myosin causes the breakup of the actin dimer to give the monomer; (*3*) actomyosin, characterized as it is by a high viscosity, must represent a new thixotropic state in which only the actin monomer participates; (*4*) *in solution*, neither the actin monomer nor the dimer combines permanently with myosin in a physical sense. Two additional points must also be stressed. First, the monomer–dimer transformation, which is slow and sometimes irreversible in the case of actin by itself, is instantaneous under the above conditions. Secondly, while ATP in "depolymerized" actin encourages the formation of monomer, in the presence of myosin, the dimer is favored. The latter contradiction is not perhaps so surprising. In a system comprising actin, myosin, ATP, and metals, any two of which interact strongly, the balance of forces, whatever their nature, may tip the scales one way or the other.

As was the case with F-actin, the interpretation of other types of study in the light of these findings is not easy, but in the present case too, considerations of viscosity, flow birefringence, light scattering, and sedimentation apply to macroscopic properties such as structure in solution. One cannot dismiss lightly a finding which for the first time in the history of reconstructed muscle systems is an *instantaneous* effect. In the last resort, we may need to think of the system in terms of long-range forces which are just as exacting in their configurational requirements as those involving physical union of enzyme and substrate. In this sense, all the work on the role of —SH groups, ion effects, and the like will lose none of its validity but will shift in emphasis. Finally, we have still to extrapolate from the effects described here in medium of high ionic strength to those in the gel state and *in situ*, where electrostatic interactions are enhanced. There would seem no doubt that in muscle itself and in the actomyosin gels used for model experiments, complex formation in a true physical sense must occur, and the behavior of models in later sections will be discussed with this assumption. It may be that all we see in solution are the specific orientational forces which cause actin and myosin *in situ* to interact in a *spatially exact* manner.

One other investigation questions the conventional interpretation of the ATP effect. Blum and Morales[125a] measured the scattered intensities at various angles and concentrations, and plotted the results by the extrapolation method of Zimm,[125b] which gives a value of particle weight independent of shape. This was the same before, during, and after the addition of ATP, though the same data analyzed by the dissymmetry method gave the impression that ATP caused a diminution of particle weight. The apparent value of this latter is of a very high order—20×10^6—and the shape functions suggested that the actomyosin particles might be considered as cylinders of finite length with respect to their diameter. For such a shape, ATP appeared to expand the system reversibly. It would seem from these results and those of Tsao that we are still a long way from any real understanding of the myosin–actin–ATP interaction.

4. Tropomyosin

a. General Properties

Tropomyosin, discovered in 1946,[126] is the most recent addition to proteins which form part of the muscle fibril. Though after isolation it

(125*a*) J. J. Blum and M. F. Morales, *Arch. Biochem. Biophys.* **43,** 208 (1953).
(125*b*) B. Zimm, *J. Chem. Phys.* **16,** 1099 (1948).
(126) K. Bailey, *Nature* **157,** 368 (1946).

has the properties of a globulin soluble at low ionic strength, it is not extracted by any simple means. Muscle mince, washed with water, is dehydrated in ethanol and in ether, and the residue is extracted with strong salt solution which slowly leaches out the tropomyosin to give a clear, viscous extract. The protein is precipitated at pH 4.6 and redissolved at pH 7, when a denatured component (possible actin) precipitates. This precipitate is removed with other impurities by adding ammonium sulfate to 40 per cent saturation, and the tropomyosin salted out from the mother liquor. The original method[127] is still the most rapid and convenient, and only one small modification has been introduced.[86]

In minced rabbit muscle, there is no evidence that water or very dilute salt solutions dissolve the protein, but solutions used for the extraction of myosin do cause some loss, and if comminution is sufficiently great, as in Hasselbach and Schneider's method[128] for the extraction of actin, then all the tropomyosin appears in the extract.[129] Tsao[129] finds in fact that it is possible to get a G-actin–tropomyosin complex out of muscle after butanol treatment, which shows a sharp viscosity drop on addition of salt, due to the depolymerization of tropomyosin (see the following), followed by a rise of viscosity as the actin becomes thixotropic. In fish muscle, Hamoir[130] found that rather acid salt solutions, unfavorable for the extraction of myosin, dissolve part of the tropomyosin, though the yield is not as great as in the author's procedure. That tropomyosin is located in the fibril was shown unambiguously by Perry,[131] who isolated it from intact fibrils freed from sarcoplasm.

Tropomyosin was discovered before the isolation of actin had been reported in England, yet it will be observed that the desiccation of the muscle which denatures the myosin component is common to both methods. The two proteins differ in many respects, and most outstanding is their behavior toward salts; these cause actin to become thixotropic and tropomyosin to fragment. This behavior may explain in part the behavior of tropomyosin to extractants, salt ions being necessary to depolymerize the aggregates as they occur *in situ;* but Hamoir's finding[130,132] that tropomyosin can occur as a complex with nucleic acid raises the possibility that salt is necessary to break the complex by a process of metathesis.

After isolation, tropomyosin can be polymerized and fragmented at will, simply by removing or adding salt.[127] When dialyzed against

(127) K. Bailey, *Biochem. J.* **43,** 271 (1948).
(128) W. Hasselbach and G. Schneider, *Biochem. Z.* **321,** 461 (1951).
(129) T.-C. Tsao, to be published.
(130) G. Hamoir, *Biochem. J.* **48,** 146 (1951).
(131) S. V. Perry, *Biochem. J.* **55,** 114 (1953).
(132) G. Hamoir, *Biochem. J.* **50,** 140 (1951).

distilled water (pH 7), the solution becomes very viscous due presumably to an end-to-end aggregation. It was first thought that the fibrils seen by drying down such solutions on a glass plate, washing off, and mounting for the election microscope, represented the state of the particles in solution. The fibrils are seen to be about 3000 A. long and 250 A. broad,[133]

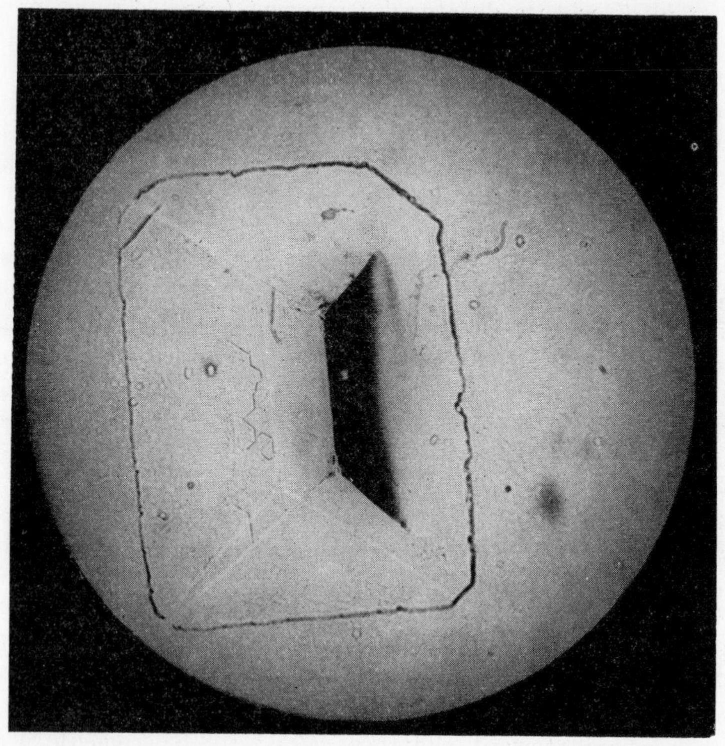

PLATE I. Large crystal of nucleotropomyosin, about 1 ml. long, prepared by Dr. T.-C. Tsao. Note the etch marks on the surface. Tropomyosin crystals which contain nucleic acid are less fragile and appear to grow larger than nucleic-free preparations.

but it is likely that the side-to-side aggregation occurs during drying (see the following). Apart from the fibrous form of tropomyosin,[127] crystals can be obtained in solutions of ionic strength 0.37 by increasing the acidity (pH 5.8–6.0) so that the solubility of the protein is diminished. These crystals are excessively fragile, and contain as much as 90 per cent of water (Plate I).

Whether by accident or design, tropomyosin has many properties in common with myosin itself, and may be thought of as a prototype of

(133) W. T. Astbury, R. Reed, and L. C. Spark, *Biochem. J.* **43**, 282 (1948).

myosin. The amino acid composition is similar, though by no means identical, some of the solubility properties (salting-in threshold), isoelectric point, and both possess an intramolecular pattern of the α-keratin type. Most significant of all, perhaps, is that by application of Sanger's method neither protein possesses appreciable amounts of N-terminal residue,[85] and taken in conjunction with certain structural evidence, it seems probable that cyclic chains rather than open chains are present.[86]

The globulin-like properties of tropomyosin are best seen at pH 5.6–6.0, the zone where crystallization occurs. An increase of salt concentration causes an increase of solubility. At pH 7, however, the protein is freely soluble in water, though above 5 per cent the solutions tend to gel. The point of minimum solubility varies with the salt concentration. In very dilute acetate buffers, it is insoluble between pH 4.5 and 5.3, but the rate of settling of the flocculus is greatest at pH 5.1, suggesting an isoelectric point in this region. On the acid side of the isoelectric point, whether in the presence or absence of salt, tropomyosin dissolves and does not suffer any loss of solubility on reneutralization; in this it resembles myosin also. It can indeed be crystallized after standing at pH 2 or 11. The salting-out threshold begins sharply at 40 per cent saturation in ammonium sulfate containing free ammonia, but the zone of precipitation is rather wide, extending up to 60 per cent.

Tropomyosin is not readily denatured. It will lose its solubility by heating at 105° in the dry state and by heating the isoelectric precipitate to 100°. At pH 7, on the other hand, it is not denatured by heat or by organic solvents. Concentrated urea solution does not modify the solubility properties, but does induce some change whereby the protein can no longer be crystallized, and in the absence of urea and salt ions forms more viscous solutions than the untreated protein at similar concentration. The fibrils present are then seen to be of indefinite length and, as a study in fibrogenesis, this phenomenon needs further investigation.[86]

b. Particle Size of Tropomyosin and Aggregation Phenomena

The relative viscosity of tropomyosin in water is very high, but the addition of quite small concentrations of salt causes a striking decrease, particularly between the values $\Gamma/2 = 0$ and 0.03; thereafter, the viscosity approaches an asymptotic value which is still much higher than that of a corpuscular protein. This behavior suggests that a disaggregation from a filamentous form to smaller asymmetric units is taking place, and that the initial aggregation is largely due to electrostatic forces weakened as the salt concentration increases.[127]

The original estimate[134] of particle size was carried out by sedimenta-

(134) K. Bailey, H. Gutfreund, and A. G. Ogston, *Biochem. J.* **43**, 279 (1948).

tion–diffusion and osmotic-pressure measurements at ionic strength 0.27, and although in the ultracentrifuge and electrophoretic apparatus the protein appeared quite homogeneous, it was later considered that these criteria were not sufficient to show that disaggregation was complete. As sedimentation proceeds, the boundary progressively sharpens, due simply to concentration effects in sedimentation; moreover, the sedimentation constant is predominantly a function of width and not of length.

A very detailed investigation[86] was then undertaken to determine by osmotic-pressure measurements (*a*) the average particle size in neutral solution at varying ionic strength, and (*b*) the particle weight in media

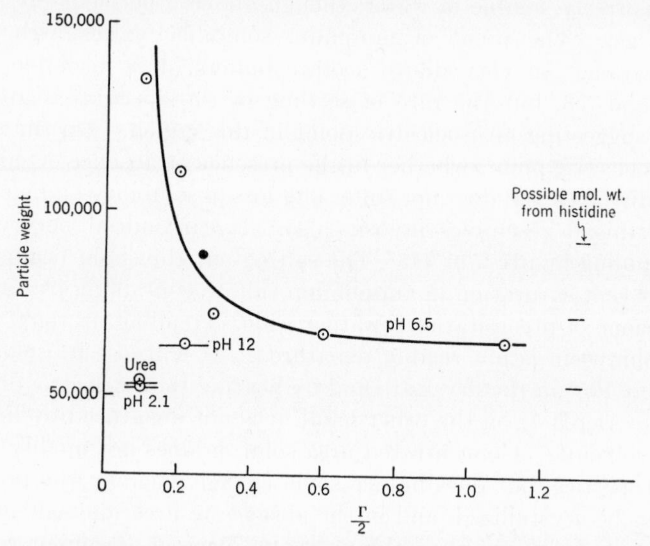

FIG. 7. Depolymerization of tropomyosin under various conditions (osmotic-pressure measurements).

likely to bring about complete fragmentation. It could not be foreseen, of course, whether the particle weight of the primary unit which shows the aggregation phenomena would differ from that found in urea solutions, i.e., whether the primary unit contains subunits which are liberated only in special, nonphysiological media.

Considering first the average particle weight as a function of ionic strength at pH 6.5, it can be seen from Fig. 7 that at $\Gamma/2 = 0.1$, it is 135,000 and falls to about half this value at $\Gamma/2 = 0.3$; thereafter, the curve flattens out and becomes asymptotic. The limiting value at $\Gamma/2 = 1.1$ (64,500) is still higher than the particle weight in acid at pH 2.1 and in 6.7 *M* urea at pH 6.5. Since in these two the particle weight is the same (53,000), it is thought that this value represents the true molecular weight of the primary particle, and that in neutral salt solution, disaggregation is never complete. In alkali (pH 12), the particle weight (61,400) is nearer the value found in salt solution.

The average particle weights in salt solutions and in urea were close enough to suggest that the primary unit does not contain subunits, and to test the hypothesis, the assay of terminal amino groups by Sanger's method was undertaken. Instead of finding one N-terminal group per mole of 53,000, as had been expected, none could be detected expect for traces of bis(dinitrophenyl)lysine, DNP-valine, and DNP-alanine, amounting *in toto* to 1 group/400,000 g. protein.[85] It is likely that these traces arise from tenacious impurities. As will be seen later, the possibility, arising directly from this observation, that the 53,000 unit is a cyclopeptide, seems possible also from other types of investigation. If subunits exist in tropomyosin, therefore, they are not manifest in the conventional depolymerizing media; and the Sanger method does not help to eliminate this rather improbable possibility. The method has been useful, nevertheless, as a check that under the acid and alkaline conditions employed for determining particle weight, no actual cleavage of peptide bonds occurs.

c. *Viscosity Studies*

In parallel with osmotic-pressure measurements, viscosity determinations have been carried out on tropomyosin dissolved in various media employed for particle-weight determination. The measurements demand a knowledge of relative viscosity under conditions of completely random orientation (i.e., under Newtonian flow) at different protein concentrations. If η_{rel} is relative viscosity and $\eta_{sp} = (\eta_{rel} - 1)$ the specific viscosity, then the intrinsic viscosity $[\eta]$ is $\lim_{C \to 0} \eta_{sp}/C$ and the viscosity increment $\nu = \frac{[\eta] \times 100}{\bar{V}}$ (C = g. protein/100 ml. of solution, and $\bar{V}$ the partial specific volume). From the viscosity increment, the axial ratio can be calculated from Simha's equation, assuming the molecule to be a prolate ellipsoid. This ratio must be corrected for an estimated hydration of 25 per cent.

Independently, the shape can be guessed from other data. The complete amino acid composition is known, and thus the average weight of an amino acid residue (116.4). For a probable molecular weight of 52,900, the total number of residues is 52,900/116.4 = 455, and since the intramolecular pattern is of the α-type, the length of a fully extended α-chain can be calculated if we can be sure of the number of amino acid residues comprising the fold of 5.1 A. The Astbury model of three was at the time the most probable and widely accepted, but a more recent model by Pauling and Corey has received considerable attention, the number of amino acids here being 3.7. From the dimension of the single α-chain, on either hypothesis, it is possible to proceed to two, three, or four chains lying side by side. A double chain, for example, can possess two configurations if the Astbury model is accepted, one in which the chains lie side by side along the side-chain direction, or one in which aggregation occurs in the backbone direction. The over-all dimensions of the two models happen to be identical (20 × 9.5 A. and 10 × 19.5 A., respectively), and the mean axial ratio would be the length (455 × 5.1/3) divided by the mean width of 14.5 A., giving a value of 26 for an unhydrated molecule. The Pauling-Corey helix would demand a lower value of 21.

In Table VII the calculated values of axial ratio are set beside those from viscosity. For the conditions under which tropomyosin exists as a monomer or primary unit (in urea and acid), the axial ratios agree better

TABLE VII
PARTICLE WEIGHT AND SHAPE OF RABBIT TROPOMYOSIN PARTICLES IN SOLUTION[86]

Medium		Average particle wt. (osmotic pressure)	Viscosity increment	Axial ratio (unhydrated) from Simha equation	Axial ratio (unhydrated) calculated (see text)
pH	Solvent and ionic strength				
6.5	Salt (0.1)	135,000	197	53	67
6.5	Salt (0.2)	111,000	141	44	55
6.5	Salt (0.3)	72,000	99	35	39
6.5	Salt (0.6)	67,000	83	32	33
6.5	Salt (1.1)	64,500	80	31	32
6.5	Urea (0.3)	53,100	74	30	26
2.1	HCl (0.3)	52,700	74	30	26
12	NaOH (0.2)	61,400	55	25	30

with the Astbury model than with Pauling-Corey, though the agreement is fairly good considering the assumptions involved in Simha's treatment. The double chain is also the most probable from the chemical finding that the molecule is a cyclopeptide. Even without this consideration, other configurations can readily be excluded. The axial ratio (unhydrated) for a single and a triple chain are about 78 and 13 (Astbury model) and 63 and 11 for the Pauling-Corey model. None of the values is remotely near that calculated from viscosity.

Legitimately, the Simha equation cannot be used to estimate the axial ratio of particles under conditions where more than one molecular species exists, so that values for media other than acid or urea have only a qualitative significance. In the salt series, for example, they serve to show that the aggregation of particles must be predominantly an end-to-end process.

While all the evidence above favors the idea of a cyclopeptide structure, recent tests to ascertain whether α-carboxyl groups occur show definitely the presence by two independent methods (carboxypeptidase and the Akabori method) that isoleucine occupies a *C*-terminal position.[135] This finding raises the question whether an NH_2 group is present but is either unreactive or blocked, or whether the molecule is cyclic but contains a branch, linked, for example, to an ω-COOH group.

d. Characterization of Tropomyosin by Other Techniques

(*1*) *Polarization of Fluorescence.* Assuming an ellipsoid of revolution, the relaxation times measured after coupling with dimethylaminonaphthalene chloride were 8×10^{-8} and 0.5×10^{-8} sec. In water at 50°, the molecules appeared to coil up to give a corpuscular particle of relaxation time 8×10^{-8} sec. and molar volume 50,000.[47] Taking the value of $\bar{V}$ as 0.71[134] and the hydration 0.31, this molecular volume corresponds to a particle weight of 52,000.

(*2*) *Light Scattering.*[136] Doty and Sanders have determined by light scattering the effect of ionic strength on the particle weight of tropomyosin at pH 6.5, under the same conditions as those employed for osmotic-pressure measurements. At $\Gamma/2 = 0.1$ the particle weight was 100,000, decreasing to 53,000–55,000 at $\Gamma/2 = 1.01$. The lowest dissymmetry obtained from light scattered at angles of 45° and 135° was 1.06, while the calculated value for a rod of length about 400 A.—the value deduced from viscosity would be about 1.05. The particle weights of 55,000 and 100,000 obtained by this method are about 20 per cent lower than the corresponding values derived from osmotic pressure. Apart from this unexplained discrepancy, the results of G. Weber and of Doty confirm that the monomer has a particle weight near 50,000, that the magnitude of the asymmetry is correct, and that the initial polymerizing process increases the asymmetry.

The aggregation process is due most probably to a charge distribution such that one end of the molecule bears a net positive charge and the other a net negative one.

(135) R. H. Locker, *Biochim. et Biophys. Acta* (In Press); see K. Bailey, *Proc. Roy Soc.* (London) **B141,** 45 (1953).

(136) Personal communication.

III. Particulate Components

The study of cell inclusions prepared by methods involving rupture of the cell and differential centrifugation has brought to light much information on the distribution of enzyme systems within the cell. Following the pioneer work[137] of Bensley and of Claude, most perhaps is known about the nuclei, mitochondria, and microsomes of the liver cell. The muscle cell, however, is difficult to break up, and has received only sporadic attention, though it is interesting to recall that the so-called "heart-muscle preparation,"[138] a favorite material for the study of cytochrome oxidase since 1929, is in reality a preparation of respiratory granules drastically damaged by the extraction procedure used. Rather severe methods[139] have also been employed to isolate the nuclei of muscle, but Robinson[140] has shown that from embryonic muscle the preparation is relatively simple.

1. Granules

It is curious that the granules first noted by Kolliker[141] in 1856 and called "sarcosomes" by Retzius[142] have been neglected until recently; this in spite of the fact that very detailed histological studies have been made upon them. In mammalian red muscle, heart muscle, and in the wing muscle of the Arthropoda they are especially plentiful, but very meager in smooth and white skeletal muscle. Holmgren[143] reported two types of granules, large ones in apposition to the A bands and small ones in the region of the I. In a later study of *Mantis*, Jordan[144] could only distinguish a segregation of the two types at "certain functional stages" especially at the midphase of contraction. He thought the sarcosomes were unrelated to the mitochondria of somatic cells since the latter in fixed preparations are filamentous, and the muscle granules round; but he did notice a bacillary type in *Mantis* muscle which to him seemed more closely related to the mitochondria of somatic cells.

The granules can readily be seen in muscle slices without fixation under the phase-contrast microscope, and Perry and Horne[145] have

(137) See A. Claude, *Advances in Protein Chem.* **5**, 423 (1949).
(138) D. Keilin, *Proc. Roy. Soc.* (London) **B104**, 206 (1928/9).
(139) M. Behrens, *Z. physiol. Chem.* **209**, 59 (1932).
(140) D. S. Robinson, *Biochem. J.* **52**, 628 (1952).
(141) A. Kolliker, *Z. wiss. Zool.* **8**, 311 (1856).
(142) G. Retzius, Muskelfibrille und Sarkoplasma Biologische Untersuchungen, Stockholm, N. F. 1.
(143) E. Holmgren, *Arch. mikroskop. Anat. Entwicklungsmech.* **75**, 240 (1910).
(144) H. E. Jordan, *Anat. Record* **16**, 217 (1919).
(145) S. V. Perry and R. W. Horne, *Biochim. et Biophys. Acta* **8**, 483 (1952).

succeeded in isolating them by homogenizing in borate buffer or in sucrose followed by differential centrifugation. There is a distinct tendency to clump in borate buffer, but in sucrose the granules remain as rounded bodies ranging in diameter from 300 to 5000 A. (Plate II) and some in pigeon breast muscle are as wide as the myofibril itself (1 μ). Like other mitochondria, the granules consist of a phospholipide–protein–

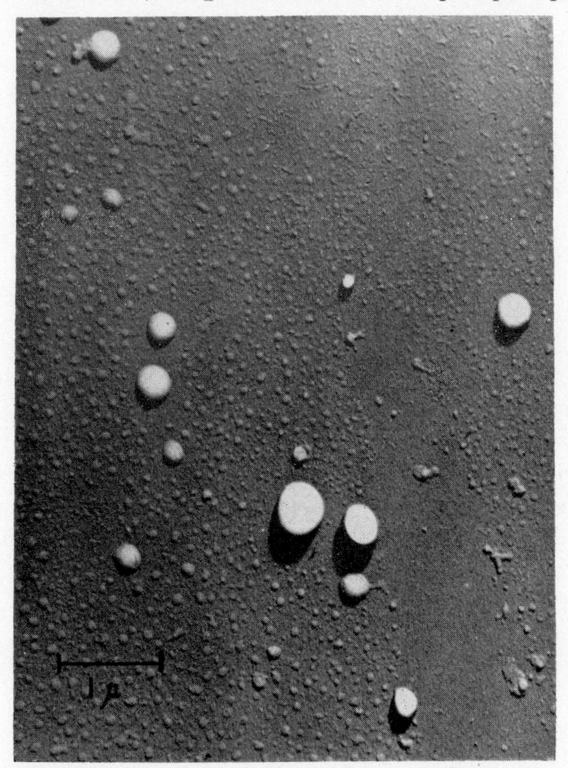

PLATE II. Rat muscle granules isolated in 30 per cent sucrose. Light fraction sedimenting after 60 min. at 19,000× *g* (S. V. Perry and R. W. Horne).[145]

nucleic acid complex, containing, in the case of the rabbit, 30–37 per cent of ether–ethanol–soluble lipide, 0.2 per cent of nonlipide P, and 10–11 per cent N; at least 65 per cent of the phosphorus could be ascribed to ribonucleic acid.[146]

Isolated granules of pigeon breast muscle have been shown to respire and to give rise to energized phosphate bonds (P/O ratio ~2).[147] For fast muscles in continuous activity, their importance must be very great indeed for the aerobic supply of ATP to the contracting myofibrils.

(146) S. V. Perry, *Biochim. et Biophys. Acta* **8,** 499 (1952).
(147) J. B. Chappell and S. V. Perry, *Biochem. J.*, **55,** 586 (1953).

A granular enzyme of great interest is the Mg-activated ATPase[59] which is discussed in the following.

2. The Isolated Myofibril

In the strictest sense the myofibril can also be regarded as a particulate component although the term is generally confined to the granules proper. Short pieces of myofibril, freed from sarcoplasm, granules, nuclei, and sarcolemma were first prepared from skeletal and heart muscle by Schick and Hass,[148,149] who sliced the frozen muscle and liberated the myofibrils after short treatment with trypsin. On standing, however, such fibrils tend to lose their morphological features, and interestingly enough, pass to a water-soluble form in which the ATPase activity is preserved. Perry[49] has employed collagenase instead of trypsin, but later[91] dispensed with all enzyme treatments and used simple homogenization. This gives a lower yield than in the freeze-slicing technique, but larger initial amounts of muscle can be used to compensate.

The fibrils possess all the details of those prepared from muscle which has first been fixed whole and then homogenized. The ultimate width under the electron microscope varies between 0.8 and 1.8 μ, and each contains between 30 and 45 filaments.[145] From the manner of splaying out, and sometimes from the elliptical appearance of the overlying Z material, there is a strong impression that the myofibril is tubular, or at least, rather loosely filled with filaments. A more detailed discussion of these features is given in sec. VII-2.

Tests applied for the presence of enzymes other than ATPase have shown them to be free of myokinase and adenylic deaminase after washing;[150] they can also be washed free of the pyruvic oxidase and succinoxidase systems, which are now generally accepted to be components of mitochondria. Myofibrils isolated by the collagenase method contain 4–7 per cent of ether–ethanol-soluble material and 0.1–0.5 per cent total phosphorus, of which ribonucleic acid plus phosphoprotein P amounts to only 0.05 per cent.[146] They contain bound nucleotide consisting predominantly of ADP (0.9, 2.7, and 0.5 μmoles of adenylic, ADP, and ATP, respectively).[91] The ADP is most probably present as the prosthetic group of actin.

3. The Granule ATPase

Kielley and Meyerhof[59] separated from myosin an Mg-activated ATPase with a pH optimum of 6.8 at 38°. The enzyme was found to be rather labile and stabilized

(148) A. F. Schick and G. M. Hass, *Science* **109,** 487 (1949).
(149) A. F. Schick and G. M. Hass, *J. Exptl. Med.* **91,** 655 (1950).
(150) S. V. Perry, private communication.

by cyanide. The purest preparations could be obtained as a pinkish pellet after high-speed centrifugation, and from the lipide content it seems likely that the enzyme had its source in microsomes. Indeed, the rate of liberation of acid-soluble P by the action of a lecithinase ran parallel with the progressive inactivation of the enzyme.[151]

Perry's granules from rabbit muscle likewise showed no activation with Ca but marked activation with Mg at 0.005 *M*; the optimal pH lay between 7.5 and 8, and both labile groups of ATP were split owing to the presence of myokinase which the most rigorous washing failed to remove entirely.[146] It does seem, however, that the enzyme per se is specific only to the terminal group, since myokinase-poor granules hydrolyze the penultimate group very slowly. It is possible to distinguish granule ATPase even from the Mg-activated ATPase of the myofibril: the first retains its activity after dispersion in *M* KCl while the second loses it and then responds to activation by Ca.

In a homogenate of muscle, the myofibrils themselves are Mg-activated, so that it is possible to derive an estimate of the contribution of the granules to the ATPase activity of the whole muscle at pH 7. This is at least 13 per cent for the rabbit psoas, and may be as high as 25 per cent. It now seems fairly certain that ATPases with activation properties rather different from myofibril ATPase—the ATPase of aqueous extracts[152] of muscle and fractions obtained from carp muscle[153]—are due to granule material.

IV. Extractability of Muscle Proteins

Deuticke[154] was the first to report that muscles which had been fatigued by stimulation, frozen, and pulverized imparted less protein to an extracting solution than those freshly extracted. The muscles were stimulated indirectly, though not to exhaustion, and the isometric tension was measured. There appeared to be a relation between tension and decrease in solubility. The stimulated muscle, on resting in oxygen, was found to acquire the solubility behavior of the control, but the recovery was a relatively slow one and not associated with the simple act of relaxation. The extracting medium in this case was phosphate $\Gamma/2 = 0.2$ mixed with ⅙th part of KI ($\Gamma/2 = 0.2$); the ionic strength is barely sufficient to extract the fibrillar protein, and the relative efficiency of the extraction may have been due to the depolymerizing action of KI. Soon after, Weber and Meyer[155] showed that in the case of rigor muscle at least, the loss of solubility which Deuticke had earlier reported was concerned solely with the myosin fraction, and Kamp[156] likewise reported in the case of fatigued muscle. Smith,[157] on the other hand, investigated

(151) W. W. Kielley and O. Meyerhof, *J. Biol. Chem.* **183,** 391 (1950).
(152) B. A. and G. F. Humphrey, *Biochem. J.* **47,** 238 (1950).
(153) N.-v. Thoai, J. Roche, and L. O. de Bernard, *Bull. soc. chim. biol.* **32,** 751 (1950).
(154) H. J. Deuticke, *Z. physiol. Chem.* **210,** 97 (1932).
(155) H. H. Weber and K. Meyer, *Biochem. Z.* **266,** 137 (1933).
(156) F. Kamp, *Biochem. Z.* **307,** 226 (1941).
(157) E. C. B. Smith, *J. Soc. Chem. Ind.* (London) **53,** 351*T* (1934).

a whole series of extracting solutions, and found that with ammonium and lithium chlorides of adequate strength, no differences could be detected in the behavior of fresh and rigor muscle 24 hr. post mortem.[158]

The most direct explanation of the results was that stimulation and rigor involve a change of state which is reflected in a loss of solubility in some salt solutions but not in all; or, in the light of recent knowledge, they involve the combination of myosin and actin to give a less soluble complex. In freshly minced relaxed muscle, the ATP acts as a specific dissociating agent; in rigor or fatigued muscle, extraction is facilitated by salts which depolymerize the complex, and it is significant that Edsall and Mehl[65] noted both with NH_4^+ and Li^+ that (acto)myosin quickly lost its double refraction of flow.

That the interpretation may not be quite so simple appears from the more recent work of Dubuisson,[159] who has studied the Deuticke effect by electrophoresis, the concentration of components being compared from the areas of peaks of defined mobility. Not only is the physiological history of the muscle of great influence, but also the treatment of physiologically normal muscle. Processes which produce shortening, such as freezing in liquid air (cf. Deuticke, above) give a lower yield of fibrillar protein than muscles cooled rather slowly to 0° before mincing. The muscle is then excised, passed through a Latapie, ground with fine sand, and extracted. Alternatively, the cooled muscle is frozen in the freezing microtome, sectioned into slices 20–30 μ thick, and extracted.

The pattern given of the myogen group of proteins which can be extracted separately by phosphate solution of low ionic strength (0.15) was investigated in some detail by Jacob[160] (Fig. 8) and later by Bosch.[161] There are about eight main peaks, though each of these must contain several components. Mobility tests[162] of certain purified proteins allow the assignation of aldolase (Baranowski's myogen A),[163] phosphoglyceraldehyde dehydrogenase and phosphoglucomutase to the *m-l* group, phosphorylases a and b to the *k*, and myoalbumin to the *h*. Physiological states which profoundly affect the extraction of the fibril protein do not alter the pattern of the myogen group, though it must be remembered that on strong contraction the adenylic acid prosthetic group of phosphorylase a tends to be split off to give the inactive phosphorylase b.[163a]

(158) E. C. B. Smith, *J. Soc. Chem. Ind.* (London) **54,** 152*T* (1935).
(159) See review by M. Dubuisson, *Biol. Revs. Cambridge Phil. Soc.* **25,** 46 (1950).
(160) J. J. C. Jacob, *Biochem. J.* **41,** 83 (1947).
(161) M. W. Bosch, *Biochim. et Biophys. Acta* **7,** 61 (1951).
(162) J. J. C. Jacob, *Biochem. J.* **42,** 71 (1948).
(163) T. Baranowski, *Compt. rend. soc. biol.* **130,** 1182 (1939).
(163*a*) G. T. and C. F. Cori, *J. Biol. Chem.* **158,** 321 (1945).

The effect of dialyzing the myogen extract against water results in the deposition of globulin-like protein (Weber's "globulin X") and simultaneously removes the electrophoretic components, i and k_2, and diminishes the concentration of all other components. Comment has already been made on the complexity of the enzymes in muscle juice, and except for the abundant enzymes of the glycolytic cycle, many others must be present in such minute amount that they can only contribute to the refractive-index gradient either by interacting with each other, or by moving with others of comparable mobility.

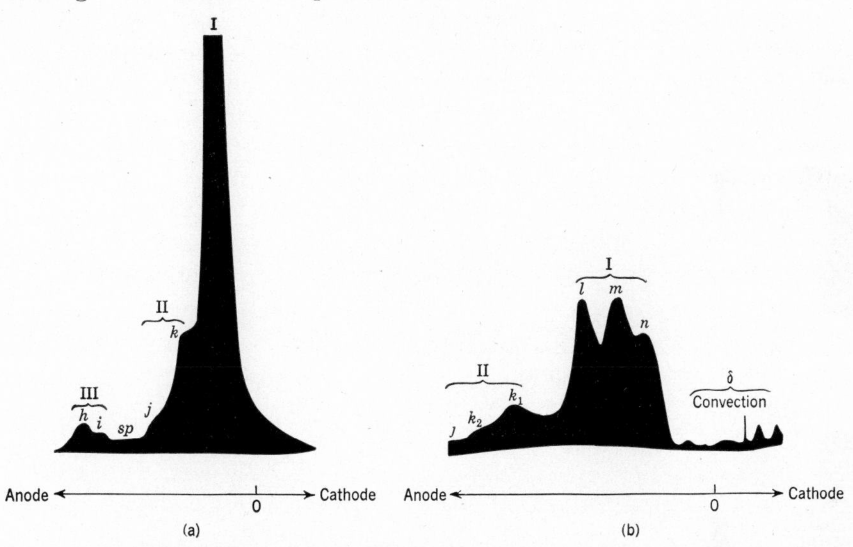

FIG. 8. Electrophoretic patterns (anodic) of extracts of rabbit muscle.[160] Γ/2 of extracting solution = 0.15. Conditions: $\frac{\Gamma}{2}$ 0.15, pH 7.5, 4.6 v./cm. (*a*) 11,000 sec. (*b*) 34,000 sec. For symbols see text.

When the ionic strength of the extracting medium is raised to 0.35 and above,[159,164] new peaks appear which have a steep refractive-index gradient, and identical with those given by the isolated myosin preparations of Weber and of Edsall. There are three components, α, β, γ the relative abundance being in the order β, α, and γ for resting muscle; there is however, more of the γ in mollusks. By fractionation with ammonium sulfate at pH 5.5, Dubuisson[165] found that the α-component is salted out completely at 27 per cent saturation, and in its properties is similar to actomyosin, while the β-component resembles Szent-Györgyi's myosin A. The γ-component has not been specifically characterized as a myosin.

(164) M. Dubuisson, *Experientia* **2,** 258 (1946).
(165) M. Dubuisson, *Experientia* **2,** 10 (1946).

In fatigued muscle[166] there is an almost complete disappearance of actomyosin (α), a large decrease in true myosin (β), and γ is no longer visible.

It might be thought that the disappearance of ATP from the muscle is largely responsible for an *in vivo* aggregation of myosin and actin which retards extraction. This is not so. While ATP hastens the rate of solution, it does not increase the final yield, except when the extracting fluid has an ionic strength above 0.5.[167]

Muscles thrown into contraction with liquid air, or into contracture with monohalogen acetate or strychnine, or into rigor show a picture somewhat different from fatigued muscle.[168,169] Actomyosin is much reduced, β-myosin disappears, and a new protein, termed contractin, which appears to be identical with γ-myosin, is found migrating faster than the myogens but less fast than β-myosin. There has not occurred under these conditions a loss of β-myosin by denaturation, because extractants such as KI and pyrophosphate which dissociate actomyosin, and also solutions of pH 9, cause its reappearance.[170] The difference in the composition of the extract has been expressed in relation to the amount of myogen present, which is constant in all circumstances:[169] e.g.,

	Per cent
normal muscle: actomyosin/myogen	11
contracture: "	6
normal muscle: contractin/myogen	0–7
contracture: "	21

These changes are not as large as they first may appear. In the most favorable conditions (long extraction time, relaxed muscle, alkaline medium) the total yield of myosin–actomyosin in a single extraction, as adopted by Dubuisson, is about 25 per cent of the total protein, as compared with a probable content of 56–60 per cent.

To explain these various effects is difficult at the present time. Weber[15] has expressed the view that for normal muscle an increase of ionic strength and the degree of comminution affect the extraction solely by creating a steep concentration gradient into the fiber particles—a process particularly important in the shorter extractions which give β-myosin. In contracture and in other abnormal states, wherever the level of ATP is diminished, a much higher ionic strength is required, if specific dissociating ions are absent, to split the F-actin–myosin complex

(166) J. J. C. Jacob, *Experientia* **3,** 241 (1947).
(167) P. Crépax, *Biochim. et Biophys. Acta* **7,** 87 (1951).
(168) M. Dubuisson, *Experientia* **4,** 11 (1948).
(169) P. Crépax, J. J. C. Jacob, and J. Seldeslachts, *Biochim. et Biophys. Acta* **4,** 410 (1950).
(170) M. Dubuisson, *Biochim. et Biophys. Acta* **5,** 489 (1950).

which is formed under such conditions. Under *all* conditions, Weber supposes that it is the surrounding muscle structures which are the main barrier, either to the diffusion of the "long F-actin filaments" or of F-actomyosin. Against this view, the Dubuisson school have shown that (*a*) ATP affects only the *rate* of extraction, (*b*) in the slow onset of rigor in myanesin-treated animals, extraction is normal although ATP has disappeared,[167] and (*c*) in bromoacetate rigor, the extract is deficient although ATP is present. (The presence of ATP under these conditions seems extraordinary.)

In the broadest sense, the Deuticke effect can certainly be interpreted as an aggregation process, and it can be influenced in several ways. First, actin and myosin can associate through lack of ATP, and by analogy with the behavior in solution, the complex possesses a higher salting-in threshold than free myosin. A relaxed muscle, freshly minced, will yield free myosin even on coarse mincing, but further comminution and stronger salt solutions will bring out large amounts of actomyosin (also tropomyosin). But homogenization must be continued to break mechanically not only the surrounding structures but to disperse further the concentrated thixotropic actin gel inside. Secondly, the extent of the interaction must be determined by physiological events, or more precisely, the molecular mechanism which accompanies contraction. Thirdly, Weber has pointed out that actomyosin aggregates exist which are "denatured" in the sense that they do not react with ATP, and Crépax and Hérion[171] find that a muscle in thaw rigor gives a normal pattern when extracted immediately, but a deficient one after 2 hr., ATP being absent in both instances. Conversely, Robinson's work[172] points to the existence of actomyosin complexes quite insoluble in salt which *do* synerese with ATP, and Perry likewise finds that isolated myofibrils do not wholly dissolve in salt solutions.[173] Lastly, the supporting structures may hinder the diffusion out to varying degrees according to the extent of the contraction; in the extreme case of thaw rigor, the ready solubility may be due to the fact that contraction is so severe that the fibrillar material is spilled out from the cut ends.

It thus seems preferable to adopt the viewpoint at present that changes in solubility are neither so complex nor so simple as the opposing viewpoints of Dubuisson and Weber suggest, and further work is desirable. On the one hand most workers (Hasselbach and Schneider, Bate Smith, Bailey) have stressed the importance of repeated extraction and comminution, and neither Deuticke nor Dubuisson employed more than

(171) P. Crépax and A. Hérion, *Biochim. et Biophys. Acta* **6**, 54 (1950).
(172) D. S. Robinson, *Biochem. J.* **52**, 621 (1952).
(173) S. V. Perry, private communication.

one. (In some methods for the preparation of fibrinogen, it is common practice to obtain by salting-out a pseudo clot which dissolves so slowly that it is often thought to be fibrin; yet a 30-sec. treatment in a blendor will disperse it at once.) While the simple physical state of the muscle should not be neglected, the experiments of Hasselbach and Schneider, discussed on pp. 995 and 1011, overemphasize its importance. Here the quantitative extraction of myosin by a buffer containing a specific dissociating agent is striking, but the further comminution, which they consider to give pure actin, gives in reality a very viscous extract, found by Tsao and Bailey[33] to contain tropomyosin, and likely to contain much particulate material which cannot be spun down.

That some very definite change occurs not only in the proteins which under set conditions become insoluble, but also in those which become soluble seems to be indicated by the appearance of "contractin," and it is unfortunate that we know so little about it: it must at least be some modification of the contractile material.

V. Estimation of Muscle Proteins

1. Adult Muscle

The first serious attempts to partition the proteins of muscle into well-defined fractions and to assess their amounts were made by Weber and Meyer[155] and later by Smith.[157,158,174] A division can broadly be made between the intracellular protein and the extracellular, the former including the contractile and the sarcoplasmic proteins, the latter the supporting structures, reticulin, collagen, and elastin, as well as extraneous tissues such as vascular material and nerve. The extracellular protein is insoluble in water and in salt solutions, but one of the main difficulties in determining its amount is that normal methods of extracting the soluble proteins leave behind a residue of intracellular protein. discussed in more detail on p. 1011. The sarcoplasmic protein can be estimated after simple extraction with very dilute salt solutions, which leave the contractile material and connective tissue behind. Most workers have preferred, however, to extract the intracellular proteins together, and then to dilute to an ionic strength in which myosin and actomyosin are insoluble. The sarcoplasmic (myogen) fraction was further divided by Weber and Meyer into an albumin portion and a globulin fraction which precipitates on dialysis to $\Gamma/2 = 0.005$ and pH 6; the precipitated material, termed "globulin X" is mostly soluble on addition of salt, and some which is insoluble may arise by denaturation of both albumins and globulins. The whole myogen complex on stand-

(174) E. C. B. Smith, *Proc. Roy. Soc.* (London) **B124,** 136 (1937).

ing in the absence of salt is relatively unstable, becomes turbid, and slowly precipitates.

The results of analysis differ according to author and method and it is useful to compare the various procedures employed.

Weber and Meyer.[155] Finely minced muscle was extracted five times with 0.6 *M* KCl, pH 8–9, and then a further 3–5 times at pH 9.5. Myosin was precipitated by dialysis at pH 7 to a KCl concentration of 0.03 *M*. Weber[15] now inclines to the view that the alkaline medium encouraged the loss of actin, which would be assessed as myogen.

Smith.[157] In this method the importance of grinding and of repeated extraction were emphasized. By extraction with MgSO ($\Gamma/2 = 1.67$) 70 per cent of the total coagulable protein was obtained in solution. The residue was autoclaved to dissolve out collagen, which could be estimated, and the final residue was digested with trypsin to bring the denatured intracellular protein into solution. Elastin was thought to be left, though this is generally stated to be dissolved by trypsin; moreover, autoclaving seems to dissolve denatured proteins which would thus be estimated as collagen. Later,[158,174] the whole method was modified substituting 7 per cent LiCl ($\Gamma/2 = 1.65$) for $MgSO_4$, and after exhaustive extraction, the residue was washed with water and extracted with 0.1 *N* HCl, which was considered to complete the extraction of intracellular material. The amount of protein N contained in the residue was comparable with that when whole muscle was directly extracted with acid. The acid-soluble material left after extracting with LiCl was partly soluble in KCl at neutral pH, and presumed to be myosin, and the insoluble portion denatured globulin X. Though in this case large errors could not have been introduced, it is worth mention that differential denaturation methods are not very reliable because of interaction between soluble and insoluble portions. Dilution of the LiCl extract to ionic strength 0.083 caused the (acto)myosin to precipitate, and this was estimated, applying a correction of 13 per cent for residual solubility.

Hasselbach and Schneider.[128] These workers were the first to attempt the direct estimation of actin. The muscle is coarsely ground and actin-free myosin plus the sarcoplasmic proteins are first extracted with 0.6 *M* KCl containing sodium pyrophosphate (pH 6–6.3). The latter reagent "dissociates" actin and myosin, and the myosin diffuses out leaving the actin associated with the stroma protein. The tissue is now homogenized in 0.6 *M* KCl, yielding a turbid, viscous extract of actin. In the presence of ATP, the isolated myosin showed only a slight drop in viscosity, indicating that very little actin is present, and the actin showed no response at all; but the addition of the one to the other caused the formation of an ATP-sensitive complex. The final residue was tested for intracellular protein by extracting with urea, and contained only 2–5 per cent of soluble protein nitrogen. The amount of actin was assessed at 13–15 per cent of the total protein, and myosin as 38 per cent. Corrected for urea-soluble protein, the total actomyosin amounts to 56 per cent. The total extracellular protein (15–17 per cent) agrees well with Smith's figures.

Tsao[129] has compared this method of extraction with others which he has devised. The myosin fraction was found to be pure, but the actin extract was so turbid and viscous that particulate matter could not entirely be spun down; moreover, it contained the whole of the tropomyosin of the muscle which was crystallized as a ribonucleic acid complex.

Robinson.[140,172] The method used by Robinson was evolved to follow the pattern of quantitative change in certain protein fractions during the development of chick muscle. The developmental aspects will be considered separately and in relation to other investigations (p. 1013). Homogenization was effected in a stainless steel vessel which itself could be used as a centrifuge tube, and extractions were repeated two or three times. Separate samples of tissue were extracted by the following solvents: (*1*) the sarcoplasmic fraction: 0.1 M KCl and phosphate, pH 7.1, $\Gamma/2 = 0.2$; (*2*) sarcoplasmic + fibrillar: 1.25 M KCl, $M/15$ K_2HPO_4, pH 8.5, $\Gamma/2 = 1.45$; (*3*) residual intracellular protein: the residue from (*2*) was extracted with 0.1 N NaOH, leaving behind the insoluble stroma protein.

Fraction 1 was diluted to ionic strength 0.1 in a medium containing 15 per cent (v/v) ethanol. Under these conditions, any myosin or actomyosin is precipitated, but none of the proteins of press-juice. Fraction 2 is similarly diluted to ionic strength 0.12 in the presence of 15 per cent ethanol. The precipitate consists of myosin, actomyosin, and desoxynucleoproteins of the cell nucleus, which in adult muscle do not contribute much to the total protein. All the extracts were sufficiently limpid to give clean separation of particular material on centrifuging, and this may help to explain why the residual intracellular fraction 3 is much higher than that found by Hasselbach and Schneider, but agrees with the findings of Herrmann and Nicholas on rat muscle. Robinson considers that this fraction is complex (see p. 1013).

TABLE VIII
PARTITION OF PROTEIN N IN SKELETAL MUSCLE

Authors	Muscle	Protein N as % Total Coagulable N					Protein N as % intracellular protein N				
		Soluble in salt									
		Sarcoplasmic (*A*)	Soluble fibrillar (*B*)	Residual intracellular (*C*)	Total fibrillar protein (*B* + *C*)	Stroma	Sarcoplasmic (*A*)	Fibrillar (*B* + *C*)	Myosin	Actin	Tropomyosin
Weber and Meyer[155]	Rabbit (white)	44	39	?		17	53	47	—	—	—
Smith[174]	Rabbit (white)	16	54	15	69	15–17	19	80	—	—	3[127]
Hasselbach and Schneider[128]	Rabbit (white)	28	52	4	56	16	37	63	44	15	—
Robinson[140,172]	Chick (1 month)	33	40	22	62	5	35	65	—	—	
Dyer, French and Snow[175]	Cod	21	70	6	76	3	22	78	—	—	

(175) W. J. Dyer, H. V. French and J. M. Snow, *J. Fisheries Research Board Can.* **7**, 585 (1950).

In Table VIII, the available data on various kinds of muscle are collected together. For rabbit and chick, the sarcoplasmic fraction comprises some 32–37 per cent of the intracellular protein, and actomyosin about 65 per cent; fish muscle contains rather more actomyosin. The largest differences are found in the amount of intracellular protein (col. 5) not extracted by strong salt solutions, but which in the present table is ultimately included in the fibrillar fraction. It should be noted that whereas the value quoted here for Smith's analysis represents only that part of the residual salt-insoluble fraction which he considered to be myosin by differential denaturation tests, those by Robinson and by Hasselbach and Schneider include the whole of the salt-insoluble fraction associated with the residue. That this does not contain denatured sarcoplasmic proteins is likely from Robinson's finding that extracts made with dilute salt give the same value as do strong salt extracts after actomyosin has been precipitated.

2. The Changes During Development of Muscle

In Plate III are shown longitudinal sections of adult fowl and of embryonic chick muscle at the 13th day of incubation. The striated fibers of the adult muscle with the elongated nuclei lying under the sarcolemma are typical; but in the developing muscle the fibers are few and are scattered in a background of reticular material (the ground substance), while the nuclei are very plentiful. At this stage they are round and lie in the middle of the fiber.

Robinson[140,172] has applied the fractionation procedure already outlined to the developing muscle of the chick, and has made an approximate assessment of the contribution of the nucleoproteins of the cell nuclei, which like actomyosin are extracted by strong salt solutions and precipitated on dilution. The bulk (80 per cent) of the desoxyribonucleic acid passes into the actomyosin extract and the remainder is left behind in the residue. The amount of protein assumed to arise from the nucleus in this extract can also be estimated from the ratio of protein to nucleic acid in the nuclei which Robinson isolated directly from young muscle and analyzed. In this way, an approximate idea of the amount of true fibrillar protein laid down at different stages could be gained.

In Table IX some of the data have been summarized to show the gross changes occurring in the prehatching period for two important groups of proteins: (*a*) the sarcoplasmic fraction consisting of the truly soluble proteins plus the particulate material which is not spun down at the speeds employed, and (*b*) the myofibril proteins together with any soluble proteins of the "ground substance." In terms of the previous fractionation scheme, this latter group is the sum of the proteins extracted

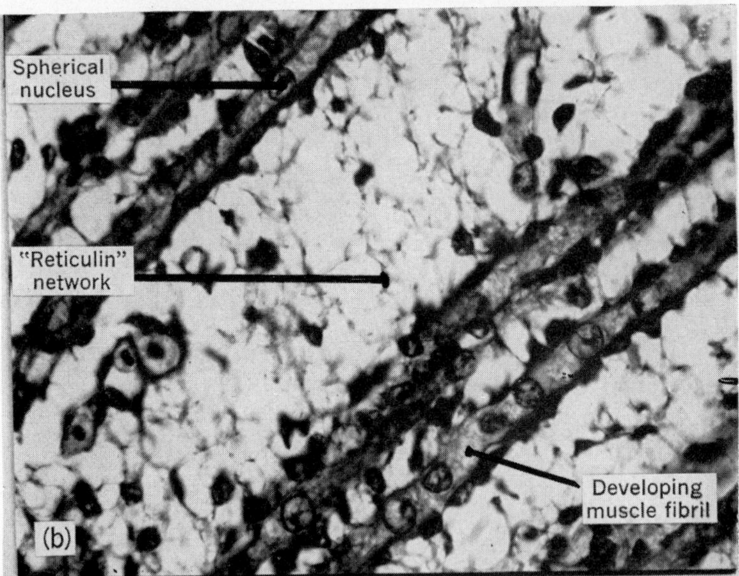

PLATE III. (*a*) A section of adult fowl muscle stained with hematoxylin and eosin. Note the striated myofibrils and elongated nuclei. The reticulin network has disappeared. Section by Mr. D. Canwell.[172] (×660).

(*b*) A section of embryonic muscle (13-day incubation) stained with hematoxylin and eosin. Note the developing muscle fibrils with centrally placed nuclei and the reticulin network (×660).

by strong salt solutions and precipitated on dilution and those extracted from the resultant residue with dilute NaOH, both corrected as outlined for nucleoproteins of the cell nucleus.

TABLE IX
THE PROTEINS OF DEVELOPING CHICK MUSCLE

Weight of embryo, g.; approx. age in days in parentheses	Sarcoplasmic group of proteins as a percentage of		Myofibrillar + soluble proteins of ground substance as a percentage of	
	Fresh weight of muscle	Total protein[a] of muscle	Fresh weight of muscle	Total protein[a] of muscle
10 (14)	2.4	58	1.9	39
20 (18)	2.8	45	3.0	44
30 (20)	3.1	38	3.8	47
Adult fowl	6.25	33	9.8	62

[a] "Total protein" includes nucleoprotein (nucleic acid plus protein).

Using a different fractionation procedure, Herrmann and Nicholas[176] have followed the development of rat muscle, and though it is difficult to make any direct comparison, very similar over-all changes of protein distribution seem to occur in the two species, and the data confirm at a biochemical level the pattern of differentiation indicated by extensive cytochemical and histological studies in the past. For a 10-g. chick embryo (14th day of incubation) the protein plus nucleic acid of the nucleic comprises no less than 66 per cent of the total protein of the fibrillar fraction, and at this stage, the true contractile material amounts to only 7 per cent of the total protein. This latter, as we have seen (Table IX) rises to over 60 per cent in adult muscle, while the *total* nucleic acid content has fallen to about 0.8 per cent of the total protein. A steady state in the amounts both of sarcoplasmic and fibrillar protein is reached for a chick weight of about 140 g., i.e., 21–28 days after hatching.

3. Changes in the Muscle Proteins During Atrophy

It is probably true to say that muscle tissue is less prone to diseases of clinical importance than any other tissue of the body, and that the most serious affliction is muscular atrophy. This can arise in a number of ways: by disuse (immobilization, neuritis, arthritis), by denervation (anatomical or by poliomyelitis), by tenotomy, or by the true primary diseases of muscle, fortunately rare, muscular dystrophy and myotonia atrophica; or in myasthenia which is often associated with hypertrophy

(176) H. Herrmann and J. S. Nicholas, *J. Exptl. Zool.* **107**, 165 (1948).

or tumors of the thymus gland. A condition brought into prominence during the London Blitz of World War II was the crush syndrome. Patients buried under debris, though relatively unhurt, later developed shock and died of renal failure.[177] This latter seems to be due to the liberation of myoglobin in the muscle, which becomes deposited in the kidney under the influence of an acid urine.

The biochemical changes undergone during atrophy have been most studied for denervation atrophy, but common to all atrophies is the initial high rate of wasting, which is due entirely to a diminution of fiber volume. In the same way, the increase in weight of a muscle induced by exercise is due entirely to an increase in the intracellular phase. For the rabbit gastrocnemius-soleus muscle which normally has a stroma component of about 15 per cent of the total muscle volume, denervation atrophy of approximately 28 days duration will produce a weight loss of 70 per cent;[178] and since the absolute volume of the stroma does not change perceptibly, the stroma content of the atrophied muscle will be 50 per cent of the muscle volume. It follows that the changes in composition during atrophy can be related either to the collagen content or to the weight loss.

Attempts to follow the change in protein pattern have been made (*a*) by exhaustively extracting with 0.5 *M* lithium chloride and subsequent precipitation of actomyosin[179] and (*b*) short extraction with 0. 6 *M* potassium chloride[180] to extract myosin A. Figure 9 shows the relative changes in the actomyosin and sarcoplasmic proteins expressed as a percentage of the muscle phase. It should be noted that the proportion of unextracted intracellular protein (white portion top) is rather high, and it is probable that the whole nonshaded portion represents the myofibrillar protein. In Fig. 10 a given fraction, first expressed as a percentage of the total intracellular protein, is worked out as a percentage of the corresponding value for the control muscle. On this basis, the amount of LiCl-extractable protein does not decline for about 10 days after denervation, but there is an immediate diminution in myosin A, much more rapid than that of actomyosin, indicating that a change of extractability rather similar to that occurring in normal fatigue is involved.

The ATPase activity of three-times precipitated myosin from atrophied muscle is significantly lower than that of the control,[180] and a

(177) E. G. L. Bywaters *in* Le Muscle, Proceedings of the Symposium at Royaumont, France, 1950, Expansion Scientifique Francaise.
(178) H. M. Hines and G. C. Knowlton, *Am. J. Physiol.* **104,** 379 (1933).
(179) E. Fischer and V. W. Ramsey, *Am. J. Physiol.* **145,** 571 (1946).
(180) E. Fischer, *Arch. Phys. Med.* **29,** 291 (1948).

relatively greater proportion remains in solution during purification. Fischer[181] has drawn an analogy between this and the reverse process which seems to occur in embryo muscle, when an ever-increasing proportion of the total ATPase activity becomes associated with the (acto-) myosin.[182,183] In neither of these cases is it certain that the granule ATPase

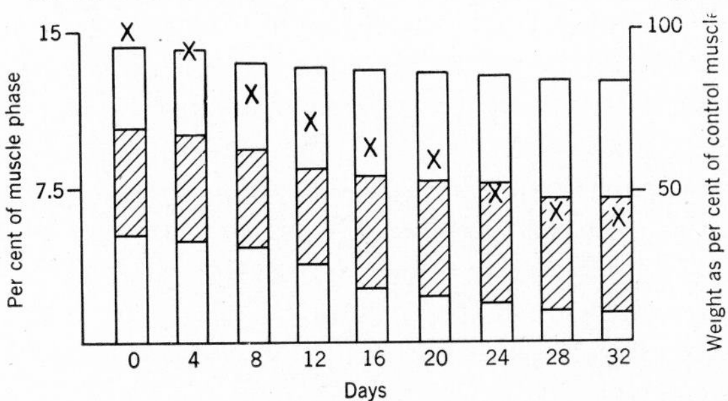

FIG. 9. Changes during denervation atrophy. Weight (×); non-collagenous protein is given by the total height of the ordinate; sarcoplasm by the cross-hatched portion; unextracted intracellular nonshaded top; myosin B nonshaded bottom. The total contractile material is represented by the sum of the last two. Adapted from Fischer.[181]

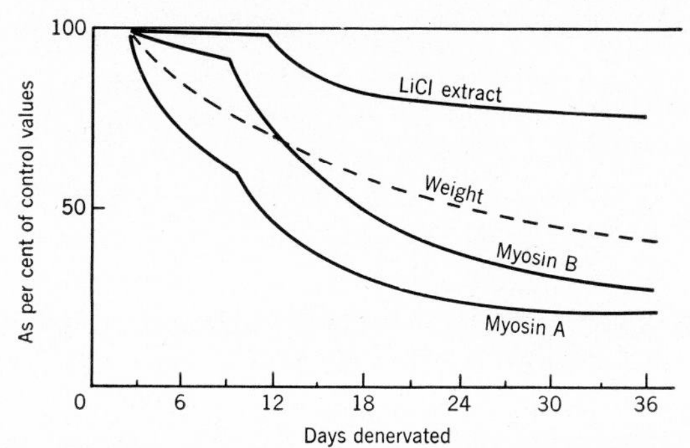

FIG. 10. Changes during denervation atrophy (rabbit gastrocnemius). Values expressed as percentage of the control values (see text). The weight represents muscle phase. From Fischer.[181]

is not involved, but the change is worthy of further study in view of the fact that trypsin can produce a water-soluble form of myosin–ATPase.[48,49]

The mean anisotropy of myosin prepared by Mommaert's method has been found to be significantly higher than that of normal muscle, and this does not seem to be due to contamination with actin.[184] If

(181) E. Fischer *in* Le Muscle, p. 280; see Ref. 177.
(182) H. Herrmann, J. S. Nicholas, and M. E. Vosgian, *Proc. Soc. Exptl. Biol. Med.* **72,** 454 (1949).
(183) D. S. Robinson, *Biochem. J.* **52,** 633 (1952).
(184) G. Schapira, J. C. Dreyfus, and M. Joly, *Nature* **170,** 494 (1952).

changes in the water-soluble portions are studied electrophoretically, there is seen to be a large relative increase in the myoalbumin fraction, particularly in tenotomized muscle.[185] Simply on the basis of its properties, the author has suggested that myoalbumin may be a mucoprotein,[186] and again, this is a conspicuous component of embryo muscle.[187] Altogether, the changes occurring in tenotomy are more severe than in denervation atrophy, and can be lessened if the contracture of the muscle is released by denervation.[181,185] Least severe are the changes caused by disuse, which produces only slight changes of protein pattern, although the initial rate of wasting is as high as in other kinds of atrophy.[181]

Of especial interest is the effect of electrical stimulation on denervated muscle. If the stimulus is sufficient to produce *maximal* contraction,[188] there is a considerable retardation of weight loss, of extractable myosin, and of metabolites such as glycogen and creatine;[189] the relative amount of myoalbumin is also much less.[185] The effect appears to be bound up with the production of tension, since the best results are given when the muscle is forcibly extended, i.e., when the contraction is isometric. For this reason, no improvement is observed in tenotomized muscle.

All these observations raise some interesting questions. What is the nature of the nervous influence on maintenance of muscle weight? What is the extra effect of contracture on wasting? Are the changes induced in atrophy a retrogression to an embryological type of muscle? What is the nature of myoalbumin? Is it to be regarded as a nonfunctional component which replaces the functional muscle proteins when they cease to play any useful role? In this connection, one may note that in the evolutionary development of electrical tissue there occurs a metamorphosis of the muscle fiber to vestigial structures at the expense of the enlargement of the motor end plate, and here, too, the cell becomes full of a mucin rich in hexosamine and phospholipide.[190]

VI. Amino Acid Composition of the Structure Proteins and Its Significance

The amino acid composition of myosin and tropomyosin was fairly complete and that of actin fragmentary until Kominz, Laki, and Hough[191]

(185) E. Fischer, R. V. Bowers, H. V. Skowlund, K. W. Ryland, and N. J. Copenhaver, *Arch. Phys. Med.* **30,** 766 (1949).
(186) K. Bailey *in* Le Muscle, p. 297; see Ref. 177.
(187) P. Crépax, reported by M. Dubuisson *in* Le Muscle, p. 22; see Ref. 177.
(188) W. H. Wehrmacher, J. D. Thomson, and H. M. Hines, *Arch. Phys. Med.* **26,** 261 (1945).
(189) See H. M. Hines *in* Le Muscle, p. 333; see Ref. 177.
(190) K. Bailey, *Biochem. J.* **33,** 255 (1939).
(191) D. R. Kominz, A. Hough, P. Symonds and K. Laki, *Arch. Biochem. Biophys.* **50,** 148 (1954); kindly communicated in advance by Dr. Laki.

undertook a new analysis of all three proteins by the Moore and Stein method. Bailey's data[127] for myosin refer to actin-containing preparations, although the amount of contamination is no more than a few per cent (see p. 958). For different (acto-)myosins (rabbit, dog, ox, chicken) there are only minor analytical differences in the limited analyses made—total N, total S, methionine, cystine, tyrosine, tryptophan—and rather greater ones between these and fish and crustacean specimens.[192] The difference that may exist in the actin content now makes these conclusions of doubtful validity, but in all probability the variations amongst individual myosins are likely to be small.

Neither myosin nor tropomyosin has a detectable prosthetic group:[127] the P content of myosin can be reduced to as little as 0.04 per cent by extracting dried myosin with strong salt solution to liberate contaminating nucleic acid (see Table X). In the case of actin, phosphorus attributable to pyro P was assessed at 0.06–0.09 per cent, representing 1 mole/40,000 reckoned as ADP and 1 mole/80,000 as ATP.[193] The actual identification of a nucleotide prosthetic group was made later: ATP for G-actin, ADP for F-actin, but the amount varies with the preparation and with the investigator.

TABLE X
SOME ANALYSIS OF TROPOMYOSIN, MYOSIN AND ACTIN

	Tropomyosin	Myosin	Actin
Total N	16.7[a]	16.7[a]	16.1[b]
			15.2[c]
P	0.02[a]	0.04–0.06	0.06–0.09[d]
			0.11[c]
S	0.80[a]	1.10[a]	1.3_6[c]
Total carbohydrate	0.2	0.2	—

[a] Bailey.[192]
[b] Tsao and Bailey.[33]
[d] Perry.[193]
[c] Mommaerts, *J. Biol. Chem.* **198,** 445 (1952).

The amino acid compositions of myosin, tropomyosin, and actin are set out in Tables XI and XII (See also Chap. 3). In the case of myosin, analytical figures for the dicarboxylic acids are not very concordant, and the estimate of free carboxyl groups by titration (Table XII) is very much higher. The detailed analysis of the titration curve of myosin by Mihalyi[25] is worthy of especial mention. The titration

(192) K. Bailey, *Biochem. J.* **31,** 1406 (1937).
(193) S. V. Perry, Ph. D. Dissertation, Cambridge, England, 1947.

TABLE XI

Analyses of Tropomyosin Actin and Myosin[a]

(Results calculated on N contents of 16.7 per cent)

	Tropomyosin			Actin[b]		Myosin			L-Meromyosin	H-Meromyosin
	Wt./100 g.	Res./10^5 g.	Res./10^5 g. (Kominz *et al.*)[a]	Wt./100 g. (Kominz *et al.*)[a]	Res./10^5 g. (Kominz *et al.*)[a]	Wt./100 g.	Res./10^5 g.	Res./10^5 g. (Kominz *et al.*)[a]	Res./10^5 g. (Kominz *et al.*)[a]	Res./10^5 g. (Kominz *et al.*)[a]
Cystine/2	0.76	6	6.5	1.34	11.2	1.4	12	8.6	5.6	10.9
Methionine	2.8	19	16	4.5	30	3.4	23	22	14	19
Tyrosine	3.1	17	16	5.8	32	3.4	19	18	12	21
Tryptophan	Nil	—	0	2.05	10	0.8	4	—	6	3
Glycine	(0.4)	(5.3)	12.5	5.0	67	1.9	25	39	24	45
Alanine	8.8	100	110	6.3	71	6.5	73	78	76	73
Valine	3.1	27	38	4.9	42	2.6	22	42	39	45
Leucine	15.6	119	95 } 124	8.25	63 } 119	15.6	119	79 } 121	85	78
Isoleucine			29	7.5	57			42	35	42
Phenylalanine	4.6	28	3.5	4.8	29	4.3	26	27	9.5	40
Proline	1.3	11	0	5.1	44	1.9	17	22	8.5	29
Serine	4.4	42	40	5.9	56	4.3	41	41	37	43
Threonine	2.9	24	28	7.0	59	5.1	43	41	38	49
Histidine	0.85	5.5	5.5	2.9	19	2.4	15.5	15	19	11.5
Arginine	7.8	45	42	6.6	38	7.4	42	41	51	29
Lysine	15.7	107	110	7.6	52	11.9	81	85	83	82
Glutamic	32.9	224 } 292	211 } 300	14.8	101 } 183	22.1	150 } 217	155 } 240	174	138
Aspartic	9.1	68	89	10.9	82	8.9	67	85	77	88
Amide N	0.89	64	64	0.92	66	1.20	86	—	—	—

TABLE XI. (*Continued*)

	Tropomyosin (Bailey)[a]	Actin[b] (Kominz *et al.*)[a]	Myosin (Bailey)[a]
Average residue weight: N partition	115.6	109	115.8
Average residue weight: Wt. residues	117.2	110	115.3
Res./10^5 g.	855	911	866
As per cent of total residues:[c]			
Free acid groups	26.6 } 45.0	12.9 } 24.9	18.0[e] } 34.1
Base groups	18.4	12.0	16.1
Polar groups[d]	63	49.3	57
OH groups	9.7	16.4	12
Amide	7.4	7.2	10

[a] Except where stated, the analyses are those of Bailey;[127] the rest were kindly communicated by Dr. K. Laki in advance of publication.[191]

[b] Other values; Tyr 4.7, Tryp 1.2, Cysteine 0.85% (Tsao and Bailey[33]); Tyr 5.5, Tryp 1.2, His 2.4, Arg 6.4, Lys 8.1% (Perry[193]).

[c] Values calculated from the data of Bailey[127] except in the case of actin.

[d] Polar groups are Arg. Lys. His. Asp. Glut. (and amides), Ser. Thre. Tyr. Cys.

[e] This figure is derived from the titration data of Dubuisson and Hamoir (see Table XII).

figures for iminazolyl and guanidino groups agree well with those from chemical analysis, but a fairly large discrepancy occurs in the lysine values. Since myosin does not appear to contain α-amino groups, whose titration range might overlap with iminazole, it is possible from the shift of pH with temperature and from calculated heats of ionization to infer that carboxyl dissociation is complete at pH 6.3 and that the iminazole groups ionize between this pH and pH 8.2. The ionization of phenolic hydroxyl groups occurs beyond pH 10.5, a fact that may suggest they are firmly hydrogen-bonded as in ovalbumin.

TABLE XII
TITRATION DATA FOR MYOSIN

	Found per 10^5 g. protein			
	Analytical		Titration	
	Kominz *et al.*[191]	Bailey[127]	Mihalyi[25]	Others
Free carboxyl	154	131	165	156[a]
Total cationic	141	138.5	150	150,[b] 150,[c] 146[d]
Histidine	15	15.5	16	
Lysine	85	81	91	
Arginine	41	42	43	

[a] M. Dubuisson and G. Hamoir, *Arch. intern. physiol.* **53,** 308 (1943).
[b] W. T. Salter, *Proc. Soc. Exptl. Biol. Med.* **24,** 116 (1926).
[c] G. E. Perlmann, *J. Biol. Chem.* **137,** 707 (1941) (metaphosphate binding).
[d] K. Bailey, unpublished.

Before further comment is embarked upon, some comments on the differences in analyses between the figures of Bailey and those of Kominz *et al.* are necessary. By the trioxalatochromiate method the presence of glycine in tropomyosin seemed doubtful, but the Moore and Stein method definitely shows some (1 per cent) to be present; by this method there appears to be *no* proline, and only a small amount (0.6 per cent) of phenylalanine. The aspartic acid figure is higher and the glutamic acid proportionately lower, such that the sum is the same as in Bailey's analysis. A fairly large discrepancy appears in the valine figure—most probably a methodological error since the same is true of myosin.

The comparisons made previously between myosin and tropomyosin are a little doubtful because of these discrepancies. However, methionine, tyrosine, valine, leucines, serine, and arginine are comparable. By Sanger's method of assay, neither protein appears to contain α-amino groups. Both are very rich in charged groups, comprising 45 per cent of the total for tropomyosin and 34 per cent for myosin. Tropomyosin, in fact, contains more groups of *mixed* charge than any protein yet analyzed. (The median value for all proteins subjected to histogram analysis is in the region of 25 per cent of the total.[194]) While the isoelectric points of both proteins are in the pH range 5–5.5, there is a large excess of acidic over basic groups in tropomyosin, and it is possible that the isoionic point has still to be accurately determined.

A noteworthy property of myosin is its low content of histidine compared with the other bases. Accepting Mihalyi's findings, one half of these groups will be ionized at physiological pH and *all other* groups will be fully ionized. Calculated per 10^5 g. protein, this means that at physiological pH there will be (from Kominz *et al.*) 154 anionic groups and 134 cationic, a net anionic charge of 20. Supposing there is a uniform distribution over the surface of the molecule, this net charge will be distributed over a segment 270 A. in length—accepting Weber's estimate, probably maximal, of 2300 A. for the molecule of 850,000 molecular weight. In other words, there would be one net negative charge in 13 A. of length in company with 7 dipoles, or 1 dipole in 2 A. of length. In reality, the charges could be anywhere on the surface of this segment, 13 A. long and about 20 A. in diameter, but the calculation shows that, apart from any other kind of intramolecular forces, the contribution of electrostatic attractions interacting on a probability basis of a few angstroms could contribute very considerably to the configuration of the molecule. If the charges are distributed on the surface of a cylinder by a random process, the first thing noticed is that the net negative charge must have a small effect in relation to the contribution of other charges; secondly, the charge density is so high that qualitatively it is difficult to imagine how anything but a major alteration of *net* charge could induce an intramolecular change of configuration, particularly when existing dipoles are reinforced by intrachain hydrogen bonds, as in the Pauling-Corey model.

Some data for the amino acid composition of Straub actin were published by Feuer *et al.*[89] Where these have been repeated in Cambridge and elsewhere, the agreement except in the case of lysine is very poor. The provisional figures kindly supplied me by Dr. K. Laki show that 96 per cent of actin nitrogen has been accounted for. There is nothing very unusual in the figures except perhaps the rather high tyrosine, methionine, and aliphatic hydroxy acids. Both basic and free acidic groups, and also total polar groups are lower than in myosin and tropomyosin. Since all three proteins possess a low histidine content (pK near 7), they must all bear at physiological pH a total charge which is near the maximal.

(194) K. Bailey, *Chemistry Industry* p. 243 (1950).

Finally, it should be noted that the two meromyosins liberated from myosin by the action of trypsin have quite different compositions. Especially striking are the differences found for proline, glycine, phenylalanine, and histidine.

VII. X-Ray and Electron Microscope Studies of Muscle and Muscle Proteins

The information that can be obtained from x-ray diffraction is of two kinds, regular patterns within a molecule and the arrangement of molecules one with another if this also is regular. The electron microscope, however, may pick out both regular and less organized parts and so is a useful adjunct to x-ray studies. The so-called wide-angle x-ray patterns are concerned with the smaller spacings, but the refinement of the technique reveals diffractions near the center of the photograph from which the larger repeating units can be calculated. We shall deal first with the wide-angle pattern, then with the electron microscopy of muscle, and conclude with small-angle patterns which in part can be correlated with electron optical data.

1. Wide-Angle Diffraction of Actomyosin

Boehm and Weber[73] were the first to show that extracted actomyosin (then called myosin), in the form of an oriented thread, gave a pattern rather like that of muscle itself. Astbury,[74] however, was the first to consider and to demonstrate the remarkable structural similarities between myosin and keratin, or rather that form of keratin ("generalized" keratin) whose elastic range is extended by the rupture of the restraining S—S bonds which link one chain to another. There is no reason to believe that the deductions which were then made about myosin are invalid because the preparations contained some actin, for the diffraction pattern of actin itself differs profoundly from that of myosin (see below).

When a myosin sol is dried down on a glass plate, the particles lie down flat, but are randomly disposed in the plane of the surface. With the beam perpendicular to the surface, the pattern is of the disoriented α-keratin type, which gives way, after stretching 1½ times, to a well-oriented α-photograph containing traces of β-keratin. It should be recalled that the α-photograph is characterized by an intense meridional arc of 5.14 A., which in Astbury's view is not composite, and which arises from a regular fold in the polypeptide chain, whose nature is still a matter of fierce argument (see Vol. I, Chap. 4). The equatorial spacing contains only the so-called side-chain spacing of 9.6 A. On extension of α-keratin to about twice its oriented length, the α-photograph has given way completely to the β: the fold is straightened out, and on the meridian a spacing representing the average length of a single amino acid residue appears (3.33 A.). Together with the side-chain spacing there appears on the equator an intense reflection at 4.65 A. (the so-called backbone spacing), whose true period is 9.3, explicable on the basis that adjacent chains run in opposite directions. It is interesting to note that a pure oriented α-photograph could be obtained by stretching myosin film in 1 per cent NaOH, which was employed to obliterate internal cohesions, but which must also have dissolved out the actin component. The points of similarity with keratin which Astbury has particularly emphasized can be discussed under four headings:

(1) The Intramolecular Transformation of α- to β-Keratin. Films of myosin when brought first of all into the oriented α-state and then stretched to about double their length give a pure β-pattern, and over this range of extension elastic recovery in water

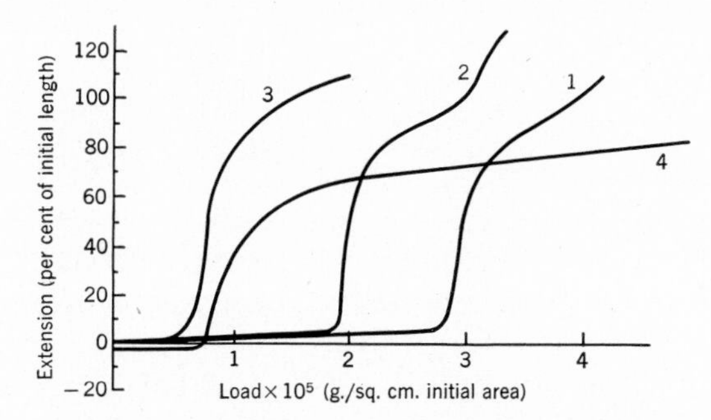

FIG. 11. Load–extension curves (1), (2), and (3) of oriented (acto-)myosin strips compared with that of "generalized" Cotswold wool (4).[74]

is complete.[74] Whole muscles, such as those of frog sartorius and retractor of *Mytilus* when freely dried, wetted, and stretched also transform from oriented α to oriented β.[194a] The stretching of undried muscle, however, does not change the α-pattern.

(2) Elasticity.[74,194b] Load–extension curves of strips of myosin follow Hooke's law to the point where the α-β transformation (as shown by x-rays) begins; thereafter, there is a sudden extension as the critical load is passed. Further loading generally elicits a third phase where intermolecular slipping is evident (Fig. 11). The first two phases in particular are characteristic of generalized keratin.

(3) The Aggregation of β-Chains. When squeezed laterally, myosin films show an interesting effect which first allowed Astbury and Sisson[195] to conclude in the case of keratin that the equatorial side-chain and backbone spacings are perpendicular to each other. (For keratin, both heating and squeezing are necessary.) In films thus treated, the α-myosin is converted to β, and the β-crystallites are so aggregated in the direction of the backbone that they roll round with the side chains perpendicular to the plane of flattening. This is proved by sending the x-ray beam parallel to, per-

(194*a*) W. T. Astbury, *Proc. Roy. Soc.* (London) **B134,** 303 (1947).
(194*b*) H. J. Woods, *J. Colloid Sci.* **1,** 407 (1946).
(195) W. T. Astbury and W. A. Sisson, *Proc. Roy. Soc.* (London) **A150,** 533 (1935).

pendicular to, and along the plane of flattening, when one obtains, respectively, (*a*) only the equatorial side-chain spacing, (*b*) only the equatorial backbone spacing, and (*c*) the side chain on the equator and the backbone on the meridian.

(*4*) *Supercontraction.* The necessary condition for supercontraction in cross-linked molecules is essentially the rupture of these links, accompanied by reactions which assist the labilization of weaker intermolecular bonds. It is not sufficient therefore in α-keratin simply to heat, but to stretch and heat such that S—S break-down occurs at a time when the steep rise of tension in the fiber favors the weakening of salt bridges. Bisulfite in the cold causes extensively the rupture of S—S bonds, but supercontraction does not occur until the fiber is heated in acid, a process which in itself does not initiate supercontraction. And again, the influence of salt bridges is shown in a deaminated fiber which supercontracts at once after treatment with bisulfite.

One of the commonest methods of producing supercontraction is to stretch in very dilute soda and to allow free contraction in water; here the same reagent breaks S—S bonds and reduces the cohesion between side chains.

In the α-type proteins which do not possess interchain covalent bonds, supercontraction can be induced by simpler methods; it is not necessary to stretch the α-form in applying heat treatment or alkali, and, in general, lower temperatures may be employed. The severity of the treatment, however, affects not only the degree of supercontraction but also very decidedly the intramolecular changes as shown by x-ray diffraction. With gentle treatment—65° for myosin films, 85° for alkali-treated keratin—supercontractions of 20 per cent do not produce any marked disorientation of the α-pattern, and analysis of the tension–extension curves shows that the restoring force K is made up predominantly of changes in internal energy $\left(\left(\frac{dU}{dl}\right)_T\right.$ component), the entropy component being only slightly positive for relaxed wool, and even negative for myosin strip[195] (see Fig. 13, sec. VIII-4). The orientation maintained in the α-pattern is taken to mean that α-type proteins are made up of chain molecules acting both in series and in parallel; the crystalline parts which give rise to a diffraction pattern are the last to be deformed under the influence of disorienting forces, and it is the less organized parts that are affected first. More violent heat treatment which can induce supercontractions of 60 per cent in oriented myosin strip, and 50 per cent in keratin, does produce a disorganization of the α-pattern, and the appearance of disoriented β. In such cases, the entropy term at these great contractions is increasingly important.

The question at once arises: Are the changes which are initiated during the first stage of supercontraction continuous or is there in the transition to the disorganized state an abrupt change? Here Rudall's experiments with epidermin are very significant.[196] By suitable heat treatment followed by stretching it is possible to pass from the disoriented β-state to one in which the crystallites are lying transversely to the direction of stretching such that an x-ray beam parallel or perpendicular to the surface but at right angles with the direction of stretching gives a backbone reflection on the meridian. Further stretching will orient these transverse chains to give the conventional β-type pattern. The interesting feature of the cross β-pattern is that it can be induced very readily in muscle simply by warming to 60°, which suggests that the first stage in the change to disoriented β is the enlarging of the small α-fold to give one much larger. Proteins having the cross β-structure can generally be brought back into the α-form by soaking in urea and stretching, and this phenomenon too

(196) K. M. Rudall, *Symposium Soc. Dyers Colourists* p. 15 (1946).

suggests that provided the backbone link is weakened the large β-folds can regain their former α-configuration.

Thus the sequence of events in supercontraction can be visualized in the simplest terms somewhat as follows: The prerequisite is that the chains shall be relatively free, so that under the influence of thermal motion they can take up new configurations. At first, the less organized parts begin to fold, a process which is resisted by the crystalline regions which give the x-ray pattern of α-keratin. There comes a stage at which they too are pulled out of alignment, and if the treatment is sufficiently drastic, the pattern is not only disoriented but gives way to β. The β-pattern is generally also disoriented, but the behavior of muscle and epidermin suggests that an orderly transition is possible through long β-folds oriented transversely to the axis. Ideally, the whole process can be regarded as a transition through various states of folding and not involving large entropy changes; when complete disorganization sets in, then the entropy term has certainly to be taken into account.

Certain quantitative considerations support these conclusions. The maximum disorientation of fully oriented long, thin particles, by calculation, can be no more than 50 per cent of the initial length; yet oriented myosin strips may contract by as much as 65 per cent and *Mytilus* muscle by 75 per cent, and, as we have seen, the first stages of the process occur without any evidence at all of disorientation. These facts would be difficult to interpret on the basis of a rubber-like elasticity, quite apart from the fact that the chain molecules which participate in the process are predominantly polar and would not be expected, by virtue of their interaction, to behave like a rubbery solid.

The general inference from all this work is that there is little evidence to invoke disorientation as having a significant part in contraction, though it is true we have come no nearer the molecular state in physiological contraction than studies of iodoacetate contracture, or of the smooth *Phascolosoma* retractor muscles which can be stretched from the shortest length over a range of 450 per cent without inducing any considerable variation from the oriented α-pattern.

2. Electron Microscope Studies of Skeletal Muscle

Electron microscope (EM) studies have in general confirmed the classical histological features observed on fixing and staining. In addition, they have brought out much fine detail at the macromolecular level of organization, and the information derived has been of rather greater value than that from small-angle diffraction, for which it is often only possible to make any interpretation at all by referring to electron optical features.

A single muscle fiber is supported in a network of fine reticular fibers which have the typical collagen banding, but which differ by their argyrophylic properties. These fibers can be considered part of the endomysium, but it is not uncommon for them to be classed as an intrinsic part of the sarcolemma itself. They can hardly be responsible for the selective permeability of muscle, or for the surface along which the action potential is propagated, and it seems advisable to reserve the name for the membrane which lies beneath the reticulin; this is quite structureless, about 0.1 μ thick, and dotted with corpuscles 400–1000 A. in diameter.[197,198] There is some evidence that beneath the membrane there is a layer of sarcoplasm which joins the myofibrils to it, and this, according to Draper and Hodge, may contain very small fibrils (100 A.) which do not possess the collagen bands. This needs reinvestigation. When the fiber is injured, the contents retract, and the clear elastic envelope, the true sarcolemma, gives negative tests for collagen, but positive for protein and lipide. It has been supposed that the contractile material at the ends of the fiber must be fused to the sarcolemma, so that tension can be exerted on the supporting structures. Barer[198] has disproved this idea by showing that retraction clots are readily produced in this region also.

The fiber contents consist of the contractile material, the sarcoplasm, particulate bodies, fat droplets, and nuclei. The contractile material is organized at two levels; when, for example, the fibers are homogenized, and especially when they are treated with collagenase, small fibrils, for the most part about 1 μ wide, can be isolated in quantity.[145,199] They appear to be the ultimate morphological unit in that they contain no sarcoplasm; but in reality they are composed of still smaller units which in the electron microscope are seen as filaments, running uniformly through both the A and the I band.

The A band in the dried state is much thicker than the I, and in it the filaments appear to be overlaid and obscured with some other material, which for want of better definition has been called the A substance (Plate IV). Its characteristic feature is that in contracted muscle it shows movement into the I band, which sometimes cannot be recognized except for the Z band which bisects it. The nature of the A substance will be discussed later on.

Draper and Hodge,[199] in their beautiful study of muscle structure, gained the impression that fibrils were in reality tubes which on drying collapse to form a ribbon, and certainly there is the appearance of one sheet underlying another. The same impression is gained from Perry and Horne's study[145] of isolated myofibrils (p. 1004). Transverse sections of muscle, however, show that the filaments are packed hexagonally in the fibril forming a solid cross section (see sec. VII-3).[200]

(197) W. M. Jones and R. Barer, *Nature* **161,** 1012 (1948).
(198) R. Barer, *Biol. Revs. Cambridge Phil. Soc.* **23,** 159 (1948).
(199) M. H. Draper and A. J. Hodge, *Australian J. Exptl. Biol. Med. Sci.* **27,** 465 (1949).
(200) C. Morgan, G. Rosza, A. Szent-Györgyi, and R. W. G. Wyckoff, *Science* **111,** 201 (1950).

PLATE IV. (*a*) Myofibrils from muscle in rigor mortis. Chromium shadowed. These fibrils show considerable contraction so that the I bands are shorter than in resting muscle (from Perry and Horne).[145]

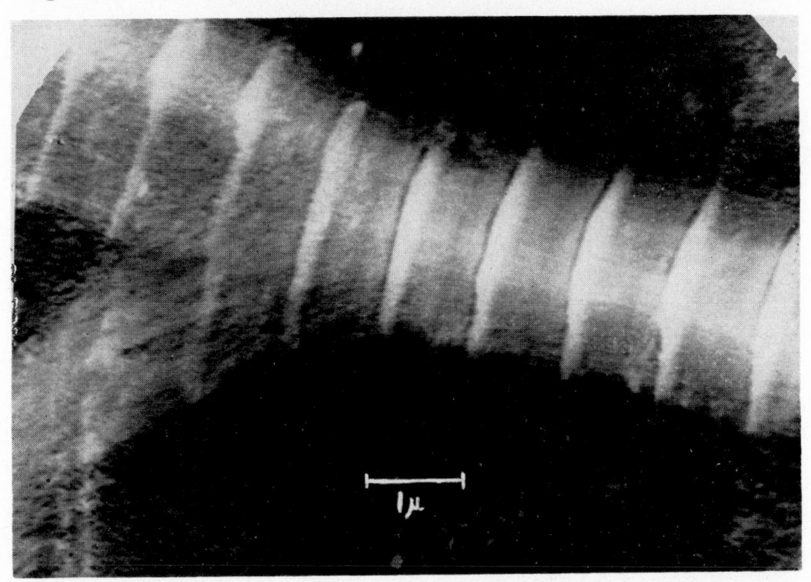

PLATE IV. (*b*) Splayed out myofibril with underlying filaments. Such preparations give the mistaken impression that the fibrils are tubelike.[145]

The Z band, which bisects the I band, possesses an ultrastructure, and in this respect differs from the M. It appears to consist of an accretion of material lying over the filaments, and after staining it can be seen to consist of two lateral subbands, 450 A. wide, enclosing three smaller bands, 100 A. wide. Draper and Hodge have observed that, on the

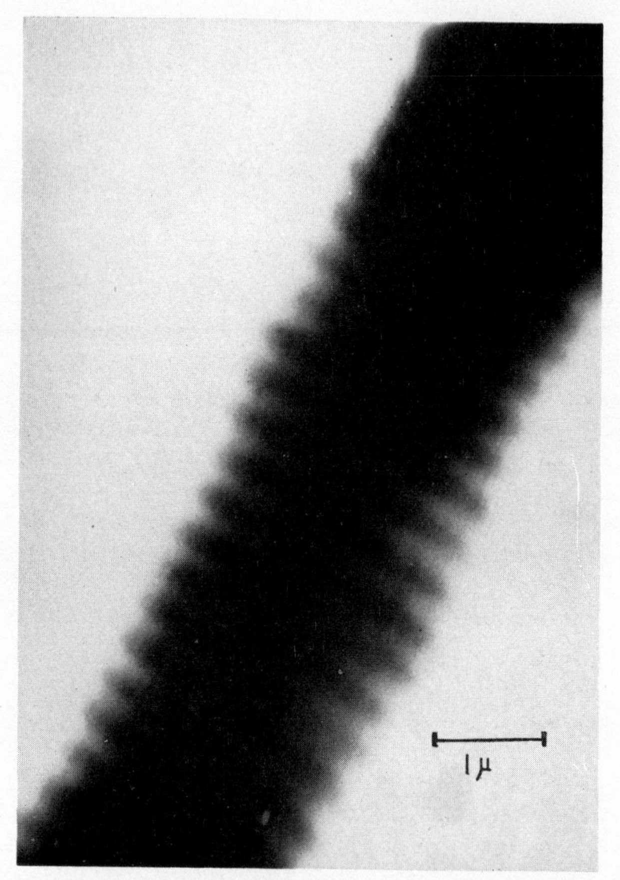

PLATE IV. (*c*) A myofibril treated with 0.0013 *M* ATP solution.[145]

inner surface of an opened-out sarcolemmal sheath, there runs a series of parallel lines, spaced 2 μ apart, which represent the remains of the Z band at the point where they are joined to the innermost membrane, and in these regions on the outside, the reticular fibers appear to reinforce these lines by running parallel; elsewhere, the network is formed by fibers running roughly 45° and 135° to the fiber axis. These observations tend to suggest that in skeletal muscle the Z band acts as a rigid cementing material overlaying the fibrils and reinforced from the outside by inelastic bands of reticulin. It may be that the bulging of a contracted *whole* fiber on either side of the Z membrane is due to these latter; but in a contracted fibril itself, the same thing is seen, as though the Z membrane were a rigid, constricting structure.

The M band, which bisects the A band, may act as a spacer for the filaments. Draper and Hodge gained this impression by finding that, on drying, the filaments were drawn together at some little distance from the M line, leaving holes at the immediate vicinity. Moreover, under the influence of bacterial enzymes, some trace of structural organization remained in the filaments only as long as the M band remained intact.

The nodose appearance of the filaments has been observed by several investigators. The periodicity along the fiber axis is about 400 A., and there is a suggestion, in Hodge and Draper's work, that as the sarcomere length diminishes, so does this repeating dimension. The width of the filament has been given as 50–250 A. (wt. average 130),[201] 150 A. for the toad,[199] and 100 A. for the rabbit.[202] Weber[15] has pointed out that the variations are probably due to the manner of staining, and according as the measurement is made for a single filament or from one filament to another. Hoffmann-Berling and Kausche[203] find in frog muscle that the interstitial space may occupy 70 A. and the actual filament 40 A., making a repeating width of 110 A.; longitudinally, they find the period to be less than do other workers, at about 250 A.

When the filaments are ashed in an intense electron beam, the residual deposit of ash showed again the 400-A. spacing, except at the Z and M bands, which are quite clear.[204] If we assume these bands consist of protein, it must be accepted that the beam really does burn off the protein and that the "ash" is not a deposit of carbonaceous material. This point is raised because the combustion actually occurs in a high vacuum. It suggests that the accompanying *gegen* ions of the protein, potassium or sodium, are also volatilized, and that any deposit, as Draper and Hodge suggest, consists of Ca and Mg. There seems no good reason to assume that these ions are bound by the contractile substance more than by the Z and M bands, and it is rather more likely that they arise from nucleotide material which tends to form fairly insoluble Ca or Mg complexes. Isolated actin itself contains such nucleotide, and so does the

(201) C. E. Hall, M. A. Jakus, and F. O. Schmitt, *Biol. Bull.* **90,** 32 (1946).
(202) G. Rosza, A. Szent-Györgyi, and R. W. G. Wyckoff, *Exptl. Cell Research* **1,** 194 (1950).
(203) H. Hoffmann-Berling and G. A. Kausche, *Z. Naturforsch.* **5b,** 139 (1950).
(204) M. H. Draper and A. J. Hodge, *Australian J. Exptl. Biol. Med. Sci.* **28,** 549 (1950).

intact isolated myofibril, in about the same quantity that one might expect from the content of actin.

To complete the picture of the muscle fiber, we may add to these newer observations other well-established facts:

(*1*) In relaxed muscle, the ultraviolet absorption of the A band compared with the I suggests that ATP is stored in the latter and tends to migrate to the A on fatigue.[205]

(*2*) The A band is more rigid than the I, so that on passive stretching of relaxed muscle the I band lengthens the more; similarly in isometric contraction, the I band may actually lengthen to accommodate the shortening of the A band.[206] In isotonic contraction, the A band contracts more than the I, but in view of the apparent reversal of striation, it is rather difficult to consider the material which formed the I band as remaining morphologically intact in relaxed and contracted muscle.

(*3*) When subjected to fracture, the weakest points transverse to the fiber are at the Z band or at the junction of the A and I.[199]

(*4*) On contraction, the total birefringence of the A band falls, attributable in equal measure to a decrease of form and intrinsic birefringence.[207,208]

It must be admitted at once that the interpretation of muscle ultrastructure in terms of known components is an exceedingly difficult task. There has been a tendency to infer that the nodose appearance of the filaments *in situ* is due to F-actin threads, overlaid, particularly in the A bands, with myosin.[202] The particulate dimension of the F-actin bead, as seen in isolated F-actin, is not very different (300 × 100 A.) from those seen in muscle, which have been variously reported as 250 × 400 A. in length and 50–150 A. in width. In support of this assumption is the fact that in the presence of dissociating agents (ATP or pyrophosphate) the myosin is shed even from coarsely minced muscle, while the long F-actin threads are trapped by the reticular network, and only further comminution liberates them.

Against this interpretation are two serious arguments. First, it is necessary to question whether structures seen in the electron microscope in dried specimens always reflect those found in solution or in the gel state. In the case of tropomyosin, Doty and Sanders[209] find that 10 molecules can polymerize lengthwise without evidence of side-to-side

(205) T. Caspersson and B. Thorell, *Acta Physiol. Scand.* **4,** 97 (1942).
(206) F. Buchtal, G. G. Knappeis, and J. Lindhard, *Skand. Arch. Physiol.* **73,** 162 (1936).
(207) V. von Ebner, *Arch. ges. Physiol. Pflügers* **163,** 179 (1916).
(208) A. von Muralt, *Arch. ges. Physiol. Pflügers* **230,** 299 (1932).
(209) Personal communication.

aggregation, although dried tropomyosin solutions reveal fibrils where this has occurred quite extensively. More relevant, still, is the probability, already discussed in sec. II-3*d*, that F-actin in solution is not a "conventional" polymer at all and we then have to explain why small-angle diffractions of F-actin dried as a film are found in muscle itself. From the available experimental details, which are often meager, it would seem that the muscle specimens so examined were themselves dried, so that a diffracting system of F-actin could arise in this way. In any case, viscosity measurements indicate that the monomeric form of actin is a thin rod, approximately 300 × 25 A., and in living muscle, this, or the dimeric form, is likely to have some special orientation with respect to the myosin particles which we know to be oriented; it too could give rise to a diffraction system.

This view, that actin in muscle is rodlike and no wider than Weber's estimate for the myosin particle, would favor the idea that the filaments consist of an intimate mixture of the two components, and not of a skeleton structure of actin threads on which the myosin is superposed. But how does this formulation accommodate other facts? For example: (*a*) the differences between skeletal and other types of muscle; (*b*) the difference in birefringence between smooth muscle and the A band of striated and between the A and the I bands; and (*c*) the behavior of muscle toward extractants. Concerning (*a*) and (*b*), we have the fact that in a contracted A band, the intrinsic birefringence can and does decrease by one half, so that the isotropy of the I band may simply exemplify less-oriented molecular states; and in comparing the A and I bands, it may be recalled first that Schmidt[210] considers the latter to be weakly anisotropic, and secondly, that this would be greatly increased if the protein content of the I band were as great as that in the A. (We are neglecting for the moment a compensating negative birefringence in the I band, against which there are rather strong arguments.)[15] Item (*c*) is harder to explain. All extracting methods employ salt solutions with the idea of first removing the myosin from the complex, and these presumably give rise to the *very concentrated* thixotropic gels of F-actin which are trapped in the reticular network; but the success of Tsao's butanol treatment[33] in liberating actin into neutral water solutions may indicate that complex formation with lipide retards its extraction, just as it retards the extraction of soluble enzymes immobilized in mitochondria.

It would be palpably unwise to speculate further on the available data, expect to comment on the movement of A substance during contraction and the "reversal of striation." Although most authors are careful not to designate the material as protein, it seems almost certain that it must

(210) W. J. Schmidt, *Z. Zellforsch. u. mikroskop. Anat.* **23**, 201 (1935).

be: Draper and Hodge's preparations were formalin-fixed and thoroughly washed, a process which must have removed extraneous ash and other substances of low molecular weight.[210a]

3. Small-Angle Diffraction (By H. E. Huxley)

The low-angle x-ray diffraction patterns given by dried specimens of muscle have been described by Astbury and his co-workers in Leeds[211] and by Bear[212,213] in America; more recently, the patterns given by wet muscle, both living and after various treatments, have been described by Huxley.[214–216]

Astbury and MacArthur observed a number of meridional or near-meridional reflections from dried frog sartorius muscle which suggested an axial period of about 54 A. Bear showed that the x-ray diagrams of dried muscles from a number of different phyla reveal two distinct types of fibril: mammalian and amphibian, containing only the "type II," and the adductor and retractor from various mollusks, the "type I." The former give the same x-ray diagram as that observed by Astbury and MacArthur,[211] while the latter contain an additional component of axial period 725 A. This was termed paramyosin and is discussed by the main author (K. B.) in the following. Bear also observed some additional reflections in the diagram from the type II fibrils which indicate that the true axial period is 410–420 A. For both types of fibril, reflections corresponding only to certain higher orders of the basic axial period appear in the diffraction diagram of the dried material. These orders obey definite selection rules for reasons which at present are not completely understood: to a first approximation, it may be said that there does exist a subperiod (145 A. for type I and 54 A. for type II fibrils) within the true axial period. A transverse period has also been observed in the dried fibrils,[217] having a value of 250–300 A. for type I, somewhat larger than that seen in the electron microscope.[218]

For type II fibrils, Bear has reported a side spacing of 115 A. A pattern very similar to that given by type II fibrils has been obtained by

(210*a*) For more recent views see Section X.
(211) I. MacArthur, *Nature* **152**, 38 (1943).
(212) R. S. Bear, *J. Am. Chem. Soc.* **66**, 2043 (1944).
(213) R. S. Bear, *J. Am. Chem. Soc.* **67**, 1625 (1945).
(214) H. E. Huxley, *Discussions Faraday Soc.* No. 11, p. 148 (1951).
(215) H. E. Huxley, Ph.D. Thesis, Cambridge University, England, 1952.
(216) H. E. Huxley, *Proc. Roy. Soc.* (London) **B141**, 59 (1953).
(217) C. M. Cannan, Ph.D. Thesis, Massachusetts Inst. Technol., Cambridge, Mass., 1950.
(218) C. E. Hall, M. A. Jakus and F. O. Schmitt, *J. Applied Phys.* **16**, 459 (1945).

Astbury *et al.*[219] from an impure sample of dried actin. The two sets of reflections are listed in Table XIII. Since Astbury has reported that dried myosin does not give any detectable low-angle reflections, it seems very probable that the meridional reflections from type II fibrils arise from the actin component alone. Confirmation of this must wait on the investigation of x-ray diagrams from actin and myosin gels and on a fuller understanding of the muscle diagram itself.

TABLE XIII
COMPARISON OF MUSCLE SPACINGS (MERIDIONAL) WITH THOSE OF ISOLATED ACTIN

Dry actin film, A.[219]	Type II fibrils, A.[213]
	58
	51
26.8	27.2
18.1	18.7
13.5	13.8
10.9	11.2
9.1	9.2
7.9	6.9
6.8	

Huxley, working with living material, has shown that striated muscle from frog and rabbit gives the same meridional reflections as were observed from dried type II fibrils (Table XIV). In addition, the lower orders of the 410–420-A. period are now visible, reflections which are extremely sharp in a meridional direction, indicating that the diffracting system repeats accurately for a distance large compared with the axial period. When living muscle is stretched by amounts up to 40 per cent, neither the spacing nor the intensity of the meridional reflections alters; even after treatment with iodoacetate and stimulated to exhaustion, or after extracting with glycerol, or after passing into rigor, the spacings are still unaltered, although here some moderate changes in intensity are found.

If the meridionally diffracting system is really due to actin alone, then the actin monomers or dimers must form long continuous structures, whether in dried F-actin, or in dried muscle or in physiological muscle. It may be that the formation of "structure" in solutions of F-actin, arising as it does without any evidence of polymerization in the accepted sense, reflects no more than the tendency of actin molecules to arrange

(219) W. T. Astbury, S. V. Perry, R. Reed, and L. C. Spark, *Biochim. et Biophys. Acta* **1,** 379 (1947).

themselves in a very regular manner. It is difficult to see how a loose association of particles could give rise to such a sharp diffraction pattern, but possibly the environment of actin in muscle is responsible for the effect.

Huxley has also observed a number of strong equatorial reflections from wet muscle. In the case of living material these indicate the presence of long molecules (filaments) arranged in a simple hexagonal

TABLE XIV

MERIDIONAL SPACINGS OF LIVING MUSCLE (FROG OR RABBIT PSOAS) AND DRIED FIBRILS TYPE II

Living sartorius muscle, A.[214–216]	Dried type II fibrils, A.[213]
410 ± 20	
210 ± 10	
136 ± 2	
105 ± 5	
83 ± 5	
67 ± 3	
59.34 ± 0.1	58
51 ± 2	51
45 ± 2	
37 ± 1	
28 ± 0.5	27.2
18.8 ± 0.4	18.7
13.5 ± 0.3	13.8
11.5 ± 0.3	11.2
9.3 ± 0.1	9.2
7.9 ± 0.2	
6.95 ± 0.2	6.9
(5.1 ± 0.05)	

array 440 A. apart, and the sharpness of the reflections indicates that the array is maintained over a considerable distance. The relative intensities of the reflections are compatible with a filament diameter of 100–150 A. (i.e., the same order of magnitude as seen in the electron microscope); the space between is occupied by material of lower and relatively uniform density—obviously, water to a large extent. The evidence is against the idea of tubelike structures and suggests that the filaments extend as a continuous hexagonal array across the myofibrils, a conclusion which has been confirmed by electron microscope studies of thin cross sections of muscle (Plate V). The appearance of fragmented material in the electron microscope in which the fibrils are seen as flat ribbons made up of filaments about 150 A. in diameter in contact is believed to be due to

the flattening effect of the large surface-tension forces present during drying.

When muscle passes into rigor, or is exhausted or glycerinated, the hexagonal array of long molecules 440 A. apart is still maintained, but the relative intensities of the reflections differ greatly from those observed on

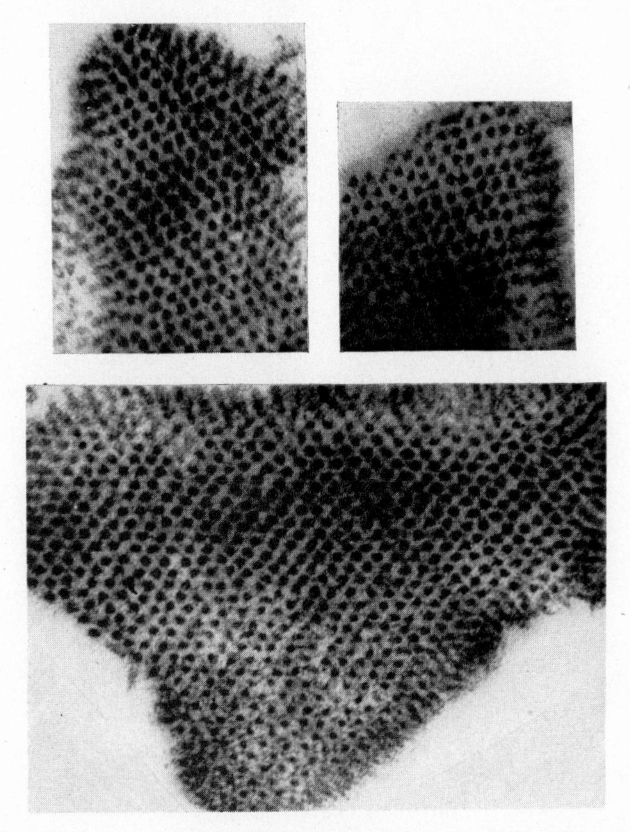

PLATE V. End-on section of rabbit psoas muscle through the H band (middle of the A band), stained with osmic and phosphotungstic acids. Magnification $\times 100,000$. This shows the continuous hexagonal arrangement of the filaments (seen here in section) within the myofibrils. In other parts of the A band a secondary array can be discerned (see Plate VI). (By H. E. Huxley.)

living material, and indicate the formation of regions of high density at specific points between the filaments where originally the density was fairly uniform (Plate VI). The observed intensity changes can be explained if material originally randomly arranged in the liquid separating the filaments becomes fixed so as to form a sixfold secondary array of long molecules around each of the filaments, and here again this has been seen in electron microscope studies of thin sections. It is believed that under these circumstances actin and myosin combine when ATP is removed from the muscle, but it is not clear which of these arrays is myosin and which is actin.

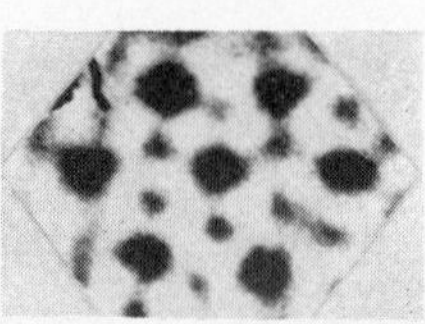

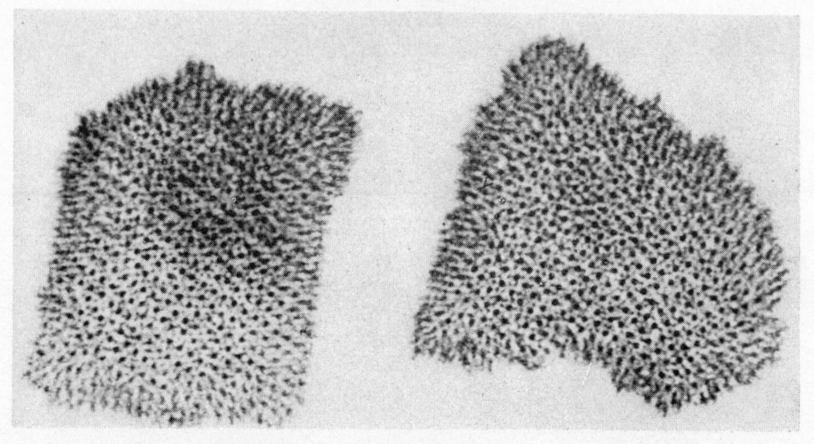

PLATE VI. Glycerinated psoas, stained with osmic and phosphotungstic acids. End-on section through the A band. Magnification ×100,000; inset ×320,000. This shows the twofold array of filaments in the A-band of muscle lacking ATP. The primary set is arranged hexagonally and the secondary set is located at the trigonal points in a sixfold array around each primary filament. (By H. E. Huxley.)

Only preliminary results are available for glycerinated fibers after contraction in ATP; these suggest that the side-by-side arrangement here resembles that found in muscle without ATP at rest length. The meridional reflections, which are weak even before contraction, have not so far been recorded satisfactorily from a contracted specimen.

4. Paramyosin

Hall, Jakus, and Schmitt[218] discovered a most interesting type of fibril in various molluskan muscles (*Venus mercenaria*, *Mya arenaria*, *Mytilus edulis*, *Ostrea virginica*, *Pecten magellenica*), and these were later termed type I fibrils by Bear. The muscles used were chiefly the adductor and foot muscles. Mollusks such as *Venus*, *Anodonta*, and *Pecten* possess both white and pink portions of muscle, of which only the tinted muscle of *Pecten* is striated. By x-ray classification, the tinted portion is most closely related in its small-angle pattern to frog sartorius and dog retractor penis, which in Bear's terminology contain the type II fibrils discussed at length above. The substance of the type II fibril has been called paramyosin.

To obtain type I fibrils, the muscle is ground in 0.3 *M* KCl and centrifuged, when the layer above the coarse debris is found to contain long, needle-shaped fibrils which dissolve completely in 0.6 *M* KCl. The fibrils range in length from 1 to 40 μ and are 200–1000 A. in width. When stained at pH 5.5 with phosphotungstic acid, they reveal in the electron microscope an extraordinary pattern of spots whose centers have the geometrical properties shown in Fig. 12. The separation of spots

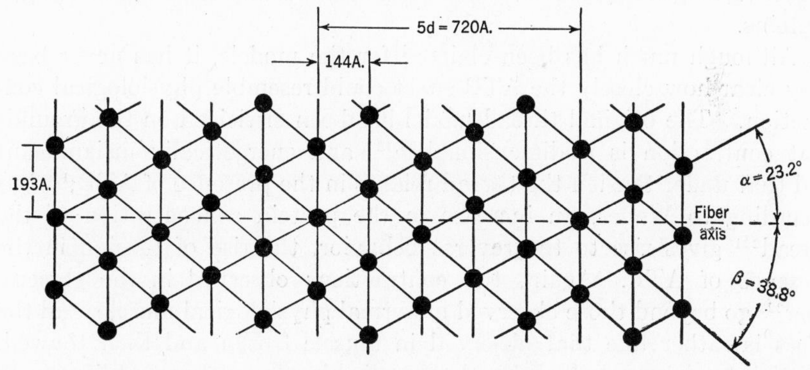

FIG. 12. Diagrammatic lattice obtained by staining the paramyosin fibril with phosphotungstic acid.[210a]

transverse to the fiber axis is 193 A. and along the axis 145 A. The angles subtended by other planes in relation to the long axis are not equal (28° and 39°), so that the full repeat along the axis is a multiple of 145, actually 720 A. In a later x-ray study, Bear found that the intense meridional spots gave a fiber axis period of 145 A., and lines indicated a true fiber period of 725, in excellent agreement with electron microscope data. The separation of spots along the layer lines suggested a lateral period of 200–325 A.

In contracted muscles, these fibrils show no apparent change, as though they do not form part of the contractile system; their solubility properties are closer to an actomyosin complex than to either protein separately. The significance of the staining areas in relation to the structure or the components of the fibril is likewise unknown. Working with *Ostrea*, the author gained the impression that paramyosin is not just one component of the adductor but the whole of it. It was found difficult to extract much soluble protein, and such as did appear gave very little viscosity response to ATP, and the ATPase activity was very feeble. Further chemical work is very desirable.

Lajtha[220] has also reported a preliminary study of the extracted actomyosin of invertebrate muscle.

VIII. Models of Muscle Contraction

Molecular studies of (acto-)myosin by Muralt and Edsall encouraged Weber to study oriented myosin threads as a model for the contractile material, and, in the same way, the newer information on actin has led to new types of model. The simplest is again the extruded actomyosin thread, but much work has been done on rabbit psoas fibers dehydrated at resting length in glycerol, which removes ATP and some of the myogen proteins.

Although much has been claimed for the models, it has never been very clear how closely the ATP effect could resemble physiological contraction. The original thread model has been criticized on the grounds that contraction is isodimensional[195,124] and energetically insignificant and that under tension the thread relaxes in the presence of ATP;[119] but according to Weber, an increase in the protein concentration of the thread[221] gives rise to the reverse behavior, the rise of tension in the presence of ATP. Again, the contractions observed in the glycerol fiber[222] go beyond those observed in normal physiological muscle, and the effect is rather like that observed in muscle frozen and then thawed. Another model we shall add is that studied by the author and Marsh, in which changes in the fibers of a minced muscle are observed as a function of time and ATP content.

1. The Behavior of Muscle Fibers after Mincing[223,224]

This study started from the familiar observation that when fresh muscle is minced it consists at first of a tacky paste from which no juice can be expressed, but after a short time, much less than 30 min., it becomes wet, and simple hand pressure suffices to squeeze out the sarcoplasm. That the change is not caused by the change of pH as glycolysis proceeds is proved by mincing in the presence of iodoacetate.

The change can be studied semiquantitatively by using a gravitational force to express the juice from a mince which has first been added to 4 parts of 0.16 *M* KCl, the volume of the fiber phase being followed

(220) A. Lajtha, *Publ. staz. zool. Napoli* **21**, 226 (1949).
(221) H. Portzehl, *Z. Naturforsch.* **6b**, 355 (1951).
(222) A. Szent-Györgyi, *Biol. Bull.* **96**, 140 (1949).
(223) K. Bailey and B. B. Marsh, *Biochim. et Biophys. Acta* **9**, 133 (1952).
(224) B. B. Marsh, *Biochim. et Biophys. Acta* **9**, 247 (1952).

during the actual centrifugation. For about 20–30 min., the volume of the fibers—on an average about 0.8 mm. long—does not change except for simple packing effects, but suddenly during a period of 2–5 min. duration the volume of the fiber phase diminishes by 20–40 per cent of its original value and then stays constant. The quicker the deposition, the greater the volume decrease, and the curve relating the two is a rectangular hyperbola, reminiscent of the classic relation between velocity and extent of shortening in whole muscle. For convenience, the phases will be called prerapid, rapid, and postrapid. The onset of the rapid phase is determined solely by the level of ATP and becomes evident when this falls to about 10–20 per cent of its original value. The addition of more ATP at the postrapid phase enables the whole cycle to be repeated.

With homogenates of rigor muscle the response to ATP is variable; sometimes the volume increase can be observed, sometimes very little volume change, and sometimes an intense and irreversible volume decrease, in extreme cases as much as 40 per cent. This latter response could always be induced at the postrapid phase by replacing the supernatant with 0.16 *M* KCl containing ATP, and the change is here so rapid that it cannot be followed, and the final volume may be no more than 10–40 per cent of the original.

The effect obtained in the absence of sarcoplasm resembles in every way the response of the thread and fiber model, while the reversal in the presence of ATP at first sight suggests that ATP can induce something akin to relaxation. A clue to the difference in behavior in the presence and absence of sarcoplasm was that the production of inorganic phosphate in the latter case was only $\frac{1}{10}$th that in the former; in other words, the very intense and irreversible shrinkage is accompanied by the rapid splitting of ATP which in whole muscle and whole brei is conserved by a factor of unknown nature (probably a protein) which inhibits myosin ATPase.

In the microscope, Marsh has observed during the rapid phase an over-all shortening and volume decrease; the reversal by a lengthening and a volume increase. The volume changes in themselves do not of course reflect the behavior of intact muscle, but it may well be that any increase of pressure involved in contraction may tend to squeeze out liquid from the cut ends of the fiber and through the injured sarcolemmal membranes. That the fiber segments are very permeable is shown by the dramatic effect of ATP in producing the immediate and irreversible form of syneresis. The relaxing effect too may arise from the straightening out of the connective tissue sheath, and is very unlikely to represent an active form of relaxation.

When the level of ATP falls and the rapid phase is over, a volume increase can be obtained only for a concentration of ATP above a threshold of about 0.07 mg. labile P/g. muscle. The addition of less than this amount, e.g., 0.02 mg./g. muscle, causes further syneresis. In the presence of the inhibiting factor, the breakdown of ATP by the fiber proceeds at the rate of 0.04 mg. labile P/min./g. muscle, and the time taken to attain the full volume increase is itself about 1 min. These figures suggest that reversibility is encouraged by a concentration of ATP which is free to dissociate the component proteins actin and myosin. In solution, Mommaerts finds that 1 mole of ATP is able to dissociate 100,000–300,000 g. of myosin in complex formation,[225] and assuming that there are 80 mg. myosin/g. muscle, the minimum amount of ATP to effect dissociation is calculated as 0.02–0.05 mg. labile P. Thus, at the time when the phase of increasing volume is over, and when the fibers begin once more to shorten, the amount of ATP remaining (approx. 0.03 mg. labile P) is barely sufficient to allow dissociation.

From the nature of the phenomena, and from the quantitative aspects, it becomes clear that ATP has two effects, one associated with its mere presence, i.e., its ability to dissociate actin and myosin, and one which depends upon its breakdown. A muscle contracts if two conditions are fulfilled: that ATP is broken down and that insufficient ATP is present to dissociate the components. In resting muscle, breakdown is slow, and the high level of ATP is more than sufficient to maintain a steady concentration at the site of breakdown. As long as ATPase–inhibitor is present, a contraction effect can only be evoked by depleting the muscle of ATP, as in contractions associated with rigor mortis (see the following), or the contraction of segments of fiber studied above. The contraction of physiological muscle must on the one hand be achieved by stimulating the rate of ATP breakdown so that the combination of actin and myosin is quicker than the relaxing action of ATP which diffuses in; the activating process itself can be regarded as a release of the inhibition by the factor which retards ATP breakdown in resting muscle. Relaxation could simply be the reverse of this process, the restoration of the inhibition and the dissociation of actin and myosin.

In the dual action of ATP we see a mechanism which could explain the great differences in the activity of various types of muscle. In smooth muscle, the cycle of contraction and relaxation is automatic, the lower ATP content suggesting that a single contraction might deplete the muscle of ATP to an extent whereby relaxation must await glycolytic resynthesis. Amongst fast muscles, the evolutionary tendency has been to raise the ATPase activity of the fibers and to localize much larger amounts of ATP in definite segments, the I band. The first development

(225) W. F. H. M. Mommaerts, *Biochim. et Biophys. Acta* **4**, 50 (1950).

increases the speed of contraction and the second encourages immediate relaxation as the A material merges with the I.

As yet the Marsh factor has not been purified, although his and subsequent work[226] have pointed to its protein character. In its absence, the rate of ATP breakdown is greatly accelerated, and the intense syneresis now observed in fiber segments is irreversible and associated with severe fibrillar disorganization; it can in fact be compared with the delta state induced by invoking large contractions by maximal stimulation.

The significance of these results are of interest in relation to the cause of rigor mortis. Erdös[227] and later Smith and Bendall[228,229] showed that the characteristic hardening of muscle is associated with a decrease in ATP content. One could assume that at low temperatures, breakdown—either by the resting fibers or by granule ATPase—is slow, and when a critical ATP concentration is reached, actin and myosin combine. But the muscle does not contract because a rather feeble tendency to do so is acting against the intact connective tissue elements. At higher temperatures (37°), however, an irreversible contraction is observed,[230] and this may be due to the greater rate of breakdown or to the destruction of the Marsh factor or both. Bendall[230] has shown that the story is rather more complex than this, and that the muscle remains flexible so long as the rate of ATP resynthesis is much higher (90 per cent) than the rate of breakdown; this probably means that a certain excess is necessary to diffuse from the glycolytic to the breakdown sites.

2. The Thread and Fiber Models

Two other models have come into extensive use, the first of which is a development of Weber's early experiments with the "myosin" thread. The discovery that such threads undergo profound change with ATP, as described in earlier sections, has reoriented entirely the manner of investigation, and Weber and Portzehl have perfected the model by greatly increasing the protein concentration from about 2.5 per cent (that of the Szent-Györgyi thread) to 6–10 per cent.[15] This is effected by incorporating glycerol into the actomyosin sol before extrusion;[221] the thread can then be dried at low temperature, the glycerol maintaining uniform protein concentration throughout and defining also the final concentration.

(226) J. R. Bendall, *Nature* **170,** 1058 (1952).
(227) T. Erdös, *Studies Inst. Med. Chem. Univ. Szeged* **3,** 51 (1943).
(228) E. C. B. Smith and J. R. Bendall, *J. Physiol.* (London) **106,** 177 (1947).
(229) E. C. B. Smith and J. R. Bendall, *J. Physiol.* (London) **110,** 47 (1949).
(230) J. R. Bendall, *J. Physiol.* (London) **114,** 71 (1951).

The glycerol fiber[222] is prepared by taking a small bundle of muscle fibers from a fresh rabbit psoas muscle, and dehydrating at post-excision length in 50 per cent glycerol at low temperatures. The ATP and some 50 per cent of the sarcoplasmic proteins are lost in the process. For use, the muscle is transferred to 15 per cent glycerol and there dissected into a few fibers 4–8 cm. in length; but for precise quantitative measurements, Weber's school has preferred to use single fibers in which the diffusion path is small enough to uphold a constant ATP concentration throughout, in spite of the ATPase activity of the fiber.

Both these models shorten with ATP and are suitable for the measurement of tension isometrically as a function of temperature and ATP concentration. The tension of a resting muscle, or of either of these models in the absence of ATP,[15,231] shows only a slight increase as the temperature is raised—about 12 per cent of that in the presence of ATP—and in all probability is due to a reversible thermoelastic equilibrium. But how can we regard the *active* state of the fibers from a thermodynamic point of view? Szent-Györgyi[114,222] has made the assumption first that the fiber–ATP model is the essential contractile machinery stripped of its excitation mechanism, and second, that the parts are made up of units ("autones") which work on the all-or-none principle, i.e., the extent of contraction as a percentage of the maximal is numerically equal to the number of units involved. This concept would seem to spring from the belief that actomyosin in the presence of ATP is unstable, loses its charge and hydration, and passes to another modification with less free energy; implicit in the theory is that the *adsorption* of ATP is responsible for contraction, and splitting for relaxation. Now it is true, as Varga[232] has shown, that there is an almost complete reversibility of tension for fibers in ATP solutions as one passes from 0° to 15° and back again; also, that the tension at a given temperature can be reached from above and below. In the absence of other evidence, this could be taken as proof of a truly reversible thermoelastic equilibrium; but it could be due also to a chemical reaction, the splitting of ATP, which is itself temperature dependent. In this case, changes of tension would arise from a steady state which shifts as the temperature changes. This view, which Weber[15] has explored in great detail, is in line with conclusions reached in the preceding section.

3. Mechanical Behavior of the Models

At 18°, a single rabbit psoas fiber (glycerol treated) develops a tension of 4 kg./sq. cm., 20 times that of an actomyosin thread, and equivalent

(231) A. Weber, *Biochim. et Biophys. Acta* **7**, 214 (1951).
(232) L. Varga, *Enzymologia* **14**, 196 (1950).

to the maximal tension found in mammals. The tension becomes zero at 20 per cent of the initial length for the fiber model and at 40 per cent for the thread.[231] If a calculation is made of the mechanical work performed by a fiber from the diminution of tension with length, a value of 3–4 $\times 10^{-2}$ cal./cc. is obtained, rather greater than that for a muscle in maximal tetanus.

A. Weber[231] has related another property of the models to physiological muscle. A frog muscle in isometric tetanus, if allowed to shorten 10 per cent, will momentarily lose its tension and then regain it—the quick-release effect of Gasser and Hill.[233] In the models, a release of 5 per cent suffices to do the same, but the recovery period is much slower.

The rigidity of the models in the absence of ATP, compared with their plasticity in its presence, can be shown in several ways. A sudden stretch will increase the tension much more in the former case, and, subsequently, the decay in tension is much less. Moreover, when at a given tension ATP is washed out of a fiber, 70 per cent of the tension remains, and now a very small decrease in length suffices to abolish the tension, and in this case no quick-release effect is observed.

4. Behavior of the Models as a Function of ATP Splitting and ATP Concentration

Strong circumstantial evidence that ATP breakdown is the necessary condition for contraction comes from the finding that there is approximately the same optimal concentration of ATP both for splitting and for development of tension. In obtaining these results, the precaution was taken of measuring initial velocity rates of splitting for models in comparable states of shortening. At 2° and 20° the ATP optima for development of tension are $10^{-2.25}$ and $10^{-1.65}$ M, respectively, and for ATP splitting $10^{-2.3}$ and $10^{-1.7}$ M; on either side the optimal, the curves are not symmetrical, but nevertheless follow the same parallel course for splitting and tension.[15,234]

In earlier sections, it has been emphasized that there is much direct evidence from inactivation or inhibition studies that decreased enzymatic activity diminishes syneresis in the gel or contraction in the model fiber. Weber and his school[15,235] have extended this direct kind of evidence by studying the effect of ATP on the models in the presence of other polyphosphates. These confer on the rigid actomyosin gel a plasticity which it would normally possess in the presence of ATP. Thus, if in an isometric experiment with a thread or fiber the ATP is washed out, 70–80

(233) H. Gasser and A. V. Hill, *Proc. Roy. Soc.* (London) **B96,** 398 (1924).
(234) A. Weber and H. Weber, *Biochim. et Biophys. Acta* **7,** 339 (1951).
(235) H. Portzehl, *Z. Naturforsch.* **7b,** 1 (1952).

per cent of the tension remains; if ATP is washed out in the presence of sodium pyrophosphate, only 30 per cent remains. If a thread is held in sodium polyphosphate, and ATP is added, tension will develop, and after washing out the latter the tension will fall. Another rather striking experiment can be carried out with an ATPase inhibitor, Salyrgan (mersalyl, an organic mercurial). A fiber is allowed to develop tension by addition of ATP; if this is then washed out, much of the tension remains even on addition of Salyrgan, which prevents the splitting of ATP but not its plasticizing effect. The further addition of ATP thus causes the *decay* of tension, an effect reversed by cysteine.

While these effects have been dealt with only summarily, they support in a general way the conclusions drawn in sec. VIII-1. A few comments may be added, however. The idea was developed there that contraction could be regarded as a competitive process between splitting and plasticizing effect, and that it occurred *before* the latter was abolished. It would be interesting to know whether polyphosphate in much higher concentration would prevent the development of tension on addition of ATP. The experiment with Salyrgan, moreover, would be opposed to the view that the ATPase centers are necessary for the formation of actomyosin, though it was noted that the plasticizing action also was affected but rather more slowly than the presumed ATPase activity. It is possible, however, that Salyrgan does not poison ATPase readily until ATP is added, when the decay of tension would be the resultant of tension increase in competition with increasing inhibition of ATPase and its accompanying effect on dissociation. The slow time course of tension decay rather supports this explanation. Incidentally, Kuschinsky and Turba[93] found that Salyrgan destroys the characteristic properties of F-actin and prevents its formation from G-actin, so that any rapid action of Salyrgan should affect the behavior of the fiber on this account alone.

5. Some Thermodynamical Considerations

For a reversibly elastic system at constant volume in which the number of particles is constant, the restoring force K is composed of two terms:

$$K = \left(\frac{\Delta U}{\Delta l}\right)_{T,p} - T\left(\frac{\Delta S}{\Delta l}\right)_{T,p} = \left(\frac{\Delta U}{\Delta l}\right)_{T,p} + T\left(\frac{\Delta K}{\Delta T}\right)_{l,p}$$

where U = internal energy and S = entropy. The magnitude of each term in relation to K tells how much of the mechanical force is due to entropy change and how much to internal energy, whether, in fact, contraction involves a randomization of particles as in rubber, or an orderly folding. In the discussion on the properties of dried actomyosin film, it was noted that the entropy contribution in a "generalized" wool

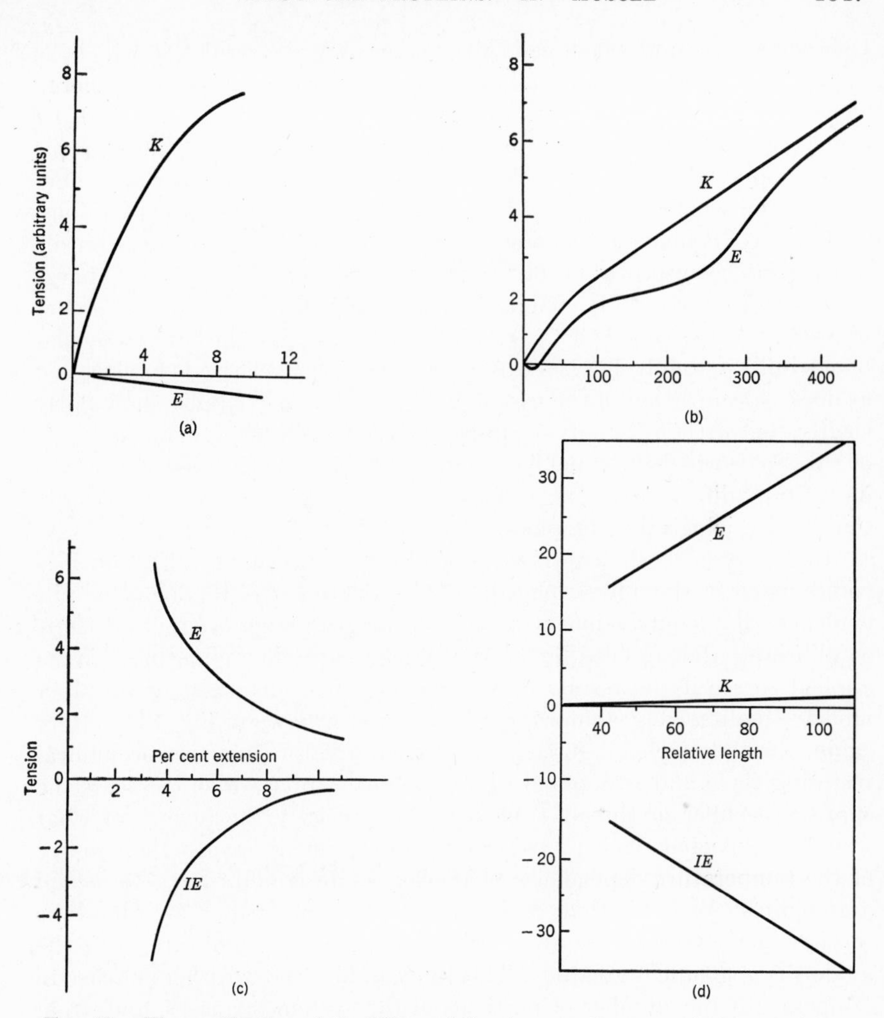

FIG. 13. The application of the Wiegand-Snyder equation to various materials:
(*a*) dried actomyosin film rewetted[195]
(*b*) rubber latex strip[199]
(*c*) undried actomyosin thread[226]
(*d*) Glycerol fiber in the presence of ATP[224]
E = entropy contribution; IE = internal energy contribution; K = mechanical restoring force.

fiber is small, and in an oriented myosin strip even slightly negative[194a,b] (Fig. 13*a*). It is instructive to compare these two cases with that of a rubber strip (Fig. 13*b*), in which the entropy contribution is large.

Such films of protein very probably differ from undried actomyosin threads by the presence of large internal cohesions built up during drying.

Experiments on undried threads give rather different results, in which, as the thread shortens, there is a gain of entropy, while the internal energy component remains negative, approaching zero at extensions of 10 per cent (Fig. 13*c*).[236] In other words, as the thread shortens, the entropic force is opposed by an extensile force which is said to arise from the net charge, and which can be modified by adsorption of cations or of anions, of which ATP holds a prominent place. The existence of an extensile force seems very probable, but that it is determined by the net charge seems unlikely (sec. VI). Morales and Botts consider that the difference between their results and those of Woods[194b] is that the intermolecular bonds built up on drying actually invert the thermoelastic behavior, just as does vulcanization in the case of rubber. These examples show fairly clearly that little information can be derived by thermodynamic analysis of this kind, when the same material, according to treatment, gives fundamentally different results, and these may change still more in a system that is physiologically complete.

In this respect, the contraction of fibers or threads in ATP solutions comes nearest the physiological, and neglecting for the moment the evidence that contraction is a steady state and not a thermoelastic equilibrium, the application of the Wiegand-Snyder equation can be applied to see if the results make sense.[234] The diagram for the fiber model—qualitatively similar for the thread—is shown in Fig. 13*d*. Here again, the thermoelastic force is very much greater than the mechanical restoring force and is opposed by an extensile force which *decreases* the shorter the fiber or thread. Since this is precisely the reverse of what would be expected, Weber[15] treats it as an argument against the validity of the temperature dependence of tension as an equilibrium. A second argument by Weber is that, unlike a perfect rubbery solid, for which the tension increases by $\frac{1}{273}$ (β) per degree rise in temperature, β in the models is very much greater. This must imply that with an increase in temperature the number of particles in the system increases, and, from all the evidence of the effect of ATP as a plasticizer, this number must be affected by the concentration of ATP, increasing to an optimum for any one temperature. On this view, the largest rise in tension should be shifted to higher temperature ranges as the concentration of ATP increases, since thermal motion should assist the ATP in diminishing cohesive forces; but in fact, as the ATP concentration diminishes, the steepest rise in tension over a temperature range of 0–20° lies in the region 0–10°.

All the evidence so far thus favors the rejection of the Wiegand-Snyder equation as having any direct applicability in the case of active

(236) M. Morales and J. Botts, *Discussions Faraday Soc.* No. 13 (1952).

models in the presence of ATP. If the driving force of contraction is derived directly from the topochemical reaction of protein and ATP (say by the formation of a very unstable S—P link) the heat derived from splitting is overwhelmingly greater than the work equivalent to give maximal tension. Some part of this heat could be consumed in deriving the configuration which characterizes the contracted state.

The heat evolved in a single twitch comprises the activation heat, which is constant and associated with the sudden change from the inactive to the active state, and the heat of shortening, which is independent of the work done and is proportional simply to the extent of shortening.[237] We have seen that for shortened fibers, the optimal ATP concentration for tension and splitting is the same, and the optimum for tension becomes less the shorter the fiber. If the same is true for splitting, then we have perhaps, as Weber suggests,[238] a dependence of heat upon shortening which might be related to the heat of shortening. Thus it seems that the inseparability of contraction and splitting, based as it is on sound experimental evidence, holds promise for the explanation of some very difficult aspects of muscle behavior.

In defining the process in more detail, and pursuing the idea of topochemical reaction, one is led at once to the historic concept of Meyer and Mark,[239,240] that changes of charge induce change of shape. While this is possible, we have already noted (sec. VI) how great is the total charge, and any modification in it must work not only in relation to electrostatic forces but to hydrogen bonds which, in the Pauling-Corey α-keratin model, for example, confer great configurational stability. However that may be, most authors,[241,242] when considering the possibility of a process such as phosphorylation, have favored the idea that it occurs as the agent of relaxation increasing the extensile force by increasing the net negative charge. Quite apart from the evidence outlined for the models, and for the independent evidence of Hill[243] that relaxation is a passive process, such a postulate does not seem necessary. The reaction could occur with —SH groups in regions of predominantly positive charge, or it might act, not by the formation of covalent links, however transient, but by deionizing an adjacent amino or imidino group. In spite of all such considerations, it cannot be emphasized too strongly, particularly in the

(237) A. V. Hill, *Proc. Roy. Soc.* (London) **B136**, 195 (1949).
(238) H. H. Weber, *Nature* **167**, 381 (1951).
(239) See K. H. Meyer, *Biochem. Z.* **214**, 253 (1929).
(240) K. H. Meyer, *Experientia* **1**, 361 (1951).
(241) J. Needham, S.-C. Shen, D. M. Needham, and A. S. C. Lawrence, *Nature* **147**, 766 (1941).
(242) J. Riseman and J. G. Kirkwood, *J. Am. Chem. Soc.* **70**, 2820 (1948).
(243) A. V. Hill, *Proc. Roy. Soc.* (London) **B136**, 420 (1949).

light of recent work, that the role of ATP in the modification of the state of the actin component is an enormous complication which will be discussed in the final section.

IX. Biological Activities Associated with the Sarcoplasm

For completeness, we must note the many individual proteins which make up the press-juice of muscle. These in many cases are not peculiar to muscle and will not be considered in detail. The procedures for purification or crystallization are to be found in the references of Chap. 1 of Vol. I.

It is likely that the entire protein of muscle juice consists of enzymes, some in quite large amounts. Triosephosphate dehydrogenase, crystallized independently by Bailey[244] and by Baranowski[245] (who were unable to specify its activity) occupies about 10 per cent of the myogen fraction,[246] creatine phosphokinase about 10 per cent,[247] phosphorylase 2 per cent, and the aldolase–isomerase (zymohexase) system 5 per cent.[248] These five enzymes together occupy about 10 per cent of the *total* muscle protein.

Two general methods have been applied to the fractionation of the proteins of muscle-juice, first by Askonas,[249] who employed organic solvents at low temperature, and by Distèche,[250,251] who precipitated various fractions at constant ionic strength by varying the pH. This procedure was repeated at different ionic strengths, and five proteins were obtained crystalline. Some enzymes (e.g., myokinase) can be partially purified by making use of their stability to acid or to heat or to both.[252]

Many enzymes in muscle are associated with the mitochondria: all those of the tricarboxylic acid cycle and those responsible for the oxidation of fatty acids. Some of these, nevertheless, are water soluble (e.g., fumarase, malic dehydrogenase), and it is not clear to what extent these are also distributed in free solution, if at all. The list given below contains nearly 50 enzyme activities which have been found to be associated with muscle tissue.

(244) K. Bailey, *Nature* **145**, 934 (1940).
(245) T. Baranowski, *Z. physiol. Chem.* **260**, 43 (1939).
(246) R. Caputto and M. Dixon, *Nature* **156**, 630 (1945); G. T. Cori, M. W. Slein, and C. F. Cori, *J. Biol. Chem.* **159**, 565 (1945).
(247) B. A. Askonas, Ph.D. Dissertation, Cambridge Univ., England, 1951.
(248) D. Herbert, H. Gordon, V. Subrahmanyan, and D. E. Green, *Biochem. J.* **34**, 1108 (1940).
(249) B. A. Askonas, *Biochem. J.* **48**, 42 (1951).
(250) A. Distèche, *Biochim. et Biophys. Acta* **2**, 265 (1948).
(251) A. Distèche, Thesis, Liége, 1951.
(252) S. P. Colowick and H. M. Kalckar, *J. Biol. Chem.* **148**, 117 (1943).

Mitochondrial (Granule) Enzymes

Aconitase	Cytochrome c
Cytochrome c reductase	Cytochrome oxidase
Diaphorase	Condensing enzyme for oxalacetate and acetyl coenzyme A (CoA)
Fumarase	Isocitric dehydrogenase
α-Ketoglutaric oxidase	Malic dehydrogenase
Oxalosuccinic decarboxylase	Pyruvic oxidase
Succinic dehydrogenase	

Glycolytic Enzymes

Aldolase, Enolase, α-Glycerophosphate dehydrogenase

The Kinases: arginine phosphokinase (invertebrates), creatine phosphokinase, fructokinase (ketohexokinase), glucokinase, myokinase, 1-phosphofructokinase, 1-phosphoglucokinase, 3-phosphoglyceric phosphokinase, pyruvic phosphokinase.

Lactic dehydrogenase, Phosphoglucomutase, Phosphoglyceraldehyde dehydrogenase, Phosphoglyceromutase, Phosphohexoisomerase, Phosphorylase, Phosphotrioseisomerase, Prosthetic-removing enzyme (for phosphorylase).

Other Enzymes

Adenosine deaminase	Leucine aminopeptidase
5-Adenylic deaminase	Lipoxidase
Amylose isomerase	Phenylalanine decarboxylase
β-Hydroxybutyric dehydrogenase	Phosphoacylase
Choline esterase	Prolinase
Glutamic dehydrogenase	Prolidase
Glutamic–pyruvic transaminase	Ribonucleic phosphorylase
Glutamic–oxalacetate transaminase	

X. Appendix and Conclusions

Since the bulk of the present chapter was written, there have appeared some important papers which may greatly influence present concepts of contraction mechanisms. These are most conveniently treated under separate headings.

1. A and I Bands During Contraction and Stretch

Mention has been made (p. 1011) of the Hasselbach-Schneider procedure[128] by which myosin and actin are selectively extracted. Both Hasselbach[253] and H. E. Huxley and Hanson[254] have followed the appearance of the myofibrils during the extraction and find that the loss of myosin causes the A band to simulate the I, except that the H zone remains rather denser than the rest. On prolonged extraction this too

(253) W. Hasselbach, *Z. Naturforsch.* **8b,** 449 (1953).
(254) H. E. Huxley and J. Hanson, *Nature* **173,** 971 (1954).

becomes less dense. It would really appear therefore that myosin is located only in the A band (cf. pp. 1032–3), whilst the actin component occupies the I and extends into the A band. (There remains the possibility that the I band could contain a nonfibrous protein which washes out from the actin filaments during preparation of the myofibrils.)

H. E. Huxley[254] now identifies the hexagonal array of filaments 110 A. wide and spaced 440 A. apart as myosin (see p. 1036). These are continuous from end to end of the A band, whilst the actin filaments, which form the secondary array, extend from the Z lines into the A band and terminate either side of the H zone. In the H zone itself, the filaments are thicker than elsewhere, and though they are assumed to consist of myosin, this has not definitely been proved.

Perhaps the most interesting advance has been made by following the relative lengths of A and I bands on stretch and after stimulation. A. F. Huxley and Niedergerke[255] have studied by interference microscopy the changes in band length of single isolated frog muscle fibers during passive stretch and isotonic contraction, whilst H. E. Huxley and Hanson[254] have followed with the phase-contrast microscope the behavior of isolated myofibrils after preparation from glycerated rabbit psoas. In passive stretch of an intact fiber or of a myofibril in the presence of plasticizer, the A bands stay at constant length whilst the I bands extend, and likewise during isotonic contraction, the A bands remain constant and only the I bands shorten, such that, at 65% of the resting length, the A bands meet at the Z line. In isometric contraction, there is little change either in A or I bands, but the H zone, ordinarily lighter at resting length, becomes more dense. When myofibrils are stretched in the "frozen" or rigor state, the actin filaments are unable to slip out of the A bands as they do in plastic stretch, and the resistance to stretch appears to be taken up by the Z material.

Huxley and Niedergerke have pointed out that if contraction can be initiated only in the areas of overlap between the actin and myosin filaments, then the tension developed isometrically would be expected to diminish linearly as the points of contact are reduced by passive stretch; this indeed is in accordance with fact.[256]

Superficially, these investigations suggest that the fundamental molecular changes, though they may be actuated by the myosin, are associated predominantly with the actin, although it is true that significant structural modifications are wrought in the A band, particularly in the H zone. Yet before we discard too readily old viewpoints, some recent work from Engelhardt's laboratory focuses once more upon myosin

(255) A. F. Huxley and R. Niedergerke, *Nature* **173,** 973 (1954).
(256) R. W. Ramsey and S. F. Street, *J. Cellular Comp. Physiol.* **15,** 11 (1940).

as the essential agent of contraction. Kafiani and Engelhardt[257] find that fibers of actin-free myosin prepared from surface films by Hayashi's method[258] will contract anisodimensionally at pH 9 in the presence of ATP, but not at all to the same extent at pH 7. Ashmarin[259] has shown that myosin itself, if stained with certain dyes, will undergo with ATP the syneresis typical of actomyosin, and it is suggested that actin modifies the charge on the myosin such that its ATPase optimum at pH 9 is diminished to 7, the pH at which most work on model systems has been carried out. In bringing back the emphasis to myosin and ATPase, it should be mentioned that Munch-Petersen[260] has provided direct experimental proof of the increase in the ADP content of muscle during the rising phase of a twitch, using tortoise muscle at 0° and at 20°.

It is clear from all the considerations discussed in the body of the chapter and above that no clear or unconflicting picture of molecular events during contraction can yet be proposed.

2. Relaxing Factors

Goodall and A. G. Szent-Györgyi[261] and Lorand[262] have considered the creatine phosphokinase system as controlling the relaxation of muscle. With the glycerinated fiber, they were able in the presence of ATP, magnesium ion, and creatine phosphate to obtain contraction effects at pH 6.6 and relaxation effects at pH 5.7. A change of pH will affect not only the rate of splitting of ATP by ATPase but will also influence the supply of ATP by the Lohmann enzyme, the formation of ATP being favored at the lower pH value. Whilst therefore the general arguments involving the dual role of ATP can be invoked in the present case, there seems to be some evidence that ATP produced actively by phosphorylation is more effective than the same concentration of ATP simply added. Perry[263] has found that a threshold concentration of ATP which just suffices to cause the intense contraction of isolated myofibrils can be reduced by 90–99 per cent if introduced via the Lohmann reaction.

An extension of Marsh's work[224] by Hasselbach and Weber[264] have shown that the Marsh factor causes relaxation of glycerated fibers under

(257) W. A. Kafiani and V. A. Engelhardt, *Doklady Akad. Nauk. S.S.S.R.* **92,** 385 (1953).
(258) T. Hayashi, *J. Gen. Physiol.* **36,** 139 (1952).
(259) I. P. Ashmarin, *Biokhimiya* **18,** 71 (1953).
(260) A. Munch-Petersen, *Acta Physiol. Scand.* **29,** 202 (1953).
(261) M. C. Goodall and A. G. Szent-Györgyi, *Nature* **172,** 84 (1953).
(262) L. Lorand, *Nature* **172,** 1181 (1953).
(263) S. V. Perry, *Biochem. J.* **57,** 427 (1954).
(264) W. Hasselbach and H. H. Weber, *Biochim. et Biophys. Acta* **11,** 160 (1953).

tension and inhibits the syneresis of actomyosin gel. In terms of the dual role of ATP, they suppose that the factor influences the ATP threshold at which relaxation occurs—10 millimolar in absence and 2 millimolar in presence of factor. Bendall,[265] however, has observed a remarkable parallelism in the behavior of Marsh factor and of preparations rich in myokinase. Using the glycerinated fiber as test system, the fiber was allowed to shorten by addition of medium containing ATP, and the response on addition of factor was then subsequently measured. At Mg concentrations between 2 and 8 millimolar, myokinase preparations caused lengthening and at the same time diminished the fiber ATPase activity. The addition of Ca ion antagonized the latter effect (as Marsh also found), and the increase of ATPase activity was then accompanied by contraction effects. Bendall inclines to the view that myokinase can fill two roles in contraction: (*a*) In a favorable ionic medium it can block the ATPase center and thus promote the relaxing action of ATP, and (*b*) in the presence of Ca ions and removal of the block, ATPase is stimulated, whilst the myokinase, acting in its own right, can maintain the ATP concentration by dismuting the ADP as it is produced by the action of ATPase.

In this latter work, there is an analogy with the reconstructed creatine phosphokinase system, in which the counterpart of the myokinase block is produced by lowering the pH to diminish ATPase activity. It seems likely, however, that Marsh factor is the one used physiologically to control the cycle of relaxation and contraction, since it could be controlled by ionic changes connected with the stimulus.

3. Conclusion

The aim of the present chapter has been to lay most stress on the historical background of modern muscle biochemistry. The subject has enlarged so dramatically in the last ten years that it is not yet possible to judge all contributions with an adequate impartiality or to predict their lasting importance. Whilst there are pointers as to the nature of the most vital mechanisms associated with contraction, the intimate details still elude us. There is general agreement, even proof, that ATP provides the energetic basis of contraction, but the nature of the energetic coupling to the contractile proteins is still a mystery. The role of ATP in relaxation is another important development, and the factors modifying it go a long way toward explaining how the active state of muscle can so suddenly be induced. But at no time in the history of muscle have molecular events seemed so contradictory and confused. The recent assignment of myosin to the A band, and the lack of change both on stretch and in

(265) J. R. Bendall, *Proc. Roy. Soc.* (London) **B142**, 409 (1954).

contraction, must for the time being cast doubts upon the classical concept that intramolecular change of configuration in the myosin component is responsible for contraction. The very nature of the interaction of actin and myosin is not at all clear, since different methods of investigation have not all led to a uniform interpretation. In retrospect, it is very striking how new techniques applied to muscle have given a definite answer to many old problems, and it is perhaps the less conservative methods of approach which will give an answer to the most vital problems, and not in the too distant future.

CHAPTER 25

Proteolytic Enzymes

BY N. MICHAEL GREEN AND HANS NEURATH*

Page

I. Introduction 1058
II. Methods of Preparation 1060
  1. Pancreatic Proteases 1061
  2. Pepsin and Pepsinogen 1064
  3. Rennin 1065
  4. Carboxypeptidase and Procarboxypeptidase 1066
  5. Intracellular Proteolytic Enzymes 1067
  6. Plant Proteases 1069
  7. Bacterial and Fungal Proteases 1069
III. Physicochemical Properties 1070
  1. Pepsinogen 1071
  2. Pepsin 1071
  3. Chymotrypsinogen 1073
    a. α-Chymotrypsinogen 1073
    b. B-Chymotrypsinogen 1075
  4. Chymotrypsins 1075
    a. α-, β-, γ- and δ-Chymotrypsins 1075
    b. B-Chymotrypsin 1078
  5. Trypsinogen 1078
  6. Trypsin 1078
  7. Carboxypeptidase 1081
  8. Papain 1081
  9. Rennin 1082
IV. Chemical Composition 1082
V. Stability 1086
  1. pH and Temperature 1086
    a. Pepsin 1086
    b. Pepsinogen 1088
    c. Chymotrypsinogen 1089
    d. α-Chymotrypsin 1090
    e. Trypsinogen 1090
    f. Trypsin 1090
    g. Carboxypeptidase 1092
    h. Cathepsins 1092
    i. Plant Enzymes 1092
  2. Irradiation 1093
  3. Pressure 1094
  4. Denaturing Agents 1094

* The authors are indebted to Miss Gretta West for her invaluable aid in the preparation of this manuscript.

Page
VI. Enzymatic Activity 1095
1. Methods 1095
2. Effect of pH on Activity 1100
a. Pepsin 1102
b. Cathepsins and Plant Proteases 1103
c. Chymotrypsin 1104
d. Trypsin 1105
e. Carboxypeptidase 1105
3. Specificity 1106
a. General Considerations 1106
b. Kinetic Principles 1109
c. Individual Enzymes 1116
*4. Transpeptidation Reactions Catalyzed by Proteolytic Enzymes 1135
5. Synthesis of Proteins and Plasteins 1143
VII. Inhibition 1144
1. Introduction 1144
2. Redox Agents and Sulfhydryl Reagents 1145
a. Activation of Plant Enzymes and Cathepsins 1145
b. Other Enzymes 1148
3. Metal Ions 1148
4. Chemical Modification 1151
a. Group Reagents 1151
b. Organic Phosphorus Compounds 1154
5. Protein Inhibitors 1159
a. Antibodies 1159
b. Other Proteins 1160
VIII. Action of Proteolytic Enzymes on Proteins 1171
1. Introduction 1171
2. Enzymatic Hydrolysis of Native and Denatured Proteins 1172
3. Products of Proteolysis 1173
a. Formation of Small Peptides Only 1173
b. Formation of Intermediate Products 1174
4. Mechanism of Proteolysis 1175
a. Experimental Results 1175
b. Theoretical Interpretations 1181
5. Limited Proteolysis 1183
a. Ill-Defined Protein Products 1183
b. Effect of Carboxypeptidase on Proteins 1187
c. Well-Defined Protein Products 1189

I. Introduction

One of the key processes of biological interest is the enzyme-catalyzed hydrolysis of peptide bonds. While the phenomenon itself has long attracted the attention of investigators, the recognition of the catalytic agents as distinct chemical components is an accomplishment belonging to more recent scientific history. Since many of the enzymes involved

* This section prepared by Y. D. Halsey.

in protein hydrolysis could be obtained in quantity, they were among the first to be isolated in a highly purified, crystalline form and to be characterized by the methods of protein chemistry.[1]

The proteolytic enzymes as a group possess many attributes of unusual interest:

1. They have the same gross chemical and physical structure as the natural substrates on which they act, thus bringing to the foreground the fundamental problem of the nature of the structural features which endow proteins with specific manifestations of biological activity. This is perhaps best illustrated by consideration of those subtle changes which occur when inactive precursors of some of these enzymes are converted into the active catalysts, a problem which in this chapter will be considered from the general viewpoint of limited enzymatic hydrolysis of proteins.

2. As a result of the pioneer work of Bergmann and co-workers,[2] synthetic substrates of well-defined structural characteristics have become available for several of the proteolytic enzymes. These have permitted a rather precise delineation of the specificity requirements of the enzyme concerned, and subsequent work has also led to a kinetic formulation of the processes involved.[3]

These studies have raised interesting problems in the field of enzyme kinetics and have contributed to efforts to elucidate the reaction mechanism and the nature of the catalytic centers of the enzymes. In this latter connection it is of interest to note that evidence has been accumulating that the number of active centers per molecule is small and perhaps not greater than one.

3. A further element of interest in these enzymes arises from their interactions with specific inhibitors. Some of them are proteins, resistant to the hydrolysis of the enzyme which they inhibit, though not to other proteases. Recently several types of synthetic inhibitors have been described, including certain dialkyl halogen phosphates (such as diisopropyl fluophosphate),[4] and structural analogs of synthetic substrates of known specificity. The former are of particular importance since they combine specifically with the active center of the enzyme to form compounds which have been isolated and analyzed. Further investigation of these compounds may well lead to the identification of the amino acid arrangement in the neighborhood of the active center.

(1) J. H. Northrop, M. Kunitz, and R. M. Herriott, Crystalline Enzymes, 2nd Ed., Columbia Univ. Press, New York, 1948.
(2) M. Bergmann, *Advances in Enzymol.* **2,** 49 (1942).
(3) H. Neurath and G. W. Schwert, *Chem. Rev.* **46,** 69 (1950).
(4) A. K. Balls and E. F. Jansen, *Advances in Enzymol.* **13,** 321 (1952).

4. Several of these enzymes are capable of mediating some stages of peptide synthesis by the route of transpeptidation.[5] These reactions will be considered herein from the viewpoint of substrate specificities, whereas their general role in protein metabolism will be considered in Chap. 26. Consideration of the physiological functions of proteolytic enzymes will likewise be deferred to the succeeding chapter.

5. With the available information on the chemical structure and enzymatic specificity of the proteolytic enzymes, their application as analytical tools in studies of protein structure has received renewed interest. Two major factors have made it possible to resume this type of investigation: (*a*) the elucidation of the specificity range of the enzymes and (*b*) the availability of chromatographic and ultracentrifugal methods to separate and identify the products of the proteolytic reaction.

The present review is limited to consideration of those proteolytic enzymes which have been sufficiently purified to warrant their characterization from the viewpoints of protein and enzyme chemistry. However, even within this restricted scope no attempt was made to achieve complete coverage. Much of the earlier literature has been adequately covered in recent reviews, notably in the monograph by Northrop, Kunitz, and Herriott,[1] the review of the crystalline pancreatic proteolytic enzymes by Neurath and Schwert,[3] and the chapters on proteolytic enzymes by Smith in *The Enzymes*,[6] and in the *Advances in Enzymology*.[6a]

The topics on the effect of proteolytic enzymes on proteins and on the natural enzyme inhibitors have hitherto not been reviewed; such overlapping with previous reviews as may be found in the discussion of other topics has been deemed advisable in the interest of clarity, consistency, and integration.

II. Methods of Preparation

The proteolytic enzymes of the pancreas and of the gastric juice are secreted in relatively high concentrations and hence are easily freed from other proteins. It is probably for this reason that the digestive proteases were among the first enzymes to be crystallized and to be unequivocally identified as proteins. The methods used in the classical work of Northrop and Kunitz in the early 1930's were essentially refinements of salt fractionation procedures, long used in the purification of proteins, and they have been lucidly described in Northrop, Kunitz, and Herriott's

(5) J. S. Fruton, Symposium sur la Biogénèse des Protéines, 2nd Intern. Congr. Biochem., Paris, 1952.
(6) E. L. Smith *in* J. B. Sumner and K. Myrbäck, The Enzymes, Academic Press, New York, 1951.
(6*a*) E. L. Smith, *Advances in Enzymol.* **12,** 191 (1951).

monograph *Crystalline Enzymes.*[1] In the following discussion, a brief outline of the fractionation procedures will be given together with a short discussion of such principles as are involved. The general principles of enzyme purification have been recently reviewed by Schwimmer and Pardee.[7]

1. Pancreatic Proteases

The stages in the isolation of chymotrypsinogen, trypsinogen and, related pancreatic proteins are shown in Fig. 1. Minced pancreas is extracted with 0.25 N sulfuric acid to give a suspension of pH 3–4 in which the zymogens are readily soluble. Under these conditions no spontaneous activation occurs. Procarboxypeptidase is unstable under such acid conditions and cannot be obtained by this method. The proteins precipitating between 0.4 and 0.7 saturated ammonium sulfate are collected, and chymotrypsinogen is isolated by crystallization from a solution of this precipitate in 0.25 saturated ammonium sulfate at pH 5. The protein from the mother liquor is further fractionated between 0.4 and 0.7 saturated ammonium sulfate, and trypsinogen is crystallized from a 25 per cent solution of this fraction at pH 8. The use of a preparative ultracentrifuge greatly facilitates the separation of the very small trypsinogen crystals from the viscous mother liquor[8] and thus avoids tedious filtration during which trypsinogen is prone to become spontaneously activated to trypsin.

Trypsin and α-chymotrypsin are obtained from their respective zymogens by activation with trypsin, as indicated in Fig. 1, and are then crystallized from concentrated solutions.

If the mother liquors from the activation of chymotrypsinogen, which themselves have low proteolytic activity, are allowed to autolyze for several weeks at 5°, pH 8, two other crystalline forms of chymotrypsin, i.e., β- and γ-chymotrypsin, can be isolated. They are similar in enzymatic properties to α-chymotrypsin and may also be prepared directly from the α-form by autolysis under the conditions just described. Chymotrypsinogen will also crystallize from salt-free solutions at pH 3[9] and near pH 9.[10] Chymotrypsinogen and α-chymotrypsin can also be recrystallized from 15–20 per cent ethanol after removal of salt by dialysis.[11] On lowering the pH of the alcoholic solution to pH 5 (chymotrypsinogen) or 4 (α-chymotrypsin), an amorphous precipitate results which becomes crystalline on standing.

(7) S. Schwimmer and A. B. Pardee, *Advances in Enzymol.* **14,** 375 (1953).
(8) F. Tietze, *J. Biol. Chem.* **204,** 1 (1953).
(9) C. J. Jacobsen, *Compt. rend. trav. lab. Carlsberg. Sér. chim.* **25,** 325 (1947).
(10) P. E. Wilcox, unpublished.
(11) M. Kunitz, *J. Gen. Physiol.* **32,** 265 (1948).

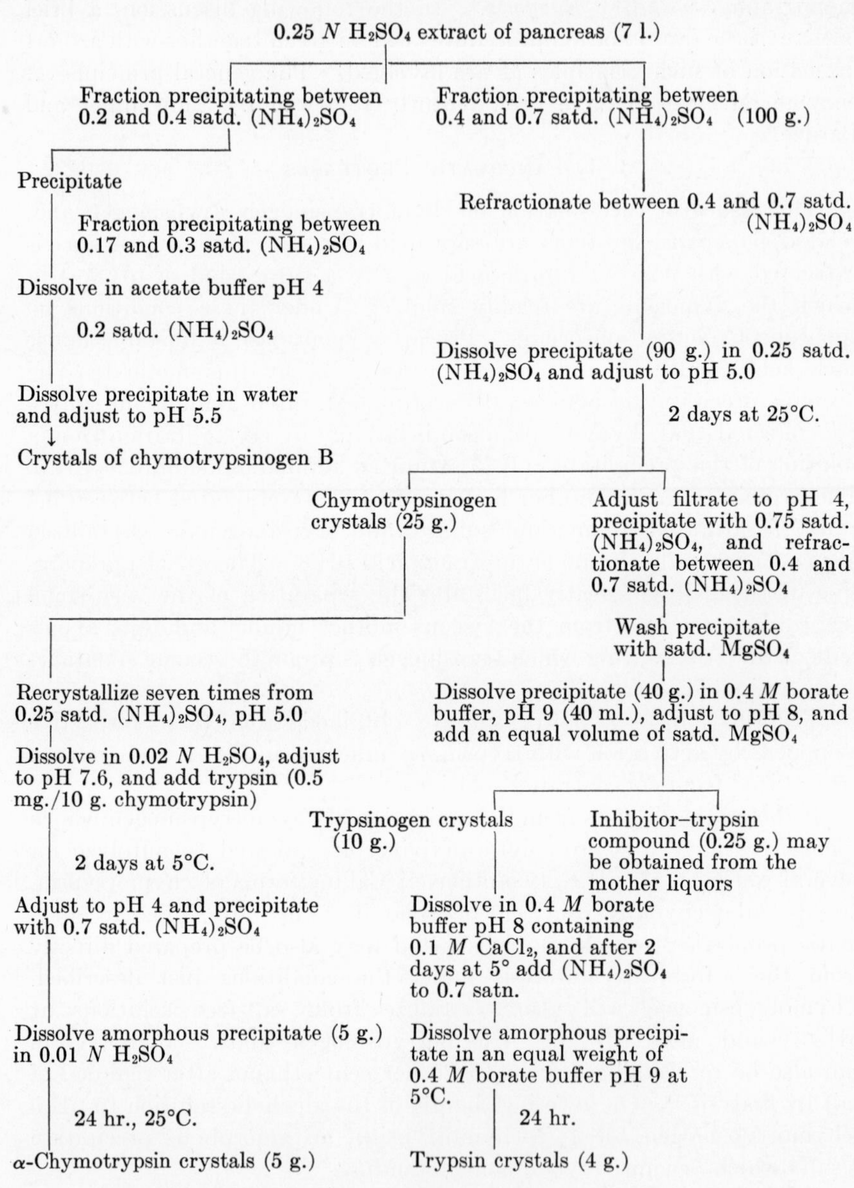

FIG. 1. Preparation of chymotrypsin, trypsin and their precursors. All weights refer to damp filter cakes.

Difficulties may arise in the preparation of trypsinogen, owing to autocatalytic activation during crystallization. While this possibility is greatly reduced by the presence of the pancreatic trypsin inhibitor in the solution from which crystallization occurs, nevertheless, some tryptic activity remains even in the presence of the inhibitor (as measured by activity toward benzoyl-L-arginine ethyl ester), the inhibitor being slowly used up as activation proceeds.[11a] The dangers of spontaneous activation are even more acute when recrystallization is attempted and the only published method for recrystallization[8] calls for the addition of diisopropyl fluophosphate (DFP) to the system.

A crystalline, enzymatically inert compound of trypsin with a low molecular weight protein has been isolated from the mother liquors remaining after trypsinogen crystallization. If this compound is allowed to stand in cold 2.5 per cent trichloroacetic acid, reversibly denatured trypsin is precipitated and activity reappears on dissolving the washed, precipitated protein in 0.02 N HCl. Reversible denaturation by trichloroacetic acid has, in fact, been used as a method for purifying both trypsinogen and trypsin.[1] The low molecular weight inhibitor may be obtained from the trichloroacetic acid filtrate by saturation with ammonium sulfate at pH 3. Crystallization is brought about with $MgSO_4$ at pH 3[12] or pH 5.5,[1] the more acid pH being preferable since it corresponds to the solubility minimum of this inhibitor.

In addition to the members of the chymotrypsinogen family already discussed, there are two other proteins which resemble them in enzymatic properties but differ in certain chemical characteristics. These are chymotrypsinogen B[13] and the corresponding active enzyme, chymotrypsin B.[14] Chymotrypsinogen B has been isolated from acid extracts of beef pancreas glands, from the fraction which precipitates initially between 0.2 and 0.4 saturated ammonium sulfate and, upon reprecipitation, between 0.17 and 0.3 saturation. Chymotrypsinogen B crystallizes when a concentrated solution is adjusted from pH 6.5 to 5.5 (resembling in this respect chymotrypsinogen α).[9] The active enzyme has milk-clotting activity and has qualitatively the same specificity toward synthetic substrates[15] as have α-, β-, and γ-chymotrypsins. The electrophoretic and solubility properties of the B proteins are, however, rather different (pp. 1075, 1078). Tryptic activation of chymotrypsinogen B (72 hours at 5°, chymotrypsinogen–trypsin ratio of 1500) yields chymotrypsin

(11a) M. Laskowski and F. C. Wu, *J. Biol. Chem.* **204**, 797 (1953).
(12) N. M. Green and E. Work, *Biochem. J.* **54**, 257 (1953).
(13) C. K. Keith, A. Kazenko, and M. Laskowski, *J. Biol. Chem.* **170**, 227 (1947).
(14) K. D. Brown, R. E. Shupe, and M. Laskowski, *J. Biol. Chem.* **173**, 99 (1948).
(15) J. S. Fruton, *J. Biol. Chem.* **173**, 109 (1948).

B which crystallizes upon dialysis against acetate buffer pH 5.5 in the form of short prisms. If the pH of the solution is initially higher than 5.5, needle-shaped crystals are obtained.

2. Pepsin and Pepsinogen[1]

Swine pepsin was first crystallized from commercial preparations which still serve as the most convenient starting material. The enzyme may also be obtained from gastric juice or from pepsinogen isolated from the gastric mucosa. Purification is accomplished by precipitation with magnesium sulfate (*ca.* 0.6 saturated) at acid pH, dissolution of the precipitate with alkali (pH less than 4), and reprecipitation by acidification to pH 2.5. Crystallization occurs when this precipitate is dissolved at pH 4, 45°, and the solution allowed to cool. This procedure has been found to lead to considerable decomposition[16] with the formation of free tyrosine and a large number of peptides. A refined modification of Northrop's procedure suggested by Philpot,[17] may therefore be preferable. Pepsin may also be crystallized from 20 per cent ethanol and is soluble in 65 per cent ethanol.[18]

The isolation of pepsinogen from the gastric mucosa consists in the extraction of the tissue with 0.45 saturated ammonium sulfate, containing sodium bicarbonate, and precipitation of the protein at 0.68 saturated ammonium sulfate. The precipitate is then adsorbed on cupric hydroxide at pH 6 and pepsinogen eluted with 0.1 *M* phosphate buffer, pH 6.8. After repeating this step, pepsinogen crystallizes from 0.4 saturated ammonium sulfate at pH 6.25.

Pepsin is produced from pepsinogen autocatalytically. Acidification of pepsinogen solutions is sufficient to bring about activation since traces of pepsin are always present. During this process, a polypeptide (molecular weight, *ca.* 5000) is split from the pepsinogen molecule, and, in solution more alkaline than pH 5.4, it combines with pepsin and causes inhibition. At pH 3.5, the inhibitor is destroyed by pepsin. The pepsin inhibitor was isolated[19] by allowing activation to proceed for a short time at pH 2, followed by denaturation of pepsin at pH 11 for 3 minutes, and precipitation of the denatured pepsin with trichloroacetic acid. After fractional precipitation of the trichloroacetic acid-soluble peptides with tungstic acid, followed by further fractionation with magnesium sulfate in the presence of trichloroacetate at pH 3,

(16) V. M. Ingram, *Nature* **167**, 83 (1951).
(17) J. St. L. Philpot, *Biochem. J.* **29**, 2458 (1935).
(18) J. H. Northrop, *J. Gen. Physiol.* **30**, 177 (1946).
(19) R. M. Herriott, *J. Gen. Physiol.* **24**, 325 (1941).

the inhibitor crystallizes. The nature of the chemical changes occurring during activation, and the properties of the inhibitor are discussed elsewhere (p. 1191). Crystalline pepsin has been obtained from other species also, e.g., cattle, salmon,[20] and tuna.[21] Pepsinogen and pepsin have also been obtained from human semen.[22,23a–e]

A few points of interest should be noted in the preparation of these proteases. Most steps are carried out with relatively large amounts of protein in order to avoid small volumes of solutions and to facilitate accurate adjustment of pH and salt concentration. The use of concentrated protein solutions (*ca.* 5–10 per cent) has proved successful in the isolation of these enzymes. In the face of the highly empirical nature of these procedures, one may await with much interest the application of a more rational approach to the determination of the most favorable conditions for protein purifications, based on solubility measurements,[24,25] as they have been applied with moderate success to the purification of the pancreatic trypsin inhibitor.[12]

The enzymes and zymogens discussed above are crystallized under conditions of pH corresponding to minimum solubility. These, incidentally, are often far removed from their respective isoelectric points as determined by the moving-boundary method (p. 1071).

3. Rennin[23a]

The original procedure for the crystallization of rennin was long and tedious,[23b] but recently two simplified versions have been published.[23c,d] In one of these,[23c] crude rennet is first saturated with sodium chloride, the precipitate is redissolved at pH 5.4, and saturated sodium chloride is then added by dialysis. When the resulting granular precipitate is redissolved in water, crystals appear on standing in the cold. The other procedure[23d,e] relies on the crystallization of rennin at pH 5.0 from dilute,

(20) E. R. Norris and D. W. Elam, *J. Biol. Chem.* **134,** 443 (1940).
(21) E. R. Norris and J. C. Mathies, *J. Biol. Chem.* **204,** 673 (1953).
(22) F. Lundquist and H. H. Seedorff, *Nature* **170,** 1115 (1952).
(23) F. Lundquist, *Acta Physiol. Scand.* **25,** 178 (1952).
(23*a*) N. J. Berridge *in* J. B. Sumner and K. Myrbäck, The Enzymes, Academic Press, New York, 1951, Vol. I. p. 1079 ff.
(23*b*) N. J. Berridge, *Biochem. J.* **39,** 179 (1945).
(23*c*) N. J. Berridge and C. Woodward, *Biochem. J.* **54,** No. 3, xix (1953).
(23*d*) R. M. De Baun, W. M. Connors, and R. A. Sullivan, *Arch. Biochem. Biophys.* **43,** 324 (1953).
(23*e*) C. L. Hankinson, *J. Dairy Sci.* **26,** 53 (1943).
(24) J. S. Falconer and D. B. Taylor, *Proc. Intern. Congr. Pure and Appl. Chem. 11th Congr., London, 1947.*
(25) J. S. Falconer, D. J. Jenden, and D. B. Taylor, *Discussions Faraday Soc.* No. **13,** 40 (1953).

dialyzed solutions of an amorphous protein preparation obtained by four consecutive precipitations with saturated sodium chloride.

4. Carboxypeptidase and Procarboxypeptidase

Carboxypeptidase is usually prepared from the exudate which is obtained when thin slices of freshly frozen beef pancreas glands are allowed to autolyze in the cold, but an alternate method involving extracts of the whole pancreatic tissue has also been described.[26] Since the enzyme is initially present as a zymogen (procarboxypeptidase),[27] it is necessary to assure activation prior to the crystallization of the enzyme. Since activation is trypsin catalyzed and since trypsin usually accumulates during autolysis, it suffices to incubate the pancreatic juice at 37°, pH 7.8 for 1 hour. The solution is then adjusted to pH 4.6 and, after ten-fold dilution with water in the cold, a euglobulin precipitate is collected which contains practically all of the active enzyme. The crystallization procedure used in this laboratory, based on the original method of Anson,[26] as subsequently modified by Putnam and Neurath,[28] Neurath, Elkins, and Kaufman,[29] and Albrecht, Neurath, and DeMaria,[30] is as follows:

The euglobulin precipitate is suspended in one-fifth of the original volume of distilled water, and inert proteins are extracted in the cold by stirring with gradual addition of freshly prepared 0.2 *M* barium hydroxide until pH 6 is reached, as measured with a glass electrode. The soluble, enzymatically inactive proteins are removed by centrifugation, and the precipitate is resuspended in distilled water and again extracted with 0.2 *M* barium hydroxide until a stable pH of 10.4 is reached. After centrifugation, the supernatant solution is carefully acidified with 1 *N* acetic acid till crystallization occurs (approximately pH 8.5) and then adjusted to pH 7.5. Repeated extractions of the euglobulin precipitate under these conditions yield additional crystalline fractions which are all combined and subjected to recrystallization. This may be done by extraction with 0.2 *M* lithium hydroxide in the cold up to pH 10.4 followed by acidification with acetic acid as described above. An alternate procedure involves the solution of the crystalline precipitate in 10 per cent lithium chloride in the cold followed by graded dialysis against 5 per cent lithium chloride, 2 per cent sodium chloride, and finally water. Although carboxypeptidase is only slightly soluble in dilute sodium chloride, Smith,

(26) M. L. Anson, *J. Gen. Physiol.* **20,** 663 (1937).
(27) M. L. Anson, *J. Gen. Physiol.* **20,** 777 (1937).
(28) F. W. Putnam and H. Neurath, *J. Biol. Chem.* **166,** 603 (1946).
(29) H. Neurath, E. Elkins, and S. Kaufman, *J. Biol. Chem.* **170,** 221 (1947).
(30) G. S. Albrecht, H. Neurath, and G. DeMaria, unpublished experiments.

Brown, and Hanson[31] have found that the contaminating proteins are even less soluble and that, therefore, extraction of crystalline carboxypeptidase with 0.2 *M* sodium chloride followed by dialysis led to more rapid purification. However, no yields were given.

With methods employed in this laboratory, the yields, though variable, were usually of the order of 1 g. of six-times recrystallized carboxypeptidase per 1.5 l. of pancreatic juice. Such preparations were completely devoid of tryptic or chymotryptic activity when tested against the corresponding synthetic substrates. When tested against carbobenzoxyglycyl-L-phenylalanine, at 25°, the apparent first-order rate constant per milligram enzyme nitrogen per milliliter was of the order of 14 (0.05 *M* substrate).

Although the zymogen has not yet been isolated in pure form, Anson[27] has described the preparation of crude fractions which, upon tryptic activation, gave enhanced carboxypeptidase activity. This problem has been taken up again in this laboratory,[30] DFP being added to the extracts of freshly frozen beef pancreas glands to suppress tryptic activity (and hence spontaneous activation of procarboxypeptidase). Using a combination of acetone precipitation and ammonium sulfate fractionation, preparations containing approximately 80 per cent procarboxypeptidase have been obtained.

5. Intracellular Proteolytic Enzymes

Purification of the intracellular proteases of animal origin (cathepsins) has received less attention and has not yet progressed very far. Although cathepsin activity can be demonstrated in a great variety of tissues (spleen, kidney, liver, muscle, and pancreas), the activity is usually of low order, and hence isolation of the respective enzymes in quantity is difficult. For this reason, techniques developed so successfully for the extracellular enzymes could not be readily applied to this group of proteases. Moreover, the cathepsins rarely occur singly but are usually found as a group of enzymes of similar properties distinguishable from one another primarily by their enzymatic specificity. The cathepsins of spleen have been studied in some detail[32,33] and classified according to their specificity. Only one of these, cathepsin C, has been separated from enzymes of different specificities and has been considerably purified.

Cathepsin A and leucinaminopeptidase (cathepsin III) were first

(31) E. L. Smith, D. M. Brown, and H. T. Hanson, *J. Biol. Chem.* **180,** 33 (1949).
(32) J. S. Fruton, G. W. Irving, Jr., and M. Bergmann, *J. Biol. Chem.* **141,** 763 (1941).
(33) H. H. Tallan, M. E. Jones, and J. S. Fruton, *J. Biol. Chem.* **194,** 793 (1952).

removed from the crude preparation by fractionation with ammonium sulfate. The fraction precipitated between 0.5 and 0.7 saturated ammonium sulfate was dialyzed against 2 per cent sodium chloride and heated to 65° for 40 minutes, causing precipitation of cathepsin B and other impurities, without destroying the activity of cathepsin C. Further purification was carried out by precipitation with ethanol in the presence of zinc acetate. This preparation had a relatively narrow specificity range, resembling that of chymotrypsin, but its homogeneity was not investigated by physical methods.

Other cathepsins have been partially purified including one from rabbit muscle[34] and a leucinaminopeptidase from beef spleen.[35]

Other intracellular proteolytic enzymes of interest which have been obtained in a partially purified state are an iminodipeptidase from swine kidney cortex, obtained in 30-fold purification, activated by Mn^{++}, and hydrolyzing L-prolylglycine and hydroxy-L-prolylglycine.[36] An enzyme of like specificity has been isolated from horse erythrocytes.[37] From the same tissue, a tripeptidase has been isolated[38] and purified from 500- to 700-fold by a procedure earlier described by Fruton *et al.*[39] for the purification of a similar enzyme from extracts of calf thymus. The procedure involved essentially repeated precipitation of the protein between 0.4 and 0.7 saturated ammonium sulfate. The enzyme is most active toward triglycine, glycylglycyl-β-alanine, L-leucylglycylglycine, and glycylglycyl-L-proline.[38]

Two proteases, which hydrolyze denatured hemoglobin optimally at pH 3.8 (protease I) and at pH 8.3 (protease II), have been isolated from pituitary extracts and partially purified. No specific activators, inhibitors, or synthetic substrates have been found for either of these two proteases.[40]

A metal-activated aminopeptidase from the particulate fraction of kidney is one of the few enzymes of this class to have been highly purified.[41] A particulate fraction was treated with 70 per cent acetone to remove soluble protein, and the enzyme was then rendered soluble by extraction with 20 per cent butanol. After removal of the organic solvent, concentration, and fractionation between 0.5 and 0.75 saturated

(34) J. E. Snoke and H. Neurath, *J. Biol. Chem.* **187,** 127 (1950).
(35) I. Krimsky and E. Racker, *J. Biol. Chem.* **179,** 903 (1949).
(36) N. C. Davis and E. L. Smith, *J. Biol. Chem.* **200,** 373 (1953).
(37) E. Adams and E. L. Smith, *J. Biol. Chem.* **198,** 671 (1952).
(38) E. Adams, N. C. Davis, and E. L. Smith, *J. Biol. Chem.* **199,** 845 (1952).
(39) J. S. Fruton, V. A. Smith, and P. E. Driscoll, *J. Biol. Chem.* **173,** 457 (1948).
(40) E. Adams and E. L. Smith, *J. Biol. Chem.* **191,** 651 (1951).
(41) D. S. Robinson, S. M. Birnbaum, and J. P. Greenstein, *J. Biol. Chem.* **202,** 1 (1953).

ammonium sulfate, a preparation with 700 times the original activity was obtained. This preparation contained 30 per cent of a phospholipide which could not be removed without inactivation of the enzyme. When lyophilized preparations were allowed to stand in the cold, a large increase in activity (up to threefold) over a period of several weeks was observed, possibly due to activation of an inactive precursor or destruction of an inhibitor. In view of the wide specificity of this enzyme, further attempts at fractionation, using ion-exchange columns, were made, but without success.

6. Plant Proteases

Crystalline proteases have been obtained from the fresh latex of papaya and fig trees, and, more recently, from commercial dried papaya latex. Procedures for obtaining the crystalline enzymes are simple, but in the case of papain the yields are very low, while the crystallization of ficin, brought about by allowing the latex to stand for several weeks at pH 5, has not been described in detail.[42] Crystalline papain was obtained from a 0.4 saturated ammonium sulfate precipitate of latex, which was dissolved in 2 per cent NaCl and allowed to stand at pH 6.5.[43] Only 2 per cent of the original activity was recovered in this way, but the crystallization of a second enzyme, chymopapain, from the mother liquors has been reported.[44]

Crystalline papain was recently obtained from a commercial preparation of dried latex.[44a] The procedure was a modification of that of Balls and Lineweaver[43] and has led to a sixfold increase in specific activity from the first extract to the recrystallized enzyme with a 2 per cent yield. Further purification was accomplished by crystallization of a mercury derivative containing one atom of mercury per two moles of enzyme protein.

A crystalline protease, asclepian, has been obtained from extracts of the root of several species of the milkweed *Asclepias*. Purification was accomplished by ammonium sulfate fractionation.[45]

7. Bacterial and Fungal Proteases

Proteases and peptidases are found in a great variety of microorganisms,[46] but only recently has any extensive purification been accom-

(42) A. Walti, *J. Am. Chem. Soc.* **60**, 493 (1938).
(43) A. K. Balls and H. Lineweaver, *J. Biol. Chem.* **130,** 669 (1939).
(44) E. F. Jansen and A. K. Balls, *J. Biol. Chem.* **137,** 459 (1941).
(44*a*) J. R. Kimmel and E. L. Smith, *J. Biol. Chem.* **207,** 515 (1954).
(45) D. C. Carpenter and F. E. Lovelace, *J. Am. Chem. Soc.* **65,** 2364 (1943).
(46) E. Maschmann, *Ergeb. Enzymforsch.* **9,** 166 (1943).
(46*a*) F. G. Lennox, *Rev. Pure and Appl. Chem.* **2,** 33 (1952).

plished. Only one fungal protease has been crystallized.[47] It was obtained by ethanol fractionation of culture filtrates of *Aspergillus oryzae*, followed by crystallization from 2.7 *M* ammonium sulfate. Kunitz[48] described a protease of a *Penicillium* species which could activate trypsinogen at pH 3–4, but no appreciable purification has been effected.

Many bacterial toxins have been shown to have proteolytic properties and often possess specific activity toward collagen (collagenases). Some of these have been partially purified by ammonium sulfate fractionation of culture filtrates. The κ[49,50] and λ[51] antigens of *Clostridium welchii* have both been found to be proteases, the κ-toxin being specific for collagen.

Similar purification methods have led to the crystallization of a protease and of its inactive precursor from the culture filtrates of a group A *streptococcus*. The precursor is activated by trypsin or autocatalytically in the presence of cysteine.[52] A collagenase from the culture filtrates of *Bacillus amyloliquefaciens* has been crystallized from 30 per cent acetone after precipitation of impurities with 40 per cent ethanol in the presence of starch.[53,54]

A protease from *Bacillus subtilis* which is responsible for the conversion of ovalbumin to plakalbumin (see Chaps. 3 and 9) has been crystallized from 12 per cent sodium sulfate after a preliminary acetone fractionation.[55]

III. Physicochemical Properties

In this section, the physicochemical properties of those proteolytic enzymes will be considered which have been obtained in a state of purity sufficiently high to render measurements of their physicochemical parameters meaningful. The values reported for some of these vary within unusually wide limits, due in part to the limited accuracy of the methods employed in earlier work, and in part due to complex monomer–polymer equilibria; however, all of these will be considered, and emphasis will be placed on those data which appear to be most reliable (Tables I and II).

(47) W. G. Crewther and F. G. Lennox, *Nature* **165,** 680 (1950).
(48) M. Kunitz, *J. Gen. Physiol.* **21,** 601 (1938).
(49) E. Bidwell, *Biochem. J.* **44,** 28 (1949).
(50) E. Bidwell and W. E. Van Heyningen, *Biochem. J.* **42,** 140 (1948).
(51) E. Bidwell, *Biochem. J.* **46,** 589 (1950).
(52) S. D. Elliott, *J. Exptl. Med.* **92,** 201 (1950).
(53) J. Fukumoto and H. Negoro, *Proc. Japan Acad.* **27,** 441 (1951); *C. A.* **46,** 7171*f* (1952).
(54) J. Fukumoto and H. Negoro, *Symposia on Enzyme Chem.* (Japan) **7,** 8 (1952); *C. A.* **46,** 8160*f* (1952).
(55) A. V. Güntelberg and M. Ottesen, *Nature* **170,** 802 (1952).

1. Pepsinogen[56]

The solubility properties of the crystalline material of Herriott approach those to be expected for a pure substance. The only value reported for the molecular weight derives from osmotic-pressure measurements (42,000 $\pm$ 3000).[56] Electrochemical characterization is limited to the determination of the isoelectric point by the microscopic cataphoresis method (pH 3.6–4.3).[56]

2. Pepsin

Crystalline swine pepsin prepared either from commercial preparations or from pepsinogen[56] is not a single protein as judged by solubility measurements. The solubility behavior of the protein resembles that of a solid solution, but at pH 5 a "more soluble fraction" (precipitated by 0.6 or 0.65 saturated magnesium sulfate) could be isolated showing constant solubility in two or more different solvents.[1]

The *electrophoretic behavior* of pepsin has been studied by several groups of investigators[57–59] using the moving-boundary method. Tiselius *et al.*[57] showed that none of four preparations obtained by different methods was quite homogeneous and that inactive components could be removed by electrophoresis, thus raising the specific activity of the active component 30–70 per cent. The residual active protein showed no isoelectric point between pH 1 and 4.6, and it was concluded, therefore, that the enzyme contained only acidic groups. These results were essentially confirmed by Herriott *et al.*[58] who found that pure pepsin, freed from nonprotein nitrogen (NPN), was negatively charged even in 0.1 N HCl. However, upon standing at 30°, significant quantities of NPN appeared, and the protein became less positively charged at pH 1.5. In a more recent study[59] it was found that crystalline pepsin as obtained by standard techniques was 96 per cent homogeneous at pH's 3.9, 5.9, and 8, but that on prolonged electrophoresis, the boundary pattern indicated heterogeneity yielding at pH 5.9 four components, two of which were active and the remaining two inactive. When subjected to paper electrophoresis at pH 5.9, crystalline pepsin reveals three components,[59a] the fastest moving one containing only pepsin, the adjacent one both

(56) R. M. Herriott, *J. Gen. Physiol.* **21,** 501 (1938).
(57) A. Tiselius, G. E. Henschen, and H. Svensson, *Biochem. J.* **32,** 1814 (1938).
(58) R. M. Herriott, V. Desreux, and J. H. Northrop, *J. Gen. Physiol.* **23,** 439 (1940).
(59) H. Hoch, *Nature* **165,** 278 (1950).
(59*a*) R. Merten, G. Schramm, W. Grassmann, and K. Hannig, *Z. physiol. Chem.* **289,** 173 (1952).

TABLE I
PROBABLE VALUES OF MOLECULAR WEIGHT OF SOME PROTEOLYTIC ENZYMES

Protein	Method	Molecular weight	Reference
Pepsinogen (swine)	Osmotic pressure	42,000 ± 3,000	56
Pepsin (swine)	S, D	35,500	17,62
	Light scattering	37,600	60
	Surface films	34,400	61
α-Chymotrypsinogen	S, D	23,000	68–70
	Light scattering	25,000	71
	Analytical	25,000	73,74
B-Chymotrypsinogen	S, D	21,600	70
α-Chymotrypsin[a]	S, D	21,600	68,79,80
	Combination with DFP or DENP[d]	25,000	85,86
	Light scattering	27,000	84
	Osmotic pressure	27,000	4
B-Chymotrypsin	S, D	23,600	70
Trypsinogen	S, D	23,700	8
Trypsin[b]	S, D	23,800	90,93
	S, D of trypsin–soybean inhibitor compound[c]	17,000	96
	Surface films	20,500	97
	Combination with DFP	20,700–24,800	90,98
Carboxypeptidase	S, D	34,300	28,31
Papain	Osmotic pressure	27,000 ± 2,000	43
	S, D	20,700	100
Rennin	S, D	40,000	106
Kidney aminopeptidase	S	75,000–80,000	41

[a] Active enzyme or the inactive dialkylphosphoryl derivatives.
[b] Active enzyme or inactive dialkylphosphoryl derivative.
[c] Assuming a molecular weight of 24,000 for soybean inhibitor. However, the values obtained for the inhibitor–trypsin combining ratio (0.81) suggest that the figures should be reversed, i.e., trypsin = 23,000 and soybean inhibitor = 18,000.
[d] diethyl-p-nitrophenylphosphate.

pepsin and gastric cathepsin, and the slowest one being enzymatically inactive. The same electrophoretic components, though in different relative proportions, are exhibited by a fraction prepared from gastric mucosa, enriched in catheptic activity (as measured by the digestion of hemoglobin at pH 3.8).

Molecular-weight determinations by a variety of methods (Table I) seem to point toward 35,000 as the most probable value.[60–62,1,17] The

(60) D. S. Yasnoff and H. B. Bull, *J. Biol. Chem.* **200,** 619 (1953).
(61) H. A. Dieu and H. B. Bull, *J. Am. Chem. Soc.* **71,** 450 (1949).
(62) J. St. L. Philpot and I. B. Eriksson-Quensel, *Nature* **132,** 932 (1933).

TABLE II
REPORTED VALUES OF ISOELECTRIC POINTS OF SOME PROTEOLYTIC ENZYMES

Protein	Method	Buffer	Isoelectric Point	Reference
Pepsinogen	Cataphoresis		3.6–4.3	56
Pepsin	Moving boundary	HCl, 0.1 N	>1.1	57,58
α-Chymotrypsinogen	Moving boundary	Glycinate, Γ/2 = 0.01	9.5	67
	Moving boundary	Glycinate, Γ/2 = 0.1	9.1	75
	Cataphoresis		5.	66
B-Chymotrypsinogen	Moving boundary	Glycinate, Γ/2 = 0.1	5.2	75
α-Chymotrypsin	Moving boundary	Veronal, Γ/2 = 0.01	8.6	67
	Moving boundary	Veronal, Γ/2 = 0.1	8.1	67
	Moving boundary	Glycinate, Γ/2 = 0.1	8.3	75
	Cataphoresis		6.0	77
	Donnan equilibrium	Acetate or phosphate 0.004 M	6.3	78
B-Chymotrypsin	Moving boundary	Acetate, Γ/2 = 0.1	4.7	75
Trypsinogen	Mixed-bed ion-exchange resins		9.3[a]	88
Trypsin	Moving boundary	Ca glycinate, 0.04 M	10.8	92
	Moving boundary	Glycinate–NaCl, Γ/2 = 0.1	10.4[b]	93
	Cataphoresis		7.0	1
	Mixed-bed ion-exchange resin		10.1[a]	88
Carboxypeptidase	Moving boundary	Cacodylate, Γ/2 = 0.1	6.0	28
Papain	Cataphoresis		<8.5	43
	Moving boundary	Veronal, glycinate, Γ/2 = 0.1	8.75	100
Kidney aminopeptidase	Moving boundary		4.6	41

[a] Isoionic point.
[b] Diisopropylphosphoryltrypsin.

results obtained by sedimentation and diffusion measurements[17] are probably most reliable and yield at pH 4–5 the following values:

$$s_{20,w} = 3.3\ S \pm 0.15,\ D_{20,w} = 9.0 \times 10^{-7},\ M = 35{,}500.$$

3. CHYMOTRYPSINOGEN

a. α-*Chymotrypsinogen*

This crystalline zymogen prepared according to the method of Kunitz[1] has been shown to conform to the solubility properties to be expected for a pure protein.[63] Provided that partial hydrolysis by contaminating

(63) J. A. V. Butler, *J. Gen. Physiol.* **24,** 189 (1940).

traces of chymotrypsin (or other proteolytic enzymes) is avoided, the protein also shows only a single component when analyzed by column chromatography on a carboxylic resin.[64,65]

The crystalline protein appears to be *electrophoretically* monodisperse when tested at pH 3.[63] Whereas Kunitz and Northrop[66] originally reported an isoelectric point corresponding to pH 5 (microscopic cataphoresis method), more recent measurements employing the moving-boundary method[67] indicate an isoelectric point above pH 9 (Table II). The heterogeneity constant (standard deviation of mobility distribution) of α-chymotrypsinogen has a low value compared to other proteins.[67]

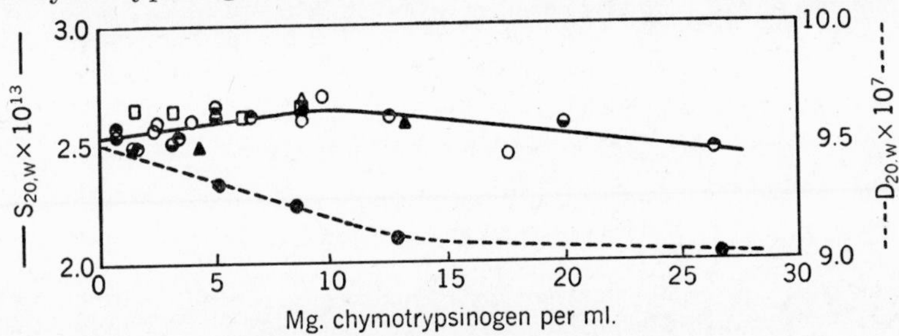

FIG. 2. Concentration dependence of the sedimentation (upper) and diffusion constants (lower curve) of chymotrypsinogen.[69]

The *molecular weight* of chymotrypsinogen has been the subject of careful studies[68,69] which yielded the following parameters: $s_{20,w} = 2.5\ S$, $D_{20,w} = 9.5 \times 10^{-7}$ (both values obtained by extrapolation to zero protein concentration), and V (partial specific volume) = 0.721. These values together with viscosity measurements correspond to a molecular weight of 23,000 as the most likely value (Table I). A slight degree of dimerization of the protein in solution is suggested by the concentration dependence of the sedimentation and diffusion constants, shown in Fig. 2. Independent measurements by Smith *et al.*[70] are in essential agreement with these data. A somewhat higher value of 25,000 has been obtained by light-scattering measurements.[71] Whereas incomplete amino acid analyses by Brand[72] correspond to a minimum molecular

(64) C. H. W. Hirs, *Federation Proc.* **12,** 218 (1953).
(65) C. H. W. Hirs, *J. Biol. Chem.* **205,** 93 (1953).
(66) M. Kunitz and J. H. Northrop, *J. Gen. Physiol.* **18,** 433 (1935).
(67) E. A. Anderson and R. A. Alberty, *J. Phys. & Colloid Chem.* **52,** 1345 (1948).
(68) G. W. Schwert, *J. Biol. Chem.* **179,** 655 (1949).
(69) G. W. Schwert, *J. Biol. Chem.* **190,** 799 (1951).
(70) E. L. Smith, D. M. Brown, and M. Laskowski, *J. Biol. Chem.* **191,** 639 (1951).
(71) F. Tietze and H. Neurath, *J. Biol. Chem.* **194,** 1 (1952).
(72) E. Brand, Abstracts of papers presented at the 112th meeting Am. Chem. Soc., New York City, Sept., 1947, p. 280.

weight of 36,800 (a value more nearly in accord with the early molecular-weight estimates of Northrop and Kunitz),[1] the analytical data of Lewis *et al.*[73] are compatible with the molecular weight of 25,000.[74]

b. B-Chymotrypsinogen

This protein is more acidic than α-chymotrypsinogen having an isoelectric point of 5.2.[75] Solubility measurements at pH 5 suggest the presence of a small amount of impurity. The molecular weight of the protein[70] appears to be of the same order as that of α-chymotrypsinogen (Tables I and II).

4. Chymotrypsins

a. α, β, γ and δ-Chymotrypsins

Although these chymotrypsins represent a group of proteins differing from one another in details of isolation, in crystal habit, in specific enzymatic activity, and in details of molecular size or electrochemical characteristics, they will be considered here as a group since they are all derived from the same precursor (α-chymotrypsinogen). Most of the detailed physicochemical measurements have been carried out with α-chymotrypsin or with its inactive dialkylphosphoryl derivatives.

None of the chymotrypsins obey the phase-rule *solubility properties* to be expected for a pure component. α-Chymotrypsin has been reported to contain about 3 per cent of inert proteins.[1] It is entirely possible that the contaminating impurities are due to autolysis products.

It has been reported that between pH 6 and 9.5, at 0.01 ionic strength, α-chymotrypsin migrates *electrophoretically* with a single symmetrical peak, with a low heterogeneity constant at the isoelectric point (pH 8.6);[67] however, a recent preliminary report[76] states that upon prolonged electrophoresis within the pH range 3.2–8.3, at 0.05 ionic strength, α-chymotrypsin yields five components, all components being enzymatically active to varying degrees. Below pH 2.6 only a single component was observed. There are wide variations in the estimates of the isoelectric point of α-chymotrypsin (Table II). Kunitz,[77] using the microscopic cataphoresis method, originally reported an isoelectric point of 6.0. However, when the moving-boundary method was applied, two groups

(73) J. C. Lewis, N. S. Snell, D. J. Hirschmann, and H. Fraenkel-Conrat, *J. Biol. Chem.* **186,** 23 (1950).
(74) H. Neurath, *in* E. S. G. Barron, Modern Trends in Physiology and Biochemistry, Academic Press, New York, 1952.
(75) V. Kubacki, K. D. Brown, and M. Laskowski, *J. Biol. Chem.* **180,** 73 (1949).
(76) R. Egan, *Federation Proc.* **12,** 199 (1953).
(77) M. Kunitz, *J. Gen. Physiol.* **22,** 207 (1938).

of investigators[67,75] found a value of 8.1–8.3 (0.1 ionic strength). More recently, a value of 6.3 was reported,[78] derived from measurements of the Donnan equilibrium in acetate or phosphate buffer, 0.004–0.005 M. While the nature of these discrepancies remains to be explained, it appears likely that the value in the alkaline range (8.1–8.3) which is closer to the value to be expected from the amino acid composition, is more nearly correct.

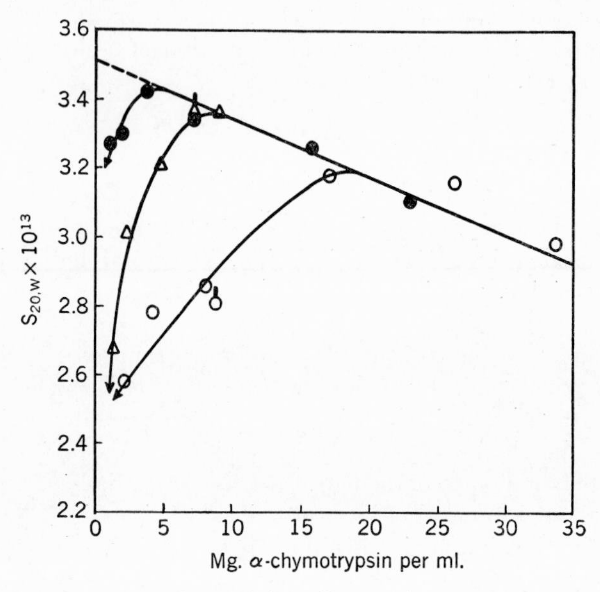

FIG. 3. Concentration dependence of the sedimentation constant of α-chymotrypsin[68] at pH 3.86 (•), pH 5.0 (△) and pH 6.20 (○). Ionic strength 0.2 in all cases.

Measurements of the concentration dependence of the *sedimentation* constant of α-chymotrypsin[68,79] have yielded a relation which is usually interpreted in terms of a monomer–polymer equilibrium (see Fig. 3). Extrapolation to zero protein concentration has yielded a value of $s_{20,w} = 2.5\ S$, which, in conjunction with an extrapolated diffusion constant of $D_{20,w} = 10.2 \times 10^{-7}$ and a partial specific volume of $V = 0.73$, corresponds to a molecular weight of 21,600. Approximately the same value was obtained from sedimentation and viscosity measurements and from diffusion and viscosity measurements.[79] Subsequent measurements by Smith and co-workers[80] are in essential agreement with this value. In higher protein concentrations, the sedimentation rate of

(78) V. M. Ingram, *Nature* **170,** 250 (1952).
(79) G. W. Schwert and S. Kaufman, *J. Biol. Chem.* **190,** 807 (1951).
(80) E. L. Smith and D. M. Brown, *J. Biol. Chem.* **195,** 525 (1952).

α-chymotrypsin approximates that to be expected for a dimer. This behavior is enhanced as the pH is decreased within the range of pH 6.28–3.86, and if the protein concentration is sufficiently high, the sedimentation rate is independent of pH. If the approximately linear concentration dependence, obtaining in high protein concentration, is extrapolated to zero protein concentration, a sedimentation constant corresponding to a molecular weight of 43,000 is found.[79] However, the data are not compatible with a simple reversible monomer–dimer equilibrium,[79] and the physical factors underlying this phenomenon are as yet indefinite. This is primarily so since no theoretical formulation exists for hydrodynamic rate processes in solutions of finite concentrations, and the interpretation just described must be considered to be largely intuitive.

It is of significance that inactive dialkylphosphoryl-α-chymotrypsin reveals the same sedimentation behavior as the active enzyme,[81,80] and that γ-chymotrypsin[68] and the product resulting from the activation of chymotrypsinogen so as to yield δ-chymotrypsin show approximately the same sedimentation behavior as does α-chymotrypsin.[82]

There are two instances in which chemical modification of α-chymotrypsin appears to destroy its ability to dimerize. One of these is oxidized diisopropylphosphoryl-α-chymotrypsin[70] and the other one the product resulting from the action of carboxypeptidase on diethylphosphoryl-α-chymotrypsin[83] (p. 1154).

While hydrodynamic measurements of the molecular weight indicate a value of approximately 22,000, somewhat higher values have been obtained by light-scattering[84] and osmotic-pressure[4] measurements (Table I). Since no experimental details have as yet been published, it is not clear whether the apparent dimerization at finite protein concentrations is responsible for these somewhat higher values. Measurements of the combination of α-chymotrypsin with diisopropyl fluophosphate and with diethyl *p*-nitrophenylphosphate[85,86] yielded a value of approximately 25,000 for the molecular weight of these inactive derivatives. Approximately the same values have been obtained in similar measurements carried out on β-, γ-, and oxidized α-chymotrypsins.[4] It has been recently reported that the analytical values for histidine, tryptophan,

(81) J. E. Snoke, *in* G. W. Schwert and S. Kaufman, *J. Biol. Chem.* **190,** 807 (1951).
(82) G. W. Schwert and S. Kaufman, *J. Biol. Chem.* **180,** 517 (1949).
(83) J. A. Gladner and H. Neurath, *J. Biol. Chem.* **206,** 911 (1954).
(84) K. J. Palmer, *in* A. K. Balls and E. F. Jansen, *Advances in Enzymol.* **13,** 321 (1952).
(85) E. F. Jansen, M. D. F. Nutting, R. Jang, and A. K. Balls, *J. Biol. Chem.* **179,** 189 (1949).
(86) B. S. Hartley and B. A. Kilby, *Biochem. J.* **50,** 672 (1952).

tyrosine, methionine, and cystine/2 in α-chymotrypsin[87] are more nearly in accord with a molecular weight of 27,000 than of 22,500.

b. *B-Chymotrypsin*

Although, as judged by solubility measurements, this enzyme is not a pure protein,[75] it is apparently electrophoretically monodisperse. In contrast to α-chymotrypsin, its isoelectric point is on the acid side of neutrality (pH 4.7, 0.1 ionic strength), and the two active enzymes appear as distinctly separated components when analyzed as mixture in the moving-boundary electrophoresis apparatus.[75] It is of interest that in the case of both the α and B proteins, the zymogen has a more alkaline isoelectric point than the active enzyme (Table II).

α- and B-Chymotrypsins do not sediment independently in the ultracentrifuge since they have approximately the same molecular weight (Table I). However, it has been reported that at pH 4, B-chymotrypsin shows a concentration dependence of the sedimentation constant characteristic of a monomer, whereas at higher pH values, up to 6.8, the concentration dependence is similar to that previously described for α-chymotrypsin.[80]

5. Trypsinogen

No solubility or electrophoretic measurements on crystalline trypsinogen have been reported. According to unpublished measurements, using mixed-bed ion-exchange resins, trypsinogen is isoionic at pH 9.3.[88]

The only molecular-weight measurements on twice-crystallized trypsinogen are those of Tietze,[8] derived from measurements of the sedimentation and diffusion rates and of the partial specific volume, carried out at pH's 3.86 and 5.0. These measurements have yielded the following values: $s_{20,w} = 2.48\ S$; $D_{20,w} = 9.7 \times 10^{-7}$; $V = 0.73$, corresponding to $M = 23{,}700$ and $f/f_0 = 1.15$. A single electrophoretic analysis at pH 4.82, 0.1 ionic strength has revealed that the preparation under investigation was essentially monodisperse, showing a small amount of a fraction of lower mobility.

6. Trypsin

In the following discussion the active enzyme and the inactive dialkylphosphoryl derivate will be considered interchangeably since both forms have been used for determination of the physicochemical properties of the enzyme. The crystalline enzyme prepared by Kunitz and Northrop[89] shows constant solubility in a saturated solution of magnesium

(87) L. Weil, S. James, and A. R. Buchert, *Arch. Biochem. Biophys.* **46**, 266 (1953).
(88) N. M. Green, unpublished.
(89) M. Kunitz and J. H. Northrop, *J. Gen. Physiol.* **19**, 991 (1936).

sulfate, pH 4.0, at 10°. This is perhaps surprising since the enzyme is subject to autolysis over a wide pH range.[1,90]

The *electrophoretic* behavior of trypsin is complicated by its instability and is affected by factors which tend to stabilize the enzyme. Thus it has been reported[91] that in the absence of stabilizing concentrations of calcium ions, trypsin was electrophoretically homogeneous between pH 4 and 5, whereas, in the presence of calcium, two boundaries appeared which were clearly separated when the calcium concentration exceeded 0.03M. Manganese and cadmium ions were similarly effective. At pH 3.5 or 5.5 a single peak appeared, even in the presence of these stabilizing ions. The faster-moving component had the mobility characteristic of trypsin in the absence of calcium, but both components were reported to be enzymatically active though their relative specific activities were not determined.[91] At pH 7.5 (which is within the optimum range of enzymatic activity), calcium ions had no initial effect on the electrophoretic pattern, but after prolonged time (20 hours or longer) in the absence of calcium, about 96 per cent inactivation occurred and the original boundary was replaced by a new stationary one which probably represented inactive protein fragments since the material could be removed by dialysis at pH 4. In the presence of calcium, only 50 per cent inactivation occurred under otherwise identical conditions and 50 per cent of the material under the initial boundary shifted toward the stationary one.[92] It is likely that these effects are related to the influence of calcium on the equilibrium between the various forms of trypsin discussed elsewhere in this chapter (p. 1091). The electrophoretic pattern of diisopropylphosphoryltrypsin (DIPT) in the absence of calcium ions is similar to that of trypsin in the presence of bivalent metal ions.[93] At pH 3.8 the protein moved as a single component; but as the pH was increased, boundary asymmetry appeared; and above pH 8, a second component comprising 7–15 per cent of the total protein was clearly separated.

Electrophoretic mobility measurements, using the moving-boundary method, yielded a value for the isoelectric point of 10.4–10.8[92,93] (Table II) which is within the same range as the isoionic point obtained with the aid of mixed-bed ion-exchange resins (pH 10.1).[88] These values, which are in accord with estimates of the amino acid composition of trypsin, are appreciably higher than the value of 7.0 obtained earlier for the isoelectric

(90) L. W. Cunningham, Jr., F. Tietze, N. M. Green, and H. Neurath, *Discussions Faraday Soc.* No. **13,** 58 (1953).
(91) F. F. Nord and M. Bier, *Biochim. et Biophys. Acta* **12,** 56 (1953).
(92) M. Bier and F. F. Nord, *Arch. Biochem. Biophys.* **33,** 320 (1951).
(93) L. W. Cunningham, *J. Biol. Chem.*, in press.

point by the microscopic cataphoresis method.[1] Accepting the higher values as being more nearly correct, it is of interest to note that in contrast to chymotrypsinogen, the activation of trypsinogen yields a product having a more alkaline isoelectric point than the parent zymogen (see p. 1192). The isoelectric point of DIPT is similar to that of active trypsin (i.e., 10.5).[93] The mobility curve is rather flat between pH 6 and 8 and shows a sharp break between pH 9 and 10.

The titration curve of trypsin has been recently determined.[94] Calcium ions shift the curve toward the acid side between pH 3.5 and 5. This has been interpreted in terms of chelation of calcium with carboxyl groups, increasing the acidity of the latter. It is not established whether this effect is functionally related to the stabilizing effect of calcium on the enzyme.

Estimates of the *molecular weight* of trypsin vary within wider limits than those of the other enzymes previously considered (Table I). While it is likely that the value of 36,500, determined by Kunitz and Northrop[89] from osmotic-pressure measurements, is too high, the variations among more recently reported values (Table I) are likely to be related to the failure of some of the investigators to recognize the influence of a variety of experimental factors on the monomer–polymer equilibrium and on the autodigestion of the enzyme. Thus it has been shown in a recent study[90] that the sedimentation behavior of trypsin varies with the concentration of the protein, with the pH, with calcium concentration, and with the age of the solution in such a manner as to suggest that when the protein is enzymatically inactive (trypsin at pH 3 or DIPT at any pH), the sedimentation rate is independent of the age of the solution and characteristic of that of a monodisperse solute. However, under conditions favorable for enzymatic activity, the sedimentation rate was found to be dependent on pH and on the age of the solution and to reveal a complex variation with protein concentration. This was ascribed in part to trypsin–trypsin interaction, (transient "enzyme–substrate" complex formation) in the course of autolytic decay. At pH 7.8 but not at pH 5, these complexities were largely alleviated by calcium ions. Similar findings have been subsequently reported by Nord and Bier[91] who have interpreted their findings to suggest that trypsin is not a homogeneous protein. Earlier sedimentation and diffusion measurements, given for a single finite protein concentration, and corresponding to $M = 15{,}100$[95] are of questionable validity.

The sedimentation constant of trypsin,[90] of DIPT[93] and of trypsinogen,[8] extrapolated to zero protein concentration, is $s_{20,w} = 2.5\ S$.

(94) J. A. Duke, M. Bier, and F. F. Nord, *Arch. Biochem. Biophys.* **40,** 424 (1952).
(95) V. G. Bergold, *Z. Naturforsch.* **1,** 100 (1946).

Together with the preliminary estimate of the diffusion constant of DIPT at pH 3.86,[93] $D_{20,w} = 9.3 \times 10^{-7}$; this yields a molecular weight of 23,800. Sedimentation and diffusion measurements of trypsin–soybean inhibitor compound[96] at finite protein concentrations have led to an estimate of 41,000 for the compound and of 17,000 for the enzyme assuming a molecular weight of 24,000 for the inhibitor. However, the values obtained for the inhibitor-trypsin combining ratio suggest that the figures should be reversed, i.e., trypsin = 23,000 and soybean inhibitor = 18,000. Measurements of the film molecular weight on 35 per cent ammonium sulfate[97] have yielded a value of 41,000, considered to represent a dimer (20,500 for the monomer).

Estimates of the minimum molecular weight, based on the phosphorus content of inactive trypsin derivatives, vary between the range of 17,500 and 25,000.[90,98,99] Thus the phosphorus content of DIPT corresponds to a molecular weight of approximately 21,000, but if allowance is made for the phosphorus content of active trypsin (0.02 per cent) this value is raised to 24,800.[90]

7. Carboxypeptidase

Of the four electrophoretic components seen in crude pancreatic exudate at pH 8.5, carboxypeptidase represents the one of next-to-lowest mobility.[28] Seven-times-recrystallized preparations of pancreatic carboxypeptidase are electrophoretically essentially monodisperse at pH 8.5, but at pH 9.3 a faster-moving component comprising 20 per cent of the protein makes its appearance. The isoelectric point of the enzyme corresponds to pH 6 (0.2 ionic strength).

The molecular-kinetic parameters of the enzyme as determined by sedimentation and diffusion measurements are as follows:[28,31] $s_{20,w} = 3.07\ S$; $D_{20,w} = 8.68 \times 10^{-7}$ corresponding to $M = 34{,}300$ (Table I). This value agrees favorably with the minimum chemical molecular weight determined by amino acid analysis.[99a]

8. Papain

The crystalline enzyme of Balls and Lineweaver[43] has an isoelectric point above pH 8.5, as determined cataphoretically. The molecular weight is 27,000 ± 2000, as determined by osmotic pressure measurements. Diffusion measurements, using the sintered-glass disk, and com-

(96) A. D. McLaren, *Compt. rend. trav. lab. Carlsberg. Sér. chim.* **28,** 175 (1952).
(97) E. Mishuck and F. Eirich, *J. Polymer Sci.* **7,** 341 (1951).
(98) E. F. Jansen and A. K. Balls, *J. Biol. Chem.* **194,** 721 (1952).
(99) B. A. Kilby and G. Youatt, *Biochim. et Biophys. Acta* **8,** 112 (1952).
(99*a*) E. L. Smith and A. Stockell, *J. Biol. Chem.* **207,** 501 (1954).

parative sedimentation measurements with lactoglobulin, in a quantity centrifuge, are in essential accord with this value.

The crystalline enzyme prepared from dried latex has been recently characterized with respect to electrophoresis, sedimentation, and diffusion.[100] The electrophoretic mobility curve revealed a broad zone of nearly constant mobility between pH 4 and 6 (ionic strength 0.1) and an isoelectric point of pH 8.75. These properties are in accord with the known acidic and basic amino acid content of the enzyme. The electrophoretic patterns suggest that the preparations were not entirely homogeneous. Papain and its crystalline mercury derivative show monomeric sedimentation behavior at pH 4. The molecular-kinetic parameters of $s_{20,w} = 2.42\ S$, $D = 10.3 \times 10^{-7}$, and $V = 0.724$ correspond to a molecular weight of 20,700, which agrees well with the minimum chemical molecular weight.

Crystalline mercuripapain shows two electrophoretic components at pH 3.9 (in the absence of cysteine and Versene, ionic strength 0.1) and two sedimenting components at pH 8, the relative amount of the heavier component decreasing with decreasing protein concentration. The sedimentation constant of the light component corresponds to that of the monomer, whereas that of the heavy one corresponds to that of a hexamer

9. Rennin

Measurement of the molecular weight of crystalline rennin by sedimentation–diffusion gave a provisional value of 40,000 ($s_{20,w} = 4.15\ S$, $D = 9.5 \times 10^{-7}$). The figure was based on relatively few measurements, owing to shortage of material.[106] Paper electrophoresis showed the protein to be isoelectric between pH 4 and 5.

IV. Chemical Composition

Data on the amino acid composition of the proteolytic enzymes are summarized in Table III. The older values have been obtained mainly by microbiological assay, whereas the more recent ones are the results of the more reliable procedure of Stein and Moore.[101] Some of the data included in Table III have not yet been published in full.

With the exception of the values for rennin and pepsin, the data given

(100) E. L. Smith, J. R. Kimmel, and D. M. Brown, *J. Biol. Chem.* **207**, 533 (1954).
(101) W. H. Stein and S. Moore, *J. Biol. Chem.* **176**, 337 (1948).
(101*a*) S. Moore and W. H. Stein, *J. Biol. Chem.* **192**, 663 (1951).
(102) G. E. Perlmann, *J. Am. Chem. Soc.* **74**, 6308 (1952).
(103) R. M. Herriott *in* W. D. McElroy and B. Glass, eds., Mechanism of Enzyme Action, Johns Hopkins Press, Baltimore, 1954.
(104) E. Brand, *in* J. H. Northrop, M. Kunitz, and R. M. Herriott, Crystalline Enzymes, 2nd ed., Columbia Univ. Press, New York, 1948.

in Table I have been corrected for the partial destruction of amino acids during hydrolysis and for incomplete hydrolysis of the more resistant peptide bonds.[106b] The published values for carboxypeptidase are generally in accord with unpublished analyses carried out in this laboratory.

In general, the amino acid compositions are in accordance with the observed isoelectric points of the proteins. Thus pepsin, with an isoelectric point below pH 1, has a very low content of basic amino acids and considerable quantities of aspartic and glutamic acids (69 residues per mole), which are, however, partially blocked by amide groups (32 residues per mole). In spite of its high isoelectric point (9.3) trypsinogen contains little arginine; however, its lysine content is high, and over two-thirds of the 34 carboxyl groups are present in the amide form.

Other noteworthy features are the single methionine residue in the pancreatic enzymes, and its complete absence from papain, a single histidine residue in papain and two in chymotrypsinogen.

Pepsin is the only purified protease to contain stoichiometric amounts of phosphorus (one equivalent per 34,000 g. pepsin). The phosphorus can be removed by treating the enzyme with intestinal or prostatic phosphatase at pH 5.6 without loss of activity.[102] The phosphorus is therefore present as a phosphate ester group. The recently purified kidney aminopeptidase[41] is reported to contain 30 per cent phospholipide, whose removal leads to inactivation, but excluding the activating metal ions of certain of these enzymes, all other proteases appear to be composed exclusively of amino acids.

The N- and C-terminal amino acids of the proteases and their precursors are discussed in detail in sec. VIII, where the available data are tabulated. It appears that pepsin, trypsin, and their respective precursors, carboxypeptidase and papain, are single peptide chains, since each possesses a single N-terminal amino acid as determined by Sanger's fluorodinitrobenzene (FDNB) reagent. The action of thioglycolic acid on pepsin supports this conclusion since it reduces the disulfide bonds without affecting the molecular weight.[103] No terminal residues could be shown in chymotrypsinogen, but two N- and two C-terminal residues are present in α-, β-, and γ-chymotrypsins.

(105) H. Fraenkel-Conrat, R. S. Bean, and H. Lineweaver, *J. Biol. Chem.* **177,** 385 (1949).
(106) H. Schwander, P. Zahler, and H. S. Nitschmann, *Helv. Chim. Acta* **35,** 553 (1952).
(106*a*) E. L. Smith, A. Stockell, and J. R. Kimmel, *J. Biol. Chem.* **207,** 551 (1954).
(106*b*) E. J. Harfenist and L. C. Craig, *J. Am. Chem. Soc.* **74,** 4216 (1952).
(107) H. Fraenkel-Conrat and H. S. Olcott, *J. Biol. Chem.* **171,** 583 (1947).
(107*a*) E. Cohen and P. E. Wilcox, unpublished experiments.
(107*b*) E. Cohen, E. W. Davie, and H. Neurath, unpublished experiments.

TABLE III

AMINO ACID COMPOSITION OF SOME PROTEOLYTIC ENZYMES[a]

Amino acid	Chymotrypsinogen				Trypsinogen[e,f]		Carboxy-peptidase[f]		Papain[f]		Rennin	Pepsin
	A^b 104	A^b 73	$A^{c,e,f}$ 107a	$B^{d,f}$	A^c 107b	B^d	A^c 99a	B^d	A^c 106a	B	A^b 106	A^b 104
Alanine	......	5.8	7.3	19	4.7	13	5.16	20	5.73	13	2.7	......
Glycine	5.3	6.3	6.6	20	6.5	21	5.06	23	8.41	23	4.5	6.4
Valine	10.1	10.3	9.9	20	7.2	15	5.58	16	8.43	15	9.9	7.1
Leucine	10.4	8.9	9.3	16	6.8	12	9.41	25	6.10	9	13.4	10.4
Isoleucine	5.7	4.9	4.8	9	6.6	12	7.65	20	6.05	9		10.8
Proline	5.9	3.8	3.8	8	3.5	7	3.66	11	5.11	9	4.6	5.0
Phenylalanine	3.6	3.8	4.1	6	2.4	4	7.16	15	3.16	4	6.8	6.4
Tyrosine	3.0	2.7	2.8	4	6.6	9	10.35	20	14.71	17	6.6	8.5
Tryptophan	5.6	......	.....	6	......	......	3.62	6	4.68	5	1.4	2.4
Serine	11.4	10.9	11.9	26	16.7	38	10.09	33	5.91	11	7.8	12.2
Threonine	11.4	10.7	10.4	20	(5.0)	(9–11)	9.21	27	3.89	7	5.6	9.6
Cystine	1.29	......	.....	...	......	......	(1.40)	(2)	4.58l	4	0.8	1.7
Cysteine	3.30[g]	......	.....	...	......	......	......	...	......	...	......	0.5
Methionine	1.22	1.11	(0.5)	(1)	0.9	1	0.44	1	0	0	0.9	1.7
Arginine	2.8	2.7	2.8	4	1.7	2	5.06	10	7.75	9	1.3	1.0
Histidine	1.23	1.16	1.1	2	2.0	3	3.47	8	0.85	1	2.8	0.9
Lysine	8.0	7.7	7.6	12	8.4	14	7.81	18	5.67	8	2.9	0.9
Aspartic acid	11.3	10.9	11.0	19	13.1	24	11.70	30	11.32	17	15.3	16.0
Glutamic acid	9.0	7.4	6.5	10	6.2	10	10.67	25	12.43	17	11.5	11.9
Amide N	1.86	......	.....	23	......	(23 ± 3)	0.95	19	1.60	19	1.4	1.6
Total	118.[h]	......	.....	...	......		117.50		114.68	178	......	115.
N	16.2	......	.....	...	......		15.1		16.1	...	......	164.
S	1.5	......	.....	...	......		......		1.22	...	......	0.94
Molecular weight		23,000			23,800		34,400		20,340		40,000	35,500
Isoelectric point		9.1			9.3[i]		6.0		8.75		4–5	<1

[a] A represents grams of amino acid/100 g. protein, and B, residues of amino acid per molecule, using the stated molecular-weight values. It should be noted that for complete recovery, the sum of the values given under col. A should be

$$100 \times \frac{(\text{average residue weight} + 18)}{(\text{average residue weight})}$$

Values in parentheses are considered to be tentative.

[b] Microbiological assay.

[c] Column chromatography.[101,101a]

[d] Based on most likely values of A, corrected to nearest integer.

[e] Most likely values obtained from 12-, 24-, 36-, 48-, and 72-hr. hydrolyzates.

[f] Values corrected for partial destruction of some amino acids and for incomplete liberation of others.

[g] It has been shown[107] that the figures for cysteine obtained by Brand were due to decomposition of cystine in the presence of tryptophan during hydrolysis. Since neither chymotrypsinogen nor pepsin contains any free sulfhydryl groups, the cysteine and cystine figures should be combined and considered as a measure of the cystine content.

[h] Including alanine from Ref.[73]

[i] Calculated from S content.

[j] Isoionic point.

V. Stability

The present discussion deals with the stability of proteolytic enzymes and their precursors as proteins; factors influencing enzyme activity will be considered in sec. VI.

Proteolytic enzymes are inactivated under conditions usually causing protein denaturation, inactivation being usually accompanied by the common manifestations of denaturation such as decreased solubility, increased susceptibility to proteolysis, etc. As with many other proteins of relatively low molecular weight, proteolytic enzymes are rather stable, many of them tolerating high concentrations of urea and elevated temperatures in favorable pH regions. In contrast to other proteins, proteolytic enzymes are susceptible to autolysis, and hence several of them are unstable at their pH optima in the absence of substrate.

Since the subject of protein denaturation has been treated elsewhere in this treatise (Vol. I, Chap. 9), only an outline of the results will be given herein.

1. pH and Temperature

Since the effects of pH and temperature are interrelated, they may be considered together in relation to the various enzymes.

a. Pepsin

Pepsin is optimally active at pH 2 and is unstable in this pH region due to autolysis. When the enzyme is allowed to stand under these conditions, or even in solutions of 2 *N* hydrogen ions, no denatured protein appears (as indicated by loss of solubility at pH greater than 3),[108] and only active pepsin and small peptides are found. Even at pH 4, where pepsin is usually crystallized, slow autolysis occurs. Thus Ingram[16] showed that on standing for 24 hours (pH 4, 45°), 40 per cent of the constituent tyrosine could be isolated in crystalline form.

Pepsin reveals maximum stability at pH 5–5.5, and at higher pH becomes highly sensitive to alkali. Steinhardt[109] studied the kinetics of the denaturation over the pH interval 6.0–6.8 and showed the rate to be approximately proportional to $(OH)^5$. At lower pH regions, the rate was higher than expected from this relation, owing to surface denaturation, and at pH higher than 6.6 the rate was lower than predicted. These observations could be explained on the assumption that five groups in the pepsin molecule underwent dissociation in this pH region and that only

(108) J. H. Northrop, *J. Gen. Physiol.* **16,** 33 (1932).
(109) J. Steinhardt, *Kgl. Danske Videnskab. Selskab. Mat.-fys. Medd.* **14,** 11 (1937).

the form which had lost five protons contributed appreciably to the rate of denaturation. From the calculated acid dissociation constant ($pK = 6.7$), it was concluded that α-amino or imidazolyl groups were involved. However, it was shown by Herriott[110] that following acetylation of pepsin with ketene, a crystalline, enzymatically active enzyme preparation could be obtained containing no α-amino nitrogen, and subject to denaturation by alkali. For this reason it is unlikely that amino groups are actually involved in the alkali denaturation. As there are only two imidizolyl groups per molecule,[104] it is difficult to correlate Steinhardt's observation with the known properties of pepsin. However, Steinhardt's theory has many attractive features, particularly as it can account for the high over-all activation energy without having recourse to improbably large changes in entropy of activation.[111] Thus from the slope and relative positions of plots of logarithm of rate constant against pH at 15 and 25°, Steinhardt deduced that of the 63.5 kcal./mole over-all activation energy, 45.2 kcal. were accounted for by the heat of ionization of the five groups and that the remaining ΔH^* for inactivation of the unstable ionic species was only 18.3 kcal./mole. This calculation would imply a heat of ionization of about 9 kcal./mole for each group, a value to be expected for amino groups but not for carboxyl groups. It is evident that a full elucidation of this process must await further experimental evidence.

More recent studies by Sturtevant *et al.*,[112,113] using a calorimetric technique, confirmed that the denaturation of pepsin between pH 6.2 and 6.8 followed first-order kinetics, but the rate was found to be dependent only on $(OH)^{1.4}$. However, if the rate was determined by measurements of enzyme activity, the reaction was no longer first order, but the time of half-inactivation was proportional to $(OH)^4$, in better agreement with Steinhardt's results. This discrepancy between heat evolution and extent of inactivation is similar to that earlier reported by Conn *et al.*[114] who measured the heat evolved on raising the pH of a pepsin solution from an initial value of pH 5–6.8 to a final value of pH 9. The higher the initial pH, the less was the heat evolved during the pH change, but the decrease in ΔH did not correspond to the decrease in activity at the initial pH. Further complexities arise from the work of Casey and

(110) R. M. Herriott, *J. Gen. Physiol.* **19**, 283 (1935).
(111) V. K. La Mer, *Science* **86**, 614 (1937).
(112) M. Bender and J. M. Sturtevant, *J. Am. Chem. Soc.* **69**, 607 (1947).
(113) A. Buzzell and J. M. Sturtevant, *J. Am. Chem. Soc.* **74**, 1983 (1952).
(114) J. B. Conn, D. C. Gregg, G. B. Kistiakowsky, and R. M. Roberts, *J. Am. Chem. Soc.* **63**, 2080 (1941).

Laidler[115] who investigated the effect of pepsin concentration on denaturation at pH 4.8 and found that at low concentrations (less than 0.04 per cent) the order of the reaction with respect to pepsin began to increase, eventually reaching a value of 5 (0.004 per cent pepsin). This effect of low protein concentration on the order of reaction was not confirmed by Buzzell and Sturtevant,[113] but their measurements were made at a higher pH (6.5). It was found in the same work that the heat of the reaction was dependent both on pH and temperature, with a maximum value with respect to pH. At 35° the maximum (48 kcal./mole) was at pH 6.4, while at 15° (15 kcal./mole) it was at pH 7.2. No simple explanation for these results could be offered. It has been subsequently pointed out[116] that the magnitude of these effects is too large to be accounted for by changes in the ionization of the two imidizolyl groups of pepsin, these being the only ones known to undergo a change of ionization in this region. It is possible that the unidentified groups postulated by Steinhardt are involved.

Partial reactivation of denatured pepsin occurs when the enzyme is allowed to stand for only a short time at pH greater than 6 and the pH is then lowered to 5.5.

The quantitative data on pepsin denaturation are thus confusing. Since different workers have used different conditions for their experiments, comparison is difficult. Moreover, as discussed above (p. 1071), pepsin is not a single protein, and the preparations used in these experiments often contained considerable amounts of nonprotein nitrogen.[113] The most that can be said in a general way is that pepsin is rapidly denatured above pH 6, the rate being dependent on (*1*) a high power of the hydroxyl-ion concentration, (*2*) the ionic strength,[109] (*3*) the nature of the buffer used,[109] and (*4*) the method of following the reaction.[113]

b. Pepsinogen

Pepsinogen is generally more stable than pepsin.[56] Thus neutral, salt-free solutions can be boiled for a limited time, and cooled without inactivation, though denatured protein can be precipitated from the hot solution (higher than 70°) by sodium chloride, indicating that reversible denaturation has occurred. Below 50° no precipitation occurs. In acid solution and in the presence of traces of pepsin, pepsinogen undergoes autocatalytic conversion to pepsin. If exposed for short periods of time to solutions more alkaline than pH 9, pepsinogen also undergoes reversible denaturation, the extent of denaturation being proportional to $(OH)^2$.

(115) E. J. Casey and K. J. Laidler, *J. Am. Chem. Soc.* **73,** 1455 (1951).

(116) K. Linderstrøm-Lang and K. M. Møller, *Ann. Rev. Biochem.* **22,** 57 (1953).

c. Chymotrypsinogen

Chymotrypsinogen is stable in acid solutions above pH 2, but the upper pH limit of stability has not been determined, being probably higher than pH 9. The protein is subject to heat denaturation which is initially reversible and shows a minimum between pH 3 and 5. Detailed studies of the denaturation as a function of pH and temperature have been carried out by Eisenberg and Schwert[117] using insolubility in glycine-HCl buffer, ionic strength 1, at pH 3 as criterion of denaturation. Between pH 2 and 3, denaturation effected by heating to 100° for 2 minutes was completely reversible, but in higher pH regions, the extent of reversal decreased, until at pH 6.5 no reversal could be obtained. The free energies and entropies of denaturation and activation of the forward reaction were determined at pH's 2 and 3 (see Vol. I, Chap. 9). The reaction rates were high, equilibrium being attained in less than 1 minute, but a reasonable fit of the points to the curve for a first-order reversible equilibrium was obtained. There are two points worthy of note pertaining to the experimental aspects of this work and of the closely related studies of Kunitz[118] involving soybean–trypsin inhibitor: (*1*) The reversibility of the process was carefully established by crystallization, solubility measurements, and activity measurements on the reversibly denatured protein. (*2*) The equilibrium position was shown to be the same whether approached from the native or the denatured side.

Quantitative results followed most closely the expression $D/N = K$ (rather than $D/N^n = K$), and the dependence of the equilibrium constant on pH followed the relation

$$\log (N/D) = 3.16(\text{pH} - 2.5)$$

It was deduced from these relations that the uptake of three protons by three particular groups of the native protein had to precede denaturation, and it was suggested that the pK of these groups was of the order of 2.5. This interpretation has since been questioned[119] on the grounds that the correct relation between pH and extent of denaturation was

$$\log (N/D) = 3.16(\text{pH} - \text{p}K_a) + \log K_d$$

where K_d is the equilibrium constant for the denaturation of the susceptible native species, ($NH^{+++} \rightleftharpoons DH^{+++}$), and that p$K_a$ can be determined only if K_d is known. Since the linear relation between log (N/D) and pH holds between pH 2 and 3, pK_a is probably less than 2 and can still be

(117) M. A. Eisenberg and G. W. Schwert, *J. Gen. Physiol.* **34,** 583 (1951).
(118) M. Kunitz, *J. Gen. Physiol.* **32,** 241 (1948).
(119) N. M. Green, unpublished.

referred to a carboxyl group. Since this group has a low heat of ionization, it is not possible to account for the high energy and entropy of activation in the same way that was done for pepsin (see p. 1086). In order to account for these high energetic constants,[117] it was suggested that denaturation was accompanied by loss of water of hydration, amounting to 70 moles of water per mole of chymotrypsinogen at pH 2 (5 per cent) and 100 moles at pH 3 (7 per cent). This loss would be accompanied by absorption of heat, increase of entropy, and lowered solubility due to decreased interaction with the solvent.

d. α-Chymotrypsin

α-Chymotrypsin has not been investigated in such detail as has its precursor. In acid solution, its behavior is rather similar.[66] It is slowly denatured on standing in 0.1 *N* HCl, losing about 20 per cent of its activity in 1 hour at 20°. Heating in 0.0025 *N* HCl leads to reversible inactivation, the denatured form being insoluble in 1 *M* NaCl at pH 2, but on cooling in salt-free solution, the activity is completely regained. Prolonged heating leads to irreversible changes (e.g., after 15 minutes at 100°, 50 per cent of the activity is not regained on cooling). In concentrated solutions at pH 7.8, activity is not lost but the enzyme is converted by autolysis into two new forms (β- and γ-chymotrypsin). No detailed investigations have been carried out on the stability of the enzyme in the alkaline pH region.

e. Trypsinogen

Trypsinogen[17] is stable in acid solution (pH 2–4) and, in fact, can be purified by precipitation with 2.5 per cent trichloroacetic acid, followed by dissolution of the washed precipitate in 0.02 *N* HCl. At more alkaline pH, the protein is less stable than chymotrypsinogen since it is more likely to contain traces of the activating enzyme, trypsin. Unfortunately, crystallization is known to occur only at pH 7.8, and recrystallization has been found impossible unless DFP or a protein inhibitor (pancreatic or soybean) is added to prevent autocatalytic activation.[90,8] In the presence of 0.6 per cent protein inhibitor, trypsinogen is stable at pH 10.5, 4°, for 48 hours, losing only 10 per cent of its potential activity.[88]

f. Trypsin

In acid solution, trypsin behaves in the same manner as chymotrypsin. In solutions more alkaline than pH 6, trypsin is much less stable, undergoing rapid autolysis. This was explained[120] on the grounds of an equilibrium between native and denatured trypsin, the reversibly

(120) M. Kunitz and J. H. Northrop, *J. Gen. Physiol.* **17**, 591 (1934).

denatured form being digested by the native form. At pH greater than 11, the concentration of native trypsin becomes so small and its activity so low that large amounts of the reversibly denatured form accumulate in solution and can be precipitated by readjustment to pH 2 in the presence of 0.5 *M* NaCl. On standing for 30 minutes, in 0.02 *N* HCl, the protein of the precipitate can be 90 per cent reactivated to trypsin,[88] thus resembling trypsin reversibly denatured by trichloroacetic acid.[89] The formation of the reversibly denatured protein is instantaneous, and on standing it changes to irreversibly denatured material, the rate of this reaction increasing with pH.

The equilibrium of trypsin with reversibly denatured protein at pH 2 has been quantitatively investigated.[121] The protein was found to be almost completely native below 40° and almost completely denatured above 50°, the heat of the reaction being 67 kcal./mole (see Vol. I, Chap. 9).

Recently, the stability of trypsin in neutral solution has received closer attention,[122,92,90] and it was found that Ca^{++} and Mn^{++} greatly reduced the autolysis in this region. Thus in the absence of Ca^{++}, trypsin is completely inactivated in 8 hours at 26°, pH 7.9; while in the presence of 10^{-2} *M* Ca^{++}, only 5 per cent inactivation occurred.[122] Similar results were reported by Bier and Nord[92] with both Ca^{++} and Mn^{++} but not with Mg^{++}, Sr^{++}, Ba^{++}, Co^{++}, or the alkali-metal cations. It was also reported that calcium salts suppressed the formation of a second peak in the electrophoretic pattern of trypsin, whereas in the absence of this ion, the peak increased in area with time at the same rate as trypsin lost its activity, the phenomenon presumably being due to the formation of digestion products. These results were interpreted in terms of an equilibrium $T_n \rightleftharpoons T_d$ postulated by Kunitz. It was suggested, therefore, that Ca^{++} combined with the native form of trypsin and shifted the equilibrium toward the left. Alternatively, however, it might be that Ca^{++} combined with the denatured form in such a way to restore the native configuration or some approximation to it. More recently, the stability of trypsin at 40° was reinvestigated by Crewther,[123] over a wide pH range (pH 3–10) and with due regard to the effects of added cations on stabilization and inactivation, respectively. Hemoglobin and casein substrates were used in this work. In agreement with some of the earlier reports (cited by Crewther), it was found that cations stabilize dilute trypsin solutions, trivalent cations being in general more effective than bivalent ones, whereas univalent ones required considerably

(121) M. L. Anson and A. E. Mirsky, *J. Gen. Physiol.* **17**, 393 (1934).
(122) L. Gorini, *Biochim. et Biophys. Acta* **7**, 318 (1951).
(123) W. G. Crewther, *Australian J. Biol. Sci.* **6**, 597 (1953).

higher concentration (2 *M* as compared to 10^{-4} *M*) to afford any measurable stabilization. At the lowest concentrations of bivalent cations used (10^{-4} *M*), and with the exception of Hg and Ni, the order of effectiveness in stabilization followed that for complex formation with 8-hydroxyquinoline or with amino acids (Be > Cu > Zn > Cd > Co > Mg > Mn > Ca > Ba > Ni > Hg). At higher concentrations (0.05 *M*) incubation with many of these ions led to inactivation which was greatest with Hg and Ni and progressively less with the other ions (Hg > Ni > Cu > Be > Zn > Cd > Co > Mn > Mg > Ba > Ca). The relative efficacy of stabilization by Ca and Mg depended on the concentration of trypsin, on pH, and on the composition of the supporting buffer. While the molecular mechanism of stabilization remains to be fully elucidated, previous hypotheses invoking autodigestion or the formation of intermediate, reversibly denatured forms, are discounted.[124]

The effects of Ca^{++} on the enzymatic activity of trypsin and on the conversion of trypsinogen to trypsin will be discussed more fully elsewhere in this chapter (p. 1149).

g. Carboxypeptidase

Carboxypeptidase is unstable below pH 6;[26,27] however, unlike the enzymes thus far discussed, it is not prone to autolyze,[125] and hence is stable at its pH optimum (pH 7.5). The upper limit of stability is pH 10.4 at 0°.[30] Pure carboxypeptidase, unlike the components of the chymotrypsin and trypsin systems, is inactivated by drying or by lyophilization.[30]

h. Cathepsins

Cathepsin C is the only member of this class of enzymes which has been extensively purified.[33] It is more stable than cathepsin B since it can be separated from the latter by heating to 65° for 40 minutes in 2 per cent sodium chloride, cathepsin B being precipitated. No further information is available on the effects of temperature or pH on the possible autolysis of these enzymes.

i. Plant Enzymes

The stability of the enzymes of this group has not been widely investigated except with respect to oxidizing and reducing agents which will be considered elsewhere (p. 1145). Crystalline papain is stable at 30° between pH 4 and 8,[126] but is rapidly inactivated below pH 2.5 and above pH 12. It would appear, therefore, that it does not undergo appreciable

(124) The original paper should be consulted for a discussion of these data.
(125) E. W. Davie and H. Neurath, *J. Am. Chem. Soc.* **74**, 6305 (1952).
(126) H. Lineweaver and S. Schwimmer, *Enzymologia* **10**, 81 (1941).

autolysis since its pH optimum is at pH 5.5.[127] Crystalline chymopapain is reported to be stable at pH 2, 10°, for several weeks.[44]

2. Irradiation

The inactivation of proteolytic enzymes by x-rays, electrons, deuterons, and especially by ultraviolet irradiation, has received some attention but has not yet added much of importance to our knowledge of these enzymes. The action of ultraviolet on proteins in general has been adequately reviewed.[128] The usual source of radiation in the ultraviolet experiments is the 2537-A. line of a high-pressure mercury-vapor lamp. In all the irradiation experiments, the enzyme activity is destroyed according to simple first-order kinetics, as would be expected if inactivation were caused by a single quantum of radiation. As pointed out by McLaren,[128] this does not mean that every quantum absorbed actually produces inactivation. Many different types of bonds can be broken in the irradiation process, and hence it is difficult to explain the phenomena in molecular terms. The quantum yield varies with the protein (e.g., trypsin 0.02, chymotrypsin 0.001, pepsin 0.00036), but it is not correlated with the aromatic amino acid content (of the three proteins, pepsin has the highest content of aromatic amino acids), nor is there any convincing correlation between efficiency of a given wavelength and the absorption spectrum of the enzyme; in general, the yield decreases with increasing wavelength.[129]

With chymotrypsin it was found that the extent of inactivation paralleled the loss of solubility in 0.5 saturated ammonium sulfate, pH 5.5.[130]

Pollard *et al.*[131] interpreted the results of deuteron and electron bombardment of dry trypsin and pepsin in terms of the target size and hence the molecular weight. For trypsin a value of 30,000 was calculated, and for pepsin, 39,000, both figures being rather high. While it is premature to judge the value of this method, it may well be useful for approximate molecular-weight determinations of impure materials whose biological activity can be measured.

The inactivation of carboxypeptidase solutions by x-rays was shown to be due to the production of OH radicals in solution.[132]

(127) S. R. Hoover and E. L. C. Kokes, *J. Biol. Chem.* **167,** 199 (1947).
(128) A. D. McLaren, *Advances in Enzymol.* **9,** 75 (1949).
(129) F. L. Gates, *J. Gen. Physiol.* **18,** 265 (1934).
(130) A. D. McLaren and P. Finkelstein, *J. Am. Chem. Soc.* **72,** 5423 (1950).
(131) E. Pollard, A. Buzzell, C. Jeffreys, and F. Forro, Jr., *Arch. Biochem. Biophys.* **33,** 9 (1951).
(132) W. M. Dale, J. V. Davies, C. W. Gilbert, J. P. Keene, and L. H. Gray, *Biochem. J.* **51,** 268 (1952).

3. Pressure

As pointed out by Curl and Jansen,[133] pressure has a twofold effect on proteolytic enzymes. At relatively low pressures (500–1000 bars), the effect is largely on the rate of the various steps of the enzymatic reaction and can be interpreted in terms of the theory of absolute reaction rates. The second effect only becomes obvious at higher pressures (greater than 5000 bars) and is related to the denaturation of the enzyme.

Chymotrypsinogen, trypsin, and chymotrypsin, in order of decreasing stability, are irreversibly inactivated by pressures of 7600 bars, when applied for 5 minutes.[133,134] Little inactivation occurs at pH less than 4 but is almost complete above pH 7. The extent of inactivation also increases with protein concentration. Inactivation is negligible below a certain critical pressure (6000 bars for trypsin, 4500 bars for chymotrypsin), but above this value it increases with increasing pressure up to a limiting value which is dependent on pH. When repeated short applications of pressure are applied to trypsin and chymotrypsin, they are found to be more effective than a single prolonged application.

Pepsin behaves somewhat differently[134] in that it shows maximum stability at pH 4 and becomes less sensitive to pressure at higher protein concentrations. In general, it is much more sensitive to pressure than the other enzymes, but it shows the same increased inactivation with several short pressings as compared to a single long one.

4. Denaturing Agents

The nonspecific effects of organic denaturing agents such as urea and synthetic detergents on the proteases have received some attention. Pepsin, trypsin, and papain are all uneffected by high concentrations of urea over limited periods of time. Thus the sedimentation constant and enzyme activity of pepsin are not influenced by 4 *M* urea or 6.5 *M* acetamide.[135] Papain is stable for 24 hours at 30° in 9 *M* urea.[126] In Anson's hemoglobin method, trypsin is assayed in the presence of 2.5 *M* urea, which apparently has little effect on activity.[136,137]

The action of soaps and detergents on proteins has been reviewed,[138] but little of the work is concerned with proteolytic enzymes. Pepsin is precipitated by sodium dodecyl sulfate below pH 2.7,[139] and trypsin is

(133) A. L. Curl and E. F. Jansen, *J. Biol. Chem.* **184,** 45 (1950).
(134) A. L. Curl and E. F. Jansen, *J. Biol. Chem.* **185,** 713 (1950).
(135) J. Steinhardt, *J. Biol. Chem.* **123,** 543 (1938).
(136) M. L. Anson and A. E. Mirsky, *J. Gen. Physiol.* **17,** 159 (1933).
(137) M. L. Anson, *J. Gen. Physiol.* **22,** 79 (1938).
(138) F. W. Putnam, *Advances in Protein Chem.* **4,** 79 (1948).
(139) F. W. Putnam and H. Neurath, *J. Am. Chem. Soc.* **66,** 692 (1944).

inhibited by low concentrations (*ca.* 10^{-5} *M*) of potassium salts of a variety of long-chain fatty acids.[140]

Most of the enzymes are stable in moderate concentrations of ethanol or methanol. Pepsin,[18] trypsin,[11] and chymotrypsinogen[11] can all be crystallized using procedures involving 15–20 per cent alcohol at room temperature. Pepsin and papain[43] can both be dissolved in 70 per cent alcohol without inactivation. The hydrolysis of synthetic substrates by trypsin[141] and chymotrypsin[142] in 30 per cent methanol or ethanol proceeds more rapidly than in aqueous solutions (see p. 1122).

VI. Enzymatic Activity

1. Methods

Since proteolytic enzymes can act on a wide variety of substrates, many different methods of estimation of enzyme activity have been devised. Methods involving protein substrates have been described in some detail in Bamann and Myrbäck's *Methoden der Fermentforschung*,[143] and those involving synthetic substrates have been briefly reviewed by Neurath and Schwert.[3] The available techniques are listed in Table IV, where those most useful for routine estimations have been *italicized.* While some of the methods are not suitable for activity measurements (e.g., dilatometry, polarimetry) they have proved useful in the investigation of the nature of proteolysis and will be considered in the section devoted to that topic (p. 1175). Protein substrates are satisfactory for most routine estimations of proteolytic activity. Gelatin, casein, hemoglobin, serum albumin, and edestin are the most commonly used substrates, the hemoglobin method of Anson[137] having probably found most frequent use. It can be used with almost any type of proteolytic enzyme (pepsin, cathepsin, trypsin), and because of its great adaptability, a brief account of the method will be given herein.

The substrate solution is prepared from purified hemoglobin by denaturation in alkaline urea (for trypsin) or in acid (for pepsin and cathepsin). After the enzyme has been allowed to act on the protein for a fixed time, 5 per cent trichloroacetic acid is added to terminate the reaction and to precipitate undigested protein. After filtration or centrifugation, the tyrosine- and tryptophan-containing peptides are estimated using the color given with the Folin reagent or by measuring

(140) R. L. Peck, *J. Am. Chem. Soc.* **64,** 487 (1942).
(141) G. W. Schwert and M. A. Eisenberg, *J. Biol. Chem.* **179,** 665 (1949).
(142) S. Kaufman and H. Neurath, *J. Biol. Chem.* **180,** 181 (1949).
(143) E. Bamann and K. Myrbäck, Methoden der Fermentforschung, G. Thieme, Leipzig, 1941.

TABLE IV[a]

Substrate	Method based on:	Quantity measured	Experimental technique	Refs.
Protein	Change in physical properties of substrate	Viscosity	Viscometer	143,144
		Optical rotation	Polarimeter	143,145
		Volume	Dilatometer	143,146,147
		Conductivity	Conductivity bridge	1,143
		Clot formation	Time measurements	143,148
	Disappearance of substrate	Turbidity		143,149
	Appearance of peptides	Tyrosine + tryptophan	*Ultraviolet absorption*	12,34,150
			Folin reagent	1,137
		Optical rotation	Polarimeter	143,151,152
		Refractive index	Refractometer	143.153
Modified proteins	Liberation of colored peptides or adsorbed dye	Color intensity	Colorimeter	50,51,154,155
Proteins, peptides or amides	Bonds broken	$—COO^-$	Acetone titration	9,156–158
		$—NH_3^+$	Alcohol titration	9,156,159
		$—NH_3^+$	*Formol titration*	160–163
		$—NH_2$	*Colorimetric ninhydrin*	164–166
		$—NH_2$	Gasometric ninhydrin	28,41,167
		$—NH_3^+ + CO_2^-$	Dilatometer	147,168
		$—NH_2$	Gasometric HNO_2	41,169
Peptides	Amino acid produced	CO_2 from tyrosine	Manometry in presence of decarboxylase	170
		Oxygen uptake due to amino acid oxidation	Manometry in presence of oxidase	171,172
Amides	Ammonia	Ammonia	Micro-Conway	173
Hydroxamides	Disappearance of substrate	Hydroxamide	Color reaction with $FeCl_3$	174

[a] Methods of choice are printed in *italics*.

TABLE IV[a] (*Continued*)

Substrate	Method based on:	Quantity measured	Experimental technique	Refs.
Esters	Disappearance of substrate	Hydroxamide from reaction of ester with hydroxylamine	Color reaction with $FeCl_3$	175–177
	Bonds broken	—COOH	*Potentiometric titration*	173

(144) E. S. Duthie and L. Lorenz, *Biochem. J.* **44,** 167 (1949).
(145) L. K. Christensen, *Compt. rend. trav. lab. Carlsberg. Sér. chim.* **28,** 37 (1952).
(146) K. Linderstrøm-Lang, *Cold Spring Harbor Symposia Quant. Biol.* **14,** 117 (1950).
(147) K. Linderstrøm-Lang and C. F. Jacobsen, *Compt. rend. trav. lab. Carlsberg. Sér. chim.* **24,** 1 (1941).
(148) J. H. Northrop, *J. Gen. Physiol.* **4,** 227 (1921).
(149) B. C. Riggs and W. C. Stadie, *J. Biol. Chem.* **150,** 463 (1943).
(150) M. Kunitz, *J. Gen. Physiol.* **30,** 291 (1947).
(151) E. Schütz, *Z. physiol. Chem.* **9,** 577 (1885).
(152) T. Winnick, *J. Biol. Chem.* **152,** 465 (1944).
(153) J. T. Groll, *Arch. néerland. physiol.* **28,** 527 (1948).
(154) C. L. Oakley, G. H. Warrack, and W. E. Van Heyningen, *J. Path. Bact.* **58,** 229 (1946).
(155) H. A. Ravin and A. M. Seligman, *J. Biol. Chem.* **190,** 391 (1951).
(156) H. O. Calvery, *in* C. L. A. Schmidt, The Chemistry of the Amino Acids and Proteins, 1st ed., C. C Thomas, Springfield, Illinois, 1938.
(157) K. Linderstrøm-Lang, *Compt. rend. trav. lab. Carlsberg. Sér. chim.* **17,** No. 4, 1 (1927).
(158) P. C. Zamecnik, G. I. Lavin, and M. Bergmann, *J. Biol. Chem.* **158,** 537 (1945).
(159) W. Grassman and W. Heyde, *Z. physiol. Chem.* **183,** 32 (1929).
(160) R. C. Sisco, B. Cunningham, and P. L. Kirk, *J. Biol. Chem.* **139,** 1 (1941).
(161) B. M. Iselin and C. Niemann, *J. Biol. Chem.* **182,** 821 (1950).
(162) D. D. Van Slyke and E. Kirk, *J. Biol. Chem.* **102,** 651 (1933).
(163) J. H. Northrop and M. Kunitz, *J. Gen. Physiol.* **16,** 313 (1932).
(164) S. Moore and W. H. Stein, *J. Biol. Chem.* **176,** 367 (1948).
(165) G. W. Schwert, *J. Biol. Chem.* **174,** 411 (1948).
(166) J. E. Snoke and H. Neurath, *J. Biol. Chem.* **181,** 789 (1949).
(167) D. D. Van Slyke, R. T. Dillon, D. A. MacFadyen, and P. Hamilton, *J. Biol. Chem.* **141,** 627 (1941).
(168) C. F. Jacobsen, *Compt. rend. trav. lab. Carlsberg. Sér. chim.* **24,** 281 (1942).
(169) D. D. Van Slyke, *J. Biol. Chem.* **9,** 185 (1911).
(170) P. C. Zamecnik and M. L. Stephenson, *J. Biol. Chem.* **169,** 349 (1947).
(171) H. Herken and H. Erxleben, *Z. physiol. Chem.* **269,** 47 (1941).
(172) E. A. Zeller and A. Maritz, *Helv. Physiol. et Pharmacol. Acta* **3,** *C*6 (1945).
(173) G. W. Schwert, H. Neurath, S. Kaufman, and J. E. Snoke, *J. Biol. Chem.* **172,** 221 (1948).

the optical density at 280 mμ.[150] The enzyme concentration is then determined from a calibration curve using a known amount of enzyme as standard. With proper care, results are reproducible to within 2–4 per cent.

Many of the other methods in which proteolysis is determined from the concentration of protein split products (see Table IV) differ from one another as to substrate, method of precipitation of undigested protein, and method of assay of reaction products, but the principles are similar. The separation into undigested material and split products is, of course, quite arbitrary,[145] but since the whole procedure is empirical, this is of relatively little consequence.

If it is desired to measure the activity of a proteinase in the presence of a peptidase, and if no specific substrate is available, it is necessary to use a method which determines the disappearance of protein rather than the appearance of peptides or the number of bonds broken. Measurements of the viscosity or of the rate of milk clotting are most convenient for this purpose; turbidimetric measurements of undigested protein have also been used. Changes in optical rotation or in volume are less readily measured, but in the hands of the Carlsberg school they have proved useful for following the course of denaturation of a protein in relation to proteolysis. In this connection it should be noted that native proteins are usually not suitable for determining proteolytic activity since the rate of hydrolysis may be greatly affected by traces of denaturing agents.[145]

A third class of methods, which can also be used with synthetic substrates, is based on the titration of the number of peptide bonds broken. These methods give more significant information about the course of the proteolytic reaction than do the empirical methods described above, but they are more tedious to use and less suited for routine measurements. They are, however, invaluable for following in detail a given proteolytic reaction.[9] In the alcohol titration,[159] titration is carried out with base, using thymolphthalein as indicator. The alcohol merely raises the pK of the indicator and allows the NH_3^+ groups to be titrated.[156] The end point is not sharp, and the high pH of the titration (pH 11–12) makes contamination by CO_2 a serious problem. The Sørensen formol titration is carried out at lower pH (pH 9), and the use of a glass electrode to determine the end point gives a satisfactory method for activity deter-

(174) B. M. Iselin, H. T. Huang, and C. Niemann, *J. Biol. Chem.* **183,** 403 (1950).
(175) S. Hestrin, *J. Biol. Chem.* **180,** 249 (1949).
(176) H. Werbin and A. Palm, *J. Am. Chem. Soc.* **73,** 1382 (1951).
(177) H. Goldenberg, V. Goldenberg, and A. D. McLaren, *Biochim. et Biophys. Acta* **7,** 110 (1951).

minations and kinetic studies.[161] This method has also been effectively applied to microscale operations.[160] Both of these methods estimate such acidic groups as give off hydrogen ions at the end point of the titration. This is in contrast to the acetone titration method of Linderstrøm-Lang,[9,157] which is carried out with acid using α-naphthyl red as indicator. The acetone serves to raise the pK of the carboxyl groups so that they can be titrated above pH 3.5, and hence the method estimates these groups and any basic groups which were uncharged at the start of the titration. If the proteins were initially isoionic, this quantity is equivalent to the total basic groups (amino, imino, and guanidino).

Methods based on the reaction of amino nitrogen with nitrous acid or ninhydrin have been used mainly to follow the hydrolysis of synthetic peptides. The gasometric method of Van Slyke for the nitrous acid[169] and ninhydrin reactions[167] is laborious and time-consuming, though when properly executed, of high accuracy. The colorimetric ninhydrin procedure[164–166] has proved more convenient in practice and has been widely used for following the kinetics of carboxypeptidase activity. When free amino acids are the end product of proteolysis, they may be estimated manometrically by introducing an excess of a second enzyme system. Thus, if the amino acid liberated is one for which a specific decarboxylase is available (e.g., tyrosine), proteolysis may be followed by determining the rate of CO_2 production.[170] The rate of oxygen consumption has been similarly used to determine D- and L-peptidase activity in the presence of D-[171] and L-[172]amino acid oxidase.

Amides, hydroxamides, and esters have been frequently used as substrates for proteolytic enzymes, and several convenient methods have been devised to follow their hydrolysis (see Table IV).

Several of the methods given above can be used to follow the hydrolysis of amide substrates, but one of the most useful is a modified form of the Conway microdiffusion technique for ammonia determination, employing boric acid in place of standard hydrochloric acid as a trapping agent.[173]

Acetyl-L-phenylalaninhydroxamide has been proposed as a chymotrypsin substrate.[178] Hydrolysis was followed either by the potentiometric formol titration or by determination of unchanged hydroxamide by its color reaction with ferric chloride. The latter method was previously adapted to follow hydrolysis of esters, the unchanged ester reacting in alkaline hydroxylamine to yield hydroxamide which can then be assayed colorimetrically.[175–177,179]

However, the most useful method for determining esterase action and,

(178) D. S. Hogness and C. Niemann, *J. Am. Chem. Soc.* **75**, 884 (1953).
(179) F. Lipmann and L. C. Tuttle, *J. Biol. Chem.* **159**, 21 (1945).

when applicable, perhaps the method of choice for proteolytic enzymes, is the continuous potentiometric titration of the liberated carboxyl groups.[173] The reaction is carried out in a dilute buffer (0.005–0.01 N), and base is added during the reaction to maintain a constant pH. The initial slope of the plot of base added against time is proportional to enzyme concentration. This method could be rendered automatic by application of the pH-stat described by Jacobsen and Léonis.[180] Ester substrates of trypsin and carboxypeptidase are hydrolyzed according to zero-order kinetics, and ester substrates for chymotrypsin are nearly so, provided that Ca^{++} is present (0.01 M).[181] The slight fluctuation in pH (less than 0.1 unit) during the reaction has no effect since the pH optima are fairly broad.

It has been shown[155] that carbonaphthoxy-L-phenylalanine is hydrolyzed by carboxypeptidase to give β-naphthol, carbon dioxide, and phenylalanine. A colorimetric assay, based on this reaction, in which the liberated β-naphthol is coupled with diazotized di-o-anisidine, has been proposed for the measurement of carboxypeptidase activity.

The actual choice of a substrate depends on several factors. Protein substrates are suitable for routine analyses since they are readily available and, especially after denaturation, are more sensitive to enzymatic hydrolysis than are most of the synthetic materials. However, they are often difficult to prepare reproducibly and they are not suited for accurate kinetic work since they present essentially a mixture of substrates. When available, synthetic compounds are preferable since the hydrolysis of these substrates follows a known course. Of possible disadvantage is the fact that some of these synthetic substrates contain aromatic amino acids and are of such low solubility in water that they can be used only in the presence of varying concentrations of methanol.[182,183] This difficulty can be alleviated, however, if polar N-acyl groups, such as nicotinyl[184] and glyceryl groups,[185] are introduced into the peptide.

2. Effect of pH on Activity

Hydrogen ions may affect enzymatic activity in many different ways. One of these is their effect on the stability of the enzyme previously considered (p. 1086). In addition, the pH exerts an effect on the ionization

(180) C. F. Jacobsen and J. Léonis, Compt. rend. trav. lab. Carlsberg. Sér. chim. **27,** 333 (1951).
(181) J. A. Gladner and L. W. Cunningham, unpublished experiments.
(182) J. E. Snoke and H. Neurath, *J. Biol. Chem.* **182,** 577 (1950).
(183) S. Kaufman, H. Neurath and G. W. Schwert, *J. Biol. Chem.* **177,** 793 (1949).
(184) H. T. Huang and C. Niemann, *J. Am. Chem. Soc.* **73,** 1541 (1951).
(185) D. G. Doherty, *Federation Proc.* **12,** 197 (1953).

of those groups on the enzyme, on the substrate, or on both, which require a particular state of ionization for enzyme–substrate combination or for substrate activation.

Quantitative interpretation of the pH effects in these terms has not progressed very far since most of the work in this field has been directed to the determination of pH optima for purposes of activity measurements or for the differentiation between two or more enzymes. The early work by Northrop[186] and by Willstätter *et al.*[187] led to the conclusion that pepsin attacked the protein cations, papain the isoelectric proteins, and trypsin the protein anions. As knowledge of protein structure advanced, more adequate hypotheses were formulated.[188–191] Thus it was suggested[189] that enzyme, substrate, and enzyme–substrate complex could all take up protons at a suitable pH, giving rise to a species which could no longer participate in the reaction. On this basis it was possible to interpret the acid branch of the pH–activity curve of histidase.[189] A study of the peptic hydrolysis of ovalbumin has been similarly carried out,[190,190a] but the complexity of the system renders quantitative interpretation difficult. The most comprehensive study of the effect of pH on enzyme activity and enzyme inhibition has been the work of Wilson and Nachmansohn[191,192] on the reaction mechanism of acetylcholinesterase. The experimental data were interpreted in terms of two dissociating groups forming part of the active center of the enzyme, with pK values of 7.2 and 9.3. General treatments of the effect of pH on enzyme activity have recently been put forward.[193,193a]

The pH optima of proteolytic enzymes fall approximately into three groups. The only enzyme active in the most acid region (less than pH 2.5) is pepsin. Between pH 3 and 7 the —SH activated proteases (cathepsins and proteases of plant origin), some mold proteases,[48] and enterokinase and rennin are optimally active. Several enzymes are active above pH 7, including trypsin, chymotrypsin, papain, carboxy-

(186) J. H. Northrop, *J. Gen. Physiol.* **5,** 263 (1922).
(187) R. Willstätter, W. Grassmann, and O. Ambros, *Z. physiol. Chem.* **151,** 286, 307 (1926).
(188) M. J. Johnson, G. H. Johnson, and W. H. Peterson, *J. Biol. Chem.* **116,** 515 (1936).
(189) A. C. Walker and C. L. A. Schmidt, *Arch. Biochem.* **5,** 445 (1944).
(190) H. B. Bull and B. T. Currie, *J. Am. Chem. Soc.* **71,** 2758 (1949).
(190*a*) H. B. Bull, *in* W. D. McElroy and B. Glass, Mechanism of Enzyme Action, The Johns Hopkins Press, Baltimore, 1954.
(191) D. Nachmansohn and I. B. Wilson, *Advances in Enzymol.* **12,** 259 (1951).
(192) I. B. Wilson, *in* W. D. McElroy and B. Glass, Mechanism of Enzyme Action, The Johns Hopkins Press, Baltimore, 1954.
(193) S. G. Waley, *Biochim. et Biophys. Acta* **10,** 27 (1953).
(193*a*) M. Dixon, *Biochem. J.* **55,** 161 (1953).

peptidase, crystalline streptococcal protease, and the metal-activated exopeptidases.[6,6a]

a. Pepsin

Using various protein substrates, Northrop[186] observed a broad pH region of peptic activity extending from below pH 2 to above pH 4 with a

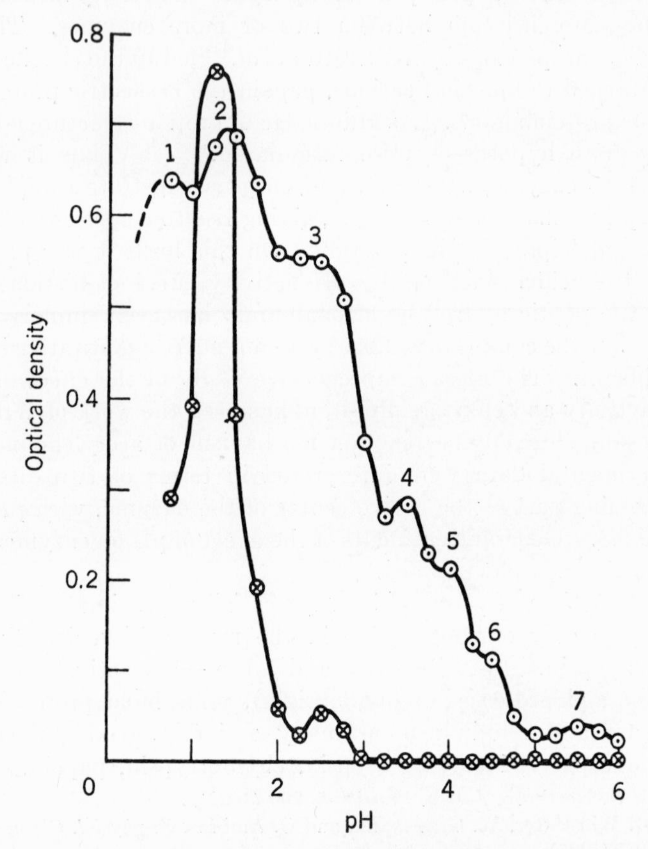

FIG. 4. The action of pepsin (⊙) and of a subfraction isolated from it (⊗) on crystalline horse hemoglobin.[196]

maximum at pH 1.8. In contrast, when synthetic substrates such as carbobenzoxy-L-glutamyl-L-tyrosine were used, maximum hydrolytic activity occurred at pH 4,[194,195] and the same pH optimum was obtained in the action of pepsin toward diphtheria antitoxin[196] and pepsin inhibi-

(194) J. S. Fruton and M. Bergmann, *Science* **87**, 557 (1938).
(195) C. R. Harington and R. V. Pitt-Rivers, *Biochem. J.* **38**, 417 (1944).
(196) C. G. Pope and M. F. Stevens, *Brit. J. Exptl. Path.* **32**, 314 (1951).

tor.[19] This discrepancy in pH optima might be attributed to the presence of more than one proteolytic enzyme in crystalline pepsin, a supposition which would be compatible with the inhomogeneity of pepsin as revealed by solubility measurements (p. 1071) and with the presence of large amounts of a catheptic type of enzyme in gastric juice.[197,198] A careful study by Pope and Stevens[196] of the effect of pH on the activity of amorphous and crystalline pepsin on hemoglobin and diphtheria antitoxin yielded the complex curves shown in Fig. 4. The many humps and points of inflection indicate the presence of several enzymes active between pH 1.5 and 6. Optimum splitting of diphtheria antitoxin [yielding the more active, heat-stable preparation (see p. 1175)] was due to an enzyme with a maximum at pH 4. While it was not possible to isolate this enzyme, a preparation was isolated with a very sharp maximum at pH 2 and which had no activity toward diphtheria antitoxin at pH 4. The relation of this active fraction to pepsin A of Herriott *et al.*,[199] which is homogeneous by solubility tests and is more active than ordinary crystalline pepsin, remains to be clarified.

Synthetic substrates have been recently synthesized which are optimally hydrolyzed at pH 2.[200] They all contain two adjacent aromatic residues (see p. 1126), and accordingly are rather insoluble in water. These substrates are hydrolyzed at higher rates than any of the synthetic substrates previously described.

b. *Cathepsins and Plant Proteases*

Cathepsin C[33] has an optimum of enzymatic activity toward glycyl-L-phenylalaninamide at pH 5 but hydrolyzes hemoglobin most rapidly at pH 3. This difference may be due to the decreased stability of hemoglobin in acid solution[201] or else may be ascribed to the presence of two distinct enzymes in this preparation. The hydrolytic action of papain is dependent on both the pH and the composition of the supporting buffer.[44a] Earlier studies,[202,44] carried out with preparations probably containing a mixture of enzymes, indicated maximum hydrolysis of protein substrates at pH 7, and of synthetic substrates at pH 5.[127] More recent studies on crystalline papain, prepared from dried latex,[44a] have revealed a broad pH maximum between pH 5.0 and 7.5, provided that cysteine and

(197) S. Buchs, Die Biologie des Magenkathepsins, S. Karger, Basle, 1946.
(198) S. Buchs, *Enzymologia* **13,** 208 (1949).
(199) R. M. Herriott, V. Desreux, and J. H. Northrop, *J. Gen. Physiol.* **24,** 213 (1940).
(200) L. E. Baker, *J. Biol. Chem.* **193,** 809 (1951).
(201) J. Steinhardt and E. M. Zaiser, *J. Biol. Chem.* **190,** 197 (1951).
(202) M. Bergmann and W. F. Ross, *J. Biol. Chem.* **111,** 659 (1935).

Versene were present, the former presumably to reduce disulfide groups, the latter to complex inactivating metal ions.

The synthetic action of papain is greatest at pH 7–8,[203,204] but this effect is probably related to the influence of pH on the ionization of the substrate rather than of the enzyme. Since only the un-ionized form of the amino group of a substrate is effective in replacement reactions, a pH optimum for replacement will depend on the pK of this group. This factor has not been always considered in studies of the effect of pH on the activity of proteolytic enzymes even though many of the substrates ionize within the pH region of enzymatic activity. In further support of this explanation, it should be noted that the papain-catalyzed synthesis of anilides has a much lower pH optimum (i.e., pH 5–6)[205] in accordance with the lower pK of aniline.

c. Chymotrypsin

Chymotrypsin hydrolyzes protein substrates with maximum velocity between pH 7 and 9.[66] Using synthetic substrates, a much sharper pH maximum is observed. Thus benzoyl-L-tyrosinamide and benzoyl-L-tyrosine ethyl ester,[183] acetyl-L-tyrosinamide,[206] and nicotinyl-L-tyrosinamide,[207] all have a sharp maximum at pH 7.8. The maxima for the hydrolysis of the corresponding tryptophan amides are relatively flat between pH 7 and 8.5.[184] Substrates containing a free amino group (e.g., L-phenylalanine ethyl ester or L-tyrosine ethyl ester) are optimally hydrolyzed at pH 6.2,[208,209] when the rate is measured by the usual procedure of titrating the liberated acid. However, when the disappearance of the ester is measured directly (by colorimetric estimation of the hydroxamic acid formed after treatment of the residual ester with hydroxylamine), the rate of the reaction apparently increases[208] and the pH optimum is shifted toward the more alkaline region (pH 7.2).[177] These results are probably due to a transpeptidation reaction (p. 1137). Oxidation and acetylation of chymotrypsin also raise the pH optimum for the hydrolysis of tyrosine ethyl ester, from pH 6.2 to 6.7,[209] but leave

(203) R. B. Johnston, M. J. Mycek, and J. S. Fruton, *J. Biol. Chem.* **185,** 629 (1950).
(204) J. S. Fruton, R. B. Johnston, and M. Fried, *J. Biol. Chem.* **190,** 39 (1951).
(205) S. W. Fox and C. W. Pettinga, *Arch. Biochem.* **25,** 13 (1950).
(206) D. W. Thomas, R. V. MacAllister, and C. Niemann, *J. Am. Chem. Soc.* **73,** 1548 (1951).
(207) H. T. Huang, R. V. MacAllister, D. W. Thomas, and C. Niemann, *J. Am. Chem. Soc.* **73,** 3231 (1951).
(208) H. Goldenberg and V. Goldenberg, *Arch. Biochem.* **29,** 154 (1950).
(209) E. F. Jansen, A. L. Curl, and A. K. Balls, *J. Biol. Chem.* **189,** 671 (1951).

the pH optimum for the hydrolysis of acetyl-L-tyrosine ethyl ester unaffected (pH 7.8). The pH optimum for the catalysis of the exchange reaction

$$\text{benzoyltyrosylglycinamide} + \text{glycinamide-N}^{15} \rightleftharpoons \text{benzoyltyrosylglycinamide-N}^{15} + \text{glycinamide}$$

is about pH 7.5,[210] but the synthesis of phenyl hydrazides of benzoylated tyrosine, tryptophan, and phenylalanine occurs maximally at pH 5–6, as measured by the over-all yield.[211]

d. *Trypsin*

The pH dependence of trypsin resembles closely that of chymotrypsin; the optimum for synthetic substrates is as broad as that for protein substrates.[212,213,173,94] As in the case of chymotrypsin, ester substrates containing a free amino group (L-arginine methyl ester) are maximally hydrolyzed at a lower pH,[208] presumably for the same reason.

e. *Carboxypeptidase*

Carboxypeptidase is the only proteolytic enzyme for which the effect of pH on k_3 and K_m has been separately determined.[3,214] While only preliminary results have been reported by one group of investigators,[3] they do show a pronounced minimum of both k_3 and K_m at pH 7.7 corresponding to a maximum of $C_{\max}$ (k_3/K_m), indicating that the minimum of K_m has a greater effect in increasing the rate than the minimum of k_3 has in decreasing it. A second maximum in both of these constants was observed at pH 8.3. Since the substrate (carbobenzoxyglycyl-L-phenylalanine) does not undergo a change of ionization in this pH range, the observed effects may be ascribed entirely to the change of ionization of groups on the enzyme. One of these is probably a positively charged group required to attract the carboxyl group of the substrate. More recently another group of investigators[214] has reported results which are at variance with those just cited. Within the relatively narrow pH range of 7.5–8.3 they failed to find any significant effect of pH on the rate of hydrolysis of carbobenzoxyglycyl-L-tryptophan and only a small monotonic increase of K_m for carbobenzoxyglycyl-L-phenylalanine.

(210) R. B. Johnston, M. J. Mycek, and J. S. Fruton, *J. Biol. Chem.* **187**, 205 (1950).
(211) W. H. Schuller and C. Niemann, *J. Am. Chem. Soc.* **74**, 4630 (1952).
(212) J. H. Northrop and M. Kunitz, *J. Gen. Physiol.* **16**, 295 (1932).
(213) M. Bergmann, J. S. Fruton, and H. Pollok, *J. Biol. Chem.* **127**, 643 (1939).
(214) R. Lumry, E. L. Smith, and R. R. Glantz, *J. Am. Chem. Soc.* **73**, 4330 (1951).

3. Specificity

a. General Considerations

The elucidation of the specificity requirements of proteolytic enzymes had to await the synthesis of synthetic substrates on the one hand and the purification of the enzymes on the other. It is, therefore, not surprising that most of the important work on this subject dates back hardly more than two decades. The synthesis of relatively simple peptides of leucine, alanine, and glycine, by Fischer's chloroacylchloride method, was soon followed by the demonstration of their hydrolysis by pancreatic juice,[215,216] and for many years these peptides were standard substrates for enzymatic studies. Willstätter, Grassmann, and Waldschmidt-Leitz used these substrates to classify proteolytic enzymes as aminopeptidases, carboxypeptidases, dipeptidases, and proteinases. The proteinases were thought not to act on small peptides since these simple substrates were not attacked by pepsin and the pancreatic proteinases. Conversely, the peptidases were thought not to attack proteins, a notion which only very recently was demonstrated to be erroneous (*vide infra*). It remained for Bergmann to widen the spectrum of available substrates by the development of the carbobenzoxy method[217] and to show that some of these substrates were hydrolyzed by the crystalline proteolytic enzymes described and prepared by Kunitz and Northrop. Thus in 1937 it was conclusively shown that a pure proteinase (chymotrypsin) was capable of hydrolyzing a small peptide such as carbobenzoxyglycyl-L-tyrosinamide.[218] The early literature dealing with the specificity of proteolytic enzymes has been adequately reviewed,[219,2] and more recently the specificity of the pancreatic proteinases[3] and of the metal-ion-activated exopeptidases[6a] has been considered in some detail.

As a point of departure, synthetic substrates which are readily hydrolyzed by purified proteolytic enzymes are listed in Table V. A structural feature common to substrates for all the enzymes considered herein is that at least one of the amino acids which contribute to the hydrolyzable bond must be of the L configuration. Depending on the enzyme in question, this amino acid may contribute the amino or the carboxyl moiety of the susceptible peptide bond. Thus intestinal leucine aminopeptidase will hydrolyze L-leucyl-D-alanine (though only

(215) E. Fischer and P. Bergell, *Ber.* **36**, 2592 (1903).
(216) E. Fischer and E. Abderhalden, *Z. physiol. Chem.* **46**, 52 (1905).
(217) M. Bergmann and L. Zervas, *Ber.* **65B**, 1192 (1932).
(218) M. Bergmann and J. S. Fruton, *J. Biol. Chem.* **118**, 405 (1937).
(219) M. Bergmann and J. S. Fruton, *Advances in Enzymol.* **1**, 63 (1941).

TABLE V

REPRESENTATIVE SUBSTRATES FOR PROTEOLYTIC ENZYMES[a]

Enzyme	Substrate	Ref.
Chymotrypsin	Nicotinyl-L-tryptophanamide	184
	Acetyl-L-tyrosine ethyl ester	82
Trypsin	Benzoyl-L-lysinamide	219
	Benzoyl-L-argininamide	220
	Benzoyl-L-arginine ethyl ester	173
Pepsin	Carbobenzoxy-L-glutamyl-L-tyrosine	219
	Acetyl-L-phenylalanyl-L-phenylalanine	200
Carboxypeptidase	Carbobenzoxyglycyl-L-phenylalanine	3
	Chloroacetyl-L-tyrosine	28
	Hippuryl-L-β-phenyllactic acid	166,221
Cathepsin A	Carbobenzoxy-L-glutamyl-L-tyrosine	32,33,222
Cathepsin B	Benzoyl-L-argininamide	32,33,222
Cathepsin C	Glycyl-L-phenylalaninamide; L-Glutamyl-L-tyrosine ethyl ester	33
Cathepsin III (leucinaminopeptidase)	L-Leucinamide	32,33
	L-Cysteine ethyl ester	35
Cathepsin IV (carboxypeptidase)	Carbobenzoxyglycyl-L-phenylalanine	32,33
Kidney acylase I	Acyl-L-α-amino acid (except aspartic acid)	223,224
Kidney acylase II	Acyl-L-aspartic acid	223,224
Papain, Ficin (Activation by —SH required)	Benzoyl-L-argininamide	225,302
	Carbobenzoxy-L-methioninamide	225
	Carbobenzoxy-L-leucinamide	225
	Carbobenzoxy-L-isoglutamine	225
	Carbobenzoxy-L-serinamide	225
	Benzoylglycinamide (papain only)	225,43
Streptococcal protease (activation by —SH required)	Benzoyl-L-argininamide	
	Carbobenzoxy-L-isoglutamine; Benzoyl-L-histidinamide	225*a*
Clostridium histolyticum (activation by —SH required)	Benzoyl-L-argininamide; L-Lysine methyl ester	225*b*

[a] More detailed specificity tables are given later.

(220) K. Hofmann and M. Bergmann, *J. Biol. Chem.* **138,** 243 (1941).
(221) J. E. Snoke, G. W. Schwert, and H. Neurath, *J. Biol. Chem.* **175,** 7 (1948).
(222) J. S. Fruton and M. Bergmann, *J. Biol. Chem.* **130,** 19 (1939).
(223) V. E. Price, J. B. Gilbert and J. P. Greenstein, *J. Biol. Chem.* **179,** 1169 (1949).

at 1/20th the rate for L-leucyl-L-alanine), but it will not attack D-leucylglycine or D-leucinamide.[226] Similarly, purified kidney aminopeptidase will hydrolyze L-alanyl-D- or -L-alanine at similar rates, but D-alanylpeptides are hydrolyzed at about 1/1000th of this rate.[41] In contrast, carnosinase will hydrolyze D-alanyl-L-histidine and β-alanyl-L-histidine, but not β-alanyl-D-histidine.[226] Pancreatic carboxypeptidase and the endopeptidases require amino acids of the L configuration to contribute to both moieties of the hydrolyzable peptide bond.[3]

This high degree of optical specificity has been widely used for the resolution of amino acids, employing the enzymes either as hydrolytic or as synthetic agents. Thus with the aid of carboxypeptidase, the chloroacetyl derivatives of racemic phenylalanine, tyrosine, and tryptophan were resolved, since only the L isomers of these peptides are hydrolyzed by the enzyme.[227] Similarly, amino acids such as tryptophan whose esters are hydrolyzed by chymotrypsin may be resolved by this enzyme.[228]

A highly purified acylase from hog kidney extracts[223,224] has proved useful for the resolution of acetyl or chloroacetyl amino acids. This enzyme could actually be fractionated into two components, one of which hydrolyzed the acylated L isomers of all naturally occurring amino acids with the exception of aspartic acid (acylase I) while the other component (acylase II) hydrolyzed acyl aspartic acids only. Optical enantiomorphs could thus be obtained with a purity of 99.9 per cent. The synthetic activity of papain and ficin have been used to resolve amino acids via their anilides[229–232] or phenylhydrazides.[320,320a]

Synthetic substrates have proved an indispensable aid for identifying the presence of several enzymes in mixtures and for following their purification. Representative examples are the characterization of the various cathepsins,[33,32] the amino peptidases[6a] or the identification of small quantities of enzymatically active impurities in preparations of the pancreatic proteolytic enzymes or their precursors.[297] Whenever an

(224) S. M. Birnbaum, L. Levintow, R. B. Kingsley, and J. P. Greenstein, *J. Biol. Chem.* **194,** 455 (1952).
(225) C. A. Dekker, S. P. Taylor, and J. S. Fruton, *J. Biol. Chem.* **180,** 155 (1949).
(225*a*) M. J. Mycek, S. D. Elliott, and J. S. Fruton, *J Biol. Chem.* **197,** 637 (1952).
(225*b*) J. D. Ogle and A. A. Tytell, *Arch. Biochem. Biophys.* **42,** 327 (1953).
(226) E. L. Smith and W. J. Polglase, *J. Biol. Chem.* **180,** 1209 (1949).
(227) J. B. Gilbert, V. E. Price, and J. P. Greenstein, *J. Biol. Chem.* **180,** 473 (1949).
(228) M. Brenner, E. Sailer, and V. Kocher, *Helv. Chim. Acta* **31,** 1908 (1948).
(229) M. Bergmann and H. Fraenkel-Conrat, *J. Biol. Chem.* **119,** 707 (1937).
(230) C. A. Dekker and J. S. Fruton, *J. Biol. Chem.* **173,** 471 (1948).
(231) H. T. Hanson and E. L. Smith, *J. Biol. Chem.* **179,** 815 (1949).
(232) T. Yoneya, *Biochem. J.* **38,** 343 (1951).

enzyme preparation possesses a wide range of specificity, the presence of more than one active component may be suspected. Thus crude preparations of papain and ficin both hydrolyze a large variety of acylated amino acid amides,[225] and it has been suggested that different enzymes present in these preparations are responsible for the hydrolysis of carbobenzoxy-L-methioninamide and of benzoyl-L-argininamide, since ficin hydrolyzes the former more rapidly than the latter, whereas the susceptibility to hydrolysis by papain is in the reverse order. A clarification of this situation, however, requires a study of the effect of purification of these enzymes on their relative activities toward these two substrates. Examples are known where a chemically homogeneous proteolytic enzyme reveals a wide spectrum of substrate specificities. For instance, the crystalline protease from group A streptococci,[52] which is electrophoretically homogeneous,[233] hydrolyzes rapidly peptides of a large number[225a] of different amino acids (see Table V). Crystalline trypsin, chymotrypsins, carboxypeptidase, and papain hydrolyze ester as well as peptide bonds. The action of carboxypeptidase on protein substrates suggests that with the exception of proline, hydroxyproline, cysteine and cystine,[234] practically all amino acids can be liberated from the C-terminal position of polypeptide chains.[235]

A proteolytic enzyme of narrow specificity can be a useful reagent for the partial degradation of a protein and the determination of its structure. Examples of such applications and the discussion of the relation between the specificity toward synthetic substrates and protein substrates will be found in a later section.

b. Kinetic Principles

In order to compare the action of one or more enzymes on one or more substrates, some quantitative measure of reaction rate is required. The early work in the field of proteolytic enzymes, notably by Bergmann and coworkers[2,236] was based on the interpretation of reaction rates by first-order kinetics at a constant and arbitrary substrate concentration of 0.05 *M*. A "proteolytic coefficient," *C*, was therefore defined as the first-order reaction constant (calculated from decimal logarithms) per milligram enzyme nitrogen per milliliter. This interpretation has been

(233) T. Shedlovsky and S. D. Elliott, *J. Exptl. Med.* **94,** 363 (1951).

(234) These four amino acids have never been found as reaction products of either protein or peptide substrates.

(235) H. Neurath, J. A. Gladner, and E. W. Davie, *in* W. D. McElroy and B. Glass, The Mechanism of Enzyme Action, The Johns Hopkins Press, Baltimore, 1954.

(236) G. W. Irving, Jr., J. S. Fruton, and M. Bergmann, *J. Biol. Chem.* **138,** 231 (1941).

criticized as being arbitrary and devoid of physical meaning[237] since according to the formulation of the kinetic process by the Michaelis-Menten theory, first-order kinetics does not apply in the general case and since the "proteolytic coefficient," in general, is a function of the substrate concentration. A more rigorous and physically meaningful analysis of the action of proteolytic enzymes on a large variety of synthetic substrates, in terms of the Michaelis-Menten theory, was proposed instead.[3,184]

(*1*) *Derivation of Equations.* Since the derivation of the kinetic equations for enzyme-catalyzed reactions has been frequently given,[3,238–240] the subject will be considered herein only in brief.

According to the Michaelis-Menten theory, a simple uninhibited enzymatic reaction may be represented by Eq. (1)

$$\underset{e}{\mathrm{E}} + \underset{a}{\mathrm{S}} \underset{k_2}{\overset{k_1}{\rightleftharpoons}} \underset{p}{\mathrm{ES}} \overset{k_3}{\rightarrow} \mathrm{E} + \mathrm{P} \tag{1}$$

where e denotes concentration of total enzyme, a that of total substrate, and p that of the enzyme–substrate complex (ES). Since the rate of formation of ES is given by Eq. (2)

$$\frac{dp}{dt} = k_1(e - p)a - (k_2 + k_3)p \tag{2}$$

and the rate of disappearance of the substrate by Eq. (3),

$$\frac{da}{dt} = -k_1(e - p)a + k_2p \tag{3}$$

the rate of the over-all reaction is given by summation of Eqs. (2) and (3), i.e.,

$$\frac{d(p + a)}{dt} = -k_3p \tag{4}$$

Since for most enzyme systems dp/dt is negligibly small as compared to da/dt (cf. Refs.[238,193]), the rate of disappearance of the substrate is given by Eq. (5)

$$-\frac{da}{dt} = k_3p = \frac{k_3ea}{K_m + a} \tag{5}$$

(237) E. Elkins-Kaufman and H. Neurath, *J. Biol. Chem.* **175**, 893 (1948).
(238) G. E. Briggs and J. B. S. Haldane, *Biochem. J.* **19**, 338 (1925).
(239) B. Chance, *Advances in Enzymol.* **12**, 153 (1951).
(240) F. M. Huennekens, *in* A. Weissberger, Technique of Organic Chemistry, Interscience Pub., New York, 1953, Chap. VIII.

and the Michaelis constant, K_m, by Eq. (6)

$$K_m = \frac{k_2 + k_3}{k_1} \tag{6}$$

It is clear that the over-all kinetics of such a reaction may be first order or zero order with respect to the substrate, depending on the relative magnitudes of K_m and a.

Integration of Eq. (5) leads to the expression

$$k_3et = 2.3K_m \log \frac{a_0}{a} + (a_0 - a) \tag{7}$$

The kinetic constants, k_3 and K_m, suffice to characterize kinetically an enzymatic system provided that the constants are obtained from measurements of initial velocities in the absence of competitive effects of any sort.[241]

In recent years, a number of cases have been described in which reaction products inhibit the hydrolysis of synthetic substrates by proteolytic enzymes. This is true, for instance, of the chymotryptic hydrolysis of *N*-acyl derivatives of aromatic amino acids, studied by Niemann and co-workers,[184,206] who found the following Eq. (8) to describe satisfactorily the course of the reaction, where K_i is the enzyme–inhibitor dissociation constant.

$$-\frac{da}{dt} = \frac{k_3\, ea}{K_m\left(1 + \frac{a_0 - a}{K_i}\right) + a} \tag{8}$$

This equation reduces to a first-order equation when $K_m = K_i$.[242,3] The kinetic constants were obtained graphically from measurements of initial reaction velocities, using the Lineweaver-Burk[243] plots. Another well-studied case of substrate inhibition is that of the tryptic hydrolysis of benzoyl-L-argininamide[3,141,213,242] which is inhibited by benzoyl-L-arginine.

(2) Effects of Temperature and Pressure on k_3 and K_m. The physical significance of k_3 and K_m has been the subject of much discussion.[3,184,244,214] The interpretation of k_3 is relatively unambiguous since k_3 represents the reaction velocity constant for the decomposition of the Michaelis complex, this complex being defined[239] as that enzyme–substrate complex whose decomposition is rate determining. It is not known whether this decom-

(241) In general the proteolytic coefficient, C, will be $\frac{k_3}{2.3(K_m + a)}$ and will tend to the value $C_{\max} = \frac{k_3}{2.3K_m}$ as a tends to zero.

(242) K. M. Harmon and C. Niemann, *J. Biol. Chem.* **178,** 743 (1949).

(243) H. Lineweaver and D. Burk, *J. Am. Chem. Soc.* **57,** 658 (1934).

(244) H. T. Huang and C. Niemann, *J. Am. Chem. Soc.* **73,** 3223 (1951).

TABLE VI

EFFECT OF TEMPERATURE ON k_3

Enzyme and buffer	Substrate	Solvent	$k_3°$,[a] sec.$^{-1}$	ΔH^*, kcal./mole	ΔS^*, E.U.	ΔF^*,[a] kcal./mole	Refs.
Chymotrypsin	Benzoyl-L-tyrosinamide	30% CH_3OH	0.37	14.0	−13.7	18.1	183
(Phosphate buffer,	Benzoyl-L-phenylalanine methyl ester	30% CH_3OH	30.0	11.9	−12.1	15.5	
pH 7.8)	Acetyl-L-tyrosine ethyl ester	30% CH_3OH	115	10.9	−12.7	14.7	
	Methyl-l-β-phenyllactate	20% CH_3OH	0.82	10.4	−24.2	17.6	
	Methyl-*d*-β-phenyllactate	20% CH_3OH	0.083	14.5	−15.1	19.0	182
	Methyl-*dl*-α-chloro-β-phenylpropionate	20% CH_3OH	0.083	14.8	−14.1	19.0	
	Methyl-β-phenylpropionate	30% CH_3OH	0.015	16.2	−12.7	20.0	
	Pepsin (denatured)[c]	Water	130	10.9	− 8.8	13.3	245
Trypsin	Benzoyl-L-argininamide	Water	0.24	16.9	− 5.0	18.4	141
(Phosphate buffer,	Benzoyl-L-arginine ester[b]	Water	15.4	10.6	−17.8	15.9	141
pH 7.8)	Sturin[c]	Water	1450	11.3	− 2.6	12.0	245
Pepsin	Carbobenzoxy-L-glutamyl-L-tyrosine	Water	0.008[d]	17.2	−10.6[e]	20.4	246
(Acetate buffer pH 4)	Carbobenzoxy-L-glutamyl-L-tyrosine ethyl ester	Water	0.0005[d]	20.7	− 4.5	22.0	246
Carboxypeptidase	Carbobenzoxyglycyl-L-phenylalanine	Water	186	15.6	+ 3.6	14.5	166
(Phosphate buffer	Carbobenzoxyglycyl-L-phenylalanine	Water	171	9.0	−18.5	14.5	214
pH 7.5)[f]	Carbobenzoxyglycyl-L-tryptophan	Water	84	9.3	−18.8	14.9	214
	Carbobenzoxyglycyl-L-leucine	Water	100	8.0	−22.8	14.8	214
	Chloracetyl-*l*-β-phenyllactic acid	Water	111	12.1	− 8.8	14.7	166

[a] Unless otherwise stated, the figures quoted all refer to measurements at 25°C. at the pH given. The assumed molecular weights were: chymotrypsin, 21,500; trypsin, 24,500; pepsin, 35,500; and carboxypeptidase, 32,000.

[b] Identical values for methyl, ethyl, isopropyl, cyclohexyl, and benzyl esters.

[c] These figures refer to 0°C. The significance of the figures is doubtful, owing to the complexity of the substrate and the unknown nature of the reactions taking place.

[d] Recalculated for 25°C. from the original data for ΔF^*.

[e] Recalculated from the figures quoted for ΔF^* and ΔH^*.

[f] The figures quoted from Ref.[214] were obtained from measurements in 0.007 *M* veronal buffer, brought to an ionic strength of 0.5 with KCl.

position involves a water molecule or whether the water is already part of the compound, but in either case the kinetics of this step will be first order since the water is present in large excess. The dimensions of k_3 will, therefore, be sec.$^{-1}$, provided that the concentration of the Michaelis complex can be expressed in moles per liter. If the molecular weight of the enzyme is unknown and its concentration is expressed in mg. protein nitrogen/ml., the units of k_3 will be moles/l./min./mg. nitrogen/ml. [cf. Eq. (5)].

The effect of temperature on k_3 has been interpreted in terms of the theory of absolute reaction rates, and free energies and entropies of activation have been calculated.[3,214] It was pointed out that caution should be exercised in the interpretation of such results since the calculated thermodynamic quantities refer to the over-all process of the decomposition of ES and include the effect of temperature on the activity of water molecules, hydrogen ions, and hydroxyl ions, and hence on the state of ionization of ES, as well as the effect of temperature on ES itself. This procedure is probably satisfactory if comparison is made between reactions carried out in the same solvent system at the same pH. Comparison of the energy constants with those of acid- or base-catalyzed reactions can only be qualitative.[6a,182] The data at present available for the effect of temperature on k_3 for a series of synthetic substrates are shown in Table VI (see also p. 142 of Ref.[3]). The exact interpretation of these results is rather speculative, and although various interesting correlations between the structure of the substrate and its effect on activation energies have been advanced, no definite conclusions can as yet be drawn. The identity of the energy constants for the tryptic hydrolysis of five different esters of benzoyl-L-arginine is noteworthy.

The effect of temperature on the enzymatic hydrolysis of proteins has been measured (cf. Refs.[245,190]), but the interpretation of the results from measurements of such complex reactions is dubious. A few representative figures for the over-all reaction are included in Table VI. It is to be noted that in these measurements the effect of temperature on the denaturation of the substrate is included.

The effect of pressure on the rate of enzyme reactions has also been interpreted in terms of the theory of absolute rates.[247] If the rate-limiting step is the decomposition of the enzyme–substrate complex, the relevant relation between pressure and velocity constant is:

$$k_3 = k_3' e^{-p\Delta V^*/RT} \tag{9}$$

(245) J. A. V. Butler, *J. Am. Chem. Soc.* **63**, 2968, 2971 (1941).
(246) E. J. Casey and K. J. Laidler, *J. Am. Chem. Soc.* **72**, 2159 (1950).
(247) K. J. Laidler, *Arch. Biochem.* **30**, 226 (1951).

ΔV^* was found to be negative for the hydrolysis by chymotrypsin[248] of tyrosine ethyl ester (−13.5 ml./mole) and casein (−13.8 ml.)[249], and of arginine methyl ester (−6 ml./mole) and β-lactoglobulin (−36 ml.)[249] by trypsin.[250] In other words, pressures of up to 500 atmospheres increased the rate of hydrolysis in these systems. [Higher pressures cause denaturation of the enzyme.] The large volume change found for the activation of the trypsin–β-lactoglobulin complex probably reflects the effect of pressure on the denaturation of this protein. In contrast with the above systems, pressure had no effect on the hydrolysis of benzoyl-L-argininamide and of benzoyl-L-arginine isopropyl ester by trypsin (i.e., $\Delta V^* = 0$).

The physical significance of K_m is less clearly defined than that of k_3 since, in general, K_m is not a true equilibrium constant but the sum of two ratios of velocity constants ($k_2/k_1 + k_3/k_1$). Accordingly, it is generally not a measure of the relative affinities of a series of substrates for a given enzyme, nor is it a valid measure of the free energy of formation of the enzyme–substrate complex. This may be readily appreciated[251] by comparing the Michaelis constants K_m and K_m', where $K_m > K_m'$, for the hydrolysis of two substrates by a given enzyme. The corresponding affinities of the substrates for the enzyme are given by k_2/k_1 and k_2'/k_1'.

Now,

$$\frac{k_2 + k_3}{k_1} > \frac{k_2' + k_3'}{k_1'} \tag{10}$$

therefore,

$$\frac{k_2}{k_1} > \frac{k_2'}{k_1'} + \left[\frac{k_3'}{k_1'} - \frac{k_3}{k_1}\right] \tag{11}$$

If the term in brackets is positive, it can be said that the first substrate has a higher dissociation constant than the second one and that its affinity for the enzyme will be less. However, the sign of the term in brackets can only be known if both k_3 and k_1 are determined, and this has not yet proved possible for any of the proteolytic enzymes. All that can be said, therefore, is that if k_3' is greater than k_3, this term is more likely to be positive than if k_3' is smaller than k_3. This is contrary to earlier impressions[3,184,166] which led to the conclusion that if $K_m > K_m'$ and $k_3' > k_3$, then k_2/k_1 is necessarily greater than k_2'/k_1'.

(248) H. Werbin and A. D. McLaren, *Arch. Biochem. Biophys.* **31,** 285 (1951).

(249) The significance of these results is doubtful, since first-order velocity constants for the hydrolysis of proteins cannot be expressed in absolute units. From the data given it is not possible to determine the quantity of protein to which these volume changes refer.

(250) H. Werbin and A. D. McLaren, *Arch. Biochem. Biophys.* **32,** 325 (1951).

(251) D. S. Hogness and C. Niemann, *J. Am. Chem. Soc.* **74,** 3183 (1952).

If k_3 is negligibly small as compared to k_2, K_m will be a true measure of the enzyme–substrate dissociation constant, whereas under all other conditions, K_m will represent an upper limit of that constant. Conversely, if k_3 greatly exceeds k_2, K_m represents simply a ratio of the reaction constants for two consecutive reactions. This latter interpretation has been accorded to the action of carboxypeptidase on the carbobenzoxyglycyl derivatives of phenylalanine, tyrosine, and leucine on the basis of what must be considered as rather tenuous arguments.[214] In contrast, it is probable that for the action of chymotrypsin on a number of amide substrates K_m represents a true dissociation constant, since the relative values of K_m for a series of substrates parallel closely the relative values of K_i (a true equilibrium constant) for a series of structurally analogous inhibitors[188,244] (see also p. 1119). While the effect of ethanol on the

TABLE VII
EFFECT OF TEMPERATURE ON K_m[a][182]

Substrate	Temperature range, °C.	ΔE, kcal./mole
Methyl-*l*-β-phenyllactate	10–32	6.3
Methyl-*d*-β-phenyllactate	10–32	11.1
Methyl-*dl*-α-chloro-β-phenylpropionate	9–25	7.6
Methyl phenylpropionate	15–25	4.4

[a] Ester substrates of chymotrypsin only.

hydrolysis of acetyl-L-tyrosinamide by chymotrypsin has also been interpreted on the basis of K_m being k_3/k_1 (k_2 being negligibly small), the same findings could also result if k_3 were negligibly small as compared to k_2.

Except for the special cases just considered, no simple interpretation can be given to the temperature dependence of K_m. Evidently if $K_m = k_2/k_1$, the temperature dependence of this quantity yields the classical thermodynamic constants for the formation of the enzyme–substrate complex. If $K_m = k_3/k_1$, the calculated Arrhenius activation energy will be the difference between two heats of activation, namely, that for the formation of the activated complex in the decomposition of ES and that for the formation of ES. The situation is therefore not very satisfactory unless an independent determination of k_1 can be made. So far this has been restricted to enzymatic reactions which can be followed spectrophotometrically.[239]

In order to indicate the magnitude of the effect of temperature on K_m, ΔE values obtained for several ester substrates for chymotrypsin are shown in Table VII.

The kinetics of inhibition of hydrolysis by structural analogs of specific substrates (e.g., reaction products, D-antipodes of substrates and other modifications) will be considered in connection with the individual enzymes. These inhibitions are usually of the competitive type though several examples of indeterminate types have been found.[141,252]

TABLE VIII
HEATS OF HYDROLYSIS OF PEPTIDES[a]

Enzyme	Peptide	$-\Delta H$, kcal./mole	Refs.
α-Chymotrypsin	Benzoyl-L-tyrosinamide	5.84	253
α-Chymotrypsin	Benzoyl-L-tyrosylglycinamide	1.55[b]	254
Carboxypeptidase	Carbobenzoxyglycyl-L-leucine	2.11	255
Carboxypeptidase	Carbobenzoxyglycyl-L-phenylalanine	2.55	253
Cathepsin C	Glycyl-L-phenylalaninamide	6.22	255

[a] The reactions were carried out in 0.05 *M* phosphate buffer, ionic strength 0.3, at 25°C.

[b] From equilibrium data $\Delta F = -0.42$ kcal./mole and $\Delta S = -3.8$ E.U./mole.

In a series of interesting studies, Sturtevant and co-workers[253,255] have measured the heat of hydrolysis of several peptide and amide substrates by chymotrypsin, carboxypeptidase, and cathepsin C. The results are listed in Table VIII and show that, as might be anticipated, the heat evolved depends more nearly on the nature of the substrate than on the enzyme. Thus the hydrolysis of amides, which yields a free ammonium ion, is accompanied by a greater heat loss than the hydrolysis of peptides which yields substituted ammonium ions.

c. *Individual Enzymes*

(1) α-Chymotrypsin. Although the action of α-chymotrypsin on an impressive series of synthetic substrates has been studied,[3,183,256–261,184,206,244,207] it has not yet been possible to draw from this work more than

(252) E. Elkins-Kaufman and H. Neurath, *J. Biol. Chem.* **178,** 645 (1949).
(253) A. Dobry and J. M. Sturtevant, *J. Biol. Chem.* **195,** 141 (1952).
(254) A. Dobry, J. S. Fruton, and J. M. Sturtevant, *J. Biol. Chem.* **195,** 149 (1952).
(255) J. M. Sturtevant, *J. Am. Chem. Soc.* **75,** 2016 (1953).
(256) J. E. Snoke and H. Neurath, *Arch. Biochem.* **21,** 351 (1949).
(257) S. Kaufman and H. Neurath, *Arch. Biochem.* **21,** 437 (1949).
(258) R. V. MacAllister, K. M. Harmon, and C. Niemann, *J. Biol. Chem.* **177,** 767 (1949).
(259) H. J. Shine and C. Niemann, *J. Am. Chem. Soc.* **74,** 97 (1952).
(260) H. T. Huang, R. J. Foster, and C. Niemann, *J. Am. Chem. Soc.* **74,** 105 (1952).
(261) H. T. Huang and C. Niemann, *J. Am. Chem. Soc.* **74,** 5963 (1952).

TABLE IX
CHYMOTRYPSIN SUBSTRATES

Substrate						
R_1 R.CO	R_2 NH—CH—CO—	R_3	K_m, $M \times 10^3$	ΔF,[a] kcal./mole	k_3,[b] $\times 10^3$	Refs.
Acetyl	L-Tryptophan	Amide	5.3	3.11	0.5	184
Acetyl	L-Tryptophan	Ethyl ester[c,d]	1.7		320	82
Acetyl	L-Tyrosine	Amide	30.5	2.08	2.4	206
Acetyl	L-Tyrosine	Amide[c]	27		3.0	82
Acetyl	L-Tyrosine	Glycinamide	30	2.09	8.9	258
Acetyl	L-Tyrosine	Hydroxamide	51		34	178
Acetyl	L-Tyrosine	Ethyl ester[g]	0.7		3060	262
Acetyl	L-Tyrosine	Ethyl ester[c,d]	32		2600	82
Acetyl	L-Phenylalanine	Amide[d]	34	2.00	0.7	260
Acetyl	L-Hexahydro-phenylalanine	Amide	27		0.6	265
Chloroacetyl	L-Tyrosine	Amide	26	2.17	4.3	259
Trifluoro-acetyl	L-Tyrosine	Amide	119	1.27	2.8	259
Nicotinyl	L-Tryptophan	Amide	2.7	3.5	1.6	184
Nicotinyl	L-Tyrosine	Amide	15.0	2.49	6.2	207
Nicotinyl	L-Phenylalanine	Amide	18	2.38	2.1	260
Benzoyl	L-Tyrosine	Amide[c,d]	42	1.89	6.5	183
Glycyl	L-Tyrosine	Amide[c]	122	1.25	4.1	183
Benzoyl	L-Phenylalanine	Ethyl ester[c,d]	6.0		390	257
Benzoyl	L-Methionine	Ethyl ester[c,d]	0.8		8.0	257
Benzoyl	Glycyl-	Methyl ester	8.5[h]		1.2	271
	L-Tyrosine	Ethyl ester[f]	Very low		700	209
	L-β-Phenyl-2-hydroxypropionic	Methyl ester[e]	11		14.4	182
	D-β-Phenyl-2-hydroxypropionic	Methyl ester[e]	16.5		1.4	182
	DL-β-Phenyl-2-chloropropionic	Methyl ester[e]	12		6.0	182
	Acetyl	p-Nitrophenol[c]	3		0.034	86
	Ethylcarbonyl	p-Nitrophenol[c]	0.0013		0.003	86

The following compounds, related to the above substrates, are not hydrolyzed by chymotrypsin: Ethyl β-phenyl-α-acetyl propionate, monoethyl malonate, *N*-benzoyl DL-phenylglycine ethyl ester, β-phenyl-2-phthalimidopropionic ethyl ester,[257] benzylmalonic diamide,[263] 2-benzamidocinnamic amide.[264]

Unless otherwise stated, all assays were performed in tris(hydroxymethyl)aminomethane hydrochloride buffer.

[a] Calculated only when the assumption that $K_m = k_2/k_1$ appears justified (see text).

[b] k_3 is expressed as moles substrate hydrolyzed/l./min./mg. enzyme N/ml.

[c] Reaction in phosphate buffer.

[d] Reaction in 30% methanol.

[e] Reaction in 20% methanol.

[f] pH 6.2.

[g] Reaction in presence of 0.083 *M* $CaCl_2$.

[h] Apparent K_m value (see original paper).

tentative conclusions concerning the nature of the active center. A representative list of substrates and of related compounds which are not hydrolyzed is given in Table IX. The substrates are of the type $R_1R_2.CHCOR_3$, where R_1 is a polar group, usually an acylamido group; R_2 is the specific amino acid side chain, usually of aromatic character; and R_3 is —NHR, —OR, $—SC_2H_5$,[266] $—NHNH_2$,[267] or —NHOH.

Substrates listed in Table IX include only those for which the kinetic constants k_3 and K_m have been determined. Other things being equal the hydrolysis rates usually decrease in the order of ester, hydroxamide, glycinamide, amide, hydrazide, and glycylglycinamide. The acyl residue of R_1 may be varied over rather wide limits and may also be substituted by other amino acids without greatly affecting the ease of hydrolysis. The nature of the specific amino acid residue in position R_2 is much more restricted, derivatives containing aromatic residues being most readily hydrolyzed. However, comparison of acetyl-L-phenylalaninamide with the corresponding hexahydro derivative suggests that the size rather than the electronic configuration of the benzene ring is of decisive importance for substrate specificity.[265] Methionine[257] and leucine derivatives[177] are also slowly attacked. Derivatives containing glutamic acid, serine, threonine, histidine, and alanine in this position do not seem to be hydrolyzed.[3]

Although the secondary peptide group may be replaced by OH, by Cl, or even entirely omitted, it cannot be replaced by $—CO_2C_2H_5$, $—CONH_2$, —COOH, or $—COCH_3$ without complete loss of susceptibility. Replacement of the secondary peptide hydrogen (as in phthalimido phenylalanine methyl ester) leads likewise to a loss of susceptibility.[257]

There are three known examples of substrates which are hydrolyzed by chymotrypsin but whose structures appear to be quite unrelated to that of the specific substrates just described. The first of these is ethyl *p*-nitrophenyl carbonate, which is slowly hydrolyzed to give *p*-nitrophenol and monoethyl carbonate.[86] The second of these substrates is *p*-nitrophenyl acetate which is hydrolyzed somewhat more rapidly (see Table IX). These substrates resemble chymotrypsin inhibitors such as diethyl *p*-nitrophenyl phosphate rather than the usual peptide or ester substrates, and further work has shown that they are hydrolyzed by a

(262) L. W. Cunningham, Jr., *J. Biol. Chem.* **207**, 443 (1954).
(263) S. Kaufman and H. Neurath, *J. Biol. Chem.* **181**, 623 (1949).
(264) M. Bergmann and J. S. Fruton, *J. Biol. Chem.* **124**, 321 (1938).
(265) R. R. Jennings and C. Niemann, *J. Am. Chem. Soc.* **75**, 4687 (1953).
(266) V. Goldenberg, H. Goldenberg, and A. D. McLaren, *J. Am. Chem. Soc.* **72**, 5317 (1950).
(267) R. V. MacAllister and C. Niemann, *J. Am. Chem. Soc.* **71**, 3854 (1949).

different mechanism.[267a] Another substrate susceptible to hydrolysis by chymotrypsin, for which no parallel exists among synthetic peptides or esters, is the peptide link between cysteic acid and serine which, according to Sanger and Thompson,[268] is opened in the A chain of oxidized insulin.

Inspection of the kinetic constants of Table IX shows clearly the large difference in the ease of hydrolysis of ester and amide substrates. This will be seen again when trypsin and carboxypeptidase substrates are considered. This difference is due both to a lower K_m and a higher k_3. The latter might be related to the more rapid acid or alkaline hydrolysis of esters as compared to amides or peptides. In view of the higher k_3 and the large differences in both constants, the lower K_m for ester substrates is most likely indicative of a higher affinity of the enzyme for esters as compared to amides, a conclusion which may be correlated with the more electropositive character of the carbonyl carbon of the ester. A higher affinity for ester substrates is also suggested by consideration of the K_i values for inhibitory esters which are usually lower than for the corresponding amides (see Table X).

Table X lists K_i values for a number of competitive inhibitors together with the free energy of binding calculated from these dissociation constants. The next to the last column of the Table (subsection I) gives the increase in binding energy, $[\Delta(\Delta F)]$, resulting from replacement of an acetyl group by a nicotinyl group. It is evident that this difference is practically constant for all compounds except the esters. If a similar calculation is carried out on the basis of the values of ΔF for structurally related substrates (L-enantiomorph of the inhibitors), derived on the assumption that $K_m = k_2/k_1$, the same difference of approximately 0.4 kcal./mole is obtained, thus providing presumptive evidence that K_m is actually a measure of enzyme–substrate affinity. This is also in accord with the low values of k_3 for amide substrates as compared with esters.[244]

The difference in free energy of binding for D and L isomers is the same for a series of compounds in which only the acid constituents of the secondary peptide group is altered (e.g., tryptophanamide, acetyltryptophanamide, nicotinyl tryptophanamide), $\Delta(\Delta F)$ amounting to about 0.4 kcal./mole). However, $\Delta(\Delta F)$ varies with the nature of the specific amino acid in position R_2 and with the hydrolyzable bond in R_3. Further regularities may be inferred from the values given in Tables IX and X, and from the original papers; however, there is as yet insufficient information for any wide generalizations.

The approach followed in the work of Niemann and co-workers is

(267*a*) B. S. Hartley and B. A. Kilby, *Biochem. J.* **56,** 288 (1953).
(268) F. Sanger and E. O. P. Thompson, *Biochem. J.* **53,** 366 (1953).

TABLE X
CHYMOTRYPSIN INHIBITORS

	Acetyl derivative		Nicotinyl derivative			
I. Acyl derivatives of:	K_i, $M \times 10^3$	ΔF, kcal./mole	K_i, $M \times 10^3$	ΔF, kcal./mole	$\Delta(\Delta F)$, nicotinyl -acetyl	Refs.
L-Tryptophan	17.5	2.39	8.8	2.81	0.42	207,184
L-Tryptophanmethylamide	4.8	3.16				207,244
D-Tryptophan	4.8	3.16				207,244
D-Tryptophanamide	2.7	3.50	1.4	3.89	0.39	207,269,184
D-Tryptophanmethylamide	1.7	3.78				207,244
D-Tryptophan methyl ester	0.09	5.52				270,207
L-Tyrosine	115	1.28	60	1.66	0.38	206,207
D-Tyrosinamide	12.0	2.62	6.2	3.01	0.39	206,207,269
D-Tyrosine ethyl ester	3.5	3.35	0.97	4.10	0.75	206,207
D-Phenylalaninamide	14	2.53	7.0	2.93	0.40	269,260
D-Phenylalanine methyl ester	2.5	3.55				260
II. Other amino acid derivatives						
L-Tryptophanamide	6.3	3.0				244,207
D-Tryptophanamide	3.2	3.4				244,207
Benzoyl-D-phenylalanine	15	2.5				3,263
Benzoyl-L-phenylalanine	29	2.11				3,263
Benzoyl-DL-methionine	46	1.83				3,263
Benzoyl-DL-phenylglycine	7.8	2.90				3,263
Benzoylglycinamide[c]	11					271
III. Miscellaneous compounds						
Tryptamine	2.5	3.55				272
Indole	0.72	4.31				273
1-Phenyl-2-acetamidobutan-3-one	7.9	2.89				263
2-Acetamidocinnamic acid	18	2.39				263
Acetanilide	10.4	2.57				261
2-Phenoxyethanol	5.8	3.05				274
2-Naphthylmethylmalonic acid	5.5	1.72				274

(269) H. T. Huang and C. Niemann, *J. Am. Chem. Soc.* **73,** 1555 (1951).
(270) H. T. Huang and C. Niemann, *J. Am. Chem. Soc.* **73,** 3228 (1951).
(271) H. T. Huang and C. Niemann, *J. Am. Chem. Soc.* **74,** 4634 (1952).
(272) H. T. Huang and C. Niemann, *J. Am. Chem. Soc.* **74,** 101 (1952).
(273) H. T. Huang and C. Niemann, *J. Am. Chem. Soc.* **75,** 1395 (1953).
(274) H. Neurath, J. A. Gladner, and G. De Maria, *J. Biol. Chem.* **188,** 407 (1951).

TABLE X. (*Continued*)

IV. Homologous series of substituted fatty acids R.$(CH_2)_n$.COOH	Ref.[a274] Free acids K_i, $M \times 10^3$	Ref.[a274] Free acids $-\Delta F$, kcal./mole	Ref.[b261] Free acids K_i, $M \times 10^3$	Ref.[b261] Free acids $-\Delta F$, kcal./mole	Ref.[b261] Acid amides K_i, $M \times 10^3$	Ref.[b261] Acid amides $-\Delta F$, kcal./mole
β-Indoleacetic acid			25	2.19		
β-(β-Indole)propionic acid	2.5	3.56	13	2.58	1.7	3.79
γ-(β-Indole)butyric acid	3.6	3.34	17	2.41		
Benzoic acid	42	1.88	200	0.96	6.6	2.98
Phenylacetic acid	42	1.88	120	1.26	10.2	2.72
β-Phenylpropionic acid	5.5	3.09	28	2.13	6.7	2.97
	4.5	3.21				
γ-Phenylbutyric acid	14	2.54	27	2.15	7.2	2.93
Cyclohexylacetic acid	86	1.46				
β-Cyclohexylpropionic acid	30	2.08				
γ-Cyclohexylbutyric acid	35	1.99				
β-(α-Naphthyl)propionic acid[d]	4.0	3.28				

[a] In 0.1 M phosphate pH 7.8. Acetyl-L-tyrosinamide as substrate.
[b] In tris(hydroxymethyl)aminomethane hydrochloride buffer pH 7.9, nicotinyl-L-tryptophanamide as substrate.
[c] Apparent K_i value (see original paper).
[d] Has been used for binding studies.[275]

essentially an extension of that originally proposed by Bergmann and Fruton[276] who assumed that the rate of hydrolysis of a synthetic substrate by a proteolytic enzyme, when expressed by the proteolytic coefficient, C, could be represented by an expression such as:

$$C = k \cdot a \cdot b \cdot c \tag{12}$$

where a, b, and c are parameters related to the various groups R_1, R_2, and R_3, present in a series of structurally related substrates. This relation was found to hold for the hydrolysis of four pairs of derivatives of phenylalanine and tyrosine by carboxypeptidase, the proteolytic coefficients of the phenylalanine substrates being 1.6–1.8 times as great as those for the tyrosine substrates. An element of uncertainty exists in this interpretation, however, since it is based on first-order constants at an arbitrarily chosen substrate concentration of 0.05 M, assuming that the substrate concentration dependence of the first-order constant is the same for the

(275) M. W. Loewus and D. R. Briggs, *J. Biol. Chem.* **199**, 857 (1952).
(276) M. Bergmann and J. S. Fruton, *J. Biol. Chem.* **145**, 247 (1942).

entire series of substrates.[3] Smith[277] reported a similar adherence to Eq. (12) when the action of carboxypeptidase on tryptophan- and glycine-containing substrates was compared, and similar adherence was reported by Dekker, Taylor, and Fruton[225] for the hydrolysis of methionine peptides by carboxypeptidase.

The three groups R_1, R_2, and R_3 are present in all substrates which are hydrolyzed by chymotrypsin at appreciable rates (β-phenylpropionic ester in which R_1 is a hydrogen atom is only very slowly hydrolyzed). In order to account for the stereochemical specificity of the enzyme and for the effects of changes in R_1, R_2, or R_3 on K_m, it was assumed, in accordance with the polyaffinity theory of Bergmann,[278] that these groups interact with corresponding areas on the enzyme surface, ρ_1, ρ_2, and ρ_3.[244]

The combination of synthetic inhibitors with the enzyme is probably less specific in that only two, or in the case of indole,[273] perhaps only one of the groups, R_1, R_2, and R_3, need to be present; in fact, such bifunctional inhibitors are often more firmly bound than are related trifunctional compounds. The only requirement for substantial inhibition appears to be an aromatic residue and a polar group separated from it by two carbon atoms.[274,261] When a bifunctional compound such as methyl hippurate serves as substrate, inhibition by a monofunctional inhibitor such as indole[273] probably occurs via the formation of a ternary complex resulting from the concurrent interaction of both the inhibitor and the specific substrate with the enzyme.

In studies involving trifunctional inhibitors, the striking anomaly was encountered that the D isomers are more firmly bound than the L isomers, D isomers of specific amide substrates being included in this observation. (The same conclusion cannot be drawn for ester substrates since k_3 is so large that K_m cannot be regarded as an equilibrium constant.) In view of the similar effects of substituents in the R_1 position on the free energy of binding for both L substrates and D inhibitors, it was argued that the same sites on the enzyme, ρ_1, ρ_2, and ρ_3 are involved in each case. This might be conceivable since the fourth position on the asymmetric carbon atom of the interacting substrate or inhibitor is occupied by a hydrogen atom, but the question remains why only L isomers of specific substrates are hydrolyzed by the enzyme. No simple explanation for this apparent paradox is as yet at hand.

(*a*) *The Effect of Methanol on α-Chymotrypsin-Catalyzed Hydrolysis.* Most of the synthetic substrates for chymotrypsin contain aromatic amino acid residues in position R_2 which render these compounds relatively insoluble in water. For this reason many of the measurements

(277) E. L. Smith, *J. Biol. Chem.* **175,** 39 (1948).
(278) M. Bergmann, *Harvey Lectures Ser.* **31,** 37 (1936).

previously alluded to have been carried out in a solvent system containing 20–30 per cent methanol. The effect of methanol on the kinetics of hydrolysis of acetyl-L-tyrosinamide has been shown to be due entirely to its effect on K_m, k_3 remaining essentially constant[142] and $1/K_m$ decreasing linearly with increasing methanol concentration. This result would be equally compatible with $k_3 \gg k_2$, or $k_2 \gg k_3$, or k_2 being independent of methanol concentration. In view of the fact that for amide substrates of chymotrypsin K_m is probably simply k_2/k_1, the second of these alternatives appears more attractive, the original interpretation of the authors[142] notwithstanding.

In order to avoid the use of methanol, strongly polar acyl groups have been introduced into the substrate molecule replacing acetyl, benzoyl, or carbobenzoxyglycyl. Nicotinic acid[184] and, more recently, glyceric acid[185,279] have been used for this purpose.

(*b*) *Number of Active Sites.* A fundamental aspect of the problem of the mode of action of proteolytic enzymes, as of any enzyme, is the number of active sites of the enzyme molecule. In the case of chymotrypsin, this problem has been attacked and solved by several experimental approaches and appears to be of particular interest in view of the diversity of structures which comply with the specificity requirements of this enzyme. All results obtained to date are in accord with a single active center for this enzyme (molecular weight 21,500).

Thus it has been found that the different manifestations of activity of chymotrypsin are all inhibited to the same extent by a given inhibitor.[85,269,86] (Similar evidence has been found for the inhibition of the amidase, esterase,[173] and proteinase[280] activities of trypsin by the soybean inhibitor). It has also been found that the kinetics of hydrolysis of a mixture of acetyl-L-tyrosinamide and acetyl-L-tryptophanamide can be accounted for by combination of both substrates with the same active site.[281] More direct evidence for the stoichiometric relation between activity and enzyme mass comes from binding studies. Thus using the specific inhibitor α-naphthylpropionic acid in equilibrium–dialysis studies, Loewus and Briggs[275] found that at pH 8 maximum binding corresponds to a protein to inhibitor ratio of 1:1. Doherty and Vaslow, using 3,5-dibromo-*N*-acetyl-L-tyrosine, containing radioactive bromine, similarly found that at pH 7.5 one mole of this substrate was maximally bound by 22,000 g. of α-chymotrypsin.[282] From the variation of the equilibrium

(279) A detailed kinetic analysis of the chymotryptic hydrolysis of glyceryl peptides will be published in the near future by Dr. D. G. Doherty (private communication).

(280) N. M. Green, unpublished experiments.

(281) R. J. Foster and C. Niemann, *J. Am. Chem. Soc.* **73,** 1552 (1951).

(282) D. G. Doherty and F. Vaslow, *J. Am. Chem. Soc.* **74,** 931 (1952).

constant with temperature, an enthalpy of binding of −5000 cal./mole was calculated. The functional relation between binding and enzymatic activation of 3,5-dibromo-*N*-acetyl-L-tyrosine by α-chymotrypsin appeared to be established by the finding that the enzyme catalyzed the carboxyl oxygen exchange with O^{18} from H_2O^{18}. This finding is in accord with previous studies of Sprinson and Rittenberg[283] who had shown that α-chymotrypsin catalyzed similar exchange reactions using carbobenzoxy-L-phenylalanine as substrate. In contrast, L-phenylalanine and α-chymotrypsin, and carbobenzoxy-L-phenylalanine without enzyme were found to be inactive in this respect. In a subsequent study, Vaslow and Doherty[284] extended these binding studies to the D-isomeric "virtual" substrate, *N*-acetyl-3,5-dibromo-D-tyrosine, to an inhibitory ketone, and to chymotrypsinogen as the binding protein. Binding occurred with a free energy change of 2–3 kcal./mole in all the systems investigated. The only significant difference between catalytically active and inactive systems was in the effect of pH on the values of ΔH and $T\Delta S$ in the neighborhood of the pH optimum. No interpretation of these results can be offered at present. In view of the apparent lack of specificity of binding and of the large number of molecules of substrate or inhibitor that are bound in pH regions below the optimum for enzyme activity, more work is required to establish the relation between the binding site and the catalytic center.

To return to the problem of the number of active sites of α-chymotrypsin, perhaps the most convincing proof for the presence of only one active center per enzyme molecule has been brought about by the stoichiometry of the reaction between chymotrypsin and diisopropyl fluophosphate (DFP). According to this work, to be discussed in more detail below (p. 1154), one mole of diisopropyl phosphate is taken up per mole of chymotrypsin and results in complete inactivation of the enzyme.

(2) β-, γ-, δ- and B-Chymotrypsins. Within the relatively restricted range of substrates which were tested, all these proteins were found to hydrolyze synthetic substrates for α-chymotrypsin. β- and γ-Chymotrypsin hydrolyze carbobenzoxyglycyl-L-tyrosylglycinamide and benzoyl-L-tyrosylglycinamide at the same rate, per milligram of enzyme nitrogen, as does α-chymotrypsin.[77] B-Chymotrypsin[15] was tested on a wider variety of substrates and found to be qualitatively similar to α-chymotrypsin. Although δ-chymotrypsin hydrolyzed both the amide and ester of acetyl-L-tyrosine, K_m was lower and k_3 higher than for the hydrolysis of the same substrates by α-chymotrypsin.[82] These latter

(283) D. B. Sprinson and D. Rittenberg, *Nature* **167**, 484 (1951).
(284) F. Vaslow and D. G. Doherty, *J. Am. Chem. Soc.* **75**, 928 (1953).

findings, however, have to be accepted with some reservation since δ-chymotrypsin has not yet been well characterized.

(3) *Trypsin.* The specificity of trypsin seems to be narrower than that of the other enzymes considered so far, only those peptide bonds being hydrolyzed in which the carbonyl group is contributed by arginine or lysine (Table XI). Derivatives of ornithine and histidine are not

TABLE XI
TRYPSIN SUBSTRATES

Substrate	K_m, $M \times 10^3$	k_3,[a] $\times 10^3$	Refs.
Benzoyl-L-argininamide	2.1	2.2	242
		3.8	141,3
Benzoyl-L-arginine ester[b]	~0.08	250	141
Benzoyl-L-arginine ethyl ester	......	280	285,286
Benzoyl-L-arginine ethyl ester[d]	......	360	285,286
p-Toluenesulfonyl-L-arginine methyl ester	......	1800	173
α-Hydroxyl-δ-guanidinovaleric acid methyl ester	......	140	287
L-Lysine ethyl ester[c]	......	110	176
L-Arginine methyl ester[c]	......	150	208

The reactions were carried out in phosphate buffer pH 7.8 in the absence of calcium ions unless otherwise stated.

[a] k_3 is expressed in moles/l./min./mg. enzyme N/ml.

[b] The following esters gave identical values for k_3: methyl, ethyl, isopropyl, cyclohexyl, benzyl, α-glyceryl.

[c] pH 5.8.

[d] In presence of 0.001 M Ca^{++} in borate buffer, pH 7.8.

attacked, and acylation of the basic side chain, such as in α-benzoyl-ε-carbobenzoxy-L-argininamide, destroys the susceptibility to tryptic hydrolysis.[288] On the other hand, omission of the α-acyl group, as in L-arginine methyl ester, while shifting the pH optimum does not destroy susceptibility to tryptic hydrolysis. Thus the secondary peptide group may be replaced by an amino group, or even by a hydroxyl group.[256] In agreement with the specificity requirements as determined by the use of synthetic substrates, it is found that trypsin readily hydrolyzes salmine,[289] polylysine,[290] and those peptide bonds in the A and B

(285) N. M. Green, J. A. Gladner, L. W. Cunningham, Jr., and H. Neurath, *J. Am. Chem. Soc.* **74,** 2122 (1952).
(286) N. M. Green and H. Neurath, *J. Biol. Chem.* **204,** 379 (1953).
(287) J. E. Snoke, unpublished experiments, cited in Ref.[3].
(288) K. Hofmann and M. Bergmann, *J. Biol. Chem.* **130,** 81 (1939).
(289) R. A. Portis and K. I. Altman, *J. Biol. Chem.* **169,** 203 (1947).
(290) E. Katchalski, *Advances in Protein Chem.* **6,** 123 (1951).

chains of oxidized insulin which contain a carbonyl group of arginine or lysine.[291,268] Hydrolysis of polylysine and of salmine liberates no free amino acids, demonstrating that trypsin cannot act on a C-terminal basic group. It will, however, attack an N-terminal basic amino acid in compounds such as lysine ethyl ester.[176] Trypsin was the first proteolytic enzyme which was shown to possess esterase activity.[173] As in the case of chymotrypsin and carboxypeptidase, substitution of the susceptible amide group by an ester group greatly increases the rate of enzymatic hydrolysis. This is shown by the very slow hydrolysis of certain esters whose amide analogs are not attacked (e.g., L-tyrosine ethyl ester),[173] by the low K_m values of ester substrates, and by the failure of benzoyl-L-arginine to inhibit the hydrolysis of esters in contrast to the strong inhibition exerted toward the corresponding amides. Kinetic data show that the nature of the alcohol group has a negligible effect on the rate of hydrolysis, indicating that the function of the enzyme is only to activate the arginine carbonyl group. The hydrolysis of benzoyl-L-arginine ethyl ester and of benzoyl-L-arginine methyl ester was accelerated in the presence of various alcohols, and the kinetics of the reaction deviated from the usual zero-order function.[141] In 16 per cent methanol, a 15 per cent increase was observed, the same concentrations of ethanol, *n*-propanol, or *tert*-butanol causing a 35–40 per cent increase in the rate.

Studies of the stoichiometry of inactivation of trypsin by diisopropylfluophosphate and related organic phosphates have shown that the binding of one mole of these reagents per mole of trypsin (molecular weight 24,000) was sufficient to cause complete inactivation, suggesting that, like chymotrypsin, trypsin possesses only one active center per molecule.

The inhibition of trypsin by structural analogs of synthetic substrates has not received much attention, and, with the exception of benzoyl-L-arginine, no synthetic inhibitor appears to be known.

(*4*) *Pepsin.* While the action of pepsin on synthetic substrates has not been measured with the same precision as has the hydrolysis of synthetic substrates by chymotrypsin, such results as have been obtained are listed in Table XII. The rates of hydrolysis are of relatively low order except for those substrates which contain adjacent aromatic residues. Moreover, only the latter substrates are optimally hydrolyzed at pH 2, the pH optimum for the action of pepsin on most proteins. All other synthetic substrates are hydrolyzed more rapidly at pH 4, which coincides with the pH optimum of a cathepsin. The most sensitive of these substrates is carbobenzoxy-L-glutamyl-L-tyrosine, which is also a typical substrate for cathepsin A, an intracellular proteinase which does not require activation by cyanide or sulfhydryl reagents.[32] A similar

(291) F. Sanger and H. Tuppy, *Biochem. J.* **49**, 481 (1951).

cathepsin-like enzyme is known to be present in gastric juice.[198,59a] It is therefore within the realm of possibility that the twofold pH optimum shown by crystalline pepsin actually reflects the presence of a contaminating cathepsin enzyme,[196] and that the slow hydrolysis of many of these substrates is actually due to this impurity. This suggestion, however, is

TABLE XII
PEPSIN SUBSTRATES[a]

Substrate	Refs.
I	
Acetyl-L-phenylalanyl-L-phenylalanine	200
Acetyl-L-tyrosyl-L-tyrosine	
Acetyl-L-phenylalanyl-L-tyrosine	
Carbobenzoxy-L-tyrosyl-L-phenylalanine	
II	
Carbobenzoxy-L-glutamyl-L-tyrosine	219,246
Carbobenzoxyglycyl-L-glutamyl-L-tyrosine	219
Carbobenzoxy-L-glutamyl-L-tyrosylglycine	
Glycyl-L-glutamyl-L-tyrosine	
Carbobenzoxy-L-glutamyl-L-phenylalanine	
Carbobenzoxy-L-methionyl-L-tyrosine	195
Methionyl-L-tyrosine	
Carbobenzoxy-L-cysteinyl-L-tyrosine	
III	
Carbobenzoxy-L-glutaminyl-L-phenylalanine	219
Carbobenzoxyglycyl-L-tyrosine	
Carbobenzoxy-L-glutamyl-L-tyrosinamide	
L-Glutamyl-L-tyrosine	
Carbobenzoxy-L-tyrosyl-L-cysteine	195
L-Tyrosyl-L-cysteine	
L-Cysteinyl-L-tyrosine	

[a] No accurate quantitative data are available for these substrates. Published figures (except Ref.[246]) refer only to per cent hydrolysis after varying times at varying enzyme concentration. The substrates have, therefore, been divided into three classes according to their approximate ease of hydrolysis (I > II > III).

contradicted by the finding that acetylation of pepsin decreases its activity toward proteins and toward carbobenzoxy-L-glutamyl-L-tyrosine to the same extent, suggesting that both activities are due to the same enzyme.[292] It is evident, therefore, that additional experimental data are required before the question can be resolved.

The most susceptible pepsin substrates possess a free terminal car-

(292) V. Hollander, *Proc. Soc. Exptl. Biol. Med.* **53**, 179 (1943).

boxyl group, conversion of this group to an amide causing a significant reduction in the rate of hydrolysis. This finding is somewhat surprising since pepsin is usually regarded as an endopeptidase. Moreover, the action of pepsin on proteins may lead to the production of free amino acids, notably of tyrosine,[16] suggesting that pepsin can act as an exopeptidase or at least that it contains exopeptidase impurities.

In contrast to its slow action on several synthetic peptides, pepsin acts rapidly on a wide variety of peptide bonds in protein substrates. The evidence on this point is largely the result of the work of Sanger and coworkers[291,268] on the peptic digestion of the A and B chains of oxidized insulin. The results are summarized in Table XIII. The nature of the

TABLE XIII

ACTION OF PEPSIN ON A AND B CHAINS OF OXIDIZED INSULIN[291,268]

Bonds containing any of the following amino acids were not attacked: serine, threonine, methionine, lysine, arginine, proline.

Slowly attacked bonds	Rapidly attacked bonds	Related bonds not attacked
Phe-val	Phe-phe[a]	Leu-val
Tyr-glu	Phe-tyr[a]	Iso-leu-val
Leu-tyr	Leu-val	Leu-cys SO_3H
Val-cys SO_3H	Leu-tyr	Val-asp
Ala-leu	Leu-glu	Val-glu
Glu-his	Glu-asp	Gly-glu
Glu-leu		Asp-glu
Gly-phe		Asp-tyr
Glu-ala		
Glu-glu		

[a] Bonds which have been shown to be split in synthetic substrates.

bonds broken was deduced from the structure of the isolated peptides. While some of these results may appear contradictory, it should be recognized that groups adjacent to those forming the peptide bonds listed in Table XIII may profoundly affect hydrolysis rates. For instance, while a leu-val bond in one part of the B chain is rapidly split, whereas a similar bond in a different portion of the chain is resistant, it should be noted that the corresponding tripeptide sequences are his-leu-val and tyr-leu-val, respectively. Since pepsin attacks the tyr-leu bond,[291] the free ammonium group of the leucyl residue prevents the subsequent hydrolysis of the leu-val bond. In the converse case, the his-leu bond is not hydrolyzed, thus rendering the leu-val bond susceptible to attack. It is worthy of note that those bonds which occur in the most susceptible synthetic substrates are also readily split in protein substrates.

(5) *Carboxypeptidase.* Specific substrates for carboxypeptidase may be represented by the symbol $R_1.R_2.CO.R_3.COOH$, hydrolysis occurring at the CO—R_3 bond. This bond may be a peptide bond if R_3COOH is an amino acid, or an ester bond if it is a hydroxy acid. The presence of a free α-carboxyl group is an absolute requirement for hydrolysis by carboxypeptidase, the corresponding amides being completely resistant.[293,277] Other structural requirements are evident from the representative list of substrates given in Table XIV, which includes only those for which complete kinetic parameters are available, and may be described as follows:

TABLE XIV

CARBOXYPEPTIDASE SUBSTRATES

Unless otherwise stated, the kinetic measurements were performed in 0.04 *M* phosphate buffer, pH 7.5, containing 0.1 *M* LiCl. k_3 is expressed as moles/l./min./mg. enzyme N/ml.

Substrate					
R_1	—R_2CO—	—R_3.COOH	K_m, $M \times 10^3$	k_3, $\times 10^3$	Refs.
Carbobenzoxy	Glycyl	L-Phenylalanine	33	2100	237
Carbobenzoxy	Glycyl	DL-Phenylalanine[a]	6.5	2230	214
Benzoyl	Glycyl	L-Phenylalanine	11	2000	166
Benzenesulfonyl	Glycyl	L-Phenylalanine	14	124	166
	Formyl	L-Phenylalanine	36	7	166
	Acetyl	L-Phenylalanine	155	2.3	166
	Chloroacetyl	L-Phenylalanine	13	137	166
Carbobenzoxy	Glycyl	L-Tryptophan[a]	5.1	1100[b]	214
Carbobenzoxy	Glycyl	L-Leucine[a]	27	1310	214
Carbobenzoxy	Glycyl	L-Leucine	93	670	294
Benzoyl	Glycyl	β-Phenyllactic acid	~0	1720	221,166
	Acetyl	β-Phenyllactic acid	13	24	166
	Chloroacetyl	β-Phenyllactic acid	~0	1300	166
	Bromoacetyl	β-Phenyllactic acid	1.6	990	166

[a] In 0.007 *M* veronal, pH 7.5, ionic strength brought to 0.5 with KCl.

[b] Previous work by the same authors, assuming zero-order kinetics gave $k_3 = 14.7 \times 10^{-3}$.[277]

The most susceptible substrates contain in position R_3 side chains of aromatic amino (or hydroxy) acids, the rate of hydrolysis decreasing in the order of phenylalanine, tyrosine, tryptophan, leucine, methionine, and isoleucine.[3,6a] In view of this wide specificity it is not surprising that derivatives of several amino acids which do not occur naturally are also hydrolyzed at appreciable rates. These include phenylglycine,[166]

(293) K. Hofmann and M. Bergmann, *J. Biol. Chem.* **134,** 225 (1940).

(294) H. Neurath and G. De Maria, *J. Biol. Chem.* **186,** 653 (1950).

p-tolylalanine, *o*-, *m*-, and *p*-fluorophenylalanine, β-2- or -3-thienylalanine and β-1- or -2-naphthylalanine.[295,6a] It is of interest that carboxypeptidase liberates also glutamic acid from polyglutamic acid (average molecular weight 10,300) though, significantly at a much lower pH (pH 5 instead of 7.5).[295a]

As in the case of trypsin and chymotrypsin, ester substrates have lower K_m and higher k_3 values than the corresponding peptides. Unfortunately, some of them are rather insoluble oils which are difficult to purify; hence they have not found much use for assay purposes. The hydrolysis of the ester bond is a convincing demonstration that the susceptible bond does not require a hydrogen atom, although in peptides, replacement of the hydrogen, as in proline or sarcosine peptides, practically eliminates enzymatic hydrolysis.[296]

In contrast to trypsin and chymotrypsin, however, the most sensitive amino acid side chain contributes the amino rather than the carbonyl group of the susceptible bond, suggesting that the activation mechanism may be different. In further contrast to the other pancreatic proteinases, carboxypeptidase is not inhibited by diisopropyl fluophosphate.[268,297] It is possible that these two phenomena are interrelated.

The nature of the amino acid which contributes the carbonyl group of the susceptible bond (R_2.CO) is of minor importance, though it is not without effect on the rate of hydrolysis. While glycine occupies this position in the most widely used substrates (such as carbobenzoxyglycyl-L-phenylalanine or hippuryl-L-phenylalanine), alanine[276] or methionine[225] do equally well, whereas glutamic acid decreases[276] and tryptophan increases[277] the rate of hydrolysis of the corresponding peptide.

While it is essential that the amino group of R_2.CO be masked, only benzoyl and carbobenzoxy groups have been tried as substituents, the former being more effective than the latter (see Table XIV). The integrity and position of the resulting (secondary) peptide bond are rather critical for rapid hydrolysis since replacement of the peptide hydrogen as in benzoylsarcosyl-L-phenylalanine[166] or separation from the hydrolyzable bond by an additional methylene group (as in carbobenzoxy-β-alanyl-L-phenylalanine)[298] reduces the rate of hydrolysis to less than 1/1000th of that of the corresponding glycyl substrates. The secondary peptide bond may be replaced, however, by a sulfonamide

(295) F. W. Dunn and E. L. Smith, *J. Biol. Chem.* **187,** 385 (1950).
(295*a*) M. Green and M. A. Stahmann, *J. Biol. Chem.* **197,** 771 (1952).
(296) M. A. Stahmann, J. S. Fruton, and M. Bergmann, *J. Biol. Chem.* **164,** 753 (1946).
(297) J. A. Gladner and H. Neurath, *Biochim. et Biophys. Acta* **9,** 335 (1952).
(298) H. T. Hanson and E. L. Smith, *J. Biol. Chem.* **175,** 833 (1948).

group, as in benzenesulfonylglycyl-L-phenylalanine[166] or entirely omitted as in carbobenzoxy-L-tryptophan or -phenylalanine.[277,276] In the latter cases, hydrolysis rates are reduced by three orders of magnitude.

When the secondary peptide group is entirely omitted, by replacement of the acylamido group by a halogen, as in chloro- or bromoacetyl-L-phenylalanine, the resulting hemipeptide will still be hydrolyzed, though at a greatly reduced rate (see Table XIV).

From the effects of structural changes on the rate of hydrolysis of substrates, interaction of the enzyme with four groupings of the substrate has been postulated.[3,6a] These are: the specific amino acid side chain in R_1, the α-carboxyl group, the susceptible bond, and the secondary peptide bond. Omission of some of these greatly diminishes substrate activity and omission of others destroys it completely.

The kinetics of hydrolysis of synthetic substrates by carboxypeptidase usually follows the integrated Michaelis-Menten equation, and one of these substrates (carbobenzoxyglycyl-L-phenylalanine) was, in fact, used to prove the applicability of this kinetic formulation to proteolytic enzymes.[237] However, the simple formulation is not applicable when inhibition by reaction products (e.g., chloroacetate)[166] or by the substrate itself (hippuryl-β-phenyllactic acid)[166] occurs. More recently, it has been reported that carbobenzoxyglycyl derivatives of L-phenylalanine and L-tryptophan, but not of L-leucine, are not only substrates but in high concentrations also inhibitors, though some of the quantitative aspects of this work[214] are at variance with observations by others,[294] and require confirmation. It has also been reported that the hydrolysis rate of carbobenzoxyglycyl-L-tryptophan is dependent on ionic strength,[214] but absolute values for this effect (which would suggest the operation of electrostatic forces in enzyme–substrate interaction) were not given.

Within the stated limitations of the structural requirements of specific substrates, the chain length of the substrate is not a contributing factor to hydrolysis by carboxypeptidase. Thus carboxypeptidase removes, at low enzyme–substrate ratios, a single amino acid from the synthetic tetrapeptide L-tyrosyl-L-lysyl-L-glutamyl-L-tyrosine,[299] whereas at higher ratios more than one equivalent of amino nitrogen was produced. In the same work, pepsin, trypsin, and chymotrypsin were also found to split one peptide bond each, as would be expected from their specificity requirements. The hydrolysis of polyglutamic acid[295a] and of certain proteins by carboxypeptidase (p. 1187) likewise demonstrates the independence of substrate hydrolysis of chain length.

In contrast to specific substrates, inhibitors for carboxypeptidase require only two interacting groups (cf. chymotrypsin). The simplest

(299) A. A. Plentl and I. H. Page, *J. Biol. Chem.* **163**, 49 (1946).

and most effective bifunctional inhibitors contain a free carboxyl group and an aromatic or heterocyclic ring, maximum inhibition occurring when the number of intervening carbon atoms is two[300,252] (see Table XV). Thus β-phenylpropionic acid is the most potent inhibitor;[252] substitution of the phenyl group by an indolyl, cyclohexyl, or naphthyl group decreases inhibitory activity in the order given.[300] In this respect,

TABLE XV
INHIBITORS OF CARBOXYPEPTIDASE

Inhibitor	K_i, $M \times 10^3$	ΔF, kcal./mole	Refs.
D-Phenylalanine	2.0	3.70	252 [a]
D-Histidine	20.0	2.33	
Phenylacetic acid	0.39	4.67	
β-Phenylpropionic acid	0.062	5.77	
λ-Phenylbutyric acid	1.13	4.04	
p-Nitrophenylacetic acid	2.5	3.57	
	K_i' [c]	$\frac{K_i' \cdot K_m}{S}$ [c]	
Indoleacetic acid	0.78	0.078	300 [b]
Indolepropionic acid	5.5	0.55	
Indolebutyric acid	33	3.3	
Benzylmalonic acid	4	0.4	
Cyclohexylpropionic acid	20	2	
Propionic acid	100	10	
Butyric acid	5	0.5	
Valeric acid	2.7	0.27	

[a] Measurements carried out in 0.04 *M* phosphate buffer, pH 7.5, containing 0.1 *M* LiCl, 25°C. The substrate was carbobenzoxyglycyl-L-phenylalanine.

[b] Measurements carried out in 0.04 *M* veronal buffer, pH 7.5, 25°C. The substrate was carbobenzoxyglycyl-L-tryptophan.

[c] K_i' was determined from the inhibitor concentration required to halve the activity at a fixed substrate concentration (0.05 *M*). Since this concentration is considerably larger than K_m (0.0052), an approximation to the true K_i may be made by multiplying by K_m/S.[301] The nature of the inhibition was not established, but it was assumed to be competitive, by analogy with the inhibitors in the upper part of the table.

(300) E. L. Smith, R. Lumry, and W. J. Polglase, *J. Phys. & Colloid Chem.* **55**, 125 (1951).

(301) J. B. S. Haldane *in* Enzymes, Longmans, Green & Co., London, New York and Toronto, 1930.

carboxypeptidase and chymotrypsin are indistinguishable from one another. Inhibition produced by benzoic acid and by carboxylic acids devoid of aromatic rings (e.g., butyric, propionic, or chloroacetic acids) was of an indeterminate type.[252]

Carboxypeptidase is also inhibited by D-amino acids, the extent of inhibition being approximately proportional to the rate of hydrolysis of the corresponding L-peptides. Thus, inhibition was found to decrease in the order; phenylalanine, histidine, alanine, isoleucine, and lysine.[252] At pH 7.5, the inhibition by L-phenylalanine was only 1/8th of that by the D isomer (and dependent on phosphate buffer concentration), but at pH 9 L was almost as effective as the D isomer at pH 7.5.[294] It was suggested that in the L isomer, the positively charged nitrogen is repelled by a positively charged group on the enzyme, thus counteracting inhibition. With the D isomer, the charged groups could be more widely separated, while at pH 9, the inhibitory species had lost its positive charge. Since D-peptides are neither hydrolyzed nor inhibitory,[29] it was concluded that the peptide group is oriented with respect to the enzyme surface in opposite direction to that in L-peptides.[3] In either case, however, the distance of separation of the nitrogen from the enzyme surface must be small since replacement of one or both of the amino hydrogens by methyl groups abolishes the inhibition by D-phenylalanine.[166]

(6) Papain and Ficin. In the past, the specificities of papain and ficin have been dealt with only in a qualitative manner (Table V). More recently, however, the substrate specificity of crystalline papain, prepared from dried latex, was reinvestigated in a more quantitative fashion.[302] Maximum hydrolytic activity was observed in the presence of cysteine and Versene, the former serving to activate by reduction, the latter by metal chelation. Under these conditions, a broad pH optimum of pH 5–7.5 was found for amidase activity, and a narrow optimum above pH 6 for esterase activity. Benzoyl-L-argininamide and tosyl-L-arginine methyl ester were the most susceptible substrates, the "proteolytic coefficient" for the hydrolysis of the amide (0.05 M) being $C = 1.2$, approximately 14 times higher than for the hydrolysis of the same substrate by crystalline trypsin. In addition, typical substrates for pepsin (carbobenzoxy-L-glutamyl-L-tyrosine), chymotrypsin (acetyl-L-tyrosinamide), carboxypeptidase (carbobenzoxyglycyl-L-tryptophan), and other peptidases (L-leucylglycine) were also hydrolyzed by crystalline papain, although at rates so much lower that the presence of adventitious enzymatic purities in amounts too low to be detected by chemical means cannot be definitely excluded.

(302) J. R. Kimmel and E. L. Smith, *J. Biol. Chem.* **207,** 515 (1954).

Exchange reactions catalyzed by papain and ficin (transpeptidation) will be considered elsewhere in this chapter (p. 1140).

(7) *Cathepsins.* The specificities of these enzymes are summarized in Table V. The esterase activity of cathepsins C and III is noteworthy. In a qualitative way, the specificities of these enzymes may be compared with those of other proteinases. Thus cathepsin A resembles pepsin; cathepsin B, trypsin; cathepsin C, chymotrypsin; cathepsin III, manganese-activated leucineaminopeptidase (pH optimum approximately pH 8); and cathepsin IV, carboxypeptidase. However, except in the case of cathepsin A, these parallels are only superficial since the cathepsins differ from their analogs in requirements for activators and in pH optima. Cathepsin C, the most highly purified of these enzymes, has a relatively narrow specificity. Thus it will not attack amides of acylated glycine, methionine, glutamic acid, or arginine, nor any of the substrates specific to the other cathepsins.

The activation and inhibition of these enzymes will be considered elsewhere in this chapter (p. 1144).

(8) *Bacterial Proteases.* The specificity of the crystalline streptococcal protease of Elliott[225a] was given in Table V. Although apparently homogeneous, it hydrolyzes derivatives of several different amino acids. This enzyme requires activation by cyanide, cysteine, or thioglycolate, and is inactivated by iodoacetate. In this latter respect, it resembles many impure bacterial proteases whose specificity has not yet been studied. Another bacterial protease which has been isolated from *Cl. histolyticum* shows the same specificity as does trypsin (see Table I).

Many bacterial toxins (Chap. 15) have collagenase activity and split only gelatin to any significant extent. If this activity is correlated with a specificity toward synthetic substrates, it might be expected that these toxins would attack peptides of proline and hydroxyproline,[50] since these amino acids are known to be present in large amounts in collagen.

(9) *Other Enzymes.* The specificity of the purified acylases of kidney, which have been used for the optical resolution of amino acids, has been discussed previously (p. 1108). More recently,[41] a related enzyme from a particulate fraction of kidney has been highly purified and its specificity investigated in detail. The specificity is rather wide since the enzyme hydrolyzes glycyl derivatives of all the naturally occurring amino acids, the D isomers, and the dehydro analogs (histidine and cystine derivatives were not tried). The rates of hydrolysis varied over wide limits, by a factor up to 70, but there was no clear correlation with the structure of the substrate. Glycine could be replaced by L-alanine without reduction in hydrolysis rates, but not by D-alanine or by chloroacetic acid Replacement of the other amino acid residue by ammonia, sarcosine, or a

dipeptide led to a considerable decrease in the rate of hydrolysis. It appears likely that the hydrolysis of the different substrates is due to a single site on a single enzyme since (*a*) a mixture of substrates was decomposed at a rate intermediate between that for the component parts, and (*b*) fractionation did not affect the relative activities toward different substrates.

4. Transpeptidation Reactions Catalyzed by Proteolytic Enzymes

By YADVIGA DOWMONT HALSEY

The replacement of one participant in a peptide bond by a related compound has been termed transpeptidation. This term will be used to include all the types of replacement reactions catalyzed by proteolytic enzymes whether the bond involved is that of an ester, amide, or peptide. Replacement reactions may be conveniently divided into two formal types.

$$RCONHR' + NH_2R'' \rightleftharpoons RCONHR'' + NH_2R' \quad (13)$$
$$RCONHR' + R''COOH \rightleftharpoons R''CONHR' + RCOOH \quad (14)$$

In the first reaction the amine component of the peptide bond is replaced by another amine, the replacement agent, and in the second reaction the carboxyl component is replaced. The reactions have been termed amine and carboxyl transfer, respectively.[303]

The catalysis of reactions of this type by proteolytic enzymes was discovered by Bergmann and Fraenkel-Conrat in the course of work on the synthesis of anilides by papain.[229,304] It was found that the synthesis of benzoylglycinanilide proceeded, in the initial stages of the reaction, nearly twice as fast from benzoylglycinamide plus aniline as from benzoylglycine plus aniline. The hypothesis was, therefore, advanced that in the former case a replacement reaction occurs rather than a simple synthesis, as in the latter case. In addition it was found that although benzoyl-L-leucine plus L-leucinanilide in the presence of papain yields benzoyl-L-leucyl-L-leucinanilide, benzoyl-L-leucine plus glycinanilide yields benzoyl-L-leucinanilide.[304] Again a replacement reaction was proposed. This reaction and similar anilide formations catalyzed by papain or ficin[305] are the only reported cases of amine transfer reactions.

Although replacement reactions have been studied extensively only within the last few years, a variety of carboxyl transfer reactions are known to be catalyzed by peptidases and proteases. However, the

(303) C. S. Hanes, F. J. R. Hird, and F. A. Isherwood, *Nature* **166**, 288 (1950).
(304) M. Bergmann and H. Fraenkel-Conrat, *J. Biol. Chem.* **124**, 1 (1938).
(305) F. Janssen, M. Winitz, and S. W. Fox, *J. Am. Chem. Soc.* **75**, 704 (1953).

identity of the enzyme catalyzing the transpeptidation reactions with that which catalyzes hydrolysis of the initial peptide bond (substrate) is better established for the reactions involving proteases.

The main examples of transpeptidation reactions apparently catalyzed by peptidases are to be found in the papers of Hanes, Hird, and Isherwood[303,306] who have studied the replacement of the cysteinylglycine moiety of glutathione by a variety of amino acids catalyzed by an extract of sheep kidney. The enzyme is not specific for glutathione; oxidized glutathione or other γ-glutamyl peptides are also capable of furnishing the initial peptide bond. Glutamine is, however, ineffective. There is, as yet, no direct proof that these reactions are catalyzed by the same enzyme which hydrolyzes the γ-glutamyl link in the substrate. The observation that eventually all of the glutathione is hydrolyzed is of doubtful significance in view of the complex enzyme system which was employed. The crude kidney extract used is known to contain a glutamine-activated enzyme which cleaves glutathione at the γ-linkage.[307] The mode of action of this enzyme is unclear since it appears likely that it does not catalyze simple hydrolysis of glutathione to glutamic acid and cysteinylglycine to any large extent.[307,308] Presumably this enzyme catalyzes replacement reactions at the γ-link of glutathione. There is at present no evidence whether there is more than one such enzyme nor as to the possible occurrence of other types of enzymes which hydrolyze glutathione in these extracts. Variations in the effectiveness of arginine in promoting glutathione reactions[303,306,308,309] in kidney preparations may be considered to indicate the presence of several enzymes.

Hanes *et al.*[306] have also noted the occurrence of carboxyl transfer when an extract of cabbage leaves is allowed to act upon a glycyl dipeptide and an amino acid. There is no proof that the hydrolytic enzymes present in this extract are actually responsible for the replacement reactions. The authors note that the presence of an acceptor amino acid appears to inhibit hydrolysis.[306] Brief reports of replacement reactions with an intestinal dipeptidase and with carboxypeptidase have also appeared.[310,311]

A number of reactions, most reasonably explained as being carboxyl

(306) C. S. Hanes, F. J. R. Hird, and F. A. Isherwood, *Biochem. J.* **51,** 25 (1952).
(307) F. Binkley and C. K. Olson, *J. Biol. Chem.* **188,** 451 (1951).
(308) P. J. Fodor, A. Miller, A. Neidle, and H. Waelsch, *J. Biol. Chem.* **203,** 991 (1953).
(309) J. H. Kinoshita and E. G. Ball, *J. Biol. Chem.* **200,** 609 (1953).
(310) I. D. Frantz, Jr. and R. B. Loftfield, *Federation Proc.* **9,** 172 (1950).
(311) P. C. Zamecnik and I. D. Frantz, Jr., *Cold Spring Harbor Symposia Quant. Biol.* **14,** 199 (1949).

transfers, have been observed to occur when α-chymotrypsin acts upon amino acid esters. Brenner and co-workers[312,313] observed that when esters of threonine, methionine, tyrosine, or phenylalanine were incubated with chymotrypsin there were obtained di- and tripeptides and insoluble polymers as well as the free amino acid expected from hydrolysis of the ester. The formation of peptides was greatest with the least rapidly hydrolyzed esters, and no peptide formation was observed with tryptophan methyl ester.[228] Free amino acids were not effective in forming peptides. When methionine esters were used,[312] it was found that there was no systematic effect of the particular alcohol residue present on the yield of peptide products obtained. The ratio of methionine in peptides to free methionine varied between 0.25 and 0.50. The pH optimum for the reaction with methionine isopropyl ester was approximately 9. In accord with the stereochemical specificity of chymotrypsin in catalyzing hydrolyses, no reaction was observed with D-methionine esters.

Peculiarities in the observed pH optima for the hydrolysis of L-phenylalanine and L-leucine ethyl ester by α-chymotrypsin[177,208] become less extreme if replacement reactions are taken into consideration. It has been found that with either of these substrates the pH optimum for hydrolysis is lower than is usual for chymotrypsin; i.e. pH 6.4–6.8 rather than pH 7.5–8.0. If the disappearance of ester is measured instead of carboxyl liberation, it is found that the pH optimum for the reaction is increased, to about 7.3 in the case of the L-leucine ester.[177] The second type of measurement would include both replacement and hydrolysis. Assuming the replacement reaction to be primarily

$$2\ \text{L-leucine ethyl ester} \rightarrow \text{L-leucyl-L-leucine ethyl ester} + C_2H_5OH$$

the disappearance of ester gives one half of the amount of substrate converted to products of the replacement reaction, as well as the amount hydrolyzed, while carboxyl liberation does not measure the replacement reaction at all. The reason for the disparity between the pH optima of the reactions just mentioned (pH 7.3) and that observed by Brenner *et al.* for L-methionine isopropyl ester (pH 9) may be, in part, the difference in conditions used in the two cases. However, the optimum pH for reaction with L-leucine ethyl ester is lower than that generally observed for synthetic substrates[314] although the conditions used were not unusual. The precipitation of L-phenylalanyl-L-phenylalanine ethyl ester during the

(312) M. Brenner, H. R. Müller, and R. W. Pfister, *Helv. Chim. Acta* **33**, 568 (1950).
(313) M. Brenner, E. Sailer, and K. Rüfenacht, *Helv. Chim. Acta* **34**, 2096 (1951).
(314) H. J. Shine and C. Niemann, *J. Am. Chem. Soc.* **74**, 97 (1952).

incubation of L-phenylalanine ethyl ester with chymotrypsin is presumably the result of the type of replacement reaction just discussed.[315]

Replacement reactions catalyzed by α-chymotrypsin have been unambiguously demonstrated by the use of isotopes.[210,204] Benzoyl-L-tyrosylglycinamide was incubated with isotopic glycinamide, and benzoyl-L-tyrosinamide was incubated with isotopic ammonia in the presence of chymotrypsin at pH 7.8. In both cases the expected carboxyl transfer reaction was observed. Replacement of the amide group or of a glycinamide group of a specific substrate for chymotrypsin by hydroxylamine has also been observed to occur to a limited extent.[210]

Little or no transamidation was found when trypsin was allowed to act upon benzoyl-L-argininamide in the presence of hydroxylamine,[210] although it is indicated by the work of Waley and Watson[316] that trypsin may catalyze carboxyl transfer reactions with the proper substrate and replacement agent. The formation of lysyllysine from L-lysyl-L-tyrosyl-L-lysine or L-lysyl-L-tyrosyl-L-leucine was noted when either peptide was acted upon by a mixture of trypsin and chymotrypsin. The authors' supposition, that the reaction is a carboxyl transfer of the *N*-terminal lysine residue to lysine or to the peptide itself, as replacement agent, is a reasonable explanation. It is likely that the replacement agent is the peptide rather than free lysine (see below). If the peptide is the main replacement agent there is no apparent necessity for the presence of chymotrypsin for the formation of lysyllysine in this system.

The catalysis of carboxyl transfer reactions by a variety of intracellular proteases has been demonstrated by Fruton and his collaborators. Highly purified preparations of cathepsin C from beef spleen have been shown to catalyze the replacement of the amide group of glycyl-L-phenylalaninamide by hydroxylamine, by amino acid amides, and by glycyl-L-phenylalaninamide.[317] At pH 5.2, no replacement by hydroxylamine was observed, while at pH 6.4 approximately one-third of the substrate which reacted yielded glycyl-L-phenylalaninehydroxamic acid rather than glycyl-L-phenylalanine. At pH 7.2, the amount of hydroxamic acid formed was decreased and a precipitate appeared. The formation of a precipitate was also observed when no replacement agent was added to the mixture of enzyme and substrate. The precipitate appeared to be a mixture of polymers formed by carboxyl transfer and had an average size corresponding to that of an octapeptide. The substrate

(315) H. Tauber, *J. Am. Chem. Soc.* **74**, 847 (1952).

(316) S. G. Waley and J. Watson, *Nature* **167**, 360 (1951); *Biochem. J.* **57**, 529 (1954).

(317) M. E. Jones, W. R. Hearn, M. Fried, and J. S. Fruton, *J. Biol. Chem.* **195**, 645 (1952).

presumably acted as replacement agent also, at least in the initial stages of the polymer formation. Roughly 30 per cent of the substrate was converted to products of replacement reactions. When amino acid amides were used as replacement agents, even larger amounts of transpeptidation occurred as shown in Table XVI. It is of interest that when L-phenylalaninamide was added as the replacement agent, hydrolysis was completely repressed and the rate of ammonia liberation, i.e., the rate of the total reaction, was decreased. In contrast, L-argininamide appeared to increase the rate of ammonia liberation. It is evident that all of the above reactions utilizing cathepsin C are complex. The replacement observed is the result of a series of reactions, as is also true of some of the replacements observed using chymotrypsin. Such complex reactions may be anticipated whenever the substrate is capable of acting as replacement agent.

TABLE XVI
REPLACEMENT REACTIONS CATALYZED BY CATHEPSIN C[317]
Substrate: glycyl-L-phenylalanine amide, 0.05 *M*

Replacement agent added	Time, hours	Ammonia liberation, μmoles/ml.	Carboxyl liberation, μmoles/ml.
None	1	15.9	11.2
	3	24.5	19.6
L-Argininamide	1	22.0	9.8
	3	27.8	12.1
L-Phenylalaninamide	1	2.3	0.0
	3	5.1	0.0
L-Isoglutamine	1	17.2	11.9
Glycinamide	1	13.8	7.1

This work with cathepsin C has been extended and polymer formation has been shown with L-alanyl-L-tyrosinamide and with glycyl-L-tyrosinamide as substrates.[318] In the former case it was found possible to recover a hexapeptide in 89 per cent yield as compared to the observed amount of replacement. The formation of considerable amounts of soluble replacement reaction products was found when glycyltyrosinamide was used.

Purified cathepsin C has also been used with papain in a two-step transpeptidation reaction; the product of the initial replacement reaction, which was catalyzed by papain, served as substrate for a second replacement reaction catalyzed by cathepsin C.[317] The amino group of the initial substrate, carbobenzoxy-L-isoglutamine, was blocked in this set of

(318) J. S. Fruton, W. R. Hearn, V. M. Ingram, D. S. Wiggans, and M. Winitz, *J. Biol. Chem.* **204,** 891 (1953).

reactions. The replacement agents were L-phenylalaninamide and L-argininamide; it appeared that papain-catalyzed replacement occurred first with L-phenylalanine as replacement agent.

Papain preparations have been used to study a number of relatively simple replacement reactions. The effect of pH on the extent of carboxyl transfer and the relation of the amino pK' of the replacement agent to the pH optimum for replacement were first demonstrated in transamination reactions catalyzed by papain.[203] A comparison of the effect of pH on replacement of the amide group of benzoylglycinamide by isotopic ammonia or hydroxylamine indicates that the optimum pH for replacement by a given enzyme depends on the amino pK_a' of the replacement agent and that the reactive form of the replacement agent is that bearing an uncharged amino group, as is also the case with synthetic reactions catalyzed by proteolytic enzymes.[319,320,320a] The data on carboxyl transfer reactions catalyzed by chymotrypsin and later work with papain are in accord with this view. The optimal pH of 9 found by Brenner *et al.*[312] is surprisingly high considering the usual optimum for hydrolysis by chymotrypsin and the fact that the replacement agent is an amino acid ester. A pH optimum of 9 for the replacement at the γ-linkage of glutathione by an unsubstituted amino acid [309] is reasonable and also corresponds to the pH optimum for glutamine-activated cleavage of the γ-linkage in similar enzyme preparations.[307] It is to be expected that when an amino acid and its ester or peptide derivative compete as replacement agents in carboxyl transfer reactions, the ester or peptide will be favored at pH values near neutrality.

Papain has been used to investigate the effect of the side chain of the replacement agent on the extent of transamidation.[321] It was found that with a given substrate, replacement agents of similar amino pK_a' frequently yield very different extents of replacement, as is shown in Table XVII. The clearest difference of this sort is seen in a comparison of glycylglycine and L-leucylglycine as replacement agents; the latter is about ten times more effective than the former. It is reasonable to suppose that the formation of an intermediate between the replacement agent and the enzyme–substrate complex occurs and that the intermediate involves an orientation, i.e., binding, of the whole molecule of the replacement agent. Papain preparations do not in general show absolute optical specificity for the replacement agent, the observed degree of optical specificity depending on the particular substrate and replacement agent and,

(319) E. Waldschmidt-Leitz and K. Kühn, *Z. physiol. Chem.* **285,** 23 (1950).
(320) E. L. Bennett and C. Niemann, *J. Am. Chem. Soc.* **72,** 1798 (1950).
(320*a*) E. L. Bennett and C. Niemann, *J. Am. Chem. Soc.* **72,** 1800 (1950).
(321) Y. P. Dowmont and J. S. Fruton, *J. Biol. Chem.* **197,** 271 (1952).

at least in some cases, on the presence or absence of alcohols.[204,317] The apparent complexity of these relations is similar to that observed for the papain-catalyzed hydrolyses and syntheses.[322–326]

Replacement reactions catalyzed by ficin and cathepsin B have been observed, including the formation of insoluble polymers by the action of ficin.[210,327] In additon, there has been a report of the catalysis of the formation of D-methionyl-D-methionine ethyl ester from D-methionine

TABLE XVII
REPLACEMENT REACTIONS CATALYZED BY PAPAIN[321]

Substrate[a]	Replacement agent	Time, hours	Ammonia liberation, μmoles/ml.	Transamidation product, μmoles/ml.	Hydrolysis product, μmoles/ml.
CGA	L-Leucine	12	24.4	1.84	14.5
	D-Leucine	10	20.2	0.50	14.8
	Glycylglycine	12	28.4	2.24	14.5
	L-Leucylglycine	10	18.6	10.7	5.2
		20	28.6	15.1	9.4
	D-Leucylglycine	9	18.6	1.10	12.3
CSA	L-Leucine	5	17.7	0.33	12.5
	L-Leucylglycine	4	12.3	2.81	6.0
	D-Leucylglycine	4	12.0	0	10.6

[a] CGA = carbobenzoxyglycinamide; CSA = carbobenzoxy-L-serinamide.

ethyl ester by liver homogenates, presumably by means of a replacement reaction.[328]

It is evident from the preceding review that carboxyl transfer reactions are of general occurrence with the common proteolytic enzymes. The catalysis of replacement reactions of this type and of hydrolysis appears to be accomplished by the same site on the enzyme.[329,330] In agreement with this view is the fact that carboxyl transfer and hydrolysis have the same optical specificities[312,203] and that approximately constant relation-

(322) M. Bergmann, L. Zervas, and J. S. Fruton, *J. Biol. Chem.* **111**, 225 (1935).
(323) O. K. Behrens, D. G. Doherty, and M. Bergmann, *J. Biol. Chem.* **136**, 61 (1940).
(324) H. B. Milne and C. M. Stevens, *J. Am. Chem. Soc.* **72**, 1742 (1950).
(325) W. H. Schuller and C. Niemann, *J. Am. Chem. Soc.* **73**, 1644 (1951).
(326) D. G. Doherty and E. A. Popenoe, Jr., *J. Biol. Chem.* **189**, 447 (1951).
(327) M. Fried, Thesis, Yale University, 1952.
(328) M. Brenner, H. R. Müller, and E. Lichtenberg, *Helv. Chim. Acta* **35**, 217 (1952).
(329) J. S. Fruton, *Yale J. Biol. Med.* **22**, 263 (1950).
(330) J. S. Fruton, Symposium sur la Biogénèse des Protéines, *Intern. Congr. Biochem. (2nd Congr. Paris, France), 1952*, p. 5.

ship of replacement to hydrolysis has been observed.[329,321] Carboxyl transfer may be considered to differ from hydrolysis only in that the second component in the reaction is an amine instead of water. Considering the great difference in concentration of replacement agent and the concentration of water in most of the systems studied, it follows that many amines must be more effective participants than is water in the reaction catalyzed by the proteolytic enzymes. At present, the lack of quantitative work on transpeptidation reactions makes it impossible to formulate any precise statement of the relationship of specificity in replacement reactions to that observed in hydrolyses.

As was noted above, very few amine transfer reactions are known. However, there appears to have been no careful attempt to investigate the possible catalysis of this type of reaction by proteolytic enzymes. It may be, as suggested by Waley and Watson,[316] that the carboxyl of the replacement agent must be uncharged for amine transfer reactions. No reaction, or very little, would then be expected at neutral pH. Although there is no indication that carboxyl transfer can occur with a substrate which is not measurably hydrolyzed in the absence of the replacement agent (but see Ref.[305]), the amine transfer reactions known to be catalyzed by papain or ficin occur at a peptide bond not appreciably hydrolyzed in the absence of a replacement agent. At present there is no obvious reason for differentiating what have been termed cosubstrate effects on the hydrolysis of normally resistant compounds[331] from an amine transfer followed by hydrolysis of the new C-terminal residue.[332] However, the amine transfers now known appear to be relatively slow reactions. In both types of reaction the compound having the bond involved in the apparent replacement differs from a proper substrate only in having an unmasked amino group. In the incubation of benzoyl-DL-phenylalanine or benzoyl-DL-tryptophan plus glycinanilide with chymotrypsin, no amine transfer reaction is observed,[305] and in this case the bond at which replacement might have been expected does not correspond to a proper substrate. No doubt the compounds effective in amine transfer with papain and ficin are not absolutely resistant to activation by the enzyme. The reactions observed are slow and probably proceed to the observed extent only because of the insolubility of the product. There is no reason to suppose that amine transfer cannot be obtained in completely soluble systems, granting that the proper reactants and conditions are used. The possibility of the occurrence of a small amount of amine transfer in papain-catalyzed reactions using a free amino acid or peptide as replacement agent has not been excluded.[321]

(331) O. K. Behrens and M. Bergmann, *J. Biol. Chem.* **129,** 587 (1939).
(332) H. Waelsch, *Advances in Enzymol.* **13,** 237 (1952).

It would appear that if the activation of a peptide (or related) bond by a given proteolytic enzyme is possible, then, in the presence of appropriate replacement agents, both amine transfer and carboxyl transfer may occur in addition to the expected hydrolysis.

5. Synthesis of Proteins and Plasteins

The resynthesis of polypeptide or protein-like material from the action of proteolytic enzymes on protein hydrolyzates is well known,[333–338] though the products, which have been termed plasteins, are not well defined. The plasteins are usually insoluble under the conditions of the experiment and precipitate out. This removal of the synthetic products from solution enables synthesis to proceed further. The molecular weight of the products generally appears to be in the neighborhood of 5000–12,000,[333,334] though the action of chymotrypsin on peptic digests of ovalbumin is stated to yield products with sedimentation constants in the region of 10–15 *S*, corresponding to molecular weights of several hundred thousand.[339] Plastein from trypsin[336] and insulin[337] solutions, respectively, were biologically inactive.

The most extensive claims for resynthesis of true proteins have been made in a series of papers by Bresler *et al.*,[340–343] in which high pressures have been used to increase resynthesis. At first sight, this action of high pressure appears unexpected since it is known that denaturation and hydrolysis of proteins are both accompanied by decrease in volume (sec. VIII). High pressure would therefore be expected to decrease the extent of synthesis. The explanation given by the author[344] is that the distance between amino acid residues is greater in a crystal of alanine than it is in a polypeptide chain. However, it is more likely[345] that electrostriction, which plays the main role in the volume change, decreases considerably

(333) H. Wasteneys and H. Borsook, *Physiol. Revs.* **10,** 35 (1930).
(334) A. I. Virtanen, H. Kerkkonen, M. Hakala, and T. Laaksonen, *Naturwissenschaften* **37,** 139 (1950).
(335) J. A. V. Butler, E. C. Dodds, D. M. P. Phillips, and J. M. L. Stephen, *Biochem. J.* **42,** 116, 122 (1948).
(336) J. H. Northrop, *J. Gen. Physiol.* **30,** 377 (1947).
(337) J. N. Haddock and L. E. Thomas, *J. Biol. Chem.* **144,** 691 (1942).
(338) H. Tauber, *J. Am. Chem. Soc.* **71,** 2952 (1949).
(339) H. Tauber, *J. Am. Chem. Soc.* **73,** 1288, 4965 (1951).
(340) S. E. Bresler, A. P. Konikov, and N. A. Selezneva, *C. A.* **44,** 10, 886*e* (1950).
(341) S. E. Bresler and N. A. Selezneva, *C. A.* **44,** 6390*d* (1950).
(342) S. E. Bresler, M. V. Glikina, and A. M. Tongur, *C. A.* **45,** 10, 273*a* (1951).
(343) S. E. Bresler, M. V. Glikina, N. A. Selezneva, and P. A. Finogenov, *C. A.* **46,** 5630*e* (1952).
(344) S. E. Bresler and M. V. Glikina, *C. A.* **43,** 704*h* (1949).
(345) K. Linderstrøm-Lang, private communication.

at higher pressures where the water is already so much compressed that the ordered state of the water molecules around the ions may actually represent a larger volume than that of the disordered molecules. The fact that the coefficient of compressibility of water changes from 46×10^{-6} at 1 atmosphere to 10×10^{-6} at 10,000 atmospheres indicates that there must be a decrease of electrostriction at higher pressures.

It is reported that the specificity of digested antigens can be restored,[340] that crystalline serum albumin can be resynthesized from tryptic and chymotryptic digests,[346] and that active insulin may be obtained from inactive proteolysis products.[342] When small amounts of digests of a foreign protein were introduced, synthesis no longer occurred. Those claims clearly require independent substantiation before they can be accepted.

VII. Inhibition

1. Introduction

Over a period of many years a wide variety of substances has been tested for their effect on proteolysis, but only more recently has the action of some of these been analyzed from the viewpoint of structural chemistry. The inhibition of proteolytic enzymes by digestion products was one of the earliest observations and was explained in terms of formation of an inactive compound between enzyme and products.[347] This interpretation has been confirmed by more recent work on synthetic substrates and inhibitors, discussed elsewhere in this chapter (Sec. VI-3). Another type of inhibition that has also been considered in this chapter is the nonspecific inactivation of proteolytic enzymes by detergents, by urea, and by other protein denaturants, and all of these have undoubtedly been responsible for many of the examples of inhibition given in the literature (*cf.* references given by Grob).[348] Other more-or-less ill-defined inhibitors of these enzymes are considered in Refs.[349–351] A decreased rate of proteolysis need not necessarily be due to enzyme inhibition or protein denaturation. Thus if proteolytic activity is being measured in the presence of an un-ionized amino compound, transpeptidation may, in fact, be a contributing factor. This will lead to a decreased rate of appearance of carboxyl groups and, if titration of carboxyl groups is used to determine the extent of hydrolysis, apparent inhibition will result.

(346) S. E. Bresler and N. A. Selezneva, *C. A.* **46,** 10, 228*g* (1952).
(347) J. H. Northrop, *J. Gen. Physiol.* **4,** 245 (1921).
(348) D. Grob, *J. Gen. Physiol.* **29,** 219 (1946).
(349) E. Kaiser and R. Hubata, *J. Am. Chem. Soc.* **72,** 4289 (1950).
(350) T. Bersin and S. Berger, *Z. physiol. Chem.* **283,** 74 (1948).
(351) M. K. Horwitt, *J. Biol. Chem.* **156,** 427 (1944).

The present discussion of inhibition will, therefore, be limited to the following topics: (*a*) inhibition (and activation) by oxidizing and reducing agents and by sulfhydryl reagents; (*b*) inhibition (and activation) by metal ions; (*c*) inhibition by chemical modification of the enzyme; and (*d*) protein inhibitors.

2. Redox Agents and Sulfhydryl Reagents

Included in this group are thiols, ascorbic acid, cyanides, "lewisite," iodine, H_2O_2, disulfides, ferricyanide, quinones, porphyrindine, iodoacetate, *p*-chloromercuribenzoate, and cupric, silver, and mercuric ions. The chemistry of action of these substances on proteins can be usually interpreted in terms of reaction with sulfhydryl or disulfide groups, though iodine also reacts with phenolic and imidazolyl groups (see Vol. I, Chap. 10). Reaction of cyanide ion with disulfide groups gives rise to —CNS and —HS,[352,353] and is not a simple reduction as is sometimes stated in the literature.

This type of reaction may be responsible for the difference in the effectiveness of cyanide and of thiols as activators for papain and cathepsin. These activators may also form complexes with metal ions such as Cu^{++}, Hg^{++}, or Ag^{+} which would otherwise inhibit the enzyme activity either by direct combination with —SH groups (Hg^{++}, Ag^{+}), or by catalyzing their oxidation to —S—S— (Cu^{++}). The other substances listed above act by oxidation, reduction, mercaptide formation, or alkylation. Proteolytic enzymes which require thiol compounds or cyanide for their action[354] include the plant enzymes and certain cathepsins.

a. Activation of Plant Enzymes and Cathepsins

Most of the plant enzymes and cathepsins require activation by sulfhydryl compounds or by reducing agents such as ascorbic acid. Cathepsin A[33,32] and the plant enzymes solanain and hurain[355,356] are exceptions. The nature of this activation process has been the subject of much controversy on account of the difficulty of purifying the enzymes (p. 1069) and because of the presence of natural activators of the thiol type[357] in the crude products which have been mostly investigated. The bulk of the evidence has been obtained from studies on papain and favors the theory that the enzymes require a free SH group for activity and that

(352) J. Mauthner, *Z. physiol. Chem.* **78,** 28 (1912).
(353) H. S. Olcott and H. Fraenkel-Conrat, *Chem. Revs.* **41,** 151 (1947).
(354) E. S. G. Barron, *Advances in Enzymol.* **11,** 201 (1951).
(355) D. M. Greenberg and T. Winnick, *Ann. Rev. Biochem.* **14,** 45 (1945).
(356) D. M. Greenberg and T. Winnick, *J. Biol. Chem.* **135,** 761 (1940).
(357) W. Grassmann, *Biochem. Z.* **279,** 131 (1935).

the function of the activator is to regenerate this group if the enzyme should be oxidized. Typical activators are H_2S, cysteine, glutathione, cyanide ions, and ascorbic acid, the sulfhydryl compounds being the most effective ones.[358,359] The activation is reversed by typical —SH reagents methyl bromide,[360] iodoacetate,[43,361] and by oxidizing agents (iodine, hydrogen peroxide,[362] quinone,[363] Cu^{++}[363]). The inhibited enzyme can be reactivated by the usual agents provided that the —SH group has been oxidized and not alkylated.[361] The early work in this field has been well reviewed by Hellermann.[364]

Crystalline papain gives no nitroprusside reaction nor is it inactivated by porphyrindin[43] suggesting a "sluggish" thiol group.[354] On the other hand, the inactivation of papain by ferricyanide[363] is indicative of a fairly reactive thiol group. Crystalline ficin likewise fails to react with nitroprusside.[358]

A series of activators of the type $ArNHOC.CH_2SH$ has been used to activate papain.[365] The relation of activator concentration to activity could be described in terms of an equilibrium such as

$$\text{ESSR} + \text{A.SH} \rightleftharpoons \text{E.SH} + \text{ASSR}$$

Once again, however, the use of crude enzyme preparations introduces an element of ambiguity into these interpretations. Moreover, such redox systems are very sluggish and equilibrium is reached so slowly[366] that the agreement between theory and experiment is indeed unexpected.

After treatment of inactive papain with methyl bromide or ethyl iodoacetate, it can still be activated with reducing agents,[360,366a] suggesting that the inactive form contains disulfide links which cannot react with alkylating agents. Papain reacts with nitroprusside after denaturation provided that it has not been treated with iodoacetate.[367]

The above mechanism of activation has been disputed and it has been suggested that, in addition to causing reduction, the activator also combines with the enzyme. The main evidence for this "coenzyme" theory

(358) T. Winnick, W. H. Cone, and D. M. Greenberg, *J. Biol. Chem.* **153,** 465 (1944).
(359) G. W. Irving, Jr., J. S. Fruton, and M. Bergmann, *J. Biol. Chem.* **139,** 569 (1941).
(360) S. E. Lewis, *Nature* **161,** 692 (1948).
(361) A. Purr, *Biochem. J.* **29,** 5 (1935).
(362) T. Bersin and W. Logemann, *Z. physiol. Chem.* **220,** 209 (1933).
(363) L. Hellerman and M. E. Perkins, *J. Biol. Chem.* **107,** 241 (1934).
(364) L. Hellerman, *Physiol. Revs.* **17,** 454 (1937).
(365) O. Gawron and K. E. Cheslock, *Arch. Biochem. Biophys.* **34,** 38 (1951).
(366) E. S. G. Barron, Z. B. Miller, and G. Kalnitsky, *Biochem. J.* **41,** 62 (1947).
(366*a*) J. F. Mackworth, *Biochem. J.* **42,** 82 (1948).
(367) A. K. Balls and H. Lineweaver, *Nature* **144,** 513 (1939).

was the loss of activity of HCN- or H_2S-activated papain or cathepsin[359,368] on dialysis or evacuation. The different activating effects of cyanide and cysteine were likewise cited in support of this hypothesis. These results are not confirmed by similar experiments on ficin,[358] and it was suggested that the inactivation on dialysis or evacuation was due to oxidation by traces of molecular oxygen.

Kimmel, Smith, and co-workers[302,369,370] have recently reinvestigated the activation of crystalline papain and mercuripapain, prepared from dried papaya latex. The mercury compound contained one atom of mercury per 43,500 molecular-weight unit, which corresponds to a dimer. As previously mentioned (p. 1104), maximum enzymatic activity was obtained in the presence of cysteine and Versene at pH 5–7. Since papain contains eight atoms of sulfur and no methionine, the active form contains at least two sulfhydryl groups; and since oxidized papain behaves in the ultracentrifuge as a mixture of monomeric and dimeric forms, it has been suggested that oxidized papain is a mixture of a monomeric intramolecular S—S form and of a dimeric form linked by an S—S bridge, as shown below:

```
papain              papain—S—S—papain
|    |
S —  S
```

The crystalline mercury complex revealed, at pH 4, the sedimentation behavior of a monomer (molecular weight 20,700), but upon electrophoresis, separation into two components was observed. At pH 8, two sedimenting components were observed, corresponding, respectively, to the monomer and to a hexamer. Mercuripapain can be fully reactivated in the presence of cysteine and Versene.

The cathepsins differ slightly among themselves in their sensitivity to inhibitors and their requirements for activation. Cathepsin A (or I) does not require activators of any sort[32,222] and is not inhibited by iodoacetate.[222] Cathepsin B (or II) is activated by cysteine but not by ascorbic acid. This enzyme is highly sensitive to inhibition by iodoacetate. Cathepsin C[33] is activated by cysteine and slightly activated by ascorbic acid, but it is less sensitive than cathepsin B to iodoacetate. Cathepsin III, a leucinaminopeptidase, is activated by cysteine or ascorbic acid and is inhibited by α,α-dipyridyl.[35] This last effect may be due to the binding of the iron which is present in this enzyme and is not removed by dialysis. Cathepsin IV, a carboxypeptidase, and a catheptic tripeptidase which splits triglycine, are activated by cysteine.

(368) G. W. Irving, Jr., J. S. Fruton, and M. Bergmann, *J. Gen. Physiol.* **25,** 669 (1942).
(369) E. L. Smith, J. R. Kimmel, and D. M. Brown, *J. Biol. Chem.* **207,** 533 (1954).
(370) E. L. Smith, A. Stockell, and J. R. Kimmel, *J. Biol. Chem.* **207,** 551 (1954).

Except for some of the experiments on cathepsins C and III, the work described above was done on only partially purified enzyme preparations.

b. *Other Enzymes*

Other proteolytic enzymes such as pepsin, trypsin, chymotrypsin, or carboxypeptidase are quite stable in the presence of oxygen even if no precautions are taken to exclude those heavy-metal ions which catalyze the oxidation of sulfhydryl groups. Furthermore, pepsin and trypsin are unaffected by "lewisite" (chlorovinyldichloroarsine)[371] and trypsin is not inhibited by *p*-chloromercuribenzoate (10^{-5} *M*),[372] indicating that none of these enzymes requires thiol groups for activity. The purported inactivation of carboxypeptidase by iodoacetate[373] requires re-examination since it is possible that this reagent may have reacted with some group on the enzyme other than a thiol group.[353] The absence of essential thiol groups in these enzymes is further supported by the finding that thiols or cyanide have no activating effects. Actually such compounds, when present in sufficiently high concentration, have been reported to inactivate trypsin and chymotrypsin,[348,374] probably due to nonspecific reduction of disulfide bonds in the protein molecule. Other workers, however, have found that reduction at lower temperatures (25° instead of 38°) and lower pH (pH 5 rather than 7.6) with 5 per cent C_2H_5SH gave no inactivation[105] based on soluble protein. The reported inactivation of carboxypeptidase by sodium sulfide and sodium cyanide is discussed below.

3. Metal Ions

Metal ions and metal-complexing agents have been found to affect proteolytic enzymes in a number of different ways, but perhaps the most striking of these is the activation of many aminopeptidases,[375] bacterial proteases, and peptidases[46] by bivalent cations such as Mg^{++}, Mn^{++}, Co^{++}, Zn^{++}, and Fe^{++}. Since few of these enzymes have been extensively purified and since the aminopeptidases have already been reviewed in detail,[6,6a] their activation by metal ions will not be dealt with further herein. More recently, a partially purified cathepsin-like enzyme from rabbit muscle[34] and two proteases of human erythrocytes[376] have been shown to be activated by Fe^{++} and Zn^{++}, respectively. In no case has the exact nature of the binding of the metal ion been fully established,

(371) E. S. G. Barron, Z. B. Miller, G. R. Bartlett, J. Meyer, and T. P. Singer, *Biochem. J.* **41,** 69 (1947).
(372) D. Grob, *J. Gen. Physiol.* **33,** 103 (1949).
(373) E. L. Smith and H. T. Hanson, *J. Biol. Chem.* **179,** 803 (1949).
(374) R. A. Peters and R. W. Wakelin, *Biochem. J.* **43,** 45 (1948).
(375) M. J. Johnson and J. Berger, *Advances in Enzymol.* **2,** 69 (1942).
(376) W. L. Morrison and H. Neurath, *J. Biol. Chem.* **200,** 39 (1953).

though it is possible that it is directly concerned with the interaction between enzyme and substrate.[377] As would be expected, these metal-ion-activated enzymes are inhibited by metal-complexing agents such as cyanide, sulfide, citrate, and Versene. It would appear that in some cases at least, the metal ion must be allowed to combine with the enzyme before the substrate is added. For instance, glycylglycine dipeptidase is activated by Co^{++},[378] but if Co^{++} and the substrate (glycylglycine) are allowed to react before the enzyme is added, the rate of hydrolysis of the peptide is greatly reduced.[379] In this case it seems unlikely that there is any correlation between complex formation with a metal and hydrolysis of the peptide by glycylglycine dipeptidase. An interesting discussion of the nature of peptidase–metal-ion complexes from a chemical viewpoint has been given by Martell and Calvin.[380]

It has been suggested[373] that carboxypeptidase is also a metal-ion-activated enzyme since it is inhibited by sodium sulfide, sodium cyanide, cysteine, phosphate, pyrophosphate, oxalate, and citrate; iodoacetate, CuCl, and lead acetate were also inhibitory. The spectrographic demonstration of the presence of magnesium in the ash of purified carboxypeptidase was taken in further support of the metal activation of this enzyme. From these observations, it was concluded that the enzyme required both magnesium and sulfhydryl groups for its activity, though no activation by added magnesium or inactivation by fluoride was found. A requirement for sulfhydryl groups would seem unlikely since no oxidation in air occurred and since mercuric chloride did not cause any inactivation. It also would be unlikely that a magnesium-activated enzyme would be inhibited by cyanide or sulfide since magnesium does not form complexes with either of these anions. The apparent inhibition by orthophosphate, citrate, oxalate, cyanide, and pyrophosphate was disputed[294] and was shown to be fully attributable to competitive inhibition by the liberated amino acid in the presence of these anions.

Following the observation that calcium ions stabilize trypsin solutions by preventing autolysis of the enzyme,[122,92] it was found[285,286] that Ca^{++} and a number of other bivalent ions (Co^{++}, Mn^{++}, Cd^{++}, Ba^{++}) caused a 25 per cent increase in trypsin activity toward the synthetic substrates benzoyl-L-argininamide and benzoyl-L-arginine ethyl ester. This activation, observed in phosphate buffer, was reversed by Versene, but the activity could not be reduced below 80 per cent of that in the pres-

(377) I. M. Klotz *in* W. D. McElroy and B. Glass, The Mechanism of Enzyme Action, Johns Hopkins Press, Baltimore, 1954.
(378) E. L. Smith, *J. Biol. Chem.* **173**, 571 (1948).
(379) J. B. Gilbert, M. C. Otey, and V. E. Price, *J. Biol. Chem.* **190**, 377 (1951).
(380) A. E. Martell and M. Calvin, Chemistry of Metal Chelate Compounds, Prentice-Hall, New York, 1952.

ence of calcium ions. Since, moreover, dialyzed trypsin contained only traces of calcium (approximately 0.1 equivalent), it is evident that calcium was not an essential ion. The results were interpreted in terms of an effect of Ca^{++} on equilibria between native and denatured trypsin, postulated by earlier workers and discussed elsewhere in this chapter (p. 1091). The experimental results are equally interpretable by the presence in crystalline trypsin of two distinct proteins, only one requiring calcium ions for its activity. Calcium ions had a similar effect on the action of α-chymotrypsin on ester and amide substrates,[285,381] but this effect was not investigated in detail. The concentration of calcium ion required for one-half activation of trypsin was approximately 4×10^{-6} *M*.

The effect of calcium ions on the tryptic digestion of proteins appears to depend on the nature of the substrate and may be either positive,[94] zero,[286] or negative.[382] The effects have been partially analyzed,[332] but the system is too complex for definite conclusions.

In addition to being activated by the above-mentioned metal ions, trypsin is completely inhibited by 10^{-4}–10^{-5} *M* Cu^{++}, Hg^{++}, and Ag^{+} salts.[286] Such inhibition of less highly purified trypsin preparations has been mentioned in the literature from time to time (e.g., by Sugai[383]), but it has only recently been investigated in detail. The inhibition is dependent on substrate concentration and is completely reversed by thioglycolate and partially reversed by Versene and by electrolytes. The effect of metal-salt concentration on the extent of inhibition cannot be explained in terms of the law of mass action as applied to reversible combination at a single site, nor does the effect of substrate concentration conform to the Lineweaver-Burk equation for competitive inhibition. From our knowledge of the binding of such cations by proteins,[384,385] it appears probable that they combine at several sites on the protein, one or more of these being concerned with the activity of the enzyme. Although these metal ions react readily with sulfhydryl groups, such cannot be involved in this case in view of the evidence already presented (p. 1148) that there is none in trypsin. Inhibition of trypsin by such ions, therefore, proceeds by a mechanism different from the inhibition of papain and cathepsins previously discussed.

Chymotrypsin is also inhibited by certain metal ions (Cu^{++}, Pb^{++}, Hg^{++})[285,381] and so is carboxypeptidase (Pb^{++}, Fe^{++}),[373,386] but these effects have not been investigated in detail.

(381) J. A. Gladner and H. Neurath, unpublished experiments.
(382) L. Gorini and L. Audrain, *Biochim. et Biophys. Acta* **9,** 180 (1952).
(383) K. Sugai, *J. Biochem.* (*Japan*) **36,** 91 (1944).
(384) I. M. Klotz and H. G. Curme, *J. Am. Chem. Soc.* **70,** 939 (1948).
(385) C. Tanford, *J. Am. Chem. Soc.* **74,** 211 (1952).
(386) E. W. Davie and H. Neurath, unpublished experiments.

4. Chemical Modification

a. Group Reagents

Chemical modification of an enzyme (as of any protein) includes all changes in which covalent bonds are broken or new groups attached to the molecule by covalent bonds. Excluding from this discussion the reactions involving the sulfhydryl group, previously considered, it appears that pepsin, trypsin, and chymotrypsin are the only proteolytic enzymes which have been modified in this way. Reagents of widely different specificity have been used, including ketene, acetic anhydride, iodine, methanolic hydrochloric acid, formaldehyde, nitrous acid, concentrated sulfuric acid, periodic acid, dialkyl halogen phosphates, and the enzyme tyrosinase. The use of these reagents in protein chemistry and their group specificity has been critically reviewed[353,387] (see also Vol. I, Chap. 10). Although some of these reagents can be made to combine specifically with only one type of group (e.g., acetic anhydride reacts under mild conditions with amino groups only), the reactions do not always go to completion. Hence, the absence of inhibition by a given reagent does not necessarily indicate that the group involved plays no part in enzyme action. Since the reactions are rarely reversible, it is not usually possible to ascertain whether an observed inactivation is due to specific blocking of a group or to some other change in the protein molecule. Certain sulfhydryl reagents (e.g., *p*-chloromercuribenzoate and methyl mercuric nitrate) are exceptions to these reservations as is the specific inhibitor diisopropyl fluophosphate (DFP). In spite of these limitations, careful work has given some interesting information.

(*1*) *Pepsins.* Among the earliest work is that of Herriott and Northrop on the effect of acetylation with ketene and of iodination on pepsin activity.[388] Derivatives obtained by partial acetylation at pH 4.5 were found to crystallize in the same form as pepsin, to possess about 60 per cent of the original activity, and to contain between 6 and 11 acetyl groups per molecule. Of these groups, three to four were combined with the three to four primary amino groups of pepsin and the rest were found to be attached to phenolic groups. When this acetylated protein was allowed to stand in 1 *N* sulfuric acid, specific activity increased to that of untreated pepsin, the acetyl content decreased to four groups per molecule, but the amino nitrogen remained constant. Acetylation of the phenolic groups was demonstrated by comparison of the color given with the Folin reagent by pepsin and by the two acetyl derivatives. The phenolic groups which had disappeared were found equivalent to the

(387) R. M. Herriott, *Advances in Protein Chem.* **3,** 170 (1947).

(388) R. M. Herriott and J. H. Northrop, *J. Gen. Physiol.* **18,** 35 (1934).

acetyl groups introduced. More extensive acetylation (e.g., 20 groups per mole) led to 90 per cent inactivation. From these results it was concluded that some of the 16 tyrosine groups of pepsin were required for its activity but that the amino groups were not essential. These results were confirmed[292] and found to hold whether a protein or the synthetic substrate carbobenzoxy-L-glutamyl-L-tyrosine was used. Similar results were obtained with carbon suboxide as acylating reagent.[389] Herriott[390] found that pepsin was also inactivated by iodination at pH 6 and that hydrolyzates of the iodinated protein contained mono- and diiodotyrosine. As would be expected for such a reaction, the titration of the tyrosine groups in iodinated pepsin occurred at lower pH than in the untreated protein. Quantitative correlation with the amount of iodine introduced was not attempted. Further evidence for the essential nature of tyrosine groups for pepsin activity has been obtained from a study of the inactivation of pepsin by HNO_2.[391] However, the evidence from this work is suggestive rather than conclusive since both iodine and nitrous acid can react with a number of different groups in a protein molecule.

A study of the action of tyrosinase on pepsin[392,393] has yielded inconclusive results since pepsin is completely inactivated at the pH optimum of tyrosinase and since such destruction of tyrosine as occurs may be due, in part at least, to nonprotein impurities.[394] Trypsin and chymotrypsin were also found to be oxidized by tyrosinase to some extent, but the observed oxidation was at least in part due to autolysis products and no inactivation was observed.

More recently,[395] the experiments on pepsin were repeated at lower pH (5.6) and although oxidation occurred, as measured by the change in ultraviolet absorption, there was no inactivation. If small quantities of a phenol were added to the oxidation system these enzymes were readily inactivated, but this was shown to be due to nonspecific oxidation by quinonoid intermediates in the oxidation of the phenol.[396]

(*2*) *Trypsin.* The inactivation of trypsin by a number of different chemical reagents has been studied extensively by Fraenkel-Conrat, Bean, and Lineweaver.[105] The effects of chemical modification on the combination of trypsin with the trypsin inhibitor ovomucoid, formed an important

(389) A. H. Tracy and W. F. Ross, *J. Biol. Chem.* **146,** 63 (1942).
(390) R. M. Herriott, *J. Gen. Physiol.* **20,** 335 (1937).
(391) J. St. L. Philpot and P. A. Small, *Biochem. J.* **32,** 542 (1938).
(392) I. W. Sizer, *J. Biol. Chem.* **163,** 145 (1946).
(393) I. W. Sizer, *J. Biol. Chem.* **169,** 303 (1947).
(394) P. Edman, *J. Biol. Chem.* **168,** 367 (1947).
(395) W. J. Haas, I. W. Sizer, and J. R. Loofbourow, *Biochim. et Biophys. Acta* **6,** 589 (1951).
(396) I. W. Sizer and C. O. Brindley, *J. Biol. Chem.* **185,** 323 (1950).

part of this investigation. One of the most notable results is the relatively slight inactivation of trypsin by acetylation of 75 per cent of the free amino groups with acetic anhydride and sodium acetate at 0°. Partial esterification, iodination, reduction, and diazo coupling of some of the imidazolyl or phenolic groups likewise affected the enzymatic activity only little. Reaction of guanidino, indolyl, or amide groups with formaldehyde under a variety of conditions (e.g., at various pH's and in the presence of acetamide or guanidine sulfate to prevent intramolecular cross-linking[397]) almost completely destroyed tryptic activity. These groups were the only ones which were markedly sensitive to blocking by chemical reagents. While treatment with concentrated sulfuric acid at −18° also inactivated trypsin, in view of the drastic nature of the treatment it is difficult to decide whether this was specifically due to blocking of the aliphatic hydroxyl groups (45 per cent of which had disappeared) or to protein denaturation.

(3) α-Chymotrypsin. Acetylation of chymotrypsin was carried out with the same method as was just described for trypsin. The resulting product had lost two-thirds of its free amino groups but retained 70 per cent of its activity toward both ester and protein substrates.[209] When chymotrypsin was oxidized with ten moles of periodic acid per mole of enzyme, at pH 6–7, 70 per cent of its esterase activity but only 35 per cent of its proteinase activity had disappeared. The oxidation product was crystallized,[209] but the nature of the groups that had been attacked could not be determined. No decrease in amino nitrogen could be detected and there was no change in absorption spectrum; nor had sulfhydryl compounds any effect on the activity of the oxidized enzyme. Although 17 per cent of nonprotein nitrogen was formed during the oxidation, the molecular weight of the product, as determined by osmotic-pressure measurements and by sedimentation in the ultracentrifuge,[80] was not significantly different from that of α-chymotrypsin. The pH optimum for the action of either acetylated or oxidized chymotrypsin on tyrosine ethyl ester was shifted from pH 6.2 to 6.5, but since the hydrolysis of this substance is probably complicated by a transfer reaction the significance of this observation is obscure.

(4) Miscellaneous Reactions. It has been claimed that pepsin, trypsin, and papain[398–401,350] are inhibited by a variety of reagents specific

(397) H. Fraenkel-Conrat and H. S. Olcott, *J. Am. Chem. Soc.* **70**, 2673 (1948).
(398) M. Bergmann and W. F. Ross, *J. Biol. Chem.* **114**, 717 (1936).
(399) O. Schales, A. M. Suthon, R. M. Roux, E. Lloyd, and S. S. Schales, *Arch. Biochem.* **19**, 119 (1948).
(400) M. Dixon and D. M. Needham, *Nature* **158**, 432 (1946).
(401) R. M. Herriott, M. L. Anson, and J. H. Northrop, *J. Gen. Physiol.* **30**, 185 (1946).

for a keto group (semicarbazide, phenylhydrazine, hydrazine, hydroxylamine, $NaHSO_3$, dimedon, Girard reagent T). However, in spite of relatively high inhibitor concentrations (0.01–0.02 *M*), the extent of inhibition was relatively low (less than 50 per cent) suggesting that in these cases inhibition was nonspecific.

Photooxidation of chymotrypsin by visible light in the presence of methylene blue leads to parallel inhibition of both protease and esterase activities.[87] The enzyme was completely inactivated when one of the two histidines and 2.4 of the six tryptophan residues had been destroyed.

A considerable amount of work has been done during the Second World War on the effect of bis(2-chloroethyl) sulfide (mustard gas) on enzyme activity. The most sensitive proteolytic enzymes were pepsin, a tissue peptidase, and a skin proteinase.[400] Chymotrypsin was also slowly inhibited under more vigorous conditions.[401] Little can be deduced from these results since the highly reactive mustard gas combines at varying rates with almost all of the reactive groups of proteins.

b. *Organic Phosphorus Compounds*

Investigations on the effect of organic phosphates of the type

$$(RO)_2\overset{\overset{O}{\uparrow}}{P}\text{—}X,\ \text{where } R = \text{alkyl, and } X = F,\ Cl,\ NO_2\text{—}C_6H_4\text{—}O^-\ \text{or}\ (RO)_2PO^-$$

was also prompted by wartime research on these toxic agents, whose action was found to be due to their ability to inhibit cholinesterase[402] at low concentration. Some of these compounds have proved to be useful as systemic insecticides and to be of great scientific interest as specific enzyme inhibitors. Besides cholinesterase, liver esterase,[403] citrus acetylesterase,[404] the following proteolytic enzymes were shown to be inactivated by these compounds: trypsin and chymotrypsin[85], and *Bacillus subtilis* protease.[405] The effect of these organic phosphates on trypsin and chymotrypsin was a logical consequence of the discovery of their esterase activity. The high purity of these proteases facilitated a partial chemical characterization of the reaction with some of these organic inhibitors. Although this work has been reviewed in another chapter of this treatise (Vol. I, Chap. 10) and elsewhere,[4] it deserves special consideration as part of the present discussion.

(402) A. Mazur and O. Bodansky, *J. Biol. Chem.* **163,** 261 (1946).
(403) E. C. Webb, *Biochem. J.* **42,** 96 (1948).
(404) E. F. Jansen, M. D. F. Nutting, and A. K. Balls, *J. Biol. Chem.* **175,** 975 (1948).
(405) D. Steinberg, *J. Am. Chem. Soc.* **75,** 4875 (1953).

Diisopropyl fluophosphate (DFP) has been used most frequently since it combines rapidly and irreversibly with chymotrypsin (and more slowly with trypsin) with a maximum velocity near the pH optimum of the enzyme. The pH dependence of the inhibition of cholinesterase by tetraethyl pyrophosphate likewise resembles that of the enzyme activity. A product of the reaction of DFP and chymotrypsin was isolated and found to crystallize in the same form as the original enzyme, but to possess less than 1 per cent of its activity.[4,297,83]

Diisopropylphosphoryl (DIP)-α-chymotrypsin contains one gram atom of phosphorus and approximately two isopropyl groups per 25,000 g. of protein.[4] Molecular-weight studies by light scattering and by osmotic pressure yielded a value of 27,000, which is probably too high as evidenced by more recent sedimentation analyses (see p. 1076). Since less than 0.25 equivalent of fluorine could be detected spectroscopically in the product, it appears that a single group of the enzyme had been substituted according to some reaction such as

$$(C_3H_7O)_2PO.F + E.H \rightarrow (C_3H_7O)_2PO.E + HF$$

The high specificity of the reaction was indicated by (*a*) the introduction of not more than one equivalent of phosphorus per mole even in the presence of a large excess of DFP; (*b*) the absence of any effect of DFP on chymotrypsinogen or on denatured chymotrypsin;[4] and (*c*) the absence of any protective effects of chymotrypsin hydrolyzates or of free amino acids[406] on the inhibition of the enzyme by DFP. β- and γ-Chymotrypsin, oxidized α-chymotrypsin, and acetylated trypsin all react with DFP in a similar manner. Similarly, crystalline inactive derivatives of α-chymotrypsin can be obtained with a number of other organic phosphates. These are given in Table XVIII together with their phosphorus contents.

From the above evidence it is clear that these inhibitors react with only one group in the enzyme molecule, which is not present in the same state in any single amino acid and which is essential for the activity of the enzyme. Since the reaction can take place between a number of enzymes of different specificities and a number of inhibitors of different structures, it is probable that the groups determining the specificity of the enzyme are not directly involved. Instead it is suggested[407] that the inhibitors react with the group that is directly responsible for the activation of the peptide or ester bond and which is similar for all DFP-inhibited enzymes (carboxypeptidase and papain which are not inhibited by DFP presum-

(406) E. F. Jansen, M. D. F. Nutting, R. Jang, and A. K. Balls, *J. Biol. Chem.* **185,** 209 (1950).

(407) N. M. Green, *J. Biol. Chem.* **205,** 535 (1953).

TABLE XVIII
INHIBITION BY ORGANIC PHOSPHORUS DERIVATIVES

Enzyme	Inhibitor[a]	Per cent P in inhibited enzyme	Molecular weight from P content	Refs.
α-Chymotrypsin	Diisopropyl fluophosphate (DFP)	0.125	24,800 (27,500)[g]	4
	Tetraethyl pyrophosphate	—		4
	Diphenyl chlorophosphate	0.127		4
	Diethyl thionofluophosphate	0.126		4
	Tetraisopropyl pyrophosphate	0.132		4
	Diethyl *p*-nitrophenyl phosphate	0.127		4
	Diethyl *p*-nitrophenyl thionophosphate			4
	Tetrapropyl dithionopyrophosphate	0.134		4
	Tetramethyl fluophosphoramide[b]	Not Inhibitory		4
	Octamethyl pyrophosphoramide[c]	Not Inhibitory		4
	Monoethyl fluophosphate	Not Inhibitory		4
β-Chymotrypsin	DFP	0.134	23,100	98
γ-Chymotrypsin	DFP	0.120	25,800	98
Oxidized α-chymotrypsin	DFP	0.120	25,700 (26,500)[g]	209
Trypsin	DFP	0.150	20,700[d]	98
		0.145	24,800[e]	90
	Diethyl *p*-nitrophenyl phosphate		17,500[f]	99

[a] Arranged in order of decreasing rate of inhibition of α-chymotrypsin.
[b] $[(CH_3)_2N]_2PO.F$.
[c] $[(CH_3)_2N)_2PO.O.PO.[N(CH_3)_2]_2$.
[d] Not corrected for traces of phosphorus in trypsin.
[e] Corrected for traces of phosphorus (0.02%) in trypsin.
[f] Determined from the nitrophenol liberated during complete inhibition of the enzyme.
[g] Molecular weight from osmotic pressure.

ably act by a different mechanism). In this connection it is also worthy of note that, whereas a zymogen such as chymotrypsinogen can presumably combine with the groupings involved in specificity (such as 3,5-dibromoacetyl-L-tyrosine) but does not react with DFP, it is likely that a peptide which is liberated during activation (p. 1195) masks the site required for combination with DFP. Further evidence for the hypothesis that different parts of the active center are responsible for the reaction with substrates on one hand and with DFP on the other, derives from the effect of calcium ions which increase the rate of hydrolysis of synthetic substrate by trypsin[286] but fail to affect the rate of reaction between trypsin and DFP.[407] It has been found, however, that when the concentration of the phosphorylating agent (diethyl *p*-nitrophenyl phosphate) is low in comparison to trypsin, calcium ions do increase the reaction between the two,[408] and alternative interpretations have been offered for these observations.[408] More experimental data are clearly needed before the problem can be properly evaluated.

Two further observations of interest in relation to the site of inhibition are the protection by the substrate (benzoyl-L-arginine ethyl ester) and by the pancreatic inhibitor[407] against the inhibition of trypsin by 10^{-2} *M* DFP. These observations confirm that DFP reacts at least with part of the active center and suggest that the pancreatic inhibitor also acts by blocking this site.

Studies of the kinetics of the reaction between α-chymotrypsin and diethyl *p*-nitrophenyl phosphate have confirmed the conclusions derived from chemical analysis of the inhibited enzyme. When the liberation of *p*-nitrophenol was followed spectrophotometrically at 400 mμ, it was found that 1 mole of *p*-nitrophenol was formed per mole of chymotrypsin inhibited.[86] The kinetics followed the first-order relation expected for a bimolecular reaction between the enzyme and excess of inhibitor. Similar results have been obtained in a parallel investigation of trypsin.[99]

More recent studies have shown that after complete reaction with diethyl *p*-nitrophenyl phosphate, the phosphorylated α-chymotrypsin was still capable of liberating nitrophenol from *p*-nitrophenyl acetate or from *p*-nitrophenyl ethyl carbonate. Since these latter esters were found to acylate phenylalanine ethyl ester, protamine, and insulin, presumably by reaction with amino groups, the reaction with inhibited, phosphorylated α-chymotrypsin was ascribed to a similar non-enzymatic mechanism.[409] In the same investigation the interesting conclusion was reached that the reaction of active chymotrypsin with *p*-nitrophenyl acetate or *p*-nitrophenyl ethyl carbonate involves a transient acylation of

(408) L. Gorini and L. Audrain, *Biochim. et Biophys. Acta* **10,** 570 (1953).
(409) B. S. Hartley and B. A. Kilby, *Biochem. J.* **56,** 288 (1954).

the active center of the enzyme followed by hydrolysis, with the liberation of acetate or carbonate.

The problem of the nature of the specific grouping on the molecule which reacts with these inhibitors should be capable of solution by careful enzymatic degradation of the DFP-inhibited enzyme. Digestion of DFP-α-chymotrypsin, labeled with P^{32}, with pepsin, and with trypsin, followed by hydrolysis with 2 *N* HCl at 100°, led to the liberation of 30 per cent of the phosphorus as serinephosphoric acid, the remainder appearing as inorganic phosphate.[410] It was suggested, however, that the phosphorus might have been transferred to serine during the acid hydrolysis. Further work, using enzymatic hydrolysis throughout and involving the isolation of phosphorus-labeled peptides should give more unequivocal results. A study of the reaction of DFP with amino acids and related compounds[411–413] showed that DFP was unusually unreactive, considering that it was an acid halide, and acylated only amines, amino acids, or phenols under alkaline conditions (e.g., in the presence of triethylamine or potassium carbonate). The only compounds with which DFP reacted under the conditions required for enzyme inhibition were catechol[412] and tyrosine (phenolic hydroxyl).[413a] These reactions are much slower than those of DFP with the enzymes, and the results cannot be regarded as evidence for the involvement of the phenolic groups of the enzyme in the reaction. Significantly, DFP was entirely unreactive toward serine. DFP-inhibited trypsin and chymotrypsin have been used as suitable substitutes for chemical and physicochemical studies of the parent enzymes under conditions under which the active enzymes are unstable and tend to autolyze (see pp. 1075, 1079). The inhibition of trypsin by DFP has likewise been a valuable tool for the isolation of trypsinogen and of procarboxypeptidase.

Although the reactions between DFP and the esterases are essentially irreversible, it has proved possible to reactivate the products of the reaction of some of the enzymes with some of the organic phosphates, using certain bases such as glycinamide, pyridine, nicotinamide, choline, and especially hydroxylamine. These experiments were first described using cholinesterase inhibited with tetraethyl pyrophosphate (TEPP) and with

(410) N. K. Schaffer, S. C. May, Jr., and W. H. Summerson, *J. Biol. Chem.* **202,** 67 (1953).

(411) T. Wagner-Jauregg, J. J. O'Neill, and W. H. Summerson, *J. Am. Chem. Soc.* **73,** 5202 (1951).

(412) B. J. Jandorf, T. Wagner-Jauregg, J. J. O'Neill, and M. A. Stolberg, *J. Am. Chem. Soc.* **74,** 1521 (1952).

(413) T. Wagner-Jauregg and B. E. Hackley, Jr., *J. Am. Chem. Soc.* **75,** 2125 (1953).

(413*a*) R. F. Ashbolt and H. N. Rydon, *J. Am. Chem. Soc.* **74,** 1865 (1952).

DFP,[414] and it was found that the *diethyl*phosphoryl derivative was reactivated 200 times as rapidly as the *diisopropyl*phosphoryl derivative (the corresponding dimethyl derivative reactivates spontaneously in aqueous solution).[415] These studies have since been extended with success to the corresponding derivatives of α-chymotrypsin.[262] It was found, however, that the reactivation reactions were much slower and did not go to completion.

5. Protein Inhibitors

Protein inhibitors may be divided into two classes, namely, (*a*) antibodies to proteolytic enzymes and (*b*) other proteins which combine specifically with certain proteases to give inactive products.

a. Antibodies

The investigation of the immunological properties of proteolytic enzymes has been devoted mainly to distinctions among various enzymes and precursors rather than to studies of the effect of antisera on enzymatic activity.[416,52] While several of the proteolytic enzymes are rather weak antigens and do not elicit precipitin production when injected by various routes, trypsin, trypsinogen, chymotrypsin, and chymotrypsinogen can be distinguished from one another by the Dale anaphylactic test,[416] and precipitins to carboxypeptidase[417] can be produced by the Freund[418] technique. Pepsin[419] and papain[420] give rise to precipitins, but those to pepsin cross-react with pepsinogen and pepsins from other species. This is possibly related to the fact that pepsin is denatured at the pH of the blood stream. This is also suggested by the finding that antisera to pepsinogen, which is stable at this pH, do not cross-react with pepsin.

Antibodies to a number of bacterial toxins and snake venoms which reveal proteolytic activity are known (e.g., the κ and λ toxins of *Cl. welchii*).[51,421] An element of ambiguity exists in the interpretation of studies of inhibition of enzymatic activity by antibodies since it is not always possible to differentiate between the effect of the antibody on the preciptiation of the enzyme on one hand and on its activity on the other. It is, however, interesting to note that, for instance, in the case of

(414) I. B. Wilson, *J. Biol. Chem.* **199,** 113 (1952).
(415) W. N. Aldridge, *Biochem. J.* **54,** 442 (1953).
(416) C. Ten Broeck, *J. Biol. Chem.* **106,** 729 (1934).
(417) E. L. Smith, B. V. Jager, R. Lumry, and R. R. Glantz, *J. Biol. Chem.* **199,** 789 (1952).
(418) J. Freund, *Am. J. Clin. Path.* **21,** 645 (1951).
(419) C. V. Seastone and R. M. Herriott, *J. Gen. Physiol.* **20,** 797 (1937).
(420) R. Haas, *Biochem. Z.* **305,** 280 (1940).
(421) E. S. Duthie and L. Lorenz, *Biochem. J.* **44,** 173 (1949).

urease,[422] catalase,[423] and papain,[420] precipitation by the antibody can occur without complete concomitant inactivation. Thus, 30–40 per cent of the original papain activity could be demonstrated in the heterogeneous system resulting from precipitation of the enzyme with antibody. Removal of the precipitate by centrifugation also removed the activity.

The purported production of trypsin inhibitors in response to trypsin injection may be illusory in view of the presence of natural trypsin inhibitors in normal serum (see p. 1165). While conflicting reports have been published, the bulk of the evidence is against any significant increase of trypsin inhibitor activity after injection of trypsin.[424] It was suggested that such positive results as had been obtained were due to inhibition of proteolysis by products formed by the action of trypsin on serum proteins.

Carboxypeptidase is the only proteolytic enzyme for which detailed studies on inhibition by antibody have been carried out.[417] A precipitate formed in the equivalence zone contained approximately equimolar proportions of antigen and antibody. Inhibition of enzyme activity by the antibody was studied in solutions sufficiently dilute to avoid visible precipitation. Under these conditions inhibition was found to be unaffected by substrate concentration. Conversely, it was found that competitive inhibition of carboxypeptidase by phenylpropionic acid, for instance, did not prevent precipitation of the enzyme–antibody complex. It was stated that inhibition was a slow reaction and its reversal on dilution rapid; however, such a situation is unlikely in view of the low apparent dissociation constant (*ca.* 10^{-7} *M*) of the antibody–antigen complex.

The evidence available to date on enzyme antibody interaction suggests that the combination of enzyme and antibody does not necessarily involve the active center of the enzyme. Such may indeed be expected on general grounds, and any relation between the combination of the enzyme with antibody and substrate, respectively, (e.g., Ref.[425]) must be considered to be purely coincidental.

b. Other Proteins

A number of proteins of relatively low molecular weight are known which combine specifically with trypsin and inhibit its activity. Some of these, listed in Table XIX, also inhibit chymotrypsin to a lesser extent, but no related proteins have been characterized which are highly specific inhibitors for chymotrypsin or any other proteolytic enzyme. Partial exceptions to this statement are (*a*) a high molecular weight polypeptide

(422) J. S. Kirk, *J. Biol. Chem.* **100,** 667 (1933).
(423) D. H. Campbell and L. Fourt, *J. Biol. Chem.* **129,** 385 (1939).
(424) D. Grob, *J. Gen. Physiol.* **26,** 405 (1943).
(425) P. C. Zamecnik and F. Lipmann, *J. Exptl. Med.* **85,** 395 (1947).

TABLE XIX
TRYPSIN[a] INHIBITORS

Inhibitor	Molecular[e] weight	Isoelectric point	Milligrams inhibitor ≡1 mg. trypsin	Enzyme protected by ester substrate[407]	K_i, $M \times 10^9$ (25°)			Inactivated by:	Not inactivated by:
					Trypsin[b]	Chymotrypsin	Plasmin[372c]		
Pancreatic (Kunitz)[89,12]	6000 (O.P.)[89]	8.7[426]	0.48[426]	+	0.3[12]	+[89]	1[d]		
	9000 (O.P.)[12]		0.27[407]		0.2[407]				
I-T compound	36,000 (L.S.)[426]	10.1[426]			0.8[d372]				
Pancreatic (Kazal)[427]		5–6 (3 components)[426]	0.39[426]						
Colostrum[428,426]		4.2[426]	0.44[426]						
I-T compound	89,000 (L.S.)[426]	7.2[426]							
Ovomucoid[105,429]	28,800 (O.P.)[431]		1.23[430]	+	5[407]	+	+	Esterification[105]	Acetylation[105]
	27,000 (S.D.)[432]		1.34[407]		1.1[d372]			Pepsin[430]	Papain[430]
					28[105]				
Soybean[433,150 434,118]	24,000 (O.P.)[433]	4.5[f150]	1.0[426]	+	0.2[407]	1000[f150]	10[d]	Pepsin[430]	Acetylation, Papain[430]
			0.8[430]						
			0.8[435]						
			0.76[407]						
I-T compound	Sedimentation data[436]								
	41,000 (S.D.)[96]	5.0[f434]							

TABLE XIX. *(Continued)*

Inhibitor	Molecular[e] weight	Isoelectric point	Milligrams inhibitor ≡1 mg. trypsin	Enzyme protected by ester substrate[407]	K_i, $M \times 10^9$ (25°)			Inactivated by:	Not inactivated by:
					Trypsin[b]	Chymotrypsin	Plasmin[c372]		
Lima bean[430]	9400 (O.P.)	4–5	0.28		0.6[d372]	+[437]	1000[d]	Acetylation	Esterification, Pepsin, Papain
Pepsin inhibitor[19]	8000 (D)	3.7[f]	Pepsin 0.16		Pepsin 70				

[a] Pepsin inhibitor included for comparison.
[b] Calculated for Mol. wt. of 24,500.
[c] Assuming molecular weight to be the same as that of trypsin.
[d] 37°.
[e] L.S. = light scattering; O.P. = osmotic pressure; S.D. = sedimentation–diffusion; and D. = Diffusion.
[f] Cataphoresis of coated collodion particles.

liberated during the activation of pepsinogen, which combines with and inhibits pepsin (p. 1191); (*b*) a crude preparation of *Ascaris* body wall which inhibits independently pepsin and trypsin; and (*c*) mammalian sera, which have been reported to inhibit chymotrypsin.[438]

Since the trypsin inhibitors have not yet been comprehensively reviewed, an outline of their occurrence and properties will be given herein. These inhibitors have been obtained from a variety of plant and animal sources, and those from pancreas (two kinds), egg white, colostrum, soybeans, and lima beans have been purified and well characterized. Other trypsin inhibitors which have not been isolated in pure form have been found in blood serum, the body wall of *Ascaris*,[439] and in a number of leguminous seeds[440] other than soybeans and lima beans.

(*1*) *Preparation and Homogeneity.* Two different inhibitors from the pancreas and those from soybean and colostrum have been crystallized. Of these, all but the pancreatic inhibitor of Kazal *et al.*[427] give crystalline compounds with trypsin, the pancreatic and soybean inhibitor compounds containing approximately equimolar quantities of inhibitor and trypsin. These compounds have been found of use in the purification of the pancreatic inhibitor of Kunitz and Northrop (from here on simply designated as pancreatic inhibitor)[89] and of the inhibitor from colostrum.[428]

The preparation of the *pancreatic inhibitor* from fresh pancreas or from the residues after insulin extraction have been considered above (p. 1063). The pancreatic inhibitor of Kazal *et al.*[427] was obtained from a later fraction of the insulin purification process (i.e., from the 15 per cent sodium chloride washings) in extremely low yield. Thus, from 1 kg. of

(426) M. Laskowski, Jr., P. H. Mars, and M. Laskowski, *J. Biol. Chem.* **198,** 745 (1952).
(427) L. A. Kazal, D. S. Spicer, and R. A. Brahinsky, *J. Am. Chem. Soc.* **70,** 3034 (1948).
(428) M. Laskowski and M. Laskowski, *J. Biol. Chem.* **190,** 563 (1951).
(429) H. L. Fevold, *Advances in Protein Chem.* **6,** 217 (1951).
(430) H. Fraenkel-Conrat, R. C. Bean, E. D. Ducay, and H. S. Olcott, *Arch. Biochem. Biophys.* **37,** 393 (1952).
(431) H. Lineweaver and C. W. Murray, *J. Biol. Chem.* **171,** 565 (1947).
(432) E. Fredericq and H. F. Deutsch, *J. Biol. Chem.* **181,** 499 (1949).
(433) M. Kunitz, *J. Gen. Physiol.* **29,** 149 (1946).
(434) M. Kunitz, *J. Gen. Physiol.* **30,** 311 (1947).
(435) A. Dobry and J. M. Sturtevant, *Arch. Biochem. Biophys.* **37,** 252 (1952).
(436) E. Sheppard and A. D. McLaren, *J. Am. Chem. Soc.* **75,** 2587 (1953).
(437) H. Tauber, B. B. Kershaw, and R. D. Wright, *J. Biol. Chem.* **179,** 1155 (1949).
(438) P. M. West and J. Hilliard, *Proc. Soc. Exptl. Biol. Med.* **71,** 169 (1949).
(439) H. B. Collier, *Can. J. Research* **19B,** 90 (1941).
(440) R. Borchers, C. W. Ackerson, and L. Kimmett, *Arch. Biochem.* **13,** 291 (1947).

pancreas glands, 1 mg. of pancreatic inhibitor is obtained as compared to 75 μg. for the inhibitor of Kazal *et al.* The crude precipitate obtained by saturating the 15 per cent sodium chloride washings with NaCl was purified by precipitation with ethanol and removal of impurities by heating to 80° for 5 minutes in the presence of 2.5 per cent trichloroacetic acid. The inhibitor crystallized from the trichloroacetic acid filtrate and gave three components on electrophoresis, all of which were active as inhibitors. The other pancreatic inhibitor (Kunitz and Northrop) gave a single boundary on electrophoresis at pH 8.7 (buffer composition was not stated).[426] Salting-out curves in phosphate buffer showed only one component.

The isolation of the trypsin *inhibitor from bovine colostrum*[428] followed a procedure similar to that developed by Kunitz and Northrop for the pancreatic inhibitor (p. 1063). After several precipitations with 2.5 per cent trichloroacetic acid, followed by ether extraction and methanol precipitation, the trypsin–inhibitor compound was prepared and crystallized from 0.5 saturated ammonium sulfate at pH 5.5. The inhibitor was regenerated from the compound using 2.5 per cent trichloroacetic acid. Prior to the formation of the compound, the yield of inhibitor was reported to be 50 mg. per gallon of colostrum. Electrophoresis of the four-times recrystallized compound and of the inhibitor gave a single peak at pH 5–5.5.[426]

Using defatted *soybean* meal as starting material, the *trypsin inhibitor* was isolated by Kunitz[433] by a series of steps which included extraction of a precipitate obtained in 80 per cent ethanol, with 0.25 N sulfuric acid. Sufficient bentonite was added to adsorb inert proteins, and additional amounts of bentonite and Hyflo Super-Cel were added to the filtrate to adsorb the inhibitor. Following elution with 10 per cent pyridine and removal of pyridine by dialysis, and of further impurities with bentonite, the inhibitor was precipitated from solution at pH 4.65, resuspended at pH 5.2, and allowed to crystallize at 35°. Recrystallization was carried out from 20 per cent ethanol at pH 5. The yield was approximately 1g./kg. of soybean meal. The trypsin–soybean inhibitor compound crystallizes readily between pH 6 and 8 from salt-free solutions of inhibitor and trypsin.[434] The solubility of the inhibitor was independent of the quantity of solid phase present.[433]

The *lima bean inhibitor*[430] was extracted and purified by a procedure very similar to that used for the soybean inhibitor, but since, under the same conditions, crystallization did not occur, further purification was carried out by precipitation with one-half saturated ammonium sulfate and fractionation between 1.3 and 1.7 M ammonium sulfate. On electrophoresis at pH 3.6, one major peak was identified with the inhibitor. The preparation appeared homogeneous in the ultracentrifuge ($s_{20,w} = 1.5\ S$).

Earlier reports[437] of the crystallization of the lima bean inhibitor were ascribed to the crystallization of an inactive protein containing adsorbed inhibitor, removable by recrystallization.[430]

Ovomucoid has been purified by two rather similar methods[432,431] based on the solubility of the protein in sodium trichloroacetate at pH 3. After removal of the other egg white proteins with trichloroacetic acid, ovomucoid was further purified by alcohol or acetone fractionation. The material was homogeneous in the ultracentrifuge[432,441] and on electrophoresis at 0.1 ionic strength (barbiturate buffer pH 8.6, containing NaCl). However, reversible boundary spreading at the isoelectric point[432] revealed marked heterogeneity which was further confirmed by electrophoretic experiments at low ionic strength [0.01, both at pH 8.6 (barbiturate) and pH 4 (acetate)][432,441] and by immunological tests using the agar diffusion technique of Oudin.[442] More detailed analysis of the electrophoretic pattern at 0.01 ionic strength in the neighborhood of the average isoelectric point (pH 4) showed the presence of five components which could not be separated by further alcohol fractionation. However, one of these components was isolated by electrophoresis–convection[441,441a] and no detectable difference was found in the activity of the major component as compared to that of the material from which it had been separated. The electrophoretic heterogeneity may account for the disagreement among published figures for the isoelectric point of the protein (e.g., pH 3.9,[432] 4.3,[443] and 4.5).[444] The three main components of the mixture showed isoelectric points of pH 4.41, 4.28, and 4.17.[441]

The *trypsin-inhibiting activity of serum* has been known for a long time and has been ascribed to various components. Its relation to antibodies to trypsin has been considered above (p. 1160). Other substances to which inhibitory activity have been attributed are a component of the albumin fraction of serum,[445,446] and proteolysis products of the serum proteins.[447] More recently, the inhibitory activity of the protein fraction of serum has been found to be predominantly in the α- and β-globulin fractions.[372] This rather confusing situation has been considered by

(441) M. Bier, J. A. Duke, R. J. Gibbs, and F. F. Nord, *Arch. Biochem. Biophys.* **37,** 491 (1952).
(441*a*) M. Bier, L. Terminello, J. A. Duke, R. J. Gibbs, and F. F. Nord, *Arch. Biochem. Biophys.* **47,** 465 (1953).
(442) L. R. Wetter and H. F. Deutsch, *Arch. Biochem.* **28,** 399 (1950).
(443) L. G. Longsworth, R. K. Cannan, and D. A. MacInnes, *J. Am. Chem. Soc.* **62,** 2580 (1940).
(444) L. Hesselvik, *Z. physiol. Chem.* **254,** 144 (1938).
(445) A. Schmitz, *Z. physiol. Chem.* **255,** 234 (1938).
(446) K. Landsteiner, *Centr. Bakt.,* **1,** *Abt.* **27,** 357 (1900).
(447) R. G. Hussey and J. H. Northrop, *J. Gen. Physiol.* **5,** 335 (1923).

Grob,[372,424] and it has been suggested that all these factors contribute in some way to the inhibitory activity of crude serum. The only factor which has been purified to any appreciable extent is the polypeptide of the albumin fraction.[445] The material was, however, obtained only in small quantities, and attempts to duplicate these observations have failed.[144] A more recent reinvestigation of the serum inhibitor by Peanasky and Laskowski[448] has revealed properties quite different from those described by Schmitz.[445] Fractional precipitation of bovine plasma with ammonium sulfate between 40 and 90 per cent saturation (pH 4) was followed by refractionation between 50 and 65 per cent saturation at pH 4.7, and 30 per cent saturation at pH 3.6. After removal of inactive protein by adsorption on bentonite, the inhibitor was obtained in 50-fold purification and 5 per cent yield. It is different from pancreatic inhibitor in having probably a higher molecular weight. It reacts with trypsin stoichiometrically in the presence or absence of electrolytes.[448] It has also been suggested that the increase in inhibitory activity of serum that often accompanies high fever and other pathologic states was due to the breakdown products of proteins which appear in the circulation under such conditions.[424]

Crude extracts of *Ascaris body* wall inhibit both trypsin and pepsin.[439] The pepsin inhibitor was precipitated by 0.3 saturated ammonium sulfate and the trypsin inhibitor by 0.7 saturated ammonium sulfate. Contaminating glycogen was removed with diastase, but no further purification was effected.

(*2*) *Properties*. The properties of the protein inhibitors are summarized in Table XIX. The inhibitors are all relatively stable, low-molecular-weight proteins, and only ovomucoid has been shown to contain non-amino acid constituents (i.e., 25 per cent carbohydrate containing mannose and glucosamine).[429] Only inhibitors of relatively higher molecular weight (soybean and ovomucoid) appear to be subject to denaturation, all others being fairly stable to extremes of pH and temperature and probably inactivated only by cleavage of covalent bonds. For example, the pancreatic and colostrum inhibitors are both unaffected by heating to 80° for 5 minutes in 2.5 per cent trichloroacetic acid; the soybean inhibitor is reversibly denatured by heat above 30° at acid or alkaline pH, and the kinetics and thermodynamics of this process have been studied in detail[118] (see Chap. 9). The *Ascaris*, colostrum, and pancreatic inhibitors dialyze slowly through cellophane membranes, but the pancreatic inhibitor is retained by suitably prepared collodion membranes.[12] The pancreatic inhibitor of Kazal *et al.* and the serum inhibitor[448] do not pass through cellophane membranes and are precipitated by 5–10 per cent trichloroacetic acid, which property clearly differentiates them from the pancreatic inhibitor of Kunitz and Northrop.

The amino acid compositions of the pancreatic inhibitor,[12] ovomucoid,[73] and the lima bean inhibitor[430] have been determined, but there is nothing to distinguish them on this basis from other proteins (see Chap. 3). The pancreatic inhibitor is devoid of histidine and tryptophan, but rich in arginine (15 per cent), whereas the lima bean inhibitor has a high cystine content (16.5 per cent). The N-terminal residues of ovomucoid[449] and of the pancreatic inhibitor[12] have been found to be single residues of alanine and arginine, respectively, and that of soybean inhibitor a single aspartic acid or asparagine residue.[450] The C-terminal amino acid of the soybean inhibitor is a single leucine residue,[450] and of ovomucoid, a single phenylalanine residue.[451]

(*3*) *Interaction with Enzymes.* The nature of the combination of the protein inhibitors with trypsin has been the subject of several important investigations. Kunitz[150] showed that the inhibition by the soybean inhibitor was directly proportional to the amount of added inhibitor and that the pure inhibitor neutralized approximately an equal weight of crystalline trypsin. The combination was too rapid to be measured and was accompanied by a decrease in the number of free amino groups as determined by formol titration, suggesting that the interaction occurred through ionic groups. These results have been recently confirmed and extended to the lima bean inhibitor and to ovomucoid.[430] Esterification of ovomucoid destroys its inhibitory activity, while acetylation of the amino groups has little effect,[105] suggesting that the carboxyl groups are involved in combination. The soybean inhibitor behaves similarly, being unaffected by acetylation, while the lima bean inhibitor shows the opposite behavior, the activity being destroyed by acetylation but not by esterification.[430] Upon combination with trypsin, 10–20 per cent of the total free amino groups of the system disappeared.[430] Although the evidence provided by these group-blocking experiments is not unequivocal, it suggests that carboxyl and amino groups are involved in the combination and that the different inhibitors react with different groups of trypsin. As evidenced by end-group analysis employing carboxypeptidase, the single C-terminal leucine residue of the soybean inhibitor is not affected by combination with trypsin.[450]

Physicochemical measurements of the combination between soybean inhibitor and trypsin are also in accord with the postulated mechanism involving interaction between ionic species. As determined by sedimen-

(448) R. J. Peanasky and M. Laskowski, *J. Biol. Chem.* **204,** 153 (1953).

(449) H. Fraenkel-Conrat and R. R. Porter, *Biochim. et Biophys. Acta* **9,** 557 (1952).

(450) H. Neurath and E. W. Davie, *Federation Proc.* **13,** 268 (1954).

(451) L. Penasse, M. Jutisz, C. Fromageot, and H. Fraenkel-Conrat, *Biochim. et Biophys. Acta* **9,** 551 (1952).

tation-velocity measurements, the inhibitor–trypsin compound dissociates into its components below pH 2.9.[436] If trypsin is replaced by DIP–trypsin, the sedimentation constant at pH 7.8 is only slightly higher than that which would be expected for complete dissociation of soybean inhibitor and DIPT.[407] The dissociation constant, calculated from the sedimentation constant, was of the order of 10^{-3} M, in contrast to 2×10^{-10} M for the system soybean inhibitor–trypsin, and might be accounted for simply by nonspecific association between oppositely charged molecules. Dilatometric experiments[96] showed that the reaction was accompanied by a volume increase of 25 ml./mole and that the rate was too fast to be measured by this technique. Assuming that the electrostriction effect is the same as for small peptides,[168] this volume change would correspond to the disappearance of two pairs of ionic groups. The heat of reaction was found to be too small for accurate measurements (less than 2 kcal./mole),[435] and hence all of the free energy of formation (*ca.* 13.8 kcal./mole) must be accounted for in terms of an entropy increase.

In contrast to the inhibitors considered above, the pancreatic inhibitor reacts with trypsin at a measurable rate ($k = 0.2 \times 10^{-6}$ l./mole/sec. at pH 7.8, 25°), the rate decreasing as the pH is lowered, until at pH 2 no appreciable reaction occurs.[89,452,280] Similarly, if the inhibitor–trypsin compound is allowed to stand in solutions more acid than pH 3, slow dissociation into the component parts occurs. At pH 2.8, the compound is 50 per cent dissociated in 10^{-5} M solution.[280] It has been suggested that this slow rate of reaction is related to the fact that in comparison to the other inhibitors, the pancreatic inhibitor has a high isoelectric point (pH 8.7). Hence, both the inhibitor and trypsin would be positively charged below pH 8.7, and any specific attraction between oppositely charged groups will be counteracted by a net repulsion between inhibitor and trypsin molecules. Trypsin is capable of hydrolyzing both Kazal's pancreatic inhibitor and ovomucoid, and hence in the presence of stabilizing concentrations of calcium ions, at pH 7 or 7.9, digestion of these inhibitors by free trypsin will lead to the progressive release of trypsin from the respective inactive compounds.[11a,408] The term "temporary inhibition" has been proposed to describe these phenomena.[11a]

There remain to be considered the kinetic investigations on the reversal of trypsin–inhibitor interaction and its relation to the active center of trypsin. Figure 5 shows the effect of increasing concentrations of the three different inhibitors on the esterase activity of trypsin.[407] It will be noted that near the equivalence point there is a departure from the linear relation between trypsin activity and inhibitor concentration, indicating

(452) N. M. Green and E. Work, *Biochem. J.* **54**, 347 (1953).

dissociation of the inhibitor–trypsin compounds. Similar results have been obtained with protein substrates,[452,372,408] and by use of much lower enzyme concentrations (10^{-9} *M*) correspondingly higher degrees of dissociation were found.[452a] The dissociation constants, calculated from curves as given in Fig. 5 are shown in Table XIX, and, considering the low levels of activity which were measured, the agreement among the values obtained by different workers is reasonably good. The nature of

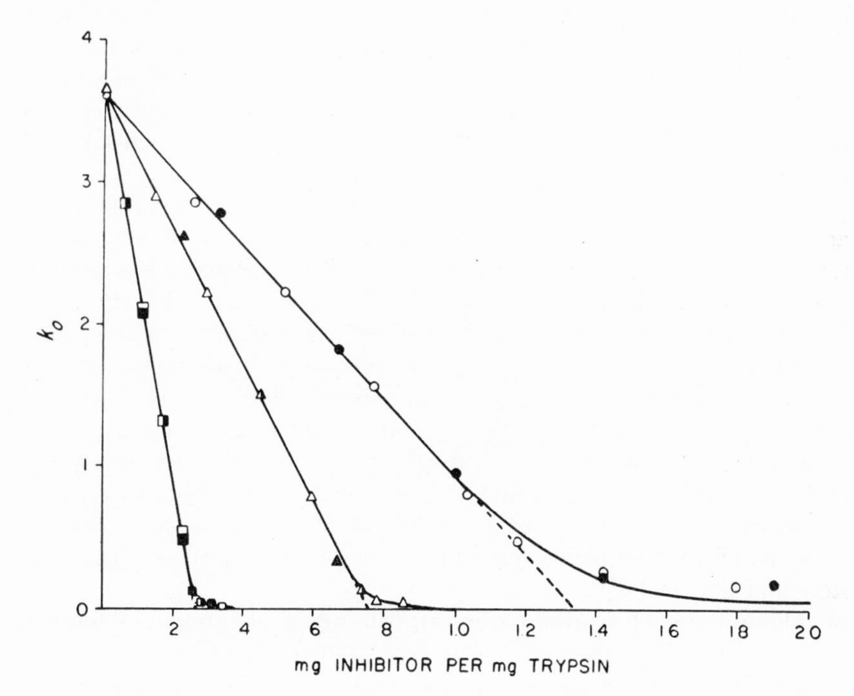

FIG. 5. Effect of pancreatic inhibitor (squares), soybean inhibitor (triangles), and ovomucoid (circles) on the esterase activity of trypsin.[407]

the substrate (protein or ester) has little effect on the measured dissociation constant, and it is evident from Fig. 5 that the substrate concentration has no effect on inhibition. It seems probable, therefore, that the inhibition is noncompetitive and has indeed so been described.[452,408] However, there are two arguments against this interpretation. In the first place, the apparent Michaelis constants of the protein substrates are so high (*ca.* 10^{-2} *M*) relative to the dissociation constants of the trypsin–

(452*a*) The results obtained for ovomucoid, using serum albumin as substrate,[408] were interpreted in terms of residual activity of the trypsin–ovomucoid complex, rather than of its reversible dissociation.

inhibitor compounds (*ca.* 10^{-10} *M*) that competitive inhibition could not be demonstrated in the usual way.[452] In the second place, although the Michaelis constant of the ester substrates is low enough to produce measurable competitive effects (*ca.* 10^{-5} *M* for benzoylarginine esters), the enzyme–inhibitor–substrate equilibrium is attained very slowly by virtue of the slow dissociation of the inhibitor–trypsin compounds. Hence, even the ester substrate does not appear to affect the inhibition unless it is added to the enzyme *before* the inhibitor.[407] In that event, the initial reaction velocity is unaffected by the inhibitor, but instead of the usual zero-order hydrolysis rates, there is a considerable falling off in reaction velocity as the inhibitor slowly displaces the substrate. The converse displacement of inhibitor by substrate can also be observed at high substrate concentration.[407]

These results show clearly that the inhibitors act by blocking the active center of trypsin. It is impossible at present to decide whether this is a relatively nonspecific effect, due to the large size of the inhibitor molecule, or whether there is some direct interaction with the active center. If the reaction were specific, then blocking of the active center by a small molecule, such as DFP, should greatly increase the dissociation constant of the inhibitor–trypsin compound. This is, in fact, the case (*vide supra*). Moreover, it was found that DIPT, when present in as high as a 30-fold excess with respect to trypsin, did not measurably affect the inhibition of trypsin by ovomucoid, soybean inhibitor, or the pancreatic inhibitor.[90,407] This evidence clearly supports the view of the close spatial relation between the site of combination of the protein inhibitors and that of DFP.

The function of several of these trypsin inhibitors presents something of a biochemical puzzle since it is unlikely that the egg white and the soybean inhibitors, for example, ever come into contact with trypsin during their normal existence. These inhibitors have not yet been found to inhibit enzymes other than trypsin and serum protease, suggesting that their enzyme-inhibiting activity may be merely a coincidence of molecular structure. The low nutritive value of soybeans, relative to their protein content, was at one time ascribed to the presence of the trypsin inhibitor. Subsequent work has cast considerable doubt on this conclusion, and it now appears that the relatively poor food value is due to a low methionine content.[453] It is possible, however, that the presence of the trypsin inhibitor aggravates the methionine deficiency by decreasing the amount that can be liberated in the digestive system.[454]

Plausible functions have been ascribed to most of the other inhibitors.

(453) H. J. Almquist, *Ann. Rev. Biochem.* **20,** 320 (1951).

(454) H. J. Almquist and J. B. Merritt, *Arch. Biochem. Biophys.* **31,** 450 (1951).

Those present in *Ascaris* enable the worm to live parasitically in the digestive juice of the host; the pancreatic inhibitor prevents premature, spontaneous activation of the pancreatic zymogens by traces of trypsin, and the colostrum inhibitor may allow the transmission of colostrum antibodies from mother to offspring by suppressing antibody digestion.[428] It is possible that the serum inhibitor is involved in the regulation of blood clotting, which is retarded by this inhibitor as well as by the pancreatic and soybean inhibitors. Apparently these inhibitors slow down the conversion of prothrombin to thrombin.[455–457]

VIII. Action of Proteolytic Enzymes on Proteins

1. Introduction

Although the proteins are the natural substrates for many proteolytic enzymes, progress in this field of research has been slow. Relatively few definitive results were obtained by early workers, due primarily to lack of experimental techniques to match the complexity of the problem. Most of the early work of value has been reviewed in papers of the Carlsberg school,[145,146,458,459] which itself has contributed richly to the subject of the action of proteolytic enzymes on proteins.

The action of proteases on proteins may be regarded from the point of view of the enzyme or of the substrate. While the former may be more germane to the subject of the present review, it has already received detailed consideration in previous sections on the behavior of proteases toward synthetic substrates. These studies have contributed much to the formulation of the kinetics of the enzymatic hydrolysis of peptide bonds and have helped to delineate the specificity requirements of some of the purified proteolytic enzymes. We may now use this information to further our understanding of the course of breakdown of a protein substrate.

The difficulties involved in the use of protein substrates are best appreciated by considering them as a mixture of constituent peptides of widely differing susceptibility to hydrolysis by a given enzyme. The similarity in structure of enzyme and substrate often leads to difficulties in differentiating between the effects of a given condition on either or both.

(455) D. Grob, *J. Gen. Physiol.* **26,** 423 (1943).
(456) M. B. Glendenning and E. W. Page, *J. Clin. Invest.* **30,** 1298 (1951).
(457) R. G. MacFarlane, *J. Physiol.* (London) **106,** 104 (1947).
(458) K. Linderstrøm-Lang, The Lane Medical Lectures. VI. Stanford Univ. Press, California, 1952.
(459) K. Linderstrøm-Lang and M. Ottesen, *Compt. rend. trav. lab. Carlsberg* **26,** 403 (1949).

This may be illustrated, for instance, by recent work on the effect of metal ions on the tryptic hydrolysis of proteins.[382] One important observation resulting from the study of proteolytic enzymes on proteins which could not have emerged from the use of synthetic substrates is that these enzymes may have effects on proteins other than the splitting of peptide or ester bonds. These hypotheses, which invoke the enzymatic denaturation of proteins, will be considered more fully in a succeeding section (p. 1176).

Prior to the introduction of chromatographic techniques, nonspecific physical or chemical methods were largely used to follow the breakdown of protein substrates. Pepsin was a most widely used enzyme, a choice that may now be considered unfortunate in view of the subsequently discovered heterogeneity of this enzyme (p. 1071). The resulting degradations of the protein were usually profound and any detailed analysis of the process was practically impossible. Nevertheless, fundamental observations of importance were made by the careful application of simple techniques.[458]

2. Enzymatic Hydrolysis of Native and Denatured Proteins

It was early recognized that native proteins were often attacked extremely slowly, or not at all, and that following denaturation by various means, the rate of proteolysis was significantly increased[460–463,150] (see Chap. 9, Vol. I). Anson and Mirsky[460] were among the first to investigate this phenomenon after they had found that hemoglobin was digested by trypsin only after it had been denatured. Partial denaturation (e.g., in 0.3 *M* salicylate) was sufficient to ensure complete degradation to low-molecular-weight peptides. This was explained by a shift in equilibrium between native and denatured hemoglobin resulting from removal of the latter by tryptic hydrolysis. The literature is replete with similar observations which include, among others, the resistance of soybean inhibitor to peptic digestion[150] and the effect of denaturation of β-lactoglobulin on hydrolysis by trypsin,[464] a system which was later studied most carefully and elegantly by Christensen.[145] It was suggested[464] that all proteins were digested by proteolytic enzymes via their denatured form. This

(460) M. L. Anson and A. E. Mirsky, *J. Gen. Physiol.* **17,** 399 (1934).

(461) F. Haurowitz, M. Tunca, P. Schwerin, and V. Göksu, *J. Biol. Chem.* **157,** 621 (1945).

(462) H. Lineweaver and S. R. Hoover, *J. Biol. Chem.* **137,** 325 (1941).

(463) F. Bernheim, H. Neurath, and J. O. Erickson, *J. Biol. Chem.* **144,** 259 (1942).

(464) K. Linderstrøm-Lang, R. D. Hotchkiss, and G. Johansen, *Nature* **142,** 996 (1938).

hypothesis may be formulated essentially by the equation in which N

$$N \rightleftharpoons D \xrightarrow{\text{enzyme}} \text{products}$$

denotes native protein and D denatured protein. Since the temperature coefficient of the rate of hydrolysis of native β-lactoglobulin by trypsin was much higher than that found for the rate of hydrolysis of the denatured form, it was suggested that the temperature coefficient of denaturation was actually being measured. In support of his hypothesis it was also found that the temperature coefficient was increased with increasing trypsin concentration; such a relation would be expected as the denatured protein becomes more rapidly removed, thus increasing the dependence of the rate on the first step of the reaction.

This hypothesis has recently received further confirmation from work on other enzyme–substrate systems.[145] It was found that, in general, the protein was more readily digested if it was unstable at the pH optimum of the enzyme. Thus, ovalbumin was readily digested by pepsin but not by trypsin, while β-lactoglobulin showed the reverse behavior, in accordance with the known susceptibility of these two proteins to denaturation at low and high pH, respectively.

3. Products of Proteolysis

a. Formation of Small Peptides Only

Another important observation to be accommodated by any general theory of proteolysis was first made by Tiselius and Eriksson-Quensel[465] who found that the products of digestion of ovalbumin by pepsin (pH 1.5, 37°) could be fractionated by ultracentrifugation or by electrophoresis into two components only, one of them being indistinguishable from the original protein and the other one having a molecular weight of the order of 1000. It thus appeared that the protein molecules were broken down one-by-one to the ultimate peptide stage without the accumulation of intermediate products. Other workers confirmed these results using different enzyme–substrate systems. Thus Haugaard and Roberts[466] studied the digestion of β-lactoglobulin by pepsin using both trichloroacetic acid and dialysis to follow the appearance of low-molecular-weight products. Both methods gave entirely concordant results showing that the ratio, amino nitrogen to nonprotein nitrogen (a measure of chain length), was constant throughout proteolysis, thus indicating a one-by-one breakdown of β-lactoglobulin molecules. These authors suggested

(465) A. Tiselius and I. B. Eriksson-Quensel, *Biochem. J.* **33**, 1752 (1939).
(466) G. Haugaard and R. M. Roberts, *J. Am. Chem. Soc.* **64**, 2664 (1942).

that the breaking of one peptide bond caused a disintegration in the internal arrangement of the peptide chains, which in turn gave rise to rapid hydrolysis of the "denatured" molecule. It was similarly found that digestion of casein with pepsin, trypsin, chymotrypsin, papain, and ficin yielded in each case products having molecular weights of approximately 600, as determined by freezing-point depression, without any marked individual characteristics.[152] A similarly narrow molecular-weight range was found by Beloff and Anfinsen[467] for the products of the action of trypsin on serum albumin and γ-globulin and of pepsin on fibrin and ovalbumin. Using the more specific fractionation method of frontal analysis on charcoal to follow the reaction, Moring-Claesson[468] confirmed that no high-molecular-weight intermediates were obtained when pepsin was allowed to act on ovalbumin. The low-molecular-weight peptides could be divided into fractions containing approximately 10, 5, 4, or 3 amino acids (as measured by the ratio of amino nitrogen to nonprotein nitrogen), smaller peptides accumulating toward the end of the reaction. All these findings tend to support the conclusion of an "all or none" reaction in which the protein molecules are broken down one-by-one to relatively small peptides.

b. Formation of Intermediate Products

There are, however, reports that indicate that under certain conditions relatively large amounts of high-molecular-weight intermediate products can be obtained. Thus Annetts,[469] using ultracentrifugal and electrophoretic analyses, showed that at 40° and pH 5, papain liberated from ovalbumin a high-molecular-weight fraction differing only slightly from the original substrate, and a fraction containing low-molecular-weight peptides. Nearly all of the substrate was converted into the high-molecular-weight component before appreciable quantities of peptide were formed. Currie and Bull,[470] using the same system as Tiselius and Eriksson-Quensel,[465] i.e., pepsin and ovalbumin, investigated the molecular-weight distribution of the resulting peptides by the monolayer technique.[471] Intermediates of high molecular weight could be detected, particularly at low enzyme concentrations, higher temperatures, and low pH (less than pH 2). It was suggested that the failure of other authors to find such intermediate products was due to the fact that proteolysis was studied at a single and unfavorable set of conditions. Some of the best examples for the accumulation of intermediate products are to be found

(467) A. Beloff and C. B. Anfinsen, *J. Biol. Chem.* **176,** 863 (1948).
(468) I. Moring-Claesson, *Biochim. et Biophys. Acta* **2,** 389 (1948).
(469) M. Annetts, *Biochem. J.* **30,** 1807 (1936).
(470) B. T. Currie and H. B. Bull, *J. Biol. Chem.* **193,** 29 (1951).
(471) H. B. Bull, *Advances in Protein Chem.* **3,** 95 (1947).

in a number of investigations by different authors on the action of pepsin, papain, and trypsin on serum γ-globulins, in particular on diphtheria antitoxin. Thus it was found[472] that digestion of the antitoxin with pepsin at pH 3.5–4 for 24 hours, 37°, yielded high-molecular-weight products which could be separated by heating to 60° in 14 per cent ammonium sulfate. This treatment precipitated approximately one-half of the product; the remaining soluble fraction possessed enhanced specific antitoxic activity. At pH 2, the true pH optimum of pepsin, low-molecular-weight products predominated.[473,474] This effect of pH on the nature of the reaction is probably due to the presence of two distinct enzymes in pepsin[196] (see p. 1071), digestion at pH 4 being due to a cathepsin-like impurity. High concentrations of crystalline trypsin and chymotrypsin were also found to cause splitting of the antitoxin at pH 3.8,[196] but not at pH 7.8.

A study of the formation of high-molecular-weight products in pepsin–antitoxin and related systems, using sedimentation in the ultracentrifuge to determine the composition of the digest, was interpreted by Petermann[474–476] in terms of the presence of unchanged γ-globulin ($s_{20,w} = 7.3\,S$), half molecules ($s_{20,w} = 5\ S$), and quarter molecules ($s_{20,w} = 3\ S$). A more exact analysis of similar data, based on the measurements of boundary spreading,[477] has shown, however, that the division into half and quarter molecules was quite arbitrary and that there was a continuous distribution of sedimentation constants from 3 to 8 S. In addition, some 20 per cent nonprotein nitrogen was found in the digestion mixture. Digestion of rabbit antiovalbumin by cyanide-activated papain has yielded a protein fraction of molecular weight approximately 40,000 which was no longer precipitated by the antigen but which was capable of inhibiting competitively the normal flocculation reaction. End-group analysis of the split antibody yielded the same *N*-terminal amino acid (alanine) as in the original globulin.[478]

4. Mechanism of Proteolysis

a. Experimental Results

Although several attempts have been made to analyze the kinetics of hydrolysis of protein substrates by pepsin or trypsin,[190,479] the reaction

(472) C. G. Pope, *Brit. J. Exptl. Path.* **20,** 132, 201 (1939).
(473) W. B. Bridgman, *J. Am. Chem. Soc.* **68,** 857 (1946).
(474) M. L. Petermann, *J. Phys. Chem.* **46,** 183 (1942).
(475) M. L. Petermann, *J. Biol. Chem.* **144,** 607 (1942).
(476) M. L. Petermann, *J. Am. Chem. Soc.* **68,** 106 (1946).
(477) J. W. Williams, R. L. Baldwin, W. M. Saunders, and P. G. Squire, *J. Am. Chem. Soc.* **74,** 1542 (1952).
(478) R. R. Porter, *Biochem. J.* **46,** 479 (1950).
(479) D. Fraser and R. E. Powell, *J. Biol. Chem.* **187,** 803 (1950).

appeared too complex for detailed interpretation. However, a more systematic study, carried out during the last 15 years by the Carlsberg group (*vide supra*) has led to a theory of proteolysis capable of explaining in a qualitative manner most of the results so far obtained.

This work was based on the assumption that the proteolytic enzymes would attack only the denatured form of a protein[464] and that the breakdown of any protein molecule went through two successive stages: namely, first, the breakdown of the internal arrangement of the polypeptide chains within the protein molecule, and second, the degradation of the unfolded, or denatured, molecule to small peptides. The first direct evidence for this has already been mentioned, namely the temperature effect on the course and extent of hydrolysis of β-lactoglobulin by trypsin. This evidence is in substantial agreement with the work of Lundgren and Williams[480,481] on the effect of papain on the denaturation of thyroglobulin. It was found that at pH 5.8 thyroglobulin gave two peaks in sedimentation analysis corresponding to two forms, termed N and α, which are in equilibrium with one another. The boundary corresponding to the α form disappeared on heating and gave way to a denatured protein (D) which sedimented at the same rate as the original N protein. The N and D form could be differentiated from one another by solubility and electrophoretic measurements. The addition of papain to the N $\leftrightarrows$ α system also led to the disappearance of the α boundary and to the complete conversion of N to D. On prolonged standing in the presence of papain, the boundary corresponding to denatured protein gave way to a broad spectrum of components of low sedimentation constant.

The problems arising from these observations may be summarized by the following questions quoted from Christensen's paper on the action of trypsin on β-lactoglobulin.[145]

"I. Is a single molecule immediately split into the ultimate number of fragments, or are the latter liberated little by little with the formation of significant quantities of high molecular intermediary products?

"II. Does the enzyme attack exclusively or preferentially molecules whose peptide chains are already uncoiled (i.e., denatured molecules)? In the latter case, the following questions arise:

"*a.* Is the uncoiled D form in equilibrium with the N form?

"*b.* Does the enzyme catalyze the N $\rightarrow$ D transformation, preceding any hydrolysis of peptide bonds?

(480) H. P. Lundgren and J. W. Williams, *J. Phys. Chem.* **43,** 989 (1939).
(481) H. P. Lundgren, *J. Biol. Chem.* **138,** 293 (1941).

"*c.* Is the uncoiling due to splitting of a few peptide bonds essential to the maintenance of the structure of the native molecule?"

In attempting to answer these questions, it is necessary to measure both the extent of denaturation and the splitting of the peptide bonds. Although the first of these quantities is difficult to define from a molecular standpoint, it may be defined operationally in a number of different ways, depending on the method used; fortunately, all lead to similar conclusions. In the study of the system, trypsin–β-lactoglobulin, the following methods were used (for a more detailed discussion, see Chap. 9):

(*1*) *Solubility.* Like most proteins, β-lactoglobulin is less soluble at its isoelectric point when denatured, and hence the process of denaturation can be followed by the appearance of insoluble protein. The precipitation technique was designed to avoid any disturbance in the equilibrium position due to removal of denatured protein, and the precipitating buffer (0.5 M $MgSO_4$, 0.8 M acetic acid, 0.4 M sodium acetate) was chosen so as to give a clear separation between native and denatured protein.

(*2*) *Change in Optical Rotation.* It has been known for some time that the optical rotatory power of proteins increases on denaturation,[482] but until recently[483,484] the method has not been extensively used in the study of the denaturation process even though the changes in rotation are quite large (*ca.* 60° for a 2 per cent protein solution) and may be followed continuously. This is a distinct advantage over the dilatometric method, in which small changes must be detected, and the precipitation method, which requires sampling.

(*3*) The dilatometric method requires careful technique since the volume increase on denaturation (approximately 5 μl. per gram of β-lactoglobulin), is small, and has to be corrected for the change in volume on mixing the protein with the denaturing agent. However, the method gives more information than the other physical methods since, for simple peptides, the volume contraction is due only to the electrostriction of the liberated ammonium and carboxylate ions. This contraction amounts to approximately 20 ml./mole peptide bonds split and can, therefore, be used as a measure of the number of bonds broken.[168,147] When native proteins are digested, much larger volume changes are observed than would be expected from the number of bonds broken, due to changes accompanying the unfolding of peptide chains. The magnitude of the difference between observed and calculated volume changes has been used to give important information on the relation between bond splitting and denaturation.[145,146]

(*4*) Viscosity measurements provide a sensitive method for detecting slight changes in hydrodynamic volume of a protein molecule and have often been used in denaturation studies.[485,486] This method was used to show that the renatured β-lactoglobulin was not appreciably different from the native form.

(*5*) Measurements of the appearance of —SH groups were less satisfactory than

(482) H. A. Barker, *J. Biol. Chem.* **103,** 1 (1933).
(483) R. B. Simpson and W. Kauzmann, *J. Am. Chem. Soc.* **75,** 5139 (1953).
(484) W. Kauzmann and R. B. Simpson, *J. Am. Chem. Soc.* **75,** 5154 (1953).
(485) H. B. Bull, *J. Biol. Chem.* **133,** 39 (1940).
(486) H. Neurath and A. M. Saum, *J. Biol. Chem.* **128,** 347 (1939).

the other methods mentioned above, since the analytical process (using potassium ferricyanide) appeared to accelerate denaturation, leading to results at variance with those obtained by the other methods.

The proteolytic reaction, using both native and denatured lactoglobulin as substrate, was followed by alcohol titration of peptide bonds broken, in parallel with one of the methods listed above.

In general, it was found that in the initial phases of the digestion larger changes in the properties measuring the extent of denaturation occurred

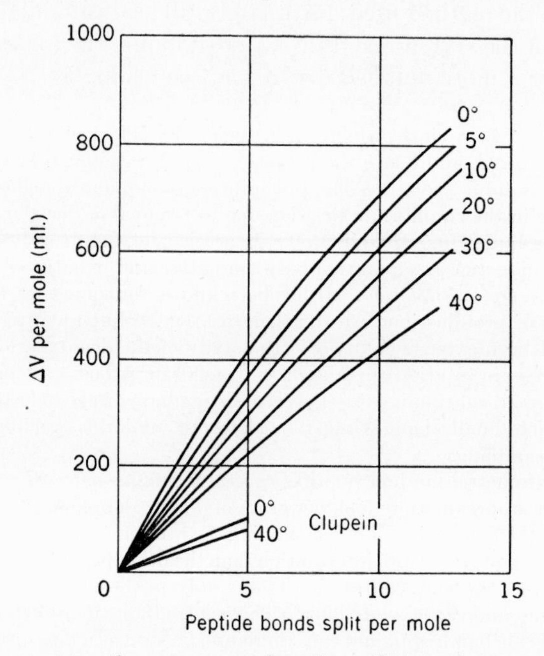

FIG. 6. Volume contraction accompanying the splitting of β-lactoglobulin (upper curves) and clupein (lower curves) by trypsin at different temperatures.[146]

than in the later stages of the reaction. This is illustrated in Fig. 6 (taken from Ref.[146]) where the change in volume is plotted against the number of peptide bonds split at different temperatures. It will be seen that at 0°C. $\Delta V/\Delta$ peptide is in the neighborhood of 100 ml./mole rather than the 20 ml./mole characteristic of the splitting of simple peptide bonds (lower two curves). The significance of these curves can best be considered in the light of two extreme theoretical possibilities illustrated in Fig. 7 and first pointed out by Tiselius and Eriksson-Quensel.[465] The upper curve shows the results to be expected if all the protein molecules were denatured by the enzyme before many peptide bonds were split. The lower curve shows what would happen if each protein molecule was completely digested in turn ("one-by-one" mechanism). The experimental curves (Fig. 6) clearly represent an intermediate case with an approach to the "one-by-one" mechanism as the temperature is raised. The lower two curves of Fig. 6 show the digestion of clupein, considered to be a representative of denatured protein, and demonstrate the much smaller volume

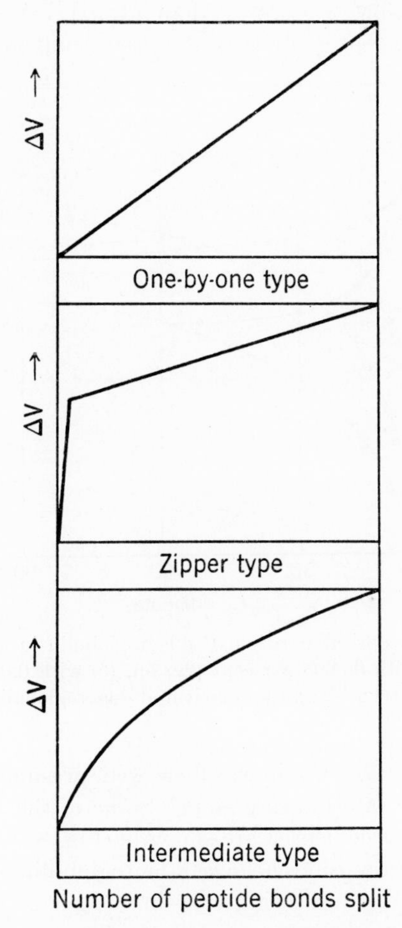

FIG. 7. Theoretical curves for the volume contraction accompanying three different postulated types of proteolysis.[458]

changes characteristic of such a process. These changes are independent of the extent of the reaction.

Similar results were found when increase in levorotation instead of volume change was used as a measure of denaturation.

Further evidence for a preliminary denaturation step in the proteolysis of the substrate derives from the study of the effects of urea and of alkaline pH, which are known to denature β-lactoglobulin, on the optical rotation of this protein. Under both conditions, the first stage of proteolysis was specifically accelerated, as indicated by an increase in the initial value of $\Delta\alpha/\Delta$ peptide (see Fig. 8). Although at pH 9.25, the alkaline pH chosen, the enzymatic activity was lower than at pH 7.4, as indicated by a decrease in the rate of proteolysis of the urea-denatured protein, the rate

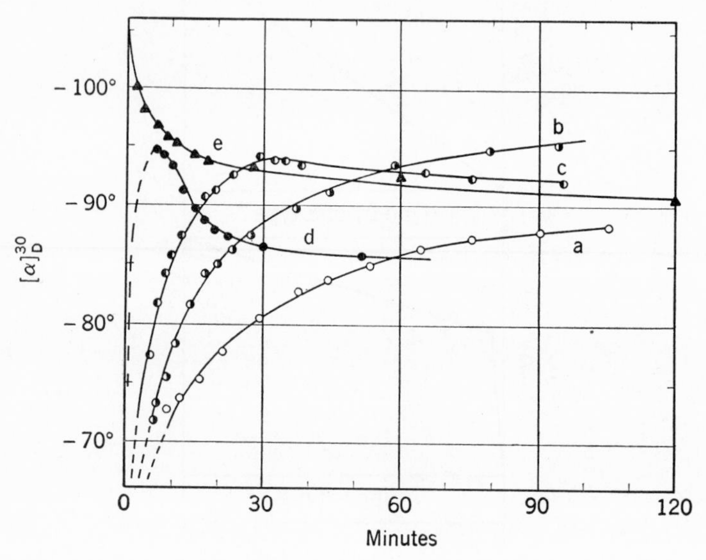

FIG. 8. Change in optical rotation of β-lactoglobulin in 19 per cent urea: (*a*) Without trypsin, (*b*) with 0.0008 per cent trypsin, (*c*) with 0.0041 per cent trypsin, (*d*) with 0.041 per cent trypsin, and (*e*) denatured β-lactoglobulin with 0.041 per cent trypsin.[145]

of digestion of the native protein was increased, presumably because denaturation became the rate-limiting step. Some of the results obtained in the presence of 19 per cent urea are shown in Fig. 8 (taken from Ref.[145]).

In the absence of trypsin (curve *a*) β-lactoglobulin is slowly denatured by urea. The effect of increasing trypsin concentrations on the optical rotation is shown by curves *b*, *c*, and *d*, the last two of these showing a rapid increase in levorotation, followed by a decrease which can be readily explained if curve *e* is taken into account. This shows that digestion of denatured β-lactoglobulin is accompanied by a slight *decrease* in levorotation. The two phases through which curves *c* and *d* pass, therefore, correspond to an initial rapid denaturation, catalyzed by trypsin (and urea), followed by a relatively slow digestion of the denatured protein. In the experiments at pH 9.25 in the absence of urea, the accumulation of denatured protein in the digestion mixture was directly shown by precipitation with the buffer mixture mentioned previously; up to 30 per cent of the total protein could be precipitated after a suitable reaction time. As determined by osmotic pressure, the precipitated protein had the same molecular weight as the native protein.

While these investigations provide partial answers to the stated questions (*vide supra*), two questions remain open for further consideration. These are: (*a*) Can trypsin attack native β-lactoglobulin? (*b*) Is the denaturation of β-lactoglobulin by trypsin a consequence of the opening of peptide bonds? While these questions are considered in the papers by Linderstrøm-Lang and co-workers,[458,146,145] there is really no experimental evidence for an unequivocal answer. One possible approach to the second of these questions may be indicated.

If trypsin can denature the protein substrate such as β-lactoglobulin without opening peptide bonds, then the denatured protein isolated from the digestion mixture should have no new end groups. Such end groups can be readily detected by Sanger's method[487] or by the use of carboxypeptidase,[488–490] and their absence would be good presumptive evidence for denaturation without peptide hydrolysis. However, the presence of new end groups would not be conclusive evidence for the converse proposition since such groups might have appeared subsequent to the initial denaturation step.

b. Theoretical Interpretations

The ideas and experimental facts considered thus far may now be integrated into a general theory of proteolysis which includes the reviewers' thoughts on the problem.

The native protein has relatively few peptide bonds exposed on its surface and, of these, only a few will be susceptible to hydrolysis by a given enzyme of relatively narrow specificity. The enzyme can attack both the native and the denatured forms, which may or may not be in equilibrium with one another, at relative rates dependent on several factors which cannot yet be dealt with experimentally. These include the relative concentrations of the native and the denatured forms (if the latter are present at all) and the relative number of susceptible bonds exposed in each form. It is likely that hydrolysis of internal peptide bonds of the polypeptide chains will lead to some disintegration of the chain structure of the protein without necessarily liberating peptide

(487) F. Sanger, *Biochem. J.* **39,** 507 (1945).
(488) A. R. Thompson, *Nature* **169,** 495 (1952).
(489) S. M. Partridge, *Nature* **169,** 496 (1952).
(490) J. A. Gladner and H. Neurath, *J. Biol. Chem.* **205,** 345 (1953).

fragments. This may, therefore, be regarded as lim ted proteolysis followed by denaturation, but this should not be confused with the reversible denaturation which occurs prior to peptide bond hydrolysis since probably different structural regions and changes are involved in each case. Similar considerations may be applied to the "reversibly denatured" form which may undergo further unfolding subsequent to proteolysis, accompanied by exposure of further susceptible bonds, by a decrease in volume, by an increase in levorotation, and by other changes. Such a scheme is outlined in Fig. 9 and accounts for variations in the course of proteolysis according to the specificity and concentration of the enzyme, the structure of the substrate, pH and temperature (the latter affecting both the extent of denaturation and the rate of hydrolysis), and the presence of denaturing agents.

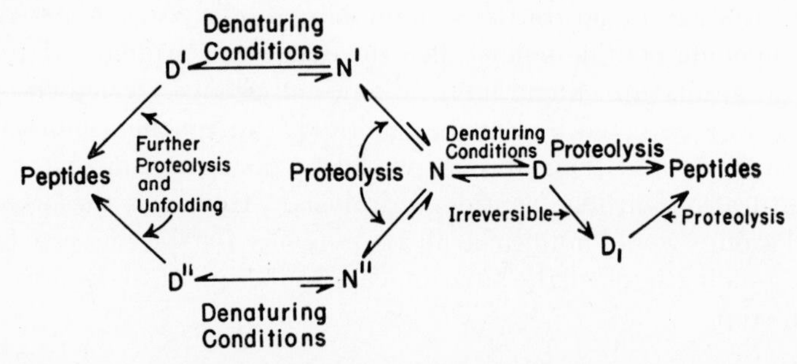

FIG. 9. Hypothetical scheme for proteolysis of a native protein.

Such a scheme is evidently flexible and capable of much *ad hoc* adjustment, and hence it is probably difficult to prove or disprove it in all of its details. However, the scheme is in harmony with present knowledge of protease action on one hand and of the structure of globular proteins on the other and can account for all experimental results without recourse to any hypothesis concerning the action of proteases as denaturing agents.

While a quantitative analysis of any far-reaching proteolytic process, which proceeds according to some scheme such as that just presented, is clearly beyond realization at the present time, the general theoretical treatment recently advanced by Linderstrøm-Lang[491] promises to be eventually applicable to such situations. In this interesting approach, Linderstrøm-Lang derived equations for the distribution of products during degradation of different types of peptides by exo- and endopeptidases, in terms of k_3 and K_m (specific rate constant and Michaelis constant), appropriate to each peptide link, and of the amount of initial substrate remaining. While it might be possible to test this theory on relatively small peptides (less than ten structural units), it cannot, as yet,

(491) K. Linderstrøm-Lang, *Abstracts, Proc. 9th Solvay Congr.*, Brussels, April 6–14, 1953.

be applied to proteins since the appropriate kinetic constants may be different from those determined for small peptides and hence are beyond experimental determination.

The reverse procedure in which the kinetic constants, K_m and k_3, are determined from the distribution of the products is equally impractical for complex molecules, and hence the theory remains at present restricted to semiquantitative correlations. The theory could be applied, however, to the relatively simple case of the stepwise degradation by carboxypeptidase of proteins composed of single peptide chains, assuming that the intrinsic rates of hydrolysis of peptide bonds are the same as in synthetic substrates.

Chemical analysis of the products resulting from the enzymatic degradation of proteins may give more unequivocal results but is probably not more revealing as far as protein structure is concerned than any of the enzymatic methods previously discussed. In contrast, analysis of processes causing more limited proteolytic changes, which can be accurately defined, has proved to be of considerable value as it relates to the structure of the protein substrate. This latter work has been greatly aided by the development of chromatographic techniques and by methods of end-group analysis.

5. Limited Proteolysis

The reactions to be considered under this heading are of the type:

$$\text{protein A} \rightarrow \text{protein B} + \text{peptides}$$

where protein B is assumed to be relatively stable to further action of the enzyme. The action of carboxypeptidase on protein substrates provides the simplest example of this type of proteolytic change, and a list of proteins whose terminal carboxyl groups have been determined in this fashion is given in Table XX. Protein B may be a well-defined protein in its own right (e.g., chymotrypsins; trypsin; fibrin; plakalbumin; meromyosins H and L) or it may be simply a resistant "core" with illdefined properties (insulin, globin, keratin, products of carboxypeptidase degradation).

a. Ill-Defined Protein Products

Investigation of the action of proteolytic enzymes on insulin[335] suggested that trypsin had no action on this protein, while chymotrypsin rapidly converted it to an inactive residue (or core) containing 50 per cent of the nitrogen and 70 per cent of the cystine of the native protein, together with several small peptides (5–6 residues per mole). The core had an average molecular weight of about 4000 and was slowly converted to nonprotein nitrogen by further action of chymotrypsin. Chromato-

TABLE XX

THE ACTION OF CARBOXYPEPTIDASE ON PROTEINS

Protein	C-Terminal amino acid	Equivalents per mole protein	Other amino acids liberated	References
Ovalbumin	None	....	None	405
α-Lactalbumin	Leu	1	None	492
β-Lactoglobulin	His	1	None	235,492
	Ileu	1		
Insulin (beef, horse, pork)	Ala	1	Leu, tyr, glu, asp, phe, lys	494,268,235,493
	Asp (NH_2)	1		
Insulin A chain	Ala	....	Lys	494,268
Insulin B chain	Asp (NH_2)	....	Cys.SO_3H, leu, tyr, glu	494,268
Ribonuclease	Val	....	....	495
Lysozyme	Leu	0.6	....	494,488
Somatotropin	Phe	2	Ala, leu, ser	497
Corticotropin A	Phe	1	Glu, Leu	498
α-Chymotrypsinogen	None	0	traces	297,490,83
B-Chymotrypsinogen	Tyr	0.6	Leu, val, gly, asp (NH_2), thr, lys	297,490,83
DIP-α-chymotrypsin	Tyr	1	Gly, ser, leu	297,490,83
	Leu	1		
DIP-β-chymotrypsin	Tyr	1	Ser, gly, asp, asp-(NH_2) (phe)	297,490,83
	Leu	1		
DIP-γ-chymotrypsin	Tyr	1		
	Leu	1		
DIP-B-chymotrypsin	Tyr	1	Ala, val, phe, gly, Asp(NH_2), lys	297,490,83
	Leu	0.5		
Trypsinogen	None	....	None	125,235,492
Acid-denatured trypsinogen	None	....	Lys, asp, ala, tyr, meth	125,235,492
DIP-Trypsin	None	....	None	125,235,492
Acid-denatured DIP-trypsin	None	0.3	Lys, asp, ala, tyr, meth	125,235,492
Carboxypeptidase	None	....	None	125
Soybean trypsin inhibitor	Leu	1	Ala, asp	235,450,492
Trypsin–soybean trypsin inhibitor compound	Leu	1	Ala, asp	235,450,492
Tobacco mosaic virus	Thre	3400	None	496

graphic analysis of the peptides[499] revealed a complex mixture (approximately 17 ninhydrin-positive spots). In the light of present knowledge of the detailed structure of the insulin molecule, the course of the reaction can be more readily understood. In view of its high cystine content, the core is probably derived from several of the constituent polypeptide chains of insulin, held together by disulfide bonds, while the small peptides are probably derived from the terminal portions of such chains which are not thus combined.[500]

When purified proteolytic enzymes were applied to the partial enzymatic degradation of the A and B chains of oxidized insulin, the very bonds were hydrolyzed by chymotrypsin and trypsin that were expected on the basis of the action of these enzymes on synthetic substrates.[291,268] Thus chymotrypsin attacked all the bonds containing the carbonyl group of an aromatic amino acid while trypsin hydrolyzed all bonds containing lysine or arginine in the same position. However, in addition, chymotrypsin was also found to attack a bond between cysteic acid and serine[268] for which no analog of synthetic origin has yet been described. In contradistinction to the findings of Butler and co-workers,[335] trypsin was found to attack also native insulin,[501] giving the same products (alanine and a heptapeptide) as were obtained from the B chain alone. Pepsin was found to reveal a much wider specificity toward the insulin substrate than had been indicated by studies with synthetic substrates. Many types of bonds, particularly those between nonpolar residues, were attacked, and the products, though complex, have been analyzed in detail (p. 1128). The action of carboxypeptidase on insulin will be considered below (p. 1187).

Using fluorodinitrobenzene and periodate titration to determine the *N*-terminal groups of the resulting peptides, Desnuelle and co-workers studied the action of pepsin,[502] trypsin, and chymotrypsin[503] on horse

(492) E. W. Davie, Thesis, University of Washington, 1954.
(493) J. Lens, *Biochim. et Biophys. Acta* **3,** 367 (1949).
(494) J. I. Harris, *J. Am. Chem. Soc.* **74,** 2944 (1952).
(495) C. B. Anfinsen, M. Flavin, and J. Farnsworth, *Biochim. et Biophys. Acta* **9,** 468 (1952).
(496) J. I. Harris and C. A. Knight, *Nature* **170,** 613 (1952).
(497) J. I. Harris, C. H. Li, P. G. Condliffe, and N. G. Pon, *J. Biol. Chem.* **209,** 133 (1954).
(498) W. F. White, *J. Am. Chem. Soc.* **75,** 4877 (1953).
(499) D. M. P. Phillips, *Biochim. et Biophys. Acta* **3,** 341 (1949).
(500) J. A. V. Butler, D. M. P. Phillips, J. M. L. Stephen, and J. M. Creeth, *Biochem. J.* **46,** 74 (1950).
(501) J. I. Harris and C. H. Li, *J. Am. Chem. Soc.* **74,** 2945 (1952).
(502) P. Desnuelle, M. Rovery, and G. Bonjour, *Biochim. et Biophys. Acta* **5,** 116 (1950).
(503) M. Rovery, P. Desnuelle, and G. Bonjour, *Biochim. et Biophys. Acta* **6,** 166 (1950).

globin and ovalbumin (see also Chap. 2, Vol. I). All three enzymes hydrolyzed bonds with alanine, phenylalanine, and serine residues at the amino end of the peptide bond more rapidly than bonds with other amino acids in the same position. This specificity, however, was only relative and, since it was independent of the enzyme used, it was probably a reflection of the chemical reactivity of the bond rather than of the specificity of the enzyme. A quantity called "the specificity index" was used to express the preferential liberation of the peptides. This quantity was defined as the fraction of bonds split containing a given amino acid divided by the fraction of the total bonds hydrolyzed. For values greater than 1, some preferential attack on bonds containing the given amino acid was indicated. As was to be anticipated from the preceding discussion, the average size of the peptides liberated at different stages of the digestion by different enzymes varied considerably. It should be noted that in these investigations those amino acids which contributed the carbonyl end of the hydrolyzed bonds were not determined.

The digestion of wool keratin by papain;[504] of ACTH "protein hormone"[505,506] and of the follicle-stimulating hormone[507] by pepsin; and of cytochrome c by pepsin, trypsin, chymotrypsin, and papain[508] have all been described, but the products have not been well characterized.

Wool keratin gives an insoluble residue of high cystine content, a number of small and large peptides, and free glycine, serine, and glutamic acid. Peptic digestion of ACTH "protein hormone" at pH 2 yielded 50 per cent of nonprotein nitrogen; all of the endocrine activity was recovered from the peptide fraction which was subsequently analyzed by paper chromatography and by displacement chromatography on charcoal,[506] one of the fractions (molecular weight about 2000) having about ten times the specific activity of the starting material. The follicle-stimulating hormone[507] reacted rather similarly toward peptic digestion at pH 4, but the reaction products have not yet been fractionated (see also Chap. 20, Vol. II).

Pepsin, trypsin, chymotrypsin, and papain all attack cytochrome c, yielding autoxidizable pigments with none of the original enzymatic activity remaining.[508] On the basis of the iron content, the product of peptic digestion had a molecular weight of approximately 2500 (as compared to 13,700 for the intact enzyme). Its absorption spectrum in neutral and

(504) S. Blackburn, *Biochem. J.* **47,** 443 (1950).
(505) C. H. Li, A. Tiselius, K. O. Pedersen, L. Hagdahl, and H. Carstensen, *J. Biol. Chem.* **190,** 317 (1951).
(506) C. H. Li and K. O. Pedersen, *Arkiv Kemi* **1,** 533 (1950).
(507) C. H. Li, *J. Am. Chem. Soc.* **72,** 2815 (1950).
(508) C. L. Tsou, *Biochem. J.* **49,** 362, 367 (1951).

alkaline solutions remained unchanged with respect to the enzyme, and the peptide combined with cyanide and carbon monoxide. However, unlike most heme pigments, it had low catalase and no peroxidase activity although it catalyzed oxidation of ascorbic acid.

More recently, the action of chymotrypsin and trypsin on aqueous solutions of silk fibroin was investigated in some detail by chromatographic methods.[509] Substrate solutions were prepared by dissolving fibroin in cupri-ethylenediamine solutions, followed by neutralization and dialysis. Chymotryptic hydrolysis led to the formation of a precipitate which differed in composition from that of the soluble fraction as well as of that of the unhydrolyzed fibroin, being composed almost entirely of glycine, serine, and alanine, with a little tyrosine. End-group analysis, based on amino nitrogen and tyrosine (assumed to occupy the *C*-terminal position), respectively, yielded a molecular weight of approximately 4000. Tryptic hydrolysis yielded insoluble products containing all the amino acids normally present in fibroin. The combined action of trypsin and carboxypeptidase on the soluble fraction proceeded continuously to the dipeptide stage. These results together with those of x-ray analysis have been correlated with "crystalline" and irregular portions within the fibroin structure.

b. Effect of Carboxypeptidase on Proteins

Although the action of carboxypeptidase is usually associated with the hydrolysis of short peptides, more recently this enzyme has come into extensive use for the stepwise degradation of protein substrates and as an analytical tool for the determination of *C*-terminal groups.

The earliest analytical use of carboxypeptidase is probably represented by the work of Grassmann and co-workers,[510] who used the enzyme to identify glycine as the *C*-terminal amino acid of glutathione. Nineteen years later, Lens[493] studied the action of carboxypeptidase on insulin and found three alanine groups per molecule (molecular weight 12,000), a result at variance with the analytical results of Sanger.[511,291] More recently, however, Harris[494] found two moles of alanine and some asparagine to be liberated by carboxypeptidase from beef insulin, and Sanger[268] reported alanine and asparagine among the main products of the action of this enzyme on the A and B fractions of oxidized insulin, in full agreement with the sequential analysis of the corresponding polypeptide chains.

(509) B. Drucker, R. Hainsworth, and S. G. Smith, *Shirley Inst. Mem.* **26,** 191 (1953).

(510) W. Grassmann, H. Dyckerhoff, and H. Eibeler, *Z. physiol. Chem.* **189,** 112 (1930).

(511) F. Sanger and H. Tuppy, *Biochem. J.* **49,** 463 (1951).

The same amino acids were liberated from beef, pork, and horse insulin, respectively.[235,492] Table XX summarizes the results of end-group analysis involving carboxypeptidase and a number of protein substrates. Among these, ovalbumin is of unusual interest since it has no *N*-terminal groups reactive toward fluorodinitrobenzene,[512] whereas the combined action of *Bacillus subtilis* protease and carboxypeptidase liberates alanine.[405] However, carboxypeptidase alone (in the presence of DFP), or in combination with other endopeptidases (chymotrypsin, trypsin) does not release alanine from ovalbumin, thus suggesting that the reaction leading to the liberation of *C*-terminal alanine requires the intervention of the same enzyme which catalyzes the ovalbumin–plakalbumin transformation. The following scheme is thought to represent this reaction sequence:

$$\text{Ovalbumin} \xrightarrow{B.\ subtilis\ \text{protease}} \text{intermediate ("open") protein} \xrightarrow{B.\ subtilis\ \text{protease}} \text{plakalbumin I, II + peptides}$$

$$\text{intermediate ("open") protein} \xrightarrow{\text{carboxypeptidase}} \text{alanine + plakalbumin-like protein}$$

Carboxypeptidase liberates rapidly and exclusively leucine from lactalbumin[492] and only isoleucine and histidine from β-lactoglobulin.[235,492] As far as is known,[125] the enzyme is without action on itself (Table XX).

It is worthy of note that the highly specific removal of threonine residues from tobacco mosaic virus[496] does not lead to inactivation of the virus and that infection by the treated virus leads to a progeny containing the normal amount of threonine. The removal of alanine from insulin does not reduce its activity, while the much slower removal of asparagine was reported to lead to a loss of biological activity.[501]

The *C*-terminal amino acids of chymotrypsinogen, of DIP-chymotrypsin, of trypsinogen and of DIPT will be considered in more detail below (p. 1195).

While the experimental details concerning the use of carboxypeptidase as an analytical tool have been considered elsewhere,[235,490] a few essential considerations may be summarized herein. It is imperative that the protein substrate be devoid of adventitious impurities of free amino acids or peptides. The conventional methods of crystallization, dialysis, and column chromatography, using ion-exchange resins, when applied with scrutiny and care, have been found to be adequate to remove these trace impurities. Perhaps even more important is the complete absence from protein substrates and carboxypeptidase of catalytic amounts of endopeptidases which are apt to lead to the production of peptides and additional

(512) P. Desnuelle and A. Casal, *Biochim. et Biophys. Acta* **2**, 64 (1948).

end groups susceptible to carboxypeptidase action, and hence to erroneous results. This experimental precaution is of particular importance when proteolytic enzymes or their zymogens are used as substrates. Diisopropyl fluophosphate, which is without effect on carboxypeptidase, proved effective in inactivating residual endopeptidase activity,[297,268] as tested by the sensitive and specific assay methods involving the use of synthetic substrates for trypsin and chymotrypsin. The use of ion-exchange resins to isolate the liberated amino acids, followed by identification by paper or column chromatography,[488,489] have proved to be sensitive and relatively simple experimental procedures.

Frequently amino acids are liberated by carboxypeptidase in nonstoichiometric quantities. In some cases this may be the result of slow and incomplete enzymatic digestion, in others of trace impurities, but the phenomenon has no been fully clarified.[83,497]

It is evident from Table XX that, when the protein substrates are taken as a group, the specificity of carboxypeptidase is very wide and wider indeed than was indicated from studies on synthetic substrates. The rate of liberation of amino acids varies over wide limits, depending on the nature of the *C*-terminal side chain and on the protein substrate. Thus, lysine is removed only very slowly,[494] while tyrosine and leucine appear rapidly,[490] and leucine is liberated much more rapidly from α-lactalbumin than from DIP-α-chymotrypsin.[492] The only amino acids which have not yet been found as reaction products are proline, hydroxyproline, cystine, and cysteine.

While, in general, it is not possible to determine unequivocally the sequence of residues following the *C*-terminal one, under certain conditions the sequence of amino acids along the polypeptide chain can be inferred from enzymatic end-group analysis. This is particularly the case if the amino acids are arranged in sequence of decreasing rates of hydrolysis, the *C*-terminal bond being hydrolyzed fastest. In many cases one has to rely on intuition as well as on chemical means of analysis to determine the amino acid sequence.

c. Well-Defined Protein Products

In a few instances, limited proteolysis has been studied in sufficient detail to describe reactant and reaction products in molecular terms. These systems are listed in Table XXI.

Since the conversion of ovalbumin to plakalbumin has been considered elsewhere in this treatise (Chaps. 2 and 9), whereas little is as yet known about the activation of procarboxypeptidase, neither of these two systems will receive further mention here. However, the activation of pepsinogen, trypsinogen, chymotrypsinogen, and fibrinogen, as well as the tryptic con-

TABLE XXI

WELL-DEFINED EXAMPLES OF LIMITED HYDROLYSIS[a]

	Protein A	Enzyme	Protein B	Peptide	References
	Pepsinogen	Pepsin	Pepsin	Pepsin inhibitor	513,19,103,514
N-Terminal	*Leu*–(ileu or leu)		(*Ileu* or *leu*)–gly	*Leu*–glu	
C-Terminal	*Ala*		*Ala*		
	Chymotrypsinogen	Trypsin + chymotrypsin (*B. subtilis* protease)[516]	α-Chymotrypsin	Peptide(s) containing lys, arg, ser, gly, asp, glu, leu, ala (val)[235,517]	77,66,9,490,235, 297,515a
N-Terminal	None detectable[490,515,515a]		*Ala*, *ileu*–val[515,515a,518]		
C-Terminal			*Leu*, *tyr*[490,83]		
	Trypsinogen[515a]	Trypsin (enterokinase, *penicillium* kinase)[520]	Trypsin	Val–$(asp)_4$–lys[521,492]	521–523, 515a,492,235
N-Terminal	*Val*[519]		*Ileu*–val[519,515a]		
C-Terminal	None detectable[125]		*Lys* (after denaturation)[125,235]		
	Procarboxypeptidase	Trypsin	Carboxypeptidase	Unknown	26,27,30
N-Terminal	Unknown		*Asp* (NH_2)–ser–thr[524]		
C-Terminal			None detectable[125]		
	Ovalbumin	*B. subtilis* protease	Plakalbumin	Ala–gly–val–asp–ala–ala	459,458,525,526
N-Terminal	None detectable		Unknown		
C-Terminal					
	Fibrinogen	Thrombin	Fibrin	Peptides A and B	527–529
N-Terminal	*Tyr* (3 or 4) *Glu* (2)		*Tyr* (3 or 4) *gly* (4)	A: *glu*, B: none detectable	
C-Terminal					
	Myosin	Trypsin	Meromyosins L + H		530,531
N-Terminal	None detectable (See Chap. 24)		L: *asp*, *gly*; H: *lys*, *ala*		
C-Terminal					

[a] Terminal groups in *italics*.

version of myosin to meromyosins, have been studied in sufficient detail to warrant special consideration.

(*1*) *The Activation of Pepsinogen.* The activation of pepsinogen to pepsin has been formulated according to the following scheme:[513,19]

$$\text{pepsinogen} \xrightarrow{\text{(pepsin)}} \text{pepsin–inhibitor compound} + \text{peptides X}$$

$$\text{pH} > 5 \Big\Updownarrow \text{pH} < 5$$

$$\text{pepsin} + \text{inhibitor}$$

$$\Big\downarrow \text{(pepsin)}$$

$$\text{peptides Y}$$

The activation itself is autocatalytic with a maximum rate at pH 2. Under certain conditions, however, it was found that pepsinogen disappeared from the activation system more quickly than could be accounted for by the appearance of peptic activity, a phenomenon which was attributed to the formation of an intermediate compound of pepsin and a polypeptide inhibitor, the latter derived from the pepsinogen molecule.

At low pH, the pepsin–inhibitor compound dissociated and the inhibitor was digested by pepsin, whereas above pH 5 little dissociation occurred and inhibition was almost stoichiometric. The inhibitor has

(513) R. M. Herriott, *J. Gen. Physiol.* **22,** 65 (1938).
(514) M. B. Williamson and J. M. Passmann, *J. Biol. Chem.* **199,** 121 (1952).
(515) M. Rovery, C. Fabre, and P. Desnuelle, *Biochim. et Biophys. Acta* **10,** 481 (1953).
(515*a*) M. Rovery, C. Fabre, and P. Desnuelle, *Biochim. et Biophys. Acta* **12,** 547 (1953).
(516) A. Abrams and C. F. Jacobsen, *Compt. rend. trav. lab. Carlsberg, Sér. chim.* **27,** 447 (1951).
(517) H. Neurath and J. A. Gladner, *Federation Proc.* **12,** 251 (1953).
(518) M. Rovery and C. Fabre, *Bull. soc. chim. biol.* **35,** 541 (1953).
(519) M. Rovery, C. Fabre, and P. Desnuelle, *Biochim. et Biophys. Acta* **9,** 702 (1952).
(520) M. Kunitz, *Enzymologia* **7,** 1 (1939).
(521) E. W. Davie and H. Neurath, *Biochim. et Biophys. Acta* **11,** 442 (1953).
(522) M. Kunitz, *J. Gen. Physiol.* **22,** 293 (1939).
(523) M. Kunitz, *J. Gen. Physiol.* **22,** 429 (1939).
(524) E. O. P. Thompson, *Biochim. et Biophys. Acta* **10,** 633 (1953).
(525) M. Ottesen and A. Wollenberger, *Compt. rend. trav. lab. Carlsberg* **28,** 463 (1953).
(526) M. Ottesen and C. Villee, *Compt. rend. trav. lab. Carlsberg* **27,** 421 (1951).
(527) L. Lóránd and W. R. Middlebrook, *Biochem. J.* **52,** 196 (1952).
(528) F. R. Bettelheim and K. Bailey, *Biochim. et Biophys. Acta* **9,** 578 (1952).
(529) L. Lóránd and W. R. Middlebrook, *Biochim. et Biophys. Acta* **9,** 581 (1952).
(530) A. G. Szent-Györgyi, *Arch. Biochem. Biophys.* **42,** 305 (1953).
(531) E. Mihályi and A. G. Szent-Györgyi, *J. Biol. Chem.* **201,** 189, 200 (1953).

been isolated as described on p. 1064. More recently Herriott and co-workers[103] have determined the *N*-terminal groups of pepsinogen, pepsin, and of the inhibitor and the *C*-terminal groups of pepsinogen and pepsin (using carboxypeptidase), with the results given in Table XXI. According to these data, pepsinogen, the inhibitor, and pepsin each contain a single peptide chain. Since the *N*-terminal sequences of pepsinogen and the inhibitor are different, it seems unlikely that the inhibitor is derived from the *N*-terminal portion of the peptide chain of the zymogen. On the other hand, the identical *C*-terminal group in pepsinogen and pepsin, i.e., alanine, renders it unlikely that the inhibitor is derived from that portion of the polypeptide chain unless the two *C*-terminal alanines represent actually different segments of the peptide chain. Evidently the completion of this work is required before a detailed structural interpretation of this proteolytic reaction can be attempted. Williamson and Passmann's analytical data, also using substitution by fluorodinitrobenzene, are in contradiction to this work, since they reported the presence of one *N*-terminal leucine (rather than isoleucine) in crystalline pepsin.[514]

The inhibitor was found to comprise the entire basic component of the trichloroacetic acid (2½ per cent) soluble fraction of the activation mixture, having a molecular weight of approximately 3000.[103]

(*2*) *The Activation of Trypsinogen.* Trypsinogen may be activated to trypsin by enterokinase (at pH 6–9), by trypsin (pH 7–9), or by a protease produced by a wild strain of *Penicillium* (pH 3–4).[520,522,523] If the activation is carried out above pH 6, a considerable amount of the potential tryptic activity is lost and an "inert protein" appears instead[522] which increases in amount as pH or temperature is increased. The formation of "inert protein" can be completely suppressed, however, by the presence of 0.01 *M* calcium ions,[532] which, as previously mentioned, are also effective in stabilizing trypsin against autodigestion[122] and in increasing the activity of trypsin toward synthetic substrates.[92,285,286] It is not known as yet to what extent these effects of calcium are interrelated. The kinetics of the activation of trypsinogen by the three enzymes has been analyzed by Kunitz[520] in terms of simple chemical reactions which, in the case of tryptic activation, may be formulated as follows:

$$\text{trypsinogen} + \text{trypsin} \rightarrow \text{trypsin} + \text{inert protein}$$
$$\text{trypsinogen} + \text{trypsin} \rightarrow 2\ \text{trypsin}$$

Satisfactory agreement with experiment was obtained without invoking the formation of an enzyme–substrate compound. In contrast to the activation by trypsin, that by the mold kinase and by enterokinase, below pH 6, followed first-order kinetics.

(532) M. R. McDonald and M. Kunitz, *J. Gen. Physiol.* **25**, 53 (1941).

Some of the results of chemical investigation of the activation process, using DIP-trypsin in place of active trypsin for purposes of end-group analysis, are shown in Table XXI. With Sanger's reagent, Desnuelle and co-workers found valine to be the single *N*-terminal group of trypsinogen, and isoleucine to occupy the same position in DIPT[519,515a] (molecular weight, about 24,000). Under the influence of carboxypeptidase,[125,235] trypsinogen as well as DIPT gave only traces of free amino acids, suggesting that in both proteins the *C*-terminal group is masked by internal ring closure or by steric inaccessibility, or else being not susceptible to the action of this enzyme. However, following acid denaturation, 0.3 mole of lysine per mole of DIPT could be identified as the only stoichiometrically significant reaction product. The replacement during activation of the *N*-terminal valine of trypsinogen by isoleucine in DIPT indicates that one or more polypeptides have been removed from this end of the single polypeptide chain. Such a

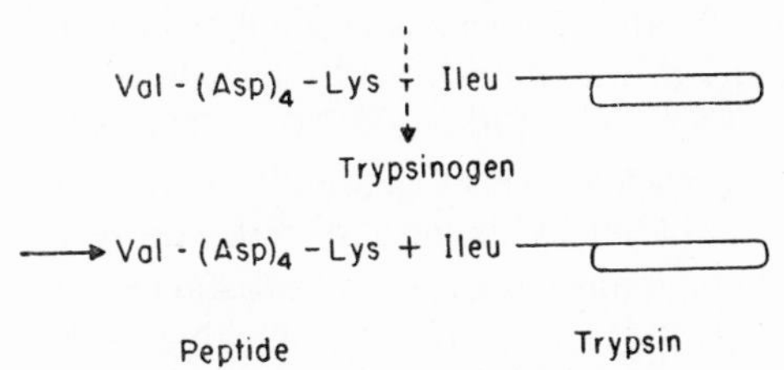

FIG. 10. Proposed scheme for the activation of trypsinogen.[492,521]

peptide of acidic properties has been identified by paper and column chromatography and by paper electrophoresis. Following acid hydrolysis, it was found to contain three amino acids in stoichiometrically significant amounts, i.e., valine, aspartic acid, and lysine, in mole ratios of 1:4:1.[521,492] Valine was identified as the *N*-terminal amino acid (Sanger reagent), whereas the *C*-terminal position of lysine is suggested by the specificity requirements of the activating enzyme, i.e., trypsin. Accordingly, the structure val–$(asp)_4$–lys has been assigned to this peptide. Significantly, the aspartic acid content of the peptide agrees with the difference in aspartic acid composition of trypsinogen and DIPT (Table III). The acidic nature of the proposed structure is in accord with the more basic isoionic point of trypsin (10.1) as compared to trypsinogen (9.3) (see p. 1078). The simplicity of this activation process, depicted schematically in Fig. 10, is in marked contrast to that found for chymotrypsinogen which yields several peptides and several active enzymes.

(3) *The Activation of Chymotrypsinogen.* The kinetics of this process has been measured by Kunitz[520] and by Kunitz and Northrop[66] and more

recently by Jacobsen.[9] Under the conditions described by Kunitz and Northrop,[66] activation is relatively slow, one part of trypsin being added to 10,000 parts of chymotrypsinogen, at pH 7.8 and 4°, and after 24 hours the crystalline product α-chymotrypsin is obtained in a yield of about 50 per cent of the original protein. The protein remaining in the mother liquor has initially little enzymatic activity; however, when allowed to autolyze (pH 8, 0–5°) proteolytic activity slowly increases to a maximum level after approximately 3 weeks. Two different crystalline products, β- and γ-chymotrypsin, can then be isolated from this solution, the two enzymes having qualitatively the same specificity as the α-form. The β- and γ-derivatives may also be obtained by autolysis of crystalline α-chymotrypsin. When higher concentrations of trypsin are used for activation (1 part per 70 parts of chymotrypsin), the activation system attains a higher specific activity,[9] approximately equivalent to 1.5 times that of α-chymotrypsin. Although this protein (δ-chymotrypsin) was never crystallized, the amorphous material, obtained under the conditions described by Jacobsen, was found[82] to hydrolyze the same substrates as α-chymotrypsin, though at different rates. The relation between the rate of activation of chymotrypsinogen and the number of peptide bonds hydrolyzed was compared by Jacobsen[9] with theoretical curves derived for several hypothetical mechanisms, and satisfactory agreement between theory and experiment was found for the following course of the activation process.

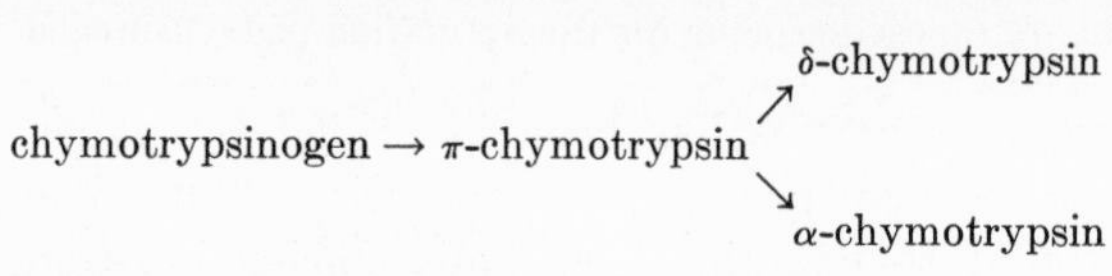

Assigning a value of 36,000 to the molecular weight of the zymogen and the enzymes, which is evidently too high (see Table I), the number of peptide bonds split per molecule of chymotrypsinogen was approximately *one*, for the conversion of chymotrypsinogen to π-chymotrypsin; *two*, for the conversion of chymotrypsinogen to δ-chymotrypsin; and *three* to *five*, with a probable value of *four*, for the conversion of chymotrypsinogen to α-chymotrypsin. However, since the molecular weight of the zymogen is probably 23,000, the above stoichiometric relations become less convincing and the detailed aspects of the postulated mechanism may require revision.

Under the conditions of rapid activation, δ-chymotrypsin is the main product of the reaction and the conversion is believed to be exclusively under tryptic control. Since it was impossible to isolate more than negligible amounts of α-chymotrypsin from the δ-activated chymotrypsinogen solutions which were stored until proteolysis had reached the same stage as in α-activated zymogen solutions, it was thought unlikely that the δ-form could be converted to the α-form. Under conditions of slow activation (as described by Kunitz and Northrop)[66] the reaction leading from the unstable π-intermediate to α-chymotrypsin predominates, and chymotrypsin participates as catalyst in this conversion.

This proposed scheme of Jacobsen has been recently subjected to more definitive experimental testing by determination of the terminal groups of chymotrypsinogen and its activation products.

Chymotrypsinogen is devoid of reactive *N*-terminal groups when tested with the Sanger reagent.[515,515a] It is likewise unreactive toward carboxypeptidase before or after denaturation by heat or acid,[297,490] and hence appears to be composed of cyclic polypeptide chains. In contrast, DIP-α-chymotrypsin contains two *N*-terminal groups per 21,500 molecular weight,[515,515a] i.e., isoleucine and alanine. The action of carboxypeptidase on DIP-α-chymotrypsin[297,490,83] suggests the presence of two *C*-terminal groups, i.e., leucine and tyrosine, though the data are also compatible with the interpretation that leucyl–tyrosine or tyrosyl–leucine forms the *C*-terminal end of one polypeptide chain and that some amino acid which is resistant to the action of carboxypeptidase occupies the *C*-terminal portion of the other polypeptide chain. Glycine, serine, and another leucine residue occupy positions internal to the *C*-terminal groups.[83] It is significant that the two *C*-terminal groups, leucine and tyrosine, conform to those to be expected when chymotrypsin is the activating enzyme, assuming the specificity requirements of chymotrypsin toward synthetic substrates and protein substrates to be the same. If this is the case, then the basic *C*-terminal group presumably created in the tryptic phase leading to the formation of π-chymotrypsin, should have been removed in the second phase when α-chymotrypsin was the reaction product. This has indeed been indicated by analysis of the protein-free filtrate remaining at various stages of the activation of chymotrypsinogen.[235,517] Acid hydrolysis of the peptides, identified by paper chromatography, yielded lysine, arginine, serine, glycine, aspartic acid, glutamic acid, leucine, alanine, and possibly valine. Assuming tentatively that these amino acids have originated from a single peptide, the minimum molecular weight would be about 1200, a value in accord with the difference in molecular weight between chymotrypsinogen and chymotrypsin (see p. 1072), and the peptide would be predominantly basic, as expected from the difference in isoelectric point between the zymogen and α-chymotrypsin (see p. 1073). A tentative scheme, incorporating these findings, is shown in Fig. 11. Figure 11 also indicates the type of end groups to be expected if the formation of δ-chymotrypsin were to proceed as suggested by Jacobsen;[9] an alternative scheme would involve the splitting of a peptide from a terminal portion of the single, open polypeptide chain of π-chymotrypsin. A paucity of experimental data, however, precludes a precise formulation of this reaction.

The DIP derivatives of β- and γ-chymotrypsin have the same *N*-terminal groups as the α-form[515,515a] even though 15 per cent of nonprotein nitrogen accompanies the formation of β- and γ-chymotrypsin. The reactivity toward carboxypeptidase is only slightly different.[83] While leucine and tyrosine are the main reaction products in all three cases, one mole of each amino acid being maximally liberated per mole of substrate, DIP-β- and γ-chymotrypsin yielded, in addition, aspartic acid, asparagine, phenylalanine, and more serine and less glycine than did DIP-α-chymotrypsin.

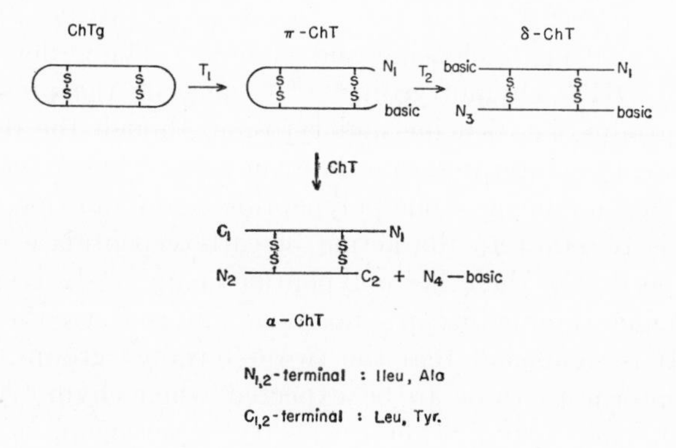

FIG. 11. Tentative scheme for the activation of chymotrypsinogen.[9,490,83]

(*4*) *General Considerations of the Activation of Zymogens.* The experimental evidence at present available points toward a common feature of the activation of precursors of proteolytic enzymes, namely the opening of peptide bonds with the concomitant liberation of one or more peptides. It is significant that in the cases thus far studied, the peptide is derived from different structural regions of the respective precursor and has a different composition in each case. In the activation of trypsinogen, which is apparently the simplest case, a predominantly acidic portion of the open amino end of the single polypeptide chain is split off, the carboxyl end remaining shielded by ring closure or being otherwise resistant to the action of carboxypeptide. In the more complex sequence of reactions attending the conversion of chymotrypsinogen to α-chymotrypsin, a predominantly basic polypeptide appears to be split off the carboxyl end of one of the polypeptide chains, whereas in the activation of pepsinogen, the inhibitory basic peptide is apparently derived from an internal segment of the peptide chain. While it is inviting to speculate that the peptide simply serves to shield the enzymatically active center of the zymogen molecule, thus suppressing enzymatic activity, it is more likely that removal of the peptide is accompanied by more profound intramolecular rearrangements which, at present, cannot be defined in molecular terms.

Since the activation of the zymogens is under control of proteolytic enzymes of relatively narrow specificity, the active enzymes may be considered as proteolytic modifications of the zymogens. However, it should be noted that in each case only a small portion of the total number of bonds conforming to the specificity requirements of the activating enzymes is hydrolyzed. For instance, trypsinogen contains approximately 14 lysine residues and 3 arginine residues, but only one of these 17 peptide bonds is hydrolyzed during tryptic activation; similarly, only 3 to 5 peptide bonds are hydrolyzed during the conversion of chymotrypsinogen to α-chymotrypsin although chymotrypsinogen contains peptide bonds contributed by about 13 aromatic residues and 27 residues of the leucine group (Table III), each of these being potentially liable to chymotryptic hydrolysis. A full elucidation of the mechanism of activation will have to take into account this high degree of selectivity of these proteolytic processes.

(5) The Conversion of Fibrinogen to Fibrin. The nature of the reaction between fibrinogen and thrombin, leading to fibrin, has been the subject of a lively controversy[533] which converged to the view that proteolysis is probably not involved since, within the limits of the experimental error, all of the fibrinogen nitrogen could be recovered as fibrin.[534] Such a conclusion, however, is of limited validity since enzymatic hydrolysis of peptide bonds does not necessarily lead to the liberation of peptides. Recent chemical investigations directed toward the determination of *N*-terminal amino acids of fibrinogen and fibrin, using the Sanger reagent, have indeed shown rather conclusively that peptide bonds are broken and that peptides are liberated during the process.[527–529] According to these results, fibrinogen contains two *N*-terminal glutamic acid residues, in contrast to four *N*-terminal glycine residues in fibrin, suggesting that during the process two terminal groups of additional peptide chains have become exposed and that a peptide has been split off from the *N*-terminal portions of the polypeptide chains of fibrinogen. However, analysis of the peptides isolated from the mother liquor of the fibrin clot by paper electrophoresis[528] showed the presence of at least two peptides, one of which contained *N*-terminal glutamic acid, as expected, while the other possessed no detectable *N*-terminal groups. Thus, whatever the structure of the second peptide may be, its apparent lack of *N*-terminal residues can

(533) T. Astrup, *Advances in Enzymol.* **10,** 1 (1950).
(534) A. G. Ware, M. M. Guest, and W. H. Seegers, *Arch. Biochem.* **13,** 231 (1947).

account for the two extra residues which appear in fibrin during the clotting process. Further implications of these findings are discussed in Chap. 21.

(6) *Meromyosins.* It has been recently discovered[530,531] that after a short period of digestion with trypsin (12 min. at 23°C., substrate to enzyme weight ratio = 140) myosin yields two new proteins, called meromyosin L and H. The two proteins were characterized by their sedimentation constants, i.e., $s_{20,w} = 2.5\ S$ for the L (or light) fraction and $s_{20,w} = 5\ S$ for the H (or heavy) fraction. These correspond to molecular weights of 96,000 and 232,000, respectively. The L fraction could be crystallized by dialysis of the reaction mixture against phosphate buffer pH 7,[530] whereas the H fraction was isolated by ammonium sulfate fractionation of the mother liquors. The molecular weights and relative proportions of the two proteins indicate that four molecules of the L-meromyosin and two molecules of the H-meromyosin are present in the intact myosin molecule. Significantly, only the H fraction possessed ATPase activity and the ability to combine with actin, whereas L-meromyosin possessed the very characteristic solubility properties of the whole myosin fraction. The two fractions differed significantly in end-group analysis see (Table XXI) and amino acid composition.

CHAPTER 26

Peptide and Protein Synthesis. Protein Turnover[1]

BY H. TARVER

Page

I. Introduction 1200
II. Synthesis of and Interaction between Peptide Bonds 1201
  1. Energetics 1201
    a. Closed System Thermodynamics 1201
    b. Open System Thermodynamics 1204
  2. Acylation of Amino Acids 1205
    a. Hippuric and *p*-Aminohippuric Acid Synthesis 1205
    b. Ornithine Synthesis 1209
    c. Acetylation of Amino Acids 1209
    d. Acetylation of Aromatic Amines 1210
  3. Synthesis of Glutathione 1211
  4. Synthesis of Glutamine (γ-Amide) 1217
  5. Glutamyl Transfer; Transamidation 1219
    a. γ-Glutamyl Transfer (from Residues Other Than Those of Amino Acids) 1219
    b. γ-Glutamyl Transfer (from True Peptide Derivatives) 1220
    c. Transamidation 1221
  6. Synthesis of Peptides 1222
  7. Conclusions 1223
III. Synthesis of Protein and Incorporation of Isotopic Amino Acids *in Vitro* ... 1224
  1. Reversal of the Proteolytic Reaction 1224
  2. Incorporation of Isotopes *in Vitro* 1225
    a. The Nature of the Reaction(s) Involved—Experimental Complications 1227
    b. The Incorporation Rate in Various Preparations 1229
    c. Stimulation of Incorporation 1236
    d. Inhibition of Incorporation 1240

(1) In this chapter use is made of the abbreviations for amino acid residues adopted by E. Brand and J. T. Edsall [*Ann. Rev. Biochem.* **16,** 223 (1947)], the first three letters of the name generally becoming the condensation. Since no confusion will arise, the same symbols are employed for the free amino acids. Other abbreviations which have been used are: P_i for orthophosphate; PP for pyrophosphate; ph, Ac, Bz, cbz, and Ph for phosphoryl, acetyl, benzoyl, carbobenzyloxy and phenyl, respectively. Also the generally used forms, TCA, RNA, DNA, ATP, ADP, and AMP are adopted.

In certain places the incorporation of DL-amino acids into protein preparations will be mentioned. By this terminology it is not meant to imply that the D-form is actually incorporated but simply that the racemic mixture was used in the investigation.

Page
e. Peptides and the Incorporation Reaction........ 1241
f. Nucleic Acids and Incorporation........ 1242
g. Anomalous Incorporation Reactions........ 1245
3. Synthesis of Specific Proteins........ 1246
a. Synthesis of Serum Albumin........ 1246
b. Ovalbumin Synthesis and a Consideration of the Mechanism of Protein Synthesis........ 1248
c. Amylase, Xanthine Oxidase, and Antibody Synthesis........ 1253
4. Incorporation of Isotopic Amino Acids into the Proteins of Tissue Cultures and Perfused Organs........ 1255
a. Tissue Cultures........ 1255
b. Perfused Livers........ 1257
5. Conclusions........ 1258
IV. Incorporation of Amino Acids *in Vivo*—Turnover........ 1259
1. Incorporation of Amino Acids into Proteins of Different Tissues........ 1262
2. Incorporation of Amino Acids into the Proteins of Cellular Components (of Liver)........ 1262
3. Turnover Data from *in Vivo* Experiments........ 1263
a. The Nature of Turnover Process and Its Measurement........ 1263
b. The Turnover of Tissue Protein and of Some Specific Proteins........ 1274
c. The Catabolism of Heterologous Protein........ 1278
d. The Interpretation of the Turnover Phenomenon........ 1280
e. Factors Affecting the Rate of Turnover........ 1282
4. Fluctuations in Enzymatic Levels........ 1283
a. Changes in Enzyme Concentrations in Animal Tissues........ 1283
b. Enzymatic Adaptation (EA)........ 1285
5. Incorporation of Foreign Amino Acids into Protein. The Pluperfect Concept of Protein Synthesis........ 1286
6. The Requirements for Protein Synthesis........ 1288
a. Amino Acids vs. Partial Breakdown Products of Plasma Protein........ 1288
b. Requirements for Energy........ 1289
7. The Magnitude of Turnover and Its Relationship to the Metabolic Rate 1290
8. Conclusions........ 1291
V. General Conclusions........ 1291
VI. Appendix........ 1292

I. Introduction

In this chapter an attempt has been made to review objectively the more recent work on the synthesis of peptide bonds, the synthesis and incorporation of isotopic amino acids into protein, and the turnover of proteins. The relationship between these phenomena and nitrogen and energy requirements for the maintenance and formation of protein as shown, for example, by the requirements for enzymatic adaptation has also been considered, somewhat briefly.

In some sections of the chapter the survey of the literature is reasonably complete (e.g., sec. II), whereas in others the subject matter has been dealt with largely by example and the discussion has been directed more toward definition of the problem. This is particularly true of the theory of protein synthesis, dealt with in sec. III-3*b*.

Little has been mentioned with respect to the effect of hormones on protein metabolism and protein turnover because the work to date does not permit of any generalizations without recourse to a very large body of supplementary material which lies beyond the scope of this treatise.

II. Synthesis of and Interaction between Peptide Bonds

In this section we shall consider the formation of true peptide bonds, involving the coupling of amino acids, and of bonds which bear a close resemblance to them. Such bonds, for example those in hippuric acids, are often formed in animals in response to "toxic agents." The "detoxication reaction" which follows is quite often simply a conjugation involving the elimination of the elements of water. The operation of such mechanisms in the intact animal has been considered in detail by Williams.[2]

1. Energetics

a. Closed System Thermodynamics

In connection with their studies on the synthesis of hippuric acid (benzoylglycine) from benzoic acid and glycine in tissue slices, Borsook and Dubnoff[3] calculated the free energies of formation of several peptide bonds. As a result they concluded that the free-energy change involved in the formation of hippuric acid is similar to that involved in the formation of the peptide alanylglycine, as shown in Table I.

TABLE I

Free Energy of Formation of Peptide Bonds in Aqueous Solution[3]

Product	Temp., °C.	$\Delta F°$, kcal./mole	K^a	Equilibrium concentration of product,[b] M
ala.gly	25	4.00	1.17	5×10^{-8}
Bz.gly	25	3.17	4.74	2×10^{-7}
Bz.gly	38	3.23	5.43	3×10^{-7}

[a] K = equilibrium constant $\times 10^3$.

[b] Concentration of product assuming the concentrations of alanine, glycine, and BzOH are 10^{-2}, 5×10^{-3}, and 10^{-2} M, respectively.

The values of K given in the table are those for the synthetic reactions calculated from the values of $\Delta F°$ by making use of the relationship

(2) R. T. Williams, Detoxication Mechanisms, John Wiley and Sons, New York, 1949.

(3) H. Borsook and J. W. Dubnoff, *J. Biol. Chem.* **132,** 307 (1940).

$\Delta F = -2.303\, RT \log K$. If reasonable assumptions are made relative to concentrations of the free amino acid and benzoic acid in the tissues involved (liver and kidney), it is possible to calculate the concentration of products existing in equilibrium with these reactants. These data are found in the last column of Table I and show that the concentration of alanylglycine in equilibrium with the free alanine and glycine in the liver at 25° is extremely low, namely $5 \times 10^{-8}\ M$. The concentration of hippuric acid in equilibrium with its precursors is likewise very low ($2 \times 10^{-7}\ M$), and, therefore, it does not appear likely that the mechanism by which peptides are formed involves equilibration between the components. However, equilibration might provide a mechanism for hippuric acid formation if some auxiliary mechanism existed whereby the product was removed from the cells. Since such a mechanism is not known, the synthesis of hippuric acid must be coupled in some manner with an exergonic reaction. This conclusion was reached by Borsook and Dubnoff.[3]

The next question that arises is whether it is justifiable to draw from data of this type a general conclusion with respect to the synthesis of peptides, both large and small. Linderstrøm-Lang[4] has approached this problem by comparing the energetic requirements for the synthesis of peptides from the dipolar ions and from the uncharged molecules. The data are shown in Table II.

TABLE II

FREE ENERGY CHANGES IN SYNTHESIS OF PEPTIDE BONDS[4]

In aqueous solution at 25°

Product	Synthesis from ions and dipolar ions		Synthesis from neutral molecules	
	$\Delta F°$, kcal./mole	$K \times 10^3$	$\Delta F°$, kcal./mole	$K \times 10^{-3}$
gly.gly	4.09	1.0	−4.98	4.5
ala.gly	4.00	1.2	−5.07	5.2
leu.gly	3.11	5.3	−5.96	23
Bz.gly	2.42	17	−4.83	3.5
Bz.gly.gly[a]	0.93	200	−4.59	2.3

[a] Synthesis from BzO^- and glycylglycine.

Inspection of this table shows that when synthesis from dipolar ions is considered, the calculated free-energy changes are all unfavorable for synthesis to an extent greater than 2 kcal. per mole. Only in the case of the benzoylation of glycylglycine is the situation more favorable. However, for the neutral reactants the energy changes are all in favor of

(4) K. V. Linderstrøm-Lang, *Stanford Univ. Publs. Med. Sci.* **6**, 93 (1953).

synthesis. Moreover, there does not appear to be any great difference in the free-energy changes except in the case of leucylglycine synthesis.

From such data as are given in Table II, Linderstrøm-Lang[4] calculated the free-energy changes associated with the coupling of glycine residues into different compounds as shown in Table III.

TABLE III
DIFFERENCES IN ENERGIES INVOLVED IN COUPLING GLYCINE RESIDUES IN DIFFERENT COMPOUNDS[4]

Reaction	ΔF°, kcal./mole
$^+$gly$^-$ + $^+$gly$^-$	4.09
$^+$gly.gly$^-$ + $^+$gly.gly$^-$	1.84
$R_1G^- + {}^+GR_2$	1.16

It is evident that ΔF° decreases with increasing distance of separation of the charges on the residues involved. Thus, the ratio of the concentration of the tetrapeptide in equilibrium with glycylglycine is much higher than the ratio of the concentration of glycylglycine in equilibrium with glycine.

The validity of these deductions is shown by the experiments of Dobry and co-workers[5] who studied the following reaction:

$$(\text{Bz-L-tyr.})^- + (\text{gly.N}^{15})^+ \rightleftharpoons \text{Bz-L-tyr.gly.N}^{15} + H_2O \qquad (1)$$

At 25° they found $\Delta F^\circ = 0.42 \pm 0.05$ kcal./mole and $\Delta H^\circ = 1.55 \pm 0.10$ kcal./mole, corresponding to $\Delta S^\circ = 0.0038$ entropy units. Hence the equilibrium is not greatly displaced by increasing the temperature to 37°. The value for ΔF° is approximately the same as that calculated from the data of Fruton and co-workers,[6] in spite of the fact that their determinations were made in 30 per cent methanol.

On the basis of these thermodynamic considerations it is clear that, for soluble peptides, the equilibrium point of the reaction:

$$^+A^- + B^- \rightleftharpoons AB^- + H_2O \qquad (\Delta F^\circ = Q_{AB}) \qquad (2)$$

is located to the left. However, where the product (AB) is insoluble the position of the equilibrium will be displaced in favor of the forward reaction, owing to the lower energy state of the product. Also, in view of the previous discussion, a further displacement of the equilibrium will occur when uncharged derivatives of the amino acids A and B are used as reactants. This is shown by the formation of various peptides from appropriate amino acid derivatives in the presence of the proper peptidase

(5) A. Dobry, J. S. Fruton, and J. M. Sturtevent, *J. Biol. Chem.* **195,** 149 (1952).
(6) J. S. Fruton, R. B. Johnston, and M. Fried, *J. Biol. Chem.* **190,** 39 (1951).

or proteases (sec. II-5,6). The possibility exists, therefore, that if small peptides are produced in significant concentrations, polymerization of the peptides might occur under favorable circumstances; the coupling of large peptides might occur even more easily.

b. Open System Thermodynamics

If, in the consideration of the synthesis of peptide bonds cognizance is taken of the fact that the biological system is open, and that we are dealing with complex equilibria involving two or more simultaneous reactions, the picture of peptide-bond synthesis becomes entirely different. Coupled reactions and energy input become of primary interest.

Such a system might operate by forming a derivative of the carboxyl group or amino group of an amino acid, yielding compounds with additional bond energy:

$$^{+}A^{-} + R_1^{+} \rightarrow {}^{+}AR_1 \qquad (\Delta F = -Q_{CO}) \tag{3}$$

$$^{+}A^{-} + R_2^{-} \rightarrow R_2A^{-} \qquad (\Delta F = -Q_{NH}) \tag{4}$$

Following reactions (3) or (4), either (5) or (6) would occur.

$$^{+}AR_1 + {}^{+}B^{-} \rightarrow {}^{+}AB^{-} + R_1^{+} \qquad (\Delta F = -Q_{CO} + Q_{AB}) \tag{5}$$

$$R_2A^{-} + {}^{+}B^{-} \rightarrow {}^{+}BA^{-} + R_2^{-} \qquad (\Delta F = -Q_{NH} + Q_{AB}) \tag{6}$$

Under these circumstances the ΔF° in favor of the forward reactions (5) and (6) would depend on the Q values for the partial reactions (3) and (4), i.e., on the energies involved in the formation of the $^{+}A \ldots R_1$ and $R_2 \ldots A^{-}$ bonds. In view of what is known concerning energy transfers in other biological systems, e.g., in the metabolism of carbohydrate, carboxyl phosphates, phosphoamides, or carboxyl coenzyme A (CoA) compounds would be given first consideration for compounds $^{+}AR_1$ and R_2A^{-}. Since the energy of carboxylphosphate is much greater than that of phosphoamide, following Lipmann,[7] the former type of amino acid derivative has generally been accorded most consideration. The high free energy of the compound would enable amino acids to be lifted from low "concentration energies" into the synthetic reaction. Apparently, the same result could be attained with the CoA types of compounds as with the carboxyl phosphates.

It is possible to liken peptide-bond synthesis still more closely to syntheses involved in carbohydrate metabolism, e.g., starch formation, by writing reactions of the following types:

$$^{+}BR_1^{-} + {}^{+}A \ldots X^{-} \rightarrow {}^{+}BA \ldots X^{-} + {}^{+}R_1^{-} \tag{7}$$

$$^{+}A \ldots X^{-} + {}^{+}R_2B^{-} \rightarrow {}^{+}A \ldots XB^{-} + {}^{+}R_2^{-} \tag{8}$$

(7) F. Lipmann, *Advances in Enzymol.* **1,** 99 (1941); *Federation Proc.* **8,** 597 (1949).

Then in the sense of Hanes and co-workers,[8] Eqs. (6) and (7) would be referred to as transpeptidations involving carboxyl transfer, whereas reactions (5) and (8) would be amide transfers when R_1 and R_2 are amino acids.

In addition, it is possible that energy might be transferred from other parts of amino acid molecules to help push synthetic reactions involving either the amino or carboxyl groups. Energy changes which occur in the formation of disulfide bonds figure prominently in this connection. However, there must exist some structure through which the energy may be transferred to the site of reaction. This might occur through a double-bond system. Since such systems do not exist in amino acids and the status of anhydropeptides is questionable, such energy transfers may not occur beyond the α-position.

It has also been suggested that energy transfers might take place through the backbone of the protein structure, or via a catalytic surface. As yet a satisfactory demonstration of the existence of such a transfer of energy is lacking. Consequently, although these mechanisms present interesting possibilities, they are at present strictly in the realm of conjecture.

2. Acylation of Amino Acids

a. Hippuric and p-Aminohippuric Acid Synthesis

Studies on hippuric acid formation *in vitro* from benzoic acid and glycine were first successfully carried out by Borsook and Dubnoff[3] using tissue slices of liver and kidney from several species. The slices were incubated with the reactants, the hippuric acid formed was extracted, and following hydrolysis its glycine content was determined by formol titration. Rather surprising species differences were found in the ability of the two tissues mentioned to carry out the synthesis. (See Table IV.)

The rates of hippuric acid formation by slices from the kidney of dogs were of the same order as those found by others in perfusion experiments.

Synthesis was completely inhibited by 1 mM cyanide; this finding together with the large yield of product was taken to substantiate the conclusion reached by thermodynamic calculations: synthesis must be coupled with exergonic reactions.

More recently it has been found that the synthesis of *p*-aminohippuric acid (PAH) occurs under conditions similar to those required for synthesis of hippuric acid.[9] The PAH formed was conveniently determined colorimetrically after preferential extraction of the *p*-aminobenzoic acid with ether. Since synthesis of PAH is completely inhibited in the

(8) C. S. Hanes, F. J. R. Hird, and F. A. Isherwood, *Nature* **166**, 288 (1950).

(9) P. P. Cohen and R. W. McGilvery, *J. Biol. Chem.* **166**, 261 (1946).

presence of benzoic acid,[10] the same enzyme system is presumably involved in both reactions. Synthesis is inhibited over 90 per cent by cyanide (1 mM), arsenite (10 mM), iodoacetate (10 mM), and fluoride (10 mM), and also in the absence of oxygen; less inhibition was obtained with azide (1 mM) or malonate (1 mM).[9]

TABLE IV
FORMATION OF HIPPURIC ACID IN LIVER AND KIDNEY SLICES FROM DIFFERENT SPECIES[3]

From 30 to 50 mg. of tissue slices incubated at 38° under 95% oxygen, 5% CO_2 in Krebs-Henseleit solution containing 0.2% glucose, 2.5 mM BzOH and 10 mM gly.

	Liver		Kidney	
	Q^a	Per cent[b]	Q^a	Per cent[b]
Dog	0	0	0.94	65
Guinea pig	0.59	76	0.41	53
Rabbit	0.59	76	0.47	59
Rat	0.35	59	0.12	18

[a] Q = amount of Bz.gly formed/hr./mg. tissue (dry weight) expressed as cu. mm. of gas at standard temperature and pressure.
[b] Per cent of added BzOH found as Bz.gly.

Cohen and McGilvery[11] and Borsook and Dubnoff[12] later found that conjugation occurred in liver homogenates, the rate of the synthetic reaction in guinea pig liver homogenate being increased in the presence of α-ketoglutarate (0.01 M) and adenylic acid (AMP) (1 mM).[12] Cohen and McGilvery[11] employed a homogenate of kidney from rats prepared in isotonic KCl and found that the mitochondrial fraction stimulated PAH formation in the presence of fumarate (2.5 mM) and potassium chloride (0.04 M). Calcium ions inhibited the reaction. A sharp optimum was found at pH 7.55. The supernatant fraction was inactive.[11,13]

Since the mitochondrial fraction as prepared by Cohen and McGilvery[14] with isotonic KCl is inevitably contaminated with nuclei, Kielley and Schneider[15] investigated the synthesis of PAH by mitochondria practically free of nuclei isolated from the livers of mice by fractionation with

(10) P. P. Cohen, *in* Sumner and Myrbäck, The Enzymes, Academic Press, New York, 1952, p. 886; P. P. Cohen, *in* McElroy and Glass, Phosphorus Metabolism, Johns Hopkins Press, Baltimore, 1951, Vol. 1, p. 630.
(11) P. P. Cohen and R. W. McGilvery, *J. Biol. Chem.* **169,** 119 (1947).
(12) H. Borsook and J. W. Dubnoff, *J. Biol. Chem.* **168,** 397 (1947).
(13) F. Leuthardt and H. Nielsen, *Helv. Chim. Acta* **34,** 1618 (1951).
(14) P. P. Cohen and R. W. McGilvery, *J. Biol. Chem.* **171,** 121 (1947).
(15) R. K. Kielley and W. C. Schneider, *J. Biol. Chem.* **185,** 869 (1950).

sucrose solutions. It was shown that the activity resided in the mitochondrial fraction only (Table V). A requirement of the system for phosphate was also demonstrated (compare Expt. 2*a* with 2*b*).

TABLE V

DISTRIBUTION OF PAH[a] SYNTHETIC ACTIVITY IN CELLULAR FRACTIONS FROM MOUSE LIVER[15]

One milliliter of homogenate (100 mg. fresh liver) or its equivalent in a medium with the following other addenda: 0.125 *M* sucrose and 0.25 m*M* PAB, 3.75 m*M* glycine, 1.25 m*M* fumarate, 1.25 m*M* glutamate; 1.25 m*M* phosphate (Expt. 2*b*), 10 m*M* KCl, 12.5 m*M* histidine (Expt. 2*a*), 0.625 m*M* $MgSO_4$, 0.25 m*M* ATP, and 10^{-3} m*M* cytochrome c at pH 7.55.

Cell fraction	Total N, mg.	Expt. 2*a*[b] PAH, μmoles	Expt. 2*a*[b] 7′P, μg.	Expt. 2*b* PAH, μmoles	Expt. 2*b* 7′P, μg.
Whole homogenate	3.26	1.61	253	2.08	323
Nuclei	0.41	0.03	55	0.03	94
Mitochondria (M)	1.02	1.15	283	1.89	305
Supernatant (S)	1.89	0.0	—	0.0	—
M + S		1.26	217	1.66	310
M + S (boiled)		1.36	269	1.76	346
			290[c]		290[c]

[a] PAH, *p*-aminohippuric acid; PAB, *p*-aminobenzoic acid.
[b] Expt. 2*a* with histidine buffer; Expt. 2*b* with phosphate buffer.
[c] 7′-P added as ATP before incubation.

Of even greater interest were the observations which showed that in those combinations in which the 7′-phosphate (presumably from adenosine triphosphate (ATP) and *not* pyrophosphate) was preserved, PAH synthesis occurred to the greatest extent. When the labile P was lost, as in the presence of nuclei, synthesis was low. The boiled supernatant was found to contain a factor which increased the rate of synthesis and helped to maintain the labile P. The nature of this material became clear from the experiments of Chantrenne,[16] who succeeded in obtaining synthesis of hippuric acid in an extract of acetone powder from rat liver. When CoA was removed from the extract by adsorption on Dowex 1, no synthesis occurred; synthetic activity appeared on repriming. Moreover, the rate of synthesis was found to be a function of CoA concentration.

The following additional observations eliminate various possible activated forms and intermediates: (*a*) Chantrenne[16] observed that benzoyl phosphate did not promote hippuric acid formation; (*b*) Cohen and McGilvery[14] found that when acetylglycine was employed, the rate of aminohippuric acid synthesis was less than when glycine was used alone

(16) H. Chantrenne, *J. Biol. Chem.* **189**, 227 (1951).

or in addition to the acetylglycine; (*c*) low rates were obtained when glycine was replaced with glyoxylate plus ammonia,[14] and (*d*) *N*-phosphoglycine does not promote PAH synthesis.[14]

It has also been observed that acetyl derivatives of various amino acids are not utilized by several species of bacteria,[17] and with some glycine peptides as substrates the synthesis was slower than with glycine itself.[13] Hydrolysis probably occurred prior to synthesis.

The results with glyoxylate and ammonia indicate that the bond is not formed by a condensation of glycine, ammonia, and glyoxalate. In fact, any condensation of nitrogen free molecules with ammonia (or other metabolically formed nitrogen compounds) and amino acid to form peptide bonds, as they occur in protein, is practically precluded by the early findings of Schoenheimer and Rittenberg and of other more recent workers. It has always been observed that the administration of amino acids labeled with N^{15} leads to the formation of tissue proteins with the highest specific activity of N^{15} in the amino acid originally introduced into the system, except for aspartic acid-N^{15}.

The nature of the active intermediate has been demonstrated by recent work of Schachter and Taggart[18] who employed synthetic benzoyl CoA in experiments where this compound was the source of the acyl radical used to synthesize hippuric acid. A close correspondence was found between the hippuric acid synthesized, the free —SH appearing (as CoA), and the disappearance of substance reacting with hydroxylamine to form a hydroxamic acid (BzCoA). The reactions involved in hippuric acid synthesis may be written as follows:

$$\text{BzOH} + \text{CoA.PP} \rightarrow \text{Bz.CoA} + \text{PP} \quad (9)$$

$$\text{Bz.CoA} + \text{gly.} \rightarrow \text{Bz.gly.} + \text{CoA} \quad (10)$$

The results of Chantrenne[16] with benzoyl phosphate are in accord with his work in non-enzymatic systems.[19] In the absence of enzymes, monobenzoyl phosphate undergoes hydrolysis at pH 7.4 at a more rapid rate than does dibenzoyl phosphate and the former does not react with glycine whereas the latter does:

$$(\text{BzO})_2\,\text{ph} + \text{gly.} \rightarrow \text{Bz.gly.} + \text{P}_i + \text{BzOH} \quad (11)$$

For further work of this type see sec. II-6.

It is of interest to note that the formation of hippuric acid may, under certain circumstances, cause a lowering of the glycine concentration in

(17) S. Simmonds, E. L. Tatum, and J. S. Fruton, *J. Biol. Chem.* **169,** 91 (1947); C. H. Eades, Jr., *J. Biol. Chem.* **187,** 147 (1950).
(18) D. Schachter and J. V. Taggart, *J. Biol. Chem.* **203,** 925 (1953).
(19) H. Chantrenne, *Nature* **160,** 603 (1947); *Biochim. et Biophys. Acta* **2,** 286 (1948).

cells. Thus Christensen and co-workers[20] observed that the α-amino N of glycine in the liver cell fell from a normal value of 32 mg. per cent to 9.0 mg. per cent in 7 to 120 min. following the feeding of benzoate (0.3 g./kg.) to guinea pigs. There was a smaller drop in extracellular glycine, but in muscle, where no hippuric acid synthesis occurred, there was little change in glycine concentration.

Synthesis of benzoylhydroxamic acid, which closely resembles that of hippuric acid, has been described by Virtanen and Berg.[21] In this system benzoic acid is utilized more readily in the presence of ATP. A reaction similar to that involved in the synthesis of glutamylhydroxamic acid occurs (page 1218).

$$\text{BzOH} + \text{NH}_2\text{OH} + \text{ATP} \rightarrow \text{BzNHOH} + (\text{ADP}) + \text{P}_i \qquad (12)$$

b. Ornithine Synthesis

The synthesis of ornithuric acids has been studied by McGilvery and Cohen[22] who employed as substrates both the α and δ-monobenzoyl-ornithines. With a sedimented fraction (mitochondria and nuclei) from the kidney of fowls they were able to show the coupling of *p*-aminobenzoic acid with either substrate. A similar preparation from liver was ineffective, but even with the kidney the rate of formation of the amino derivatives of ornithine was much slower than was that of hippuric acid formation with an analogous liver preparation from the rat. The system in the fowl required a high substrate concentration in order to attain maximum velocities of coupling (about 10 m*M*), being in this respect similar to the PAH synthesizing system. In most respects, the requirements for synthesis were similar to those for the synthesis of the hippurates.

c. Acetylation of Amino Acids

The classical case of acetylation occurs with cysteine, or more probably with its derivatives in which the sulfur is blocked by a conjugation with an aromatic compound. Other acetylation reactions of amino acids have been described in the literature,[2] but since none of these have been studied *in vitro* they will not be considered here. However, using the S^{35}-labeled substrate, Gutman and Wood[23] showed that *S*-benzyl-homocysteine is acetylated in rat liver and kidney slices. Liver slices appeared to be more active than kidney slices.

(20) H. N. Christensen, J. A. Streicher, and R. L. Elbinger, *J. Biol. Chem.* **172,** 515 (1948).
(21) A. I. Virtanen and A. M. Berg, *Acta Chem. Scand.* **5,** 909 (1951).
(22) R. W. McGilvery and P. P. Cohen, *J. Biol. Chem.* **183,** 179 (1950).
(23) H. R. Gutman and J. W. Wood, *J. Biol. Chem.* **189,** 473 (1951).

Recently Stadtman, Katz, and Barker[24] described a cyanide (or azide)-induced acetylation of amino acids and proteins which occurs in a dried-cell preparation of *Clostridium kluyverii*. Acetyl phosphate (AcOph) or other acyl phosphates transfer acyl groups resulting in the formation of acylated amino acids or acylated protein (either endogenous protein or added protein such as egg albumin). The reaction, or some step in the reaction sequence, is enzymatic since heating the preparation abolishes the acetylation, or rather it decreases the deacylation of the AcOph, which is the variable actually measured. It was shown by isolation that with glycine or leucine present as acceptors and AcOph-C^{14} as donor the corresponding acylated amino acids appeared. Most D- and L-amino acids, except the dicarboxylic acids, were quite effective as acceptors,[25] and many aliphatic but not aromatic amines could act as acceptors. Therefore, the reaction differs from the acetylation of the aromatic amines which occurs in animal tissues.[26] It is also of interest to note that no acetylation of glycine occurred with free acetate plus ATP in the presence of cyanide (0.1 *M*). The reaction has been formulated as follows:[25]

$$\text{AcOph} + \text{CoA} \xrightarrow{\text{phosphotransacetylase}} \text{Ac.CoA} + \text{P}_i \quad (13)$$

$$\text{Ac.CoA} + \text{gly.} \rightarrow \text{Ac.gly.} + \text{CoA} \quad (14)$$

Reaction (13) is not reversible in the system under consideration, and no hydrolysis of the acetylglycine was observed. Evidently the reaction is analogous to the benzoylation of glycine except for the participation of acetyl phosphate rather than of ATP.

d. Acetylation of Aromatic Amines

Since the pioneering investigation of Lipmann[26] who demonstrated that cell-free extracts of pigeon liver were capable of catalyzing the acetylation of sulfanilamide (sulfa), and that a soluble cofactor was required in the reaction, this type of coupling reaction has received considerable attention. However, it should be noted that Klein and Harris[27] had previously demonstrated the same reaction in rabbit liver slices. A coupling with the respiratory mechanism of the cells was also observed. Lipmann[26] found the acetylating system of pigeon liver to reside in soluble components. Some type of specific stoichiometric coupling between the synthesis of the acetyl derivative and the breakdown of ATP could be demonstrated when the breakdown of ATP was inhibited by fluoride.

(24) E. R. Stadtman, J. Katz, and H. A. Barker, *J. Biol. Chem.* **195,** 779 (1952).
(25) J. Katz, I. Lieberman, and H. A. Barker, *J. Biol. Chem.* **200,** 417, 431 (1953).
(26) F. Lipmann, *J. Biol. Chem.* **160,** 173 (1945).
(27) J. R. Klein and J. S. Harris, *J. Biol. Chem.* **124,** 613 (1938); J. S. Harris and J. R. Klein, *Proc. Soc. Exptl. Biol. Med.* **38,** 78 (1938).

Consequently, a reaction of the following type was indicated:

$$\text{ATP} + \text{AcOH} + \text{sulfa} \rightarrow \text{Ac.sulfa} + \text{ADP} + \text{P}_i \quad (15)$$

Although Lipmann at first thought that acetyl phosphate was involved as an intermediary in reaction (15), it was later found that the active agent was a coenzyme which could activate acetyl groups at either the carbonyl or the methyl end of the molecule.[28] The coenzyme, CoA, has been found to be widely distributed in animal and plant tissues, for example in liver and in bacteria of the *Clostridium* group, by an assay involving the acetylation of sulfa.[29]

Further information concerning the function of CoA in various systems will be found in the reports of several symposia.[30] In the light of these more recent observations,[31] it appears that reaction (15) should be formulated as follows:

$$\text{ATP} + \text{CoA} \rightarrow \text{CoA.PP} + \text{AMP} \quad (16)$$
$$\text{CoA.PP} + \text{AcOH} \rightarrow \text{Ac.CoA} + \text{PP} \quad (17)$$
$$\text{Ac.CoA} + \text{sulfa} \rightarrow \text{Ac.sulfa} + \text{CoA} \quad (18)$$

where CoA.PP and PP indicate CoA pyrophosphate and pyrophosphoric acid, respectively.

In addition to the normal type of acylation, it is possible to observe under the proper conditions the succinylation of sulfa.[32] In the presence of the specific enzyme CoA, α-ketoglutarate, and other accessory factors, succinyl CoA is formed and from this intermediate the succinyl group can be transferred to form the derivative of the sulfa drug. The behavior of this system is shown in Fig. 1, adapted from Sanadi and Littlefield.[32]

Acylation of sulfanilamide like that of glycine is mediated by CoA. Whether CoA is likewise required for the benzoylation of ornithine (sec. II-2*b*) and of glutamine in the human[2] is not certain.

3. Synthesis of Glutathione

The synthesis of glutathione (GSH) which involves the synthesis of one normal peptide bond and of one bond involving the γ-carboxyl group

(28) F. Lipmann and N. O. Kaplan, *J. Biol. Chem.* **162,** 743 (1946); F. Lipmann, N. O. Kaplan, O. G. Novelli, L. C. Tuttle, and B. M. Guirard, *J. Biol. Chem.* **167,** 869 (1949); F. Lipmann, *Harvey Lectures Ser.* **44,** 99 (1948–49).

(29) N. O. Kaplan and F. Lipmann, *J. Biol. Chem.* **174,** 37 (1948).

(30) McElroy and Glass, Phosphorus Metabolism, Johns Hopkins Press, Baltimore, 1951, Vol. 1; 1952, Vol. 2; *Federation Proc.* **12,** (3) (1953).

(31) F. Lipmann, M. E. Jones, S. Black, and R. M. Flynn, *J. Am. Chem. Soc.* **74,** 2384 (1952).

(32) D. R. Sanadi and J. W. Littlefield, *J. Biol. Chem.* **193,** 683 (1951); *ibid.* **201,** 103 (1953); *Science* **116,** 327 (1952); S. Kaufman *in* McElroy and Glass, Phosphorus Metabolism, Johns Hopkins Press, Baltimore, 1951, Vol. 1, p. 370; H. R. Mahler, *ibid.*, 1952, Vol. 2, p. 286.

s of paramount interest. The first *in vitro* studies were carried out by Bloch and Anker[33] using glycine-N^{15} as a labeling agent in order to detect incorporation in the absence of net synthesis; later in the same laboratory the work was continued with glycine-1-C^{14} and glutamate-C^{14}.[34–37] In the earlier of these studies there may or may not have been net synthesis of GSH, and carrier was added in different amounts in order to trap the newly formed GSH. In the later studies[36] a net synthesis was generally demonstrated; however, with or without net synthesis the basic requirements of the system were substantially similar.

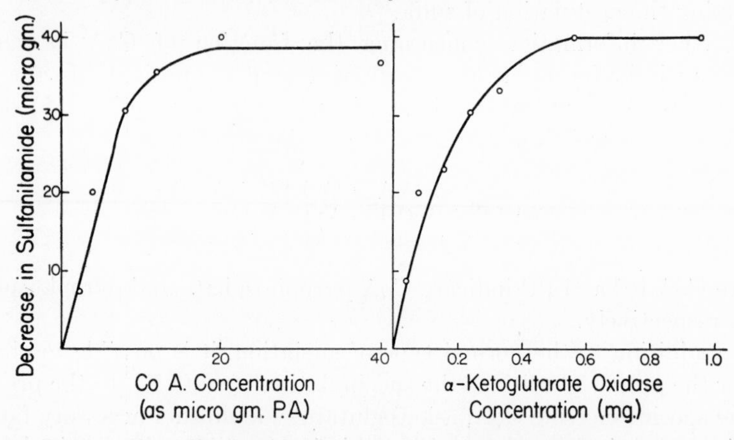

FIG. 1. Decrease in sulfanilamide (increase in acetyl derivative) as a function of CoA and α-ketoglutarate oxidase concentration.[32]

The general procedure used in this work[34–37] was as follows: A preparation of pigeon, or sometimes rat liver, in the form of slices or homogenate (pigeon liver), homogenate fraction, or extract was incubated with labeled amino acid, generally glycine-1-C^{14}. After the incubation period (usually one hour at 37°), the GSH present was isolated in the form of the Cd salt with the aid of carrier, added either before or after the period of incubation. The precipitated material was purified through the cuprous salt.

When partially purified liver extracts were employed and GSH carrier was added prior to incubation, the apparent synthesis was inversely pro-

(33) K. Bloch and H. S. Anker, *J. Biol. Chem.* **169,** 765 (1947).
(34) K. Bloch, *J. Biol. Chem.* **179,** 1245 (1949).
(35) R. B. Johnston and K. Bloch, *J. Biol. Chem.* **179,** 493 (1949).
(36) R. B. Johnston and K. Bloch, *J. Biol. Chem.* **188,** 221 (1951); J. E. Snoke and K. Bloch, *J. Biol. Chem.* **199,** 407 (1952); K. Bloch, J. E. Snoke, and S. Yanari *in* McElroy and Glass, Phosphorus Metabolism, Johns Hopkins Press, Baltimore, 1952, Vol. 2, p. 82.
(37) J. E. Snoke, S. Yanari, and K. Bloch, *J. Biol. Chem.* **201,** 573 (1953).

portional to the amount of carrier added for small amounts of carrier. With larger amounts of carrier the apparent synthesis became independent of the amount added. This shows that in some experiments GSH was being broken down as well as synthesized; not all the GSH synthesized was trapped. The apparent synthesis was calculated from the following relationship:

$$\text{GSH} = \frac{C_2(\text{GSH}_1 + \text{GSH}_2)}{C_1} \tag{19}$$

where C_1 = specific activity of labeled atom in added precursor
C_2 = specific activity of labeled atom in isolated GSH (e.g. glycine-C^{14})
GSH = GSH formed
GSH_1 = GSH originally present in preparation
GSH_2 = GSH added.

In order to avoid possible erroneous interpretations when relationships of this sort are employed, several rather obvious facts should be emphasized: (*a*) if the precursor is present in or is formed in the system, then C_1 is not the specific activity of the precursor; C_1 may undergo a change during the experiment. This can lead to gross errors, for example, if labeled glutamic acid is employed, because in many systems glutamic acid is formed at rapid rates (or else its specific activity becomes rapidly reduced, particularly if the labeling agent is N^{15}, e.g., by transamination); and (*b*) $\text{GSH}_1 + \text{GSH}_2$ may change significantly during the period of the experiment. It would appear that in the particular experiments with which we are dealing these considerations are not of such significance as to alter the interpretation of the results.

In pigeon liver homogenates the GSH formed amounted to 0.2–0.4 mg./g. liver/hr. Braunstein and co-workers[38] found a net synthesis of 1–5 mg./g. rat liver/hr.

The requirements for the synthesis of GSH by pigeon liver homogenate are shown in Table VI.

It will be noted that the conditions under which this synthesis proceeds most rapidly closely resembles those which are optimal for the formation of the hippurates. Either ATP, or metabolites, which are oxidized with the appearance of energy in phosphate bonds, are required for best results; respiratory poisons have the opposite effect. Experiments (6) and (7) show that insufficient glutamic acid and cysteine are present in the homogenate to allow maximum synthetic rates, and indicate further that the added GSH is not broken down at a sufficiently rapid rate to maintain optimum concentrations of these precursors. Interest is added to these results when the depressing effect of ammonium ion is noted. This most probably is due to a competition for the available

(38) A. E. Braunstein, G. A. Shamshikova, and A. L. Jaffe, *Biokhimiya* **13,** 95 (1948); *C. A.* **42,** 8302 (1948).

TABLE VI

T OF METABOLITES AND INHIBITORS ON THE INCORPORATION OF GLYCINE INTO GSH IN PIGEON LIVER HOMOGENATES[34]

mogenization was carried out in the medium: 1.6 g. of homogenate incubated in gen for 1 hour at 37° in 20 ml. of medium with the following composition: carrier GSH 3.8 m*M*, labeled gly 16 m*M*, glu 10 m*M*, cySH 3 m*M*, KCl 30 m*M*, $MgSO_4$ 2.4 m*M*, and phosphate buffer 50 m*M*; final pH 7.4.

Conditions		Concentration, m*M*	Activity in GSH, counts/min.
Control		—	100
Added:	ATP	0.30	313
	Succinate	10	264
	Fumarate	10	250
	Ammonium chloride	10	77
Removed:	Glutamic acid	—	25
	Cysteine	—	33
Added:	Malonate	10	39
	2,4-Dinitrophenol	0.1	105
	2,4-Dinitrophenol	0.4	10

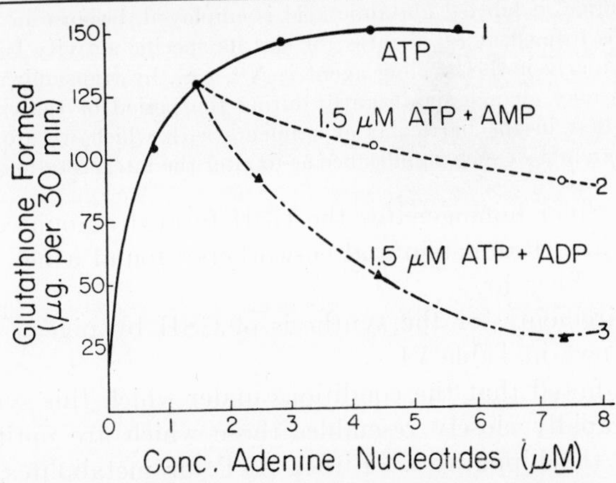

FIG. 2. Glutathione formation as a function of the concentration of nucleotides in the system. Inhibition of synthesis by high concentrations of ATP, ATP + AMP, or ATP + ADP (Bloch, Snoke, and Yanari).[36]

glutamic acid, between the GSH-synthesizing system and that forming glutamine (sec. II-4). In pigeon liver homogenates this system is very active.

As with other anabolizing systems, the generation of ATP *in situ* (or of energy-rich phosphates in general) is more efficient in promoting synthesis than is ATP itself, particularly when the latter is used in high concentrations. Then inhibitory effects supervene, as shown in Fig. 2.

By suitable purification procedures,[37] the system which catalyzes the reaction between cysteine and glutamic acid may be completely removed from the one which catalyzes the coupling of glutamylcysteine with glycine. Coincidentally, the hydrolytic enzymes specific for GSH are eliminated.

The progress of the coupling reaction with a partially purified system in the presence of several possible substrates is shown in Fig. 3.

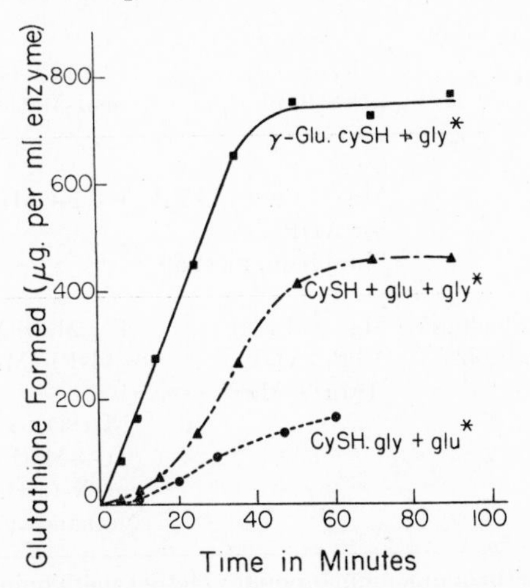

Fig. 3. Glutathione formation as a function of the type of precursor present. Precursors: γ-glutamylcysteine, cysteinylglycine, or free amino acids only. Indicators of synthesis: labeled glycine and glutamic acid (Bloch, Snoke and Yanari).[36] All precursors had their amino acids in L-form.

It is obvious from the results of these particular experiments that the glutamylcysteine-synthesizing component was not removed completely by any means. (No salt fractionation was employed in this case.)

During the synthesis, the liberation of inorganic phosphate is stoichiometrically equivalent to the amount of GSH formed and there is an equivalent fall in ATP and rise in adenosine diphosphate (ADP).[39] No inorganic pyrophosphate appears, and there is no evidence for the involvement of a cofactor such as CoA.

The condensing enzyme (synthetase B) is specific for glycine, insofar as it has been investigated, although at high concentrations there is a reaction in which hydroxylamine replaces glycine.

$$\text{glu.cySH} + NH_2OH + \text{ATP} \xrightarrow{\text{syn.B}} \text{glu.cySH.NHOH} + \text{ADP} + P_i \qquad (20)$$

Arsenate has no effect on the reaction.

(39) J. E. Snoke, *J. Am. Chem. Soc.* **75**, 4872 (1953).

TABLE VII
ENZYMES CONCERNED WITH GLUTAMINE METABOLISM

Enzyme	Source	Activators	Inhibitors	References
Glutaminase (hydrolytic)	Plant, bacteria	Cl, CTM[a]	*p*-Benzoquinone	40,41
		Arsenate, sulfate	Bromosulfalein	42,43
	Kidney, brain			41,42
	Liver glu I	pH 7–8, pyruvate	D- and L-Glutamate	43,44
	Liver glu II	pH 8–9, phosphate	DL-α-Methylglutamate	45,46
Glutamyl transferase	Bacteria	Cu^{++}		43
	Plant	Mn^{++}(Mn^{++}), ATP or ADP	F^-, *p*-CMB[a]	47
	Animal	Phosphate, arsenate		43
Synthetase[b]	Plant, bacteria	Mg^{++}(Mn^{++})	F^-, MetSO,[a] *p*-CMB	43,49
	Liver, brain	ATP, (ADP), (Mn^{++})Mg^{++}	(ADP) (Mn^{++}), F^-,	48.49
			MetSO,[a] *p*-CMB[a]	50
			DL-α-Methylglutamate	45
			DL-N-(α-Glutamyl)-ethanolamine	

[a] CTM, cetyltrimethylammonium bromide; MetSO, methionine sulfoxide; *p*-CMB, *p*-chloromercuribenzoate.

[b] According to Elliott,[51] the transferase enzyme is probably the same as the synthetase because even during extensive purifications the activities run parallel.

(40) C. A. Zittle *in* Sumner and Myrbäck, The Enzymes, Academic Press, New York, 1951, Vol. 1, p. 932.
(41) H. A. Krebs, *Biochem. J.* **29,** 1951 (1935); *ibid.* **43,** 51 (1948).
(42) J. P. Greenstein and F. M. Leuthardt, *Arch. Biochem.* **17,** 105 (1948); M. Errera, *J. Biol. Chem.* **178,** 483 (1949); M. Errera and J. P. Greenstein, *ibid.* **178,** 495 (1949).
(43) H. Waelsch, *Advances in Enzymol.* **13,** 237 (1952); H. Waelsch *in* McElroy and Glass, Phosphorus Metabolism, Johns Hopkins Press, Baltimore, 1952, Vol. 2, p. 109.
(44) A. Meister and S. V. Tice, *J. Biol. Chem.* **187,** 173 (1950); A. Meister, H. A. Sober, S. V. Tice, and P. E. Fraser, *ibid.* **197,** 319 (1952).
(45) N. Lichtenstein, H. E. Ross, and P. P. Cohen, *Nature* **171,** 45 (1953).
(46) N. Lichtenstein, H. E. Ross, and P. P. Cohen, *J. Biol. Chem.* **201,** 117 (1953).
(47) P. K. Stumpf and W. D. Loomis, *Arch. Biochem.* **25,** 451 (1950); P. K. Stumpf, W. D. Loomis, and C. Michelson, *ibid.* **30,** 126 (1951); C. C. Delwiche, W. D. Loomis, and P. K. Stumpf, *Arch. Biochem. Biophys.* **33,** 333 (1951).
(48) J. F. Speck, *J. Biol. Chem.* **179,** 1387, 1405 (1949); *ibid.* **168,** 403 (1947).
(49) W. H. Elliott, *Nature* **161,** 128 (1948); W. H. Elliott and E. F. Gale, *ibid.* **161,** 129 (1948).

More recently, Snoke[39] has prepared the coupling enzyme (synthetase B) in highly purified form from yeast and found that it catalyzes the transfer of phosphate from ATP to ADP. Presumably in this process the enzyme becomes phosphorylated as shown by experiments with labeled ADP.

$$\text{Syn.B} + \text{ATP} \rightarrow \text{Syn.B.ph} + \text{ADP} \quad (21)$$
$$\text{Syn.B.ph} + \text{ADP}^* \rightarrow \text{Syn.B} + \text{ATP}^* \quad (22)$$

Hence, it may be assumed that prior to coupling, the enzyme (or its prosthetic group) becomes phosphorylated. It is also evident that reaction (22) explains the inhibition of the GSH synthesis by ADP (Fig. 2).

The results of experiments in which labeled glutamate was used to investigate the incorporation of glutamic acid indicated the requirements to be similar to those for the reaction just considered. However, since the enzyme(s) required (synthetase A) has not been purified, it cannot be catagorically stated whether the reaction involved is of type (20).

These results of Bloch and co-workers are of the utmost importance because they show:

(*1*) The synthesis of GSH proceeds stepwise through a dipeptide to a tripeptide.

(*2*) The synthetase is not related to a hydrolytic enzyme.

(*3*) The synthesis of a true peptide bond does not require CoA activation.

The first of these conclusions is of paramount importance if it is permissible to extend the analogy further. If no template mechanism is involved in the synthesis of a tripeptide of a very specific structure, then why should one be required for the synthesis of a longer peptide chain?

A mechanism for the synthesis of glutamylcysteine is given in sec. II-4*a*.

4. Synthesis of Glutamine (γ-Amide)

The present picture with respect to the synthesis of glutamine, the γ-amide of glutamic acid, is somewhat confused owing to the existence of a variety of enzymatic activities, which have not been definitely associated with well-characterized entities, and which, in some cases, may be identical. A synopsis of the situation is given in Table VII.

There is a simple hydrolytic enzyme, glutaminase, which breaks down glutamine to ammonia and glutamic acid, although in animal tissues there appear to be two such enzymes with this characteristic, glutaminases

(50) W. H. Elliott, *Biochem. J.* **49,** 106 (1951).
(51) W. H. Elliott, *J. Biol. Chem.* **201,** 661 (1953).

I and II, which differ in their pH optima and in their activation characteristics.[41,42] In addition, there is an enzymatic activity which is referred to as glutamyl transferase. Finally, the enzyme or enzyme system has been described with which we are most immediately concerned, namely the conjugating enzyme for glutamine (glutamine synthetase).

The latter enzyme differs from the glutamyl transferase in the activator requirements. However, there is no concordance among several workers in the field as to the activator for the transferase nor in the interpretation of the data (see Table VII and footnote).

The rapid synthesis of glutamine *in vitro* was first described by Krebs,[41] as occurring in a wide variety of animal tissues, with the exception of liver. However, liver homogenates or extracts of acetone powder are capable of glutamine synthesis as shown by Leuthardt and co-workers,[52] by Speck,[48] and by Elliott,[50] the disagreement between the results being due to the lack of penetration of glutamate into the cells.[53,54] Glutamine is also synthesized by a purified extract from green peas.[51] The synthesis depends on the maintenance of respiration,[48] and is stimulated in homogenates by ATP and Mg ion.[48,50,52] There is inhibition by fluoride ion (6.4×10^{-2} M).[52] There is a nonspecific inhibition by pyruvate which is due either to removal of glutamate by transamination or of glutamine by the activation of glutaminase II.[54]

With acetone powders of pigeon liver[48] or of sheep brain,[50] synthesis closely resembles that involved in the formation of GSH from glycine and glutamylcysteine; the preparation of the enzymatic extract is similar, activation effects with ATP and Mg ion bear a close resemblance to each other, phosphate equivalent to glutamine formed is released from ATP, K ion is required, and under the proper conditions synthesis proceeds anaerobically. Fluoride ion is inhibitory, so although most of these effects were discovered in relation to glutamine synthesis, further details will not be given here. No CoA is needed for glutamine synthesis.[48]

The investigation of glutamine synthesis was greatly facilitated by the discovery that ammonia could be replaced by hydroxylamine,[48,50] with the formation of a hydroxamic acid which may be determined photometrically. The reaction in this case may be written:

$$\text{glu} + \text{ATP} + \text{NH}_2\text{OH} \xrightarrow[\text{synthetase}]{\text{glu-NH}_2} \text{glu-NHOH} + \text{ADP} + \text{P}_i \qquad (23)$$

(52) F. Leuthardt and E. Bujard, *Helv. Med. Acta* **14,** 274 (1947); J. Frei and F. Leuthardt, *Helv. Chim. Acta* **32,** 1137 (1949).

(53) P. P. Cohen and M. Hayano, *J. Biol. Chem.* **166,** 239 (1946).

(54) P. P. Cohen *in* Sumner and Myrbäck, The Enzymes, Academic Press, New York, 1952, Vol. 2, p. 897.

Hydrazine may be substituted for hydroxylamine but not other organic bases. However, it was found by Speck[48] that L-cysteine would serve as an acceptor, or rather that it catalyzed the release of inorganic phosphate from ATP in the presence of glutamine and the enzyme. The reaction proceeded at about one third the rate of the reaction with ammonia. This is interesting because the product of the reaction should be glutamylcysteine, the precursor required for the second step in the synthesis of GSH (sec. II-3).

5. Glutamyl Transfer; Transamidation

a. γ-Glutamyl Transfer (from Residues Other Than Those of Amino Acids)

During the last several years there has been described an enzyme-catalyzed transfer of the glutamyl group of L-glutamine to hydroxylamine or to other ammonia molecules (isotope studies), a reaction which presumably occurs as follows:

$$\text{glu-NH}_2 + \text{H}_2\text{NR} \xrightarrow{\text{transferase}} \text{glu-NHR} + \text{NH}_3 \tag{24}$$

where R = H, OH, or CH_3.

Stumpf and co-workers[47] and Waelsch and co-workers[43] have investigated the transfer reaction as it occurs in plants and in animals and bacteria, respectively. Unlike the glutamine synthesis, the transfer requires no exergonic coupling.[43] The enzyme in bacteria (*Proteus vulgaris*) can react with asparagine, although the analogous one in plants (pumpkin seedlings) and animal tissues is specific for glutamine.[55] There is no catalysis of reaction between benzamide and hydroxylamine, as observed in liver extracts by Virtanen and Berg,[21] and other amides are also likewise inactive.[43] Cupric ion activates the bacterial enzyme, whereas Mn ion is required for the plant enzyme. Cupric ion has no effect on glutaminase, and since purification of bacterial transferase also led to a change in ratio between glutaminase and transferase activity, and in the same preparation there was no glutamine synthetase activity, it appears that these three enzymes are different. Also, synthetase but not glutaminase is present in extracts of acetone powders from both plant and animal tissues (compare Elliott[51]).

The plant enzyme[47] and the enzyme occurring in *Neurospora crassa*[43] and in animal tissues[43] require, in addition to the factors already considered, phosphate or arsenate; after dialysis, a requirement for ATP or ADP makes its appearance.

(55) N. Grossowicz, E. Wainfan, E. Borek, and H. Waelsch, *J. Biol. Chem.* **187,** 1 (1950).

b. γ-Glutamyl Transfer (from True Peptide Derivatives)[55a]

It has been found by Hanes and co-workers[56] that extracts of mammalian kidney and pancreas catalyze the transfer of the glutamyl group from GSH to a number of different amino acids such as valine and phenylalanine, to form new peptides, e.g.,

$$\text{glu.cySH.gly} + \text{phe} \rightarrow \text{glu.phe} + \text{cySH.gly} \quad (25)$$

More recently it has been shown that GSH is not the sole donor but that other glutamyl peptides can serve in the same capacity[57] (Table VIII).

TABLE VIII
EXAMPLES OF γ-GLUTAMYL GROUP TRANSFER[57]

Donor[a]	Acceptor[b]	New Peptide[c]
GSH	phe	glu.phe
glu.glu	phe	glu.phe
glu.glu	gly	glu.gly
glu gly	phe	glu.phe
glu.Ph	gly	glu.gly

[a] The donors used were either γ glutamyl dipeptides, GSH or GSSG.
[b] Of the 11 amino acids tested, 9 acted as acceptors, arginine and proline failing.
[c] New peptides were isolated on chromatograms and identified by comparison and hydrolysis.

The relationship of these reactions to the glutamyl transfers discussed in the previous section is obvious. These reactions all follow the type:

$$\text{GSH} + \text{glu} \rightarrow \gamma\text{-glu.glu} + \text{cySH.gly} \quad (26)$$

Hendler and Greenberg[58] were unable to demonstrate transfers in similar systems, perhaps because cofactors were eliminated in the preparation of the materials used. If it proves possible to demonstrate the conversion of a γ-peptide into an α-peptide, perhaps by ring formation, the γ-glutamyl transfer reactions will assume greater importance; but since it is unlikely that the energy in a peptide bond at one end of a molecule can be shifted down to the other to form a new bond with rupture of the old, reactions of this type and of the type shown below are somewhat improbable:

$$\text{glu.gly} + \text{gly} \rightarrow \text{glu.} + \text{gly.gly} \quad (27)$$

(55a) See Chap. 25.
(56) C. S. Hanes, F. J. R. Hird and F. A. Isherwood, *Nature* **166,** 288 (1950); *Biochem. J.* **51,** 25 (1952).
(57) C. S. Hanes, G. E. Connell, and G. H. Dixon *in* McElroy and Glass, Phosphorus Metabolism, Johns Hopkins Press, Baltimore, 1952, Vol. 2, p. 95.
(58) R. W. Hendler and D. M. Greenberg, *Nature* **170,** 123 (1952).

c. *Transamidation*[55a]

The phenomenon of transamidation appears first to have been observed by Bergmann and Fraenkel-Conrat[59] in the reaction:

$$\text{Bz.gly-NH}_2 + \text{Ph-NH}_2 \xrightarrow{\text{papain}} \text{Bz.gly-NH-Ph} + \text{NH}_3 \tag{28}$$

It was postulated that the reaction occurred directly, without hydrolysis to hippuric acid, to form the less soluble anilide, because there was a more rapid formation of the anilide from hippurylamide than from hippuric acid.

TABLE IX
EXAMPLES OF TRANSAMIDATION

Enzyme	Donor	Acceptor	New Peptide	Ref.
Papain	cbz.gly-NH_2	L-phe-NH_2	cbz.gly-L-phe-NH_2	59
Liver extract	Bz-NH_2	NH_2OH	Bz-NHOH	21
Cathepsin C (beef spleen)	gly-L-phe-NH_2	L-α-glu-NH_2	gly-L-phe-L-α-glu-NH_2	60,61
Cathepsin B (beef spleen)	Bz-L-arg-NH_2	NH_2OH	Bz-L-arg-NHOH	62,63
Glycyl enzyme (cabbage leaves)	gly.gly	L-met	gly-L-met	56
Chymotrypsin (C)	Bz-L-tyr-NH_2	$N^{15}H_3$	Bz-L-tyr-$N^{15}H_2$	63
Chymotrypsin	L-met-O-iso-Pr[a]	L-met	L-met-L-met	64
C + trypsin	L-lys-L-tyr-L-lys	(L-lys)	L-lys-L-lys	65

[a] iso-Pr = isopropyl.

Since this observation was made, many other reactions involving similar amide transfers have been observed with a wide variety of proteolytic enzymes, and the phenomenon has assumed considerable importance.[66,67] Examples of various types are given in Table IX.

(59) M. Bergmann and H. Fraenkel-Conrat, *J. Biol. Chem.* **119,** 707 (1937).
(60) M. E. Jones, W. R. Hearn, M. Fried, and J. S. Fruton, *J. Biol. Chem.* **195,** 645 (1952).
(61) Y. P. Dowmont and J. S. Fruton, *J. Biol. Chem.* **197,** 271 (1952).
(62) R. B. Johnston, M. J. Mycek, and J. S. Fruton, *J. Biol. Chem.* **185,** 629 (1950).
(63) R. B. Johnston, M. J. Mycek, and J. S. Fruton, *J. Biol. Chem.* **187,** 205 (1950).
(64) M. von Brenner, H. R. Müller, and R. W. Pfister, *Helv. Chim. Acta* **33,** 568 (1950); *ibid.* **35,** 217 (1952).
(65) S. G. Waley and J. Watson, *Nature* **167,** 360 (1951).
(66) S. W. Fox and M. Winitz, *Arch. Biochem.* **35,** 419 (1952).
(67) F. Janssen, M. Winitz, and S. W. Fox, *J. Am. Chem. Soc.* **75,** 704 (1953).

It will be noted that one of the transfers, that observed by von Brenner and co-workers,[64] is of a mixed type in that the donor is an ester rather than an amide. The possibility that such reactions as this might proceed is evident from the work of Schwert, Neurath, Kaufman, and Snoke[68] who showed that trypsin catalyzes the hydrolysis of the methyl ester of α-benzoyl-L-arginine. Von Brenner and co-workers[64] noted similar reactions to the one shown in Table IX with other esters of methionine and with the same ester of L-threonine, also with the esters of D-methionine. In all cases several products, dipeptides and higher polymers as well as peptide esters, were found in the reaction mixtures.

It is noteworthy that reaction (28) comes to equilibrium from equimolar concentrations when it reaches to within 6 per cent of completion.[69] From this it may be calculated, according to Borsook and Deasy,[70] that the ΔF is -5 kcal. at 37°, thus differing greatly from that for hippuric acid formation of approximately 3 kcal.

At the present time none of these reactions has been investigated to a sufficient extent to support any claim as to their importance or otherwise in the biological sphere. It is possible that by the proper choice of conditions, greater transfers than have yet been observed in soluble systems will be demonstrated.

In 1939 Behrens and Bergmann[71] discovered the so-called cosubstrate effect. It was found, for instance, that although gly-L-leu did not hydrolyze in the presence of papain, when Ac-DL-phe.gly was also present hydrolysis occurred. The assumption was made that there was a tetrapeptide intermediate formed and that this tetrapeptide broke down stepwise to give in turn leucine and glycine. From what has preceded, it is more likely that a transfer is involved, as suggested by Waelsch.[43]

$$\text{Ac-DL-phe.gly} + \text{gly-L-leu} \rightarrow \text{Ac-DL-phe.gly-L-leu} + \text{gly} \quad (29)$$
$$\text{Ac-DL-phe.gly-L-leu} \rightarrow \text{Ac-DL-phe.gly} + \text{L-leu} \quad (30)$$

As yet there is no proof of either mechanism.

6. Synthesis of Peptides

Examples of peptide synthesis have been mentioned in sec II-5; however, the catalysis of the synthesis of insoluble peptides by proteolytic

(68) G. W. Schwert, H. Neurath, S. Kaufman, and J. E. Snoke, *J. Biol. Chem.* **172,** 221 (1948).
(69) E. Waldschmidt-Leitz and K. Kühn, *Z. physiol. Chem.* **285,** 23 (1950).
(70) H. Borsook and C. L. Deasy, *Ann. Rev. Biochem.* **20,** 209 (1951).
(71) O. K. Behrens and M. Bergmann, *J. Biol. Chem.* **129,** 587 (1936).

enzymes was discovered by Bergmann and co-workers[72] about the same time the transamidation reaction was uncovered. The following example may be given:

$$\text{Bz.leu} + \text{leu-NHPh} \rightarrow \text{Bz.leu.leu-NHPh} \tag{31}$$

Since then a large number of peptides have been synthesized by similar types of reactions. It is evident that these syntheses occur because of product insolubility, but it remains to be shown that they are of significance in any living system. Further information concerning the factors concerned in the operation of such synthetic systems as these will be found in publications of Fox and collaborators[66] and others,[69,73] to which the reader is referred for further insight into the literature.

It is possible that the reactions which have led to synthesis of peptide bonds in the hands of Chantrenne[74] are of more significance from the biosynthetic point of view because in this case the driving force is not that provided by the insoluble nature of the products but is rather that due to differences in the energy content of the bond ruptured and the bond formed. This is shown in the following example:

$$\begin{array}{l}\text{CH}_2\text{NHcbz} \\ | \\ \text{COOPO}_3\text{HPh}\end{array} + \text{gly} \rightarrow \begin{array}{l}\text{CH}_2\text{NHcbz} \\ | \\ \text{COgly}\end{array} + \text{PhH}_2\text{PO}_4 \tag{32}$$

Unfortunately, since no enzymatic intervention is required in these syntheses, they will remain type reactions of questionable import. At least the Bergmann–Fraenkel-Conrat–Fruton type reactions require this intervention, and in general the products are predictable from the known specifications of the hydrolytic enzymes involved. (See, however, Refs. 66 and 67.)

Note should be taken of the change of peptide AB into BA, observed by Sanger and Thompson.[75]

7. Conclusions

From the thermodynamic point of view it is necessary to conclude that the net synthesis of bonds to form small peptides occurs only when energy is supplied by coupling with some exergonic reaction. The formation of γ-glutamyl peptides may be exceptional. However, the coupling of long peptides may occur without such a large endergonic requirement, so that,

(72) M. Bergmann, J. S. Fruton, and H. Fraenkel-Conrat, *J. Biol. Chem.* **119,** 35 (1937); M. Bergmann and H. Fraenkel-Conrat, *J. Biol. Chem.* **124,** 1 (1938); M. Bergmann, *Ann. N. Y. Acad. Sci.* **45,** 409 (1944).

(73) E. Waldschmidt-Leitz and K. Kuhn, *Ber.* **84,** 381 (1951).

(74) H. Chantrenne, *Pubbl. staz. zool. Napoli (Suppl.)* **23,** 1 (1951); *Biochim. et Biophys. Acta* **4,** 484 (1950); *Nature* **164,** 576 (1949).

(75) F. Sanger and E. O. P. Thompson, *Biochim. et Biophys. Acta* **9,** 225 (1952).

pending clarification of the energetic relationships, little of a more definite nature can be said concerning such couplings.

It appears also that transamidation may be of considerable significance, but transamidation from primary peptide amides with the consequent release of ammonia is open to question till such time as the existence of such primary amides in biological systems is demonstrated. Further, unless conditions are found such that, in the presence of a proteolytic enzyme, the velocity of the transamidation is favored above that of the purely hydrolytic reaction, the significance of the general phenomenon will rest uncertain. The importance of γ-glutamyl transfers is open to serious question because of the obscure relationship between the synthesis of α- and γ-peptide bonds and the questionable interconvertability of such bonds.

Granting that the primary synthesis of peptide bonds requires an energetic coupling, the mechanism of the coupling remains obscure. For the synthesis of the pseudo-peptide bond in which an amino acid is acylated with a simple carboxylic acid, it appears, from the investigations with hippuric acid synthesis, that CoA is definitely involved. However, for the synthesis of GSH and glutamine no CoA coupling appears to be obligatory, and ATP is the effecting agent. This leaves the precise mechanism in a penumbra (otherwise known as a smog).

III. Synthesis of Protein and Incorporation of Isotopic Amino Acids *in Vitro*

1. Reversal of the Proteolytic Reaction

Starting with the work of Danilewski and others (reviewed by Wastenays and Borsook)[76] and continuing with Virtanen,[77] Tauber,[78] and Bresler and co-workers,[79] to mention a few of the more recent endeavors, attempts have been made to reverse the action of proteolytic enzymes by changes in the reaction system composed of proteolytic enzyme–partial digestion products. The equilibrium may be shifted toward synthesis by changing the concentration of the water, by decreasing the charge on the amino groups, or by increasing the pressure on the system.

(76) H. Wastenays and H. Borsook, *Physiol. Revs.* **10,** 110 (1930).
(77) A. I. Virtanen, *Makromol. Chem.* **6,** 94 (1951); A. I. Virtanen, H. Herkkonen, M. Hakala, and T. Laaksonen, *Naturwissenschaften* **37,** 139 (1950); F. R. Williams and A. I. Virtanen, *Acta Chem. Scand.* **7,** 285 (1953).
(78) H. T. Tauber, *J. Am. Chem. Soc.* **73,** 4965 (1951).
(79) S. E. Bresler, M. V. Glikima, and A. M. Tongur, *Doklady Akad. Nauk S.S.S.R.* **78,** 543 (1951); *C. A.* **45,** 10273 (1951); S. E. Bresler and N. A. Selezneva, *Doklady Akad. Nauk S.S.S.R.* **84,** 1013 (1952); *C. A.* **46,** 10228 (1953).

The insoluble materials which are formed in concentrated solutions containing peptides have generally been referred to as plasteins; but these substances, though they may be polymers of amino acids, do not possess the physical and biological properties of the original proteins from which they were derived. In many of the earlier investigations, the insolubility of the products did not solely arise from the formation of peptide bonds. Rather, disulfide bond formation occurred, with the resultant formation of large from small peptides.[80,81] It is possible that in some other cases esters and other types of linkages were also formed.

It is apparent that nonspecific polymers may be formed depending on the reactivities and concentrations of the various free carboxyl, amino, sulfhydryl, and other groups present in the system. Nevertheless, it has been claimed by Bresler and co-workers[79] that when proteolysis is interrupted at an early stage and the mixture is brought to a proper pH on the alkaline side of neutrality, then, in the presence of sucrose, the application of a pressure of 6000 atm. for some hours will result in the synthesis of significant amounts of specific protein. Obviously, the probability of reattaining the original structural configuration of the protein is very low unless the degree of hydrolysis is extremely small or the protein split products act as a template.[82] Consequently, it is very difficult to comprehend positive results which have been confirmed by immunological and electrophoretic methods, and by the return of specific biological activity (insulin resynthesis), as well as by the less specific demonstrations of decrease in free amino nitrogen, etc.[79]

2. Incorporation of Isotopes in Vitro

It was not until the advent of suitable isotopes, such as the radioactive sulfur and carbon, that it became possible to approach the problem of the synthesis of protein *in vitro* any more directly than has been indicated in the previous sections. Soon after sulfur with high specific activity became available, this element in methionine and cystine became

(80) H. H. Strain and K. Linderstrøm-Lang, *Enzymologia* **5**, 86 (1938).

(81) M. E. Maver and C. Voegtlin, *Enzymologia* **6**, 219 (1939).

(82) In this connection it is of interest to note that the action of carboxypeptidase on Fraction B of insulin results in the removal of alanine from its terminal position.[83] This reaction occurs at a rate ten times as fast as the removal of asparagine, the amino acid which appears in next highest concentration to the alanine. The product after the removal of the terminal alanine from insulin retains the original biological activity of the hormone.[84] Under the proper experimental conditions, therefore, it may be possible to both remove and replace the alanine. This reaction is under investigation.

(83) J. I. Harris, *J. Am. Chem. Soc.* **74**, 2944 (1952).

(84) J. I. Harris and C. H. Li, *J. Am. Chem. Soc.* **74**, 2945 (1952).

the first tool used to demonstrate protein synthesis in slices.[85] Work with C^{14}-labeled amino acids followed very shortly thereafter,[86–88] and it became possible to demonstrate the incorporation of isotopes into protein from precursors other than amino acids. Thus an incorporation of labeled carbon dioxide into rat liver proteins via the dicarboxylic amino acids[89,90] and of acetate into the proteins of liver[90] and bone marrow[91] proved to be demonstrable.

During this early period, investigations were made of the synthesis of protein in tumor[92–95] and embryonic tissues,[92,95] as well as in a variety of other tissues such as bone-marrow cells,[96] intestinal sections,[87] diaphragm,[97] and nonproliferating bacteria.[98] Homogenates of liver and other tissues also were employed in order to simplify the system.[88,95,99–101] In addition, the effects of inhibitors[87,94,96,102,103] and activators[95,104]

(85) J. B. Melchior, Dissertation, University of California, Berkeley, 1946; J. B. Melchior and H. Tarver, *Arch. Biochem.* **12,** 309 (1947).

(86) I. D. Frantz, Jr., R. B. Loftfield, and W. W. Miller, *Science* **106,** 544 (1947).

(87) T. Winnick, F. Friedberg, and D. M. Greenberg, *Arch. Biochem.* **15,** 160 (1947).

(88) F. Friedberg, T. Winnick and D. M. Greenberg, *J. Biol. Chem.* **171,** 441 (1947).

(89) C. B. Anfinsen, A. Beloff, A. B. Hastings, and A. K. Solomon, *J. Biol. Chem.* **168,** 771 (1947).

(90) C. B. Anfinsen, A. Beloff, and A. K. Solomon, *J. Biol. Chem.* **179,** 1001 (1949).

(91) R. Abrams, J. M. Goldinger, and E. S. G. Barron, *Biochim. et Biophys. Acta* **5,** 74 (1950).

(92) P. C. Zamecnik, I. D. Frantz, Jr., R. B. Loftfield, and M. L. Stephenson, *J. Biol. Chem.* **175,** 299 (1948).

(93) E. Farber, S. Kit, and D. M. Greenberg, *Cancer Research* **11,** 490 (1951).

(94) S. Kit and D. M. Greenberg, *Cancer Research* **11,** 495, 500 (1951).

(95) T. Winnick, *Arch. Biochem.* **27,** 65; **28,** 338 (1950).

(96) H. Borsook, C. L. Deasy, A. J. Haagen-Smit, G. Keighley, and P. H. Lowy, *J. Biol. Chem.* **186,** 297 (1950).

(97) H. Borsook, C. L. Deasy, A. J. Haagen-Smit, G. Keighley, and P. H. Lowy, *J. Biol. Chem.* **186,** 309 (1950).

(98) J. B. Melchior, M. Mellody, and I. M. Klotz, *J. Biol. Chem.* **174,** 81 (1948); J. B. Melchior, O. Klioze, and I. M. Klotz, *J. Biol. Chem.* **189,** 411 (1951).

(99) T. Winnick, I. Moring-Claesson, and D. M. Greenberg, *J. Biol. Chem.* **17,** 441 (1947); T. Winnick, F. Friedberg, and D. M. Greenberg, *J. Biol. Chem.* **175,** 117 (1948); T. Winnick, I. Moring-Claesson, and D. M. Greenberg, *J. Biol. Chem.* **175,** 127 (1948).

(100) H. Borsook, C. L. Deasy, A. J. Haagen-Smit, G. Keighley, and P. H. Lowy, *J. Biol. Chem.* **179,** 689 (1949); R. Schwett and H. Borsook, *Federation Proc.* **12,** 266 (1953).

(101) H. Borsook, C. L. Deasy, A. J. Haagen-Smit, G. Keighley, and P. H. Lowy, *J. Biol. Chem.* **184,** 529 (1950).

(102) I. D. Frantz, Jr., P. C. Zamecnik, J. W. Reese, and M. L. Stephenson, *J. Biol. Chem.* **174,** 773 (1948).

(103) M. V. Simpson and H. Tarver, *Arch. Biochem.* **25,** 384 (1950).

became the foci of specific investigations. The uptake into liver slices from different strains of rats was also examined.[105]

These investigations have been reviewed by several authors.[106–111] The purpose of the following discussion will be to indicate some of the difficulties encountered and to make an appraisal of this type of experimentation.

a. The Nature of the Reaction(s) Involved—Experimental Complications

Most workers in this field have carried out tissue slice or homogenate experiments employing the usual techniques. At the beginning of the experiments labeled amino acids, acetate, or bicarbonate with high specific activity have been added to the medium, then after a suitable period of time, the incubation has been stopped, and the protein in the preparation has been precipitated, washed, and its radioactivity determined. In some cases the protein has been hydrolyzed and specific amino acids isolated and their activities or the activities of their specific carbon atoms have been measured.

It is quite necessary, first of all, to demonstrate that the appearance of radioactivity in the protein is actually due to the formation of peptide bonds, that the radioactive material is not simply adsorbed on the protein, and that no bonds other than peptide are involved. This is easier said than done. The first step in this direction necessitates the development of an adequate washing procedure for the protein. The most satisfactory one for most purposes involves the employment of several changes of trichloroacetic acid (TCA), cold and hot, followed by organic solvents to remove lipide. (The formation of labeled lipide material occurs quite freely when glycine is used in liver preparations.[112])

(104) E. A. Peterson, T. Winnick, and D. M. Greenberg, *J. Am. Chem. Soc.* **73,** 503 (1951).
(105) R. Rutman, E. Dempster, and H. Tarver, *J. Biol. Chem.* **177,** 491 (1949); R. J. Rutman, *Genetics* **36,** 54 (1951).
(106) H. Borsook, C. L. Deasy, A. J. Haagen-Smit, G. Keighley, and P. H. Lowy, *Federation Proc.* **8,** 589 (1949); H. Borsook, *Physiol. Revs.* **30,** 206 (1950).
(107) H. Borsook, *Fortschr. Chem. org. Naturstoffe* **9,** 292 (1952); H. Borsook and C. L. Deasy, *in* Bourne and Kidder, Biochemistry and Physiology of Nutrition, Academic Press, New York, 1953, Vol. 1, p. 188; H. Borsook, *in* Advances in Protein Chemistry, Academic Press, New York, 1953, Vol. 8, p. 128.
(108) D. M. Greenberg, F. Friedberg, M. P. Schulman, and T. Winnick, *Cold Spring Harbor Symposia Quant. Biol.* **13,** 113 (1948).
(109) P. C. Zamecnik and I. D. Frantz, Jr., *Cold Spring Harbor Symposia Quant. Biol.* **14,** 199 (1949).
(110) I. D. Frantz, Jr. and P. C. Zamecnik, *in* Youmans, Symposia on Nutrition, II, Plasma Proteins, Thomas Publ. Co., Springfield, 1950, p. 94.
(111) H. Tarver, *Ann. Rev. Biochem.* **21,** 301 (1952).
(112) T. Winnick, E. A. Peterson, and D. M. Greenberg, *Arch. Biochem.* **21,** 235 (1949).

Such washed protein preparations from incubations with carboxyl-labeled amino acids contain little of any radioactivity which can be liberated by treatment with ninhydrin.[95,113] Moreover, when the labeled protein is successively treated with a series of proteolytic enzymes, pepsin, trypsin–chymotrypsin, carboxypeptidase, and erepsin, the amount of radioactivity which is released by ninhydrin gradually increases.[112,113] Similarly, it has been found that almost none of the radioactive glycine, leucine, or lysine incorporated by the protein of rabbit reticulocytes had its amino group free.[114]

When cobaltous ion is present, the uptake of a large part of the radioactivity from L-leucine and glycine by a sedimented fraction of a guinea pig liver homogenate is apparently due to the formation of a cobalt–amino acid–protein complex because the ninhydrin treatment releases most of the activity.[106] In the absence of cobalt, no such release occurs.

Thus the indications are that the TCA washing procedure quite efficiently removes free amino acid. However, under certain conditions difficulties may arise. For instance, Winnick[95] found that when the TCA-washed proteins from incubations of homogenates of the livers from adult rats with glycine-1-C^{14} were treated at 26–28° for 2 hours with 0.5 N sodium hydroxide, the reprecipitated protein had lost a great deal of its original radioactivity. Such losses did not occur when alanine-1-C^{14} was the labeling agent.

Another type of difficulty may arise owing to the binding of labeled materials by non-peptide bonds. This is particularly well illustrated by the early results of Melchior and Tarver[115] with DL-cystine-S^{35}. Cystine was taken up by slices, probably by disulfide-bond formation, so that it became difficult to measure the relatively small fraction of the incorporation due to peptide-bond synthesis. In this particular case it has proved feasible to wash out such cystine by treating the slice protein with a reducing agent, the most useful of which is 2-mercaptoethanol.[104,116] This particular type of interference assumes additional importance because methionine prepared by synthetic methods may be contaminated with homocystine,[117] and it appears that there may be traces of the same contaminant in this amino acid prepared by microbiological methods.[118]

Furthermore, it is possible for tissue preparations to form labeled products from the amino acid used. These products may be bound by the protein. Such interference is exemplified by the results of Canellakis and Tarver,[119] who found that, with a mitochondrial preparation from the liver of the rat, methionine is broken down to form methyl mercaptan which binds to the protein. With liver slices this difficulty does not arise.

It has been assumed that if activity is removed by reductive treatment, then that part removed is of little significance for the incorporation process concerned. However, this is not necessarily the case. It is possible that fragments of a protein

(113) E. A. Peterson and D. M. Greenberg, *J. Biol. Chem.* **194,** 359 (1952).
(114) H. Borsook, C. L. Deasy, A. J. Haagen-Smit, G. Keighley, and P. H. Lowy, *J. Biol. Chem.* **196,** 669 (1952).
(115) J. B. Melchior and H. Tarver, *Arch. Biochem.* **12,** 301 (1947).
(116) H. Tarver, *in* F. W. Hoffbauer, Liver Injury, *Trans. 7th Conf. Josiah Macy, Jr. Foundation,* 1948.
(117) H. Tarver, *Advances in Biol. and Med. Phys.* **2,** 281 (1951).
(118) M. Tabachnick, Dissertation, University of California, Berkeley, 1953.
(119) E. S. Canellakis and H. Tarver, *Arch. Biochem.* **42,** 387 (1953).

molecule are removed, in some cases, by this type of treatment and that the part removed may be of great importance, since it is possible to consider the formation of peptides which are thus bound to a protein as a mechanism for the stabilization of such peptides or of anchoring such peptides to template structures.

Another necessary requirement when amino acids like lysine, glutamic acid, and aspartic acid are used is to show that the bond formed involves either the α-amino or α-carboxyl groups, assuming that in all the proteins occurring in the system these amino acids are bound in this fashion only. This becomes obligatory when it is found, as in the case of the lysine-ϵ-C^{14} experiments of Borsook and co-workers,[100,101] that the amino acid behaves in an apparently anomalous fashion (see sec. III-2*g*).

It is also evident that it is advisable to employ amino acids which have been resolved into the D- or L-components. The difficulties of interpretation which otherwise may arise are illustrated by those encountered by Simpson and Tarver[103] in experiments with DL-methionine-S^{35}, diluted with either D- or L-methionine. As the concentration of the unlabeled amino acid was increased, the increase in rate of uptake of amino acid was less when the D-amino acid was used as a diluent. Although this result was taken to indicate an inhibition of uptake of L-methionine by D-methionine, the precise interpretation of the data is difficult because nothing is known concerning the rate of conversion of D- to L-amino acid in the system, the effect of keto acid, and the rate of metabolism of the L-amino acid.

From many of the labeled amino acids which have been used there may be formed others also strongly labeled; for instance, from methionine-S^{35}, cystine-S^{35} may arise; and from glycine-C^{14}, serine-C^{14}. Consequently, if in experiments carried out with such amino acids under different conditions, the nonspecific uptake of radioactivity alone is measured, comparison of the results may be of little value because the different conditions may have had different effects on the secondary transformations, and hence on the total uptake observed. Unfortunately, many workers completely ignore these possibilities.

b. The Incorporation Rate in Various Preparations

In this section and elsewhere *the incorporation of amino acids will be expressed generally in terms of micromoles of amino acid per gram of protein, the calculations being made on the assumption that there is no significant dilution of the labeled amino acid added by amino acid already in the preparation, and that the specific activity remains constant over the period of the experiment.* Calculated in this way, the incorporation rates of DL-methionine-S^{35} observed by Melchior and Tarver[85] with liver slices from rats varied between 0.3 and 2.4 micromoles/g. in 2 hours.[120] In later experiments, with lower concentrations of methionine, incorporations of 0.4–0.6 micromoles/g. was observed;[103] and with nonproliferating *Escherichia coli*, an incorporation of 3.6 micromoles of sulfur per gram of protein was observed in a 6-hour period.[98,121] Zamecnik and co-workers,[92] using DL-alanine-1-C^{14} as labeling agent, found incorporation of the order of 1.7 micromoles/g. of protein in 3.5 hours (alanine, 4.5 micromoles/ml. medium). The results obtained by Borsook and co-workers, who used several different tissues, bone marrow from rabbits,[96] diaphragm from rats,[97] and reticulocytes from rabbits,[114] with a variety of labeled amino acids, are summarized in Table X.

TABLE X

INCORPORATION OF LABELED AMINO ACIDS INTO PROTEIN OF DIFFERENT TISSUES[96,97,114]

Tissues incubated in Krebs'–bicarbonate under 95% oxygen–5% carbon dioxide at 37° for 1 hour. Labeled amino acid, 1 micromole/ml.

	Rat diaphragm[a]	Rabbit[b] bone marrow[a]	Rabbit reticulocytes[a]
Glycine	0.08	1.1	0.78
L-Histidine[c]	—	—	1.08
L-Leucine	0.11	2.0	0.70
L-Lysine	0.12	1.6	0.99

[a] Data in micromoles per gram protein per hour. The incorporation figures include all the labeled material taken up, i.e., as the amino acid presented and as derived amino acid. In the case of glycine and bone marrow probably only about 27 per cent of the incorporation was due to glycine as such. With leucine and lysine, the corresponding figures were 75 and 87 per cent.[96]

[b] Data taken from Fig. 2.[96]

[c] It has also been demonstrated that the L-histidine-N^{15}-imidazole is taken up by the protein of the reticulocytes from the duck [D. Shemin, I. London, and D. Rittenberg, *J. Biol. Chem.* **183,** 757 (1950)].

From the figures and those mentioned in the preceding paragraph it is evident that each tissue must have a different rate of amino acid incorporation and that the rates are individual for each amino acid when they

(120) The conditions in these experiments were as follows: Krebs-bicarbonate solution under 95 per cent oxygen and 5 per cent CO_2; methionine concn., 2.6–11 m*M*. The data of Table III[85] were utilized, and it has been assumed that the livers contained 17 per cent protein. The figures are for methionine incorporation only (cystine not included).

(121) In the experiments with slices (Table III),[103] the DL-methionine concn. was 0.27 m*M*, in those with *E. coli* (Table I),[98] 0.08 m*M*. The assumption is made that the liver slices contained 36 micromoles of methionine in the form of protein per gram net weight, and that in the *E. coli* experiments 8 mg. of protein was present.

are applied in equal concentrations. It is also evident from the work of Zamecnik and co-workers[92,109] that the rates of incorporation by different types of liver from the same species may vary greatly, as shown (see also Rutman *et al.*)[105] in Table XI.

TABLE XI
RELATIVE RATES OF INCORPORATION OF ALANINE INTO VARIOUS TYPES OF LIVER SLICES FROM RATS[109]

Slices incubated in Krebs'–phosphate under 100% oxygen at 37° for 3.5 hours; DL-alanine-1-C^{14} concentration, 4.5 micromoles/ml. Hepatoma induced with *p*-dimethylaminoazobenzene.

	Rate[a]		Rate
Normal control liver	38	Fetal liver	179
Hepatoma	255	Nonmalignant part of liver with hepatoma	91
Regenerating liver	91		

[a] In these experiments nearly all of the incorporation is due to the alanine itself.[109] Rates in arbitrary units.

Similar observations have been made with homogenates of fetal and adult liver tissue.[95] However, it must not be concluded from these data that tumor tissues in general show higher rates of incorporation than similar or homologous normal tissues; indeed, the opposite situation may sometimes prevail. For instance, Kit and Greenberg[94] found the incorporation of glycine-2-C^{14} into the cells of the Gardner lymphosarcoma grown in C3H mice to be about 70 per cent less than the incorporation into normal spleen, in oxygen and with glucose absent. With glucose in the medium both tissues showed about the same incorporation rate for glycine. When DL-alanine-1-C^{14} was employed, both in the presence and absence of glucose, the tumor tissue showed a lower incorporation rate than did the normal spleen (40–70 per cent). The reader is referred to the literature[122–124] for a more complete discussion of incorporation into tumor tissues.

At this point it is well to draw attention to the fact that cells show a considerable ability to concentrate amino acids and, moreover, this ability is more pronounced in fetal tissues than in those from adult animals, more pronounced in regenerating than in normal liver,[125] and

(122) C. Heidelberger, *Advances in Cancer Research* **1,** 274 (1952).
(123) P. C. Zamecnik, R. B. Loftfield, M. L. Stephenson, and J. M. Steele, *Cancer Research* **11,** 592 (1951).
(124) V. R. Potter and P. Siekevitz, *in* McElroy and Glass, Phosphorus Metabolism, Johns Hopkins Press, Baltimore, 1952, Vol. 2, p. 682.
(125) H. N. Christensen and J. A. Streicher, *J. Biol. Chem.* **175,** 95 (1952); H. N. Christensen, J. T. Rothwell, R. A. Sears, and J. A. Streicher, *ibid.* **175,** 101 (1948).

that it is especially noteworthy in one type of tumor tissue, the Ehrlich mouse ascites tumor.[126] This being the case, it is manifestly possible to observe increased rates of incorporation, under certain conditions, due to modifications in the amino acid concentration in cells rather than due to effects on the incorporation reaction as such.

It has been found that when tissues are homogenized, the rate of amino acid incorporation is greatly reduced. Thus the incorporation of DL-methionine by liver homogenates was 20 per cent or less than that observed with intact slices, the actual rate depending largely on the nature of the homogenization process; the incorporation by the homogenized cells of the Gardner lymphosarcoma was only about 30 per cent of that of the intact cells;[93] ruptured bone marrow cells incorporated no glycine, leucine, or lysine;[96] homogenized rat diaphragm incorporated the three amino acids just mentioned at one quarter the rate of the intact tissue.[97] However, in spite of the lower rates of incorporation by this type of preparation, homogenates and their fractions separated by differential centrifugation possess such obvious advantages for investigational purposes that their properties have been examined in some detail.

With such preparations a difficulty arises in that they tend to lose activity at a considerable rate, whereas with preparations consisting of intact cells, particularly those supplied with the proper substrates to maintain respiration or glycolysis, the rate of incorporation is maintained fairly constant for 3–5 hours, e.g., liver,[103] diaphragm,[97] and bone-marrow cells.[96] In homogenates the incorporation in the second hour is almost invariably much less than in the first hour, and even before this, activity may have fallen off considerably.[95,100,113,127] This being the case it is obvious that the production of a homogenate fraction exhibiting significant activity will depend, to a large extent, on the methods and the rapidity with which the necessary manipulations are carried out. Even with tissue slices, aging for any significant length of time results in the loss of considerable activity.[103] Similar losses of activity were found when spleen cells were aged at 38°.[93] It is not clear whether this loss of activity is due to the breakdown of essential factors in the preparations or whether the loss is due to the passage of necessary cofactors, including protein, into the external medium.

The actual rates of incorporation into homogenates of liver and their fractions prepared in several different ways are shown in Tables XII and XIII.

(126) H. N. Christensen and T. R. Riggs, *J. Biol. Chem.* **194,** 57 (1952); H. N. Christensen, T. R. Riggs, H. Fischer, and I. M. Palatine, *ibid.* **194,** 198 (1952).

(127) P. Siekevitz, *J. Biol. Chem.* **195,** 549 (1952).

TABLE XII

INCORPORATION OF AMINO ACIDS INTO THE PROTEINS OF LIVER PREPARATIONS (HOMOGENATES)

All preparations from the livers of rats or guinea pigs. Data expressed as micromoles amino acid incorporated at 37° per gram protein per period. Concentrations of amino acids, millimolar; all C^{14}-labeled in position indicated. Time in hours.

Homogenate preparation	Labeled amino acid	Concentration	Period	Incorporation	References and footnotes to Table
Fetal, rat	DL-ala-1	10	1	1.1	[b] 95
Adult, rat	DL-ala-1	10	1	0.2	[b] 95
Sediment (S), rat	gly-2	1.2	1	0.04–0.06	[c] 113
S + supernatant		1.2	1	0.37	[c] 113
Adult, guinea pig	L-lys-6	14.3	2	1.8	[d] 100
S, guinea pig	L-lys-6	14.3	2	3.3	[d] 100
SI, guinea pig	gly-1 or (L-lys-6)		4 or (2)	0.52[a] (4.1)	[e] 101
SII, guinea pig				0.4 (3.2)	[e] 101
SIII				0.075 (0.9)	[e] 101
Adult, rat	DL-leu-1	0.23	0.5	0.08	[f] 128

[a] In the experiments with glycine a considerable fraction of the incorporation may be attributed to amino acids other than glycine (e.g., serine). With the others this type of complication is minor. Values in brackets refer to incorporations of L-lys-6.

[b] Homogenate prepared in KCl–$KHCO_3$ medium (1:1) at pH 7.5. All operations in connection with homogenization done in the cold in this and other cases below.

[c] Homogenate in above medium (1:4) but nuclei and debris removed by 5 min. at 680 × *g*. Sedimented by 2700 × *g* for 30 min. Incubations at pH 7.4–7.5.

[d] First line; homogenate in HCO_3^- free Ringer's solution containing Ca ion, 3 m*M*. Incubation at pH ~ 6. Second line: homogenate in Kreb's–HCO_3^- and centrifuged approximately as in footnote *c*. Incubation at pH ~ 7.3.

[e] Homogenate in isotonic sucrose or Kreb's–HCO_3^- (1:2). SI, unbroken cells and debris; SII, mitochondria (?); SIII, microsomes. Incubations at pH 7.5.

[f] Homogenate in 0.3 *M* sucrose buffered with phosphate and containing: $KHCO_3$, nicotinamide, glucose, pyruvate, and Mg ion. Erythrocytes and nuclei but no intact liver cells present.

There appears to be a considerable lack of concordance in the data, which may arise from any one of several reasons: (*1*) different techniques used in the preparation and fractionation of the homogenate, (*2*) incorporation of derived amino acid besides the one introduced, and (*3*) anoma-

(128) P. C. Zamecnik, *Federation Proc.* **12,** 295 (1953); see also P. C. Zamecnik and E. B. Keller, *J. Biol. Chem.* **209,** 337 (1954).

lous behavior of the amino acid used, as with L-lysine-1-C^{14} or -6-C^{14}.[100,101] This behavior of lysine will be discussed in sec. III-2*g*.

TABLE XIII

INCORPORATION OF ALANINE INTO THE PROTEINS OF VARIOUS FRACTIONS OF A RAT LIVER HOMOGENATE[127]

Homogenate fractions prepared and incubated in media having the composition given in Fig. 4. Esterification of phosphate determined in 5-min. period in separate flasks which in addition to the components noted contained the following substances (micromoles/ml.): glucose, 20; NaF, 10; plus hexokinase.
Values refer to changes per half hour.

System	Oxygen consumption, micro atoms	Phosphate esterified, μmoles	Incorporation, μmoles/g.
Homogenate	18.7	24.0	0.0119
Mitochondria (M′)	4.1	25.2	0.0014
Microsomes (pH 5)(M″)	0.2	0.	0.0012
Supernatant (pH 5)(S)	0.4	0.	0.0004
M′ + M″	7.0	24.6	0.0112[a]
M′ + S	4.9	24.6	0.0017
M′ + M″ + S	9.4	22.8	0.0047

[a] In the absence of α-ketoglutarate, in a comparable experiment the incorporation was 0.0015. The maximum rate of incorporation amounted to 0.072 micromoles alanine/g. protein/hr. during the first 10 min. of incubation (0.006% of the alanine replaced per hour).

From the various lines of evidence presented, in particular by Winnick,[95] by Peterson and Greenberg,[113] and by Siekevitz,[127] it is reasonably certain that amino acids are actually incorporated into the proteins from the homogenates by peptide-bond formation. Besides glycine, Peterson and Greenberg[113] observed the incorporation of the amino acids DL-serine-3-C^{14}, DL-phenylalanine-3-C^{14}, and DL-leucine-2-C^{14}, although Borsook and co-workers[101] failed to find incorporation of L-leucine-1-C^{14} into homogenate sediments I and III when the sucrose method of fractionation was employed.

An approach to the question as to which fraction of the homogenate contains (synthesizes?) the protein was found by Siekevitz,[127] by incubating a whole homogenate of liver with alanine and determining the incorporation into the different fractions by subsequent centrifugal separation. The results are shown in Fig. 4.

It is clear that the microsomal fraction incorporates amino acid at a much greater rate than does any other fraction, and that the supernatant exhibits the least activity. This is noteworthy because in the experiments of Peters and Anfinsen,[129] in which the proteins of liver slices from chickens were labeled by incubation in bicarbonate-C^{14}-containing media, the fraction with the highest specific activity was present in the supernatant medium. When the soluble material from the chicken livers was subjected to fractionation with alcohol, the protein with the highest specific activity corresponded in properties to albumin. No specific separation of microsomes was attempted.

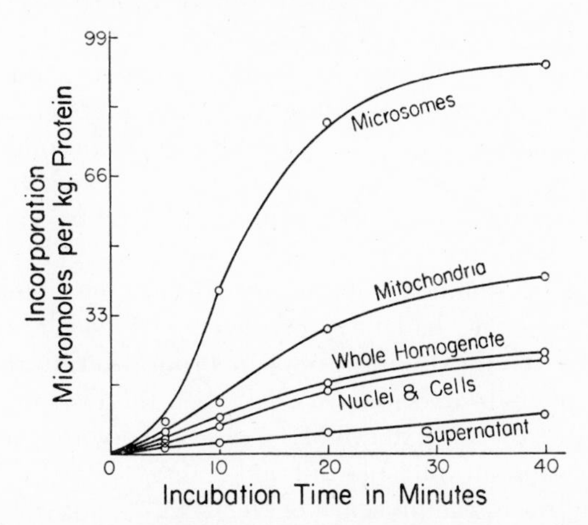

FIG. 4. Incorporation of alanine into the proteins of the various fractions of rat liver homogenate.[127] One milliliter of a 1:10 rat liver homogenate made in isotonic sucrose incubated at 37° in oxygen in a total volume of 2 ml. with other components in concentrations as follows: α-ketoglutarate 20 mM, AMP-5 1.6 mM, $MgCl_2$ 5 mM, phosphate pH 7.5 7.5 mM, K ion 55 mM, DL-alanine-1-C^{14} 0.9 mM. After incubation, fractions were separated by modification of the method of W. C. Schneider and G. H. Hogeboom [*J. Biol Chem.* **183**, 123 (1950)]. Microsomes were separated at 45,000 × *g* for 60 min. at 4°.

The results of Siekevitz[127] are in agreement with those of others who have labeled liver fractions by the administration of labeled amino acid to intact animals[130] (see also sec. IV-2). Nonetheless, when the other type of experiment was carried out by Siekevitz, that is, when centrifugal fractionation was done first and incubations subsequently, the results were quite different as shown in Table XIII.

In this case mitochondria and microsomes were labeled to about an equal extent. Maximum labeling was attained when the two fractions were remixed. These results should be compared with those obtained in PAH synthesis (Table V).

(129) T. Peters, Jr. and C. B. Anfinsen, *J. Biol. Chem.* **182,** 171 (1950).
(130) T. Hultin, *Exptl. Cell Research* **1,** 376 (1950).

c. *Stimulation of Incorporation*

(*1*) *Stimulation by Providing Energy Sources.* It was found by Winnick[95] that the homogenate prepared from fetal liver could be largely inactivated by dialysis in the cold for a short period. The preparation was partially reactivated when Mg ion; a mixture of L-amino acids; and ATP, ADP, AMP, or L-glycerophosphate were added, the last substance being somewhat less effective than the adenylic derivatives. Similarly, Peterson, Winnick, and Greenberg[104] noted that the sedimented fraction from the liver of rats could be activated in much the same fashion with respect to the incorporation of DL-serine-3-C^{14}, DL-phenylalanine-3-C^{14}, DL-leucine-2-C^{14}, or glycine-2-C^{14}. These results were confirmed in a more complete investigation[113] with the liver sediment using glycine-2-C^{14}, and by Kit and Greenberg[131] with threonine-1,2-C^{14} and DL-valine-2-C^{14}.

In contrast, Siekevitz[127] found no activation by ATP although previously,[132] using a sediment containing both microsomes and mitochondria, such an effect had been observed. However, the conditions which stimulated the maximum rate of incorporation by the recombined microsome–mitochondrial system existed when there was rapid oxidation of α-ketoglutarate (i.e., Mg ion, AMP-5, and phosphate present) coupled with the generation of phosphate-bond energy.[127] By preincubation of the mitochondria in the presence of the α-keto acid, there appeared in the supernatant a *nonspecific* factor which stimulated the incorporation reaction in the absence of oxygen. This factor did not correspond in its properties to either an activated amino acid or to ATP.

(*2*) *Stimulation by Changes in Concentration of Amino Acid*(*s*). Borsook and co-workers claim that, for bone marrow cells[96] and for diaphragm,[97] the logarithm of the fraction of the maximum rate of incorporation for any one of several amino acids is proportional to concentration. However, there appears to be no theoretical justification for anticipating a relationship of this type on the basis of simple kinetic theory. The same workers[96,97] observed inhibitions of incorporation at high amino acid concentrations. Others[103,110] have found with DL-methionine-S^{35} or DL-alanine-1-C^{14} in liver slices that with increasing concentration of amino acid the increase in incorporation rate gradually decreased. Moreover, with the former amino acid the curve appeared to show a break, possibly indicating the participation of two processes. However, over a smaller range of concentration of DL-methionine-S^{35} Melchior and coworkers[98] found, with resting *E. coli*, incorporation and concentration to be linearly related. By making some assumption with respect to

(131) S. Kit and D. M. Greenberg, *J. Biol. Chem.* **194,** 377 (1952).
(132) P. Siekevitz and P. C. Zamecnik, *Federation Proc.* **10,** 246 (1951).

Michaelis' constants, these authors demonstrated that such a linear relationship should exist over a limited concentration range, even if two processes were involved. Others[93] have found a similar relationship using Gardner lymphosarcoma cells and glycine-2-C^{14} as the indicator of incorporation.

More recently Gale and Folkes,[132a] in studies on the incorporation of L-glutamic acid and DL-phenylalanine-3-C^{14} into the proteins of *Staphylococcus aureus*, have shown that incorporation rate—concentration relationships—behave in a manner predictable from the Michaelis-Menten theory; that is, a linear relationship exists between the reciprocal of the incorporation rate and the reciprocal of the concentration. The same authors found the incorporation of phenylalanine to be inhibited in a competitive way by *p*-chlorophenylalanine, and non-competitively by chloroamphenicol, aureomycin, and terramycin (sec. III-2*d*).

Be these relationships between incorporation rate and concentration as they may, their interpretation in systems of metabolizing cells is fraught with the utmost difficulty because of the concentrating powers for amino acids which are possessed by both animal and bacterial cells, and also on account of the uninvestigated effects of the concentrations of the amino acid in the external medium on the concentration of others within the cells.[125,126,133,134] Even with mitochondria the relationship between medium concentration and "particle concentration" is not at all clear. It is evident that at low concentrations of amino acid added to the medium, the incorporation rates in cellular systems will fail to obey simple enzymatic kinetics, unless the data refer to *actual amino acid concentrations and specific activities at the reaction sites.*

It has been found that in a sedimented system incorporation is stimulated by the addition of a suitable amino acid mixture containing methionine, arginine, aspartic acid, glutamic acid, proline, and alanine which does not comprise all the amino acids known to be present in tissue protein and only two of those considered essential.[95,104,113] Moreover, the precise effect of adding different concentrations of this mixture on the incorporation of glycine-2-C^{14} depends on other factors such as the presence of citrate. These observations are difficult to interpret with any degree of certainty. Borsook and co-workers[114] have also observed effects of added amino acids on the incorporation rates of the individual amino acids, glycine, histidine, leucine, and lysine, with the reticulocyte

(132*a*) E. F. Gale and J. P. Folkes, *Biochem. J.* **55**, 721, 730 (1953).
(133) H. N. Christensen, M. K. Cushing, and J. Streicher, *Arch. Biochem.* **23**, 106 (1949).
(134) E. F. Gale, *Biochem. J.* **48**, 286, 290 (1951); E. F. Gale and T. F. Paine, *ibid.* **48**, 298 (1951).

system. Stimulation of incorporation by the "essential" amino acids histidine, phenylalanine, and valine in low concentrations were observed, but in view of the afore-mentioned concentration differences between amino acids inside and outside cells—the interpretation of these data is likewise difficult.

It is quite clear that if net protein synthesis with respect to one particular protein occurs, then all the amino acids required for that protein must be supplied. In the systems studied it is not known whether the "non-essential" amino acids are synthesized, but it is known that many essential acids can be supplied by the catabolism of one or another protein in the system.[135] Consequently, it cannot be concluded whether net synthesis of any protein occurs. Nor do the data necessarily support some exchange type mechanism involving minimal disruption of peptide chains.

One important finding with respect to incorporation is that when more than one labeled amino acid is presented, the incorporation is additive. This has proved to be the case with the intact cell systems of bone marrow,[96] reticulocytes,[114] and diaphragm[97] as well as for the homogenate sediment system from the livers of guinea pigs.[101] This is of importance because it indicates that other important factors are involved in incorporation besides energy input.

(*3*) *Stimulation by Other Factors.* It is of interest to note that in addition to amino acids there appears to be some incorporation-stimulating heat-stable factor found in extracts of plasma, animal tissues, and yeast,[114] which is neither amino acid nor inorganic salt. Since the factor is present in plasma, the incorporation by reticulocytes was in general significantly greater in plasma than in saline medium. Whether any loss of the protein component from reticulocytes, such as occurs particularly with liver,[136] has anything to do with this effect is not immediately clear. Unfortunately the effect of the factor on the respiration and glycolysis in the system was not determined.

The relationship, if any, between this factor and factors described by Kutsky[137] and by Rosenberg and Kirk,[138] which stimulate growth of tissue cultures, is not clear.

(135) D. M. Greenberg, L. A. Miller, and R. Sills, *Proc. Soc. Exptl. Biol. Med.* **82,** 206 (1953).
(136) C. Hoch-Ligeti and H. Hoch, *Brit. J. Exptl. Path.* **31,** 138 (1950).
(137) R. J. Kutsky, *Proc. Soc. Exptl. Biol. Med.* **83,** 390 (1953); R. J. Kutsky and M. Harris, *Anat. Record* **112,** 419 (1952); M. Harris and R. J. Kutsky, *Growth* **17,** 147 (1953).
(138) S. Rosenberg and P. L. Kirk, *Science* **117,** 566 (1953); *Arch. Biochem. Biophys.* **44,** 226 (1953); *J. Gen. Physiol.* **37,** 231, 239 (1953).

Totter and co-workers[139] have made observations concerning the incorporation of glycine-1-C^{14} into the proteins and phospholipides of homogenates made from the livers of chickens. In livers from folic acid-deficient chicks there was a notable reduction in the rates of both processes, but little or no effect was observed when folic acid was added to the deficient system. Evidently the deficient homogenate metabolized glycine less readily than did the control from the chicks fed folic acid. Unfortunately it is not known what percentage of the activity in the protein was due to glycine and what to serine or other amino acid, in the normal as compared with the deficient homogenate–protein.

Various effects of hormones on incorporation rates have been described. Thus Sinex and co-workers,[140] using hemidiaphragms from rats have shown that insulin (0.6 units/ml.) has a distinct effect on the uptake of alanine-1-C^{14} from a bicarbonate saline solution. The greatest effects (av. +57 per cent) were found in the absence of added glucose or pyruvate. These two substances also caused a decrease in uptake by the tissue in the absence of insulin (−14 and −54 per cent with glucose and pyruvate). The pyruvate effect could be due, as the authors suggest, to the dilution of the labeled alanine. Krahl[141] made rather similar observations by showing that the slices of liver from alloxan-diabetic rats were less able to incorporate glycine-1-C^{14} and phenylalanine-3-C^{14} than those from normal rats. The incorporation rate in the deficient slices were brought to normal by the *in vitro* addition of glucose and insulin (0.1 unit/ml.). Diaphragms from deficient rats also showed reduced rates of incorporation which were rendered normal by the addition of glucose to the medium. A depression in the rate of incorporation into glutathione was also observed in the diabetic animals. Most of these effects appear to be rather nonspecific.

Krahl[141] also found that fasting in itself would cause a reduction of glycine incorporation into both liver and diaphragm. This reduction was abolished simply by the provision of glucose.

It has also been observed by Melchior and Halikis[142] that there is a decrease in the rate of incorporation of DL-methionine-S^{35} into the pituitaries of lactating female rats when compared with nonlactating females. Male pituitaries incorporated the same amount per gram of

(139) J. R. Totter, B. Kelley, P. L. Day, and R. R. Edwards, *J. Biol. Chem.* **186,** 145 (1950).
(140) F. M. Sinex, J. Macmullen, and H. B. Hastings, *J. Biol. Chem.* **198,** 615 (1952).
(141) M. E. Krahl, *J. Biol. Chem.* **200,** 99 (1953).
(142) J. B. Melchior and M. N. Halikis, *J. Biol. Chem.* **199,** 773 (1952).

protein but less per gland since the size in the male is less. The high rate of incorporation into this gland is also noteworthy.

d. Inhibition of Incorporation

Among the first observations made with respect to the incorporation of alanine was that inhibition occurred under anaerobic conditions.[86] This observation has been confirmed by all more recent work (cf. Kit and Greenberg)[94] although the effects of anaerobiosis are somewhat variable as indicated in Table XIV.

TABLE XIV
EFFECTS OF INHIBITORS ON INCORPORATION OF LABELED AMINO ACIDS INTO PROTEINS OF BONE MARROW (BM) AND RETICULOCYTES (R)[96,114]
Incorporation expressed as per cent of that without inhibitor

Amino acid	Glycine		L-Leucine		L-Lysine	
Inhibitor[a]	BM	R[b]	BM	R	BM	R
Anaerobiosis	0	67	0	43	0	100
Arsenite	0	9	0	12	0	9
Arsenate	4	56	23	87	21	86
Azide	17	63	23	95	32	91
2,4-Dinitrophenol		11		12		21
Fluoride		0		9		0
Hydroxylamine		16		11		32
Iodoacetate		40		51		76

[a] Concentration of inhibitors: arsenite, arsenate, azide, 2,4-dinitrophenol, 0.001 *M*; fluoride, hydroxylamine, 0.02 *M*; iodoacetate, 0.002 *M*.

[b] R, reticulocytes from rabbits injected with phenylhydrazine. Bone marrow also from rabbits.

Siekevitz[127] observed that incorporation of alanine in the microsome system proceeded anaerobically in the presence of the nonspecific factor.

In general it has been observed that inhibitors of respiration and phosphorylation—in particular 2,4-dinitrophenol—have a detrimental effect on the incorporation reaction.[94,102] Similar effects have been noted in homogenate systems;[113,127] so that there appears to be a direct connection between the maintenance of respiration or of glycolysis[94] or rather of the generation of phosphate-bond energy, and incorporation. However, the same agents also have a detrimental effect on the concentration of amino acids by reticulocytes[143] and other cells.[144]

(143) T. R. Riggs, H. N. Christensen, and I. M. Palatine, *J. Biol. Chem.* **194,** 53 (1952).

(144) E. Negelein, *Biochem. Z.* **323,** 214 (1952).

Many other substances besides those noted in Table XIV, malonate,[95] and heavy metals also have been found to act as inhibitors. This is of some interest because the inhibitors of some types of peptidases are heavy metals.[145] Borsook and co-workers[96] showed inhibitions in the incorporation of leucine, lysine, and glycine into bone-marrow cells at concentrations of Co^{++}, Mn^{+3}, and Cu^{++} as low as 1 mM. Similar inhibitions have been observed in homogenate systems.[113]

e. Peptides and the Incorporation Reaction

It is attractive to speculate, as numerous authors have done, that peptides are intermediates in the biological synthesis of proteins; in fact, it is rather difficult to conceive a satisfactory mechanism in which peptides are not involved in one way or another. Most of the work directed toward the demonstration of such a relationship has been carried out in the microbiological field, and Gale and Van Halteren[146] have shown that peptides of glutamic acid are excreted into the medium by one bacterium. But, it appears from recent work in Snell's laboratory[147,148] that there are a number of hitherto largely unrecognized pitfalls in this type of research. Growth stimulation was not obtained with any specific peptides, and effects observed were either minor or could be attributed to products of hydrolysis. Virtanen and Nurmikko[149] also came to the latter conclusion with respect to the utilization of peptides by *Leuconostoc mesenteroides*, and much the same position has been taken by other workers. For a review of the effects of peptides in bacterial growth, the reader should consult Fruton and Simmonds.[150]

In the field of warm-blooded animals, Hoberman and Stone[151] have shown that the observed incorporation rates of glycine-N^{15} from various peptides into hippuric acid could best be accounted for on the basis of hydrolysis prior to coupling. It is also of interest to note that glycine from γ-glutamylglycine is less readily incorporated into protein than is glycine under similar conditions.[58] Other examples of a similar type were mentioned in connection with GSH and hippuric acid formation.

(145) E. L. Smith, *Advances in Enzymol.* **12,** 191 (1951); E. L. Smith and R. Lumry, *Cold Spring Harbor Symposia Quant. Biol.* **14,** 168 (1949).
(146) E. F. Gale and M. B. Van Halteren, *Biochem. J.* **50,** 34 (1951).
(147) H. Kihara, W. G. McCullough, and E. E. Snell, *J. Biol. Chem.* **197,** 783 (1952); H. Kihara and E. E. Snell, *ibid.* **197,** 791 (1952); H. Kihara, O. A. Klatt, and E. E. Snell, *ibid.* **197,** 801 (1952).
(148) V. J. Peters, J. M. Prescott, and E. E. Snell, *J. Biol. Chem.* **202,** 521, 533 (1953).
(149) A. I. Virtanen and V. Nurmikko, *Acta Chem. Scand.* **5,** 681 (1951).
(150) J. S. Fruton and S. Simmonds, *Cold Spring Harbor Symposia Quant. Biol.* **14,** 55 (1949).
(151) H. D. Hoberman and D. Stone, *J. Biol. Chem.* **194,** 383 (1952).

Human subjects largely excrete intravenously administered peptides in the urine.[152] However, peptides are found in normal blood,[153] and the rejection of peptides may simply mean that they are not readily taken up by cells. In this connection it can be said that many workers who have obtained negative effects on the growth of bacteria by adding peptides to the medium have failed to show that the peptides actually penetrated into the cells.

Borsook and co-workers[154] after incubating guinea pig liver homogenate with leucine-1-C^{14} and an amino acid mixture corresponding in composition to casein, found a peptide fraction labeled with leucine in the TCA-soluble material. This peptide material was isolated using a starch column, and the same peptide, as identified by its position in the column chromatogram, was found in normal guinea pig liver. Upon hydrolysis, it was found to yield all the usual naturally occurring amino acids with the exception of hydroxyproline, cystine, cysteine, and valine. These authors claim that a similar peptide fraction exists in a wide variety of products such as plasma; livers from a representative group of animals; and heart, kidney, and spleen from the guinea pig; and even in Witte's peptone. It has recently been shown that there are at least four components in the peptide material.[155]

f. Nucleic Acids and Incorporation

It has long been a favorite speculation that there is some more or less direct connection between protein synthesis and nucleic acids. The subject has been reviewed extensively, for example in the *Cold Spring Harbor Symposia* series,[156] and elsewhere[157–163] (see also Chap. 12). In

(152) H. N. Christensen, E. L. Lynch, D. G. Decker, and J. H. Powers, *J. Clin. Invest.* **26,** 849 (1947); H. N. Christensen, E. L. Lynch, and J. H. Powers, *J. Biol. Chem.* **166,** 649 (1946).
(153) H. N. Christensen and E. L. Lynch, *J. Biol. Chem.* **163,** 741 (1946); *ibid.* **166,** 87 (1947).
(154) H. Borsook, C. L. Deasey, A. J. Haagen-Smit, G. Keighley, and P. H. Lowy, *J. Biol. Chem.* **174,** 1041 (1948); *ibid.* **179,** 705 (1949).
(155) I. G. Fels and A. Tiselius, *Arkiv. Kemi* **3,** 369 (1951).
(156) *Cold Spring Harbor Symposia Quant. Biol.* **12** (1947).
(157) J. Brachet, *Ann. N. Y. Acad. Sci.* **50,** 861 (1950).
(158) D. Mazia *in* E. S. G. Barron, ed., Modern Trends in Physiology and Biochemistry, Academic Press, New York, 1952, p. 77.
(159) T. O. Caspersson, *Symposia Soc. Exptl. Biol.,* **1,** 127 (1947); *idem,* Cell Growth and Cell Function, W. W. Norton, New York, 1950.
(160) P. C. Caldwell and C. Hinshelwood, *J. Chem. Soc.* **1950,** 3156.
(161) H. Chantrenne, *Second Symposium Soc. Gen. Microbiol.,* The Nature of Virus Multiplication, Cambridge, 1953.
(162) J. N. Davidson, The Biochemistry of the Nucleic Acids, John Wiley and Sons, New York, 1950.

spite of all the extensive material available, any very specific experimental demonstration of a connection between either desoxyribonucleic acid (DNA) and ribonucleic acid (RNA) and protein synthesis appears to be lacking. However, recently the approach has become somewhat less oblique and some specific examples will be mentioned here.

The fact that the microsome fraction which contains the main part of the cytoplasmic RNA has been demonstrated to be intimately concerned with amino acid incorporation is very suggestive.[127] The experimental findings have been confirmed by much work performed *in vivo* so that it appears clear that the protein in the microsome fraction is replaced at a rapid rate (sec. IV-2). It also proved possible for Holloway and Ripley[164] to show, in experiments with reticulocytes taken from rabbits poisoned with phenylhydrazine, that the incorporation of L-leucine-1-C^{14} was a function of RNA content of the cells. However, the relationship is not linear and, moreover, RNA has to attain a significant level before any effect on incorporation rate is seen. DNA showed much less increase with increase in incorporation rate. (See also Allfrey and co-workers.[164a])

The experiments of Gale and Folkes[165] with bacterial growth are of interest. In some of their experiments a good correlation was obtained between the rate of protein synthesis in *Staphylococcus aureus* and the nucleic acid content of the cells. When cells were incubated with glucose and amino acids the addition of a mixture of purines and pyrimidines caused a simultaneous increase in the nucleic acid and in the rate at which the organisms formed protein. On the other hand in the presence of chloramphenicol, aureomycin, or terramycin, protein synthesis was inhibited but the effect on nucleic acid synthesis was the opposite; bacitracin inhibited both syntheses to the same extent.

An entirely different approach to this problem has been taken by Hakim and Happold.[166] Using the phenomenon of enzymatic adaptation (EA), they found that in *E. coli* adaptation to tryptophan, by the formation of tryptophanase, may be related to a rise in DNA and total nucleic acid, and, moreover, DNA addition speeded up the adaptive process. An approximate proportionality between the rate of adaptation and DNA concentration was reported, whereas RNA had no effect. Presumably EA must be related to changes in the genetic constitution of *E. coli* and not to some process which is not so linked. Since EA

(163) N. A. Eliasson, E. Hammarsten, P. Reichard, S. Äquist, B. Thorell, and G. Ehrensvard, *Acta Chem. Scand.* **5,** 431 (1951).

(164) B. W. Holloway and S. H. Ripley, *J. Biol. Chem.* **196,** 695 (1952).

(164*a*) V. Allfrey, M. M. Daly and A. E. Mirsky, *J. Gen. Physiol.* **37,** 157 (1953).

(165) E. F. Gale and J. P. Folkes, *Biochem. J.* **53,** 483, 493 (1953); see also *Nature* **173,** 1223 (1954).

(166) A. A. Hakim and F. C. Happold, *Biochem. J.* **52,** xxv (1952); **53,** xxxvi (1953).

involves protein synthesis (formation of enzymes) the question of the relationship between DNA and the synthesis of protein appears to be close.

Nonetheless, Brachet and Chantrenne[167] have carried out experiments which make the connection appear rather remote. Measurements were made of the incorporation of amino acids from bicarbonate-C^{14} into the nucleated and nonnucleated parts of an alga, *Acetabularia mediterranea*, at different times after the parts had been separated. When the algae were sectioned the nucleated parts grew into normal individuals, but the nonnucleated parts had a more limited regenerative capacity. Results are shown in Table XV.

TABLE XV
INCORPORATION OF AMINO ACIDS INTO SECTIONS OF AN ALGAE WITH OR WITHOUT NUCLEUS[167]

Time since sectioning, days	Incorporation time, hr.	Specific activity ratio[a]	Time since sectioning days	Incorporation time, hr.	Specific activity ratio
9	38	1.01	23	24	0.61
11	26	1.02	33	24	0.78
16	24	.92	38	27	0.70

[a] Specific activity (S.A.) of carboxyl groups of proteins released by ninhydrin following hydrolysis. Ratio of S.A. non-nucleated to S.A. nucleated. Labeling by incubation with bicarbonate-C^{14}.

It is evident that the nonnucleated sections are capable of incorporating amino acids with carboxyl labels at a normal rate for about 16 days following the sectioning, but that, thereafter, a slow falling off occurs.

In the case of virus reproduction in *E. coli*, phage T2, it appears from the work of Hershey and Chase[168] that the nucleic acid (DNA) and not the protein part, or at least the sulfur-containing part, is the one intimately concerned with the propagation. Presumably the protein envelope is solely concerned with the adsorption of the virus to the surface of the bacteria prior to infection. Hence, the nucleic acid may be looked at as the template for further phage production, both nucleic acid and protein (?) parts.

Winnick and co-workers[169,170] have also made observations on the relationship between nucleic acid and protein synthesis in tissue cultures (sec. III-4*a*).

(167) J. Brachet and H. Chantrenne, *Nature* **168,** 950 (1951).
(168) A. O. Hershey and M. Chase, *J. Gen. Physiol.* **36,** 39 (1952).
(169) H. W. Gerarde, M. Jones, and T. Winnick, *J. Biol. Chem.* **196,** 69 (1952).
(170) T. Winnick, *Texas Repts. Biol. and Med.* **10,** 452 (1952); H. W. Gerarde, M. Jones, and T. Winnick, *J. Biol. Chem.* **196,** 51 (1952).

g. Anomalous Incorporation Reactions

The most striking case of this type is that described by Brunish and Luck[171] in which the incorporation of glycine-1-C^{14}, DL-alanine-1-C^{14}, DL-phenylalanine-1-C^{14}, and DL-lysine-2-C^{14} into the desoxypentose nucleohistone (DNH) prepared from liver in the fibrous or nonfibrous state was employed. The DNH, or in many experiments the histone alone, was incubated with one of the above amino acids in a Krebs bicarbonate medium or in distilled water, and uptakes of amino acid were measured. In all experiments, with histone and DNH, increased rates of uptake were found when the temperature was increased—up to 100°. Rates were directly proportional to amino acid concentrations and to time of reaction. The effect of the temperature on the incorporation was not on the histone as such because preheating this component alone interfered *completely* with subsequent incorporation. Incorporation proved to be pH dependent in the range 5.2–9.9, increasing with increasing pH. The amino acid once bound was not removed by dialysis, weak base, ninhydrin treatment (carboxyl-labeled amino acids), or the usual TCA washing procedures.

The crux of this matter is that the process is evidently non-enzymatic. The optical specificity of the reaction also remains to be determined.

The second type of anomalous reaction is that resulting in the uptake of L-lysine by the two systems of Borsook and co-workers.[100,101] In these systems D-lysine is not taken up. These two lysine systems are particularly characterized by their pH optima and the differences between them are summarized in Table XVI.

From these results it would appear that neither system functions according to expectations as outlined in the preceding discussion. The lack of effects of dinitrophenol in *B*, and of anaerobiosis in *A*, together with the pH optimum of 6.3 for *A* puts both these systems in a class by themselves. Since lysine has a second reactive amino groups, and heat-treated proteins in some cases become nutritionally lysine deficient, the mode of lysine binding is in need of investigation. Until this is done it does not appear profitable to speculate further on the anomaly.[172]

It is obvious that one of the chief stumbling blocks with respect to the

(171) R. Brunish and J. M. Luck, *J. Biol. Chem.* **197,** 869 (1952).

(172) It may be mentioned that some years ago peculiar results of this same general type were obtained with DL-tyrosine-2-C^{14} in experiments done in this laboratory under the direction of Prof. D. M. Greenberg, and that some of the published results with DL-phenylalanine are rather anomalous (see Peterson and Greenberg,[113] and Gerarde and co-workers;[162] the phenylalanine incorporation is phenomenally high). The possibility exists that these effects are due to self radiolysis of the amino acids concerned and the reaction of the products with the protein.

interpretation of the experimental work noted in this section (III-2) arises from the ill-defined nature of the material "synthesized."

TABLE XVI
DIFFERENCES IN PROPERTIES OF THE TWO SYSTEMS RESPONSIBLE FOR THE INCORPORATION OF LYSINE BY LIVER PREPARATIONS FROM GUINEA PIGS[100,101]

		System A^a	System B
pH optimum		6.1	7.3
Requirements	Ca ion	+	$\pm^c$
	O_2	—	+
Inhibition, %	Fluoride	100	±
	Arsenite	50	±
	Azide[b]	<50	±
L-Lysine concn.		Linear change	Independent—depends on total lysine in system

[a] A is the whole homogenate, B is the sediment from the dilute homogenate (1:15) separating at 2500 × g.
[b] Approximately similar effects obtained with arsenate, cyanide, and dinitrophenol.
[c] ± indicates that no definite effect was observed.

3. SYNTHESIS OF SPECIFIC PROTEINS

a. *Synthesis of Serum Albumin*

One of the earliest attempts to find a system which synthesized a specific protein was that of Zamecnik and co-workers[173,173a] who investigated the production of silk by injecting labeled alanine or glycine into larvae of the silk moth. Labeled coccoons were eventually formed, but the process does not appear to have been investigated further. In the following year Peters and Anfinsen[129] found that slices prepared from chickens' livers incorporated amino acids, from bicarbonate-C^{14} in the medium, into the albumin fraction with higher specific activity than any of the insoluble or fractions of the soluble protein separated by alcohol. In the ultracentrifuge and in the electrophoretic apparatus the most radioactive fraction also behaved like serum albumin. It was later shown by the extremely sensitive immunological method that the protein isolated from the medium was actually serum albumin and that a net synthesis of the protein could be demonstrated.[174] The average net synthesis amounted to 0.12 mg./g. liver/hr. Assuming 17 per cent protein in the liver, this is equivalent to 0.71 mg. albumin/g. protein/hr., or

(173) P. C. Zamecnik, R. B. Loftfield, and M. L. Stephenson, *Science* **109,** 624 (1949).
(173*a*) C. B. Anfinsen, *J. Biol. Chem.* **185,** 827 (1950). This paper is concerned with the synthesis of labeled ribonuclease.
(174) T. Peters, Jr. and C. B. Anfinsen, *J. Biol. Chem.* **186,** 805 (1950).

for a chicken with a 45-g. liver, 5.4 mg. albumin/chicken/hr. or 129 mg./day. If the *total* exchangeable albumin in a 2.5-kg. chicken is assumed to be 5000 mg.; this leads to a calculated turnover time of 39 days and a half life of 27 days, somewhat longer perhaps than would be anticipated from the results with other species (Tables XX and XXI). (Regarding *total* exchangeable albumin, see secs. IV-3*a* and *b*.)

In some experiments a simultaneous synthesis and incorporation of radioactivity was shown. The radioactivity of the serum albumin isolated after a 4-hour incubation was such as to indicate a replacement of 37.5 per cent of the glutamic and aspartic acids with labeled material; on the assumption that there is no turnover of the preformed albumin, the authors calculated that 63 per cent of the dicarboxylic acids were replaced. Again the figures indicate a net production of plasma protein *in vitro* such as can be demonstrated *in vivo* by plasmapheresis and nutritional depletion methods.[175]

Studies were made of the factors necessary for the maintenance of the incorporation rate in the system, with the following results: inhibition in the absence of oxygen and in the presence of cyanide, azide, iodoacetate, dinitrophenol, arsenate, and fluoride. One-hundredth molar malonate had a negligible effect; either Ca or Mg ion was required for the maximum rate of synthesis.

The question of the mechanism of the protein synthesis which occurs in this case has been broached by the study of Peters.[176] It was shown that there was a lag in the incorporation of the label from bicarbonate-C^{14} into albumin which presumably is not attributable to the time required to build up the necessary concentrations of labeled dicarboxylic acids because there is apparently no similar lag in the incorporation into the total liver protein. Similar results were also obtained when glycine-1-C^{14} replaced bicarbonate-C^{14}. There was likewise a lag in the appearance of substance with the physical-chemical and immunological properties of albumin. Therefore, according to Peters, there exists in liver slices some precursor which is not easily washed out, and which is converted by an "all at once" process into serum albumin.

First, with regard to the lack of knowledge of the specific activity of the precursor: this is largely circumvented by a reverse isotope experiment in which labeling occurred for 60 min.; then the medium was changed to one without bicarbonate-C^{14}, and the experiment was continued. More labeled albumin was produced in the second 60-min. period, presumably from the precursor. The only question which remains is whether the 5-min. washing procedure between the two intervals served to wash out all the

(175) S. C. Madden and G. H. Whipple, *Physiol. Revs.* **20,** 194 (1940); S. C. Madden *in* Youmans, Symposia on Nutrition, II, Plasma Proteins, Thomas Publ. Co., Springfield, 1950, p. 62.

(176) T. Peters, Jr., *J. Biol. Chem.* **200,** 461 (1953).

labeled precursors—the dicarboxylic acid and their amides together with such small peptides as may have been present. The lag observed in these experiments should be compared with the lag observed by Siekevitz[127] in the uptake of alanine-1-C^{14} by the microsome–mitochondria fraction of liver (Fig. 4). The second weakness in these experiments is that the conclusion rests on comparisons between incorporation into total liver protein—a large pool of miscellaneous proteins—and incorporation into a small individual pool of albumin.

b. Ovalbumin Synthesis and a Consideration of the Mechanism of Protein Synthesis

Anfinsen and Steinberg in a series of studies[177] have attempted to probe more deeply into the mechanism of protein synthesis. Experiments were conducted in much the same way as those done on serum albumin synthesis, except that oviduct tissue replaced the liver slices. Again, the labeling agent generally used was bicarbonate-C^{14} with incubation times from 1 to 6 hours. Activities in protein were ninhydrin stable. The ovalbumin, isolated in crystalline form after the incubation, showed only the A_1-A_2 complex in the electrophoretic pattern (see Chap. 16), but contained some free amino acid which could be removed by dialysis.

When ovalbumin is treated with an enzyme from *Bacillus subtilis* it is broken down into a large fragment, plakalbumin (PA), together with several smaller fragments which have the constitution: ala.ala, ala.gly.val.asp., and ala.ala.ala.gly.val.asp. Some glutamic acid is also split off.[178] The hexapeptide and its split products (apart from the glutamic acid) will be referred to henceforward as HP. In the first of these experiments the aspartic acid components of the PA and HP were the focus of attention, since both become labeled from the bicarbonate-C^{14} in the incubation process. When subjected to analysis, the specific activities of the aspartic acid moiety in PA and HP were found to be different, the HP-asp being about 2.4 times (1.3–3.5, depending on the time of incubation?) as active as the PA-asp.

In other experiments, labeling was done with both bicarbonate and with alanine-1-C^{14}.[177] The labeled ovalbumin was degraded using either the bacterial enzyme or by digestion with crystalline pepsin. Ala.ala was isolated from the first type of degradation by separation on Dowex–formate[179] with further purification by paper chromatography; the peptide

(177) C. B. Anfinsen and D. Steinberg, *J. Biol. Chem.* **189,** 739 (1951); *ibid.* **199,** 25 (1952).

(178) K. Linderstrøm-Lang and M. Ottesen, *Nature* **159,** 807 (1947); *Compt. rend. trav. lab. Carlsberg. Sér. chim.* **26,** 403 (1949); C. A. Villee, K. Linderstrøm-Lang, and M. Ottesen, *Federation Proc.* **9,** 241 (1950).

(179) S. Moore and W. H. Stein, *J. Biol. Chem.* **176,** 367 (1948).

mixture resulting from pepsin digestion was fractionated on paper. The specific activities of both alanine residues in the ala.ala were determined by appropriate methods, as well as those of the alanine in the various peptides. In addition, specific activities of the PA-glu and that freed by the enzymatic digestion were determined.

Briefly, all these specific activities turned out to be somewhat different, except those in ala.ala, whether the labeling proceeded via the amino acid or bicarbonate.

Deductions Concerning the Mechanism of Protein Synthesis Made from (These) Experiments. Since the synthesis of protein is such an intriguing subject, it is not surprising that there is more speculation and less experimentation than is usual in the scientific field, the present effort being an exceptional one. For more complete reviews and speculations in this connection the reader is referred to the following literature.[180–184] Many authors have allowed themselves considerable latitude in speculating on the role of nucleic acids in the process of protein synthesis,[183,160] but on the whole there does not appear to be too much profit in these flights of fancy. Here the subject will be treated somewhat differently with the principal objective of definition leaving out as many of the complicating factors as possible. Having done this it is hoped that the validity of the present approach will be either substantiated or that other methods will be demonstrated as necessary. The assumption has been made that a partial solution is possible in the foreseeable future. By Occam we are admonished not to multiply hypotheses until the simplest has been disproved. However, with regard to protein synthesis the designation of the simplest is not easy and there appears to be no particularly valid reason to discard any one of a large number, or at least only a few of the wilder variety.

It is possible to formula several different mechanisms for protein synthesis all of which have approximately equal validity. Two principal types of mechanism have been proposed: i.e., (*1*) template and (*2*) peptide.

By the *template mechanism* it is generally understood that amino acids are accumulated on a template until a sequence is established corresponding to that in the protein (or large peptide). Then by a *Deus ex machina* transformation, protein is built up. By the *peptide mechanism* it is understood that a mixture of peptides is synthesized and that the protein pattern is built up by polymerization.

(180) J. H. Northrop, Synthesis of Proteins, Conference on the Chemistry and Physiology of Growth, Princeton University, 1946; J. H. Northrop, M. Kunitz and R. M. Herriott, Crystalline Enzymes, 2nd Ed., Columbia University Press, New York, 1948.

(181) M. Stacey, *Quart. Revs.* (London) **1,** 179 (1948).

(182) F. Haurowitz, Chemistry and Biology of Proteins, Academic Press, New York, 1950.

(183) A. L. Dounce, *Enzymologia* **15,** 251 (1952).

(184) P. N. Campbell and T. S. Work, *Nature* **171,** 997 (1953).

It is convenient to subdivide these mechanisms as follows into:

1*a* Template simultaneous	2*a* Peptide large
1*b* Template serial	2*b* Peptide small
	2*c* Peptide serial

By these terms the following is meant:

(1*a*) Template simultaneous: as defined above; (1*b*) Template serial: template mechanism in which the peptide is build up step by step from either end while part at least is held in place on the template structure. No free peptides appear in the medium unless the mechanism breaks down. (2*a*) Peptide large: the units used in the synthesis are large, i.e., the protein pattern is already well established in the peptide (the large peptides used are probably not common to other proteins in contrast to the small peptides); (2*b*) Peptide small: the protein is built up by the polymerization of relatively small peptides (2–3 units); and (2*c*) Peptide serial: peptides are built from small to large either serially or by transamidation.

Besides these mechanisms which obviously may overlap, other reactions have been proposed in order to account for the appearance of isotopic amino acids in protein molecules. It may be assumed (*1*) that in a given case there is net protein synthesis, or (*2*) a preformed protein is degraded just sufficiently at some specific point to allow the incorporation of the isotopic label. For instance, a terminal amino acid may be removed by a carboxypeptidase type of reaction or by transamidation. In the former case the original terminal amino acid may be replaced subsequently either by the operation of the same enzyme or of some synthetic system. This may be classed under 1*b* or 2*c*. Alternatively, two peptide bonds may be ruptured and a new amino acid molecule introduced or a peptide may be inserted by transamidation. This type of incorporation may be accommodated under type 2*a* plus 1*b* (or 2*c*), although actually the mechanisms are not identical because the sources of the large peptides are different, one being formed by synthesis and the other by degradation (concerning utilization of protein split products see p. 1241). However, for the sake of brevity these differences will not be dealt with further. Having gone this far we can then say that the incorporation of isotopes into protein may involve any one of the five mechanisms, or any combination of them. Furthermore, *protein synthesis may simply be defined as any process involving one of these mechanisms or combinations.* Although this obviously does not fit with what is generally thought of as protein synthesis, yet, having a general term to encompass all possible processes is of obvious advantage. *In effect they may be indistinguishable except in special cases.*

With this aspect of the matter defined, it is possible to extend the basic concepts to include nucleic acid, carbohydrate, or any other substance which may eventually end up in the unique and subtle entity designated "pure protein."

One of the main advantages of the template idea is that it may be evoked at any stage in the process of synthesis to fill in and banish the unseemly vacuum until such time as new knowledge makes its revocation (or substantiation) possible. By this it is not necessarily implied that the template mechanism is not probable, rather the reverse, but enzymatic specificity in some other form can nearly always be induced to substitute in the same role.

With respect to protein synthesis, in general if the protein under consideration contains, like insulin, several distinct peptide chains, then a template mechanism may be postulated for each chain.

The experimental approach of Anfinsen and Steinberg,[177] therefore, is an attempt to justify discarding one of the two main mechanisms. Needless to say, it is almost impossible to see how a template mechanism as defined by 1*a* could possibly be entertained considering the almost insuperable kinetic and thermodynamic requirements of such a postulated mechanism. Such a mechanism requires that the amino acid residues of a given type in a protein be laid down simultaneously. Hence, if these residues all come from the same homogeneous pool *at the same time* they would all have the same specific activity if labeled. There appears to be no actual experimental demonstration that the dicarboxylic acid precursors are maintained at the same specific activity throughout the period of the experiments on ovalbumin synthesis. If this is not the case, the whole argument falls down, so this possibility will not be entertained here.[185] The experiments show without a shadow of a doubt that the specific activities are different. Hence, mechanism 1*a* may be discarded, unless it is assumed that amino acids are bound to a template for a sufficient length of time so that the pool specific activity changes before the template is completely filled. [In this connection the argument concerning SS-bound peptides is of interest (sec. III-2*a*).]

In order to clarify the next step in the argument, Fig. 5 should be consulted.

From the experiments under consideration there is information concerning the product ovalbumin and the amino acid precursor (specific activity but not concentration). In between there is a large *terra incognita*. Moreover, the relationship between the rates of anabolic processes (↑) and catabolic processes (↓) is not known. Consequently, a considerable field is left in which to speculate, the more so because net synthesis was not demonstrated in this case as with serum albumin.

(185) C. E. Dalgliesh, *Nature* **171,** 1027 (1953).

It is obvious that if the degradation of ovalbumin, or of any other labeled protein in the system, is occurring due to the activity of catheptic enzymes, then any small peptides which may be synthesized will be diluted by similar peptides arising from the catabolic process. The larger the peptide the less chance there will be for dilution. Consequently, until such time as more definite information is available concerning the relative rates of catheptic action, of synthetic action, and of relative size of the peptides, and lastly, of the pool sizes of these peptides together with their rates of formation and the protein pool size, not too much can be gleaned from these results concerning the relative importance of mechanisms *2a*, *2b*, *2c*, or *1b*, or of a mechanism involving

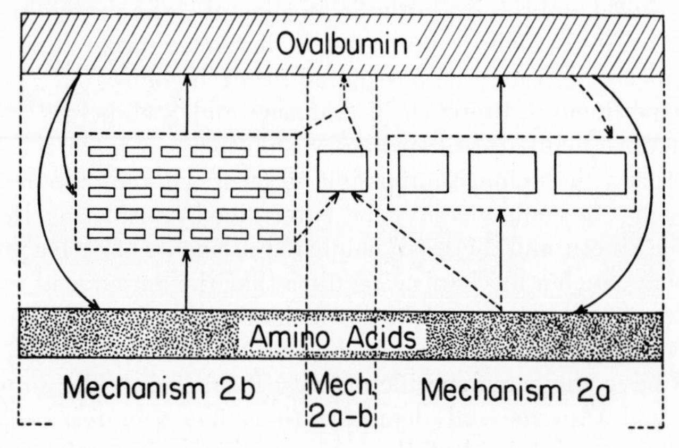

FIG. 5. Illustration of possible protein synthetic mechanisms *2a* and *2b* and a combined system *2a-b*.

template synthesis of large (and small) peptides. The possibility that amino acids in various loci are exchanged at different rates is covered by mechanism *2b* plus *1b* or *2c* if the intermediate peptides are looked upon as being of large size (or very large and small *2a*, *2b* plus *1b* or *2c*).

The question whether catheptic enzymes are involved in the synthetic reaction is of major importance. On the basis of some experiments of a negative nature, Loftfield and co-workers[186] have concluded that peptidases are not a necessary component. They showed that although the peptides DL-alanylglycine, DL-α-aminobutyrylglycine, and DL-leucylglycine were quite readily hydrolyzed by the enzymes in a suspension of rat liver homogenate at pH 8.2, the foreign peptide at a rate intermediate between the other two, yet DL-α-amino butyrate was "incorporated into the protein to only a very slight extent, if at all," when incubated with

(186) R. B. Loftfield, J. W. Groves, and M. L. Stephenson, *Nature* **171,** 1025 (1953).

slices from the livers of rats. The question as to the incorporation of foreign amino acids is dealt with in sec. IV-4.

The conclusion that can be made regarding the experiments of Anfinsen and Steinberg is that it will be necessary to garner information concerning the numerous unknowns in the system or else to modify the approach in some manner. In view of the success of Bloch and co-workers in investigating the synthesis of GSH and considering the preceding treatment of the problem, it is obvious that the most logical approach must involve the following steps: (*1*) the attainment of synthesis of a sufficient magnitude in a homogenate system, (*2*) the removal or inactivation of the catheptic enzymes, and (*3*) the isolation of peptides of the same sequences from the protein after incubating the homogenate system with peptides with multiple labels, or (*4*) the demonstration of the incorporation of large peptides (obtained by degradation of the protein being synthesized?). Until such experiments are performed, interpretations will remain more subjective than objective.

c. *Amylase, Xanthine Oxidase, and Antibody Synthesis*

(*1*) *Amylase.* In a series of papers Hokin[187] has shown, by enzymatic methods, the *in vitro* synthesis of a specific protein, pancreatic amylase. Slices from the pancreatic tissue of pigeons, pretreated with carbamylcholine to deplete the tissue of as much as possible of the preformed enzyme and zymogen, were incubated at 40° in a medium containing a supplement of a casein hydrolyzate and glucose. The amylase was determined by conventional methods before and after incubation.

TABLE XVII

IN VITRO SYNTHESIS OF PANCREATIC AMYLASE BY TISSUE FROM PIGEONS[187]

Saline medium with glucose and a supplemental casein hydrolyzate, at 40°, aerobically.

Incubation time, min.	Enzyme units found			Synthetic rate[a] units/hr.
	Medium	Tissue	Total	
0	0	59	59	—
30	16	48	64	10
60	21	59	80	32
120	31	75	106	26
180	31	117	148	42

[a] Synthesis expressed as units according to B. W. Smith and J. H. Roe [*J. Biol. Chem.* **179,** 53 (1949)], in the interval between adjacent measurements. Average rate 20 units/hr. or 5 mg. amylase/g. dry weight/hr.

(187) L. E. Hokin, *Biochem. J.* **48,** 320 (1951); *ibid.* **50,** 216 (1951); *Biochem. et Biophys. Acta* **8,** 225 (1952).

The progress of the synthesis with time is shown by the data in Table XVII.

The formation of the amylase is clearly demonstrated, and since it does not occur in the absence of oxygen the interpretation of the results cannot be complicated by the formation of enzyme from zymogen. Synthesis was also shown to be inhibited by cyanide, dinitrophenol, and iodoacetate. The rate of synthesis increased with the addition of a mixture of the following ten amino acids: tryptophan, threonine, valine, tyrosine, lysine, leucine, isoleucine, histidine, phenylalanine, and arginine.

An excretion of the enzyme into the medium resulted from the addition of carbamylcholine or of eserine plus acetylcholine. These agents caused an increase in RNA phosphorus turnover which did not occur when protein synthesis itself was stimulated by other means. They had no effect on protein synthesis as such.

(*2*) *Xanthine Oxidase.* In a similar manner, Dhungat and Sreenivasan[188] were able to demonstrate the synthesis of xanthine oxidase *in vitro* using slices of liver tissue from protein-depleted rats. The medium used to obtain the best results contained inactivated horse serum, glucose, oxygen, bicarbonate, methionine, glycine, and riboflavin. When riboflavin alone was added, the protein part of the enzyme did not appear to be synthesized. Noteworthy is the fact that in these experiments D-amino acid oxidase did not show any increase.

(*3*) *Antibody.* Ranney and London[189] showed the *in vitro* formation of the antibody to Type III pneumococcus by demonstrating the incorporation of labeled carbon from glycine-1-C^{14} into this specific protein. Slices of liver and spleen from rabbits, actively forming antibody due to pretreatment with the antigen, were incubated in a suitable medium containing the labeled glycine. In control experiments similar slices from the tissues of animals not generating antibody were incubated with preformed antibody and glycine. Following the incubation the antibody was precipitated with the specific polysaccharide antigen, and it was found that none of the labeled glycine had been introduced into the preformed antibody in the control experiments but that the label was found in the antibody obtained from the tissues of the antigen-forming animals, as shown in Table XVIII.

The absence of activity in the preformed antigen in the control experiments shows that that found in the others could not have been due to contamination by adsorption or some similar process. Nor was there any exchange of glycine residues between the preformed protein and the labeled glycine in the medium.

(188) S. B. Dhungar and A. Sreenivasan, *J. Biol. Chem.* **197,** 831 (1952).
(189) H. M. Ranney and I. M. London, *Federation Proc.* **10,** 562 (1951).

TABLE XVIII
INCORPORATION OF LABELED AMINO ACID INTO ANTIBODY IN VITRO[189]

	Total antibody found		Specific activity of precipitate[a]	
Tissue	Control, mg.	Experimental, mg.	Control,[b] counts/min.	Experimental, counts/min.
Spleen	4.4	0.25	0	630
Liver	25.	4.6	2	93
Kidney	6.4	1.9	8	7

[a] Different amounts of glycine-1-C^{14} were added to each flask except the first pair (spleen 10 mg.), so no direct comparison of activities among the tissues or between control and experimental pairs, appears to be possible.

[b] Control refers to tissues taken from animals not synthesizing specific antibody. The synthesis of other proteins has been reported more recently.[189a]

4. INCORPORATION OF ISOTOPIC AMINO ACIDS INTO THE PROTEINS OF TISSUE CULTURES AND PERFUSED ORGANS

a. Tissue Cultures

Hull and Kirk[190] studied the growth of tissue cultures in relation to the uptake of labeled phosphate by DNA, but it remained for Winnick and co-workers[169,170,191] to investigate protein synthesis in cultures. Before considering these experiments the differences between these tissue culture studies and tissue slice–homogenate experiments will be considered: (*1*) The tissue-culture experiments are long-term experiments for the most part, two weeks or more, as compared with hours or less for homogenates; (*2*) the culturing is done in media which are changed frequently, including the addition of fresh labeled amino acid (constant isotope level in the pool); (*3*) the type of cell changes during culturing from that typical of the original tissue to one composed mainly of fibroblasts; (*4*) there may be vast changes in the amount of protein in the system, including increases in extracellular protein. Since these differences prevail it is impossible to interpret this type of experiment from the same point of view as that taken in the previous section. In culture experiments the specific activity attained by protein in the long period will be determined by the specific activity of the labeled precursors, and the protein specific activity will asymptotically approach a specific activity determined by the amino acid specific activity unless "dead" protein is present. However, owing to the fact that the protein pool

(189*a*) M. V. Simpson and S. F. Velick, *J. Biol. Chem.* **208,** 61 (1954); M. Heimberg and S. F. Velick, *ibid.*, **208,** 725 (1954).

(190) W. Hull and P. L. Kirk, *J. Gen. Physiol.* **33,** 335 (1950).

(191) M. D. Francis and T. Winnick, *J. Biol. Chem.* **202,** 273 (1953).

changes in both size and type, the mathematical treatment of the data become rather involved.

The experiments[169,191] showed that there was a net increase in protein in the cultures maintained in 25 per cent embryo extract (EE); e.g., lung cultures increased tenfold in protein content in 50 days, the increase being approximately linear with time. With heart muscle there was a lag period of some 10 days, following which a similarly large increase in protein formation occurred, but no corresponding increase in nucleic acid was found in any case. The net protein formation was reduced at 5° and there was inhibition by the following reagents: fluoride, 5 m*M*; cyanide or azide, 1 m*M*; dinitrophenol, 0.5 m*M*; colchicine, 0.15 m*M*; and malonate, 10 m*M*. It is of interest to note that if fluoride or dinitrophenol was added after the 20th day of culturing, it was ineffective in preventing the further increase in protein content, the medium containing EE.

As cultures growing in EE over a period of 6 days aged, an increasingly greater percentage of the building material came from the protein in the EE and decreasingly less from the free amino acids present, the actual values noted being from 4 to 29 per cent per day. Does this suggest that the proteins of embryo extract are more readily attacked by the fibroblasts?

When EE containing labeled protein (methionine-S^{35} or phenylalanine-3-C^{14} labeling *in vivo*) was used in the culturing medium, the specific activity of the protein in the culture increased very rapidly with time. If labeled EE was used and in addition the proper unlabeled amino acid was added, to provide a diluent, then the apparent uptake of the labeled amino acid in the EE–protein was lowered. These results might be interpreted to mean that utilization of the protein proceeded after breakdown to amino acid if it were not for the various complications, noted by the workers:[191] (*1*) inhibition of the cultures by the addition of the unlabeled amino acid, (*2*) lack of equilibration between the unlabeled amino acid and free amino acid released from the EE, and (*3*) synthesis of utilizable peptides or activated amino acid derivatives from the unlabeled amino acid. Other experiments done with doubly labeled EE have the same inherent stumbling blocks in the way of clear interpretation.

The effect of a series of amino acid analogs was also tested, and it was found that DL-methoxinine, DL-ethionine, DL-β-3-thienylalanine, and DL-*o*-fluorophenylalanine at concentrations of the order of 1 m*M* would significantly inhibit net protein formation. The analogs also blocked the incorporation of the corresponding amino acids (DL-methionine-2-C^{14} and DL-phenylalanine-3-C^{14}) into the protein. The fluorinated analog and ethionine also inhibited the incorporation of DL-alanine-1-C^{14}. The concentrations of the labeled amino acids were around 0.04 m*M*.

These experiments are of considerable interest in relation to the mechanism of protein synthesis, and confirm the effect of the same analogs on growth in intact animals,[192,193] and the effect of DL-ethionine on the incorporation of DL-methionine-S^{35}

(192) D. W. Woolley, A Study of Antimetabolites, John Wiley and Sons, New York, 1952.

(193) J. D. Lewis and M. D. Armstrong, *J. Biol. Chem.* **188,** 91 (1951).

and of glycine-1-C^{14} into the proteins of the tissues from intact rats and mice.[194] It had been shown previously that DL-ethionine inhibits the uptake of DL-methionine-S^{35} by liver slices, an effect reversed by the addition of extra DL- or L-methionine (1:1).[195] Likewise, ethionine can be shown to inhibit the uptake of glycine-1-C^{14}, serine-3-C^{14}, and alanine-1-C^{14} by liver slices.[196] However, in many cases L-methionine in concentrations of the same order as the ethionine used in the last mentioned experiments (10–30 mM) had similar inhibitory effects. In connection with these experiments as with those of Francis and Winnick, there are difficulties in interpretation of the same type as those mentioned in sec. III–2*c*(*2*) and III–2*d* with respect to the effect of amino acid concentration and of inhibitors on incorporation reactions. The reality of such effects is shown by the data of Levy and co-workers[197] who have shown that in animals treated with ethionine the concentration of several amino acids in the liver undergoes a significant increase. The analogs may have had effects on precursor concentration or specific activity at the site of incorporation so that until more is known concerning these factors, the data cannot be interpreted in any satisfactory manner.

Winnick[169] and co-workers[191] observed that the conditions requisite for the production of protein with the highest specific activity due to incorporation of labeled amino acids, such as DL-alanine-1-C^{14}, glycine-1-C^{14}, and DL-phenylalanine-3-C^{14} from the medium, were not those which resulted in optimum growth and maintenance of nucleic acid concentration. Because the nature and size of the protein pool is varying and only protein specific-activity values are given, the proper interpretation of the data is uncertain. In any event the percentage of the various amino acids in the protein of the cultures which was replaced was high, amounting to more than 70 per cent in 9 days when glycine was used under the best conditions for incorporation.

b. Perfused Livers

Miller and co-workers[198] carried out very interesting experiments on plasma protein synthesis in rats by using the liver perfusion technique. The livers used were taken from fasting rats and were perfused with whole blood generally containing the DL-form of lysine-6-C^{14} together with variable amounts of glucose and amino acids.

Rates of protein synthesis were calculated by making the following assumptions: (*1*) All the activity in the proteins was in the form of L-lysine; (*2*) D-lysine was not utilized for synthesis; (*3*) the plasma proteins all contained 7 per cent lysine; and (*4*) incorporation results from protein synthesis. The second assumption is justified by the observation that D-lysine-6-C^{14} was hardly utilized at all by the perfused liver over a period of 6 hours.

(194) M. V. Simpson, E. Farber, and H. Tarver, *J. Biol. Chem.* **182,** 81 (1950).
(195) M. Simpson, Dissertation, University of California, Berkeley, 1949.
(196) M. V. Simpson, J. Rutman, and H. Tarver, unpublished data.
(197) H. M. Levy, G. Montanez, E. A. Murphy, and M. S. Dunn, *Cancer Research* **13,** 507 (1953).
(198) L. L. Miller, D. G. Bly, M. L. Watson, and W. F. Bale, *J. Exptl. Med.* **94,** 431 (1951).

On the basis it was found that the isolated liver synthesized globulin more rapidly than albumin and that fibrinogen was formed more slowly than either. The rate of synthesis gradually fell off over a period of 6 hours. Maximum rates of incorporation were observed when mixtures of essential and non-essential amino acids together with glucose were supplied. Leaving out either the glucose or non-essential amino acids resulted in reduced incorporation rates, and the omission of the essential amino acids caused the rates to decline very considerably. It was shown that when the rate of incorporation fell off owing to the lack of amino acids, the rates could be increased again by adding more of the amino acid mixture to the perfusion fluid. The maximum rate observed amounted to approximately 4.2 mg. *plasma protein*/g. liver protein/hr., considerably higher than the rate of *albumin* synthesis observed by Peters and Anfinsen[129] in chicken liver, namely 0.7 mg. Abdou and Tarver[199] in experiments on intact rats, in which the degradation of *plasma protein* was actually measured, calculated a rate of formation of 5.2 mg./hr./200-g. rat, or assuming 1.3 g. *protein* per rat liver this amounts to 4 mg. *plasma protein*/g. liver protein/hr. Since this rate is grossly underestimated owing to the fact that no account has been taken of the extravascular *plasma protein*, which introduces a factor of 3 or more in the calculation, the maximum rate observed *in vitro* with slices of liver from chickens is considerably less than that observed with livers from rats *in vivo*. The rate of plasma protein synthesis in the perfused liver may be likewise underestimated because it is assumed in the calculation of the results that the labeled lysine is undiluted.

It is clear from these experiments that the incorporation reaction is directly tied up with the utilization of both essential and non-essential amino acids.

5. Conclusions

It has been demonstrated without any reasonable doubt that incorporation of amino acids into proteins in homogenates or homogenate fractions represents, in major part, incorporation by the synthesis of peptide bonds or protein synthesis; also, net protein synthesis has been demonstrated to occur *in vitro*. Whether all the incorporation reactions observed involve the formation of peptide bonds is uncertain.

It may be concluded that the process of incorporation resembles in its energetic requirement the processes described earlier where the synthesis of isolated peptide or pseudo-peptide bonds to give a soluble product was considered. Energy must be supplied.

In addition, some kind of a cofactor and/or other soluble factor(s)

(199) I. A. Abdou and H. Tarver, *Proc. Soc. Exptl. Biol. Med.* **79,** 102 (1952).

may be involved in the synthetic process as well as "microsomes," or some combination of these small particles with larger ones of the mitochondrial type. Beyond this, the relationship between the synthesis and peptides, the RNA protein of the microsomes, and DNA of the nuclei is preponderantly in the realm of speculation.

It is clear also from the summation effect with several labeled amino acids that the incorporation problem is much more than one of energetic in-feed. Other important factors must be operative.

Since activation of homogenate preparations and reticulocytes requires a limited number of amino acids only, it would appear superficially that net protein synthesis is not involved but for the fact that amino acids can be provided by the catabolic processes occurring simultaneously in the preparations. With respect to amino acid requirements only, the *in vitro* synthesis of amylase and the synthesis of plasma protein by the perfused liver, which proceed best when all the amino acids are supplied, behave in a manner such as would be predicted from studies of a nutritional type in the intact animal.[200,201]

In general, the systems studied are not entirely satisfactory for the following reasons: The precursor system is not defined nor is the relationship between the catabolic and anabolic system known. In many cases it is not known whether a single protein or a large heterogeneous mixture is being synthesized.

Little can be gleaned concerning the mechanism of protein synthesis from any of the experiments noted except by the employment of wishful thinking.

IV. Incorporation of Amino Acids in Vivo—Turnover

1. Incorporation of Amino Acids into Proteins of Different Tissues

Since the pioneering investigations of Schoenheimer and co-workers with DL-tyrosine-N^{15},[202] with ammonium citrate-N^{15},[203] and especially those with L-leucine-N^{15},[204] all carried out by feeding the labeled materials to intact adult rats for several days, it has been apparent that different

(200) P. R. Cannon, C. H. Steffee, L. J. Fraser, D. A. Rowley, and P. C. Stepto, *Federation Proc.* **6,** 390 (1947).

(201) E. Geiger, *J. Nutrition* **36,** 813 (1948); *Science* **108,** 42 (1948); *ibid.* **111,** 594 (1950).

(202) R. Schoenheimer, S. Ratner, and D. Rittenberg, *J. Biol. Chem.* **127,** 333 (1939).

(203) D. Rittenberg, R. Schoenheimer, and A. S. Keston, *J. Biol. Chem.* **128,** 603 (1939).

(204) R. Schoenheimer, S. Ratner, and D. Rittenberg, *J. Biol. Chem.* **130,** 703 (1939).

tissues take up labeled amino acids into their proteins at quite different rates, so that when examined at a short interval after cessation of administration the specific activities of the proteins of different tissues are found to be widely divergent. These and more recent experiments carried out by administering single or multiple doses of variously labeled amino acids by different routes have led to rather similar results. They have been reviewed elsewhere[107,117,205–207] and will not be dealt with in detail here. In brief, the tissues which rapidly attain high specific activity are intestinal mucosa, liver, kidney, spleen, pancreas, and bone marrow. Other tissues like skin, muscle, brain, and erythrocytes attain only relatively low activities.

The significance of observations of this type is limited for several reasons.

(*1*) The relative activities of the different tissues after administration of labeled amino acid varies with the time at which they are compared. A week after the initial labeling the specific activity of protein from the intestinal mucosa may be lower than that of muscle although initially the situation is quite the reverse.[208] The shift which occurs in the relative activities of plasma albumin and globulin in dogs following the administration of lysine-6-C^{14} also provides a good illustration of this type of change.[209]

(*2*) One tissue may incorporate amino acid into protein which is retained *in situ*, whereas another may incorporate amino acid into protein which may then be lost from the tissue in the form of secretion, e.g., albumin from the liver, enzymes from the glands secreting into the stomach and intestine, hormones from various endocrine glands, and collagen which may be deposited extracellularly.

(*3*) Some cells have a finite lifetime which largely determines the incorporation into their protein, e.g., erythrocytes.

(*4*) The protein in one tissue may become labeled by transfer of label from another tissue, e.g., muscle may be labeled by transfer of labeled amino acid (protein) from liver and other internal organs.[210]

(205) R. Schoenheimer, The Dynamic State of Body Constituents, Harvard Univ. Press, Cambridge, Mass., 1942.

(206) B. Vennesland, *Advances in Biol. and Med. Phys.* **1**, 45 (1948).

(207) H. Tarver *in* Greenberg, The Amino Acids and Proteins, Thomas Publ. Co., Springfield, 1951.

(208) F. Friedberg, H. Tarver, and D. M. Greenberg, *J. Biol. Chem.* **173**, 355 (1948).

(209) L. L. Miller, W. F. Bale, C. L. Yuile, R. E. Masters, G. H. Tishkoff, and G. H. Whipple, *J. Exptl. Med.* **90**, 297 (1949).

(210) D. Shemin and D. Rittenberg, *J. Biol. Chem.* **153**, 401 (1944).

(5) If small doses of labeled amino acid are given, the dilution of the material in one tissue may be greater than in another owing to the different concentrations[211,212] and rates of metabolism of amino acids in different tissues. Thus, apparent rates of incorporation will not always bear the same relationship to the true rates.

(6) Erroneous ideas as to the relative rates of amino acid incorporation by tissues may be obtained because different tissues concentrate injected amino acid to different degrees.[213]

(7) Tissue protein is a heterogenous mixture of species, some of which may become highly labeled, whereas others may not be labeled to any significant extent. Consequently, although two tissues may attain the same specific activity in their protein, the reasons for the equality may be different. The one may contain much of a protein with an intermediate rate of incorporation, the other, a little protein with a high rate of incorporation giving at some time interval similar specific activities.

From the metabolic point of view it is not the specific activity which is attained by the protein which is of paramount significance, but rather the product of the specific activity and the total amount of the protein labeled, i.e., the rate of incorporation per unit time (replacement rate), or the rate of protein synthesis.

In view of these considerations, it is obvious that isolated observations on the specific activity of tissue proteins, or even of individual proteins, are of little value unless bolstered by other factual information.

The *efficiency* with which a labeled amino acid is utilized for protein synthesis depends on several factors such as the nature of the amino acid and its labeling,[214] the route of administration, and the size of the dose or doses which will determine what fraction of the dose is catabolized and thus lost,[215,216] and the dietary state,[117] age, and hormonal balance of the animal under investigation. If, for instance, protein labeling is attempted with a small dose of glutamic acid-C^{14} or -N^{15}, then the specific activity of the administered labeled material will be greatly reduced by the large amount of glutamine–glutamate in tissues,[217] that is, as compared with most other amino acids, and by the presumably high rate of formation or transamination of this amino acid.

(211) P. E. Schurr, H. T. Thompson, L. M. Henderson, J. N. Williams, Jr., and C. A. Elvehjem, *J. Biol. Chem.* **182,** 39 (1950).
(212) J. D. Solomon, C. A. Johnson, A. L. Sheffner, and O. Bergeim, *J. Biol. Chem.* **189,** 629 (1951).
(213) J. Awapara and H. N. Marvin, *J. Biol. Chem.* **178,** 69 (1949).
(214) H. Borsook, C. L. Deasy, A. J. Haagen-Smit, G. Keighley, and P. H. Lowy, *J. Biol. Chem.* **187,** 839 (1950).
(215) D. B. Sprinson and D. Rittenberg, *J. Biol. Chem.* **180,** 707 (1949).
(216) D. B. Sprinson and D. Rittenberg, *J. Biol. Chem.* **180,** 715 (1949).
(217) P. B. Hamilton, *J. Biol. Chem.* **158,** 397 (1945).

2. Incorporation of Amino Acids into the Proteins of Cellular Components (of Liver)

Because of the difficulties attending the interpretation of data concerning the incorporation of amino acids into the total protein of tissues, it is of interest to examine the results which have been obtained with respect to incorporation into the cellular structures of a single tissue following the administration of labeled amino acids. In these investigations some of the previously noted complications are circumvented. The data obtained by various workers are shown in Table XIX.

TABLE XIX

Relative Specific Activities of the Proteins of Subcellular Components of Liver after Administration of Labeled Amino Acids

Relative activity of microsomal protein = 100.

All amino acids labeled with C^{14} as indicated except cystine-S^{35}.

References and footnotes to Table	b 219	c 127	d 218	e 214	e 214	e 214	e 214
Labeling agent	DL-leu-1	DL-ala-1	DL-cySS	gly-1	L-his-2-imid	L-leu-1	L-lys-1
TL[a]	53	23	77	—	—	—	—
N	—	20	84	47	48	54	45
M	14	36	71	50	39	26	55
S	39	7	74	57	39	42	55

[a] TL, total liver; N, nuclei; M, mitochondria; S, supernatant.

[b] Liver of rats fractionated by the sucrose method 15 min. after the intravenous injection of leucine. The value used for the microsomes is the average of Mic-1 and Mic-i.

[c] Homogenate of liver from rats incubated *in vitro* with alanine for 15 min. Data taken from Fig. 4.

[d] Liver of rats fractionated by method of Ada,[220] nuclei separated by method of Marshak,[221] 8 hours after the injection of cystine into the peritoneal cavity.

[e] Liver of mice fractionated 30 min. after the intravenous injection of amino acids as indicated.

Except in the experiments of Lee and co-workers,[218] in which an incorporation period of 8 hours was employed, the livers from the animals were fractionated 15–30 min. after the amino acid administration to give nuclei, mitochondria, microsomes, and supernatant. In order to permit comparison among the results, the data are all calculated to a relative

(218) N. D. Lee, J. T. Anderson, R. Miller, and R. H. Williams, *J. Biol. Chem.* **192,** 733 (1951); N. D. Lee and R. H. Williams, *ibid.* **200,** 451 (1953).

(219) E. B. Keller and P. C. Zamecnik, *Federation Proc.* **10,** 206 (1951).

(220) G. L. Ada, *Biochem. J.* **45,** 422 (1949).

(221) A. Marshak, *J. Gen. Physiol.* **25,** 275 (1941).

specific activity of 100 in the microsomal protein. It is seen that with some exceptions there is a reasonable agreement in the data from various laboratories. In all experiments the microsomal protein has the highest activity, lower values being found for both the nuclear and mitochondrial protein. The results with the *in vitro*[127] labeling are also in fair accord with those obtained *in vivo*. Again it may be noted that the microsomal protein is associated with more RNA and DNA than are the other fractions.[219]

The values for the various fractions obtained by Lee and co-workers are not so widely separated as those obtained in other laboratories. The longer labeling period employed by these workers probably results in a more complete equilibration in the system, the differences in the rates of replacement in the various fractions becoming masked. Thus, it is evident that unless something is known about the change of specific activity with time, the interpretation may be difficult. At any rate the importance of the microsomes is amply demonstrated.

3. Turnover Data from in Vivo Experiments

a. The Nature of Turnover Process and Its Measurement[222]

The administration to an animal in nitrogen and sulfur balance, as measured by the daily excretion of these substances, of a labeled amino

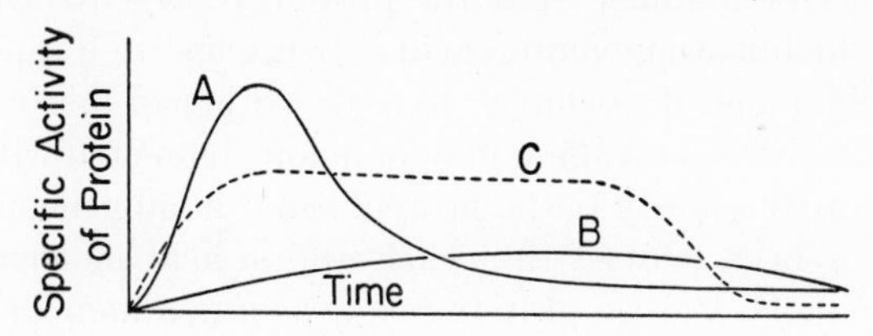

Fig. 6. General types of specific activity (S.A.)–time curves obtained following the administration of a single dose of labeled amino acid. Curve *A*, rapid turnover; curve *B*, slow turnover; curve *C*, finite life curve.

acid in small dosage or of another compound capable of labeling protein leads to labeling curves of a typical form, that is, when isotopic concentrations in protein (specific activities) are measured at different time intervals. The nature of these curves became evident following the early work with DL-methionine-S^{35},[223,208] and to be sure it had been evident since the classical experiments of Shemin and Rittenberg with glycine-N^{15}.[210] The curves found in these studies, and since confirmed by all subsequent work by others using different labeling agents, are shown in Fig. 6. They

(222) A. K. Solomon, *Advances in Biol. and Med. Phys.* **3,** 65 (1953); J. M. Reiner, *Arch. Biochem. Biophys.* **46,** 53, 80 (1953).

(223) H. Tarver and L. Morse, *J. Biol. Chem.* **173,** 53 (1948).

will be referred to henceforth as turnover curves determined by method D, the direct method.

As illustrated, there are two main types of curves, A and B, where curves of type A are typical of the tissues with rapid turnover such as liver, kidney, spleen, and glandular tissues in general, and where curves of type B are more typical of such tissues as muscle, cartilage, skin, and erythrocytes. However, the last-mentioned cells should be placed in a special category illustrated by the dotted curve C. In this curve the specific-activity scale is greatly magnified, and it is drawn on the assumption that the protein in these cells has a finite life, that is "protected" protein. What the turnover rate of the same protein (globin) outside the cells may be, is an open question, but certainly it will not behave as indicated by curve C.

If curve A is considered first and attention is directed toward the descending portion, it is found that if the data are plotted with the specific activity (i.e., concentration of isotopic material in protein), on a logarithmic scale, then for a period of several days a linear relationship with time is found.[117] Later the curve deviates from linearity and approaches the abscissa. The interpretation of this portion of curve A necessitates making certain assumptions which have limited validity.

First it must be assumed that the protein represents a homogeneous system from which isotopic amino acid is being lost in a random manner, i.e., the protein molecule behaves like any other stable chemical compound, and that there is no aging phenomenon connected with the protein molecule. Quite obviously tissue protein is not homogeneous, and, moreover, it has not been proved that each amino acid molecule of a given kind, e.g., glycine, in a given protein reacts as if in a homogeneous system relative to the other molecules of glycine. The second assumption made is that the protein all becomes labeled instantaneously in the same way and that then the labeled amino acid immediately disappears from the pool. This is manifestly not true.

Having made these two assumptions the problem of correlating the descending part of curve A with theory is the relatively simple one of isotope dilution in a homogeneous system of labeled protein.[222] If the specific activity of the amino acid in the protein or of the protein generally is denoted by C_0 at the beginning and C_t after time t, then

$$k = \frac{2.3}{t} \log \frac{C_0}{C_t}$$

or in a given system

$$t \propto -\log C_t$$

Here k is the specific velocity constant. If the time with which we concern ourselves is that for half of the reaction to be completed, $t_{1/2}$, then

$$t_{1/2} = \frac{\ln 2}{k} = \frac{0.69}{k}$$

Then if the size of the pool of labeled protein is P and T_T is the time required to remove an amount of protein equivalent to the size of the pool, i.e., the *turnover time*, then

$$T_T = t_{1/2} \times \frac{1}{\ln 2} = 1.44 t_{1/2} = \frac{1}{k}$$

and the amount of protein formed per unit time, the replacement rate (R), becomes

$$R = P/T_T$$

Hence the semilogarithmic plot of C_t against time will be a straight line from which $t_{1/2}$ may be determined. Knowing the pool size (P), the value of R may be calculated.

When we are dealing with metabolic problems it is of much greater importance to have a knowledge of R rather than simply that of $t_{1/2}$ or T_T. In other words pool size is of pre-eminent importance. Without this knowledge, deductions such as have been made by many authors[224–227] with regard to protein synthetic rates may be very erroneous. This point cannot be overemphasized.

As the descending curve A is followed, it is found that with increasing time the deviation from linearity of the semilogarithmic plot of specific activity against time becomes greater and greater. Several causes for this deviation may be suspected:

(*1*) The protein is not homogeneous or does not behave homogeneously—this is clearly the case if mixed proteins from a tissue are taken;

(*2*) the concentration of the isotope in the pool is not zero so that relabeling of the protein from either its own breakdown products or from elsewhere in the system occurs;

(224) H. Tarver and W. O. Reinhardt, *J. Biol. Chem.* **167,** 395 (1947).

(225) L. W. Kinsell, S. Margen, H. Tarver, J. McB. Frantz, E. K. Flanagan, M. E. Hutchin, G. O. Michaels, and D. P. Mc Callie, *J. Clin. Invest.* **29,** 238 (1950).

(226) V. C. Kelley, M. R. Ziegler, D. Doeden, and I. McQuarrie, *Proc. Soc. Exptl. Biol. Med.* **75,** 153 (1950).

(227) H. Kreiger, C. A. Hubay, W. D. Holden, and W. T. Hancock, *Proc. Soc. Exptl. Biol. Med.* **78,** 402 (1951).

(*3*) there is some physical discontinuity in the system, e.g., as between plasma proteins in the blood circulation, lymph, interstitial fluids, and cells;

(*4*) the labeling process is complex, e.g., methionine is changed to cystine and the two labels behave differently both in labeling and in relabeling.

The existence of curves of type *B* must be attributed largely to the operation of relabeling complications, as Shemin and Rittenberg[210] originally showed with respect to muscle protein labeling. The degree to which relabeling takes place is extremely difficult to assess because there are two types of relabeling which may, and no doubt do, occur. Tissue protein may break down and relabeling may take place intracellularly, or label from the catabolic process in one tissue, such as liver, may be transferred to another like muscle.[210] Evidently in liver, in which we have a tissue with a strategic location, with high concentrating power for amino acids and with a high replacement rate for the protein, the relabeling is not too significant at the beginning of the experiment but becomes increasingly significant as the specific activity of the amino acid in the liver approaches that in the muscle, skin, and elsewhere.

It should not be necessary to state that with respect to the turnover process the rising part of curve *A* must have the same meaning as the falling, albeit the interpretation of the rising part is complicated by factors of a type different from those which have been discussed. Before considering these factors it will be convenient to deal with another experimental method, the constant pool method, *CP*, by which it is possible to determine turnover rates and which demonstrates the relationship between the rising and descending parts of the curve. Madden and Gould[228] determined the turnover rate of fibrinogen in the dog by using both methods *D* and *CP*. To carry out the experiments by method *CP*, dogs were given labeled amino acid (L-methionine-S^{35} or yeast labeled with S^{35}, presumably in methionine and cystine residues) with the food throughout the period, thus keeping the precursor pool at approximately the same level of activity. In the other experiments, a single dose of the labeled material was administered (method *D*). Specific activities of fibrinogen were determined at intervals by appropriate methods. When these results are plotted as indicated in Fig. 7 two linear relationships are found, the one obtained while the isotopic amino acid concentration in the fibrinogen is steadily rising and the other while it is falling.

The average half life obtained, using several dogs, turned out to be the same by either method, namely 4.2 days. Preliminary results in humans indicated a half life of 5.6 days. By this experimental technique the incorporation and loss of isotope from the protein are shown to be related phenomena.

The factors other than the rate of protein synthesis which modify the rising part of curve *A* (Fig. 6) are as follows: (*1*) When the labeled amino acid is injected or given orally, time will be required to reach equilibrium with the free amino acid in the tissue, and the labeled amino acid will be diluted by that already present; (*2*) if peptides are the precursors of protein, then there will be another equilibration delay. The operation of

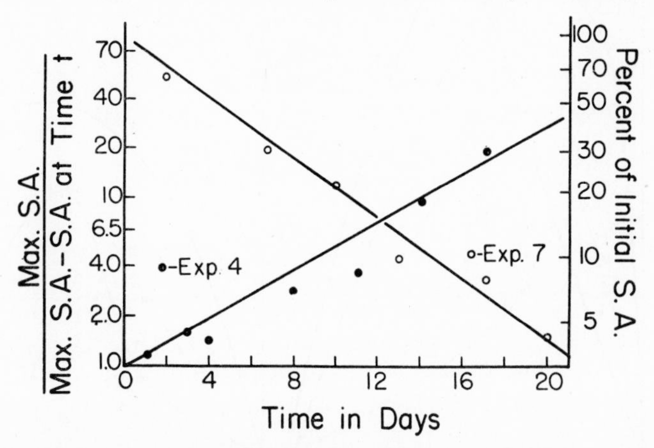

FIG. 7. Determination of the half life of fibrinogen in the dog by two methods.[228]

Expt. 7: right-hand scale, direct method, *D*, by giving a dog a single dose of yeast-S^{35}. (Presumably the yeast contained only cystine and methionine as the protein-labeling agents in significant amount.)

Expt. 4: left-hand scale, constant pool method, *CP*, by feeding a dog methionine-S^{35} in the diet. Half lives found, 4.2 and 4.6 days. Maximum S.A. is the maximum specific activity attained by the fibrinogen after reaching equilibrium with the labeled material fed.

these factors is shown in the data obtained by Larson[117] and others.[229,230] There will be a lag in the attainment of the maximum rate of labeling, but if the dose of labeled amino acid is high enough, the free unlabeled amino acid in the pool becomes relatively insignificant, and the maximum rate of incorporation observed will be the true rate of protein synthesis under the conditions. (It is assumed that the dose of amino acid does not affect a change in the rate of protein synthesis.)

This provides another method for the determination of turnover *in vitro*, a modification of method *CP*. Values calculated from the data of

(228) R. E. Madden and R. G. Gould, *J. Biol. Chem.* **196,** 641 (1952).

(229) D. R. Sanadi and D. M. Greenberg, *Proc. Soc. Exptl. Biol. Med.* **69,** 162 (1948).

(230) G. Solomon and H. Tarver, *J. Biol. Chem.* **195,** 447 (1952).

Borsook and co-workers,[214] in which relatively large doses of the amino acids indicated were injected into groups of mice which were then sacrificed at short intervals thereafter, are shown in Fig. 8.

It is noted that following the injection of any one of the amino acids employed, the maximum rates of incorporation into the visceral proteins

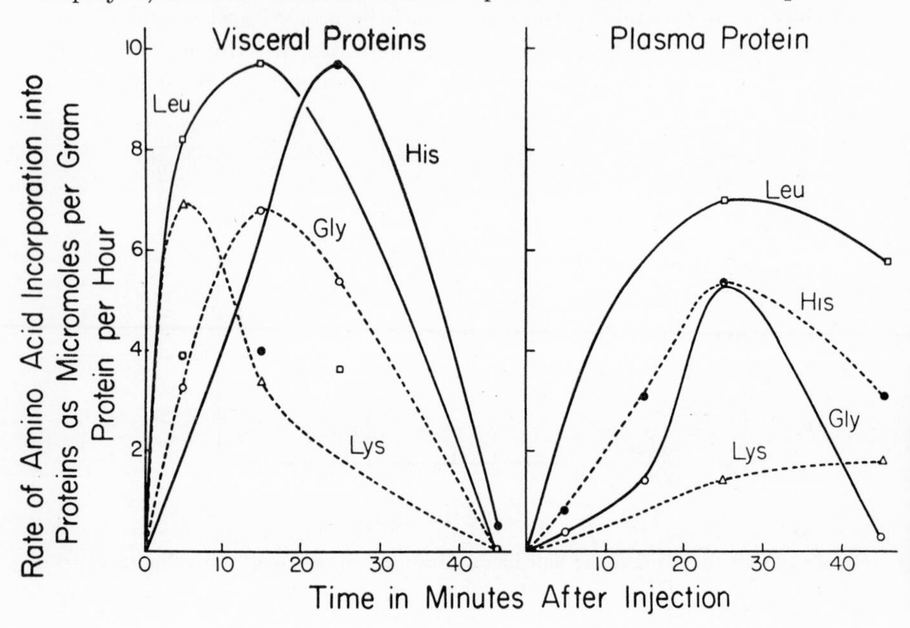

Fig. 8. Rate of incorporation of the amino acids glycine, histidine, leucine, and lysine into the visceral and into the plasma proteins of intact mice as a function of time.[214]

Amino acids were injected into tail vein of a series of 20–30-g. mice in doses, in micromoles, as follows: glycine 9, histidine 9, leucine 15, and lysine 10. Animals were sacrificed at six intervals after injection. The points on the curves are the rates of incorporation at the midpoints of the intervals assuming linear changes in rates with time. In the case of the plasma protein, the measurements were made on the whole blood protein. The values given were calculated assuming the plasma protein represents 30 per cent of the total protein, and assuming the rest to be unlabeled. The plasma protein curve following the feeding of leucine is rendered uncertain owing to the lack of concordance in the experimental data. There may be a lag in the initial uptake as observed with glycine.

occur 10–20 min. after the intravenous injection. The maximum rate for plasma protein is somewhat delayed, which appears reasonable. The actual values (taken from the curves) are as follows for the visceral and plasma proteins, respectively, glycine 6.9, 5.4, leucine 9.7, 7.0, lysine 6.9, 1.8, histidine 9.7, 5.4 micromoles/g. protein/hr. These figures should be compared with those calculated from the replacement rate of plasma protein in rats given by Tarver.[111] The values are: leucine 5.1, lysine 1.6 and histidine 1.3 micromoles/g. protein/hr.[231] It is of considerable interest to note that, apart from the value for histidine, the agreement is such as might be expected because there is no assurance that the doses of amino acid used by Borsook and co-workers[214] were sufficient to meet the requirement of flooding the system, and because mice and rats are compared. Actually the dose of histidine was relatively low yet the incorporation rate was higher than that calculated from the replacement rate.

Finally, turnover may be determined by the method adopted by Fink and co-workers,[232] which may be referred to as the Hevesey method, *H*. This involves the labeling of the protein in a donor animal, removing the protein from the donor, and transferring it to the recipient; protein specific activities are then measured at intervals in these animals. The object of this procedure is to obviate the relabeling such as may occur in method *D*. In the experiments of Fink and co-workers, the protein was labeled by administering large doses of lysine-ϵ-N^{15} to dogs, a relatively inefficient procedure with respect to the utilization of the labeled compound; later, similar experiments from the same laboratory were carried out with lysine-6-C^{14}.[209] More recently the same type of work has been carried out in rats in which the donor plasma protein was produced by administration of either serine-3-C^{14} or C^{14}-labeled *Rhodospirillum rubrum* to rats.[233,199] For convenience the particulars of the rat data will be dealt with here. The recipient animals were all males, and whether the label used was serine or *R. rubrum*, the data fell on the curve given in Fig. 9.

Considering the upper plasma protein curve, it is found that there is a linear portion which extends for 2–3 days from the beginning of the experiment. After that time the curve assumes a different slope which is maintained for at least another 4 days. The extrapolated portion of this curve is shown as the broken line labeled "Metabolic Curve." Thus there appears to be a complex situation which can be described by three curves, although this is somewhat an arbitrary analysis of the data, in

(231) The assumptions made in arriving at these figures are as follows: In the plasma protein (rat) there is, leucine 7.1, lysine 7.0, and histidine 2.1 per cent.[212] The rat synthesizes 5.2 mg. plasma protein/hr., and has a total plasma protein of 555 mg. For leucine the calculation becomes:

$$\text{micromoles amino acid incorporated/g. protein/hr.} = \frac{5.2}{0.555} \times \frac{71}{131} = 5.1$$

(232) R. M. Fink, T. Enns, C. P. Kimball, H. E. Silberstein, W. F. Bale, S. Madden, and G. H. Whipple, *J. Exptl. Med.* **80**, 455 (1944).

(233) I. A. Abdou and H. Tarver, *J. Biol. Chem.* **190**, 769 (1951).

this instance at least. Before going further it should be mentioned that in Hevesey experiments this type of curve had been found by all other workers. Inspection of the tables in the following part reveals why the slowest component is assigned to the operation of the catabolic phase

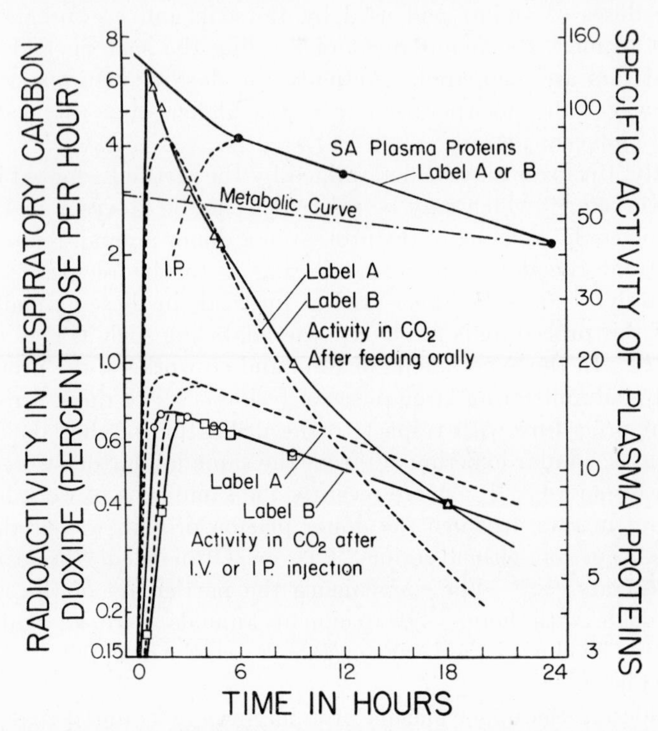

FIG. 9. Upper solid and dotted curve: rate of loss or appearance in the circulation of rats of labeled plasma protein administered intravenously, solid; or intraperitoneally, dotted curve. Dotted curve (Metabolic Curve): portion of same curve extrapolated toward $t = 0$ from $t = 72$ to $t = 168$ hours, i.e., curve representing loss of label from plasma protein due to its metabolism.[199]

Pairs of curves (△, ○, □) indicating rate of appearance of C^{14} from the labeled plasma protein in the respiratory carbon dioxide, either after oral or after intravenous or intraperitoneal injection. Label *B* refers to plasma protein biologically labeled by feeding boiled *R. rubrum;* label *A* by administering serine-3-C^{14} to donor rats. All plots are semilogarithmic.

of the turnover process. The half lives derived from this part of the curve approximate those found by methods *D* and *CP*. The precise interpretation of the fast component is less certain. However, it is clear that it is concerned with mixing phenomena between the different compartments in the system, blood, interstitial fluid, lymph, and cells. The passage of plasma protein into the lymphatic system was demonstrated by Cope and Moore[234] and others.[235,236] Moreover, Wasserman and Meyerson[237] showed in dogs, by means of tracer studies with the iodinated proteins, that there is an actual migration of albumin and globulin from the blood vascular system into the lymph and back into the circulation. The albumin traveled about 1.6 times faster than the globulin, apparently owing to different blood capillary permeabilities, not because of different lymph capillary permeabilities. The same type of circulation has been shown to exist in rats and dogs by others,[238–240] and, without the benefit of isotopes, in rabbits by Courtice and Steinbeck.[241]

It is found that biological labeling led to the production of plasma protein with different concentrations of isotopes in the several factions. This is well illustrated by the data of Miller and co-workers,[209] London,[242] and by the earlier work of Schoenheimer and co-workers.[243] The existence of these differences in specific activities between the biologically labeled albumin and globulin leads to difficulties in the unequivocal interpretation of data when activities in total plasma protein are measured.

It is generally conceded that mixing time within the blood vascular system takes only a matter of minutes; therefore, if two fast components exist, it may be assumed that the one is concerned with the transfer into interstitial fluid and lymph, and the other slower one with the transfer into cells, other than the liver cells, since it appears likely that the liver must be extremely permeable to albumin at least. The possibility exists, however, that albumin is formed at the boundary of the liver cell, so that the protein is not readily transferred back into the cell. Yet this appears to be unlikely because the C^{14}-labeled plasma proteins when injected are rapidly broken down, as shown by the data on respiratory carbon dioxide in Fig. 9. It is reasonable to suppose, as Madden and Whipple[175] have done, that plasma protein passes into cells generally, although in view of the passage of the plasma proteins into interstitial fluid and lymph this is a little more difficult to prove with or without isotopes than

(234) O. Cope and F. D. Moore, *J. Clin. Invest.* **23,** 241 (1944).
(235) H. Krieger, W. D. Holden, C. A. Hubay, M. W. Scott, J. P. Storaasli, and H. L. Friedell, *Proc. Soc. Exptl. Biol. Med.* **73,** 124 (1950).
(236) S. P. Masouredis and L. R. Melcher, *Proc. Soc. Exptl. Biol. Med.* **78,** 264 (1951).
(237) K. Wasserman and H. S. Meyerson, *Am. J. Physiol.* **165,** 15 (1951); *Cardiologia* **21,** 296 (1952).
(238) I. A. Abdou, W. O. Reinhardt, and H. Tarver, *J. Biol. Chem.* **194,** 15 (1952).
(239) L. L. Forker, I. L. Chaikoff, and W. O. Reinhardt, *J. Biol. Chem.* **197,** 625 (1952).
(240) F. W. McKee, W. G. Wilt, Jr., R. E. Hyatt, and G. H. Whipple, *J. Exptl. Med.* **91,** 115 (1950).
(241) F. C. Courtice and A. W. Steinbeck, *Australian J. Exptl. Biol. Med. Sci.* **28,** 161, 171 (1950); *J. Physiol.* (London) **114,** 336, 355 (1951).
(242) I. M. London, *in* Youmans, Symposia on Nutrition, II, Plasma Proteins, C. C. Thomas Publ. Co., Springfield, 1950, p. 72.
(243) R. Schoenheimer, S. Ratner, D. Rittenberg, and M. Heidelberger, *J. Biol. Chem.* **144,** 541 (1942).

appears superficially to be the case.[244,245,245a] The hypothesis deserves consideration that each tissue cell may have its individual permeability toward albumin (and globulin) and that this permeability may be related to cellular protein metabolism.

In view of these considerations, definite causes are not assigned for the existence of the fast components in the die-away curves for injected plasma protein until such time as the appropriate experiments have given some clearer indication as to the realities of the situation.

It is of interest to note here that by means of labeled material it can be shown that the plasma protein moves rapidly from the peritoneal cavity into the circulation.[240,233,241] There also exists a placental transmission of globulins[246] and, in general, plasma proteins often readily penetrate cellular barriers.

Owing to the difficulties attending turnover studies by the Hevesey method, which requires relatively large doses of labeled amino acids for the donor animals, numerous investigators have turned to a modification of the method by employing proteins labeled *in vitro* with I^{131}. Thus it is possible to take a pure specimen of albumin and iodinate with a trace amount of I^{131} to produce a highly labeled protein, the label being stable to dialysis and migrating with the protein in the ultracentrifuge. The protein appears to behave in a fashion closely resembling natural albumin in at least some metabolic systems as shown by the data of Sterling.[247] This worker labeled human albumin in the fashion indicated and injected the material into rabbits in large enough doses so that he was able to determine the human albumin in the rabbit serum by both immunological and radioactive methods. When the concentration of the human albumin in rabbits was followed over a period of 14 days, results in all cases similar to those shown in Fig. 10 were obtained. It is noted that the rate of disappearance is precisely the same in both sets of measurements. In contrast to these results, when albumin was labeled with Cr^{51}, a more rapid rate of loss of the radioactive label was observed.[247,248] Other workers have arrived at essentially the same conclusions from similar data[237,240,249] and, in fact, proteins labeled in this fashion have found extensive application in the field of immunology (Chap. 22) and in blood-volume studies. Thus, approximately the same half lives have been found with proteins, e.g., albumin, whether they are labeled *in vitro* with I^{131} or *in vivo* (Tables XX and XXI).

(244) I. A. Abdou and H. Tarver, *J. Biol. Chem.* **190,** 781 (1951).
(245) N. Kaliss and D. Pressman, *Proc. Soc. Exptl. Biol. Med.* **75,** 16 (1950).
(245*a*) D. Gitlin and C. A. Janeway, *Science* **120,** 461 (1954).
(246) S. G. Cohen, *J. Infectious Diseases* **87,** 291 (1951).
(247) K. Sterling, *J. Clin. Invest.* **30,** 1228 (1951).
(248) It has also been found by Margen and Lange, working in the author's laboratory, that denatured albumin behaves in an entirely abnormal manner, disappearing at an extremely rapid rate.
(249) L. R. Melcher and S. P. Masouredis, *J. Immunol.* **5,** 393 (1951).

However, it must be noted that Berson and co-workers[250] have brought forward evidence which suggests that iodinated albumin preparations may behave differently depending on the degree of iodination, and that the rate of catabolism of this type of albumin changes with time, the half life steadily increasing. This perhaps shows that all the albumin-I^{131} molecules are not behaving in the same manner, owing to different degrees of iodination. In this connection it may be observed that the

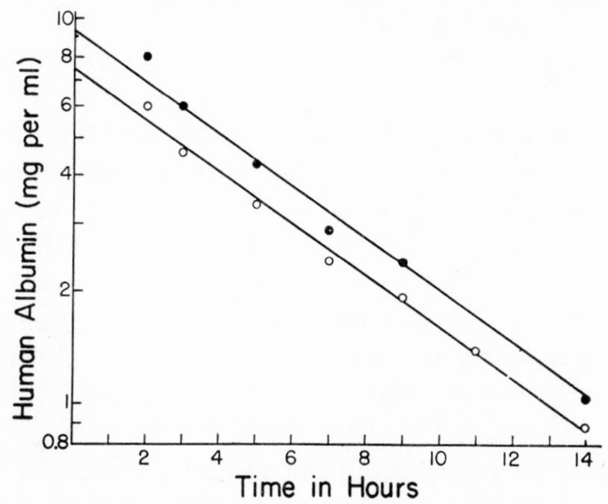

FIG. 10. Rate of disappearance of I^{131}-labeled human albumin and unlabeled human albumin from the circulation of the rabbit.[247] ○, I^{131} measurements; ●, immunochemical determinations.

iodinated protein should behave in a different manner from the normal protein when subjected to electrophoresis because of the effect of iodination on the strength of the phenolic hydroxyl group of tyrosine. Moreover, it might be anticipated that iodination would change the rate of enzymatic attack on the protein (see further discussion after Table XXI).

The behavior of erythrocyte protein presents an entirely different problem from that which has been under consideration. These cells have a finite life, under a given set of conditions, this being determined by factors not especially investigated from a biochemical point of view. The life of the cells in various conditions has been determined by several workers[251–255] and will not be dealt with further.

(250) S. A. Berson, R. S. Yalow, S. S. Schreiber, and J. Post, *J. Clin. Invest.* **32**, 746 (1953).
(251) D. Shemin and D. Rittenberg, *J. Biol. Chem.* **166,** 627 (1946).
(252) I. M. Landon, D. Shemin, R. West, and D. Rittenberg, *J. Biol. Chem.* **179**, 463 (1949).
(253) A. Neuberger and J. S. F. Niven, *J. Physiol.* (London) **112,** 292 (1951).
(254) W. F. Bale, C. L. Yuile, L. DeLa Vergne, L. L. Miller, and G. H. Whipple, *J. Exptl. Med.* **90,** 315 (1949).
(255) D. S. Amatuzio and R. L. Evans, *Nature* **171,** 797, 798 (1953).

To summarize, protein turnover studies may be made *in vivo* by any one of the following methods:

(*1*) *Direct Method* (*D*). Determination of isotopic concentration in protein at intervals after injection of labeled material.

(*2*) *Hevesey Method* (*H*). Transfer of labeled protein to recipient animal, followed by the measurements.

(*a*) Labeled protein prepared biologically.

(*b*) Protein labeled *in vitro*.

(*3*) *Constant Pool Method* (*CP*). Maintain precursor pool of constant specific activity and measure rising activities.

(*a*) Multiple dose or constant infusion method.

(*b*) Single large dose method in short-term experiments.

It is also possible to calculate half lives of various proteins from experiments carried out *in vitro*.[107] Such calculations show the essential agreement between the *in vitro* and *in vivo* determinations and provide proof that what is being measured *in vitro* is not some type of artifact due to adsorption or related phenomena.

b. *The Turnover of Tissue Protein and of Some Specific Proteins*

From the data of several groups of workers, it is possible to calculate the half lives and rates of replacement of the mixed proteins in various tissues, but such measurements do not appear to be of major importance, and only liver protein will be used as an example. Thus from the data of Schoenheimer and co-workers[204] the liver protein of rats has a $t_{1/2}$ of 6 days, Shemin and Rittenberg[210] give 5–6 days, and Tarver[207,208] estimates the value at 4 days. All these estimates are probably low and somewhat subjective, largely depending on the period over which the turnover is estimated. It is also clear from the data of Solomon and Tarver[230] that the rate of turnover of liver protein depends on the protein content of the diet which influences the state of the organ.

Data with respect to specific proteins, apart from those on the proteins of plasma, are more scarce. Bidinost[256] has compared the rates at which the whole muscle protein, myosin and actomyosins I and II take up N^{15} or C^{14} from glycine-N^{15} or -1-C^{14} or deuterium (D) from enriched water with the rates at which the same labels are taken up by internal organ (less liver) and liver proteins. It was found that the internal organ proteins of rats *in vivo* took up the labeled materials 4–7 times as fast as the whole muscle protein, that actomyosin I incorporated the labels about half as fast as actomyosin II, and that actin was even less active than actomyosin I. Liver was more active than the internal organs.

The turnover of collagen has also received considerable attention at

(256) L. E. Bidinost, *J. Biol. Chem.* **190,** 423 (1951).

the hands of Dziewiatkowski,[257] Bostrøm,[258] Robertson,[259] and Neuberger and co-workers.[260] The first-mentioned author[257] used sulfate-S^{35} to label articular cartilage of suckling rats, and showed the sulfate to be present in purified chondroitin sulfate. In rats of this age the uptake appeared to be quite rapid and probably largely occurred because of the increase in amount of the material and because of the turnover of procollagen. The rate of disappearance of the label was increased in rats injected with thyroxine. Bostrøm[258] found it possible to label the chondroitin sulfate of adult (costal cartilage) with S^{35} and found that the material had a $t_{1/2}$ of 17 days. The results are complicated, however, owing to the fact that the experimental animals grew 10 per cent during the period of the experiment. It was shown that in cartilage an enzyme system exists which is capable of catalyzing the "exchange" of sulfate from the chondroitin sulfate. Robertson[259] examined the incorporation of glycine-N^{15} into the collagen and non-collagenous proteins of liver, muscle, skin, bone, and lung of adult guinea pigs in various dietary states by feeding the labeled amino acid for a week or more. He found that ascorbic acid had no effect on the incorporation and that the non-collagenous protein of all the organs was more active than the collagen by a factor of at least two. However, the collagen always became labeled. Essentially similar results were obtained earlier by Neuberger and co-workers[260] by feeding adult rats glycine-2-C^{14}. These authors found the activity in the glycine of the collagen of tendon to be very low but significant. In young rats, skin and tendon collagen was more active but the collagen of bone and liver showed higher activities.

Since approximately one-third of the body protein is comprised of extracellular collagen[260] and the turnover of this material is extremely slow in the adult animal, it is evident that not all the protein component of the animal is in a state of dynamic equilibrium. These experiments also show that it is extremely difficult to obtain a proper estimate of turnover rates of proteins that exhibit but slight activities. This is particularly true in experiments with adult rats in which slow persistent growth exists.

More satisfactory turnover data have been obtained with respect to the blood proteins, and most of the important literature is summarized in Table XX, in which figures for $t_{1/2}$ are given. It is indeed unfortunate

(257) D. Dziewiatkowski, *J. Biol. Chem.* **189,** 187, 717 (1951).
(258) H. Bostrøm, *J. Biol. Chem.* **196,** 477 (1952); H. Bostrøm and B. Mansson, *ibid.* **196,** 483 (1953).
(259) W. V. B. Robertson, *J. Biol. Chem.* **197,** 495 (1952).
(260) A. Neuberger, J. C. Perrone, H. G. B. Slack, *Biochem. J.* **49,** 199 (1951); A. Neuberger and H. G. B. Slack, *ibid.* **53,** 47 (1953).

TABLE XX

TURNOVER OF PLASMA PROTEINS IN DIFFERENT SPECIES

Protein and species[a]		Labeling agent and method[b]		Interval, days	$t_{1/2}$, days	Reference
TP	(Rat)	DL-met-S^{35}	(D)	1–8	2.6	208
"	"	DL-met-S^{35}	(D)	1–8	4.4	223
"	"	DL-ser-3	(H)	3–7	3	233
"	"	*R. rubrum*-C^{14}	(H)	3–7	3[c]	199
"	"	L-met-S^{35}	(D)	2–6	3.5	261
"	(Dog)	DL-met-S^{35}	(D)	2–8	5.2–6.4	117
"	"	DL-lys-N^{15}	(H)	1–5	5.4	232
"	"	DL-lys-6	(H)	1–5	5.0	209
"	"	DL-met-S^{35}	(H)		5.0	117
"	"	DL-met-S^{35}	(D)	2–15	6.6	262
"	(Hum.)	DL-met-S^{35}	(D)	1–9	9.2	117
"	"	gly-N^{15}	(D)	4–9	10.0	263
"	(Hum. ♂)	gly-N^{15}	(D)	—	7.	242
"	(Hum. ♀)	gly-N^{15}	(D)	—	9.	242
A	(Rat)	L-met-S^{35}	(D)	2–12	5.	261
"	(Dog)	DL-lys-6	(H)	1–5	6.9	209
"	(Hum. ♂)	gly-N^{15}	(D)	—	20	242
"	(Hum. ♀)	gly-N^{15}	(D)	—	20	242
G	(Rat)	L-met-S^{35}	(D)	2–6	3.0	261
G	(Dog)	DL-lys-6	(H)	1–5	3.3	209
G-α_1	(Rat)	L-met-S^{35}	(D)	1–5	3.0	261
G-α_2	"	"	(D)	1–5	2.0	261
G-β	"	"	(D)	1–5	3.0	261
G-γ	"	"	(D)	1–5	3.2	261
G-$\beta + \gamma$	(Hum. ♂)	gly-N^{15}	(D)	—	12.	242
G-γ	"	"	(D)	—	19.	242
G-$\beta + \gamma$	(Hum. ♀)	"	(D)	—	9.	242
G-γ	"	"	(D)	—	18.	242
F	(Dog)	S^{35}	(D, CP)	—	4.2	228
F	(Hum.)	S^{35}	(D)	—	8.1	228

[a] TP, total plasma protein; A, albumin; G, globulin; F, fibrinogen.

[b] Labeling with C^{14} in position indicated unless other label is given.

[c] Replacement rate of 16–31 mg./200 g. rat/hr.

that data for actual replacement rates are not calculated in all the experiments in which method *H* is employed. The other methods do not permit making this calculation unless assumptions or auxiliary determinations of pool size are made.

From the data in Table XX it is evident that the Hevesey method

(261) A. Niklas and W. Maurer, *Biochem. Z.* **323**, 89 (1952).

(262) L. L. Forker and I. L. Chaikoff, *J. Biol. Chem.* **196**, 829 (1952).

(263) D. Rittenberg, *Cold Spring Harbor Symposia Quant. Biol.* **13**, 173 (1948).

(H) of turnover measurement generally gives lower values than the direct method. This is reasonable since this method is not open to such serious interference by the relabeling phenomenon. The data are in fair accord in showing that the albumin turns over at a much slower rate than any one of the globulin fractions; that fibrinogen turns over quite rapidly; and that γ-globulin probably has a relatively low rate of turnover although its turnover rate is not as low as that of albumin. These data should be compared with those obtained by Dixon and co-workers[264,265] with respect to the turnover of homologous iodinated γ-globulin and albumin in several species (Table XXI).

TABLE XXI

TURNOVER OF HOMOLOGOUS-I^{131} PROTEINS IN DIFFERENT SPECIES[264,265]

Species	Albumin: No. of subjects and method of fractionation[a]	Albumin: Time interval, days	Albumin: Half life, days	γ-Globulin: No. of subjects and method of fractionation[a]	γ-Globulin: Time interval, days	γ-Globulin: Half life, days
Cow	3 A	2–30	20.7	4 A	6–27	21.2
Human	11 A	7–28	15.0	14 A	4–21	13.1
Monkey				10 S	4–24	6.6
Dog	9 E	7–28	8.2	8 A	7–17	8.0
Rabbit	9 A	4–22	5.7	9 A	4–21	4.6
Guinea pig				13 S	4–13	5.4
Mouse	30 E	3–7	1.2	27 S	1–5	1.9
Rabbit[b]	8 A		4.5			

[a] S = salt, A = alcohol, E = electrophoresis. All subjects given extra KI.

[b] Rabbits made hyperthyroid by the daily injection of thyroxine 0.1 mg./kg. It is not known whether these rabbits were in caloric and N equilibrium or not.

It is evident from the figures that there is fair agreement between the data for albumin half lives whether biologically labeled or iodinated albumin is used in the investigation, although it appears that the iodinated protein has a somewhat longer half life. This may be because of the difference in the time intervals over which the measurements are made. This is brought out by comparison with the results of Sterling.[247] Although Dixon and co-workers[265] give a $t_{1/2}$ of 15.0 days for the iodinated albumin in the human subject over the interval 7–28 days after the initial injection, Sterling finds in 21 adult male subjects a $t_{1/2}$ of 10.5 ± 1.5 days

(264) F. J. Dixon, D. W. Talmage, P. H. Maurer, and M. Deichmiller, *J. Exptl. Med.* **96,** 313 (1952).

(265) F. J. Dixon, P. H. Maurer, and M. P. Deichmiller, *Proc. Soc. Exptl. Biol. Med.* **83,** 287 (1953).

over a period of 2–14 days after the injection, a significantly lower value (see additional comment on page 1273). The data of Sterling show that the replacement rate of albumin in the normal human subject amounts to 16.9 g. per day for six subjects averaging 71.6 kg., and that the total exchangeable albumin, i.e., the blood vascular plus extravascular albumin, is 2.25 times that in the vascular system alone. In pathological states such as cirrhosis the replacement rate was found to be abnormal.[266,267]

The data in Table XXI show very clearly that the larger the animal the slower the turnover of the plasma proteins. From what has been said before it would be anticipated that there should be some relationship between basal metabolism and replacement rate, rather than between basal metabolic rate (BMR) and $t_{1/2}$ as previously suggested.[117]

c. *The Catabolism of Heterologous Protein*

In connection with the turnover of homologous albumins and globulins in different species, it is of interest to examine the half lives of heterologous proteins. Such studies as these have been largely carried out from the immunological point of view, and many authors have based theories of protein synthesis largely on the results of immunological investigations.[182] Since the injection of foreign proteins and other specific substances (antigens) into animals causes, in many instances, the synthesis of specific γ-globulins (antibodies), it has always appeared tempting to suggest that the foreign protein is part of some template mechanism.

Dixon and Maurer[268] gave large doses of crystalline bovine serum albumin (BSA) to rabbits over a period of 6 weeks so that the BSA in the circulation of the animals amounted to about one quarter of the total albumin. This caused a rise in the total circulating globulin and a fall in the rabbit albumin (RA) in circulation, and the animals became immunologically unresponsibe to the BSA. When the injection of the BSA was discontinued, the BSA in the circulation of the rabbits fell to about one-fourth of the injection level in 2 weeks. The rates of metabolism of both the BSA and RA were investigated over the period of the injection by injecting either BSA-I^{131} or RA-I^{131}, and it was found that half lives were, respectively, 4.0 and 4.1 days, whereas the normal half life of the RA was 5.8 days (cf. Table XXI). Evidently the foreign protein was catabolized at the same rate as the homologous protein. Gitlin and co-workers[269] had previously found that the rabbit catabolized large doses of BSA and human albumin at the same rate as it catabolized its

(266) K. Sterling, *J. Clin. Invest.* **30,** 1238 (1951).
(267) M. P. Tyor, J. K. Aikawa, and D. Cayer, *Gastroenterology* **21,** 79 (1952).
(268) F. J. Dixon and P. H. Maurer, *Federation Proc.* **12,** 441 (1953).
(269) D. Gitlin, H. Latta, W. H. Batchelor, and C. A. Janeway, *J. Immunol.* **66,** 451 (1951).

own albumin. It appears also that, in the experiments of Dixon and Maurer,[268] the large amount of foreign protein did not have any great effect on the replacement rate of the rabbit's own albumin.

The effect of the immunological response on the disappearance of heterologous protein as compared with normal protein is shown in Table XXII, summarized from Dixon and co-workers[270] and others as indicated in the footnotes.

TABLE XXII

RATE OF LOSS OF NORMAL AND HETEROLOGOUS PROTEIN FROM THE CIRCULATION[270]

Protein	Subject	Label[a]	Period	Rate of loss injected protein
R-γ-G[b]	R	I^{131}	0–4	Normal
			4–	Normal
B-γ-G	R	I^{131}	0–4	Normal
			4–	Very rapid[c]
B-γ-G	R[d]	I^{131}	0–2	Very rapid
B-γ-G	R[e]	I^{131}	0–2	Normal
			2–	Very rapid
B-γ-G	R[f]	I^{131}	0	Very rapid

[a] The stability of the I^{131}-label in the antigen has been demonstrated;[269,271,272] see also Sterling.[247]

[b] R-γ-G, rabbit γ-globulin; B, bovine.

[c] The disappearance rate of the B-γ-G from the tissues approximately paralleled its loss from the circulation. Antibody appeared in the circulation after virtually all antigen had disappeared.[273] However the disappearance rate of azo proteins from the tissues is quite slow.[274] They appear to be laid down in mitochondria.[275]

[d] Actively sensitized with antibody in circulation.[276]

[e] Previously actively sensitized but without antibody present in the circulation (anamnestic).

[f] Rabbit passively immunized by injection of antibody. Following the injection, B-γ-G immediately falls precipitously.

(270) F. J. Dixon, S. L. Bukantz, G. J. Dammin, and D. W. Talmage, *Federation Proc.* **10**, 553 (1951).
(271) H. N. Eisen and A. S. Keston, *J. Immunol.* **63,** 71 (1949).
(272) W. C. Knox and F. C. Endicott, *J. Immunol.* **65,** 523 (1950).
(273) C. V. Z. Hawn and C. A. Janeway, *J. Exptl. Med.* **85,** 571 (1947).
(274) H. Krause and P. D. McMaster, *J. Exptl. Med.* **90,** 425 (1949).
(275) C. F. Crampton, H. H. Reller, and F. Haurowitz, *Proc. Soc. Exptl. Biol. Med.* **80,** 448 (1952); F. Haurowitz, C. F. Crampton, and R. Sowinski, *Federation Proc.* **10,** 560 (1951).
(276) F. J. Dixon and D. W. Talmage, *Proc. Soc. Exptl. Biol. Med.* **78,** 123 (1951); D. W. Talmage, F. J. Dixon, S. C. Bukantz, and G. J. Dammin, *J. Immunol.* **67,** 243 (1951).

It may be noted that Schoenheimer and co-workers[277] found the half life of the antibody to *Pneumococcus*, Type III and of normal globulin in the rabbit to be 14 days. This is much longer than that estimated in the experiments given in Table XXII, and appears to be due to the experimental errors in these early investigations. Glycine-N^{15} was not incorporated into passive antibody injected into a rabbit.[278] This finding has recently been confirmed by Bulman and Campbell.[279] Thus the antibody protein, itself, did not stimulate the synthesis of more antibody, nor was glycine exchanged in and out of the protein.[279a]

From Table XXII it appears that the immunological response causes a rapid disappearance of protein from the circulation, and that the injection of specific antibody has a similar effect. The antigen–antibody complex is probably taken up by cells. However, in animals which are rendered immunologically unresponsive by the injection of massive doses of protein (albumin), foreign protein may be removed from the circulation at a more normal rate.[268,269]

d. The Interpretation of the Turnover Phenomenon

In the preceding sections information relative to a process which has been loosely labeled "turnover" has been given. It is now necessary to define more precisely what is meant by this process, and to this end it will be convenient to consider serum albumin as a specific example. It has been seen that there appear to be two more or less separate and distinct systems, one for the synthesis of protein and one for its degradation. Therefore, the simplest way in which the turnover phenomenon can be regarded is to assume that it results from the simultaneous operation of these two processes. Moreover, it may be assumed also, for the sake of simplicity, that when there is net protein synthesis, the synthetic component of the turnover process is in the ascendent and vice versa. In other words, *no separate system is responsible for protein synthesis or breakdown apart from the two already involved in the turnover process.* More complex hypotheses appear to be unnecessary at the present time.

Suppose, then, in a given biological system there is a modification so that the rate of protein synthesis increases (or decreases), then, unless there is an increase (or decrease) in cell numbers or some other factor influences the system, presumably the system will come to a new state of equilibrium by responding with an increase (or decrease) in the rate of

(277) R. Schoenheimer, S. Ratner, D. Rittenberg, and M. Heidelberger, *J. Biol. Chem.* **144,** 545 (1942).

(278) M. Heidelberger, H. P. Treffers, R. Schoenheimer, S. Ratner, and D. Rittenberg, *J. Biol. Chem.* **144,** 555 (1942).

(279) N. Bulman and D. H. Campbell, *Proc. Soc. Exptl. Biol. Med.* **84,** 155 (1953).

(279*a*) J. H. Humphrey and A. S. McFarlane, *Biochem. J.* **57,** 186, 195 (1954).

degradation. If the primary effect is on the rate of degradation, the synthetic system will show a secondary response. On the basis of these assumptions it is possible to predict what effect a change in the rate of synthesis will have on pool sizes, half lives, and replacement rates when a new state of equilibrium is reached. However, even if the system does not attain a new equilibrium, the variables will change in the same direction although the magnitude of the changes will not be the same. Hence, if observations are made of pool sizes, half lives, and replacement rates in a system under different conditions, it is possible to predict from the direction of the observed changes whether the initial modification affected the synthetic rate or the degradation rate. This is shown in Table XXIII.

TABLE XXIII

THE INTERPRETATION OF OBSERVED CHANGES IN POOL SIZE, HALF LIFE AND REPLACEMENT RATE

Change in system			
Pool size	Half life	Replacement rate	Cause
+	−	+	Synthetic rate greater[a]
−	+	−	Synthetic rate less[a]
+	+	−	Degradation rate less
−	−	+	Degradation rate greater

[a] When the primary change is in the synthetic system, the initial effect observed before the pool size has changed significantly and before there has been a secondary response of the degradation system will be greater than the effect observed at equilibrium.

If the secondary change in the degradative system occurs slowly so that the pool size increases greatly, the effect on the half life may be the opposite to that shown.

One example from the table will clarify the argument. If the synthetic rate for albumin increases, before the pool size changes significantly, it is clear that the rate of dilution of the labeled pool will be observed to increase, that is, the half life of the protein will decrease, whereas the replacement rate will increase. Eventually, the pool size will increase and the effect on the half life will be reduced. The situation, therefore, is described by the first line in the table. This change (probably) occurs when animals are moved from a low protein to a high protein diet.[230] The other interpretations appearing in Table XXIII may be deduced in like manner.

In the case of albumin it is probable that the protein is being continually synthesized, but it is highly improbable that this continual synthesis takes place at a constant rate, because the liver is exposed to a variable concentration of amino acid entering the organ from the intestinal tract and elsewhere. Consequently, the rate of albumin synthesis probably fluctuates quite widely and, with this varying rate, the pool size will show corresponding changes unless the degradation rate immediately responds to changes in synthetic rate. This is improbable.

The simplest concept of turnover is one in which the rate of degradation of protein exactly equals the rate of synthesis. This is illustrated by the diagram (Type I) of Fig. 11. Protein is built into the pool in a given amount in time A, and a corresponding amount is removed over the same period, now labeled B. The turnover is constant and continuous. The more complicated situation is depicted as Type 2 in Fig. 11. Here the pool size undergoes an increase during a period A', and later in a period B' reverts to normal size. In this situation the removal of labeled amino acid from the pool will proceed at a different rate from that observed in the ideal system of Type 1, and, in fact, it is highly improbable that the ideal situation exists unless the synthetic process is

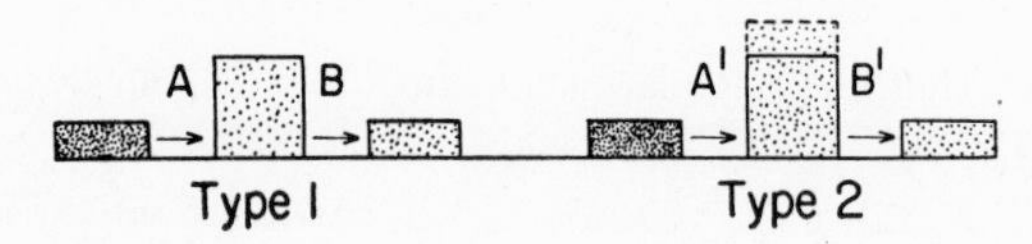

FIG. 11. Illustration of simple turnover process Type 1 and more complex process Type 2.

quite directly coupled with the breakdown process. Simpson[280] has investigated this possibility by labeling the liver protein with either DL-methionine-S^{35} or DL-leucine-3-C^{14}. The rate of release of the methionine or leucine from the protein was then studied in slices of the organ, and it was found that the release was inhibited by agents such as anaerobiosis, cyanide, and dinitrophenol. Thus the process of protein degradation was inhibited by the same agents as affect the synthesis, although the agents have no such specific effect on the catheptic enzymes. The further investigation of the phenomenon by the employment of a specific protein would be of considerable interest.

It may be concluded that it is a matter of some experimental difficulty to distinguish between systems of Type 1 and Type 2. *Moreover, it is not easy to exclude the possibility that isotope is only introduced into protein under conditions where there is a simultaneous net synthesis.*

e. Factors Affecting the Rate of Turnover

As mentioned in the previous section, it is probable that an increase in the protein content of the diet elicits an increase in the turnover of

(280) M. V. Simpson, *J. Biol. Chem.* **201,** 143 (1953).

tissue and plasma proteins.[230,280a] However, the data of Hubay and co-workers[281] are not in agreement with this concept.

Numerous authors have injected labeled amino acids into animals in different states of hormonal balance and have measured uptakes into various proteins at one interval following the dose. From the comparison of the results they have arrived at somewhat premature conclusions as to the way in which the turnover has been changed. It is obvious that erroneous conclusions may be reached from data of this sort if the shape of the turnover curve is altered by the change in conditions, or if the distribution of amino acid between different tissues is affected by the hormone, or if the synthesis of all proteins in a given cell is not influenced in the same direction. Consequently, it does not appear profitable to pursue the details of these investigations.

4. Fluctuations in Enzymatic Levels

a. Changes in Enzyme Concentrations in Animal Tissues

Elsewhere the fluctuations which occur in the total protein, and in particular of the cytoplasmic protein of liver and also of the enzymatic composition of liver and other tissues, have been emphasized.[207,282,283] Such changes have been shown to extend to the cellular fractions of liver separated by centrifugation.[284] More recently, changes in the enzymatic makeup of the liver and other tissues have been described more fully in the following communications which give an introduction to the new literature: arginase,[285] xanthine oxidase,[286] alkaline phosphatase,[287] rhodanese,[288] adenosinetriphosphatase (ATPase),[288] and tryptophan

(280*a*) H. L. Steinbock and H. Tarver, *J. Biol. Chem.* **209,** 127 (1954).

(281) C. A. Hubay, H. Kreiger, W. D. Holden, and T. Hancock, *Proc. Soc. Exptl. Biol. Med.* **78,** 402 (1951).

(282) H. W. Kosterlitz and R. M. Campbell, *Nutrition Abstracts & Revs.* **15,** 1 (1945/46); K. M. Henry, H. W. Kosterlitz, and M. H. Quenouille, *Brit. J. Nutrition* **7,** 51 (1953).

(283) R. Y. Thompson, F. C. Heagg, W. C. Hutchison, and J. N. Davidson, *Biochem. J.* **53,** 460 (1953).

(284) E. Muntwyler, S. Seifter, and D. M. Harkness, *J. Biol. Chem.* **184,** 181 (1950); *Proc. Soc. Exptl. Biol. Med.* **75,** 46 (1950).

(285) J. Mandelstam and J. Yudkin, *Biochem. J.* **51,** 681 (1952).

(286) G. Litwack, J. N. Williams, and C. A. Elvehjem, *J. Biol. Chem.* **201,** 261 (1953).

(287) O. Rosenthal, J. C. Fahl, and H. M. Vars, *J. Biol. Chem.* **194,** 299 (1952).

(288) O. Rosenthal, C. S. Rogers, H. M. Vars, and C. C. Ferguson, *J. Biol. Chem.* **185,** 669 (1950).

peroxidase.[289] Changes also occur during the process of regeneration[290] with sex[291] and hormonal balance.[289,292,293]

The variations which occur in xanthine oxidase and in tryptophan peroxidase are extremely noteworthy. For instance, according to Westerfeld and Richert,[294] the rat is born without detectable traces of xanthine oxidase whereas the adult animal on an appropriate diet has abundant stores. The situation with respect to tryptophan peroxidase is more or less similar.[289] Following the administration of tryptophan to the adult rat or guinea pig, the peroxidase increases tenfold in 3–5 hours. It is of interest to note that this formation of the peroxidase is partially inhibited by DL-ethionine.[295] Liver catalase also falls following the administration of ethionine.[296]

More recently, Lee and Williams[297] have made a definite correlation between the formation of tryptophan peroxidase and the turnover of the protein in the cytoplasmic structures of the liver from rats. During the period in which the enzymatic activity was increasing, there was a loss of protein from the mitochondrial fraction, with a simultaneous increase in the proteins of a microsomal fraction. At the same time, changes in the turnover of the proteins in the several fractions of the liver were observed. There apparently was an increased rate of turnover in the mitochondrial protein and a decrease in the microsomal protein.

Also noteworthy is the behavior of the hemoglobin in *Daphnia*, which responds by an increase when there is an oxygen deficiency in the water in which the animals are maintained.[298] Although this is not the response of an enzyme to a substrate, yet the phenomenon appear to be related to those just considered.

It is pertinent to note here that Lee and Williams[299] found that following the injection of DL-cystine into intact rats the uptake into the various cellular structures of the liver varied with dietary, hormonal, and other conditions under which the animals were maintained prior to treatment.

(289) W. E. Knox, *Brit. J. Exptl. Path.* **32,** 462 (1951); *Science* **113,** 237 (1951).
(290) O. Rosenthal, C. S. Rogers, H. M. Vars, and C. C. Ferguson, *J. Biol. Chem.* **189,** 831 (1951); O. Rosenthal, H. M. Vars, C. S. Rogers, and J. C. Fahl, *Acta Unio Intern. contra Cancrum* **7,** 386 (1951).
(291) R. M. Campbell and H. W. Kosterlitz, *J. Endrocrinol.* **6,** 308 (1950).
(292) H. Fraenkel-Conrat and H. M. Evans, *Science* **95,** 305 (1942).
(293) C. M. Sego and S. Roberts, *J. Biol. Chem.* **178,** 827 (1949).
(294) W. W. Westerfeld and D. A. Richert, *J. Biol. Chem.* **184,** 163 (1950).
(295) N. D. Lee and R. H. Williams, *Biochim. et Biophys. Acta* **9,** 698 (1952).
(296) W. Bollag and E. Gallico, *Biochim. et Biophys. Acta* **9,** 193 (1952).
(297) H. Lee and R. H. Williams, *J. Biol. Chem.* **204,** 477 (1953).
(298) H. M. Fox and E. A. Phear, *Proc. Roy. Soc.* (London) **B141,** 179 (1953); H. M. Fox, E. A. Phear, and B. M. Gilchrist, *Nature* **171,** 347 (1953).
(299) N. D. Lee and R. H. Williams, *J. Biol. Chem.* **200,** 451 (1953).

It is evident that these changes in protein content of organs and tissues, and the fluctuations in their enzymatic constitution must be related to the turnover of the proteins; in fact it appears that the lability of the body protein and hence of the enzymatic activity of tissues must represent one mechanism which allows the organism to cope with changes in its external environment.

b. *Enzymatic Adaptation* (*EA*)[300–303]

In bacteria and other lower organisms, the adaptation to new substrates occurs to a quite astonishing extent, although the process appears to be, generally, of a somewhat different order from that which occurs in the vertebrate animal. In some cases the enzymatic activity may be potential in all the cells of a given bacterium, in others it can perhaps only arise by mutation or by clone variation with selection.[303] Whatever the precise situation may be in a given case, it is evident that large amounts of new enzymatically active protein appears in the bacteria in a short space of time when they are adapting to a new substrate. The question therefore arises as to the source of this protein.

Halvorson and Spiegelman[304] found that during the process of adaptation of *Saccharomyces cerevisiae* to maltose, the free amino acids in the bacteria decreased very greatly in concentration. Moreover, both the processes of adaptation and growth were inhibited by many amino acid analogs, particularly by *p*-fluorophenylalanine, and were reversed by the corresponding amino acids. In addition, the process was inhibited by respiratory poisons.[305] The requirement for amino acids and energy to promote the synthesis of adaptive enzymes is confirmed and extended by work from other laboratories.[306,307] Consequently, it must be concluded that the synthesis of the enzyme proceeds from free amino acids and not from larger (preformed) components. Essentially the same conclusions have been reached by Monod and co-workers[307a] using a different organism and methods.

(300) J. Monod and M. Cohn, *Advances in Enzymol.* **13,** 67 (1952).
(301) S. Spiegelman, *in* Sumner and Myrbach, The Enzymes, Academic Press, New York, 1951, Vol. I, part 1, p. 267; *Symposia Soc. Exptl. Biol.* **2,** 286 (1948); *Cold Spring Harbor Symposia Quant. Biol.* **11,** 256 (1946).
(302) R. Y. Stanier, *Ann. Rev. Microbiol.* **5,** 35 (1951).
(303) J. Yudkin, *Nature* **171,** 541 (1953); *Biol. Revs. Cambridge Phil. Soc.* **13,** 93 (1938).
(304) H. Halvorson and S. Spiegelman, *J. Bact.* **64,** 207 (1952).
(305) S. Spiegelman, J. M. Reiner, and R. Cohnberg, *J. Gen. Physiol.* **31,** 27 (1947).
(306) M. J. Pinsky and J. L. Stokes, *J. Bact.* **64,** 145 (1952).
(307) D. Ushiba and B. Magasanik, *Proc. Soc. Exptl. Biol. Med.* **80,** 626 (1952).
(307*a*) J. Monod, A. M. Pappenheimer, and G. Cohen-Bazire, *Biochim. et Biophys. Acta* **9,** 648 (1952).

5. Incorporation of Foreign Amino Acids into Protein. The Pluperfect Concept of Protein Synthesis

Levine and Tarver[308] attempted to show in experiments with ethionine-$C_2{}^{14}H_5$ that this "unnatural" amino acid is incorporated into the proteins of rats, presumably in positions which would normally have been occupied by methionine. This study grew out of the well-know inhibitions of growth, and presumably of protein synthesis, which result when suitable amino acid analogs, such as ethionine, are applied to appropriate systems.[192,193,309] Insofar as ethionine is concerned, it had been shown that this analog is capable of reducing the uptake of DL-methionine-S^{35} and of glycine when administered to rats or mice.[310]

The ethionine was apparently incorporated into the proteins of various tissues, and of the blood, in such a manner that it proved impossible to remove it by any of the generally effective washing procedures. That the analog existed as such in association with the protein was shown by its reisolation with the aid of carrier from hydrolyzates of various proteins and the conversion of the ethyl group to *S*-ethylisothiourea picrate. The activity in the protein invariably carried over into the derivative. Consequently, unless the amino acid analog was bound to the protein in some fashion, as yet undetermined, then the ethionine must have been introduced into the protein by peptide-bond formation. The possibility remained that the ethyl group was transferred from the amino acid into the preformed protein or peptide precursor. Such a mechanism appears highly unlikely since the label from methyl mercaptan-S^{35} or -C^{14} is not introduced into protein to any significant extent.[311]

Various other amino acid analogs have been investigated, such as DL-norleucine-3-C^{14} and DL-norvaline-3-C^{14},[312] L-α-aminoadipic acid-6-C^{14},[313] and α-amino-*n*-butyric acid,[186] without any significant amount of the labeled material being found in the protein. Possible explanations of the several negative results are: (*1*) the definition of "significant amount"; (*2*) the analogs are too readily metabolized, i.e., their concentration is not maintained high enough for a sufficient length of time within the cells; (*3*) the amino acid which they might replace is in high concen-

(308) M. V. Levine and H. Tarver, *J. Biol. Chem.* **192,** 835 (1951).

(309) W. W. Ackermann, *J. Exptl. Med.* **93,** 337 (1951); G. C. Brown and W. W. Ackermann, *Proc. Soc. Exptl. Biol. Med.* **77,** 367 (1951).

(310) M. Simpson, E. Farber, and H. Tarver, *J. Biol. Chem.* **182,** 81 (1950).

(311) E. S. Canellakis and H. Tarver, *Arch. Biochem. Biophys.* **42,** 446 (1953).

(312) M. ul Hassan and D. M. Greenberg, *Arch. Biochem. Biophys.* **39,** 129 (1952); M. ul Hassan, Dissertation, University of California, Berkeley, 1951.

(313) H. Borsook, C. Deasy, A. J. Haagen-Smit, G. Keighley, and P. H. Lowy, *J. Biol. Chem.* **187,** 839 (1950).

tration or has a high rate of formation so that the analog is in an unfavorable competitive position; or else (*4*) the analogs do not bear a close enough resemblance to any specific amino acid.

Actually the concept that the foreign amino acid ethionine may be introduced into protein becomes less surprising if the following considerations are taken into account. S-CH_3-Homocysteine is a normal constituent of protein. Are S-CD_3-homocysteine, S-CT_3-homocysteine, and S-C_2H_5-homocysteine incorporated? If the incorporation of no foreign amino acid occurs, it would mean the enzymatic specificity in the areas of protein synthesis is absolute. This would appear to be highly unlikely.

Work and Work[314] also entertained the idea that the foreign amino acids might become incorporated into proteins in biological systems, and the concept of Synge[315] that families of closely related proteins may exist is rather similar. The same hypothesis, that an abnormal structure may be synthesized, was introduced into the nucleic acid field by Kidder and Dewey[316] as a result of experiments which demonstrated the great inhibition of growth of *Tetrahymena geleii* which followed their exposure to low concentration of 8-azaguanine. More recently, the incorporation of the same analog into nucleic acid has actually been demonstrated.[317–319a] Other analogs have also been incorporated in nucleic acids, e.g., 2,6-diaminopurine[320] and bromouracil[321] *in vivo*.

If the point of view that "foreign" amino acids may be incorporated into protein is adopted, it is possible to propose some rather interesting questions such as the following: Is the mechanism of thyroxine (or triiodothyronine) action one in which this amino acid is actually incorporated into general body protein? The administration of labeled thyroxine results in its appearance in liver protein where it is found in the nucleus, mitochondria, as well as in the centrifugal supernatant.[322] Another possibility is that certain types of proteins exist, e.g., immune

(314) T. S. Work and E. Work, The Basis of Chemotherapy, Interscience, New York, 1948, p. 227.
(315) R. L. M. Synge, *Cold Spring Harbor Symposia Quant. Biol.* **14,** 191 (1950).
(316) G. W. Kidder and V. C. Dewey, *J. Biol. Chem.* **179,** 181 (1949).
(317) M. R. Heinrich, V. C. Dewey, R. E. Parks, Jr., and G. W. Kidder, *J. Biol. Chem.* **197,** 199 (1952).
(318) J. H. Mitchell, Jr., H. E. Skipper, and L. L. Bennett, Jr., *Cancer Research* **10,** 647 (1950).
(319) H. G. Mandel and P. E. Carlo, *Intern. Congr. Biochem., 2nd Congr., Paris,* 1952, p. 475; H. G. Mandel, P. E. Carlo, and P. K. Smith, *J. Biol. Chem.* **206,** 181 (1953).
(319*a*) R. E. F. Matthews, *Nature* **171,** 1065 (1953).
(320) H. E. Skipper, *Cancer Research* **13,** 545 (1953).
(321) F. Wegand, A. Wacker, and H. Dellweg, *Z. Naturforsch.* **7b,** 19 (1952).
(322) E. A. Carr, Jr. and D. S. Riggs, *Biochem. J.* **54,** 217 (1953).

globulins, in which the structure is less rigidly fixed so that small changes in amino acid composition can occur, leading to a condition in which in a given "pure protein" the amino acid composition represents the statistical average from a variety of unique species. It is assumed that immune α-globulins are not simply normal α-globulins with the chains refolded.[323]

Nothing need be said in this context with respect to the self-evident relation between these speculations with regard to the incorporation of foreign amino acids and the theories of inhibition, competitive and otherwise. In this connection the presence of "peculiar" amino acids in the antibiotics may be noted: the case of penicillin is well known; others also exist.[324]

Whether L-methionine sulfoximine is incorporated into protein deserves more thorough investigation.[325]

6. The Requirements for Protein Synthesis

a. Amino Acids vs. Partial Breakdown Products of Plasma Protein

Various authors, in particular Madden and Whipple,[175] have speculated that plasma proteins might give rise to tissue protein without complete breakdown to amino acids. Unfortunately, neither the experiments of Yuile and co-workers[326] nor those of Abdou and Tarver[244] can be interpreted as giving any satisfactory support to this concept. However, it is evident from the work of Anfinsen and Steinberg[177] that, in the chicken, ovalbumin may be synthesized from amino acids at a rapid rate without the necessity of supplying serum albumin.

There is more definite information with respect to the synthesis of the casein in milk from the experiments of Campbell and Work[327] and of Barry.[328] The experiments of the former authors were performed on lactating rabbits by the administration of either DL-valine-4,4′-C^{14} or DL-lysine-2-C^{14}. The specific activities of these amino acids were determined at intervals in the casein, and whey proteins and also in the plasma and erythrocyte proteins of the animals. The results are shown graphically in Fig. 12.

It is evident from these results that the specific activity of the casein and whey protein in the intervals shortly following the injection of the

(323) L. Pauling, *J. Am. Chem. Soc.* **62,** 2643 (1940).
(324) G. G. F. Newton and E. P. Abraham, *Nature* **171,** 606 (1953).
(325) J. S. Roth, A. Wase, and L. Reiner, *Science* **115,** 236 (1952).
(326) C. L. Yuile, B. G. Lamson, L. L. Miller, and G. H. Whipple, *J. Exptl. Med.* **93,** 539 (1951).
(327) P. N. Campbell and T. S. Work, *Biochem. J.* **52,** 217 (1952).
(328) J. M. Barry, *J. Biol. Chem.* **195,** 795 (1952).

labeled material is far above that in the plasma protein. Hence, it is not possible that the casein is derived from the plasma protein. The experiments of Barry[328] were carried out in goats in which the proteins were labeled with either DL-lysine-1-C^{14} or DL-tyrosine-1-C^{14}. Activities in the carboxyl groups in hydrolyzates of the casein and plasma protein of the animals following the administration of these amino acids were determined. By far the highest activities were found in the casein. It was also shown by perfusion experiments carried out with halves of cow's udders that the casein was formed from amino acids added to the

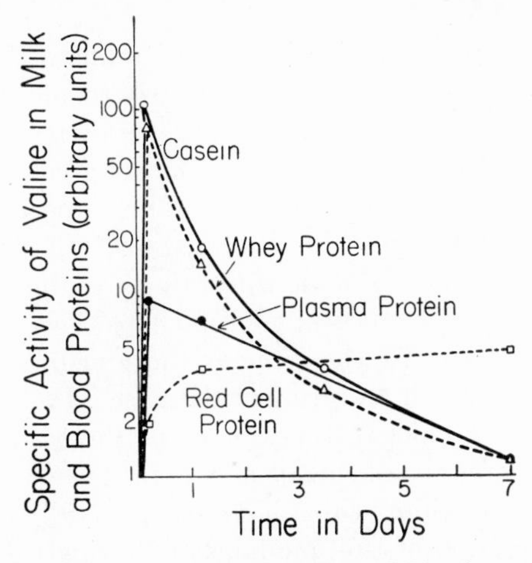

FIG. 12. Specific activity of valine in milk and blood proteins from a rabbit at different time intervals after the administration of valine-$C^{14}H_3$: the relationship between the synthesis of casein and of the blood proteins.[327]

perfusing blood. Similarly, the phosphorus in the casein appears to be derived from the blood inorganic phosphate.

Campbell and Work[327] also labeled plasma protein with glycine-2-C^{14} and found that the labeled protein did not give rise to significant labeling in the casein. It must be concluded from the results of these experiments on casein synthesis that the protein is synthesized, at least largely, from the free amino acids supplied by the blood stream.

b. *Requirements for Energy*

It is amply demonstrated from a large body of data of a nutritional type that the storage of protein can be affected by the provision of extra calories either in the form of fat or carbohydrate when the dietary nitrogen is maintained at a constant level.[329] Although the results are clear,

(329) H. N. Munro, *Physiol. Revs.* **31,** 449 (1951); H. N. Munro and D. J. Naismith, *Biochem. J.* **54,** 191 (1952).

the interpretation of the data is less certain, but in view of what has been found with respect to protein and peptide synthesis *in vitro*, it may be assumed that the effect is due, at least in part, to the provision of the necessary energy to support the synthetic process. In addition, because of the presence of excess substrate, there may be a conservation of amino acids or a synthesis of extra non-essential amino acids.

7. The Magnitude of Turnover and Its Relationship to the Metabolic Rate

Following the work of Sprinson and Rittenberg[216] and of Hoberman,[330] various authors[331–334] have attempted to measure the rate of protein turnover in animals and human subjects by measuring the excretion of N^{15} following the administration of either glycine-N^{15},[216,330,332,334] yeast-N^{15},[331] or L-aspartic acid-N^{15}.[333,334] This experimental approach has been criticized by Russell[335] and by Tarver[111,207] on account of the various assumptions which are made with respect to the biological system in order to apply the mathematical treatment to the experimental data. Many of the assumptions are the same as those made in order to apply the simple kinetic approach to turnover data (sec. IV-3*a*), but some are of greater scope. The chief of these involves the assumption that the total protein in the organism behaves as a homogeneous system.

In spite of the objections which have been raised, there is no doubt that the rate of excretion of isotopic labels following the administration of labeled amino acids is related in some complex manner to the process of protein synthesis–degradation, to the turnover, which is occurring in the organism. It is, therefore, of interest to see what may be concluded from the actual figures which have been deduced by Pietro and Rittenberg[334] regarding the replacement rate of the nitrogen in the protein of two human subjects of about 70 kg. Values of between 39 and 92 g. of N per day were found for the rate, and if an average figure of 63 g. of N per day is assumed, i.e., 4.5 moles of N per day, and it is also assumed that in the synthesis of each peptide bond one energy-rich phosphate is consumed, and that three energy-rich phosphates arise per atom of oxygen, then the oxygen required to maintain the daily replacement rate

(330) H. D. Hoberman, *Yale J. Biol. and Med.* **22,** 341 (1950); *J. Biol. Chem.* **188,** 797 (1951).
(331) A. G. C. White and W. Parson, *Arch. Biochem.* **26,** 205 (1950).
(332) P. D. Bartlett and O. H. Gaebler, *J. Biol. Chem.* **196,** 1 (1952).
(333) H. Wu and S. E. Syndermann, *J. Gen. Physiol.* **34,** 339 (1950).
(334) A. S. Pietro and D. Rittenberg, *J. Biol. Chem.* **201,** 445, 457 (1953).
(335) J. A. Russell, *Ann. Rev. Physiol.* **13,** 327 (1951).

would amount to

$$\frac{4.5}{3} \times \frac{22.4}{2} = 17 \text{ l./day}$$

This is about 4 per cent of the total oxygen consumption. Figures of the same order of magnitude may be calculated from the estimated half lives and amounts of the various proteins in the animal as obtained by direct turnover measurements.

The calculation shows that there is ample energy available from the basal metabolism to maintain the normal replacement rate for protein even on the assumption that the synthesis of each peptide bond requires one energy-rich phosphate. If the energy required to maintain the normal replacement rate for protein is added to that needed to maintain the replacement rate for other body constituents, it would appear probable that a large fraction of the metabolic energy is involved in turnover phenomena.

8. Conclusions

From the experimental work dealt with in this section it is clear that there is a wide spread in the half lives and replacement rates of the various proteins in animal tissues. The most satisfactory determinations have been made on plasma proteins whose turnover has been measured by several different methods. No satisfactory evidence has been adduced to show that plasma proteins form tissue proteins prior to extensive breakdown. The turnover phenomenon appears to be related to the rapid changes which occur in the enzymatic makeup of such tissues as liver. It is possible that foreign amino acids are incorporated into tissue proteins, so that the synthesis of protein may not proceed at all times to give absolutely specific products.

V. General Conclusions

It appears, both from the theoretical point of view and from experiments carried out *in vivo* and *in vitro*, that energy must be supplied in order to synthesize peptide bonds and protein. No precise mechanism for the coupling process has been discovered, although there are suggestions that various cofactors are necessary, and work with glutathione synthesis suggests the involvement of enzyme systems, quite apart from the hydrolytic ones, cathepsins.

However, it may be that the proteolytic enzymes are necessary to the process of equilibration among amino acids and peptides, although the importance of peptides to the synthetic process is not actually demonstrated. In particular, it is not clear whether small or large peptides are needed in synthesis or if some template mechanism is operative; in general, the evidence available to date does not support any specific operative mechanism for protein synthesis.

There is a rapid turnover of many proteins which is probably related to the rapid changes in the enzymatic constitution which occur in some tissues and to the production of proteinaceous secretions in others. It is likely that diurnal changes in rates of protein formation, in particular of plasma protein formation, occur, resulting in conservation of amino acids by removing them from the catabolic phase. Much the same half lives and replacement rates have been found for several specific proteins with a variety of methods involving the labeling of different amino acids. This suggests that the turnover phenomenon observed with respect to the amino acid constituents of proteins does not result from the exchange of amino acids in and out of the molecule as a whole, nor out of large peptides, but rather that the synthesis–degradation of proteins involves the formation of either free amino acids or at least of small peptides which come to equilibrium rapidly.

To maintain the turnover it is no doubt necessary to supply energy, and a significant fraction of the basal metabolism may represent heat evolved owing to the persistence of this process. From the study of turnover phenomena in different systems, it is possible to deduce whether changing from one to another involves a change in rate of protein synthesis or of degradation.

VI. Appendix

Concerning the General Applicability of the Turnover Concept

Monod and co-workers[336,337] have recently carried out experiments with *E. coli* with the object of determining whether there is turnover of the total protein and, more specifically, of the protein formed by induction with β-galactoside, the adaptive enzyme β-galactosidase. Experiments of two general types were carried out.

(*a*) The total protein in the organisms was labeled by growing them a media containing sulfate-S^{35}, both cystine and methionine becoming marked.[338] The labeled sulfate was then removed and the inducer was added to the medium so that the protein-enzyme was formed in the presence of unlabeled sulfate and hence of unlabeled methionine and cystine. The enzyme was isolated by a combination of chemical fractionation

(336) D. Hogness, *in* Adaptation in microorganisms, *3rd Symposium Soc. Gen. Microbiol. Cambridge,* p. 126 (1953).

(337) M. Cohn and J. Monod, *5th Intern. Congr. Microbiol., Rome* (1953).

(338) D. B. Cowie, E. T. Bolton, and M. K. Sands, *Arch. Biochem. Biophys.* **35,** 140 (1952).

and immunological technic. A determination of its specific activity was then made.

(*b*) The enzyme formed by induction and the total protein were both labeled and the dilution of the label was measured in both fractions following growth in the absence of labeling agent. The results in either case indicated that the general protein pool contributed no amino acid to the galactosidase nor did the enzyme contribute to the protein pool. Protein of both types proved to be essentially stable, that is, little or no turnover was demonstrable. Moreover, neither free cystine nor free methionine exchanged with the protein-bound amino acid, otherwise the labeling of the protein would have changed with time as the degree of labeling of the free amino acid pool changed.

Since the results of these experiments are perfectly clear and, moreover, are in accord with previous work of Monod and others[300–302] it becomes quite necessary to re-evaluate the situation in the mammalian organism to see if intracellular protein stability might also obtain there. The foregoing review was written on the premise, explicitly stated (sec. IV-3d), that there is a persistent breakdown of protein in the organism, presumably both intracellular and intercellular. Obviously, intercellular protein breakdown occurs in the mammal. Enzymes, hormones and blood-lymph proteins are formed in specific tissues and are broken down elsewhere, perhaps even in the cells of origin. The question remains, however, as to whether specific intracellular protein undergoes breakdown or not; or, reforming the question, are (intracellular) cathepsins active in the viable cell? More broadly still, is the individual peptide bond stable in the biological system?

Insofar as the reviewer is aware, the literature contains little evidence which cannot be interpreted just as easily from the point of view of "no intracellular turnover" as from the point of view of the "general dynamic state" postulated by Schoenheimer. The turnover observed in tissues containing cells which do not secrete protein, may simply be due to the replacement of such cells, necessitated by the "wear and tear" postulated by classical workers with the mechanistic point of view.

Folin[339] gave the name endogenous metabolism to the metabolism of nitrogen involved in the wear and tear process. This is clearly a "no turnover" concept. Concomitantly, in the fed animal, other nitrogen appearing in the urine was named exogenous nitrogen. Implicitly there was an identification of at least part of that fed with that which appeared in the urine. Presumably, the nitrogen and carbon of most of the fed protein was immediately converted to non-protein products without going through a protein stage. Obviously, isotopic data show this to be

(339) O. Folin, *Am. J. Physiol.* **13,** 66, 117 (1905).

untrue; first, endogenously formed and exogenously supplied amino acids are in general indistinguishable, and, second, the rapid turnover of the plasma protein suggests that a large part of the ingested nitrogen passes through the stage of plasma protein prior to catabolism. Assuming the validity of the Monod hypothesis, this latter pathway assumes great importance, the pathway which is, in effect, the main component of the "continuing metabolism" of Borsook and Keighley.[340]

There is little evidence bearing on the activity of cathepsins in intact cells, but during the last few years it has become increasingly clear that the enzymatic activities exhibited by cells depend on their state of organization. When the cell structure is in any way disrupted, activities which were formerly negligible come to the fore. In this connection the behavior of the energy-transfer systems of various types are of particular interest. There is an obvious activation of the phosphate-transfer system in skeletal muscle following stimulation. More specifically, the activity of the mitochondrial system depends first, on its age[341] and its tonicity,[342] APT-ase activity increasing as the preparation ages;[339,341] second, on the availability of Ca-ion;[343] and third, on the presence of phosphate acceptors.[344,345] In another field recent work of Putnam and Hassid[346] has shown that in leaf cells the activity of invertase is masked under some conditions, although the macerated tissue contains copious amounts of the enzyme. Examples of the same general type might be multiplied.

Thus, in the animal cell either one of two situations may exist. Either, the cathepsins are inactive intracellularly, in which case they might be activated at the time of cell division or at the time of cell dissolution. In the latter eventuality, the enzymes would serve the purpose of breaking down specific intracellular proteins and thus preventing their release into the organism generally. Or, the cathepsins may be active normally but only at cell surfaces or following excretion into intercellular fluid. Thus plasma protein and other proteins might be broken down intercellularly in the interstitial fluid or lymphatic channels. There appears to be no specific information which might be enlightening with respect to these questions.

(340) H. Borsook and G. Keighley, *Proc. Roy. Soc. (London)* **B118,** 488 (1933).
(341) W. W. Kielley and R. K. Kielley, *J. Biol. Chem.* **191,** 485 (1951).
(342) E. C. Slater and K. W. Cleland, *Biochem. J.* **53,** 557 (1953).
(343) E. C. Slater and K. W. Cleland, *Nature* **170,** 118 (1952).
(344) H. A. Lardy and H. Wellman, *J. Biol. Chem.* **195,** 215 (1952).
(345) V. R. Potter and R. O. Recknagel, *in* McElroy and Glass, Phosphorus Metabolism, Johns Hopkins Press, Baltimore, 1951, vol. I, p. 377.
(346) E. W. Putnam and W. Z. Hassid, in press.

To be more specific with respect to the plasma protein and to the albumin in particular, recent work has shown that in the normal adult male rat on a stock diet the rate of replacement is about 0.34 g. of albumin per day per rat. The protein consumption of such an animal is of the order of 2 g.[347] So the possibility exists that a considerable fraction of the amino acid taken in actually enters the continuing metabolism of the total plasma protein, since the contribution of the globulins to the picture has been ignored. This is also suggested by the work of Denton and Elvehjem[348] on differences between the amino acid concentrations in portal and radial blood of dogs following a protein-containing meal.

It therefore becomes evident that considerable reinterpretation is possible on the basis of the Monod hypothesis. Here, only one more example will be dealt with, and this concerns the introduction of foreign amino acids into protein (sec. IV-5). The incorporation of labeled ethyl from ethionine into the proteins of rats[308] has been confirmed by Stekol and co-workers,[349] and there has been observed a considerable incorporation of this foreign amino acid into the protein of *Tetrahymena*, grown in a synthetic medium containing both L-ethionine-C^{14}-ethyl and methionine-unlabeled.[350] There is also a report that *p*-fluorophenylalanine is likewise incorporated into the protein or peptides of *L. arabinosus*,[351] and in this connection it should be mentioned that such an incorporation had been predicted on the basis of "inhibition" experiments.[352] Evidently *L. arabinosus* is capable of adapting to the analog of phenylalanine and, under certain conditions, the analog actually stimulates growth. Thus the possibility exists that the incorporation of foreign amino acids may be of quite general occurrence, and that perhaps D-amino acids are actually incorporated into both normal and neoplastic proteins.[353]

Assuming that foreign amino acids are directly incorporated into proteins under the proper conditions, it is of interest to call attention

(347) F. Ulrich, H. Tarver, and C. H. Li, *J. Biol. Chem.*, in press.

(348) A. E. Denton and C. A. Elvehjem, *J. Biol. Chem.* **206**, 449 (1954).

(349) J. A. Stekol, S. Weiss, E. I. Anderson, and B. Hsu, *Federation Proc.* **13**, 304 (1954). However, the interpretation of the data by these workers is that the ethyl is exchanged (transferred) into the protein or peptide subsequent to the incorporation of methionine, that is, ethyl is exchanged for methyl *in the protein.*

(350) D. Gross unpublished data. Thanks are due to Prof. Kidder for his assistance in this project.

(351) R. S. Baker, J. E. Johnson, and S. W. Fox, *Federation Proc.* **13**, 178 (1954).

(352) D. E. Atkinson, S. Melvin, and S. W. Fox, *Arch. Biochem. Biophys.* **31**, 205 (1951).

(353) D. Burke and R. J. Winzler, *Ann. Rev. Biochem.* **13**, 493 (1944); J. A. Miller, *Cancer Research* **10**, 65 (1950); P. Boulanger and R. Osteux, *Biochim. et Biophys. Acta* **5**, 416 (1950).

to one particular aspect of the data presented by Levine and Tarver,[308] this is the lability of the label once this is introduced into the protein. The labeled amino acid is very rapidly lost from the protein. This means, presumably, that if ethionine is incorporated in the proteins generally, these proteins are abnormal in their vulnerability with respect to catheptic action. Since there is no reason to assume the mere replacement of an ethyl group for a methyl group in methionine would have any such outstanding effect on peptide bond stability, it must be assumed that the catheptic enzymes are more active in tissues which have been exposed to ethionine. The cells are perhaps disorganized. That disorganization actually occurs under the influence of ethionine is exemplified by the production of fatty livers under the influence of this agent,[354] the cells becoming infiltrated with globules of fat. There is, also, a complete breakdown of the acinous tissue in the pancreas following treatment with ethionine.[355] A similar breakdown occurs with a virus infection in mice.[356] It has also been observed that one or two days after treatment with ethionine there is an increase in the rate of incorporation of labeled amino acid into the total protein of pancreatic tissue and also an increase in the rate of formation of amylase, trypsinogen, and chymotrypsinogen by the tissue.[356] This is understandable because if catheptic action is increased in a tissue the rate of turnover should increase, provided that the synthetic system is still capable of making an appropriate response. Hence, the incorporation rate should be enhanced; enzymes and virus may be expected to increase, as observed.

To summarize, it is evident that the "no turnover" hypothesis of Monod should be given careful consideration—at least as an alternative to the hypothesis of the "dynamic state" of Schoenheimer.

(354) E. Farber, M. V. Simpson, and H. Tarver, *J. Biol. Chem.* **182,** 91 (1950).
(355) R. C. Goldberg, I. L. Chaikoff, and A. H. Dodge, *Proc. Soc. Exptl. Biol. Med.* **74,** 869 (1950); E. Farber and H. Popper, *ibid.* **74,** 838 (1950).
(356) E. Farber, H. Sidransky, and E. D. Kilbourne, *Federation Proc.* **13,** 428 (1954).

Author Index

The numbers in parentheses are footnote numbers and are inserted to enable the reader to locate a cross reference when the author's name does not appear at the point of reference in the text.

A

Abderhalden, E., 281, 850, 1106
Abdou, I. A., 1258, 1270(199), 1271, 1272, 1276(199, 233), 1288
Abel, J. J., 624, 632, 635
Abelin, A., 657
Abitz, W., 917
Abonyi, I., 267, 272(542)
Abraham, E. P., 462, 1288
Abramowitz, A. A., 620
Abrams, A., 372, 1190(516), 1191
Abramson, H. A., 636, 828
Acher, R., 660
Ackermann, D., 43
Ackermann, W. W., 1286
Ackerson, C. W., 1163
Ada, G. L., 1262, 1269
Adair, G. S., 289, 292, 302, 307, 340, 448, 702, 705, 828, 949, 973, 997(86), 998 (86), 1000(86)
Adair, M. E., 292, 307, 340, 448, 680, 702, 705, 828
Adams, D. H., 132
Adams, E., 1068
Adams, M., 784
Adams, M. H., 101, 149, 153
Adams, R., 565
Addison, Thomas, 596
Addoms, R. M., 591
Adler, E., 162, 163, 166, 171(160), 172, 179(182), 184, 187, 188, 189(276), 190, 191, 192(151), 203, 204, 205 (319), 206(319), 207(289), 211, 249, 573
Agner, K., 348
Ahrens, E. H., Jr., 637
Aikawa, J. K., 1278
Åkeson, Å., 276, 277, 428, 429
Albaum, H. G., 39
Albers, H., 163
Albert, A., 605
Alberty, R. A., 458, 464(106), 678, 679 (71), 718, 720, 741(248), 1074, 1076 (67)
Albrecht, G. S., 1066, 1067(30), 1092(30), 1190(30)
Albritton, E. C., 731, 754
Alderton, G., 442, 443, 449, 456, 459, 462, 463(23), 464(23, 24), 465(24), 468(23), 470(24), 473(24), 474, 475 (158), 476, 478, 479, 482, 483(158)
Aldrich, T. B., 615, 623, 625(154), 626 (154), 627(154), 628(154), 630(154)
Aldridge, W. N., 1159
Alexander, B., 730, 753
Alexander, J., 790
Alexander, P., 860, 863, 872, 874, 875
Alfert, M., 54
Alfthan, M., 559
Allen, D. W., 302, 303, 308, 320
Allen, F. W., 20, 32
Allen, T. H., 140, 149
Allfrey, V., 52, 1243
Allfrey, V. G., 18(65), 19
Allsopp, A., 529, 562
Almquist, H. J., 1170
Alsberg, C. L., 480
Altman, R., 5
Altmann, K. I., 292, 1125
Altschul, A. M., 252, 282, 300
Alvarez-Tostado, C., 412
Amantea, G., 282
Amatuzio, D. S., 1273
Ambache, N., 372, 373
Ambros, O., 1101

Volume II part A—pages 1 to 661
Volume II part B—pages 663 to 1296

Ambrose, E. J., 854, 855, 857, 858, 870 (32), 891, 899(92, 93), 900, 901, 920, 938, 941(32)
Ames, R. G., 634
Ames, S. R., 153, 277
Amos, A. J., 493
Anderson, C. G., 367
Anderson, E. A., 458, 464(106), 720, 1076 (67)
Anderson, E. I., 1295
Anderson, J. T., 59, 1262
Anderson, L., 190
Anderson, T. F., 86, 95, 98, 99(321), 100, 103(340), 105, 107, 108, 109, 110 (340)
Andersson, B., 184, 207(249)
Andersson, K. J. I., 636, 640
Andregg, J. W., 78
Andrewes, C. H., 88
Anfinsen, C. B., 169, 171, 188, 189, 190, 588, 673, 1174, 1184(495), 1185, 1226, 1235, 1246, 1248, 1251, 1253, 1258, 1288
Angrist, A. A., 731
Anker, H. S., 1212
Annetts, M., 1174
Anselmino, K. J., 613
Anson, M. L., 286, 288, 289, 293, 331(48), 343, 452, 963, 1066, 1067, 1091, 1092 (26, 27), 1094, 1096(137), 1153, 1154 (401), 1172, 1190(26, 27)
Anton, H., 357
Appenzeller, H., 235
Åqvist, S., 1243
Aragona, C., 326
Arai, K., 191
Arbogast, R., 98, 104
Archer, R., 463
Archibald, R. M., 568, 569, 667
Ardenne, M. V., 970, 986
Armstrong, D. M. G., 918
Armstrong, M. D., 1256, 1286(193)
Armstrong, S. H., Jr., 667, 677, 696, 739 (67), 807
Arndt, U. W., 45
Arnold, V., 452, 963
Aronoff, S., 562
Arrhenius, S., 496, 497(51), 821(286), 832
Arthington, W., 558, 559, 563(179)
Arthus, M., 355
Aschaffenburg, R., 432, 737
Aschheim, S., 618
Ascoli, A., 5
Ash, L., 612, 615
Ashbolt, R. F., 1158
Ashmarin, I. P., 1053
Ashworth, J. N., 680, 685, 725(77)
Asimov, I., 148, 154(99)
Askonas, B. A., 1050
Astbury, W. T., 30, 33, 766, 854, 861, 862, 863, 864(65–67), 866, 868, 870, 872, 876, 887, 888, 889, 890, 891(33), 892, 893, 896, 897, 902, 907, 910, 937, 944, 948, 971, 972(74), 992, 1024, 1025, 1026, 1034, 1035, 1040(195), 1047 (194a, 195)
Astrup, T., 754, 1197
Astwood, E. B., 615, 616
Atkinson, D. E., 1295
Atno, J., 734
Attwood, A. M. P., 381
Aubel-Lesure, G., 793, 816, 819(253)
Audrain, L., 1150, 1157, 1168(408), 1169 (408), 1172(382)
Auhagen, E., 124, 259
Austin, J. H., 295
Austin, M., 320
Avery, G. S., Jr., 586
Avery, O. T., 55(181–183), 56, 775
Awapara, J., 564, 1261
Axelrod, B., 276
Ayer, D. E., 617

B

Babcock, S. M., 429
Babers, F. H., 775
Bach, S. J., 126
Bachawat, B. K., 170, 171, 210(212)
Bacher, J. E., 20, 32
Bachman, C., 618, 619, 620
Backus, R., 71, 99
Bader, R., 316
Baddiley, J., 966
Baer, E., 190
Bagdy, D., 877, 880
Bahn, A., 850
Bahnel, E., 809
Bahr, G., 926
Bailes, R. H., 306

Bailey, A. E., 500(85), 501
Bailey, C. H., 493
Bailey, K., 379, 381(199), 492, 493, 502, 507(14), 728, 729, 766, 876, 877, 879, 880, 887, 958, 963, 964, 965, 966, 967, 973, 979(33), 980, 989, 994, 995, 996, 997, 998(86), 999(85), 1000(86), 1001, 1009, 1010, 1018, 1019, 1021, 1022, 1023, 1033(33), 1040, 1050, 1190(528), 1191, 1197(528)
Bailey, L. H., 500(76), 501
Bailey, W. T., 115
Bain, J. A., 443, 449, 456, 458, 484
Bain, J. M., 519
Baker, D. L., 135
Baker, L. E., 1103, 1107(200), 1127(200)
Baker, R. S., 1295
Baker, R. W., 328
Balassa, G., 333
Bald, J. G., 78, 552
Baldwin, E., 128, 129, 343
Baldwin, R. L., 346, 678, 679(71), 798, 1175
Bale, W. F., 292, 673, 1257, 1260, 1269, 1271(209), 1273, 1276(209)
Ball, E., 162, 166, 218(185), 221(185), 222(185), 224(397), 225, 226, 227, 228, 229(397), 230(397), 430, 1136, 1140(309)
Ballantyne, M., 462, 464
Ballou, G. A., 672, 685(46)
Balls, A. K., 276, 466, 551, 1059, 1069, 1077, 1081, 1093(44), 1095(43), 1103 (44), 1104, 1107(43), 1117(209), 1123 (85), 1146, 1153(209), 1154, 1155, 1156(4, 98, 209)
Baló, J., 948
Bamann, E., 1095, 1096(143)
Bamford, C. H., 852, 857, 858(49), 868 (23), 870(23), 890, 891, 896(23), 899(211), 902
Banga, I., 37, 40, 207, 216, 262, 516, 948, 949, 958, 965, 967, 977(17), 986
Banting, F. G., 635
Baranowski, T., 169, 171, 191, 1006, 1050
Barbieri, J., 519
Barbu, E., 908
Barcroft, J., 281, 299, 300, 301(21), 310 (21), 311(21), 317(21), 319
Barer, R., 872, 1028
Barger, G., 563, 654
Barkan, G., 317, 327, 699
Barker, C., 500(93), 501
Barker, G. R., 7
Barker, H. A., 263, 1177, 1210
Barlow, J. L., 98, 100(341)
Barltrop, J. A., 582
Barnes, B. A., 707, 719
Barnes, J. M., 360
Barnett, B., 381
Barnum, C. P., 59
Barr, D. P., 706
Barr, M., 351, 821
Barrien, B. S., 543, 548(130), 549(130), 590
Barron, C., 761
Barron, E. S. G., 141, 172, 174(217), 177, 186, 189, 197, 209(217), 211(217), 217, 233(281), 262, 271(217), 275 (217), 297, 298(109), 300(109), 964, 1075, 1145, 1146, 1148, 1242
Barry, J. M., 433, 1288, 1289
Bartlett, G. R., 233, 271, 1148
Bartlett, P. D., 1290
Bartlett, S., 737
Bartner, E., 414, 427(87), 433(87), 712, 718(222)
Bartz, Q. R., 608
Bass, L. W., 7, 10
Bassett, R. B., 605
Bassham, J. A., 523, 581(50)
Baster, E., 555(172c), 594
Batchelder, A. C., 672, 676, 677(59, 60), 685(45)
Batchelor, W., 707
Batchelor, W. H., 1278, 1279
Bates, R. W., 602, 603, 638
Bath, J. D., 854
Battelli, F., 166, 179(179)
Battersby, S. R., 641
Bauman, E., 650(292), 651
Baumberger, P., 328
Baur, E., 119
Bawden, F. C., 61, 64, 65(212), 67(212), 68, 71, 72, 79, 80, 82, 83, 84, 85, 120, 790
Baybutt, R. B., 903
Bayliss, J. H., 373, 374(161a, 161b)
Bayliss, W. M., 596
Baylor, M. R. B., 95

Bazemore, A. W., 617
Beach, E. F., 291, 391
Beadle, G. W., 585
Beale, R. N., 17
Bean, R. S., 468, 474, 693, 1083, 1148 (105), 1152, 1161(105, 430), 1162 (430), 1163, 1165(430), 1167(430)
Bear, R. S., 869, 870(90, 91), 908, 910, 912, 914, 915(269), 916, 917(269), 918, 919, 932(269), 933(300), 934, 935, 936(269), 939, 940, 941, 942, 944(269), 1034, 1035, 1036(213), 1039
Beard, D., 90, 91, 95, 96, 98(329), 99 (329), 100(329), 101(330), 103, 437, 471(8), 472(8), 473(8)
Beard, J. W., 90, 91, 95, 96, 98(329), 99 (329), 100(329), 101(330), 103, 437, 471(8), 472(8), 473(8, 155), 810
Beaven, G. H., 320, 781, 855, 872
Beckel, A. C., 500(83), 501
Becker, E. L., 799, 802
Beeman, W. W., 78
Beesley, M. B., 658
Beevers, H., 136
Behnke, J., 776, 801, 831(89)
Behrens, M., 8, 1002
Behrens, O. K., 649, 650, 1141, 1142, 1222
Beinert, H., 165, 168(174), 225, 265(174), 278
Belfanti, S., 375
Bell, F. K., 635
Bell, F. O., 33, 897, 937
Bellamy, W. D., 573
Beloff, A., 1174, 1226
Belozersky, A., 8
Belter, P. A., 490
Benaglia, A. E., 360
Bénard, H., 281
Bendall, J. R., 1043, 1047(226), 1054
Bender, A. E., 235, 243, 245
Bender, M., 403, 1087
Bender, M. M., 503(103), 504
Bendersky, J., 374
Benhamon, N., 293
Bennett, E., 531
Bennett, E. L., 1108(320, 320a), 1140
Bennett, J. P., 594
Bennett, L. L., 616
Bennett, L. L., Jr., 1287
Bennhold, H., 697
Bensley, R. R., 38, 516, 542(13)
Benson, A. A., 523, 562, 581
Bentley, J. A., 546, 560(140, 140a), 573 (140, 140a)
Bentley, R., 185, 223, 258(257)
Benz, F., 161
Beraldo, W. T., 384
Berenblum, I., 770
Berg, A. M., 571, 1209, 1219, 1221(21)
Berg, D., 811
Berg, G., 377
Bergeim, O., 1261
Bergell, P., 1106
Berger, E., 775
Berger, J., 1148
Berger, L., 134
Berger, S., 1144, 1153(350)
Berggren, R. E. L., 403
Bergmann, M., 648, 850, 928, 1059, 1067, 1096(158), 1097, 1102, 1103, 1105, 1106, 1107(32, 219), 1108, 1109, 1111 (213), 1117(264), 1118, 1121, 1122, 1125, 1126(32), 1127(219), 1128, 1130, 1131(276), 1135, 1141, 1142, 1145(32), 1146, 1147, 1153, 1221, 1222, 1223
Bergold, G., 62, 92, 93, 970, 1080
Bergström, S., 276, 277
Bering, E. A., 884
Bernal, J. D., 68, 69, 73(222, 223), 75, 76
Bernard, H., 809
Bernheim, F., 235, 1172
Bernheimer, A. W., 369, 370, 371
Bernstein, S. S., 291, 391
Berridge, N. J., 409, 1065
Berry, W. E., 529, 575(87)
Bersin, T., 1144, 1146, 1153(350)
Berson, S. A., 1273
Bertrand, G., 135
Berulava, I. T., 493(41), 494
Bessey, O. A., 230, 231(416)
Bessman, S. P., 564
Best, C. H., 258, 635
Best, R. J., 61
Betke, K., 319
Bettelheim, F. R., 379, 381(199), 879, 729, 1190(528), 1191, 1197(528)

Beveridge, J. M. R., 944
Bevilacqua, E. B., 503(103), 504
Bevilacqua, E. M., 503(103), 504
Bezer, A. E., 346, 358(4), 791
Bhattacharya, D. B., 375
Bickle, A. S., 546, 559, 565(191a), 573 (140a)
Bidinost, L. E., 1274
Bidwell, E., 367, 383, 1070, 1096(50, 51), 1134(50), 1159(51)
Bier, M., 453, 454, 468, 1079, 1080, 1091 (92), 1105(94), 1149(92), 1150(94), 1165, 1192(92)
Bigwood, E. J., 230
Billimoria, M. C., 531
Bing, F. C., 328
Binkley, F., 386, 588, 1136, 1140(307)
Biosotti, A., 605
Bird, G. W. G., 789, 809(147), 814(147)
Birkinshaw, J. H., 221, 224(384), 257 (384), 258(384)
Birköfer, L., 247, 291
Birnbaum, S. M., 1068, 1096(41), 1107 (224), 1108(41), 1134(41)
Birnie, J. H., 625
Bischoff, F., 620, 623
Bischoff, G., 376
Biscoe, J., 916
Biserte, G., 497, 510
Bishop, L. R., 490, 495, 498, 499, 510, 511 (11)
Bishop, M. N., 613(88), 614
Bjerrum, J., 692
Bjørneboe, M., 811
Black, L. M., 85
Black, S., 180, 181, 182, 183, 1211
Blackburn, S., 874, 1186
Blackwood, J. H., 480
Blake, C. H., 942
Blaker, R. H., 716, 777
Blanchard, M., 224(395, 400), 225, 235 (395, 400), 236(395), 237(395, 400)
Blanchard, M. H., 283, 636, 639(199)
Blanchard, M. L., 212
Blank, F., 530
Blatt, W. F., 726
Bliss, A., 178
Blix, G., 701, 702, 706, 736, 737
Bloch, B., 149
Bloch, K., 286, 1212, 1214, 1215, 1228(35)
Block, R. J., 291, 489, 511(134), 512, 522, 533, 539(112), 885
Blum, J. J., 994
Blumenthal, D., 766
Bly, C. G., 673
Bly, M. L., 1257
Boardman, N. K., 522
Boas, M., 469
Bock, R. H., 168, 261(198), 264(198), 268(198), 269(198), 275(198), 276 (198)
Bock, R. M., 170(209), 171, 203, 224 (403), 225, 255(403), 256(403), 257 (403), 273, 274(554)
Bodansky, O., 314, 1154
Bode, G., 376
Bodine, J. H., 140, 149
Boehm, G., 970, 1024
Boehm, R., 347
Boell, E. J., 149
Bohr, Ch., 300, 308
Boivin, A., 53, 56, 374, 386
Bokelman, E., 183
Bolduan, O. E. A., 916, 919, 933(300), 934, 935
Bollag, W., 1284
Bolling, D., 489, 511(134), 512, 533, 539(12)
Bolton, E. T., 1292
Bonath, K., 284
Bonjour, G., 1185
Bonner, J., 539, 542, 545, 546, 548, 550, 551, 552, 553, 554, 556
Bonnichsen, R., 167, 168(192), 169, 171, 172(192), 173, 174, 175, 176, 177, 178(192), 200(192)
Bonot, A., 693
Booth, V. H., 226, 230(406)
Boquet, P., 354, 355(57), 378, 382(57)
Borasky, R., 912
Borchers, R., 1163
Bordet, J., 94, 730, 756, 820, 831
Bordner, C. A., 152
Borek, E., 571, 573(238), 1219
Borell, U., 605
Bormsen, H., 811
Boroff, D. A., 373
Borrero, L. M., 670
Borrows, E. T., 655

Borsook, H., 162, 319, 1143, 1201, 1202, 1205, 1206, 1222, 1224, 1226, 1227, 1228, 1229, 1230(96, 97, 114), 1232(96, 97, 100), 1233(100, 101), 1234(101), 1236, 1237(114), 1238(96, 97, 101, 114), 1241(96), 1242, 1245(100, 101), 1246(100, 101), 1260(107), 1261, 1262(214), 1268(214), 1274(107), 1286, 1294
Bortels, H., 592
Bosch, M. W., 1006
Boss, W. R., 625
Bostrøm, H., 1275
Boswell, F. W., 95
Boswell, J. G., 528, 581
Bosworth, A. W., 396, 409(28)
Bot, G. M., 543
Bott, M. J., 925
Bott, P. A., 675
Bottazzi, F., 314
Botts, J., 836, 1048
Boulanger, P., 16, 217, 220, 221(365), 248(365), 1295
Bourquelot, E., 135
Bovarnick, M. R., 133, 234(34)
Bovet, D., 376
Bovet, F., 376
Bowen, H. E., 362
Bowen, W. J., 325, 978
Bowers, R. V., 1018
Bowes, J. H., 918, 920, 921, 923, 926, 943, 946(368)
Bowin, A., 671
Bowling, J. D., 589, 590(274), 591(274)
Bowman, W. M., 843
Boxer, G. E., 617
Boyd, W. C., 758, 759, 761, 762(10), 769, 775, 776, 778, 779, 782(113), 783, 786, 789, 790(133), 797, 799, 800(186), 801, 805, 807, 809, 814 (149, 150), 817, 818, 821, 825, 826, 827, 828, 829, 831, 832, 833(62), 835 (62), 837
Boyer, P. D., 197, 198, 200, 201, 202, 278, 672, 685(46), 815, 816
Boynton, D., 566
Brachet, J., 50, 51, 58, 59, 280, 1242, 1244
Brackett, F. S., 309
Brackman, J. A., Jr., 261
Braganca, B. M., 375, 378(175a)
Bragg, W. L., 294, 295, 324, 898, 900
Brahinsky, R. A., 1161(427), 1163
Brakke, M. K., 85
Brambell, F. W. R., 739
Brand, E., 423, 427(113), 682, 686, 687 (111), 712(110), 1074, 1082, 1084 (104), 1087(104), 1199
Brann, L., 323
Branson, H. R., 894, 897(225), 901(225), 939
Braun, A. C., 533
Braun, C., 789, 814(147d)
Braunstein, A. E., 1213
Brawerman, G., 18(71), 19
Bregoff, H. M., 564
Breinl, F., 790, 810
Brenchley, W. E., 592
Brenner, M., 1108, 1137, 1141
Bresler, S. E., 926, 933, 936, 1143, 1144, 1224, 1225(79)
Brewster, J. F., 493
Bricas, E., 950
Brice, B. A., 420, 448, 465(57)
Bridgman, W. B., 1175
Briggs, C., 737
Briggs, D. R., 425, 497, 500(77), 501, 506, 1121, 1123
Briggs, G. E., 1110
Briggs, G. P., 593
Brill, R., 851, 854, 855
Brimley, R. C., 522
Brindley, C. O., 1152
Brink, N. G., 613, 614, 615(87), 617
Brinkmann, R., 319
Briot, M. A., 353
Brodie, A. F., 254
Brömel, H., 165, 172, 192(173), 224(396), 225, 233, 234(396)
Brohult, S., 340, 505
Bronfenbrenner, J., 97, 102
Brooks, J., 312, 314
Brother, G. H., 502
Brown, A., 668(61), 672, 676, 677(59, 60), 685(45), 705(61), 713(61), 720(61), 726(61), 791, 877, 924, 933(323)
Brown, A. P., 283, 780
Brown, C. J., 890
Brown, D. H., 776, 801(81), 831(88), 835 (88)

Volume II part A—pages 1 to 661
Volume II part B—pages 663 to 1296

Brown, D. M., 12, 19(35), 21, 23, 164, 610, 1067, 1074, 1075(70), 1076, 1077 (80), 1078(80), 1081(31), 1082, 1147, 1153(80)
Brown, G. C., 1286
Brown, G. L., 926
Brown, H. J., 382
Brown, K. D., 1063, 1075, 1076(75), 1078(75)
Brown, L., 852, 868(23), 870(23), 890, 891(23, 210), 896(23), 902(210)
Brown, R., 519, 520, 521(25), 530, 531
Brown, R. A., 711, 712, 713, 717, 805
Brown, R. K., 707, 719(193)
Browne, F. L., 396
Browne, J. S. L., 618
Browning, C., 756
Bruce, H. M., 615
Bruckner, V., 949, 950
Bruhn, J. M., 618
Brumberg, E. M., 51
Brunelli, E., 184
Brunish, R., 41, 1245
Brunius, E., 178
Bryan, W. R., 90
Buchert, A. R., 1078, 1154(87)
Buchs, S., 1103
Buchtal, F., 990, 1032, 1040(119)
Buckley, E. S., Jr., 666
Budka, M. J. E., 667, 677, 739(67)
Bücher, T., 201, 312, 326
Buehler, H. J., 372
Buell, M. V., 259
Bürger, M., 649
Bürker, K., 281
Bugbee, Y. P., 615, 623, 625(154), 626 (154), 627(154), 628(154), 630(154)
Bujard, E., 526, 571(74), 1218
Bukantz, S. L., 1279
Bull, H. B., 419, 448, 453, 516, 645, 1072, 1101, 1113(190), 1174, 1175(190), 1177
Bulliard, H., 885
Bullock, M. W., 261
Bulman, N., 761, 777, 793(27), 794, 816, 818, 1280
Bunding, T. M., 613(89), 614, 615(89)
Bunge, G., 480
Burdel, A., 344
Burgen, A. S. V., 372
Burk, D., 306, 406, 1111, 1295
Burk, N. F., 400, 448, 453(52), 492, 507 (22)
Burn, J. H., 625
Burnet, F. M., 89, 109, 116, 811, 813, 828
Burnett, W. T., Jr., 278
Burns, J. H., 299
Burow, R., 396
Burrell, R. C., 591
Burrows, W., 385
Bursa, F., 292, 765, 949
Burström, H., 531, 592
Burton, K., 133, 134(35), 162, 222, 224 (390), 234(35), 245
Burton, R. M., 181, 182(235)
Butement, F. D. S., 768, 790
Butler, J. A. V., 30, 33(89), 647, 648, 1073, 1074, 1112(245), 1113, 1143, 1183(335), 1185
Butterfield, E. E., 295
Buxton, C. L., 738
Buzzell, A., 1087, 1088, 1093
Bywaters, E. G. L., 324, 1016

C

Cable, R. S., 403
Cadden, J. F., 135
Cain, C. K., 221, 257, 619
Calaby, J. H., 989
Caldwell, P. C., 58, 1242, 1249(160)
Califano, L., 343
Calkins, E., 149, 152(103), 153(103)
Callan, H. G., 59
Calmette, A., 349, 783
Calvery, H. O., 475, 477(163), 1097
Calvin, M., 306, 523, 562, 581, 582, 1149
Cameron, J. W., 701, 711(177), 718(177), 730(177), 741(177)
Camien, M. N., 373
Camis, M., 300
Campbell, B., 416, 433(90)
Campbell, D. H., 708, 709, 710(200), 712, 715, 716, 722(200), 761, 776, 777, 784, 793(27), 794, 801(88), 802, 804, 814, 816, 818, 829, 830, 831 (88), 835(88), 1160, 1280
Campbell, E. D., 647
Campbell, G. F., 445, 456, 466(38), 471 (38), 475, 476, 480(160), 482, 495, 500(74, 94), 501, 504, 506

Campbell, J. M., 546, 547
Campbell, P. N., 411, 588, 1249, 1288, 1289
Campbell, R. M., 1283, 1284
Canellakis, E. S., 1228, 1286
Cann, J. R., 449, 711, 712, 713, 717, 805, 808
Cannan, C. M., 1034
Cannan, R. K., 407, 423, 424, 427, 437, 438(11), 439, 441(11), 443, 445, 446, 447(34), 449(11), 453(34), 456(11), 461, 464(11), 465(11), 467(11), 471 (11), 473, 1165
Cannon, P. R., 1259
Cantoni, G. L., 371
Canwell, D., 1014
Capen, R. G., 500(76), 501
Caplin, S. M., 528, 529, 530
Caputto, R., 126, 169, 196, 198(10), 1050
Carlisle, C. H., 73, 75, 76
Carlo, P. E., 1287
Carney, A. L., 693
Caroline, L., 306, 459
Carpenter, D. C., 400, 402, 1069
Carpenter, F. H., 232
Carpenter, L. M., 400, 402
Carr, E. A., Jr., 1287
Carr, P. H., 95
Carrére, M., 94
Carroll, W. R., 988
Carson, G. P., 120
Carstensen, H., 615, 1186
Carter, C. E., 12, 15, 17, 21(33), 32, 966
Carter, D., 874, 875
Cartland, G. F., 621, 622, 623(142, 145)
Cartwright, G. E., 672
Cary, E. G., 788
Casal, A., 1188
Caselli, P., 343
Casey, E. J., 1088, 1112(246), 1113, 1127(246)
Caspary, E. A., 362, 877
Caspersson, T., 7, 8, 28(19), 50, 51, 58, 59, 521, 1032, 1242
Castañeda-Agulló, M., 500(91), 501
Castoro, N., 529
Catch, J. R., 559
Catchpole, H. R., 621
Catron, D., 412
Cayer, D., 1278
Cazal, P., 789, 814(147a)
Cecil, R., 224(404), 225, 258(404), 420
Ceithaml, J., 204
Chaikoff, I. L., 656, 1271, 1276, 1296
Chain, E., 377, 378(191), 385
Chalopsis, H., 314
Chalmers, J. R., 655
Chambers, D. C., 686
Champétier, G., 944, 945
Chance, B., 125, 167, 168, 176, 178, 200, 209, 315, 1110, 1111(239), 1115(239)
Chandler, C. A., 800, 842(195)
Chantrenne, H., 47, 59, 1207, 1208, 1223, 1242, 1244
Chanutin, A., 736, 740(354)
Chappell, J. B., 1003, 1004(147)
Chargaff, E., 14, 17, 18, 19, 31, 33, 475, 476
Chase, J. H., 810
Chase, M., 104, 113, 1244
Chatterjee, A. K., 377
Chaudhuri, D. K., 375
Chaudhuri, R. P., 83
Chauvet, J., 660
Cheng, L. H., 809
Chenowith, M. B., 347
Cheo, C. C., 550, 551
Cherbuliez, E., 397
Chernoff, A. I., 319
Chernoff, L. H., 500(69), 501
Cheslock, K., 165, 168(175), 263(175), 265(175), 266(175), 268(175), 269, 273(175), 274(175), 1146
Chèvremont, M., 885
Chibnall, A. C., 291, 489, 507, 519, 521 (24), 523, 524, 525, 527, 532, 543, 544, 549, 560, 561, 565, 567, 570, 572, 578(24), 585, 639, 642
Chick, H., 405, 453
Chittenden, R. H., 396
Chopra, R. N., 381
Chou, T. C., 263
Chouaiech, M. S., 331
Choudhury, J. K., 574, 575(253)
Chow, B. F., 338, 339(349), 599, 600, 601, 607, 624, 633, 638, 643(220), 717, 722 (243), 723, 782
Christensen, H. N., 671, 1209, 1231, 1232 (125, 126), 1237, 1240, 1242

Christensen, L. K., 425, 426, 1096(145), 1097, 1098(145), 1171(145), 1172, 1173(145), 1176(145), 1177(145), 1180(145), 1181(146)
Christensen, L. R., 382, 731, 732(327)
Christian, W., 126, 133, 161, 162, 163 (139), 165(8), 169, 171, 172, 186, 192 (8), 193, 194, 195, 201, 214, 217, 219, 220(375), 221(32, 375), 224(265), 225, 233, 234(375), 246(265), 247, 248(32, 265), 312
Christiansen, G. S., 526, 531, 562(68), 573, 586(68)
Christiansen, J., 307
Christie, R., 368
Chu, F., 840
Chu, H. P., 349, 350(33), 367(33)
Ciereszko, L. S., 605, 606(49), 607(49)
Ciotti, M., 199, 200(307), 215
Circle, S. J., 502
Ciuca, M., 94
Ciusa, W., 161
Claesson, L., 619
Clapp, F. L., 721
Clark, A. M., 978
Clark, C. L., 359
Clark, E. M., 95
Clark, G. L., 95, 916
Clark, W. M., 162, 329, 331(304), 332
Clarke, E. G. C., 349, 358(25), 359
Clarke, F. A., 809
Clarke, P. H., 349, 366(30)
Claude, A., 46, 47(144), 51, 516, 542(14), 1002
Clayton, J. C., 655
Cleland, K. W., 1294
Cline, J., 109
Clough, H. D., 649
Clutton, R. F., 762, 771, 775(32)
Coca, A. F., 785, 787
Cochran, W., 570, 901
Coghill, R. D., 7, 791
Cohen, B., 370
Cohen, C., 939, 941
Cohen, E., 1083, 1084(107a, 107b)
Cohen, H., 873
Cohen, P. P., 525, 573(65), 587, 589 (269, 270), 732, 1205, 1206, 1207, 1208(14), 1209, 1216, 1218
Cohen, S., 724
Cohen, S. G., 738, 1272
Cohen, S. S., 18, 19, 32, 67, 73, 98, 103 (340), 104, 105, 106, 110(340), 111, 112, 214
Cohen-Bazire, G., 1285
Cohn, E. J., 283, 403, 405, 407(54), 419, 420(54), 636, 639(199), 666, 677, 678, 680, 681(68), 682(68), 683, 684, 686(68), 700, 725(77), 732, 753, 790, 807
Cohn, M., 439, 444(14), 713, 1285, 1292, 1293(300)
Cohn, W. E., 12, 13, 14, 15, 21(34), 24, 27, 164, 966, 970
Cohnberg, R., 1285
Cole, A. G., 437, 456(3), 459(3), 466(3), 472, 473, 484, 782
Cole, H. H., 621
Coleman, D., 857, 890
Coleman, R. G., 559, 564(191c)
Collier, H. B., 1163, 1166(439)
Collip, J. B., 608, 618, 635
Colowick, S. P., 134, 164, 173(167), 175 (167), 199, 200(307), 208(167), 215, 216(359), 1050
Colson-Guastalla, H., 314
Colwell, C. A., 800
Comar, C. L., 278
Comline, R. S., 416
Commoner, B., 552, 554, 555, 556(172)
Conant, J. B., 303, 314, 328(192), 338, 339(349), 341
Condliffe, P. G., 612, 1184(497), 1185, 1189(497)
Cone, W. H., 1146
Conn, E. E., 162, 173
Conn, J. B., 289, 1087
Conn, J. C., 817, 818
Connell, G. E., 1220
Connors, W. M., 1065
Conrad, R. M., 437
Consden, R., 522, 527(44), 643, 873, 877
Conway, B. E., 33
Cook, W. H., 490
Coombs, R., 724
Coombs, R. R. A., 756, 757, 785
Coon, J. M., 625
Coon, M. J., 202, 203(316)
Coons, A. H., 670, 693
Cooper, G. R., 90, 685

Cope, O., 1271
Copenhaver, N. J., 1018
Corey, R. B., 295, 324(99), 570, 853, 868, 890, 893(25), 894, 895, 896(227), 897 (225), 900, 901(225, 238), 903, 904, 907, 916, 926, 939
Cori, C. F., 126, 134, 167, 169(9), 195 (188), 196, 197, 198, 199(188, 189), 200(9, 189), 201(9, 189), 649, 969, 1006, 1050
Cori, G. T., 126, 167, 169(9), 195(188), 196, 197, 198, 199(188, 189), 200 (9, 189), 201(9, 189), 1006, 1050
Cori, O., 185
Corner, G. W., 602
Corran, H. S., 221, 224(386), 226(386), 227(386), 230(386), 231(386), 249, 250, 431
Cortis-Jones, B., 614
Corwin, A. H., 334
Coryell, C., 312, 315, 329, 330(306), 331, 333, 334(206, 317), 335
Cosby, E. L., 276
Cosslett, V. E., 76
Cotchin, E., 737
Coulon, M. H., 712
Coulson, E. J., 434
Coulthard, C. E., 221, 224(384), 257 (384), 258(384)
Courtice, F. C., 1271, 1272(241)
Cowan, P. M., 915
Cowgill, R. W., 332
Cowie, D., 1292
Cox, H. R., 614
Coy, N. H., 414, 427(87), 433(87), 719
Craig, L. C., 628, 637, 640(212), 641, 646, 1083
Craigie, J., 85
Cramer, F., 522
Crammer, J. L., 452, 659, 693
Crampton, C. F., 670, 810, 818, 1279
Crandall, M. W., 324
Crandell, D. D., 359
Crane, F. A., 531, 534(101), 536(101), 590 (101)
Crawford, E. J., 230, 231(416)
Creech, H. J., 775
Creeth, J. M., 640, 648, 682, 1185
Crépax, P., 1008, 1009, 1018
Crewther, W. G., 1070, 1091
Crick, F. H. C., 33, 34, 901, 903, 907, 941
Crook, E. M., 367, 544
Crooke, A. C., 613, 614
Cross, R. J., 498
Crumb, C., 352
Crowe, M. O'L., 362
Crowe, S. J., 608
Crowfoot, D., 82, 419, 640
Crowther, C., 411, 413, 433(76)
Cruz, W., 317
Csáky, T. Z., 95, 96(327), 98(327), 99 (327), 101(327), 103, 472, 473(155)
Csonka, F. A., 493, 500(86), 501
Cubin, H. K., 453
Culhane, K., 635
Cunningham, B., 1096(160), 1097, 1099 (160)
Cunningham, L. W., Jr., 1079, 1080(93), 1081(90, 93), 1090(90), 1091(90), 1100, 1117(262), 1118, 1125, 1149 (285), 1150(285), 1156(90), 1159 (262), 1192(285)
Curl, A. L., 1094, 1104, 1117(209), 1153 (209), 1156(209)
Curme, H. G., 1150
Currie, B. T., 419, 1101, 1113(190), 1174, 1175(190)
Curtis, G. M., 731
Curtis, R. M., 738
Cushing, H., 608
Cushing, M. K., 1237
Cushney, A. R., 347
Custer, J. H., 392, 393(12), 397(12), 398 (12), 399, 400(12), 410(38), 424, 425, 434
Cuthbertson, W. R., 874

D

Dainty, M., 969, 986(66)
Dale, H. H., 484, 625
Dale, W. M., 1093
Dalgliesh, C. E., 588, 854, 890(33), 891 (33), 1251
Dalling, T., 738
Dalton, H. R., 139, 150, 151
Daly, M. M., 18, 19, 40, 41(124), 1243
Dammin, G. J., 1279
Damodaran, M., 408, 566, 572

Danielli, J. F., 517
Danielsson, C. E., 492, 494(17, 18), 495 (16), 496(17, 18), 497(17, 18), 503 (17, 18, 107, 109), 504, 506(17), 508 (110), 509, 510(16), 511(137), 512
Danilewsky, A., 958
Darling, R. C., 304, 328
Darlington, C. D., 120
Darmon, S. E., 854, 890(33), 891(33)
Darrow, R., 320
Das, N. B., 184, 187, 188(276), 189(276), 190, 232, 238(428)
Datta, N., 141, 142(89)
Davidson, C. S., 712, 741(219)
Davidson, J. N., 16, 17, 44, 48, 53, 1242, 1283
Davie, E. W., 1083, 1084(107b), 1092, 1109, 1150, 1167, 1184(125, 235, 450, 492), 1185, 1188(125, 235, 492), 1189 (492), 1190(125, 235, 492, 521), 1193 (125, 237, 492, 521), 1195(235)
Davies, J. V., 1093
Davies, W. L., 390
Davis, B., 697
Davis, B. D., 425, 791, 794(9), 795, 796, 843
Davis, H. F., 949
Davis, N. C., 1068
Davis, S. G., 534, 537
Davisson, E. O., 170, 171, 210(212)
Davoll, H., 629, 630
Davson, H., 517
Dawson, C. R., 135, 136(46), 137(46), 138, 140(72), 141, 142, 143, 144, 145, 146(72, 91, 96), 147, 148, 149, 152, 153, 154, 155, 156
Dawson, I. M., 86, 87(281)
Day, A. J., 375
Day, F. P., 261
Day, P. L., 1239
Dayhoff, M. O., 448, 740
De, S. S., 347, 366, 371, 375, 377, 381
Dean, H. R., 757, 826, 828
Dean, R. F. A., 390, 518
Deanesly, R., 651
Deasy, C. L., 1222, 1226, 1227, 1228, 1229 (100, 101), 1230(96, 97, 114), 1232 (96, 97, 100), 1233(100, 101), 1234 (101), 1236(96, 97), 1237(114), 1238 (96, 97, 101, 114), 1241(96), 1242, 1245(100, 101), 1246(100, 101), 1261, 1262(214), 1268(214), 1286
De Baun, R. M., 1065
de Bernard, L. O., 1005
De Busk, B. G., 132, 261, 266, 269, 270 (503)
Decker, D. G., 1242
de Duve, C., 324
Deffner, M., 257
de Garilhle, M., 465
Deichmiller, M. P., 669, 1277
Dekker, C. A., 11, 48(32), 559, 1107(225), 1108, 1122, 1130(225)
de Kock, P. C., 136
Delange, L., 730
Delaunay, A., 671
DeLa Vergne, L., 292, 1273
De Lawder, A. M., 649
Delbrück, M., 94, 95, 97, 106, 108, 109, 110, 111, 114, 115
Del Campillo, A., 202, 203(316), 212, 213 (345, 348), 265, 274(534)
Delezenne, M. C., 365
Delfs, E., 618
Della Monica, E. S., 424, 425, 431
Dellweg, H., 1287
Delory, G. E., 324
Delwiche, C. C., 553, 555, 571
DeMaria, G., 1066, 1067(30), 1092(30), 1120, 1121(274), 1122(274), 1131 (294), 1133(294), 1190(30)
de Mars, R. I., 113
DeMe'o, R. H., 141
Demerec, M., 94, 95, 97
Dempster, E., 1227
den Dooren de Jong, L. E., 116
Dénes, G., 950
Denes, K., 397, 400(36)
Denis, P. S., 677
Dent, C. E., 527, 559(79), 561, 562, 563 (79, 204), 565, 583(79)
Denton, A. E., 1295
Denton, R., 757
Denys, J., 788
Denz, F. A., 372
DeRenzo, E. C., 230
Derksen, J. C., 919
de Robertis, E., 947
de Rooy, A., 876
Derrien, R., 315

Derrien, Y., 288, 652, 653
Dersch, F., 341
Desnuelle, P., 247, 291, 292(83), 452, 682, 1185, 1188, 1190(515, 515a, 519), 1191, 1193(515a), 1195(515, 515a), 1196(515, 515a)
Desreux, V., 678, 1071, 1103
Dethier, V. G., 137
Deuticke, H. J., 1005, 1006
Deutsch, A., 17, 990, 1040(119)
Deutsch, H. F., 394, 395, 439, 442, 443, 444(14), 449, 456, 458, 464(16), 465, 466, 467, 468, 482, 484, 672, 698(43), 713, 718, 737, 739, 741(248), 761, 770, 1161(432), 1163, 1165
Devi, S. G., 522, 556(46)
Dewan, J. G., 162, 166, 168, 182(177), 187, 221, 224(386), 226(386), 227 (386), 230(386), 231(386), 249, 963
Dewan, J. W., 431
Dewar, M. J. S., 818
Dewey, T. L., 533
Dewey, V. C., 1287
Deysher, E. F., 408
Dhéré, Ch., 337, 338(342, 346), 339, 343, 344
d'Herelle, F., 93, 97, 106
Dhungar, S. B., 1254
Diamond, L., 757
Diamond, L. K., 785, 803, 822(127)
Dibben, H. E., 33
Dickens, F., 162, 186, 214, 215(272), 372
Dickinson, S., 766, 887, 971, 972(74), 1025(74)
Dickman, S., 197
Dickson, G. T., 655
Dieckmann, M. R., 412
Diehl, H., 306
Diemair, W., 142, 144(93)
Dietrich, H., 376
Dietz, P. M., 555, 556
Dieu, H. A., 1072
Digaud, A., 493
Dilks, E., 713
Dill, L. V., 135
Dillon, E. S., 95, 96(327), 98(327), 99 (327), 101(327), 103, 437, 471(8), 472(8), 473(8)
Dillon, M. L., 95, 96(327), 98(327), 99 (327), 101(327)
Dillon, R. T., 1096(167), 1097, 1099(167)
Dills, W. L., 432
DiMola, F., 453
Dingemanse, E., 608
Dingle, J. H., 800, 842(195)
Dintzis, H., 681, 683(83)
Dirnhuber, F., 316
Dirscherl, W., 263, 270(522)
Dische, Z., 735
Dissoway, C. F. R., 115
Distèche, A., 1050
Ditmars, R. L., 355
Dixon, F. J., 669, 1277, 1278, 1279
Dixon, G. H., 1220
Dixon, H. B. F., 614
Dixon, M., 124, 126, 162, 166, 169(10), 172, 173(149), 179, 180, 196, 198(10), 226, 231(407), 1050, 1101, 1153, 1154(400)
Dobry, A., 1116, 1161(435), 1162, 1168 (435), 1203
Doctor, N., 141, 142(89)
Dodds, E. C., 647, 1143, 1183(335)
Dodds, M. L., 138
Dodge, A. H., 1296
Dodson, R. W., 335
Doeden, D., 1265
Doermann, A. H., 111, 115
Doerner, H., 387
Doerr, R., 782
Doherty, D. G., 13, 164, 1100, 1123, 1124, 1141
Doisy, E. A., 221, 257, 618, 619
Dole, V. P., 667
Done, J., 559, 560(191), 564, 567
Doniger, R., 17, 18, 19
Donohue, J., 853, 893(25), 900, 901, 904
Dornberger, K., 73
Doty, P. M., 71, 639, 640
Doudoroff, M., 266
Dougherty, T. F., 810
Douglas, C. G., 299, 307, 311(119)
Dounce, A. L., 277, 587, 1249
Dovey, A., 411
Dowmont, Y. P., 1140, 1141(321), 1142 (321), 1221
Drabkin, D. L., 281, 282, 283(27, 33), 295, 298(105), 299, 324

Draper, M. H., 1028, 1030, 1031, 1032 (199), 1047(199)
Dresel, E. I. B., 281
Drew, R., 348
Dreyfus, J. C., 1017
Drinker, C. K., 669
Drosdoff, M., 593
Drucker, B., 851, 857, 1187
Dubnoff, J. W., 1201, 1202, 1205, 1206
DuBois, D., 316
DuBois, K. P., 134, 140
Dubos, R., 349, 374(24), 425, 761, 799
Dubuisson, M., 959, 961, 978, 1006, 1007, 1008, 1009, 1022
duBuy, H. G., 120, 548
Ducay, E. D., 9, 48(30), 463, 469, 470, 471(147, 148), 1161(430), 1162(430), 1163, 1165(430), 1167(430)
Duckert, F., 380
Duke, J. A., 468, 1080, 1105(94), 1150 (94), 1165
Duncan, C. W., 432
Duncan, J. T., 815, 828, 829
Dunn, E. E., 381, 382
Dunn, F. J., 141, 142(92), 143, 144(92)
Dunn, F. W., 1130
Dunn, M. S., 17, 373, 1257
Du Pan, R. Martin, 738
Duran-Reynals, F., 385
Duthie, E. S., 380, 385, 1096(144), 1097, 1159, 1166(144)
Duvick, D. N., 562
du Vigneaud, V., 624, 626(156), 628, 629, 630, 631, 632, 634, 638, 647, 660, 661 (337)
Dworetzky, M., 378
Dyckerhoff, H., 1187
Dyer, H. M., 624, 626(156)
Dyer, W. J., 1012
Dykeshorn, S. W., 602
Dziewiatkowski, D., 1275

E

Eades, C. H., Jr., 12?8
Eagle, H., 381, 382(208, 209), 724
Eakin, R. E., 439, 459, 469
Earland, C., 860, 872, 874, 875
Earle, I. P., 412
Eaton, M. D., 361, 767
Eaton, S. V., 590
Ebihara, T., 142, 144(94), 148(94)
Ecker, E. E., 771, 840
Eckert, E. A., 437, 471(8), 472(8), 473(8)
Eddy, C. A., 638
Edelhoch, H., 223, 246(394), 254, 316, 691, 692(123), 709, 804
Eder, H. A., 667, 706
Edlbacher, S., 316
Edman, P., 1152
Edsall, G., 381, 711
Edsall, J. T., 405, 407(54), 419, 420(54), 691, 692(123), 726, 727(286), 728, 730, 753, 874, 877, 880, 881(153), 954, 958, 960, 962(12), 963(13), 964, 969, 970, 988, 1006, 1007, 1040, 1199
Edwards, M. B., 538, 562(118)
Edwards, R. R., 1239
Eeg-Larsen, N., 454, 767
Egan, R., 1075
Eggleston, L. V., 262
Eggman, L., 545, 547
Ehrenberg, A., 326, 332(288)
Ehrensvard, G., 1243
Ehrich, W. E., 753
Ehrlich, G., 891
Ehrlich, P., 358, 730, 757, 881, 882(160)
Eibeler, H., 1187
Eichholz, A., 471
Eiger, I. Z., 148, 149
Eijkman, C., 948
Eirich, F., 1081
Eirich, R., 322
Eisen, H. M., 768
Eisen, W., 777, 801
Eisenberg, M. A., 1089, 1090(117), 1095, 1111(141), 1112(141), 1116(141), 1125(141), 1126(141)
Eklund, H. W., 372
Ekström, D., 496
Elam, D. W., 1065
Elbinger, R. L., 1209
Elder, J. H., 618
Elek, S. D., 368, 369(123), 799, 827
Eley, D., 323
Elford, W. J., 88, 674
Eliasson, N. A., 1243
Elkes, A. C., 289
Elkins, E., 1066, 1133(29)

Elkins-Kaufman, E., 1110, 1116, 1129 (237), 1132(252), 1133(252)
Elks, J., 655
Ellenbogen, E., 639, 640(225), 709
Elliott, A., 852, 854, 855, 857, 858(49), 864, 868(23), 870(23, 32, 81), 872 (81), 890, 891, 896(23, 81), 899(93), 902, 920, 938, 941(32)
Elliott, H. A., 707
Elliott, L., 204
Elliott, R. G., 926
Elliott, S. D., 1070, 1107(225a), 1108, 1109, 1130(225), 1134, 1159(52)
Elliott, W. H., 526, 571(73, 73a) 1216, 1217, 1218, 1219
Ellis, E. L., 97, 106
Ellis, J. W., 854
Elmes, P. C., 18, 29
Elmore, D. T., 11, 48(32)
Elöd, E., 860, 861, 873
Elowe, D. G., 278
Elson, D., 17
Elvehjem, C. A., 384, 1261, 1283, 1295
Emde, H., 376
Emmelin, N., 564
Emmens, C. W., 608, 651
Enders, J. F., 753
Endicott, F. C., 768, 1279
Engel, F., 396
Engel, W., 257
Engelhardt, V. A., 964, 965, 968, 990, 1053
Engley, F. B., 374
Engstrom, A., 50
Enns, T., 1269
Enselme, J., 150
Epstein, R., 161
Epstein, S., 818
Erdmann, J. G., 334
Erdös, T., 971, 983, 987(77), 1043
Erickson, J. O., 290, 400, 685, 722, 765, 786, 963, 1172
Ericksson-Quensel, B., 281, 293(10), 294 (10)
Eriksen, T. S., 190
Erikson, R. O., 54
Eriksson, I. B., 505
Eriksson-Quensel, I., 648, 817, 1072, 1173, 1174, 1178
Erlenmeyer, H., 775
Erway, W. F., 140
Erxleben, H., 1096(171), 1097, 1099(171)
Esselen, W. B., Jr., 534, 537
Essex, H. E., 378
Etili, L., 818
Euler, H. v., 162, 163, 164(155), 166, 171 (160), 172, 173(155), 178, 179(182), 184, 187, 188, 189(276), 190, 191, 192 (151), 203, 204, 205(319), 206(319), 207(289), 249, 250, 573
Evans, E. A., Jr., 11, 207, 211(334), 212, 641
Evans, F. R., 408
Evans, H. J., 278, 592
Evans, H. M., 598, 599, 600, 602(17), 603, 604(33), 606(50), 607(50), 608, 609, 613, 620, 621, 622, 623, 1284
Evans, J. S., 622, 623(145, 146)
Evans, R. L., 1273
Everard, B. A., 381
Everett, J. E., 190
Eversole, W. J., 625
Evertzen, A., 646

F

Fabre, C., 682, 1190(515, 515a, 518, 519) 1193(515a, 519), 1195(515, 515a), 1196(515, 515a)
Fagraeus, A., 673, 811
Fahey, J. L., 731
Fahl, J. C., 1283, 1284
Fairley, D., 41
Fairley, J. L. 17
Falconer, J. S., 1065
Fankuchen, I., 68, 69, 73(222, 223), 419, 420(100), 516, 890
Fano, U., 94, 95, 97
Fantl, P., 381
Farber, E., 1226, 1232(93), 1237(93), 1257, 1286, 1296
Farnworth, A. J., 863
Farnsworth, J., 1184(495), 1185
Farr, L., 283
Farrant, J. L., 905, 907
Farthing, A. C., 890
Fassbender, W., 357
Faure, M., 712
Fauré-Fremiet, E., 918, 944(292), 945
Fawcett, D. W., 666

Feeney, R. E., 452, 459, 460
Feldberg, W., 357, 375, 377, 384, 564
Felix, K., 41, 45
Fell, H. B., 51
Fell, N., 791
Fellers, C. R., 534, 537
Fels, I. G., 1242
Felton, L. D., 717, 719, 807, 808
Fenner, F., 811, 813
Fenton, E. T., 362
Ferguson, C. C., 1283, 1284
Ferguson, J. H., 382, 729, 753
Ferguson, J. K. W., 308
Fernholz, M. E., 357
Ferrebee, J. W., 666
Ferrel, R. E., 647, 695
Ferri, C., 948
Ferry, J. D., 636, 639(199), 674, 726, 728, 729, 730, 877, 881, 882(160), 883 (175, 176), 884, 923, 925(176)
Ferry, R. M., 282
Fettig, B., 377
Feuer, G., 978, 979(89), 982, 983, 1023
Feulgen, R., 8
Fevold, H. L., 437, 442, 443, 445, 449(24), 456, 459, 462, 463(23), 464(23, 24), 465(24), 468(23), 470(24), 473(24), 474, 475(158, 159), 476, 478, 479, 482, 483(159), 598, 1161(429), 1163, 1166(429)
Fiala, S., 460
Field, C. W., 346, 358(5)
Fierce, W. L., 613(91, 91a), 614, 615 (91), 617(91, 91a)
Fiess, H. A., 692
Fildes, P., 112
Filitti-Wurmser, S., 789, 793, 816, 819
Fink, K., 656
Fink, R. M., 656, 1269
Finkelstein, H., 91
Finkelstein, P., 1093
Finks, A. J., 500(78, 79), 501
Finnerty, J., 617
Finogenov, P. A., 926, 933(340), 936 (340), 1143
Fischer, E., 849, 1016, 1017, 1018, 1106
Fischer, F. G., 258
Fischer, H., 45, 284, 285(40), 336, 1231, 1232(126), 1237(126)
Fischer, H. O. L., 190
Fisher, A. M., 636, 645
Fisher, H., 113
Fisher, H. F., 162, 173
Fisher, J. D., 613(89), 614, 615(89)
Fishman, J. B., 608, 610
Fishman, M. M., 548, 549
Fisk, A. A., 295
Fitch, A., 647
Fitzpatrick, T. B., 137, 149, 152(103), 153(103)
Fixl, J. O., 855, 919
Flanagan, E. K., 1265
Flavin, M., 1184(495), 1185
Fleckenstein, A., 377
Fleisher, J. A., 604
Fleming, A., 462
Fletcher, W. E., 33
Flick, J. A., 757
Florkin, M., 336, 341, 342(358), 665, 698 (6)
Flory, P. J., 836
Flynn, R. M., 1211
Fodor, P. J., 1136
Fogg, G. E., 592
Folch-Pi, J., 753
Folin, O., 1293, 1294(339)
Folkers, K., 613, 614, 615(87), 617
Folkes, J. P., 1237, 1243
Fontaine, Th. D., 500(75, 85), 501, 505, 506(117)
Forbes, W. H., 301
Forker, L. L., 1271, 1276
Forro, F., Jr., 1093
Forsham, P. H., 613(90), 614, 615(90)
Forster, W., 365
Forsythe, R. H., 437, 438, 439, 442, 444 (12), 471(12), 473
Foster, G. C., 651
Foster, J. F., 412, 437, 438, 439, 442, 444 (12), 448, 471(12), 473, 490, 496, 728, 877, 881
Foster, R. A. C., 112
Foster, R. J., 1116, 1117(260), 1120(260), 1123
Foster, S. B., 923
Fothergill, L. D., 800, 842(195)
Foulkes, E. C., 317
Fournier, P., 493
Fourt, L., 784, 799, 1160
Fourt, P. C., 799

Fowden, L., 534, 537(113), 539, 540(113), 559, 560(191), 564, 567
Fox, H. M., 280, 281, 335, 336(327), 1284
Fox, M., 874
Fox, S. W., 1104, 1135, 1142(305), 1221, 1223(66), 1295
Fraenkel-Conrat, H., 9, 48(30), 346, 347, 365, 381(6), 445, 447(36), 449(36), 458, 459, 460, 463(36), 465, 467(36), 468, 469, 470(36), 471(147, 148), 477 (36), 600, 603, 606(50), 607(50), 623, 647, 693, 695, 723, 781, 1075, 1083, 1084(73), 1085(73, 107), 1108, 1135, 1145, 1148(105, 353), 1151(353), 1152, 1153, 1161(105, 430), 1162 (430), 1163, 1165(430), 1167, 1221, 1223, 1284
Fraenkel-Conrat, J., 365, 605, 606, 607 (50), 647
Frampton, V. L., 437
France, W. G., 497
Franchi, C. M., 947
Francis, G. E., 483
Francis, M. D., 1255, 1256(191), 1257
Franck, J., 548
Frank, K. W., 358
Franke, W., 184, 211(252), 257, 276
Frankel, S., 564
Franklin, R. E., 33, 34
Franks, W. R., 775
Frantz, I. D., Jr., 1136, 1226, 1227, 1230 (92), 1231(92, 109), 1236(110), 1240 (86, 102)
Frantz, McB. J., 1265
Franz, E., 849
Fraser, A. M., 624, 626
Fraser, D., 401, 1175
Fraser, L. J., 1259
Fraser, P. E., 1216
Fraser, R. D. B., 902, 911, 914(270), 919 (270), 920(270), 928(270), 931(270), 933(270), 936(270), 939, 941(270, 304, 362), 942(270), 944(270)
Frazer, A. C., 289, 707
Frederic, J., 885
Fredericq, E., 442, 449(22), 455, 466, 467, 468, 637, 640(211), 641, 964, 1161 (432), 1163, 1165(432)
Fredericq, L., 337
Frei, P., 219
French, D., 496, 874
French, H. V., 1012
Frenkel, S. Ya., 926, 933(340), 936(340)
Freud, J., 608
Freudenberg, K., 647
Freund, J., 793, 797, 800(169), 1159
Freundlich, H., 825, 833, 834
Frey-Wyssling, A., 530
Fried, M., 1104, 1138, 1139(317), 1141, 1203, 1221
Friedberg, F., 1226, 1227, 1260, 1263 (208), 1276(208)
Friedell, H. L., 1271
Friedell, R. W., 412
Friedemann, U., 350
Friedrich-Freksa, O., 854
Fromageot, C., 463, 465, 467, 630, 639, 950, 1167
Fronsacq-Collin, M., 499, 509(66)
Frosch, 60
Fruton, J. S., 559, 648, 1060, 1063, 1067, 1068, 1092(33), 1102, 1104, 1105, 1106, 1107(32, 33, 219, 225, 225a), 1108, 1109, 1111(213), 1116, 1117 (264), 1118, 1121, 1124(15), 1126 (32), 1127(219), 1130, 1131(276), 1134(225a), 1138, 1139, 1140, 1141, 1142(321, 329), 1145(32, 33), 1146, 1147, 1203, 1208, 1221, 1223, 1241
Fuerst, R., 564
Fugitt, C. H., 695
Fujio, O., 770
Fukomoto, J., 1070
Fuld, M., 165, 168(175), 263(175), 265 (175), 266(175), 268(175), 273(175), 274(175), 854
Furuhata, T., 789

G

Gaby, W. L., 221, 257
Gaddum, J. H., 655
Gaebler, O. H., 1290
Gale, E. F., 212, 1216, 1237, 1240, 1243
Gallagher, T. F., 626, 631
Gallico, E., 1284
Galston, A. W., 548
Galvin, J. A., 462, 464
Gamble, J. L., Jr., 669

Gansser, E., 292
Garbade, K. H., 201
Gard, S., 89, 887
Gardiner, S., 701
Garen, A., 109
Garrod, M., 948
Gasser, H., 1045
Gates, F. L., 1093
Gaudino, M., 669
Gäumann, E., 559
Gaunt, R., 625
Gaunt, W. E., 768
Gaw, H. Z., 66
Gawron, O., 1146
Gay, F., 756
Gay, H., 49
Gayer, J., 377
Geiger, A., 319
Geiger, E., 1259
Geiger, J. W., 349, 374(24)
Geiling, E. M. K., 624, 632(157), 635, 638, 645, 649
Geiman, Q. M., 666
Gellart, M., 639, 640(227)
Gelotte, B., 987
George, P., 316, 327
Gerard, R. W., 664
Gerarde, H. W., 1244, 1245, 1255(169, 170), 1256(169)
Gerber, I. E., 775
Gergely, G., 966, 968, 976, 979, 1017(48)
Gergely, J., 266, 274(540)
Gerheim, E. B., 382
Gerischer, W., 186, 187(271)
Gerkhardt, H., 377
Gerlough, T. D., 638, 719
German, B., 307
Gerngross, O., 917, 924
Gerpe, M. C., 493
Gersdorff, C. E. F., 500(78, 79, 88, 90, 92) 501
Gest, H., 113
Ghosh, B. N., 366, 375, 377, 381, 834
Gibbs, C. B. F., 649
Gibbs, R. J., 453, 468, 1165
Gibson, D. M., 170, 171, 210, 699
Gibson, J. G., 2d, 666, 677
Gibson, S. T., 672, 685(45)
Gilbert, C. W., 1093
Gilbert, J. B., 1107, 1108, 1149
Gilbert, L., 33
Gilbert, S. G., 593
Gilchrist, B. M., 1284
Gilchrist, M., 327
Gilder, H., 334
Gillespie, J. M., 522, 556(45)
Gilmour, D., 989
Ginko, R., 625
Girard, P., 796
Giri, K. V., 522, 556(46), 558
Giroud, A., 885
Gitlin, D., 639, 640(225), 669, 671, 686, 693(107), 706, 709, 712, 741(219), 804, 1272, 1278, 1279
Gladner, J. A., 1077, 1100, 1108(297), 1109, 1120, 1121(274), 1122(274), 1125, 1130, 1149(285), 1150, 1155 (83, 297), 1181, 1184(83, 235, 297, 490), 1188(235, 490), 1189(83, 297, 490), 1190(83, 235, 297, 490), 1192 (285), 1193(235), 1195(83, 235, 297, 490, 517), 1196(83, 490)
Gladstone, G. P., 346, 349(26), 374
Glantz, R. R., 1105, 1111(214), 1112 (214), 1113(214), 1129(214), 1131 (214), 1159, 1160(417)
Glaser, R. W., 91, 92
Glass, B., 1082, 1083(103), 1101, 1109, 1149
Glassman, H. N., 349
Glendening, M. B., 647, 1171
Glenn, J. T., 352
Glenny, A. T., 351, 821, 828
Glick, D., 50
Glikina, M. V., 1143, 1144(342), 1224, 1225(79)
Glock, G. E., 186, 214(272), 215(272)
Go, Y., 889
Goddard, D. R., 131, 575, 874
Godeaux, J., 964, 988
Godfrid, M., 619
Goebel, W. F., 386, 715, 717, 719, 722 (243), 723, 775, 799, 805, 816
Gofman, J. W., 702, 707
Göksu, V., 1172
Gold, N. I., 18
Goldberg, R. C., 1296
Goldberg, R. J., 802, 804, 836, 839

Golden, F., 736
Goldenberg, H., 1098, 1099(177), 1104, 1105(208), 1118, 1125(208), 1137 (177)
Goldenberg, V., 1098, 1099(177), 1104, 1105(208), 1118, 1125(208), 1137 (177)
Golder, R. H., 420
Goldie, H., 767
Goldinger, J. M., 262
Goldman, D. S., 225
Goldschmidt, S., 854
Goldstein, G., 39
Goldstein, M., 870
Goldsworthy, N. E., 842
Goldwater, W. H., 423, 427(113)
Golovchenko, T. V., 499
Gonell, H. W., 910
Gonon, W., 863
Goodall, M. C., 1053
Goodman, D. S., 692
Goodman, M., 713
Goodner, K., 782, 784, 799, 800
Gopalkrishnan, K. S., 558
Gordon, A. H., 221, 224(386, 398), 225, 226(386), 227(386), 230(386), 231 (386, 398), 431, 522, 527(44), 643, 873, 921
Gordon, F. H., 719
Gordon, H., 1050
Gordon, J., 788, 840
Gordon, R. S., Jr., 699
Gordon, S., 628, 660
Gordon, W. G., 393, 403, 417
Gordon, W. S., 738
Gorini, L., 1091, 1149(122), 1150, 1157, 1168(408), 1169(408), 1172(382), 1192(122)
Gortner, R. A., 498
Gosling, R. G., 29, 33(87), 34
Goss, H., 621
Gosting, L. J., 718, 741(248)
Gough, D., 874
Gould, R. G., 708, 1266, 1267(228), 1276 (228)
Gowdy, R. A., 95, 96(327), 98(327), 99 (327), 101(327)
Grabar, P., 761, 767, 770, 924
Grady, A. B., 616
Graffin, A. L., 203, 204(321), 205(321), 206(321)
Graham, C. E., 872, 920, 946(99)
Graham, D. M., 702(197), 707
Graham, W. R., Jr., 431
Gralen, M., 307
Gralén, N., 347, 365(10), 490, 492, 498, 781, 860
Granick, S., 334, 543, 548
Grassmann, W., 914, 1071, 1096(159), 1097, 1098(159), 1101, 1127(59a), 1145, 1187
Gray, B. A., 522
Gray, C. H., 362
Gray, L. H., 1093
Graydon, J. J., 368
Greco, A. E., 39
Green, A. A., 126, 282, 283, 702, 706, 707 (191)
Green, C., 17, 18, 19
Green, D. E., 128, 131, 162, 165, 166, 167, 168, 170(209), 171, 190, 191(290), 192(177), 202, 203, 211, 221, 224 (386, 395, 400), 225, 226(386), 227 (386), 230, 231(398), 232, 235(395, 400), 236(395), 237(395, 500), 249, 250, 254, 255, 256(403), 257(315, 403), 259, 261, 262, 265(174), 270 (508), 271(508), 272(508), 274(518, 519), 276(514, 519), 278, 431, 573, 963, 1050
Green, F. C., 463
Green, M., 319, 1130, 1131(295a)
Green, N. M., 1063, 1065(12), 1078, 1079 (88), 1080(90), 1081(90), 1089, 1090 (88, 90), 1091(88, 90), 1096(12), 1123, 1125, 1149(285, 286), 1150 (285, 286), 1155, 1156(90), 1157(286, 407), 1161(12, 407), 1166(12), 1167 (12), 1168, 1169(407, 452), 1170(90, 407, 452), 1192(285, 286)
Green, R. H., 86
Greenbaum, K. W., 710
Greenberg, D. M., 400, 448, 453(52), 647, 695, 708, 710(200), 716(200), 722 (200) 794, 1145, 1146, 1220, 1226, 1227, 1228, 1231, 1232(93, 113), 1233 (113), 1234, 1236, 1237(93), 1238,

1240, 1241(113), 1245, 1260, 1263 (208), 1267, 1276(208), 1286
Greene, B., 791
Greene, R. D., 414, 427(87), 433(87), 511 (135, 136), 512, 712, 718(222)
Greenspan, F. S., 608
Greenstein, J. P., 37, 289, 290, 400, 452, 958, 963, 964, 1068, 1096(41), 1107 (224), 1108(41), 1134(41), 1216, 1218(42)
Greep, R. O., 599, 600, 601(13), 607, 624, 633, 638, 643(220)
Gregg, D. C., 152, 154(122), 157, 159, 817, 818, 1087
Gregory, F. G., 521, 528, 576, 577, 578, 579, 582, 591
Grevile, G. D., 203, 967
Griese, A., 161, 162(137), 186(137)
Griffee, J. W., Jr., 607
Griffith, A. B., 335, 336, 343, 344
Grignolo, A., 944
Grimmer, W., 432
Grisolia, S., 205, 207, 211(333), 212(333)
Grob, D., 1144, 1148, 1160, 1161(372), 1162(372), 1165(372), 1166, 1169 (372), 1171
Grobbelaar, N., 565
Groh, J., 397, 400
Grönwall, A., 394, 423, 811
Groll, J. T., 1096(153), 1097
Gross, C. R., 566
Gross, D., 1295
Gross, J., 656, 657, 694, 913, 921, 922, 927, 928, 929, 931, 936, 944, 945, 947
Gross, M., 775
Gross, P. M., Jr., 701, 711(177), 718(177), 722, 730(177), 741(177)
Grossberg, A. L., 776, 801(88), 831(88), 835(88)
Grossberg, D. B., 373
Grosscurth, G., 308
Grossman, W. I., 161
Grossowicz, N., 571, 573(238), 1219
Grote, I. W., 615, 623, 625(154), 626 (154), 627(154), 628(154), 630(154)
Groves, J. W., 1252, 1286(186)
Groves, M. L., 392, 393(12), 397(12), 398 (12), 399, 400(12), 401, 402(44), 403 (44), 404(44), 405, 406(57), 407, 410 (38), 422, 423(110), 426
Grueter, F., 602
Grunbaum, A., 768
Guba, F., 880
Güntelberg, A. V., 448, 455, 1070
Günther, G., 166, 179(182), 184, 187, 188 (276), 189, 190, 191, 203, 204, 205 (319), 206(319), 207(289), 249, 250
Gürber, A., 677, 680(65)
Guest, M. M., 1197
Guest, P., 594
Guggenheim, M., 560, 563(194)
Guggisberg, H., 404
Guirard, B. M., 1211
Gulland, J. M., 7, 21, 32, 33, 384, 626
Gunsalus, I. C., 265, 269, 274(534), 573
Gurd, F. R. N., 639, 640(225), 689, 692, 702, 704, 705(185), 706(185), 707, 709, 719, 753
Gurin, S., 618, 619, 620, 622
Gustavson, K. H., 922, 923, 944, 945
Gutfreund, H., 448, 636, 639, 640, 689, 997, 1001(134)
Gutfreund, K., 726, 728(289), 877
Guthe, K. F., 303
Gutman, A. B., 672, 733
Gutman, E. B., 733
Gutman, H. R., 1209
Gutmann, A., 116
Gutsche, A. E., 583
Gutter, F. J., 448, 449(62), 454(62)
Guy, R., 786
Guyon, L., 943
György, P., 217, 471

H

Haagen-Smit, A. J., 1226, 1227, 1228, 1229(100, 101), 1230(96, 97, 114), 1232(96, 97, 100), 1233(100, 101), 1234(101), 1236(96, 97), 1237(114), 1238(96, 97, 101, 114), 1241(96), 1242, 1245(100, 101), 1246(100, 101), 1261, 1262(214), 1268(214), 1286
Haas, E., 124, 127, 186, 187(269), 221, 222, 224(17, 266, 267, 388), 246(17), 248, 249, 250, 251, 336
Haas, R., 1159, 1160(420)
Haas, W. J., 1152
Haberman, S., 786
Hackley, B. E., Jr., 1158

Haddock, J. N., 1143
Haddox, C. H., 559
Hagdahl, L., 615, 1186
Hagerman, J. A. S., 708
Haggith, J. W., 868
Hahn, G., 357
Hainsworth, R., 1187
Hakala, M., 525, 558, 1143, 1224
Hakim, A. A., 1243
Haldane, J., 310, 327(173)
Haldane, J. B., 299, 311(119)
Haldane, J. B. S., 1110, 1132
Haldane, J. S., 299, 307, 311(119)
Hale, C. W., 385
Hale, J. H., 363, 379, 380
Halford, R. S., 870
Halikis, M. N., 1239
Hall, B. V., 675
Hall, C. E., 503, 728, 878, 882, 883, 913, 917, 926(289), 940(289), 970, 971, 983, 986, 1031, 1034, 1038
Hall, D. A., 949
Hall, F. G., 319
Hall, J. L., 327, 636
Halliburton, W. D., 36, 336
Halvorson, H., 1285
Halwer, M., 420, 448, 465(57)
Hambleton, A., 842
Hamer, D., 41
Hamilton, P. B., 667, 1096(167), 1097, 1099(167), 1261
Hamilton, R. C. M., 328
Hamm, A., 276
Hammarsten, E., 7, 28(19), 55, 1243
Hammarsten, O., 36, 395, 396, 725
Hamoir, G., 17, 48, 961, 995, 1022
Hamsik, A., 285, 288(43)
Hanania, G., 327
Hanby, W. E., 353, 852, 857, 858(49), 868 (23), 870(23), 890, 891(23, 210), 896 (23), 899(92, 211), 902(210), 949
Hancock, W. T., 1265, 1283
Handler, P., 235
Hanes, C. S., 526, 571(75, 76), 587(75, 76), 1135, 1136, 1205, 1220, 1221 (56)
Hankinson, C. L., 1065
Hannig, K., 1071, 1127(59a)
Hansen, R. G., 412, 413, 427, 433(82), 737, 792
Hanson, E. A., 543, 548(130), 549, 590 (130)
Hanson, H. T., 1067, 1081(31), 1108, 1130, 1148, 1149(373), 1150(373)
Hanson, J., 1051, 1052
Hanut, C. J., 381
Happey, E., 875, 876, 890, 891, 899(211)
Happold, F. C., 136, 1243
Hardcastle, S. M., 281
Harden, A., 432
Hardy, W. B., 701, 725
Harfenist, E. J., 637, 638, 640(212), 641, 646, 1083
Hargreaves, A. B., 381
Harington, C. R., 635, 647, 654, 655, 656, 657, 658, 659(329), 762, 771, 775(32), 1102, 1127(195)
Harkins, W. D., 799
Harkness, D. M., 1283
Harkness, R. D., 930
Harmon, K. M., 1111, 1116, 1117(258), 1125(242)
Harper, G. J., 368
Harrington, W. F., 637, 640(211b)
Harris, A., 791, 794(9), 795, 843
Harris, G., 559, 562
Harris, I. F., 5, 6, 358, 500(72), 501
Harris, J. I., 642, 643, 647, 1184(494, 496, 497), 1185, 1187, 1188(496, 501), 1189(494, 497), 1225
Harris, J. L., 463
Harris, J. S., 1210
Harris, M., 695, 1238
Harris, R. J. C., 17
Harris, S. A., 21
Harris, T. N., 381, 382(208)
Harrison, D. C., 184, 185
Harrop, G. A., Jr., 186
Hart, G. H., 621
Harte, R. A., 775
Harting, J., 182, 195(236, 239), 200, 201 (236, 239), 202(239)
Hartley, B. S., 1077, 1117(86), 1119, 1123 (86), 1157
Hartley, H., 515
Hartley, P., 484
Hartree, E. F., 135, 222, 224(392, 393), 235, 257(393), 258(392, 393), 281, 299, 315, 316, 332(205a)
Hartridge, H., 282, 304, 309

Haselbach, C., 863
Hass, G. M., 1004
Hasse, K., 249
Hasselbalch, H. A., 311
Hasselbach, W., 995, 1009, 1010, 1011, 1012, 1051, 1053
Hassid, W. Z., 1294
Hasson, M., 667
Hastings, A. B., 217, 303, 307, 309, 314, 335(131), 1226, 1239
Haugaard, G., 446, 498, 734, 1173
Haurowitz, F., 279, 281, 282, 283, 285, 286, 288, 289, 292, 295(19), 297(19), 298, 303(36), 304(36), 305, 306(38), 312(30, 114), 313, 314, 315, 316(209), 318, 319(28, 140), 320, 321(196), 323, 325(140), 326(140), 327, 329, 330 (62), 331(62, 231), 332, 335(36), 336 (2), 453, 518, 533(22), 587, 620, 670, 708, 759, 760, 762, 764, 765, 773, 790, 791, 801, 807, 810, 814, 818, 841, 949, 1172, 1249, 1278(182), 1279
Hauschildt, J. D., 622, 623(146)
Hausmann, W., 637
Havemann, R., 308, 316
Hawkins, J. A., 308
Haworth, R. D., 533
Hawn, C. van Z., 728, 882, 1279
Hawrylewicz, E. J., 607
Hayaishi, O., 223, 246(394), 254(394)
Hayano, M., 1218
Hayashi, T., 1053
Hayes, J. F., 404, 405(51)
Haynes, L. J., 21
Haynes, R., 649
Hays, E. E., 613(89), 614, 615(89)
Heagg, F. C., 1283
Hearn, W. R., 1138, 1139, 1141(317), 1221
Hearon, J., 306
Hedon, E., 634
Hegetschweiler, R., 849
Heidelberger, C., 1231, 1271
Heidelberger, M., 282, 307, 309, 346, 358 (4), 453, 651, 653, 671, 709, 713, 714, 715, 725, 765, 766, 780, 782, 783, 784, 793, 794(1), 795, 796, 797, 800(169), 801, 803, 807, 808, 809, 816(42, 43, 44), 817, 820, 821, 822, 824, 825, 829, 830(197), 831, 832, 834, 840, 841, 842, 1280
Heilmeyer, D., 295, 327(106)
Heinrich, M. R., 1287
Heinz, E., 991
Heiwinkel, H., 162
Hektoen, L., 437, 456(3), 459(3), 466(3), 472, 473, 484
Hele, P., 266, 274(540)
Hellerman, L., 133, 234(34), 1146
Hellström, H., 162, 163, 164(155) 171 (160), 172(160), 173(155)
Hellström, V., 190, 191(289), 207(289), 249
Helson, V. A., 349
Hemming, W. A., 739
Hemmingsen, A. M., 635
Hempel, W., 396
Hems, B. A., 655
Henbest, H. B., 546, 560(140), 573(140)
Hendee, E. D., 362
Henderson, L. J., 308, 698
Henderson. L. M., 792, 1261
Hendler, R. W., 1220
Hendriks, R. H., 523, 539(52), 540(52)
Henly, A. A., 613, 614
Henn, G., 306
Henri, M. V., 354
Henriques, O. M., 308
Henry, K. M., 1283
Henschen, G. E., 1071
Henseleit, K., 566
Henze, M., 337, 343
Heppel, L. A., 24, 224(399), 225, 226, 227, 229(399), 230(399), 252, 965
Herbain, M., 261
Herbert, D., 261, 270(508), 271(508), 272 (508), 295, 369, 370, 1050
Heringa, G. C., 931
Herion, A., 1009
Herken, H., 1096(171), 1097, 1099(171)
Herkkonen, H., 1224
Hermann, K., 917, 924
Hermann, L., 918
Hermann, V. Sz., 968
Herne, R., 41
Hernler, F., 337
Herold, L., 613
Herriott, R. M., 98, 100(341), 348, 678, 1059, 1060, 1061(1), 1063(1), 1064, 1071, 1072(1), 1079(1), 1080(1), 1082, 1083(103), 1084(104), 1087

(104), 1088(56), 1096(1), 1103, 1151, 1152, 1153, 1154(401), 1159, 1162 (19), 1190(19, 103, 513), 1191, 1192, 1249
Herrmann, H., 1012, 1015, 1017
Hershey, A. D., 97, 102, 104, 113, 114, 115, 833
Hershey, A. O., 1244
Herzog, R. O., 849, 851, 864, 915
Hess, K., 851
Hesselvik, L., 1165
Hestrin, S., 1098, 1099(175)
Hetrick, J. H., 431
Heubner, W., 313, 314
Hevesy, G., 55
Hewitt, E. J., 592
Hewitt, L. F., 371, 680, 686, 734, 735, 767
Hewson, W., 666
Heyde, W., 1096(159), 1097, 1098(159)
Heyl, J. T., 677
Heyman, U., 190
Heyroth, F. F., 340
Heytler, P., 230
Hicks, C. S., 298
Hier, S. W., 872, 920, 946(99)
Higgins, H. G., 404, 405(51)
High, L. M., 875
Highberger, J. H., 921, 922, 927, 928, 929
Hijman, A. J., 559
Hilditch, T. P., 500(93), 501
Hill, A. V., 301, 302(126), 1045, 1049
Hill, D. W., 408
Hill, J. M., 786
Hill, R., 285, 288, 289(42), 305, 324, 325, 330(42), 331(42)
Hill, T. L., 836
Hiller, A., 317, 328, 667
Hiller, D. D., 283
Hilliard, J., 1163
Hilmoe, R. J., 24, 965
Hines, H. M., 1016, 1018
Hink, J. H., Jr., 717
Hinshelwood, C., 58, 1242, 1249(160)
Hipp, N. J., 392, 393(12), 397(12), 398 (12), 399, 400(12), 401, 402, 403(44), 404, 405, 406(57), 407, 410(38), 422, 423(110), 426
Hird, F. J. R., 526, 571(75, 76), 587(75, 76), 1135, 1136, 1205, 1220, 1221(56)
Hirs, C. H. W., 522, 1074
Hirsch, J., 270
Hirschmann, D. J., 445, 447(36), 449(36), 463(36), 465(36), 467(36), 468(36), 470(36), 477(36), 1075, 1083(73), 1085(73), 1167(73)
Hirsh, E. F., 808
Hisaw, F. L., 598, 620
Hoagland, C. L., 86, 87, 88(275)
Hoberman, H. D., 1241, 1290
Hoch, H., 319, 320, 709, 736, 740(354), 781, 802, 1071, 1238
Hoch-Ligeti, C., 1238
Hocker, H., 247
Hocking, C. S., 728, 877
Hodge, A. T., 1028, 1030, 1031, 1032 (199), 1047(199)
Hodgkin, D. Crowfoot, 640
Högberg, B., 162, 619
Hoerr, N. L., 542
Hoffbauer, F. W., 1228
Hoff-Jørgensen, E., 168
Hoffman, E., 270
Hoffman, F., 613
Hoffman, O. D., 391
Hoffman, W. F., 493, 498
Hoffmann-Berling, H., 1031
Hofmann, K., 1107, 1125, 1129
Hofmann, U., 914
Hofmeister, F., 443, 445, 459(26)
Hofstee, B. H. J., 230, 231(415)
Hogeboom, G. H., 47, 149
Hogness, D. S., 1099, 1114, 1117(178), 1292
Hogness, T. R., 127, 166, 221(17), 224 (17), 246(17), 250, 282, 297, 298 (109), 300(109)
Hokin, L. E., 1253
Holden, C., 54, 575
Holden, H. F., 285, 288, 289(42), 298, 330(42), 331, 366, 375
Holden, W. D., 1265, 1271, 1283
Hole, N. H., 756, 757
Holiday, E. R., 15, 320, 781, 855, 872
Hollander, A., 350
Hollander, V., 1127
Hollies, N., 699
Holloway, B. W., 1243
Holm, G. E., 408
Holman, R. T., 276, 277, 278(564)

Holmberg, C. G., 135, 140(49), 158, 159, 160, 700, 701, 728
Holmes, H. L., 560
Holmgren, E., 1002
Holmgren, H., 605
Holt, L. B., 362, 363
Holter, H., 409
Holton, F., 991
Homan, D. M., 221, 257
Homans, J., 608
Homiller, R. P., 642
Hook, A. E., 95, 96, 98(329), 99(329), 100 (329), 101(330)
Hooker, S. B., 758, 759, 762(10), 769, 775, 776, 778, 782(113), 783, 801, 805, 807, 827, 828, 829, 831
Hoover, R. D., 658
Hoover, S. R., 286, 288(49), 405, 428, 856, 1093, 1103(127), 1172
Hopkins, F. G., 134, 430, 443, 452
Hopkins, S. J., 723, 770
Hoppe-Seyler, F., 282, 308, 309(167), 313
Hoppe-Seyler, G. K. F., 476
Horecker, B. L., 127, 185, 213, 214, 215 (351, 357), 220, 221(17), 224(17, 377, 399), 225, 226, 227, 229(399), 230 (399), 246(17), 250, 251, 298, 309, 328(113)
Horn, M. J., 505, 874
Horne, R. W., 1002, 1003, 1004(145), 1028, 1029, 1030
Horowitz, N. H., 222, 235, 245, 566
Horsfall, F. L., Jr., 676, 782, 784, 799, 800
Horwitt, M. K., 1144
Hotchkiss, R. D., 14, 56, 57, 425, 816, 1172, 1176(464)
Hottle, G. A., 372, 478, 481
Hough, H., 1018, 1020(191), 1021, 1022, 1023
Houssay, B. A., 374, 375, 383(172), 384 (172), 605
Howard, H. W., 120
Howard, J. G., 371
Howat, G. R., 408
Howe, P. E., 412, 413, 737
Howell, E. R., 294(101), 295
Howell, W. H., 623
Howitt, F. O., 636, 857
Hoyle, L., 89
Høyrup, M., 421, 441, 444, 445(31), 456 (19), 457(19), 459(19)
Hsu, B., 1295
Huang, H. T., 1096(174), 1098, 1100, 1104, 1107(184), 1110(184), 1111, 1114(184), 1115(244), 1116, 1117 (184, 207, 260, 271), 1119(244), 1120, 1121(261), 1122(244, 261, 273), 1123 (184, 269)
Hubata, R., 1144
Hubay, C. A., 1265, 1271, 1283
Hubbard, R., 174, 177(225)
Huddleson, I. F., 834
Hudson, C. S., 185
Hudson, R. F., 874
Hüfner, G., 282, 292, 299, 309, 312
Huennekens, F. M., 167, 1110
Hug, E., 374
Hughes, A., 365
Huggins, C., 685, 721(105)
Huggins, C. G., 191
Huggins, M. L., 893, 897(221), 899, 938
Hughes, A. F., 51
Hughes, W., 162, 192(151)
Hughes, W. L., Jr., 666, 678, 679, 680, 681(68), 682(68), 683, 685, 686(68), 687, 689, 690(116), 691, 692(116, 123), 693, 694(134), 700, 725(77), 790, 807
Hugon de Scoeux, F., 122
Hugounenq, L., 480
Huiskamp, W., 36, 46
Hull, R., 425
Hull, W., 1255
Hulme, A. C., 519, 520, 558, 559, 563 (179), 574
Hultin, T., 41, 59, 1235
Human, M. L., 108
Humel, E. J., 649
Hummel, F. C., 291
Humphrey, B. A., 1005
Humphrey, G. F., 1005
Humphrey, J. H., 370, 1280
Hunt, G. E., 562
Hunter, M., 690, 695(121)
Huseby, R. A., 59
Hussey, R. G., 1165
Hutchin, M. E., 1265
Hutchinson, A. O., 390

Hutchinson, M. C., 752
Hutchinson, W. C., 1283
Huxley, A. F., 1052
Huxley, H. E., 903, 941, 952, 1034, 1036 (214–216), 1037, 1038, 1051, 1052
Hyatt, R. E., 1271, 1272(240)

I

Ikawa, M., 776, 801(88), 831(88), 835(88)
Ikeda, C., 776, 801(88), 831(88), 835(88)
Ilin, G. S., 500(87), 501
Ingalls, E. N., 295, 315, 319
Ingbar, S. H., 322
Ingelman, B., 503(107), 504, 699, 811
Ingle, D. J., 614
Ingram, V. M., 1064, 1076, 1086, 1128 (16), 1139
Innerfeld, I., 731
Inoue, S., 359
Irvine, D. H., 316
Irving, E. A., 943
Irving, G. W., Jr., 500(75), 501, 505, 506 (117), 624, 626, 631, 1067, 1107(32), 1109, 1126(32), 1145(32), 1146, 1147
Isbell, H. S., 185
Iselin, B. M., 1096(161, 174), 1097, 1099 (161)
Isherwood, F. A., 526, 571(75, 76), 587 (75, 76), 1135, 1136, 1205, 1220, 1221(56)
Ishii, M., 557
Ishii, T., 161
Isliker, H., 716, 808
Iso, K., 30
Itano, H. A., 321, 322, 781
Ivánovics, G., 949
Iwanowski, D., 60
Iyengar, N. K., 376, 381

J

Jaaback, G., 572
Jackson, D. S., 922
Jackson, E. M., 384
Jackson, H., 136
Jackson, S., 911, 912(270), 919(270), 920 (270), 928(270), 931(270), 933(270), 936(270), 941(270), 942(270), 944 (270)
Jackson, S. Fitton, 914, 926, 928, 930 (279), 946
Jacob, F., 110, 111, 117, 118
Jacob, J. J. C., 1006, 1007(160), 1008
Jacobs, H. R., 262
Jacobs, S. L., 767
Jacobs, W. A., 6, 8, 27
Jacobsen, C. F., 425, 696, 1096(147, 168), 1097, 1100, 1168(168), 1177(147, 168), 1190(516), 1191
Jacobsen, C. J., 1061, 1063(9), 1065(9), 1096(9), 1098(9), 1099(9), 1190(9), 1194(9), 1195, 1196(9)
Jacobson, W., 943
Jacquot-Armand, Y., 789, 816, 819(253)
Jäger, B. V., 708, 712(202), 723, 791, 792, 799, 1159, 1160(417)
Jaeger, W., 377
Jaffe, A. L., 1213
Jaffe, W., 161
Jagannathan, V., 168, 264, 273, 274, 275 (197)
Jagendorf, A. T., 542, 543, 545, 546, 547, 548(124), 553
Jaggi, M. P., 908
Jaisle, F., 987
Jakus, M. A., 913, 917, 926(289), 940 (289), 970, 971, 983, 986, 1031, 1034, 1038
James, S., 1078, 1154(87)
James, W. F., 30, 33(89)
James, W. O., 136, 559
Jameson, C., 412
Jancke, W., 849, 851, 864, 915
Jandorf, B. J., 1158
Janeway, C. A., 669, 677, 693, 742, 753, 1272, 1278, 1279
Jang, R., 1077, 1123(85), 1154(85), 1155
Jansen, E. F., 1059, 1069, 1077, 1093(44), 1094, 1103(44), 1104, 1117(209), 1123(85), 1153(209), 1154, 1155, 1156(4, 98, 209)
Janssen, F., 1135, 1142(305), 1221
Janssen, S., 605
Janzen, R., 308
Jaques, L. B., 880
Jarvinen, H., 525, 558
Jeannerat, J., 855
Jeener, R., 47, 48, 58, 59
Jeffreys, C., 1093

Jelinek, V. C., 617
Jenden, D. J., 1065
Jennings, R. K., 721
Jennings, R. R., 1117(265), 1118
Jenrette, W. V., 37
Jensen, E. V., 685, 721(105)
Jensen, H., 641, 645, 648, 649(276i)
Jensen, R., 496, 511(44), 521
Jermyn, M. A., 522, 556(45)
Johansen, G., 1172, 1176(464)
Johns, C. O., 493, 500(69, 78, 88), 501
Johns, R. G. S., 709, 802
Johnsen, V. L., 500(83), 501
Johnson, A. H., 498
Johnson, C. A., 1261
Johnson, E. A., 15
Johnson, F. H., 306
Johnson, G. H., 1101, 1115(188)
Johnson, J. E., 1295
Johnson, J. L., 565
Johnson, K. G., 875
Johnson, L. R., 761
Johnson, M. J., 1101, 1115(188), 1148
Johnson, P., 492, 503(20), 504(101), 505, 506(119), 972, 983, 984, 987
Johnson, T. B., 7, 658, 659
Johnston, R. B., 1104, 1105, 1138(204, 210), 1140(203), 1141(203, 204), 1203, 1212, 1221, 1228(35)
Joly, M., 908, 1017
Joncich, M. J., 437
Jones, C. B., 875
Jones, D. B., 500(78, 79, 86, 90, 92), 501, 505, 782(115), 783, 874
Jones, E. R. H., 546, 560(140), 573(140)
Jones, E. W., 592
Jones, F. S., 432, 822
Jones, F. T., 464
Jones, H. B., 702(197), 707
Jones, L. R., 221, 257
Jones, M., 1244, 1245(169), 1255(169, 170), 1256(169)
Jones, M. E., 1067, 1092(33), 1107(33), 1138, 1139(317), 1141(317), 1145 (33), 1147(33), 1211, 1221
Jones, P. E. H., 390
Jones, T. S. G., 522, 559
Jones, W., 8, 20
Jones, W. M., 589, 1028
Jonsen, J., 842
Jonxis, J. H. P., 319
Jope, E. M., 288, 299, 331(63), 333, 872
Jordan, D. O., 7, 32, 33
Jordan, H. E., 1002
Jordan, L. D., 948
Joselow, M., 144, 145, 146(96)
Joseph, J., 373
Josepovits, G., 968
Joslyn, M. A., 562
Jowett, M., 182
Jukes, T. H., 437, 476, 477, 481, 484, 485(4)
Jung, F., 312, 356
Junge, J. M., 194(207), 196(207)
Juni, E., 270
Jutisz, M., 463, 467, 642, 1167

K

Kabat, E. A., 346, 358(4), 713, 714, 722 (226), 782, 787, 791, 793, 794, 795, 796(172), 797, 800(170), 801, 809 (172), 811, 820, 821, 822(172), 825, 843
Kaesberg, P., 78, 728, 917
Kaeske, H., 317
Kafiani, W. A., 1053
Kahnt, F. W., 161
Kaiser, E., 1144
Kalckar, H. M., 227, 230, 231, 967, 968, 1050
Kaleita, E., 230
Kaliss, N., 1272
Kalman, A., 603
Kalmanson, G., 97, 102
Kalnitsky, G., 1146
Kamen, M. D., 113
Kamin, H., 233
Kaminski, M., 767, 770
Kamm, O., 615, 623, 625(154), 626(154), 627, 628, 630
Kamp, F., 1005
Kapeller-Adler, R., 258
Kaplan, M. H., 670
Kaplan, N. O., 164, 173(167), 175(167), 181, 199, 200(307), 208(167), 215, 216(359), 1211
Karczag, L., 124, 270
Kardos, E., 397, 400(36)
Karel, L., 359

Kari, S., 559
Karlberg, O., 466
Karrer, P., 161, 162(140, 141), 219, 235, 564
Karush, F., 716, 776, 777, 801, 818
Kaspers, J., 326
Kass, E. H., 322
Kassanis, B., 79, 83, 120
Kassell, B., 423, 427(103), 686, 687(111)
Katagiri, H., 493
Katchalski, E., 691, 692(123), 889, 1125
Katchman, B., 264, 273, 274(554)
Kato, E., 899
Katsoyannis, P. G., 660
Katz, J., 1210
Katz, J. R., 876, 919
Katz, L., 570
Katz, S., 726, 728(289), 977
Katzman, P. A., 221, 257, 618, 619
Kauder, 677
Kaufman, J., 264, 265(533), 275(533)
Kaufman, P., 740
Kaufman, S., 212, 213(348), 1066, 1076, 1077, 1095, 1096(173), 1097, 1099 (173), 1100, 1104(183), 1105(173), 1107(82, 173), 1112(183), 1116, 1117 (82, 183, 257, 263), 1118(257), 1120 (263), 1123(142, 173), 1124(82), 1125(173), 1126(173), 1133(29), 1194 (82), 1211, 1222
Kaufmann, B. P., 49
Kausche, G. A., 1031
Kauzmann, W., 685, 696(103), 1177
Kay, H. D., 408, 431, 437, 475, 476, 477 (162), 480, 481, 483, 484, 485(4)
Kayaloff, 354
Kazal, L. A., 1161(427), 1163
Kazenko, A., 1063
Kearney, E. B., 134, 160(40), 162, 164, 165, 173(170), 190, 203(170), 211 (170), 218(150), 221, 222, 223(150), 224(389), 225(389), 234(378), 237 (40, 389), 238, 239, 240, 241, 242, 243 (40, 380, 389), 244, 248(150), 275 (170), 383
Keene, J. P., 1093
Kegeles, G., 372, 448, 449(62), 454(62)
Keighley, G., 1226, 1227, 1228, 1229(100, 101), 1230(96, 97, 114), 1232(96, 97, 100), 1233(100, 101), 1234(101), 1236 (96, 97), 1237(114), 1238(96, 97, 101, 114), 1241(96), 1242, 1245(100, 101), 1246(100, 101), 1261, 1262(214), 1268(214), 1286, 1294
Keilin, D., 124, 135, 136, 137(1), 139, 140, 151, 152(77), 153(58), 154, 157, 158, 222, 224(392, 393), 231(407), 235, 257(393), 258(392, 393), 281, 289, 299, 303, 305, 310(130), 313, 315(141a), 316, 322, 325(130), 331, 332(205a), 700, 1002
Keith, C. K., 258, 1063
Kekwick, R. A., 224(402), 225, 247(402), 443, 714, 734, 752, 796, 798(180, 181), 877
Kellaway, C. H., 353, 356, 357, 374(68), 375, 376(68), 384
Keller, E. B., 1233, 1262, 1263(219)
Kelley, B., 1239
Kelley, H., 390, 391(4), 397(4), 411(4), 412(4), 416(4), 427(4)
Kelley, V. C., 1265
Kelly, F. C., 926, 946
Kelly, M. F., 785
Kendal, L. P., 136, 152
Kendall, E. C., 650, 654
Kendall, F. E., 678, 680, 681(74), 709, 715, 782(112), 783, 800, 801, 803, 807, 808, 809, 824, 829, 830(197), 832, 834
Kendrew, J. C., 299, 305, 319, 324, 898, 900, 903
Kennedy, J. W., 113
Kent, P. W., 352
Kent-Jones, D. W., 493
Kenten, R. H., 918, 920, 923, 943, 946 (368)
Kenyon, A. E., 531
Kerby, G. P., 95, 96(327), 98(327), 99 (327), 101(327)
Kerkkonen, H., 1143
Kershaw, B. B., 1162(437), 1163, 1165 (437)
Kertész, D., 152
Keston, A. S., 427, 768, 1259
Khorazo, D., 462, 466
Khym, J. X., 13, 27
Kibat, D., 276

Kibrick, A., 407, 423(58), 445, 446, 447 (34), 453(34)
Kidd, F., 520
Kidder, G. W., 1287
Kido, I., 618
Kielley, R. K., 1206, 1259, 1263(15), 1294
Kielley, W. W., 367, 968, 1004, 1005, 1294
Kies, M. W., 276
Kiese, M., 304, 314, 317
Kiesel, A., 8, 500(70), 501
Kihara, H., 1241
Kilbourne, E. D., 1296
Kilby, B. A., 1077, 1081, 1117(86), 1119, 1123(86), 1157
Kimball, C. P., 649, 1269
Kimmel, J. R., 1069, 1082, 1083, 1084 (106a), 1103(44a), 1107(302), 1133, 1147
Kimmett, L., 1163
Kimura, F., 102
King, E. J., 327
King, F. E., 559, 565(188)
King, H. H., 327
King, T. J., 559, 565(188)
Kingsley, R. B., 1107(224), 1108
Kinoshita, J. H., 1130, 1140(309)
Kinsell, L. W., 1265
Kirk, E., 1096(162), 1097, 1099(160)
Kirk, J. S., 784, 808, 1160
Kirk, P. L., 1096(160), 1097, 1238, 1255
Kirkwood, J. G., 711, 712, 713, 717, 805, 808, 1049
Kirkwood, S., 564
Kistiakowsky, G. B., 289, 817, 818, 1087
Kit, S., 1226, 1227(94), 1231, 1232(93), 1236, 1237(93), 1240
Kitagawa, M., 559
Kittel, H., 329
Kjelgaard, N., 110, 117(380), 227, 231
Klatt, O. A., 1241
Kleczkowski, A., 764, 786, 790
Klein, D. E., 327
Klein, J. R., 233, 235, 1210
Kleinschmidt, W. J., 815, 816
Kleinzeller, A., 966, 969, 986(66)
Kleispool, R. J. C., 565
Klemm, O., 854
Klemperer, F., 141
Klenow, H., 185, 227, 231
Kline, D. L., 732
Kline, L., 474
Klingmüller, G., 304
Klioze, O., 1226, 1230(98), 1236(98)
Klobusitzky, D., 356
Klotz, I. M., 692, 693, 1149, 1150, 1226, 1230(98), 1236(98)
Knappeis, G. G., 990, 1032, 1040(119)
Kniazuk, M., 313, 314(185)
Knight, B. C. J. G., 366, 784
Knight, C. A., 17, 65, 66, 87, 88, 552, 1184(496), 1185, 1188(496)
Knight, S. G., 245
Knowlton, G. C., 1016
Knox, G. A., 370
Knox, W. C., 768, 1279
Knox, W. E., 161, 259, 262, 276(514), 1284
Kobal, H., 559
Kobert, R., 315, 316(210), 336
Koch, P. K., 374
Kocher, V., 1108, 1137(228)
Kocsis, J. J., 613(89), 614, 615(89)
Kodama, S., 397, 849
Koechlin, B. A., 459, 699
Koenig, H., 235
Koenig, V. L., 682, 728
Koerner, J. F., 28
Koger, M., 658
Kohler, G. O., 658
Kokes, E. L. C., 1093, 1103(127)
Koller, F., 380
Kolliker, A., 1002
Kominz, D. R., 700, 731, 1018, 1020, 1021, 1022, 1023
Kon, S. K., 737
Konikov, A. P., 1143
Koning, C. J., 430
Konz, W., 356, 375(67)
Korey, S., 991
Korkes, S., 183, 184, 185, 212, 213(345, 348), 265, 274(534)
Korn, A. H., 405, 428, 856
Kornberg, A., 126, 164, 165, 187, 188, 204(14), 205, 206, 207, 211(333), 212, 213(342)
Kossel, A., 5, 35, 37, 39(106), 40(106), 42(106), 43(106), 49
Kosterlitz, H. W., 1283, 1284

Koulumies, R., 789, 814(147b)
Kovács, J., 950
Kovács, K., 950
Kozloff, L. M., 96, 98(331), 102(331), 103, 104(352), 105(352), 111
Krahl, M. E., 1239
Kramer, H., 649, 943
Kratky, O., 851, 854, 855, 858, 910, 916, 919, 931, 936
Kraus, R., 820
Krause, H., 1279
Kraybill, H. R., 589
Krebs, E. G., 126, 193, 194, 195, 196 (207), 230, 710, 784
Krebs, H. A., 232, 233(427), 234(425), 235, 236, 243, 245, 350, 383(36), 566, 573, 1216, 1218(41), 1230, 1233
Kreiger, H., 1265, 1271, 1283
Krejci, L., 496, 497(47), 503(108), 504, 666, 721
Krekels, A., 45
Krimsky, I., 167, 168(194), 177(94), 182 (194), 198, 199, 200, 201(194), 202, 1068, 1107(35), 1147(35)
Krogh, A., 635, 673, 674
Kroner, H., 377
Krotkov, G., 569
Krueger, A. P., 96
Krüger, F., 281, 317, 318(228)
Krueger, R. C., 152, 154(24), 713
Krüpe, M., 786, 789, 814(147c, 147d)
Krukenberg, C. F. W., 336, 341
Kubacki, V., 1075, 1076(75), 1078(75)
Kubo, H., 280, 322(6)
Kubowitz, F., 124, 136, 140, 141, 144, 149, 150, 152(57), 162(57), 170, 171, 210, 261, 270, 271, 272(509), 341, 784
Kuehl, F. A., Jr., 613(88), 614, 617
Kühn, K., 1140, 1222, 1223
Kühnau, J., 493
Kühne, W., 957
Küster, W., 284, 314
Kuhn, R., 217, 219, 220(365), 221(365), 247, 248(365), 323, 452
Kuhns, W. J., 787
Kunitz, M., 20, 194, 225, 358, 359(79), 1059, 1060, 1061(1), 1063(1), 1070, 1072(1), 1074, 1075, 1078, 1079(1), 1080(1), 1082, 1084(104), 1087(104), 1089, 1090(66), 1091(89), 1095(11), 1096(1, 150, 163), 1097, 1098(150), 1101(48), 1104(66), 1105, 1124(77), 1161(89, 118, 150, 433, 434), 1163, 1164, 1166(118), 1167, 1168(89), 1172(150), 1190(66, 77, 520, 522, 523), 1192, 1193, 1194, 1195, 1249
Kunkel, H. G., 629, 719
Kuntzel, A., 919
Kuriyama, S., 851
Kurnick, N. B., 50
Kuschinsky, G., 979, 989, 990(94, 95), 991, 1046
Kusmin, S., 287
Kutsky, R. J., 1238
Kuttner, A., 413, 738
Kuzin, A. M., 814
Kyes, P., 759
Kylin, E., 697
Kym, J. X., 164

L

Laaksonen, T., 1143, 1224
Lack, C. H., 382
Lackey, M. D., 548
Lagsdin, J. B., 796, 797, 798(179)
Laidlaw, P. P., 625
Laidler, K. J., 210, 1088, 1112(246), 1113, 1127(246)
Laine, T., 563, 572
Lajtha, A., 1040
Laki, K., 700, 726, 730, 731, 879, 880, 881, 978, 984, 988, 1018, 1020(191), 1021, 1022, 1023
Laland, S. G., 18, 19
Lalaurie, M., 789, 814(147a)
Lamanna, C., 346, 349, 359, 372, 374(2a)
Lamb, W. G. P., 52
La Mer, V. K., 1087
Lamm, O., 448, 449(60), 496, 497(48)
Lampl, H., 772
Lamson, B. G., 1288
Lamy, F., 731
Lancaster, E. R., 280
Lancaster, R., 335
Landing, B. H., 671
Landis, E. M., 667, 669, 673, 675
Landolt, R., 972, 983, 984, 987
Landow, H., 787

Landsteiner, K., 317, 346, 352(1), 484, 708, 709, 714(210), 722(203a), 761, 762, 763, 765(30), 766(30), 767, 768 (30), 769, 772, 773, 774(30), 775, 780, 782(114), 783, 789, 805, 820, 831, 1165
Landua, A. J., 564
Landwehr, G., 730
Lane-Claypon, J. E., 432
Lang, E. H., 319
Lang, J. M., 872
Langenbeck, W., 260, 263(496, 497, 498), 270(496, 497), 332
Langmuir, I., 825, 833, 834
Lanni, F., 437, 471(8), 472(8), 473(8, 155), 708, 710(200), 715, 716(200), 722(200), 829
Lanni, Y. T., 472, 473(155)
Lardy, H. A., 190, 1294
Larionow, L. T., 51
Larmour, R. K., 498
Larny, R., 712
LaRosa, W., 443
Laskowski, M., 258, 415, 416(88, 89), 737, 877, 1063, 1074, 1075(70), 1078 (75), 1161(426, 428), 1163, 1164(426, 428), 1166, 1168(11a), 1171(428)
Laskowski, M., Jr., 415, 416(88, 89), 728, 737, 1161(426, 428), 1164(426, 428), 1171(428)
Laties, G. G., 548
Latta, H., 1278, 1279
Lauffer, M. A., 70, 71, 72, 73, 83, 87, 88, 423
Laughton, P. M., 678, 679(71)
Laurell, C. B., 135, 140(49), 158, 159, 160, 699, 700, 701
Laurie, R. A., 44
Lausten, A., 474, 475(159), 478, 482(159), 483(159)
Lavin, G. I., 86, 1096(158), 1097
Lawler, H. C., 660
Lawrence, A. S. C., 969, 986(16), 1049
Laws, W. D., 497
Layne, E. C., 564
Lea, D., 73, 91(242)
Leach, S. J., 570
Leavenworth, C. S., 493, 518, 524, 526, 532(67), 566, 570
Leblond, C. L., 656, 885
Le Breton, E., 358, 359
Leclef, J., 788
Le Clerc, J. A., 500(76), 501
Ledebt, S., 365
Ledingham, J. C. G., 85
Leduc, E. H., 670
Lee, H., 1284
Lee, K. H., 782
Lee, N. D., 59, 1262, 1284
Lee, S. B., 95
Lee, W. A., 32
Legge, J. W., 281, 305, 313(18), 316, 317 (220)
Legler, R., 235
Leglise, F., 724
Lehmann, E., 860
Lehmann, H., 967
Lehmann, J., 396
Lehmann, W., 410
Lehninger, A. L., 131, 166, 203
Lehoult, Y., 56
Lejeune, G., 499, 509(66)
Leland, J. P., 651
Leloir, L. F., 259
Lemberg, R., 281, 313(18), 316, 317
Le Messurier, D. H., 353
Lemétayer, E., 788
Lemley, J. M., 258
Lennon, B., 605
Lennox, F. G., 1069, 1070
Lens, J., 636, 642, 646, 647, 1184(493), 1185
Leonard, B. R., 78
Leonard, S. L., 598
Léonis, J., 1100
Lepow, I. H., 362
Lerman, J., 656
Lermann, S. G., 716
Lerner, A. B., 137, 149, 152(103), 153 (103)
Lesh, J. B., 613(89, 90, 91), 614, 615(89, 90, 91), 617(91)
Leslie, I., 53
LeStrange, R., 522
Leuthardt, F., 526, 571(74)
Leuthardt, F. M., 1206, 1208(13), 1216, 1218(42)
Levene, P. A., 6, 7, 8, 10, 21, 27, 31, 408, 445, 467, 479, 480, 485

Lever, W. F., 707, 719(193), 880, 881(153)
Levin, L., 606, 621
Levine, M. V., 1286, 1295(308), 1296
Levinthal, C., 113
Levintow, L., 1107(224), 1108
Levitt, M. F., 669
Levy, A. L., 630
Levy, E., 827
Levy, H., 137, 459, 574, 575
Levy, H. M., 1257
Levy, M., 360, 368, 369(123), 850, 851
Lewin, J., 681
Lewis, J. C., 445, 447, 449, 463, 465(36), 467, 468, 470, 477, 1075, 1084(73), 1085(73), 1167(73)
Lewis, J. D., 1256, 1286(193)
Lewis, J. H., 484, 758, 759(12), 766, 782 (115), 783
Lewis, L. A., 702, 706, 707, 741
Lewis, P. S., 453
Lewis, S., 141, 142(91), 143(91), 146(91), 153
Lewis, S. E., 1146
Lewisohn, R., 665
l'Héretier, Ph., 122
Li, C. F., 49
Li, C. H., 347, 365, 424, 599, 600, 601, 602, 603, 604, 608, 609, 610, 612, 613, 614, 615, 616, 617, 620, 621, 622, 623, 643, 647, 659, 693, 724, 1184(497), 1185, 1186, 1188(501), 1189(497), 1225, 1295
Li, C. P., 809
Lichtenberg, E., 1141
Lichtenstein, N., 1216
Liddle, G. W., 616
Liebecq, C., 317
Lieberman, I., 1210
Liebig, R., 230, 231(418)
Likhniskaya, I. I., 320
Lilienfeld, L., 36, 37, 46
Lindan, O., 534, 537(114), 540(114)
Lindberg, J., 860
Linderstrøm-Lang, K., 397, 400, 410(34), 425, 445, 446, 448, 454, 455, 586, 696, 767, 1088, 1096(146, 147, 157), 1097, 1099, 1143, 1171, 1172, 1176 (464), 1177(146, 147), 1178(146), 1179(458), 1181(146, 458), 1182, 1190(458, 459), 1202, 1203, 1225, 1248
Lindgren, F. T., 702(197), 707
Lindhard, J., 1032
Lindhorst, T. E., 554
Lindley, H., 570, 872, 874
Lindquist, F., 811
Lindsay, A., 133, 234(34)
Lindsley, C. H., 383
Lineweaver, H., 286, 288(49), 449, 466, 468, 474, 693, 1069, 1081, 1083, 1092, 1094(126), 1095(43), 1107(43), 1111, 1146, 1148(105), 1152, 1161(105, 431), 1163, 1172
Ling, S. M., 453, 456
Linggood, F. V., 348
Lipmann, F., 181, 185, 214, 260, 262, 263, 264, 266, 408, 479, 480(170), 587, 589(271), 672, 784, 1099, 1160, 1204, 1210, 1211
Lipton, M. A., 262
Liquori, A. N., 321, 322
Liret, N., 59
Lis, H., 511(137), 512
Lisbonne, M., 94
Lissetzky, S., 656
Little, K., 943
Little, R. B., 412, 413
Littlefield, J. W., 168, 261(198), 263, 264 (198), 265, 268(198), 269(198), 270 (521), 274(521), 275(198, 521, 536, 537), 276(198), 1211, 1212(32)
Litwack, G., 1283
Liubimova, M. M., 964, 967
Livermore, A. H., 628
Livingood, J. J., 636, 639(199)
Livingstone, B. J., 881
Llewellyn Smith, M., 368
Lloyd, D. J., 856, 918(43)
Lloyd, E., 1153
Locke, A., 808
Locker, R. H., 1001
Lockhart, E. E., 249
Lockwood, W. H., 316, 317(220)
Lockwood, W. W., 647
Lodge, A. L., 380
Loeb, L., 605
Loeffler, 60
Loeliger, A., 380
Loeser, A., 605

Loewus, F., 173
Loewus, M. W., 1121, 1123
Loewy, A. G., 730, 880
Loftfield, R. B., 1136, 1226, 1230(92), 1231(92), 1240(86), 1246, 1252, 1286 (186)
Logan, J. B., 614
Logan, M. A., 936, 944(357), 946(357)
Logemann, W., 1146
Lohmann, K., 259, 261(493)
Lomax, R., 907, 944
London, E. S., 6
London, I. M., 286, 287, 1254, 1255(189), 1271, 1273, 1276(242)
Long, C. N. H., 613
Long, E. R., 352
Long, J. A., 608
Longsworth, L. G., 437, 438, 439, 441, 448, 449, 455, 456(11), 461, 464, 466 (11), 467(11), 469, 470, 471(11), 473, 484, 738, 1165
Lontie, R., 507, 719, 722, 728
Loofbourow, J. R., 1152
Loomis, E. C., 730, 731
Loomis, W. D., 571, 573(241), 1216, 1219
Loomis, W. E., 548
Loose, L., 530
Lorand, L., 379, 381(199), 729, 730, 731, 879, 880, 1053, 1190(527, 529), 1191, 1197(527, 529)
Lorenz, F., 257
Lorenz, L., 380, 1096(144), 1097, 1159, 1166 (144)
Loring, H. S., 17, 67
Lo-Sing, 314
Louran, M., 796
Lovelace, F. E., 1069
Lovell, R., 737
Lovett-Janison, P. L., 139, 142, 143(81)
Low, B., 636, 682, 903
Lowell, F. C., 760
Lowenthal, J. P., 359
Lowry, O. H., 230, 231(416)
Lowry, P. H., 1226, 1227, 1228, 1229 (100, 101), 1230(96, 97, 114), 1232 (96, 97, 100), 1233(100, 101), 1234 (101), 1236(96, 97), 1237(114), 1238 (96, 97, 101, 114), 1241(96), 1242, 1245(100, 101), 1246(100, 101), 1261, 1262(214), 1268(214), 1286
Lowther, A. G., 874
Lucas, C. C., 872, 944
Lucas, R. E., 593
Luck, J. M., 41, 672, 685(46), 1245
Ludwig, B. J., 136, 138, 148(63), 152(75)
Ludwig, W., 657
Lüers, H., 493
Luescher, E., 716
Luetscher, J. A., 678
Luft, R., 613
Lugg, J. W. H., 507, 532, 533, 534, 536 (106), 538(106), 539, 540, 541, 542, 544, 547, 550, 556
Lumry, R., 1105, 1111(214), 1112(214), 1113(214), 1129(214), 1131(214), 1132, 1159, 1160(417), 1241
Lund, E., 445, 446
Lundberg, W. O., 278
Lundgren, H. P., 619, 653, 791, 794(4), 795, 801, 825, 875, 1176
Lundquist, F., 1065
Lundsgaard, E., 197, 963
Lundsteen, E., 410
Luria, S. E., 95, 99, 108, 112, 113, 114
Luttgens, W., 261, 270, 271, 272(309)
Lutwak-Mann, C., 166, 174, 177(224), 179, 180
Lu Valle, J. E., 131
Lwoff, A., 110, 116, 117, 118
Lyman, C., 262
Lynch, E. L., 1242
Lynch, V. H., 523, 581(50)
Lynen, F., 181, 184, 202, 203, 204(322), 205, 206(322), 211(252), 254, 255, 263, 264, 358
Lyon, T. P., 702(197), 707
Lyons, W. R., 598, 602, 603, 604(33), 621

M

Maaløe, O., 98, 113
MacAllister, R. V., 1104, 1111(206), 1116, 1117(206, 207, 258), 1118, 1120 (206, 207)
MacArthur, C. G., 632, 633
MacArthur, I., 867, 869, 901, 907, 1034
McCaig, J. D., 490
McCalla, A. G., 489, 490, 492, 498, 504, 508

McCallie, D. P., 1265
MacCallum, W. G., 60, 85(205)
McCance, R. A., 138, 390, 433
McCarthy, E. F., 299, 320, 706
McCarty, M., 55(181–183), 56, 57, 672, 784
McCay, C. M., 569
Macchi, M., 261
McClaughry, R. T., 731
McClean, D., 385
McClement, W. D., 79
McCoy, R. H., 561
McCrea, J. F., 89
McCullough, W. G., 1241
McDonald, H. J., 522, 539(46a)
McDonald, M. R., 49, 194, 225, 358, 359 (79), 1192
MacDonnell, L. R., 452
McElroy, O. E., 372
McElroy, W. D., 278, 592, 1082, 1083 (103), 1101, 1109, 1149
McEwen, M. B., 933
MacFadyen, D. A., 1096(167), 1097, 1099 (167)
McFarlane, A. S., 86, 87(281), 701, 706, 1280
Macfarlane, M. G., 349, 366, 367, 784
Macfarlane, R. G., 381, 382, 383, 1171
McFarlane, W. D., 328
McGee, F. M., 578
MacGillivray, R., 533
McGilvery, R., 587, 589(269, 270), 1205, 1206, 1207, 1208(14), 1209
McGirr, J. L., 710
McGrew, R. V., 303
McHargue, J. S., 593
Macheboeuf, M., 451, 702, 703, 704(179, 183), 706, 753
McIlwain, H., 162
MacInnes, D. A., 437, 438(11), 439, 441 (11), 448, 449(11), 456(11), 461, 464 (11), 466(11), 467(11), 471(11), 473, 740, 1165
Mackay, M. E., 752
McKee, F. W., 1271, 1272(240)
McKee, H. S., 523, 560(59), 567
McKie, M., 116
McKinney, H. H., 551
Mackintosh, J., 373, 374(161a, 161b)
Mackor, E. L., 58
Mackworth, J. F., 1146
McLaren, A. D., 1081, 1093, 1098, 1099 (177), 1104(177), 1114, 1118, 1137 (177), 1161(96, 436), 1163, 1168 (96, 436)
McLauchlan, T. A., 95
Maclennan, J. D., 383
MacLeod, C. M., 56, 370, 382, 731, 732 (327)
Macleod, J. J. R., 635
McMaster, P. D., 1279
McMeekin, T. L., 391, 392(9), 394(9), 397(12), 398(12), 399, 400(12), 401, 402(44), 403(44), 404(44), 405, 406, 407, 410(38), 419, 420, 421, 422, 423 (110), 424, 425, 426, 428, 429(12), 433, 434(9), 444, 678, 680(73), 683 (73), 686, 700, 728, 734, 790
Macmullen, J., 1239
McMurtrey, J. E., Jr., 589, 590(274), 591 (274)
MacPherson, C. F. C., 449, 453, 455, 822
McQuarrie, I., 1265
McShan, W. H., 601
MacVeigh, I., 558
Macy, I. G., 291, 390, 391, 397(4), 411 (4), 412(4), 416(4), 427(4)
Madden, R. E., 1266, 1267(228), 1276 (228)
Madden, S. C., 671, 809, 1247, 1269, 1288
Magasanik, B., 17, 31, 1285
Magdalena, A., 605
Mager, A., 41
Mahadevan, S., 566
Mahdihassan, S., 8
Mahler, H. R., 170(209), 202, 203, 219, 222(370), 223(370), 224(370, 403), 225, 246, 252, 253, 254, 255, 256, 257 (403, 472), 278, 1211
Maier-Hüser, H., 630
Main, E. R., 808
Maitra, S. R., 377
Makino, K., 7
Malkiel, S., 759, 770, 825
Mallette, M. F., 135, 138, 141, 152, 153, 154, 155, 156
Malmstrom, B. G., 50
Mandel, H. G., 1287
Mandelstam, J., 1283
Mangun, G. H., 673
Mann, R. L., 500(77), 501, 506

Mann, T., 136, 139, 140(58, 77), 151, 152 (77), 153(58), 154, 157, 158, 700
Manske, R. H. F., 560
Manski, W., 842
Mansson, B., 1275
Marceau, F., 954
Margen, S., 1265
Margolis, D., 566
Marion, L., 564
Maritz, A., 237, 238(444), 243, 244, 1096 (172), 1097, 1099(172)
Mark, H., 852, 856, 890
Mark, H. J., 692
Markham, R., 13, 14, 15, 16, 17, 18, 19, 20, 22, 23(36), 24(80b, 80c), 26(75), 28, 67(60), 72(47), 73, 74, 76, 77, 82, 91(242), 120
Markin, L., 759
Markley, A. L., 430
Marko, A. M., 930
Marks, H. P., 635
Marks, J., 368, 369(128)
Marrack, J. R., 708, 709, 764, 776, 777, 790(34), 791, 800(34), 801, 802, 806 (34), 817, 818, 820, 826(34), 828(34), 829, 830
Marriott, R. H., 856, 918(43)
Mars, P. H., 415, 416(89), 1161(426), 1164(426)
Marsh, B. B., 1040, 1041, 1047(224), 1053, 1054
Marshak, A., 16, 17, 19, 105, 1262
Marshall, C. R., 312
Marshall, J., 282
Marshall, L., 406, 728
Marshall, M. E., 482, 672, 698(43)
Marshall, P. G., 475, 477(162), 480, 481, 483, 620
Marshall, W., 312
Martell, A. E., 1149
Martiensien, E. W., 771
Martin, A. J. P., 522, 527(44), 643, 873, 921
Martin, A. V. W., 911, 914(220), 919 (270), 920(270), 928(270), 931(270), 933(270), 936(270), 941(270), 942 (270), 944(270)
Martin, C. J., 405, 453
Martin, L. F., 551
Martius, C., 263
Marvin, H. N., 1261
Marwick, T. C., 868, 870
Maschmann, E., 383, 1069, 1148(46)
Mason, H. L., 922
Mason, J. H., 738
Masouredis, S. P., 1271, 1272
Massoro, W. J., 569
Masters, R. E., 1260, 1269(209), 1271 (209)
Mathews, M. B., 161
Mathies, J. C., 1065
Mathieu, L., 978
Matoltsty, G., 992
Matoltsy, A. G., 944
Matthews, R. E. F., 71, 77, 1287
Maurer, P. H., 669, 765, 766, 816(42, 43, 44), 1277, 1278, 1279
Maurer, W., 1276
Mauthner, J., 1145
Mautz, F. R., 673
Maver, M. E., 39, 1225
May, L. M., 322
May, S. C., Jr., 1158
Maybury, R., 691, 692(123)
Mayer, M., 784, 793, 794, 796(172), 803 (124), 809(172), 817, 822(172), 825, 842, 843
Mayer, M. M., 725
Mayerson, H. S., 1271, 1272(237)
Maynard, J. T., 776, 801(88), 831(88), 835(88)
Mazia, D., 1242
Mazur, A., 338, 1154
Mazureke, C., 712
Mead, T. H., 771
Mecham, D. K., 463, 475, 477(164), 478, 479(164), 480(164), 875
Medveczky, A., 768
Meerwein, H., 219
Meeuse, B. J. D., 575
Mehl, J. W., 735, 736(346), 956, 969, 988, 1006
Mehler, A. H., 205, 207, 211(333), 212, 213(342)
Meiklejohn, G. T., 140
Meisinger, M. A. P., 613, 614, 615(87), 617
Meister, A., 170(210), 171, 208, 209, 570, 1216
Meitina, R. A., 990

Melcher, L. R., 1271, 1272
Melchior, J. B., 1226, 1228, 1229, 1230 (85, 98), 1236(98), 1239
Meldrum, N. U., 289
Melin, M., 680, 701, 702, 704, 705(185), 706(185), 711(177), 718(177), 725 (77), 730(177), 741(177)
Mellander, O., 390, 392, 393(1, 10), 396, 397, 398, 408(1), 427(1, 10), 428(1)
Mellody, M., 1226, 1230(98), 1236(98)
Mellon, E. F., 405, 428, 856
Mellors, R. C., 673
Melnick, J. L., 261
Melville, J., 566, 567
Melvin, S., 1295
Mendel, L. B., 358, 500(72), 501, 515
Mendes, T. G., 336
Menefee, S. G., 391
Meneghini, M., 553, 555
Menke, W., 543
Mercer, E. H., 850, 856, 860, 875, 886, 887, 904, 905, 907, 946
Mercer, F. L., 554
Merejkowski, 343
Merino, A. S., 500(81), 501
Merkelbach, O., 298, 328(112)
Mernan, J. P., 878
Merrill, P., 554, 605
Merritt, J. B., 1170
Merten, R., 1071, 1127(59a)
Mesrobeanu, L., 374, 386
Meyer, C. E., 561
Meyer, D., 467, 642
Meyer, F., 397
Meyer, J., 1148
Meyer, K., 462, 466, 467, 471(139), 473, 474, 734, 921, 956, 1005, 1010, 1011, 1012
Meyer, K. H., 852, 853, 854, 855, 856, 863, 889, 915, 948, 1049
Meyer, P., 673
Meyer, R. K., 601
Meyerhof, O., 190, 201, 367, 965, 968, 1004, 1005
Miall, M., 969, 986(66)
Michaelis, L., 131, 134, 217, 218(369), 306, 335, 874
Michaelis, M., 211
Michaelis, R., 221, 224(384), 257(384), 258(384)
Michaels, G. O., 1265
Micheel, F., 356, 376, 387
Michel, H. O., 312
Michel, J., 293
Michel, R., 651, 652, 653(295), 656, 659, 945
Michelson, A. M., 966
Michelson, C., 571, 573(241), 1216
Middlebrook, W. R., 379, 381(199), 729, 873, 879, 880, 1190(527, 529), 1191, 1197(527, 529)
Miège, E., 493(37), 494
Miescher, F., 3, 4, 5, 38, 40, 43(1), 45, 480
Miettinen, J. K., 559, 562
Mihalyi, E., 730, 880, 881, 961, 962, 976, 977, 1019, 1022, 1190(531), 1191, 1198(531)
Mii, S., 170(209), 171, 202, 203, 224(403), 225, 254(315), 255, 256(403), 257 (315, 403)
Miles, A. A., 110, 349, 350(32), 367(32), 387
Miles, E. M., 349, 350(32), 367(32)
Millar, F. K., 523, 539(52), 540(52), 587
Miller, A., 1136
Miller, B., 765
Miller, E., 890
Miller, E. G., 946
Miller, G. L., 78, 88, 420, 636, 640
Miller, H. K., 570
Miller, J. A., 1295
Miller, L. A., 1238
Miller, L. L., 292, 565, 673, 1257, 1260, 1269(209), 1271, 1273, 1276(209), 1288
Miller, O. N., 191
Miller, R., 59, 1262
Miller, S. G., 726, 727(286)
Miller, W. H., 154, 157, 159
Miller, W. W., 1226, 1240(86)
Miller, Z. B., 1146, 1148
Millerd, A., 548
Millikan, G. A., 305, 325, 338(138)
Milne, H. B., 1141
Mims, V., 564
Minkowski, O., 634
Minkowski, V., 634
Mirsky, A. E., 18, 19, 37, 38, 39, 40, 41 (124), 42(116), 44, 46, 49, 50(116),

51, 52, 53, 54, 286, 288, 289, 331(48), 343, 452, 963, 1091, 1094, 1172, 1243
Mirsky, I. A., 645
Mirsky, R. A., 625
Misani, F., 18, 19
Mishuck, E., 1081
Mitchell, J. H., Jr., 1287
Mitchell, R. F., 878
Mittelman, D., 683
Mizushima, S., 899
Mönch, J., 276
Mohammad, A., 449, 463, 465(68), 471 (68)
Mohammed, A. H., 354
Mohr, R., 45
Moisio, T., 559
Møller, K. M., 1088
Molnár, F., 978, 979(89), 982(89), 983 (89), 1023(89)
Mommaerts, W. F. H. M., 640, 647, 881, 961, 967, 968(58), 972, 978, 979, 980, 987, 988, 1019, 1042
Monod, J., 111, 1285, 1292, 1293(300)
Monroe, R. A., 278
Montanez, E. A., 1257
Montanez, G., 1257
Montreuil, J., 16
Moon, H. D., 615
Moore, D. H., 224(401), 225, 235(401), 319, 348, 449, 453, 455, 606, 634, 738, 765, 791, 794(9), 795, 816(43), 843, 970
Moore, F. D., 1271
Moore, S., 522, 629, 686, 687(112), 723, 850, 1082, 1085(101, 101a), 1096 (164), 1097, 1099(164), 1248
Moos, A. M., 721
Morales, M. F., 836, 994, 1048
Morawetz, H., 683
Morel, A., 480, 769
Morel, J., 924
Morell, D. B., 226, 227, 228, 229, 230 (405)
Morena, J., 343
Morgan, C., 1028
Morgan, E. J., 134, 430
Morgan, F. G., 368
Morgan, R. S., 373, 374(161, 161a, 161b)
Morgan, V. E., 326
Morgan, W. T. J., 386
Morgenroth, J., 783
Mori, T., 445, 467
Moring-Claesson, I., 1174, 1226
Moriyama, L., 359
Morner, C. T., 441, 466(21), 943
Morris, C. J. O. R., 613, 614
Morris, H. J., 474
Morris, M., 403
Morris, P., 614
Morrison, K. C., 667, 677, 739(67)
Morrison, P. R., 726, 727, 728, 882, 883 (175), 884
Morrison, R. I., 559, 565
Morrison, W. L., 1148
Morse, L., 1263, 1276(223)
Morton, J. A., 804
Morton, R. K., 965, 980
Moskowitz, M., 610
Mosley, V. M., 88
Moss, J. A., 921, 926
Moulder, J. W., 207, 211(334)
Moulé, Y., 358, 359
Mourant, A. E., 785
Moutte, M., 288
Moyer, A. W., 614
Moyer, L. S., 548, 549
Mudd, S., 790
Müller, D., 172, 257
Müller, F. H., 849
Müller, H. R., 1137, 1141, 1221, 1222(64)
Mueller, J. H., 561
Mueller, J. M., 629
Mugibayasi, N., 493
Muhlbock, O., 703
Muir, H. M., 286, 287, 292, 930
Muir, P., 756, 840
Muir, R. D., 221, 257
Mukerji, B., 376, 381
Mulder, E. G., 592
Mulder, G. J., 395, 515
Mulford, D. J., 680, 725(77)
Mullen, J. E. C., 431
Muller, H. J., 812
Munch, R. H., 297, 298(109), 300(109)
Munch-Petersen, A., 368, 990, 1040(119), 1053
Munday, B., 352
Munk, L., 267, 272(542)
Munoz, J., 799

Volume II part A—pages 1 to 661
Volume II part B—pages 663 to 1296

Munro, H. N., 1289
Muntwyler, E., 673, 1283
Murlin, J. R., 649
Murphy, E. A., 1257
Murray, C. D., 309
Murray, C. W., 449, 466, 468, 1161(431), 1163
Murray, G. R., 650
Muscatine, N. A., 100
Mustacchi, P. O., 917
Mustakallio, K. K., 697
Mutsaars, W., 768, 769
Muus, J., 693
Mycek, M. J., 1104, 1105, 1107(225a), 1108, 1134(225a), 1138(210), 1140 (203), 1141(203), 1221
Mydans, W. E., 341
Mynors, L. S., 724
Myrbäck, K., 525, 754, 922, 1060, 1065, 1095, 1096(143), 1102(6), 1206

N

Nachmansohn, D., 1101
Naef-Roth, S., 559
Nageotte, J., 925, 943
Nagler, F. P. O., 367
Naismith, W. E. F., 505
Najjar, V. A., 126, 195(11), 784
Nanninga, L., 725
Narayanan, K. G. A., 566
Nason, A., 137, 278, 592, 593
Nebbia, G., 161
Needham, D. M., 166, 192(177), 963, 969, 986(66), 1049, 1153, 1154(400)
Needham, J., 436, 437, 466, 484, 485, 969, 986(66), 1049
Neel, J. V., 781
Negelein, E., 124, 165, 169, 171, 172, 173, 186, 187(269, 271), 192(173), 224 (396), 225, 233, 234(396), 312, 336, 1240
Negoro, H., 1070
Nehari, V., 555
Neidle, A., 1136
Neil, J. C., 96, 98(331), 102(331)
Neilands, J. B., 167
Neill, J. M., 307, 328
Neish, A. C., 543, 548, 593(151)
Nelson, C. D., 569
Nelson, F. E., 95
Nelson, J. M., 135, 136, 138, 139, 142, 143(81), 148(63, 78), 150, 151, 152, 153, 154(122), 432
Nelson, J. W., 621, 622, 623(142, 145), 796
Nemetschek, Th., 914
Nesbitt, L. L., 500(84), 501
Netter, H., 308
Neuberg, C., 124, 270
Neuberger, A., 185, 223, 258(257), 286, 287, 292(56a), 293, 445, 452, 561, 568, 647, 659, 693, 735, 930, 1273
Neufeld, E. F., 215, 216(359)
Neuman, R. E., 936, 944(356, 357), 946 (356, 357)
Neumann, A., 5
Neumeister, R., 465
Neurath, H., 70, 72, 90, 290, 400, 453, 637, 639, 640, 647, 685, 722, 762, 765, 786, 793, 807, 963, 1059, 1060, 1066, 1067(30), 1068, 1074, 1075, 1077, 1079, 1080(90), 1081(28, 90), 1083, 1084(107b), 1090(90), 1091(90), 1092 (30), 1094, 1095, 1096(28, 34, 166, 173), 1097, 1099(166, 173), 1100, 1104(183), 1105(3, 173), 1107(3, 28, 166, 173), 1108(3, 297), 1109, 1110 (3), 1111(3), 1112(166, 182, 183), 1113(3, 182), 1114(3, 166), 1115 (182), 1116, 1117(182, 183, 257, 263), 1118(257), 1120, 1121(274), 1122 (3, 274), 1123(142, 173), 1125, 1126 (173), 1129, 1130, 1131(3, 166, 294), 1132(252), 1133(3, 29, 166, 252, 294), 1148, 1149(285, 286), 1150, 1155(83, 297), 1156(90), 1157(286), 1167, 1172, 1177, 1181, 1184(83, 125, 235, 297, 450, 490), 1188(125, 235, 490), 1189(83, 297, 490), 1190(30, 83, 125, 235, 297, 490, 521), 1192(285, 286), 1193(125, 235, 521), 1195(83, 235, 297, 490, 517), 1196(83, 490), 1222
Neutelings, J., 647
Nevraeva, N. A., 814
Newcomer, E. H., 543
Newell, J. M., 700, 790
Newmark, P., 552, 554(160)
Newton, G. G. F., 1288
Newton, N., 556

Newton, W. H., 626
Nichol, J. C., 713
Nicholas, D. J. D., 278, 592
Nicholas, J. S., 1012, 1015, 1017
Nichols, A. V., 702(197), 707
Nichols, J. B., 448, 449(63)
Nicolet, B. H., 409
Niedergerke, R., 1052
Niederland, T. R., 191
Nielands, J. B., 200(193), 208, 209
Nielsen, H., 1206, 1208(13)
Niemann, C., 656, 850, 1096(161, 174), 1097, 1098, 1099(161), 1100, 1104, 1105, 1107(184), 1108(320, 320a), 1110(184), 1111, 1114, 1115(244), 1116, 1117(178, 184, 206, 207, 258, 259, 260, 265, 271), 1118, 1119(244), 1120, 1121(261), 1122(244, 261, 273), 1123, 1125(242), 1137, 1140, 1141
Nightingale, G. T., 523, 590, 591
Niklas, A., 1276
Nitsch J. P., 529, 532(92), 542(92)
Nitschmann, H., 404, 410, 1083, 1084 (106)
Niven, J. S. F., 1273
Nixon, H. L., 120
Noble, E. C., 635
Nocito, V., 221, 224(395, 400), 225, 235 (395, 400), 236(395), 237(395, 400), 259(382), 573
Noda, H., 926, 928
Noordam, D., 79
Nord, F. F., 257, 453, 454, 468, 1079, 1080, 1091(92), 1105(94), 1149(92), 1150(94), 1165, 1192(92)
Norris, E. R., 230, 1065
North, A. C. T., 911, 914(270), 915, 919 (270), 920(270), 926, 928(270), 931 (270), 933(270), 936(270), 939, 940 (270, 362), 940(270), 944(270), 946
Northen, H. T., 586
Northrop, J. H., 96, 98, 101, 102(336), 293, 401, 678, 719, 794(6), 795, 798, 799, 805, 1059, 1060, 1061(1), 1063 (1), 1064, 1071, 1072(1), 1074, 1075, 1078, 1079(1), 1080(1), 1082, 1084 (104), 1086, 1087(104), 1090(66), 1091(89), 1095(18), 1096(1, 148, 163), 1097, 1101, 1102, 1103, 1104 (66), 1105, 1143, 1144, 1151, 1153, 1154(401), 1161(89), 1165, 1168(89), 1190(66), 1193, 1194, 1195, 1249
Norton, S., 713
Nossal, P. M., 213
Novelli, G. D., 264
Novelli, O. G., 1211
Novick, A., 115
Nowakowski, S., 320
Nunheimer, T. D., 95
Nurmikko, V., 1241
Nuttall, G. H., 780
Nutting, G. C., 420, 448, 465(57)
Nutting, M. D. F., 1077, 1123(85), 1154, 1155
Nygaard, A. P., 200, 201(309), 375, 378 (175b)

O

Oakley, C. L., 349, 366(30), 367, 383, 1096(154), 1097
Obermayer, F., 760, 761, 765(20)
O'Brien, J. R. P., 288, 331(63), 333
Ochoa, S., 203, 204, 205, 206(320, 321, 323), 207, 211(333), 212, 213, 260 (347), 261(347), 262, 264, 265, 274 (534)
O'Connell, R. A., 875
Odin, L., 736
Oesper, P., 201
Ofner, P., 162, 173(146)
Ogle, J. D., 1107(225b), 1108
Ogston, A. G., 224(404), 225, 258(404), 420, 432, 640, 997, 1001(134)
Ogur, M., 54
O'Kell, C. C., 828
Olcott, H. S., 475, 477(164), 478, 479 (164), 480(164), 647, 695, 1083, 1085 (107), 1145, 1148(353), 1151(353), 1153, 1161(430), 1162(430), 1163, 1165(430), 1167(430) .
Oleson, J. J., 614
Olitzki, L., 374
Olitsky, P. J., 715
Oliver, G., 623
Olofsson, B., 875, 886
Olsen, G. A., 590
Olsen, N. S., 649
Olson, C. K., 1136, 1140(307)
Olson, J. A., 169, 171, 188, 189, 190

Oncley, J. L., 449, 465, 639, 640(225), 668, 676, 699, 700, 701, 702, 704, 705 (61, 185), 705(185), 709, 711(177), 713(61), 718, 719, 720(61), 726(61), 730(177), 741(177), 753, 790, 791, 807, 877, 924, 933(323), 986
O'Neill, J. J., 1158
Onoe, T., 482
Onslow, M. W., 152, 489, 490
Oppenheimer, E. H., 60, 85(205)
Orcutt, M., 822
Orcutt, M. L., 413, 737
Orekhovich, K. D., 926, 944(334)
Orekhovich, V. N., 926, 944(334)
Orgel, G., 39
Orlando, A., 119
Orr, J. H., 382
Orth, H., 284, 285(40)
Osato, S., 288
Osborn, C. M., 625
Osborne, T. B., 5, 6, 358, 399, 400, 410, 433(73), 443, 445, 456, 466(38), 471 (38), 475, 476, 480(160), 482, 488, 489, 490(1), 491, 493, 495, 500(25, 72, 74, 82, 94), 501, 504, 506, 515, 516, 517, 518, 534(18)
Osborne, T. S., 782(116), 783
Oster, G., 70, 71, 72
Osteux, R., 1295
Oswald, A., 650
Otey, M. C., 1149
Otis, A. B., 149
Ott, P., 170, 171, 210, 784
Ottesen, M., 454, 455, 767, 1070, 1171, 1190(459, 525, 526), 1191, 1248
Ouchterlony, Ö., 799
Oudin, J., 362, 799
Ouellet, C., 523, 581(50)
Overman, O. R., 391
Overend, W. G., 18, 19, 49
Owades, P., 571
Owen, R. D., 552
Owren, P. A., 754

P

Page, E. W., 1171
Page, I. H., 702, 706, 707, 741, 1131
Païc, M. M., 793, 794(8), 795
Paine, T. F., 1237
Painter, E. P., 358
Painter, H. M., 396
Palatine, I. M., 1231, 1232(126), 1237 (126), 1240
Pallade, G. E., 47
Palm, A., 1098, 1125(176), 1126(176)
Palmer, A. H., 393, 407, 419, 423(58), 445, 446, 447(34), 453(34)
Palmer, J. W., 462, 466, 734
Palmer, K. J., 462, 464, 1077
Palmer, W. W., 651, 653
Pan, S. Y., 602
Panier, A., 507
Panum, P., 677
Panzer, F., 277
Papendick, A., 284
Papkoff, H., 615
Pappenheimer, A. M., 349, 353(23b), 767, 785, 787, 791, 793(164), 794(4, 5), 795, 798, 801, 803(130), 804, 824, 825, 1285
Pappenheimer, A. M., Jr., 314, 361, 362, 363, 710, 711, 713, 754
Pappenheimer, J. R., 670, 671, 674(40)
Pardee, A. B., 716, 776, 777, 798(94), 1061
Parfentjev, I., 790
Parker, E. A., 916
Parkes, A. S., 615, 651
Parkinson, G. G., 139, 148(78), 153(78)
Parks, R. E., Jr., 1287
Parmelee, C. M., 95
Parnas, J., 166, 167, 179(180)
Parnas, J. K., 822
Parrish, R. G., 972
Parson, W., 1290
Partridge, S. M., 386, 522, 949, 1181, 1189(489)
Passmann, J. M., 1190(514), 1191, 1192 (514)
Pasternak, R. A., 570
Patterson, E. L., 261
Paul, K.-G., 125
Paul, M., 165, 168(175), 263(175), 265 (175), 266(175), 268(175), 273(175)
Pauli, W., 307, 308
Pauling, L., 295, 303, 304, 315, 320, 321, 322(251), 324(99), 329, 330(306), 331, 333, 334(206), 335, 709, 714 (210), 774, 776, 790, 798(94), 801,

807, 813, 814, 829, 830, 831(88), 835, 890, 893, 894, 895, 896(227), 897 (225), 900, 901(225, 238), 903, 904, 907, 939, 1288
Pauly, H., 724
Pautrizel, R., 724
Payne, R. W., 615, 616
Peacock, N., 860
Peacocke, A. R., 32
Peanasky, R. J., 1166, 1167
Pearsall, W. H., 530, 531, 532
Peck, R. L., 1095
Pedersen, K. O., 224(402), 225, 247(402), 293, 306, 339(148), 340, 392, 393, 401, 402, 417, 418, 429, 499, 503 (105), 504, 505(105), 600, 608, 610 (67), 613, 615, 640, 652, 653(295), 701, 702(232), 714, 728, 737, 791, 794(1, 2, 10), 795, 796, 831, 1186
Pekarek, E., 647
Pekelharing, G. A., 665
Pembroke, R. H., 738
Pénasse, L., 467, 642, 1167
Pencharz, R. I., 621
Penfold, B. R., 570
Pennell, R. B., 834
Pennington, D., 439, 469
Pensky, J., 127, 261, 263(500, 501), 264, 265(501), 270(501), 271, 272(500, 501), 273(500)
Perez, J. J., 712
Perkins, M. E., 8, 370, 1146
Perlman, E., 386
Perlmann, G. E., 401, 448, 449, 450, 451, 455, 672, 740, 1022, 1082, 1083(102)
Perrings, J. D., 682, 728
Perrone, J. C., 287, 292(56a), 1275
Perry, S. V., 966, 968, 976, 978, 979(91), 987, 989, 992, 995, 1002, 1003, 1004, 1005(146), 1009, 1017(49), 1019, 1021, 1028, 1029, 1030, 1035, 1053
Perry, W., 615
Persky, H., 252
Perutz, M. F., 281, 294, 295, 319, 322, 324, 335, 421, 867, 872, 896(84), 898, 900, 901, 941
Peskett, G. R., 434
Petat, J. M., 150
Petermann, M. L., 713, 722, 791, 793 (164), 794(5), 795, 798, 1175
Peters, R. A., 262, 1148
Peters, T., 673, 1235, 1246, 1247, 1258
Peters, V. J., 1241
Petersen, W. E., 416, 433(90)
Peterson, E. A., 1227, 1228, 1232(113), 1233(113), 1234, 1236, 1237(104), 1240(113), 1241(113), 1245
Peterson, W. H., 1101, 1115(188)
Petrie, A. H. K., 527, 532, 585, 589
Pettinga, C. W., 1104
Pettko, E., 978, 979(89), 982(89), 983 (89), 1023(89)
Pevsner, D., 967
Pfeiffer, H. H., 919
Pfister, R. W., 1137, 1141(312), 1221, 1222(64)
Phear, E. A., 1284
Philip, B., 860
Philipp, K., 318
Philippi, E., 337
Phillips, D. M. P., 647, 648, 649(274), 1143, 1183(335), 1185
Phillips, H., 874
Phillips, P. H., 412, 413, 427, 433(82), 737, 792
Phillips, R. A., 667
Philpot, J. St. L., 227, 348, 431, 769, 1064, 1072, 1152
Phinney, J. I., 483, 485(182)
Phisalix, M., 353, 354
Pick, E. P., 760, 761, 765(20)
Pickels, E. G., 86, 549, 637, 640
Picken, L. E. R., 946
Pickles, M. M., 804
Pierce, J. G., 628, 629, 630, 631(169, 178), 632
Pierce, J. V., 261
Pietro, A. S., 1290
Pijper, A., 889
Pillai, R. K., 192
Pillemer, L., 348, 362, 368, 373, 725, 752, 771, 840
Pilling, J., 382
Pincus, S. N., 443
Pinsky, M. J., 1285
Pirie, A., 943
Pirie, N. W., 47, 61, 64, 65(212), 67(212), 68, 71, 72, 79, 80, 82, 84, 85, 110, 347, 387, 542, 543, 545, 775

Pitt-Rivers, R., 651, 656, 657, 658, 694, 1102, 1127(195)
Plass, M., 203, 205(319), 206(319)
Plentl, A. A., 1131
Plescia, O. J., 711, 802
Plimmer, R. H. A., 476, 480
Plomley, K. F., 404, 405
Plotnikova, N. E., 926, 944(334)
Polgase, W. J., 1108, 1132
Polis, B. D., 424, 425, 428, 429, 431, 434, 965
Polis, E., 398, 429
Pollard, E., 1093
Pollard, J. K., Jr., 558, 562(182), 563 (182), 564(182)
Pollinger, A., 322
Pollister, A. W., 37, 38, 39, 40(118), 42 (116), 43, 44, 46, 50(116), 54
Pollock, J. R. A., 559
Pollok, H., 1105, 1111(213)
Polson, A., 96, 102, 105, 109(351), 373, 412, 448, 449(60), 496, 497(48), 503 (106), 504
Pon, N. G., 1184(497), 1185, 1189(497)
Ponticorvo, L., 286
Pope, C. G., 362, 790, 1102, 1103, 1127 (196), 1175
Popenoe, E. A., 632, 660, 1141
Popjak, G., 706
Popper, H., 1296
Porath, J. O., 617
Porter, K., 728, 882, 914, 931(278)
Porter, P. M., 828
Porter, R. M., 416, 433(90)
Porter, R. R., 317, 321(226), 428, 446, 458, 467, 636, 714, 724(235), 765, 814, 1167, 1175
Portier, P., 353, 757
Portis, R. A., 1125
Portzehl, H., 958, 959, 960(15), 970, 1033, 1040, 1043, 1044(15), 1045
Post, J., 1273
Posternak, S., 408, 480
Posternak, T., 480
Potter, J. S., 51
Potter, R. L., 427, 792
Potter, V. R., 134, 246, 1231, 1294
Potts, A. M., 626, 631
Pouradier, J., 924
Powell, J. F., 559
Powell, R. E., 401, 1175
Powers, J. H., 1242
Powers, W. H., 141, 142(91), 143(91), 146 (91, 97), 147, 148
Powilleit, E., 786
Prakke, F., 919
Prasad, A. L. N., 522, 556(46)
Prášek, E., 769
Pratt, A. W., 913
Pratt, M. I., 933
Preer, J. R., 121
Prescott, J. M., 1241
Pressman, D., 724, 765, 774, 776, 777, 801 (88), 808, 829, 830, 831(88), 835(88), 1272
Preston, C., 520, 521, 526, 528, 529, 531 (35), 568, 569, 574(35), 575(39, 40, 87), 580(39), 589, 591(40)
Prévot, A. R., 373
Prianischnikov, D. N., 565
Price, S. A., 368
Price, V. E., 1107, 1108, 1149
Price, W. C., 77, 78, 549, 902
Price, W. H., 107, 969
Pricer, W. E., Jr., 126, 164, 188, 204(14), 205(14), 206
Prideaux, E. B. R., 636
Privat de Garilhe, M., 630
Proescher, F., 665
Proom, H., 564
Prosser, C. L., 665
Prudhomme, R. O., 761
Pryor, M. G. M., 946
Pucher, G. W., 515, 518, 524, 525, 526, 532(67), 553(6), 566, 570, 587(6)
Puck, T. T., 109
Pugh, C. E. M., 152
Pullman, M. E., 161, 164, 173(167), 175 (167), 208(167)
Purnell, M. A., 826, 828
Purr, A., 1146
Putnam, E. W., 1294
Putnam, F. W., 95, 96, 98(331), 102(325, 331, 333), 103, 104(352), 105(352), 111, 290, 346, 374(2a), 400, 497, 963, 1066, 1081(28), 1094, 1096(28), 1107 (28)
Putterman, S., 370

Q

Quastel, J. H., 162, 182, 375, 378(175a)
Quensel, O., 492, 495, 496(15), 497(50), 505
Quibell, T. H., 175, 177(226)

R

Racker, E., 166, 167, 168(194), 169(201), 171, 172(201), 177(194), 178(184), 179, 180, 181, 182, 183, 198, 199, 200, 201(194), 202, 1068, 1107(35), 1147 (35)
Raben, W. S., 615, 616
Rabideau, G. S., 538, 539, 562(118, 119)
Rabinowitch, B., 639, 640(227)
Rabinowitch, E. I., 548(152), 549
Race, R. R., 785, 803, 804, 822(127, 128)
Radhakrishnan, A. N., 558
Radin, N. S., 287
Rafferty, J. A., 319
Rafter, G. W., 194(207), 196(207)
Raistrick, H., 221, 224(384), 257(384), 258(384), 411, 413, 433(76)
Ralph, P. H., 729
Ram, J. S., 522, 556(46)
Ramachandran, B. V., 408
Ramakrishnan, C. V., 266, 274(540)
Ramamurti, T. K., 529, 575(87)
Ramasarma, G., 141, 142
Ramaswamy, R., 566
Ramdas, K., 566
Ramon, G., 788, 808, 826
Ramsdell, G. A., 396
Ramsdell, P. A., 162, 226
Ramsey, R. W., 1052
Ramsey, V. W., 1016
Randall, J. T., 911, 912, 914, 915, 918 (272), 919, 920, 922(272), 926, 928, 931(270), 933, 936, 939, 941(270, 362), 942, 943(272), 944(270), 946
Randall, S. S., 656
Ranney, H., 781, 1254, 1255(189)
Raper, H. S., 136
Rapkine, L., 134, 177, 197(37, 227), 198, 963
Rapkine, S. M., 177, 197(227)
Rapp, H. J., 843
Rappaport, C., 100
Ratner, B., 413, 738
Ratner, S., 221, 224(395, 400), 225, 235 (395, 400), 236(395), 237(395, 400), 259(382), 573, 671, 1259, 1271, 1274 (204), 1280
Ratzenhofer, M., 931
Raub, A., 564
Rauch, K., 258
Rauen, H., 205
Ravin, H. A., 1096(155), 1097, 1100(155)
Rawlinson, W. A., 339, 341(351), 363
Rawson, R. W., 605
Ray, N. E., 671
Ray, P., 355
Ray, R. C., 381
Raynaud, M., 373
Rebeyrotte, P., 702(183), 703, 704(183)
Recknagel, R. O., 1294
Record, B. R., 714, 752, 796(180, 181)
Redfield, A. C., 337, 338(344), 339(345), 340(344)
Reed, G. B., 349, 382, 569
Reed, L. J., 132, 261, 266, 269, 270(503), 558, 563(178)
Reed, R., 946, 949, 987, 996, 1035
Rees, A. L. G., 905
Rees, M. W., 447
Reese, J. W., 1227, 1240(102)
Reese, M. C., 642
Reguera, R. M., 789, 814(150)
Reichard, P., 1243
Reichert, E., 181, 263
Reichert, E. T., 283, 780
Reifer, I., 566, 567
Reimer, F. C., 590
Reinbold, B. von, 280, 312, 313, 343(5)
Reineke, E. P., 433, 651, 655, 656, 657, 658
Reiner, J. M., 1263, 1285
Reiner, L., 319
Reinert, M., 357
Reinhardt, W. O., 614, 616, 1265, 1271
Reinhart, R. W., 459
Reinke, J., 548
Reiser, H. G., 731
Reiss, M., 327, 620
Reithel, F. J., 221, 257
Reitz, H. C., 647, 695
Reller, H. H., 1279
Remmert, L. F., 732

Renaux, E., 94
Renkin, E. M., 670
Renkonen, K. O., 789, 809(148), 814(148)
Renold, A., 613(90), 614, 615(90)
Ressler, C., 660, 661(337)
Retzius, G., 1002
Reumuth, H., 860
Rhees, R. C., 448
Rice, F. E., 430
Rich, A. R., 709
Richards, A. N., 675
Richards, F. J., 559, 564(191c), 576, 589
Richert, D. A., 230, 701, 711(177), 718 (177), 730(177), 741(177), 1284
Richet, C., 757
Richet, M. C., 353
Richmond, J. E., 292
Richter, D., 152
Richter, J. W., 617
Rickless, P., 530
Riddell, C. B., 605
Riddle, O., 474, 602, 603
Rideal, E. K., 492, 505(21)
Rigdon, H., 319
Riggs, A., 303, 319
Riggs, B. C., 1096(149), 1097
Riggs, D. S., 665, 1287
Riggs, T. R., 671, 1231, 1232(126), 1237 (126), 1240
Riker, A. J., 583
Riley, D. P., 45, 295
Rimington, C., 324, 363, 408, 622, 735
Ringel, S. J., 874
Ripley, S. H., 1243
Ris, H., 40, 41(124), 44, 49, 51, 53, 54
Riseman, J., 1049
Rittenberg, D., 286, 287, 515, 553(6), 587 (6), 671, 1124, 1259, 1260, 1261, 1263 (210), 1266, 1271, 1273, 1274, 1276, 1280, 1290
Ritter, H., 614
Ritthausen, H., 572
Rivers, T. M., 86, 87, 88(275)
Robbins, K. C., 880
Robbins, W. J., 558
Robbins, W. R., 590, 591
Robb-Smith, A. H. T., 383, 922
Roberts, C. W., 660
Roberts, E., 564
Roberts, E. A. H., 136, 462, 562
Roberts, E. C., 221, 257
Roberts, G. L., 453
Roberts, H. E., 416
Roberts, R. M., 289, 817, 818, 1087, 1173
Roberts, S., 1284
Robertson, J. S., 669
Robertson, R. N., 519
Robertson, W. V. B., 1275
Robinson, C., 900, 901, 925
Robinson, D. S., 1002, 1009, 1011, 1012, 1013, 1017, 1068, 1096(41), 1108(41), 1134(41)
Robinson, E., 520, 531
Robinson, E. S., 136, 824
Robinson, J. R., 370
Robinson, M. E., 138, 152, 323
Robinson, R., 462
Rocha e Silva, M., 384, 842
Roche, A., 292, 336, 337(334), 340, 342
Roche, J., 279, 280, 288, 292, 294(1), 317 (1), 324, 326, 331, 332(1), 335, 336 (1, 327), 337(1, 334), 338(1), 339(1), 340, 341, 342, 343, 651, 652, 653 (295), 656, 659, 945, 1005
Rochow, E. G., 665
Rockland, L. B., 562
Rodbart, M., 369
Rodenberg, S. D., 552, 554(160)
Rodwell, A. W., 360
Roe, E. M. F., 17
Roe, J. H., 1253
Roedig, A., 258
Roettig, L. C., 731
Rogers, C. S., 1283, 1284
Rogers, H. J., 385
Roman, J., 924
Romanoff, A. J., 436, 437, 474(2)
Romanoff, A. L., 436, 437, 474(2)
Romans, R. G., 636
Ronzoni, E., 196
Root, W. S., 328
Rose, C. S., 471
Rose, R. C., 490, 498
Rose, W. C., 512, 561
Rosen, G. V., 54
Rosenberg, H. R., 216, 217, 260(360)
Rosenberg, S., 1238
Rosenberg, S. D., 555(172c), 594
Rosenberg, Th., 619
Rosenblum, H., 432

Rosendal, K., 370
Rosenfeld, G., 384
Rosenfeld, M., 624, 632
Rosenfeld, S., 381
Rosenthal, O., 1283, 1284
Ross, A. F., 82, 83
Ross, H. E., 360, 1216
Ross, W., 286, 288(47)
Ross, W. F., 1103, 1152, 1153
Rossen, J., 564
Rossenbeck, H., 8
Rossi, A., 324, 326
Rostorfer, H. H., 319, 327
Roth, J. S., 1288
Roth, L. J., 148, 154
Rothen, A., 453, 458, 470, 516, 599, 624, 633, 638, 643(220), 722, 794(7), 795, 816
Rothstein, M., 565
Rothwell, J. T., 1231, 1232(125), 1237 (125)
Rotman, R., 115
Rouelle, H. M., 542
Roughton, F. J. W., 299, 300, 301, 304, 305, 308, 309, 328
Rouiller, Chas. A., 624, 632(157), 635
Rous, P., 810
Routh, J. I., 873
Roux, R. M., 1153
Rovery, M., 291, 292(83), 682, 1186, 1190 (515, 515a, 518, 519), 1191, 1193 (515a, 519), 1195(515, 515a), 1196 (515, 515a)
Rowe, L. W., 615, 623, 625(154), 626 (154), 627(154), 628(154), 630(154)
Rowen, J. W., 165
Rowland, S. J., 391
Rowlands, I. W., 622
Rowlands, W. T., 739
Rowles, S. L., 966
Rowley, D. A., 1259
Rozsa, G., 971, 983, 1028, 1031, 1032(202)
Rubinstein, J., 381
Rudall, K. M., 863, 866, 876, 885, 886, 896, 902(187), 905, 946, 1026
Rudy, H., 219
Rüfenacht, K., 1137
Rueff, L., 202, 203(314)
Rugo, H. J., 869, 870(91), 908
Ruhland, W., 568
Ruska, H., 99, 101(344), 882, 884(171)
Russ, E. M., 706
Russ, V. K., 782
Russell, H. L., 429
Russell, J. A., 608, 1290
Rusznyak, St., 697
Rutman, R. J., 1227, 1231, 1257
Ryan, F. J., 423, 427(113)
Rydon, H. N., 353, 949, 1158
Ryland, K. W., 1018

S

Sabin, F. R., 810
Sabine, J. C., 234
Sacks, M. G., 320
Saenz, A. C., 715
Saetren, H., 52
Säverborn, S., 492, 495(16), 496, 503 (109), 504, 510
Sager, D. S., 431
Saha, N. N., 889
Sahli, E., 327
Sahyun, M., 636
Saidel, L. J., 686, 687(111)
Saidel, W. H., 423, 427(113)
Sailer, E., 1108, 1137
St. Faust, E., 355
St. George, R. C. C., 304
St. John, J. L., 590
Salaman, M. H., 784
Salaman, R. L. N., 119
Salázar, V. P., 500(91), 501
Sallans, H. R., 498
Salo, T. P., 916, 919, 934(301), 935
Salter, W. T., 654, 693, 962, 1022
Sanadi, D. R., 168, 261(198), 263, 264 (198), 265, 268(198), 269(198), 270 (521), 274(521), 275(198, 521, 536, 537), 276(198), 1211, 1212(32), 1267
Sandegren, E., 496
Sandelin, A. E., 396
Sanders, G. P., 431
Sandman, R. P., 191
Sandor, G., 706
Sands, M. K., 1292
Sanger, F., 317, 321(226), 428, 561, 568, 612, 630, 641, 642, 643, 644(254a), 645, 648, 728, 850, 873, 1119, 1126, 1128(268, 291), 1130(268), 1181,

1184(268), 1185(268, 291), 1187, 1189(268), 1223
Sanger, R., 785, 804, 822(127)
Sanigar, E. B., 666
Santos Ruiz, A., 500(80, 81, 89), 501
Sanz Muños, M., 500(89), 501
Sarkar, B. B., 377
Sarkar, N. K., 219, 222(370), 223(370), 224(370), 246(370), 252(370), 253, 254, 377
Saroff, H. A., 684, 692, 693
Sasaki, R., 500(71), 501
Saum, A. M., 70, 1177
Saunders, F. J., 621
Saunders, W. M., 798, 1175
Savur, G. R., 500(73), 501
Sawamura, S., 500(71), 501
Sax, K. B., 54
Sayers, G., 613, 614
Sayers, M. A., 614
Scallet, B. L., 497
Scatchard, G., 668(61), 672, 676, 677(59, 60), 685, 688, 696, 705(61), 713(61), 720(61), 726(61), 791, 877, 924, 933 (323)
Schaad, J. A., 916
Schachman, H. K., 423, 637, 640(211b)
Schachter, D., 1208
Schade, A. L., 137, 306, 459, 574, 575
Schaffer, N. K., 1158
Schaffner, G., 533
Schales, O., 317, 564, 699, 1153
Schales, S. S., 564, 1153
Schantz, E. J., 372
Schapira, G., 1017
Schauenstein, E., 855, 858, 919, 931
Scheier, H., 203, 204(322), 205, 206(322)
Scheinberg, H., 728, 877
Scheinberg, I. H., 696
Schellman, J., 696, 697
Scheraga, H. A., 728, 877, 878
Schermerhorn, L. G., 590, 591
Schick, A. F., 1004
Schick, B., 710
Schipitzina, G., 8
Schlegel, F. McM., 306
Schlenk, F., 160, 162, 163, 164(155), 165 (162), 173(155), 184
Schlesinger, M., 61, 97, 105, 109
Schlockerman, W., 260
Schmid, K., 707, 719(193), 735, 736(347), 928
Schmidt, C. F., 299
Schmidt, C. L. A., 533, 1096(156), 1097, 1098(156), 1101
Schmidt, G., 103, 943
Schmidt, G. M. S., 82
Schmidt, K., 326
Schmidt, W. J., 948, 1033
Schmitt, F. O., 516, 913, 917, 921, 922, 926(289), 927, 928, 929, 940, 950, 970, 1031, 1034, 1038
Schmitz, A., 1165, 1166(445)
Schmitz, H., 376
Schneider, G., 995, 1009, 1010, 1011, 1012
Schneider, W. C., 47, 1206, 1259(15), 1263(15)
Schneiter, W., 377
Schoenbach, E. B., 338, 339(349)
Schoenheimer, R., 324, 515, 553(6), 587 (6), 671, 1259, 1260, 1271, 1274, 1280
Schönig, S., 377
Schörmuller, A., 479, 480(170)
Schopfer, W. H., 560
Schou, L., 523, 581(50)
Schou, M., 571
Schramm, G., 959, 970, 972(21), 1043, 1071, 1127(59a)
Schreiber, D. L., 555
Schreiber, S. S., 1273
Schroeder, W. A., 322, 463
Schryver, J., 39
Schuber, D., 948
Schubert, H., 322
Schubert, M. P., 217, 218(369)
Schütz, E., 1096(151), 1097
Schütz, F., 316
Schuller, W. H., 1105, 1141
Schulman, J. H., 289
Schulman, M. P., 339, 1227
Schulman, S., 78
Schultz, J., 51, 58, 59
Schultz, E. L., 707, 719(193)
Schultze, E., 519, 524, 527, 529, 560(61), 561, 565
Schulz, F., 288
Schurr, P. E., 1261
Schuster, P., 259, 260, 261(493)
Schwander, H., 1083, 1084(106)

Schwartz, A. W., 731
Schwartz, I. L., 669
Schwarzacher, W., 307
Schwarzenbach, G., 161, 665
Schweet, R. S., 165, 168(175), 263(175), 264, 265(175), 266(175), 268(175), 269, 273(175), 274(175, 197, 554), 275(197), 1226, 1229(100), 1233 (100), 1234(101), 1245(101), 1246 (101)
Schweigert, B. S., 277
Schwerin, P., 323, 814, 1172
Schwert, G. W., 497, 1059, 1060, 1074, 1075(68), 1077, 1089, 1090(117), 1095, 1096(165, 173), 1097, 1099 (165, 173), 1100, 1104(183), 1105(3, 173), 1107(3, 82, 173), 1108(3), 1110 (3), 1111(3, 141), 1112(141, 183), 1113(3), 1114(3), 1116(3, 141), 1117 (82, 183), 1118(3), 1120(3), 1123 (173), 1124(82), 1125(141, 173), 1126 (141, 173), 1129(3, 221), 1131(3), 1133(3), 1194(82), 1222
Schwimmer, S., 1061, 1092, 1094(126)
Scott, D. A., 635, 636, 645
Scott, D. B. M., 214
Scott, F. H., 476
Scott, H. M., 437
Scott, M. W., 1271
Scriban, R., 497, 510
Scripture, P. N., 593
Seagram, H. L., 17
Seale, B., 564
Sealock, R. R., 634, 638
Sears, R. A., 1231, 1232(125), 1237(125)
Seastone, C. V., 1159
Sebelien, J., 411
Seedorff, H. H., 1065
Seeds, W. E., 29, 33(87), 70, 926, 946
Seegers, W. H., 730, 731, 1197
Seegmiller, J. E., 181, 214
Seemann, C. v., 336
Segal, H. L., 197, 198, 200, 201, 202
Segesser, A. v., 316
Sego, C. M., 1284
Sehra, K. B., 376, 381
Seibert, F. B., 352, 734, 796
Seibert, M. V., 352
Seifter, S., 840, 1283
Sekora, A., 854, 855, 858, 916
Seldeslachts, J., 1008
Selezneva, N. A., 1143, 1144, 1224, 1225 (79)
Seligman, A. M., 669, 1096(155), 1097, 1100(155)
Sell, H. M., 593
Selye, H., 608
Semmett, W. F., 393, 403, 417
Sen, P. K., 521, 528, 576, 578, 579, 582, 591(38)
Sendroy, J., 309, 328
Senti, F. R., 419
Seraidarian, K., 967, 968(58)
Serenyi, V., 397, 400(36)
Serrano, J. P., 500(80), 501
Sevag, M. G., 784, 843
Severens, J. M., 95
Shack, J., 329, 331(304)
Shäfer, E. A., 623
Shahani, K. M., 391
Shakrokh, B. K., 693
Shamshikova, G. A., 1213
Shapiro, B., 264
Sharp, D. G., 90, 91, 95, 96, 98(329), 99 (329), 100(329), 101(330), 346, 374 (2a), 437, 471(8), 472(8), 473(8)
Shaw, J. T. B., 849
Shedlovsky, T., 599, 602, 1109
Sheffner, A. L., 1261
Shemin, D., 286, 287, 686, 687, 1260, 1263 (210), 1266, 1273, 1274
Shen, S.-C., 969, 986(66), 1049
Shepard, C. C., 102, 478, 481
Sheppard, E., 1161(436), 1163, 1168(436)
Sherry, S., 382, 729, 878
Shine, H. J., 1116, 1117(259), 1137
Shinn, L. A., 409
Shipley, R. A., 649
Shmukler, J. H., 424, 429, 431, 434
Shooter, E. M., 492, 503(20), 505(20, 21)
Shore, W. S., 570
Shorey, E. C., 527, 561
Short, W. F., 221, 224(384), 257(384), 258(384)
Shugar, D., 198
Shulman, S., 682, 726, 728, 729(290), 877, 881, 882(160)
Shumaker, J. B., Jr., 713
Shupe, R. E., 1063

Shurman, M. M., 917
Sidransky, H., 1296
Sidwell, A. E., 282, 297, 298(109), 300 (109)
Siegal, M., 777
Siegel, B. H., 878
Siegert, M., 493
Siekvitz, P., 1231, 1232, 1234, 1235, 1236, 1240, 1243(127), 1262(127), 1263 (127)
Sifferd, R. H., 638
Signer, R., 7, 28(19)
Sikorski, J., 860, 863
Silberschmidt, K., 119
Silberstein, H. E., 1269
Siley, P., 769
Sills, R., 1238
Silva, R. B., 452
Simanouti, T., 899
Siminovitch, L., 110, 117, 118, 198
Simmonds, S., 1208, 1241
Simms, H. S., 31, 432
Simonovits, S., 333
Simpson, M. E., 598, 599, 600, 602(17), 603, 606(50), 607(50), 608, 609, 613, 620, 621, 623
Simpson, M. V., 1227, 1229, 1230(103), 1232(103), 1236(103), 1255, 1257, 1282, 1286, 1296
Simpson, R. B., 685, 1177
Sinclair, A. T., 490
Sinex, F. M., 1239
Singer, J. J., 545, 547
Singer, S. J., 552, 709, 713, 802, 816
Singer, T. P., 127, 128, 134, 160(40), 162, 164, 165, 172, 173(170), 174(217), 177, 189, 190, 203(170), 209(217), 211(217), 218(150), 221, 222, 223 (150), 224(389), 225(389), 233, 237 (40, 389), 238, 239, 240, 241, 242, 243(40, 380, 389), 244, 248(150), 261, 263(500, 501), 264, 265(501), 267, 268(501), 270(501), 271, 272(500, 501, 543), 273(500), 275(170, 217), 383, 964, 1148
Sinha, A. C., 377
Sinn, L., 649
Sinsheimer, R. L., 28
Sisco, R. C., 1096(160), 1097, 1099(160)
Sisson, W. A., 893, 897(200), 1025, 1026, 1040(195), 1047(195)
Sizer, I. W., 922, 1152
Sjögren, B., 416, 499, 503(104), 504, 613, 636
Skipper, H. E., 1287
Skoog, F., 560
Skowlund, H. V., 1018
Skrimshire, G. E. H., 221, 224(384), 257 (384), 258(384)
Slack, H. G. B., 1275
Slade, H. D., 370
Slater, E. C., 1294
Slein, M. W., 126, 134, 167, 169(9), 195 (188), 196, 197, 198, 199(188), 200 (9), 201(9), 1050
Sloan, R., 390, 391(4), 397(4), 411(4), 412(4), 416(4), 427(4)
Slobodiansky, E., 850, 851
Slotin, L., 212
Slotta, K. H., 346, 365, 374, 381(6), 781
Smadel, J. E., 86, 87, 88(275)
Small, P. A., 348, 769, 1152
Smellie, R. M. S., 16, 17, 48, 53
Smelser, G. K., 606
Smiles, J., 324
Smiley, W. G., 502
Smith, A. K., 490, 500(83), 501, 502
Smith, A. U., 651
Smith, B. W., 1253
Smith, C. A., 320
Smith, C. B., 593
Smith, D. E., 724
Smith, E., 791
Smith, E. C., 953
Smith, E. C. B., 1005, 1006, 1009, 1010, 1011, 1012, 1043
Smith, E. L., 392, 394, 411, 412, 414, 416 (11), 427(17, 87), 433(17, 77, 87), 511 (135, 136), 512, 549, 610, 672, 708, 712, 718, 719, 723, 736, 791, 792, 799, 922, 1060, 1067, 1068, 1069, 1074, 1075 (70), 1076, 1077(80), 1078(80), 1081 (31), 1082, 1083, 1084(99a, 106a), 1102(6, 6a), 1103(44a), 1105, 1106 (6a), 1107(302), 1108, 1111(214), 1112(214), 1113(6a, 214), 1122, 1129 (6a, 214, 277), 1130, 1131(6a, 214, 277), 1132, 1133, 1147, 1148, 1149,

1150(373), 1153(80), 1159, 1160 (417), 1241
Smith, G. F., 546, 560(140), 573(140)
Smith, I. P., 605, 608(44)
Smith, J. D., 13, 14, 15, 16, 17, 18, 19, 20, 22, 23(36), 26(75), 28, 67(60), 72 (47), 73, 74, 77(47, 75), 82, 91(242), 92(68), 120, 322
Smith, K. M., 77, 78, 79, 93, 118
Smith, L. A., 362
Smith, L. D., 721
Smith, L. de S., 383
Smith, M. Llewellyn, 349
Smith, M. M., 379, 380
Smith, P. E., 598, 605, 608(44)
Smith, P. K., 1287
Smith, S. G., 849, 851, 857, 1187
Smith, T., 412, 413
Smith, T. O., 589
Smith, W., 363, 379, 380
Smith, W. H., 519
Smolens, J., 784
Smoluchowsky, M. V., 108
Smyrniotis, P. Z., 185, 213, 214, 215(351, 357)
Smyth, E. M., 734
Smythe, C. V., 217, 218(369)
Snapper, I., 313, 768
Snell, E. E., 439, 459, 469, 489, 792, 1241
Snell, N. S., 9, 48(30), 445, 447(36), 449 (36), 463(36), 465(36), 467(36), 468 (36), 469, 470(36), 471(147, 148), 477(36), 1075, 1084(73), 1085(73), 1167(73)
Snellman, O., 971, 983, 987
Snoke, J. E., 1068, 1077, 1096(34, 166, 173), 1097, 1099(166, 173), 1100, 1105(173), 1107(166, 173), 1112(166, 182), 1113(182), 1114(166), 1115 (182), 1116, 1117(182), 1123(173), 1125, 1126(173), 1129(166, 221), 1130 (166), 1131(166), 1133(166), 1148(34), 1212, 1214, 1215, 1217, 1222
Snow, J. M., 1012
Sober, H. A., 1216
Socquet, I. M., 210
Solmssen, U., 161
Solomon, A. K., 1226, 1263
Solomon, G., 1267, 1274, 1281(230), 1283 (230)
Solomon, J. D., 1261
Somers, G. F., 128
Sommer, H. H., 391
Sommers, S. C., 767
Sonderhoff, R., 354
Sonneborn, T. M., 120, 121
Soodak, M., 266
Sørensen, M., 417, 418, 432, 433(95), 439, 445(13), 446, 451, 456(13), 458, 467 (13), 471(13), 472, 473, 734
Sørensen, S. P. L., 417, 421, 432, 433(95), 441, 444, 445, 446, 447, 448, 451, 455 (75), 456(19), 457(19), 459(19), 680
Soret, A., 298
Sortor, H. H., 412
Sowinkski, R., 818, 1279
Spark, L. C., 992, 1035
Sparrow, A. H., 685
Speakman, J. B., 862, 863
Speck, J. F., 526, 571(70–72), 1216, 1218, 1219
Spencer, E. L., 553
Spencer, E. Y., 498
Sperling, E., 221, 259(383)
Spicer, D. S., 1161(427), 1163
Spicer, S. S., 979, 984, 988, 991
Spiegel-Adolph, M., 765
Spiegelman, S., 1285, 1293(301)
Spies, J. R., 686
Spoehr, H. A., 578
Spohn, A. A., 474
Spooner, E. T. C., 71, 79
Sprinson, D. B., 1124, 1261, 1290
Spychalski, R., 499
Squire, P. G., 798, 1175
Srb, A. M., 566
Sreenvasan, A., 500(73), 501, 1254
Sreerangachar, H. B., 140
Stacey, M., 49, 352, 467, 1249
Stack, A. H., 361
Stack-Dunne, M. P., 614
Stadie, W. C., 133, 234(33), 308, 327, 1096(149), 1097
Stadtman, E. R., 183, 264, 266, 1210
Stahmann, M. A., 1130, 1131(295a)
Stainer, R. Y., 1285, 1293(302)
Stamm, A. J., 492, 503(13)
Standfast, A. F. B., 221, 224(384), 257 (384), 258(384)
Stanier, W. M., 877

Stanley, W. M., 32, 61, 66, 67, 72, 73, 82, 87, 88, 91, 92, 518, 552
Stansly, P. G., 170(209), 171, 203, 225
Stare, F. J., 217, 248(367)
Starling, E. H., 596, 673
Starr, M. P., 887, 888
Stary, Z., 862
Staub, A., 649, 650
Staudinger, M., 971
Stedman, E., 43, 44, 50, 51, 339
Steele, J. M., 1231
Steelman, S. L., 607
Steensholt, G., 563
Steere, R. L., 67, 71
Steffee, C. H., 1259
Stehle, R. L., 624, 626
Stein, M., 625
Stein, W. H., 464, 522, 569, 629, 686, 687 (112), 723, 850, 928, 946, 1082, 1085 (101, 101a), 1096(164), 1097, 1099 (164), 1248
Steinbeck, A. W., 1271, 1272(241)
Steinberg, A., 665
Steinberg, D., 446, 455(44), 588, 1154, 1184(405), 1188(405), 1248, 1251, 1253, 1288
Steinberg, R. A., 558, 576, 583, 589, 590, 591(274), 592, 593
Steinbock, H. L., 1283
Steiner, R. F., 640, 730, 881, 984
Steinhardt, J., 293, 307, 695, 1086, 1088 (109), 1094, 1103
Steinman, H., 138, 148(70)
Stekol, J. A., 1295
Stent, G. S., 109
Stephen, J. M. L., 647, 648, 1143, 1183 (335), 1185
Stephenson, M., 212
Stephenson, M. L., 1096(170), 1097, 1099 (170), 1226, 1227, 1230(92), 1231 (92), 1240(102), 1246, 1252, 1286 (186)
Stephenson, N. R., 258
Stepka, W., 523, 527, 559(79), 561, 562, 563(79), 565, 572, 581(50), 583(79)
Stepto, P. C., 1259
Sterling, K., 669, 1272, 1278, 1279
Stern, H., 52
Stern, J. R., 202, 203(316), 264, 265, 274 (534)
Stern, K. G., 39, 261, 316, 647, 791
Stern, L., 166, 179(179)
Sternberg, H., 322
Sternberger, L., 765, 808
Sternberger, L. A., 722, 724
Sterne, M., 373
Steudel, H., 6, 43, 45
Stevens, C. M., 141
Stevens, H., 434
Stevens, M. F., 362, 1102, 1103, 1127 (196), 1175(196)
Stevenson, J. W., 349
Steward, F. C., 489, 520, 521, 522, 523, 525, 526, 527, 528, 529, 530, 531(35), 532, 533(47), 534(36, 47, 80, 81), 535 (80, 81), 537(36, 47), 538(80, 81), 539 (52, 80, 81), 540(52), 542, 558, 559, 560(48, 189), 561, 562, 563(47, 182, 204), 564(182), 565, 567, 568, 569, 570, 574(35), 575(39, 40, 66, 86, 87), 578, 579, 580(39, 66), 581, 583, 586 (66), 587, 589, 591(40)
Stewart, C. P., 140, 430
Stewart, H. C., 289
Still, J. L., 221, 259(383)
Stitt, F., 312, 315, 333, 334(206, 317)
Stockell, A., 1081, 1083, 1084(99a, 106a), 1147
Stocken, L. A., 966
Stockmayer, W. H., 836
Stöver, R., 970, 972(68)
Stoker, M., 18, 19
Stokes, A. M., 649
Stokes, A. R., 33, 70
Stokes, J. L., 1285
Stokstad, E. L. R., 261
Stolberg, M. A., 1158
Stollerman, G. H., 370
Stone, D., 559, 1241
Stone, J. D., 89
Storaasli, J. P., 1271
Stotz, E., 141, 142, 262
Stout, P. R., 521, 528, 575(39), 580(39)
Straessle, R., 690, 693, 694(134)
Strain, H. H., 1225
Straub, F. B., 126, 170, 171, 172, 208, 209, 211, 221(218), 223(218), 224

(218), 225, 233, 235(430), 246(218), 249, 250, 251, 977, 978, 979, 982, 983 (89), 986, 989, 1023(89)
Strauss, K., 854
Strecker, H., 169, 171, 183, 184, 185, 188, 189
Street, A., 861, 862(66), 864(66), 892, 896, 897(66)
Street, H. E., 489, 507, 523, 524, 525, 526, 530, 531, 532, 542, 558, 560(56, 57), 561, 568, 575(66), 578, 579, 580(65), 586(66)
Street, S. F., 1052
Streicher, J. A., 1209, 1231, 1232(125), 1237
Streng, O., 756
Stricker, P., 602
Strisower, B., 702(197), 707
Strong, L. E., 680, 685, 699, 725(77), 807
Stumpf, P. K., 232, 262, 274(518, 519), 276(519), 571, 573, 1216, 1219
Sturgis, S. H., 18
Sturtevant, J. M., 1087, 1088, 1116, 1161 (435), 1163, 1168(435), 1203
Subrahmanyan, V., 224(398), 225, 231 (398), 261, 270(508), 271(508), 272 (508), 1050
Süllman, H., 276
Sugai, K., 1150
Sugita, T., 899
Sullivan, R. A., 1065
Sullivan, T. J., 733
Summerson, W. H., 149, 152(103), 153 (103), 1158
Sumner, J. B., 128, 200, 201(309), 276, 277, 349, 351(34), 375, 378(175b), 522, 525, 566(42), 754, 784, 808, 922, 1060, 1065, 1102(6), 1206
Sumner, R. T., 277
Suominen, H. S., 496
Surgenor, D. M., 699
Sutermeister, E., 396
Sutherland, E. W., 649
Sutherland, G. B. B. M., 854, 890(33), 891(33)
Suthon, A. M., 1153
Svedberg, T., 281, 293, 294(10), 336, 340, 347, 365(10), 400, 402, 416, 417, 418, 448, 492, 495(16), 496, 497(47, 50), 499, 503(13, 104, 105, 108), 504, 505, 510(16), 636, 682, 781, 791, 817
Svedmyr, A., 89
Svennerholm, L., 736
Svensson, H., 449, 701, 702, 706(178), 718, 737, 753, 1071
Swain, R. E., 498
Swan, J. M., 660
Swann, M. M., 946
Swarm, R. L., 320
Swedin, B., 258
Sweeney, L., 666
Swenson, P. A., 58
Swenson, T. L., 466
Swift, H., 54
Swingle, S. M., 776
Sykes, G., 221, 224(384), 257(384), 258 (384)
Symonds, P., 731, 1018, 1020(191), 1021, 1022, 1023
Syndermann, S. E., 1290
Synge, R. L. M., 558, 560, 563(181), 643, 921, 1287
Szafarz, D., 48, 59
Szára, St., 877
Szent-Györgyi, A., 135, 207, 216, 231, 516, 958, 965, 968, 969, 976, 977, 978, 979, 980, 983, 986, 989, 990, 1007, 1028, 1031, 1032(202), 1040, 1043, 1044, 1053, 1190(530, 531), 1191, 1198(530, 531)
Szilard, L., 115
Szyszka, G., 374

T

Tabachnick, M., 1228
Tabareŝ, L. G., 493
Tabor, H., 258
Taborda, A. R., 384
Taborda, L. C., 384
Tachibana, T., 359
Tadakoro, T., 142
Tager, H., 380
Taggart, J. V., 1208
Tahmisian, T. N., 149
Takahashi, W. N., 553, 554, 557, 572
Takamatsu, Y., 857
Takasugi, N., 142
Tallan, H. H., 464, 1067, 1092(33), 1107 (33), 1145(33), 1147(33)

Talmage, D. W., 1277, 1279
Tamm, I., 676
Tanford, C., 453, 570, 688, 689(117), 692, 693(117), 1150
Tang, F. F., 88
Tang, Y., 320
Tangl, F., 396
Tanko, B., 267, 272(542)
Tapley, T. F., 685, 721(105)
Tappel, A. L., 278
Tarpley, W. B., 135, 136(46), 137(46), 138, 142, 148(46)
Tartar, H. V., 590
Tarver, H., 1226, 1227, 1228, 1229, 1230 (103), 1231(105), 1232(103), 1236 (103), 1257, 1258, 1260, 1261(117), 1263, 1264(117), 1265, 1267, 1269, 1270(199), 1271, 1272, 1274, 1276 (199, 223, 233), 1278(117), 1281 (230), 1283, 1286, 1288, 1295, 1296
Tasman, A., 362
Tatum, E. L., 1208
Tauber, H., 1138, 1143, 1162(437), 1163, 1165(437)
Tauber, H. T., 1224
Taurins, A., 291
Taurog, A., 656
Tayeau, F., 724
Taylor, A. R., 90, 91, 95, 96, 98(329), 99 (329), 100(329), 101(330), 103(335)
Taylor, D. B., 1065
Taylor, D. S., 334
Taylor, G. L., 448, 680, 828
Taylor, H. E., 56, 57
Taylor, H. L., 680, 699, 725(77)
Taylor, H. S., 899
Taylor, J. F., 126, 167, 195(188), 199 (188), 303, 314, 320, 326, 335(131)
Taylor, S. P., 629, 1107(225), 1108, 1122
Teague, D. M., 291, 391
Teel, H. M., 608
Tekman, S., 453
Telkkä, A., 697
Temple, R. B., 870, 899(93)
Templeman, W. G., 576, 589
Ten Broeck, C., 761, 1159
Teng-Yi Lo, 493(39), 494
Tennent, H. G., 30, 32(90)
Tenow, M., 983
Teorell, T., 807, 827, 831, 836, 837(274)
Teply, L. J., 223, 246(394), 254(394)
Terminello, L., 1165
Terry, P., 737
Tesar, W. C., 614
Tetsch, C., 354
Tewkesbury, L. B., 658, 659
Thannhauser, S. J., 103
Thayer, P. S., 222
Theorell, H., 125, 167, 168, 171(192), 172 (192), 173, 174(192), 175, 176, 177, 178(192), 186, 187, 200(192, 199), 219, 220, 225(266), 247, 248, 276, 277, 303, 324, 325, 326(279), 331, 332 (288), 334, 428, 429
Thimann, K. V., 526, 531, 562(68), 573, 586(68)
Thoai, N.-v., 1005
Thoday, D., 543
Thomann, H., 979, 990(94)
Thomas, A. W., 923
Thomas, D. W., 1104, 1111(206), 1116 (206, 207), 1117(206, 207), 1120(206, 207)
Thomas, J., 230
Thomas, L. E., 1143
Thomas, M. D., 523, 539(52), 540(52)
Thompson, A. R., 463, 1181, 1184(488), 1189(488)
Thompson, D. L., 608
Thompson, E. O. P., 642, 643, 644(254a), 645, 1119, 1126(268), 1128(268), 1130(268), 1184(268), 1185(268), 1187(268), 1189(268), 1190(524), 1191, 1223
Thompson, F. C., 788
Thompson, H. T., 1261
Thompson, J. F., 521, 522, 523, 525, 526, 527, 529, 530, 532(92), 533(47), 534 (47, 80, 81, 136), 535(80, 81), 537 (36, 47), 538(80, 81), 539(52, 80, 81), 540(52), 542, 558, 559, 560(48, 189), 562, 563(47, 182, 204), 564(182), 565, 567, 569, 570, 579, 581, 583
Thompson, R., 462, 466
Thompson, R. E., 635, 637(196)
Thompson, R. Y., 1283
Thompson, S. Y., 737
Thomson, J. D., 1018
Thomson, R. Y., 53
Thorell, B., 1032, 1243

Thornton, H. G., 592
Thunberg, T., 211
Thung, T. H., 79
Thurlow, S., 124, 226(3)
Tice, S. V., 1216
Tietze, F., 637, 639, 640, 1061, 1074, 1078, 1079, 1080(4, 90), 1081(90), 1090(8, 90), 1091(90), 1156(90)
Tillet, W. S., 382
Tippelt, H., 377
Tiselius, A., 449, 615, 648, 677, 701, 702, 706(178), 737, 794(3), 795, 796, 797, 1071, 1173, 1174, 1178, 1186, 1242
Tishkoff, G. H., 1260, 1269(209), 1271 (209), 1276(209)
Tislowitz, R., 613
Tissières, A., 139, 140, 157
Titchen, D. A., 416
Tittsler, R. P., 431
Todd, A. R., 12, 19(35), 21, 23, 164, 966
Todd, E. W., 369, 370, 371
Todrick, A., 452, 964
Toennies, G., 642
Tolbert, N. E., 523, 581(50)
Tomlin, S. G., 943
Toms, B. A., 857
Tong, W., 656
Tongur, A. M., 1143, 1144(342), 1224, 1225(79)
Torquati, T., 559
Totter, J. R., 278, 1239
Tracy, A. H., 1152
Tracy, P. H., 391, 431
Traub, F. B., 350
Trease, G. E., 524
Treffers, H. P., 784, 793, 797, 800(169), 803(124), 817, 825, 842, 1280
Trethewie, E. R., 375
Trevan, J. W., 381, 635
Trikojus, V. M., 651
Trippett, S., 660, 661(337)
Tristram, G. R., 41, 42, 290, 291(76), 326 (76), 332(76), 445, 447, 516, 533, 639, 850
Trogus, C., 851
Troll, W., 729, 878
Trotter, I. F., 852, 868(23), 870(23), 890, 891(23, 210), 896(23), 902(210)
Trotter, W. R., 657
Trout, G. M., 432
Trpinac, P., 177, 197(227)
Truland, L. T., 812
Tsao, T. C., 958, 959, 963, 964, 965, 972 (18), 973, 975(18), 977, 979(33), 980, 984, 985, 994, 995, 996, 997(86), 998 (86), 1000(86), 1001(47), 1010, 1011, 1019, 1021, 1033(33)
Tsou, C. L., 1186
Tsuboi, K. K., 31
Tsuboi, M., 899
Tsui, C., 593
Tsumaki, T., 306
Tuft, L., 759
Tullis, J. L., 666, 702, 703, 711, 753
Tunbridge, R. E., 949
Tunca, M., 1172
Tung, T. C., 190
Tuppy, H., 643, 644(254a), 645, 660, 1126, 1128(291), 1185(291), 1187
Turba, F., 979, 990(94, 95), 991, 1046
Turner, A. H., 673
Turner, A. W., 360
Turner, C. W., 433, 651, 655, 657, 658
Turner, J. F., 519
Turner, R., 286, 288(47)
Turner, R. A., 629, 630, 631(178)
Turpeinen, K., 598
Turschin, T. W., 589
Tustanovsky, A. A., 926, 944(334)
Tuttle, L. C., 1099, 1211
Twort, F. W., 93
Tyler, F. H., 672
Tyor, M. P., 1278
Tytell, A. A., 1107(225b), 1108

U

Udenfriend, S., 427, 564
Uhrovits, A., 768
ul Hasson, M., 1286
Ulrich, F., 1295
Umbreit, W. W., 573
Underwood, J. C., 562
Urion, E., 499, 509(66)
Uroma, E., 707, 719(193)
Urquhart, J. M., 692, 693
Ushiba, D., 1285
Utter, M. F., 212
Uyei, N., 608

V

Vahlquist, B., 738
Vaidyanathan, C. S., 558
Valko, E., 308, 863
Vallee, B. L., 666
Van Abelee, F. R., 647
Vanamee, P., 914, 931(278)
VanBruggen, J. T., 221, 257
Vand, V., 901
Vandenbelt, J. M., 730, 731
Vandendriessche, L., 703
van der Scheer, J., 484, 614, 773, 775, 782 (114), 783, 796, 797, 798(179), 805, 809, 831
van der Want, J. P. H., 79
van der Wyk, A. J., 863
Vandevelde, A. J. J., 430
van Dyke, H. B., 599, 600, 601(13), 602, 607, 624, 632, 633, 634, 638, 643(220)
Van Halteren, M. B., 1241
van Heyningen, W. E., 112, 346, 349(26), 350, 367, 374, 383, 385(89), 1070, 1096(50, 154), 1134(50)
Van Klaveren, K. H. L., 289
van Ramshorst, J. D., 362
Van Slyke, D. D., 283, 307, 308, 317, 328, 493, 667, 1096(162, 167, 169), 1097, 1099(167, 169)
Van Slyke, L. L., 396, 409(28)
Van Veen, A. G., 559
Van Vunakis, H., 682, 715
van Wagtendonk, W. J., 121, 122
Varga, L., 992, 1044
Vargha, L., 216
Varin, R., 410
Vars, H. M., 1283, 1284
Vaslow, F., 1123, 1124
Vassel, B., 500(84), 501
Vaughan, A. C. T., 368, 369(128)
Vaughan, J. H., 787
Veiga Salles, J. B., 212, 213(343)
Vegezzi, G., 343
Velick, S. F., 166, 182, 195, 196, 197, 198, 199, 200(189, 299), 201(189, 236, 239), 202(239), 1255
Vellard, J., 374
Velluz, L., 261
Vendrely, C., 53, 55(171)
Vendrely, R., 53, 55(171), 56
Venet, A. M., 924
Venkastesan, T. R., 566
Vennesland, B., 162, 173, 204, 205, 207, 211(334), 212, 260, 262, 276(514), 1260
Venning, E. M., 618
Verney, E., 361
Vernon, L. P., 219, 222(370), 223(370), 224(370), 246(370), 252(370), 253, 254, 562
Vichess, P., 724
Vickery, H. B., 291, 489, 515, 518, 523, 524, 525, 526, 527, 532, 533, 542, 553 (6), 561, 566, 569, 570, 587(6), 885
Vieil, H., 324
Vilbrant, C. F., 30, 32(90)
Villafañe, A. L., 757
Ville, J., 315
Villee, C. A., 454, 1190(526), 1191, 1248
Vinograd, M., 493
Virtanen, A. I., 322, 525, 558, 559, 562 (183a), 563, 571, 572, 593, 1143, 1209, 1219, 1221(21), 1224, 1241
Vischer, E., 14, 17, 18, 19
Visco, S., 493(38), 494
Voegtlin, C., 1225
Vogel, H. J., 16, 17, 19
Volkin, E., 13, 14, 24, 27, 32, 164, 636
von Brenner, M., 1221, 1222
von Ebner, V., 1032
von Fürth, O., 956, 957, 958(7)
von Magnus, P., 89
von Mering, J., 634
von Muralt, A. L., 954, 960, 969, 1032, 1040
von Mutzenbecker, P., 657, 658
von Susich, G., 863
von Weimarn, P. P., 857
Voorhees, C. G., 493
Vosgian, M. E., 1017

W

Wacker, A., 1287
Wada, M., 559
Wadsworth, A. B., 362
Waelsch, H., 286, 288, 330(62), 331(62), 568, 570, 571, 573(238), 1136, 1142, 1150(332), 1216, 1219, 1222
Wagman, J., 39

Wagner, R., 822
Wagner-Jauregg, T., 205, 217, 221, 1158
Wagreich, H., 138
Wainfan, E., 571, 573(238), 1219
Wainio, W. W., 184, 185
Waitkoff, H. K., 872, 920, 946(99)
Wakelin, R. W., 1148
Wakeman, A. J., 399, 400, 518, 524, 526, 532(67), 566
Wakil, S. J., 170(209), 171, 203, 225
Waksman, B. H., 842, 922
Walaszek, W. F., 613(89), 614, 615(89)
Wald, G., 174, 177(225), 303, 339
Waldschmidt-Leitz, E., 1140, 1222, 1223
Waley, S. G., 642, 1101, 1110(193), 1138, 1142, 1221
Walker, A. C., 1101
Walker, D. M., 737
Walker, E., 452, 964
Walker, J. B., 559
Walkley, J., 532
Wall, M. E., 591
Wallace, K. R., 371
Wallner, L., 849
Walter, C., 665
Walti, A., 1069
Wang, T. Y., 555(172c), 594
Wang, Y. L., 303, 310(130), 322, 325(130)
Warburg, O., 124, 125, 126, 133, 161, 162, 163, 165(8), 169, 171, 172, 186, 192 (8), 193, 194, 195, 201, 214, 217, 219, 220(375), 221(32, 375), 224(265), 225, 233, 234(375), 246(265), 247, 248(32, 265), 312, 336, 664
Ward, A. G., 924
Ward, S. M., 86
Ward, W. H., 9, 48(30), 442, 449(24), 456, 459, 462, 464(24), 465(24), 470(24), 471, 473(24), 479, 794, 875
Ware, A. G., 1197
Warner, R. C., 393, 397, 398, 399(16), 400 (16), 419, 420, 421, 429, 441, 444, 449 (20), 457, 458, 459, 460, 500(75), 501, 505, 506(117), 699
Warrack, G. H., 349, 366(30), 383, 1096 (154), 1097
Warren, M. E., 360
Warren, W. J., 916
Warwick, A. J., 559, 565(188)
Wase, A., 1288
Wassen, A., 169, 171, 174
Wasserman, K., 1271, 1272(237)
Wasserman, P. W., 645
Wassermann, A., 783
Wasteneys, H., 1143, 1224
Watanbe, I., 30
Waters, J. W., 943
Watson, C. C., 496, 497(51)
Watson, J., 642, 1138, 1142, 1221
Watson, J. D., 33, 34, 98, 108, 113, 908
Watson, J. M., 531
Watson, M., 946
Watson, R., 589
Watson, R. H., 324
Waugh, D. F., 635, 637, 682, 684, 731, 881, 907, 908
Weare, J. H., 678, 680(68), 681(68), 682 (68), 683(68), 686(68)
Webb, B. H., 408
Webb, E. C., 965, 967(43), 1154
Webb, M., 18, 19
Webb, R. A., 826, 828
Webb, T. J., 313, 314(185)
Weber, A., 683, 1044, 1045, 1048(234)
Weber, G., 682, 724
Weber, H., 1045, 1048(234)
Weber, H. H., 956, 958, 959, 960(15), 962, 968, 970, 971, 972, 974, 984, 986, 988 (84), 990, 991, 995, 1005, 1007, 1008, 1009, 1010, 1011, 1012, 1024, 1031, 1033(15), 1043, 1044, 1045, 1048, 1049, 1053
Weber, I., 441, 449(20), 457, 458, 459, 460, 699
Wedum, A. G., 775
Wegand, F., 1287
Wegman, T., 647
Wehmer, C., 561
Wehrmacher, W. H., 1018
Weibull, C., 887, 888, 889
Weidinger, A., 931
Weil, A. J., 721, 842
Weil, L., 1078, 1154(87)
Weil-Malherbe, H., 263
Weimar, H. E., 735, 736(346)
Weimer, R. J., 635, 637(196)
Weinstein, B. R., 432
Weiss, S., 1295
Weissberger, A., 1110
Weisz-Tabori, E., 204, 206(323)

Weiziger, J., 317
Weld, J. T., 369
Welecki, S., 820
Welker, W. H., 282
Weller, R. A., 544, 547
Wellman, H., 1294
Wells, H. G., 410, 433(73), 484, 758, 760, 782(115, 116), 783
Wells, H. Gideon, 709
Wells, I. C., 322
Wensinck, F., 374
Wentzel, L. M., 373
Wenzel, C. C., 500(68), 501
Werbin, H., 1098, 1099(176), 1114, 1125 (176), 1126(176)
Werle, E., 564
Werner, I., 736
Wessely, L., 202, 203(314)
West, C., 520
West, P. M., 1163
West, R., 287, 1273
Westall, R. G., 558, 563(180)
Westerfeld, W. W., 230, 262, 276(514), 1284
Westerman, A., 619
Westheimer, F. H., 162, 173
Wetmore, R. H., 529, 532(92), 542(92)
Wetter, L. R., 439, 444(14), 449(16), 464 (16), 465, 504, 761, 770, 1165
Wetterlow, L. H., 712, 741(219)
Wetzel, K., 568
Weygand, F., 219, 247
Wheatley, A. H. M., 162
Whewell, C. S., 862
Whipple, A., 671
Whipple, G. H., 292, 671, 809, 1247, 1260, 1269, 1271, 1272(240), 1273, 1276 (209), 1288
Whitby, L. G., 219
White, A., 291, 475, 477(163), 603, 607, 613, 647, 810
White, A. G. C., 1290
White, J. C., 18, 29
White, W. F., 613(89, 91, 91a, 91b), 614, 615(89, 91), 617, 1184(498), 1185
Whitehead, H. R., 788, 840
Whitfield, P. R., 24
Whiting, G. C., 528, 581
Whitney, I. B., 278
Whittaker, V. P., 132
Whittier, E. O., 396
Whittlestone, W. J., 632
Wibaut, J. P., 559, 565(191a)
Wichmann, A., 416
Widdowson, E. M., 433
Wieland, H., 230, 231(418), 356, 375(67)
Wieland, O., 202, 203(314)
Wieland, T., 183
Wieland, U., 358
Wiener, A. S., 710, 756, 785, 788, 803 (129), 822(129)
Wiesner, B. P., 620
Wiggans, D. S., 1139
Wilcox, P. E., 695, 1061, 1083, 1084 (107b)
Wildman, S. G., 542, 543, 545, 546, 547, 548(124), 550, 551, 552, 553, 554, 556
Wilhelmi, A. E., 608, 610
Wilkins, M. H. F., 29, 33(87), 70
Wilkinson, G., 312
Wilkinson, G. R., 926, 946
Wilkinson, S., 559
Williams, A. H., 592
Williams, C. M., 733
Williams, C. W., 363
Williams, F. R., 1224
Williams, G., 537
Williams, H. H., 291
Williams, J. N., Jr., 384, 1261, 1283
Williams, J. W., 496, 497(51), 503(103), 504, 653, 720, 790, 791, 794(4), 795, 796, 798, 801, 802, 804, 825, 924, 933 (323), 1175, 1176
Williams, M. B., 657
Williams, R. C., 67, 69, 71, 99, 887, 888
Williams, R. E. O., 368
Williams, R. F., 589
Williams, R. H., 59, 1262, 1284
Williams, R. J., 459, 469
Williams, R. T., 1201, 1209(2)
Williamson, M. B., 433, 1190(514), 1191, 1192(514)
Willstätter, R., 322, 1101
Wilmarth, W. K., 306
Wilson, A. T., 432
Wilson, D. W., 618, 619, 620
Wilson, H. C., 314
Wilson, H. R., 33
Wilson, I. B., 1101, 1159

Wilson, James W., 493(40), 494
Wilson, P. W., 524, 525, 558(63), 572
Wilson, T. H., 162
Wilt, W. G., Jr., 1271, 1272(240)
Windsor, E., 533
Winegarden, H. M., 319
Winikoff, D., 651
Winitz, M., 1135, 1139, 1142(305), 1221, 1223(66)
Winnick, T., 1096(152), 1097, 1145, 1146, 1174(152), 1226, 1227, 1228(95, 104, 112), 1231(95), 1232(95), 1233(95), 1234, 1236, 1237(95), 1226(95), 1244, 1245(169), 1255, 1256(169, 191), 1257
Wintersteiner, O., 635, 636, 645
Wintrobe, M. M., 672
Winzler, R. J., 735, 736(346), 1295
Wishart, G. M., 480
Wittenberg, J., 287
Wittler, R. G., 368, 373
Woelfflin, R., 944
Woiwood, A. J., 348, 564
Wolfe, H. R., 713
Wolfe, M., 592
Wolfers, D., 230
Wolff, K., 354
Wollaston, W. H., 561
Wollenberger, A., 454, 767, 1190(525, 526), 1191
Wollman, E. L., 109, 111; 118
Wolpers, C. H., 882, 884(171), 913, 947
Wolvekamp, H. P., 305
Wonder, D. H., 622
Wood, D. J., 562
Wood, G. C., 949
Wood, J. G., 527, 543, 548(130), 549 (130), 576, 585, 590
Wood, J. W., 1209
Woodbury, L. A., 614
Woodin, A. M., 875
Woodruff, H. B., 95
Woodruff, L. M., 672, 685(45)
Woods, E. F., 522, 556(45)
Woods, H. J., 860, 861, 862(67), 863, 864 (65, 67), 897(67), 1025, 1047(194b), 1048
Woods, M. W., 120, 548
Woods, N. W., 554
Woodward, C., 1065
Woolley, D. W., 439, 449(18), 469, 470, 533, 1256, 1286(192)
Woolley, J. M., 467
Work, E., 533, 534, 537(114), 540(114), 1063, 1065(12), 1096(12), 1161(12), 1166(12), 1167(12), 1168, 1169(452), 1170(452), 1287
Work, T. S., 588, 1249, 1287, 1288, 1289
Wormall, A., 723, 761, 768, 769(59), 770, 788, 840
Wormell, R. L., 876
Wosilait, W. D., 137
Wrede, H., 260
Wright, B. A., 919
Wright, C. I., 234
Wright, E. A., 373, 374(161)
Wright, G. G., 721
Wright, G. Payling, 371, 373, 374(161, 161a, 161b)
Wright, N. C., 408
Wright, R. D., 1162(437), 1163, 1165 (437)
Wright, R. R., 784
Wróblewski, A., 396
Wu, F. C., 1063, 1168(11a)
Wu, H., 293, 328, 453, 456, 782, 809, 1290
Würz, A., 854, 858
Wulff, H. J., 169, 171, 172(200), 173
Wurmser, R., 789, 793, 816, 819(253)
Wyatt, G. R., 7, 14, 15, 17, 18, 19, 92(68), 93, 104
Wyckoff, R. W. G., 78, 82, 85, 88, 90, 92, 93, 96, 98(332), 105, 516, 520(11), 522(11), 549, 796, 797, 798(179), 809, 868, 913, 916, 926, 928, 983, 1028, 1031, 1032(202)
Wyman, J., Jr., 281, 295, 302, 303, 307, 308, 315, 319, 320, 331, 335, 447

Y

Yakel, H. L., 890
Yalow, R. S., 1273
Yamada, H., 559
Yamade, M., 555(172c), 594
Yanari, S., 1212, 1214, 1215
Yang, E. F., 293
Yang, J. T., 490

Yasnoff, D. S., 1072
Yemm, E. W., 571, 578
Yenson, M. M., 323, 841
Yoffey, J. M., 669
Yoneda, M., 348
Yoneya, T., 1108
Yoshida, H., 135
Youatt, G., 1081, 1157(99)
Youmans, G. P., 800
Young, E. G., 468, 471, 473, 485(182)
Young, F. G., 614
Young, R. A., 943
Yphantis, D. F., 682
Yudkin, J., 1283, 1285
Yui, N. H., 490
Yuile, C. L., 292, 675, 1260, 1269(209), 1271(209), 1273, 1276(209), 1288
Yuill, M. E., 762, 771, 775(32)

Z

Zachariades, M. P.-A., 925
Zacharius, R. M., 527, 534(80), 535, 536 (116), 537(80), 538(80), 559, 560 (116, 189), 562(116, 189), 563(116), 565, 567
Zahn, H., 849, 854, 858, 859, 860, 861, 872 (55), 873(55), 899, 939
Zahler, P., 1083, 1084(106)
Zaiser, E. M., 307, 1103
Zamecnik, P. C., 784, 1096(158, 170), 1097, 1099(170), 1136, 1160, 1226, 1227, 1230, 1231, 1233, 1236, 1240 (102), 1246, 1262, 1263(219)
Zamenhof, S., 18, 19, 31
Zapp, J. A., Jr., 133, 234(33)
Zarudnaya, K., 262, 274(518, 519), 276 (519)
Zatman, L. J., 372
Zeller, E. A., 237, 238, 243, 244, 258, 259, 357, 365(72), 376(72), 383(72), 384, 1096(172), 1097, 1099(172)
Zerban, K., 142, 144(93)
Zerfas, L. G., 126, 162
Zervas, L., 1106, 1141
Zeynek, R. v., 313, 315
Ziegler, M. R., 1265
Ziff, M., 970
Zill, L. P., 122
Zimm, B. H., 71, 994
Zimmer, A. J., 554
Zinoffsky, O., 286
Zinsser, H. H., 320, 788
Zittle, C. A., 431, 1216
Zondek, B., 618
Zuckerman, R., 17
Zweig, G., 522

Subject Index

A

ACTH, see Adrenocorticotropic hormone, of anterior pituitary
ADP, see Adenosine diphosphate
ATP, see Adenosine triphosphate
ATPase, see Adenosinetriphosphatase
A substance, of muscle, 1028, 1043
  nature of, 1033–1034
Abrin, 357, 360
Absorption spectra,
  of hemoglobin and its derivatives, 295–299
Acetyl-L-phenylalaninehydroxamine,
  as substrate for chymotrypsin, 1099
Acetyl phosphate,
  role in amino acid acetylation, 1211
Acetylcholinesterase,
  activity, mechanism of, 1101
Achroglobins, 344
Acids,
  fatty, reaction with human serum albumin, 678
Aconitin,
  lethal dose, 347
Actin, 977–994, see also Actomyosin
  ATP sensitivity, 982
  amino acid composition, 979, 1018, 1019, 1020–1021, 1023
  estimation, 1011
  isolation, 977, 986
  molecular weight, 979
  nucleotide in, 1031
  phosphorus in, 978
  physicochemical properties, 977–979
  polymerization, 979, 981, 983
  preparation, 979–982
  purification, 980
  reaction with myosin, 977, 986–988
    preparation and, 982
    role in muscle contraction, 957
  thixotropic form, reactivity of, 993
F-Actin,
  action of salyrgan on, 1046
  nature of, 984–986
  nucleotide content, 978
  preparation, 983
  ultrastructure of muscle and, 1032, 1033
  viscosity, 982
G-Actin,
  conversion to F-actin, effect of ions on, 977–978
    inhibitors of, 979
    mechanism, 978, 979, 983
  dimerization, 984ff.
  globular nature of active, 983
  magnesium in, 978
  molecular weight, 983
  nucleotide content, 978
  prosthetic group of, 1019
  reaction with tropomyosin, 995
(Acto)-myosin, see Actomyosin, natural
Actomyosin, 986ff., see also Actin
  action of ATP on, 988–994
    in gel form, 990–992
    nature of, 989, 992–994
    in solution, 988–990
  artificial, 959
    electron microscope studies on, 986–987
    nature of, 986
    proportion of components in, 987
  contraction, 991ff.
  dissociation by ATP, effect of metals on, 990
  natural, 959
    composition, 1019
      species differences in, 1019
    intramolecular structure, 970
    molecular size and shape, 970
    physicochemical constants, 961
    polydispersity, 970
    sedimentation, 970

reaction with ATP, role in muscular contraction, 957
solubility, 960
stability, ATP and, 1044
sulfhydryl content of, 964
turnover rates, 1274
viscosity, 993
x-ray diffraction pattern of, 1024
Acylase(s), kidney,
resolution of amino acids by, 1108, 1134
substrates for, synthetic, 1107
Acylcoenzyme A,
formation, 130
Adenine (6-aminopurine),
in nucleic acids, 5, 10
occurrence in plants, 560
Adenosine diphosphate,
in actin, 978, 979
in myofibrils, 978, 979
as prosthetic group of F-actin, 1019
Adenosinetriphosphatase,
action on inosine triphosphate, 966
activity, effect of ions on, 967–968
in myosin, 964ff.
in animal tissues, 1283
inhibition by salyrgan, 1046
mitochondrial, 968
of muscle granules, 1004–1005
activation by Mg, 1004–1005
isolation, 1005
role in amide synthesis in plants, 571
Adenosine triphosphate, in actin, 978
action on actomyosin, 988–994, 1042
in gel form, 990–992
mechanism, 989, 992–994
in solution, 988–990
on myosins, 960, 977
breakdown, acetylation of aromatic amines and, 1210
destruction by snake venoms, 384
glutamine synthesis and, 526, 1218
in myofibrils, 978
peptide bond synthesis and, 1224
as prosthetic groups of G-actin, 1019
role in muscular contraction, 1042, 1043ff., 1054
in muscular relaxation, 1044
in protein biosynthesis, 587
synthesis of, 966
viscosity response of myosins to, 960
Adenylic acid(s), 5
in actin, 978
effect on incorporation of amino acids into proteins of tissue homogenates, 1230
in myofibrils, 978
Adrenaline, 596
action, 597
chemical constitution, 597
source, 597
Adrenocortical hormones,
release of antibodies from lymphocytes and, 810
Adrenocorticotropic hormone, of anterior pituitary, 596ff., 613–618
action of pepsin on, 1186
activity, 597, 604, 613
assay, 614–615
essential groups for, 618
chemical constitution, 597
inactivation, 617
isolation, 616–617
polypeptide nature, 613, 614, 617
purification, 615–616, 617
source, 597
Agglutination,
of bacteria, 820, 821–822
zones in, 821–822
Agglutinins, 783, see also Antibodies, agglutinating
anti-B(β), 793
molecular weight, 793
thermodynamic values, effect of genotype on, 819–820
Alanine,
incorporation into proteins of tissue preparations, inhibitors of, 1240
removal of terminal, from insulin, 1225
role in insulin activity, 1188
β-Alanine,
occurrence in plants, 559, 562, 563
DL-Alanine-1-C^{14},
incorporation into desoxypentose histone, 1245
into proteins of tissue preparations, 1229
concentration and, 1236
rates of, 1230

Alanylglycine,
formation, free energy of, 1201
Albumin(s),
carbohydrate-containing (McMeekin), 734, 737
in cereals, 493
circulation, 1271
of dicotyledons, 500, 504, 506
effect of ripening on, 509
half life, 1272, 1278
iodinated, catabolism, 1273
turnover, 1277
labeled with I^{131}, 1272
of peas, amino acid composition, 511
plasma, isolation, 744, 747, 748
physicochemical properties, 740
therapeutical application, 742
serum, 677–698
antigenicity, 759
species differences in, 779
cellular agglutination and, 757
complex of human, with fatty acids, 678
composition, 686–687
crystallization of, 680–681
definition of mammalian, 677–678
dimerization, 681
thiol groups and, 689–692
effect of urea on, 685
formation, site of, 673
heterogeneity, 678–680
interaction with other proteins, 684
intracellular distribution, 671
iodinated, 693, 694
in milk, 434
molecular size and shape, 675, 682
molecular weight, 676, 682, 777
physicochemical properties, 681–684
physiological functions, 697–698
as protein reserve in fasting animal, 672
reaction with anions, 696–697
of bovine, with horse antibody, 804
with rabbit antibody, 805
with mercury, 681
physiological function and, 697
reactive groups in, 689–697
resynthesis from proteolysis products, 1144
solubility, 683
species differences in, 680
stability, 684–686
structure, 682, 685–686, 687
sulfur in, 686, 687
synthesis in liver preparations, 1246–1248, 1258
inhibitors, 1247
mechanism, 1247
net production, 1246
precursor, 1247–1248
titration curve, 688–689
tryptophan content, species differences in, 686
turnover, 672, 1247
valency, 777
Alcohol dehydrogenase(s),
chemical properties, 168, 175
horse liver, inhibition, 174, 177
isolation, 174
oxidation of vitamin A by, 178
purified, properties, 169, 174
occurrence, 171
yeast, chemical properties, 172
homogeneity, 171
inhibition, 172
isolation, 171
purified, properties, 169
Alcohols, action on proteolytic enzymes, 1095
Aldehyde dehydrogenase(s), 178–183
liver, action mechanism, 181
assay, 179
inhibition, 180
purification, 178
yeast, action mechanism, 181
inhibition, 181
isolation, 180, 181
Aldehyde oxidases, 220
horse liver, purification, 232
pig liver, isolation, 231
purified, 224
Aldolase,
in cow milk, 428, 431
in muscle proteins, 1006
Alfalfa,
bulk proteins of, amino acid composition, 539
Alfalfa mosaic virus, 82–83
isolation, 83

properties, 83
transmission, 82
Algae,
α,ϵ-diaminopimelic acid from, 533
incorporation of amino acids into, 1244
Alkaloids,
occurrence in plants, 560, 561, 573
Allantoic fluid, mucinase activity, 89
Allantoin,
occurrence in plants, 560
Allergy,
antibodies and, 787
Amandin, 501
elementary analysis, 491
molecular constants, 503
molecular weight, in urea solution, 507
occurrence, 503
Amanitin, 357
polypeptide nature, 358
Amides, see also individual compounds
formation, carbohydrates and, 527
hydrolysis by proteolytic enzymes, heats of, 1116
in proteins, 572–573
synthesis in plants, 570–572
ATP and, 571
protein synthesis and, 571
Amine transfer reactions, 1142
Amines, aromatic, see also individual compounds
acetylation, 1210–1211
in animal tissues, 1210–1211
acylation, coenzyme A and, 1211
diazotized, preparation of conjugated proteins with, 771–773
D-Amino acid oxidases,
mammalian kidney, homogeneity, 233
inactivation, 234
inhibitors for, 234
isolation, 233
prosthetic group of, 133
purification, 224
specificity, 234
occurrence, 235
L-Amino acid oxidases,
of *Neurospora crassa*, 224, 245
occurrence, 237
purification, 224
rat kidney, activity, 236
isolation, 235
purification, 224, 235
of snake venoms, 237, 383
inactivation, 239–243
inhibitors for, 243
prosthetic group of, 133, 221
purification, 224, 238
role in venom proteolysis, 383
specificity, 237, 243
Amino acid oxidases, 220, 232–245
action mechanism, 130
Amino acids,
acetylation, 1209–1210
in *Clostridium kluyverii*, 1210
in actin, 1018, 1019, 1020–1021
as activators of pancreatic amylase synthesis *in vitro*, 1254
acylation, 1205–1211
CoA and, 1224
analogs of, effect on growth, 1286
on protein synthesis, 1286
antigenic specificity, 774
aromatic, antigenicity of proteins and, 760–761
in bacteriophages, 105
concentration by cells, 671, 1231–1232
cobalt complexes, combination with oxygen, 306
in collagen, 912, 920–921
decarboxylation, 563, 564
products of, occurrence in plants, 563–564
in egg yolk protein, 477
in epidermin, 886
essential, for animal nutrition, 515
daily allowance for man, 512
esters of, action of α-chymotrypsin on, 1137
in fibrinogen, 877
in follicle-stimulating hormone, 601
foreign, incorporation into tissue proteins, 1291, 1295
formation, carbohydrates and, 527
by transamination, 525
free,
effect of enzymatic adaptation on, in *Saccharomyces cerevisiae*, 1285
in glucagon, 649
in growth hormone, 611
in hemocyanins, 338
in human chorionic gonadotropin, 620

in immune proteins, 712, 792
incorporation into algae, 1244
of isotopes into, 1226
inhibition of carboxypeptidase by D-, 1133
in insulin, 638–639
species differences, 646
in keratins, 872–873
labeled, determination of protein turnover with, 1263, 1266, 1267ff.
incorporation into proteins of cellular components of liver, 1262–1263
incorporation into proteins of perfused livers, 1258
incorporation into proteins of tissue cultures, 1255–1257
effect of amino acid analogs on, 1256
inhibitors of, 1256
incorporation into proteins of tissue homogenates, 1234
activators of, 1236–1240
inhibitors of, 1240–1241
mechanism, 1234, 1258
incorporation into proteins of tissue preparations, 1234ff.
activities of different tissues, 1260
amino acid concentration and, 1236–1238
anomalous reactions, 1245–1246
effect of homogenization on, 1232
effect of hormones on, 1239
factors responsible for, 1258–1259
factors stimulating, 1238–1240
inhibition of, 1240–1241
lability of label, 1296
mechanism, 1250
peptides and, 1242
rate of, 1229–1235
incorporation into proteins *in vitro*, 1229–1246
utilization for protein synthesis in live tissue, 1261
in lactogenic hormone, 604
in meromyosins, 1020–1021
metabolism in ripening grain, 508
in milk proteins, 427, 428
in myosin, 1018, 1019, 1020–1023
in ovalbumin, 447
oxidation, inhibition by cyanide, 130
in oxytocin, 629
"peculiar," of antibiotics, 1288
in plants, 527, 558–559
in bulk proteins of, 532
in growing points of, 529–530
in leaf proteins, 549
role in protein biosynthesis, 527, 582–583
in soluble and protein nitrogen fractions, 583–585
in pregnant mare serum gonadotropin, 622
protein synthesis from, 1288ff.
in proteolytic enzymes, 1082–1085
resolution by proteolytic enzymes, 1108
in seed proteins, 510–512
in serum albumin, 686–687
in silk fibroin, 850–851
in structure proteins, 847
of muscle, 1018–1024
synthesis of casein in mammary gland from, 433
in thyroglobulin, 653
in tobacco mosaic group of viruses, 66
in tropomyosin, 1018, 1019, 1020–1021
in vasopressin, 630, 631
Amino groups,
free, activity of ACTH and, 618
of growth hormone and, 609
terminal, in serum albumin, 687
α-Aminoadipic acid,
as protein hydrolytic product, 533
γ-Aminobenzoylphenylaminoacetic acids,
serological differentiation of, 775
γ-Aminobutyric acid,
occurrence in plants, 558, 562, 563, 582, 583
distribution, 563
origin of, 563, 564
p-Aminohippuric acid,
synthesis of, 1207
activators of, 1207
inhibitors of, 1206
phosphate requirement for, 1207

Aminopeptidase(s), 1106
  kidney, isoelectric point, 1073
    molecular weight, 1072
  metal activation of, 1148
  substrates for, synthetic, 1108
Ammonia,
  formation, carbohydrates and, 527
  physiological role, 572
    in nitrogen and protein metabolism of plants, 525
Amphibians,
  collagen in skin of, 944
  poisons in skin glands of, 353, 354
Amygdalin,
  occurrence in plants, 560
Amylase, of cow milk, 428, 430
  in egg yolk, 481
  pancreatic, synthesis *in vitro*, 1253–1254
    amino acids and, 1254, 1259
    inhibitors of, 1254
Androgens, 596
  action, 597
  sources, 597
Anemias,
  hemoglobins in, 321, 322
Angiosperms,
  seed proteins of, 488–489
Anhydrohemoglobin, 283
  formation, 306
Anilides,
  synthesis by papain, 1135
Annelids,
  respiratory proteins of, 341
Anserine,
  in muscle, 956
Antibiotics,
  "peculiar" amino acids in, 1288
Antibodies, 708–725, 782–814
  see also individual antibodies
  "active patch" of molecule, 807, 812
  agglutinating, action of, 783
    blocking antibodies and, 785, 786
  allergic, action of, 787
  amino acid arrangement in, 814
  to animal's own proteins, 709, 758
  artifacts, 713, 714, 721
  to artificially conjugated proteins, 805
  avidity of, 820–821
  blocking, 785–786, 803
    agglutinating antibodies and, 785, 786
  chemical properties, 790–805
  chicken plasma, 713
  in colostrum, 390
  combination with antigen, 776, 782
    see also Antibody-antigen reactions
  complement-fixing, action of, 784
  composition of, 791, 792
  Coombs, 785–786
  definition, 708–710
  denaturation, 790
  determinants of, chemical nature, 806–807
  effects of enzymes on, 790–791, 798–799
  electrophoretic mobility, 796–798
  estimation, quantitative, 809
  formation, 711, 714, 809–814
    enzymes and, 813
    in multiple myeloma patients, 711
    sites of, 810, 811
    theories of, 811–814
  functions of, 782–789
  globulin nature of, 782, 786, 790
  γ-globulins and, 710, 711, 713, 714, 782, 786
  halogenation, 790
  heterogeneity, 710–715, 805
    determination, 799
  human, end groups in, 715
  incomplete, 709, 803–804
    in blood-grouping sera, 822
    molecular weight, 803
    valence of, 804
  interaction with enzymes, 1159–1160
  iodination, effects of, 724
  isoelectric points, 796
  to β_1-lipoprotein, 706
  lytic, action of, 784
  mammalian, 713
    molecular weights, 713
  molecular shape, 793–796
  molecular structure, 790
  molecular weight of, 791, 793, 794
  multivalency of, 776
  mutual valence of antigen and, 830, 832
  natural, 788–789
    classes of, 788

formation, 789
plant, 789, 809, 814
neutralizing, action of, 783–784, see also Antitoxins
nomenclature, 782
polymerization with other proteins, 721
precipitating, 783
proteolytic activity, 1159
purification of, 715–720, 807–809
release from lymphocytes, adreno-cortical hormones and, 810
role of placenta in transference of, from human mothers to infant, 413
serological differences between normal globulin and, 786
for silk, 763
species differences between, 799–800
specific combining groups in, 786
number of, 829
specificity, 708
stability, 713
differences in, 799
synthesis, 1278
"third-order," 786
valence of, 800–805
Antibody,
to Pneumococcus, type III, half life, 1280
synthesis *in vitro*, 1254–1255
Antibody-antigen reactions, 708–710, 814–840
see also Precipitation reaction, serological
effect of electrolytes on, 819–820
equations for, 832–839
forces involved, 815–816
heat of reaction, 815, 816, 817–818
liberation of water molecules in, 818
mathematical treatments, 832–839
equations for, 833ff.
rate of reaction, 817
requirements for, 708–709
similarity between enzyme reactions and, 843–844
stages, 814–815
first, 815–820
second, 820–840
Anticoagulants, 382
Antigens, 757–782, see also individual antigens
administration, 793
artificial, preparation, 776
reaction with antipneumococcus horse sera, 816
specificity of antisera to, 772
chemical modifications, 722
of *Clostridium welchii*, protease nature, 1070
definition, 757, 758
determinant groups of, 760, 844
nature of, 763–765
enzymes and, 844
heterogeneity, determination, 799
mutual valence of antibody and, 830, 832
nature of, 708
phagocytosis of particulate, 671
proteins as, 757
requirements for, 759–763
reaction with antibodies, 708, 710, 776, 782
see also Antibody-antigens reactions
requirements for, 708–709
structure of complexes formed, 709
retention in body, 762
specificity, polar groups and, 764, 815
species, 761
toxic, of gram-negative bacteria, 385–387
structure, 386–387
valency, 777
molecular weight and, 777ff.
Anti-γ-globulin, formation, 786
Antihemocyanin,
reaction with Busycon hemocyanin, heat of, 817
Antiovalbumin,
purification, 715
rabbit, action of papain on, 1175
iodination, 724
physical properties, 794
rabbit γ-globulin and, 714
skin-sensitizing activity of, 787
stability, 721, 722
Antipertussis plasma,
human, fraction II of, 717
Antiplasmin, 382
Antipneumococcus globulins,
horse, denaturation of type I, 723
stability, 722

purification, 715
Antipneumococcus serum antibody,
horse, effect of pepsin on, 794
electrophoretic mobility, 797
heterogeneity, 799
molecular weight, 71
physical properties, 791, 794, 797, 800
reaction with artificial antigens, 816
Anti-Rh serum,
blocking antibodies in, 822
Antisera,
rabbit, species specificity, 780
Antitoxins,
action of, 783–784
combination with toxins, 820
Apoenzymes, 165–168
action mechanism, 166
activation of substrate and, 131
compound formation between coenzyme and, 167
definition, 128
stabilization, 134
Apple,
bulk protein of, amino acid composition, 537, 538
protein synthesis in, 519–520
Apyrases,
action, 967
Arachin,
amino acid composition, 511
extraction, 505
legumin and, 506
molecular constants, 503, 505
occurrence, 503
Arginase,
in animal tissues, 1283
occurrence in plants, 566
Arginine,
in myosin, 1022
occurrence in plants, 560, 561
role of, 567
Ascaris,
body wall of, pepsin inhibitor in, 1166
trypsin inhibitor in, 1163
Asclepain,
isolation, 1069
Ascorbic acid,
enzymatic oxidation, 137, 143, 144, 145, 146, 147, 148, 157, 159
Ascorbic acid depletion test,
for adrenocorticotropic activity, 614
Ascorbic acid oxidases,
activity, determination of, 143
mechanism, 137
chemical properties, 141
copper-protein bond in, 144
distribution, 135
inactivation, 140, 146
isolation, 141
purification, 142
role of copper in, 139
specificity, 138
stability, 146
Asparagine,
in edestin, 572
occurrence in plants, 559, 560, 561, 562
metabolic relation to glutamine, 567–569
physiological role, 524, 525–526, 530, 568, 569
protein metabolism and, 566
relative content of, 567–568
role in insulin activity, 1188
structure, 526, 569, 570
physiological behavior and, 569–570
in tobacco mosaic virus, 572
Aspartic acid, biosynthesis, 572–573
in myosin, 1022
occurrence in plants, 561
in bulk proteins, 535, 536, 537, 538
role in nitrogen metabolism, 525
D-Aspartic oxidase,
of rabbit tissues, 259
Aspergillus oryzae,
α-aminoadipic acid from, 533
protease from, 1070
Atrophy,
muscular, biochemical changes in, 1016
etiology, 1015–1016
Atropine,
lethal dose, 347
Auxins,
formation, zinc and, 593
occurrence in plants, 560
role in protein biosynthesis, 573, 586
Avidin, 48, 439, 469–471
activity, antibiotin, 469
essential groups for, 471

complexes with nucleic acid, 469, see also Avidin NA
forms of, 469–471
inactivation, 471
isolation, 469
physical constants, 449
stability, 471
Avidin A, 469
composition, 470
preparation, 470
Avidin NA, 469
composition, 470
desoxyribonucleic acid content, 470
preparation, 470
Avidin XA, 469
composition, 470
glycoprotein in, 469
preparation, 470
Azides,
preparation of conjugated proteins with, 771
p-Azophenylarsonic acid antibody, from rabbit serum, purification, 716

B

Bacillus anthracis,
toxins of, 367
virulence, role of protein component in, 353
Bacillus cereus,
toxins of, 367
Bacillus megatherium,
lysogenic, effect of ultraviolet irradiation on, 117, 118
release of phage by, 116
Bacillus mycoides,
toxin of, 367
Bacillus subtilis,
enzyme effecting conversion of ovalbumin to plakalbumin in, 454, 455, 1070, 1248
Bacitracin,
inhibitory effect on nucleic acid and protein synthesis in *Staph. aureus*, 1243
Bacteria,
agglutination, 821–822
salts and, 820
decarboxylation of amino acids by, 563
effect of virus infection on synthesis of cell constituents in, 110ff.
elastase in, 948
enzymatic adaptation in, 1285
glutamine as growth factor for, 569
gram-negative, polymolecular antigenic toxins of, 385–387
identity with O-antigens, 385
growth, peptides and, 1241, 1242
lysogenic, 94, 116–118
bacteriophage production in, 117–118
latent viruses in, 62
relationship between bacteriophages and, 93–94
spreading factors in culture filtrates of invasive, 385
toxins of, see Toxins, bacterial and individual toxins
virulence of pathogenic, role of proteins in, 352–353
Bacteriophages, 62, 93–122, see also Coliphages
assay of infective particles, 96–97
chemical composition, 102–105
desoxyribonucleic acid in, 10
electron microscopy of, 99–101
formation within bacterial cell, 110–114
genetics, 114–116
genetic recombinations, 114–115
mutations, 114
phenotypic variations, 115–116
isolation, 97–98
multiplication of, 105–110
adsorption onto host, 107–110
effect of ions, 109–110
kinetics of, 108–109
specificity of, 109
lytic cycle, 106–107
physical properties, 96, 99–102
diffusion, 102
sedimentation, 101–102
stability, 101
production in lysogenic bacteria, 117–118
purification, 97–98
relation between bacteria and, 93–94

size and shapes of, 95, 99ff.
Baikiain,
occurrence in plants, 559, 565
structure, 565
Barley,
germination,
effect on seed proteins of, 509–510
globulins of, 495
amino acid composition of β-globulin, 511
components of, 495
in seeds, 495
protein content, 493
protein formation in ripening grains of, 508
Beans,
proteins of, anti-blood groups A activity of, 789
Bee venom, 356–357
components of, 357
dehydrogenase inhibitor of, 378
histamine in, 357
pathologic effects of, 356
Beet seed globulin,
molecular constants, 503
Benzoylglycine, see Hippuric acid
S-Benzylhomocysteine,
acetylation in liver and kidney, 1209
Benzylhydroxamic acid, synthesis, 1209
Betaines,
occurrence in plants, 561
Biliverdin, 316
Birds,
collagen in skin of, 944
Bis(2-chloroethyl) sulfide, see Mustard gas
Blood,
barrier between maternal and fetal, in placenta, 737
clots, see also Fibrin
formation, 876
structures, 884
types of, 880, 883–884
urea soluble, 880, 881
clotting components of, 725–732, see also Fibrinogen Prothrombin, Plasminogen
clotting, mechanism of, 379
coagulation, prevention of, 665
defibrination, 666
hemoglobin content, species differences in, 281
plasma, see Plasma, blood
serum, see Serum
types of, protein variations as explanation of, 680
Blood groups,
isoagglutin of human, 788
Blood pressure,
effect of posterior pituitary hormones on, 623
Bone marrow,
incorporation of labeled amino acids into proteins of, 1240
inhibitors of, 1240, 1241
of vertebrates, formation of hemins in, 286
of porphyrins in, 286
Bones,
composition, 846
Boron,
effect on nitrogen metabolism in plants, 592–593
Bothropotoxin, 356
Botulinus type A neurotoxin,
lethal dose, 346
molecular weight, 346
properties of purified, 346
Botulism, 360
Bradykinin, 384
Bradykininogen, 384
Broad bean mottle virus, 83–84
properties, 83
Brucella melitensis,
toxic antigen of, structure, 387
Brucine,
antibodies for, 775
5-Butyl picolinic acid,
occurrence in plants, 559
"Byssokeratin,"
sources, 911

C

CA, see Cenapse, acide
CoA, see Coenzyme A
"C reactive component," of McCarty, 672
Cadaverine,
occurrence in plants, 563

Calcium,
activation of myosin by, 967, 968
effect on nitrogen metabolism in plants, 590–591
on trypsin, 1149, 1150
protein synthesis and, 579, 580
role in blood clotting, 665
Canavanine,
occurrence in plants, 559
Capillaries,
leakage of proteins from, 669
Carbobenzoxy-L-glutamyl-L-tyrosine,
as substrate for cathepsin A, 1126
for pepsin, 1102, 1126
Carbobenzoxy proteins,
immunological behavior, 768
Carbohydrate(s), see also individual compounds
in egg white globulin, 473, 474
in α-globulins, 678
in gonadotropins, 601, 607, 620, 622, 623
microbial, antigenicity of, 757
occurrence in plants, 560
in ovalbumin, 445, 446
in ovomucoid, 466–467
in proteins, 737
role in protein synthesis, 527–528
Carbon dioxide,
fixation, role in protein metabolism, 578
protein biosynthesis and, 575
Carbon monoxide hemoglobin, 308–312
absorption spectrum, 296, 297, 298, 309
human, solubility, 283
molecular weight, 309
oxyhemoglobin acid, 309
photochemical dissociation, 311–312
stability, 283, 309
Carbonaphthoxy-L-phenylalanine,
action of carboxypeptidase on, 1100
Carbonylhemoglobin, see Carbon monoxide hemoglobin
Carboxyhemoglobin,
magnetic properties, 333
Carboxyl phosphates,
as energy source in peptide bond synthesis, 1204
Carboxylase, 124
of pig heart, composition, 276
of wheat germ, isolation, 271
purification, 271
yeast, isolation, 270
purification, 270
α-Carboxylases, 270–272
reaction pattern, 265
Carboxymyoglobin, 325–326
Carboxypeptidase(s), 1106, 1129–1133
action on itself, 1184, 1188
on proteins, 1109, 1183, 1184, 1187–1189
activation by metal ions, 1149
activity, pH and, 1102, 1105
amino acid composition, 1083, 1084
determination of C-terminal groups with, 1187ff.
effect of carbonaphthoxy-L-phenylalanine, 1100
of irradiation on, 1093
electrophoretic behavior, 1081
hydrolysis of peptides by, heats of, 1116
immunological behavior, 1159
inhibitors, 1131–1133, 1150
antibody as, 1160
isoelectric point, 1073, 1084
molecular weight, 1072, 1084
physicochemical properties, 1081
preparation, 1066–1067
resolution of amino acids by, 1108
removal of terminal alanine from insulin by, 1225
specificity, 1129
stability, 1092, 1148
structure, 1083
substrates for, synthetic, 1105, 1107, 1112
hydrolysis of, 1129ff.
kinetics, 1131
mechanism, 1131
structural requirements, 1129, 1130–1131
temperature and, 1112
Carbylamine hemoglobins, 312
Carcinoma,
prostate, serum acid phosphatase and, 733
Carnitine, in muscle, 956

Carnosine, in muscle, 956
Carrot,
  bulk protein of, amino acid composition, 537, 538
Cartilage,
  enzyme system in, 1275
  fibrils other than collagen in, 914
α-Casein, 392, 397
  density, 406
  molecular weight, 402
  preparation, 398–399
  separation from β-casein, 397, 398
    from milk protease, 429
  terminal amino groups in, 428
β-Casein, 392, 397
  density, 406
  preparation, 398–399
  separation from α-casein, 397, 398
  terminal amino groups in, 428
γ-Casein, 392
  density, 406
  preparation, 399
Casein(s), 395–411, see also individual members
  action of proteolytic enzymes on, 1174
  components, 392, 397
    combining capacity of, 403–404
    distribution, 401
    interaction between, 400–401
    separation of, 397–400
  composition, 396–397
    amino acid, 427
    species differences in, 397, 410–411, 427, 428
  heat coagulation, 408
  hydrolysis by chymotrypsin, effect of pressure on, 1114
  iodinated, isolation of thyroxine from, 657
  isolation of phosphopeptones from, 408–409
  in milk, 390, 392
    effect of lactation on, 394
    synthesis of, 1288–1289
  molecular shape, 404–405
  physical properties, 401–408
  preparation, 395
  synthesis in mammary gland, 433
    in milk, 1288–1289
Castor bean,
  poisonous protein of, immunization against, 757
Catalase,
  in cow milk, 428, 432
  in egg white, 474
  in egg yolk, 481
  in plasma, origin of, 672
Catecholase, 135
  purification, 153
Cathepsin A, 1126, 1145, 1147
  isolation, 1067
  pepsin and, 1134
  substrates for, synthetic, 1126
Cathepsin B,
  activators, 1147
  as catalyst of replacement reactions, 1141
  inhibitors of, 1147
  trypsin and, 1134
Cathepsin C,
  activators, 1147
  activity, pH optima, 1103
  as catalyst of replacement reactions, 1138–1139
  chymotrypsin and, 1134
  hydrolysis of peptides by, heats of, 116
  inhibitors of, 1147
  isolation, 1067
  specificity, 1134
Cathepsin III, see also Leucineaminopeptidase
  activators of, 1147
  inhibitors of, 1147
  manganese-activated leucineaminopeptidase and, 1134
  nature of, 1147
Cathepsin IV, see also Carboxypeptidase
  carboxypeptidase and, 1134
  nature of, 1147
Cathepsin(s),
  activators of, 1145
  activity, cellular, 1294
    determination, 1095
    pH optima, 1101
  homogeneity, 1103
  preparation, 1067–1069
  role in protein synthesis, 1252, 1294
  specificities, 1134

classification based on, 1067
stability, 1092
substrates for, synthetic, 1107
applications of, 1108
Cells,
agglutination, substances responsible for, 756, 757
concentration of amino acids by, 1231–1232
inhibitors of, 1240
enzymes of, activity, 1294
growth, ribonucleic acid and, 58
liver, incorporation of labeled amino acids into proteins of, 1262–1263
permeability for plasma proteins, 1271–1272
cellular protein metabolism and, 1272
plasma, formation of immune globulin in, 673
as plasma protein reservoirs, 670
uptake of peptides by, 1242
Cenapse, acide, 704, 706
composition, 702
molecular weight, 704
properties, 704
Cereals,
globulins of, 496ff.
isolation, 496–497
molecular constants, 496
gluten content, 495
proteins of, 493, 495
effect of storage on, 509
factors influencing, 494–495
formation in ripening grains, 508–509
seed, molecular constants, 496
Ceruloplasmin, 158, 700–701
copper content, 701
enzymatic activity, 701, 732
hemocuprin and, 700
nature of, 732
physicochemical properties, 740
purification, 700
Chaetopodes,
chlorocruorin in, 335
Chalazae,
composition, 473
Chitin, 846
Chlorella vulgaris,
bulk protein of, amino acid composition, 537, 538
Chlorocruorin, 335–336, see also Hemoglobin, invertebrate
absorption spectra of, and derivatives, 336
amino acid composition, 336
carbon monoxide affinity, 336
iron content, 335
isolation, 335
molecular weight, 336
occurrence, 280, 335
oxygen affinity, 335
prosthetic group of, 336
protein component, 336
Chlorophyll,
compound with protein, 548, 549
molecular weight, 549
extraction from chloroplasts, 548, 550
formation, iron and, 593
Chloroplasts,
extraction of chlorophyll from, 548, 550
nature of, 548
proteins of, 543, 544, 548–550
photosynthesis and, 54
Choleglobin, 316
Cholesterol,
inhibition of bacterial hemolysins by, 370
in plasma lipoproteins, 702, 703, 704
Choline,
occurrence in plants, 573
Cholinesterase,
effect of organic phosphates on, 1154
Chondroitin sulfate, 921
half life, 1275
Chromium,
labeled, use in protein turnover studies, 1272
salts, effect on collagen, 923
Chromosin, 38
Chromosomes,
composition, 51, 52
role of nucleic acids in self-duplication of, 521
structure proteins in, 950
Chromosomin, 44

Chymopapain,
  stability, 1093
α-Chymotrypsin,
  action on amino acid esters, 1137
  activity, effect of acetylation on, 1153
    of organic phosphorus compounds on, 1156, 1157
    of oxidation on, 1153
  as catalyst of replacement reactions, 1137, 1138, 1140
  components of, 1075
  conversion of chymotrypsinogen to, 1190, 1193–1196
    enzymes effecting, 1190, 1194
    mechanism, 1194
  dialkylphosphoryl derivatives,
    physical properties, 1077
  homogeneity, 1075
  hydrolysis of peptides by, heats of, 1116
  isoelectric point, 1073, 1075–1076
  molecular weight, 1072, 1077, 1078
  preparation, 1061, 1062
  purity, 1075
  sedimentation constant, concentration and, 1076–1077
  stability, 1090
  substrates for, synthetic, 1117, 1118
    hydrolysis, 1116, 1118
      methanol and, 1122–1123
    specificity, structure and, 1118
β-Chymotrypsin,
  formation, 1194
  isolation, 1061
  substrates for, synthetic, 1124
δ-Chymotrypsin, 1194
  formation, 1194
  substrates for, 1124
γ-Chymotrypsin,
  formation, 1194
  isolation, 1061
  sedimentation behavior, 1077
  substrates for, synthetic, 1124
B-Chymotrypsin,
  activity, 1063
  electrophoretic behavior, 1078
  isoelectric point, 1073, 1078
  molecular weight, 1072
  preparation, 1063–1064
  sedimentation behavior, 1078
  substrates for, synthetic, 1124
Chymotrypsin(s),
  action on insulin, 1183, 1185
    on A chain of oxidized, 1119
    on peptic digests of ovalbumin, 1143
    on silk fibroin, 1187
    of tyrosinase on, 1152
  active site of, 1123, 1124
  activity, 1109
    determination of, 1099
    effect of organic phosphorus compounds on, 1154, 1155, 1156
      mechanism, 1155, 1157
    of pressure on, 1094
    of ultraviolet radiation on, 1093
    pH optima, 1101, 1104–1105
    stoichiometric relation between enzyme mass and, 1123
  as catalyst of replacement reactions, 1142
  immunological behavior, 1159
  inactivation of follicle-stimulating hormone by, 601
  inhibitors of, 1123, 1150
    proteins as, 1160, 1163
  molecular weight, 1123
  oxidized, molecular weight, 1077
  photooxidation, 1154
  physicochemical properties, 1075–1078
  precursor of, 1075
  reaction with diisopropyl fluophosphate, 1124
  resolution of amino acids by, 1108
  specificity, stereochemical, 1137
  stability, 1148
  substrates for, synthetic, 1104, 1107, 1112
    hydrolysis of, 1112ff.
      inhibitors of, 1111, 1118, 1120, 1121, 1122
      kinetics of, 1119
      pressure and, 1114
      temperature and, 1112
  terminal residues in, 1083
α-Chymotrypsinogen,
  homogeneity, 1074
  isoelectric point, 1073, 1074
  molecular weight, 1072, 1074, 1075
  as precursor of chymotrypsins, 1075
  solubility behavior, 1074

B-Chymotrypsinogen,
isoelectric point, 1073, 1075
isolation, 1062, 1063
molecular weight, 1072, 1075
tryptic activation, 1063
Chymotrypsinogen,
activity, effect of pressure on, 1094
amino acid composition, 1083, 1084
conversion to chymotrypsin, 1190, 1193–1196
enzymes effecting, 1190, 1194
possible mechanism, 1194, 1195–1196
denaturation, kinetics of, 1089–1090
immunological behavior, 1159
isoelectric point, 1084
isolation, 1061, 1062
molecular weight, 1084, 1194
physicochemical properties, 1073–1075
stability, 1089
structure, 1195
terminal residues in, 1083
Circulation,
peripheral, physiology of, 673–674
Citrulline,
occurrence in plants, 559
Clostridia, see also individual members
toxins of, 361, 366ff.
nature of, 366
Clostridium, botulinum,
neurotoxins of, 372–373
type A toxin, 372
type B toxin, 372
species variations in toxicity of, 349
type D toxin, 373
Clostridium histolyticum,
protease from, 1134
substrates for, synthetic, 1107
Clostridium kluyverii,
acetylation of amino acids and proteins in, 1210
Clostridium septicum,
hemolysin of, 369
Clostridium welchii,
antigens, protease nature, 1070
collagenase of, 383
toxins of, 360, 366ff.
activity, 367
α-toxin, neutralization, 350
κ-toxin, specificity for collagen, 1070
Clupein,
action of trypsin on, 1178, 1179
Coagulants,
in snake venoms, 381
of staphylococci, 379–380
Coagulase factor, 379
prothrombin and, 380
Cobra hemolysin,
lethal dose, 346
molecular weight, 346
properties of purified, 346
Cobra venom,
cardiotoxic component of, 377
inhibitor of acetylcholine synthesis in, 377
zinc content of, 378
Coconut,
globulin of, 499
plant growth factor in milk of, 529
Coenzyme,
compound formation between apoenzyme and, 167
definition, 128
Coenzyme I,
effect of snake venoms on, 378
Coenzyme III,
differentiation from DPN, 165
isolation, 164
replacement of DPN by, 165
structure, 164
Coenzyme A,
acetylation of aromatic amines and, 1211
distribution, 1211
peptide bond synthesis and, 1210, 1224
role in *p*-aminohippuric acid synthesis in liver, 1207ff.
Coenzyme A dehydrogenases, fatty acyl, see Fatty acyl coenzyme A dehydrogenases
Cofibrin, 879
Coliphage(s), see also Bacteriophages
nature of, 61
T-series of, 95–96
properties, biological, 94
physical, 95, 96, 99ff.
Collagen(s), 909–946, 1010, see also individual compounds
amino acid composition, 847, 873, 920–921

definition, 910
electron microscope studies on, 910, 933
fish, see also Ichthyocol, Elastoidin
stability, 945
formation *in vivo*, 930–931
immunological behavior, 922
mammalian, absorption spectra, 919–920
amino acid composition, 912
birefringence, 919
configuration, 919
extensibility, 917
fibrils, electron microscope studies on, 912–914
fine structure, 912ff.
fibrils morphology, 912ff.
properties, 912–923
physical, 917–920
stabilizing bonds in, 945
swelling, 918
thermal contraction, 918, 920
x-ray diffraction patterns, 914–917
effect of hydration on, 918–919
of stretching on, 915
in mammalian and other vertebrate tissues, 943–945
manufacture of gelatin from, 923
molecular dimensions, 933
molecular weight, 921
mucopolysaccharides and, 942
occurrence, 847, 911
physiological role, 909
polypeptide chain, possible configuration, 932, 936–942
polysaccharides in, 921, 922
precursor of, 930
reaction with proteolytic enzymes, 922
regenerated, collagen and, 933
regeneration reactions, 928–930
tissue genesis and, 930
secreted, 910, 945–946, see also Ovokeratin, Byssokeratin
sources, 911
soluble, see also Procollagen
Nageotte's, 925–926
specificity of *Clostridium welchii* κ-toxin for, 1070
stability, 944
structure, 921
theories of, 931–942
Ambrose-Elliott and related structures, 938
Astbury structure, 937–938
helical structures, 939–941
Pauling-Corey structure, 939
protofibrillar theory, 932
Randall's structure, 933–934, 941–942
unidimensional, evidence for, 932–933
tanning of, 922–923
turnover rates, 1274–1275
x-ray diffraction patterns, 848, 910, 933
effect of drying on, 934, 935
Collagenase(s),
isolation from *Bacillus amyloliquefaciens* filtrates, 1070
sources, 922
Colostrum, 412–416
amino acid composition of globulin fraction, 427
antibody content of, 390
biological importance, 412, 413
immune globulin of, 412–413, 416
molecular weight, 414
origin of, 416
protein components of, 412–414, 416
"regression" milk and, 433–434
trypsin inhibitor in, 414–416, 1163
isolation, 1164
vitamins in, 412
Complement,
components, 839–840
chemical nature, 840
purification, 840
fixation, 841–844
theories of, 842–844
metal activation, 725
quantitative determination, 841
sources, 840
hemolytic titer of, 841–842
Conalbumin, 439, 456–462
acid modification reaction of, 461
crystallization, 699
in egg yolk, 482
homogeneity, 458
isolation, 440, 441, 442
metal-binding by, 459–461
binding sites, 460

molecular weight, 458
physical constants, 449
preparation, 456–458
properties, 456, 458
reaction with copper, 460
with iron, 460
stability, 456
transferrin and, 698
Conarachin,
components of, 505
isolation, 505
molecular constants, 503
molecular dissociation, 505
occurrence, 503
vicilin and, 506
"Congestin," 353
Conglutinin, 756, 757
action, 756
Contractin, 1008, 1009
Copper,
as component of oxidases, 593
effect on fibroin, 857, 858
on nitrogen metabolism in plants, 593
in flavoproteins, 278
in hemocyanins, 337
in plasma, 700
reaction with conalbumin, 460
Copper enzymes, 135–160, see also individual enzymes
action mechanism, 137
biological function, 136
chemical properties, 141
distribution, 135
oxido-reduction, 132
role of copper in, 139
Copper proteins, 135–160, see also Copper enzymes
biological function, 135
distribution, 135
in milk, 432
Coproporphyrin, 287
Coproporphyrin III,
secretion by diphtheria bacillus, 362
Corals,
dibromotyrosine from, 533
Corn,
α-aminoadipic acid from, 533
protein content, 493
Cornein,
sources, 911
Corpus luteum,
as source of progesterone, 597, 598
Corticoids, 596
action, 597
sources, 597
Corticotropin, see Adrenocorticotropic hormone, of anterior pituitary
Corticotropin B, 617
Corynebacterium,
α,ϵ-diaminopimelic acid from, 533
Corynebacterium diphtheriae,
toxin of, 360, see also Diphtheria toxin
Creatine phosphokinase,
role in muscle relaxation, 1053
separation from myosin, 969
Cresolase, 135
purification, 153
Crotalidae, 355
Crotalotoxin, 355
Crotalus terrificus,
venom, see Crotoxin
Crotin, 357, 360
Crotoxin,
activity of, 365
components of, 366
inactivation, 365
lethal dose, 346
molecular weight, 346
nature of, 347, 365
properties of purified, 346
Crush syndrome, 1016
Cryptagglutinoids, 786
Cucumber virus 4, 64
amino acid composition, 66
Curcin, 357
Cyanide,
inhibition of amino acid oxidation by, 130
Cyanide hemoglobin, 312
Cyanomethemoglobin, 315
Cysteine,
acetylation, 1209
in actin, 979
effect on cobra neurotoxin, 376
inactivation of crotoxin by, 365
DL-Cysteine,
uptake into liver cells, 1284

Cystine,
in keratin, 873, 874
keratin stability and, 873–874
occurrence in plants, 561
relation to methionine in plant bulk proteins, 539–540
Cytochrome b,
diphtherial, diphtheria toxin and protein moiety of, 363
Cytochrome c,
action of proteolytic hormones on, 1186–1187
reoxidation of pyridine nucleotides by, 218
Cytochrome reductases, 220, 245–254, see also Diphosphopyridine nucleotide cytochrome reductase and Triphosphopyridine nucleotide cytochrome reductase
autoxidability, 222
diaphorase and, 278
isolation, 246
specificity, 223
Cytochromes, 324
Cytolysins, bacterial, 383
Cytoplasm,
desoxyribonucleic acid in, 9
synthesis of proteins in, 58–59
Cytosine (2-hydroxy-6-amino-pyrimidine),
in nucleic acids, 10
occurrence in plants, 560

D

DFP, see Diisopropyl fluophosphate
DNA, see Desoxyribonucleic acid
DPN, see Diphosphopyridine nucleotide
DPT, see Diphosphothiamine
Daphnia,
hemoglobin in, response to lack of oxygen, 1284
Dehydroascorbic acid,
from ascorbic acid by enzymatic oxidation, 137
Dehydrogenase inhibitors,
of animal venoms, 357, 377–379
toxic effects, 378
Dehydrogenases, 127
action mechanism of, 127
substrate specificity, 133, 165
Desoxypentose nucleohistone,
incorporation of labeled amino acids into, 1245
Desoxyribonuclease,
action on bacterial transformation factors, 56
Desoxyribonucleic acid(s),
in avidin NA, 470
bacterial transformation factors and, 55, 56, 57
in bacteriophages, 102–103, 104, 105
formation of, 111ff.
function of, 113, 114
in cellular nucleus, 53ff.
biological function, 52–55
linkage between histone and, 49
species differences in, 53
composition, 18
species differences in, 54
degradation, 19
enzymatic, 27–28
effect on amino acid incorporation into proteins of tissue preparations, 1243
in fish sperm nucleoproteins, 45
in plants, 8
in polyhedral insect viruses, 62, 92, 93
protein biosynthesis and, 812, 1243
virus reproduction and, 1244
2-Desoxy-D-ribose,
in nucleic acids, 10
2-Desoxyribose, 6
Desoxyuridine, 11
Detergents,
action on proteins, 1094
on proteolytic enzymes, 1094, 1095
Detoxication reaction, mechanism of, 1201
Deuteroporphyrin, formation, 285
Diacetyl mutase,
identity with pyruvic acid oxidase from pigeon breast muscle, 274
Diamine oxidases, 258
α,γ-Diaminobutyric acid,
occurrence in plants, 559
α,ϵ-Diaminopimelic acid,
as protein hydrolytic product, 533
Diaphorases, 220, 245–254
autoxidizability, 222

cytochrome reductase and, 278
isolation, 246
prosthetic group of, 133
Straub's, 249
isolation, 250
purification, 224
Dibromotyrosine,
as protein hydrolytic product, 533
Dicotyledons,
albumins of, 504
cytoplasmic protein in leaves of, 547, 548, 549–550
globulins of,
components, 503–504
molecular constants, 503
physicochemical analysis, 503–507
protein composition, 500–503
seed proteins, 500
amino acid composition, 510ff.
Dihydropyridine nucleotides,
ultraviolet absorption spectra, 163
Dihydroxyphenylalanine,
occurrence in plants, 559
Dihydroxyphenylalanine oxidase, 135
Diiodotyrosine,
in gorgonin, 745
in spongin, 745
3,5-Diiodotyrosine, 656
as precursor of thyroxine, 656
Diisopropyl fluophosphate,
chymotrypsin derivative, 1124, 1184, 1195, 1196
action of carboxypeptidase on, 118, 1195, 1196
structure of, 1195, 1196
inhibition of proteolytic enzymes by, 1059, 1126, 1155ff., 1189
trypsin derivative, 1126
6,7-Dimethyl-9-(D-1′-ribityl) iso-alloxazine, see Riboflavin
Dipeptidase(s), 1106
intestinal, replacement reactions with, 1136
1,3-Diphosphoglyceric acid,
from glyceraldehyde 3-phosphate, 129
Diphosphopyridine nucleotide, see also Coenzyme I
enzymatic synthesis, 165
isolation, 163
oxidation-reduction potential, 162
replacement by coenzyme III, 165
structure, 163, 164
Diphosphopyridine nucleotide cytochrome reductases, 252–254
inactivation, 253
isolation, 252, 254
purification, 224
Diphosphopyridine nucleotide pyrophosphatase,
in snake venoms, 378, 384
Diphosphopyridine nucleotide specific isocitric dehydrogenases. See Isocitric dehydrogenases, Diphosphopyridine nucleotide specific
Diphosphothiamine, 259, 260
Diphtheria,
number of deaths in the United Kingdom in 1950, 361
Diphtheria antitoxin,
action of proteolytic enzymes on, 798, 1102, 1103, 1175
concentration in plasma, 740
horse, electrophoretic mobility, 797, 798
molecular weight, 791
physical properties, 794
Diphtheria toxin, 361–363
action, possible mechanism, 347, 363
lethal dose, 346
molecular weight, 346, 777
production, iron and, 362–363
mode of, 362
properties of purified, 346
protein moiety of diphtherial cytochrome b and, 363
purity, 362
toxicity, species variations in, 349
toxoidation, 348
valency, 777
Dipicolinic acid,
occurrence in plants, 559
Disaccharides,
as haptens, 775
Diseases,
caused by immune response to animal's own protein, 709
Dismutations, 166
Djenkolic acid,
occurrence in plants, 559
Dodecyl sulfate,
reaction with β-lactoglobulin, 425, 427

Dopa,
see Dihydroxyphenylalanine
Dopa oxidase, see Tyrosinase
Drosophila,
CO_2 sensitivity of, 122
Drugs, idiosyncracy to, 351–352

E

Echinoderms,
venoms of, 354
Edema,
etiology, 674
therapy, 674
Edestan, 507
decomposition, 507
formation, 507
Edestin, 501
amino acid composition, 511, 572
denaturation, 507
elementary analysis, 491
isolation, 502
molecular constants, 503
occurrence, 503
water of crystallization, 503
Egg proteins, 435–485, see also Egg white, Egg yolk and individual compounds
composition, species variations in, 484–485
Egg white, 437–474
antibacterial activity, 459
composition, 436
electrophoretic behavior, 437–438
enzymes in, 474
globulins in, 473–474
nature of, 473
layers of, 437
nature of, 436
proteins of, 439ff., see also individual compounds
in egg yolk, 482
globulin components, 471, 473–474
ovomucin content, 471
physical constants, 449
separation, 440ff.
trypsin inhibitor in, 1163
see also Ovomucoid
"Egg white injury," in rats,
avidin and, 469
Egg yolk, 474–483
composition, 436
egg white proteins in, 482
enzymes of, 481
proteins of, 474ff., 482–483, see also individual compounds
amino acid composition, 477
Elapidae, 355
neurotoxins of, 374–377
Elapine neurotoxin, 374–377
activity, 376
components of, 376
effect of cysteine on, 376
Elastase, 948
action on elastin, 948
mucase nature of, 949
occurrence, 948
Elastin, 943, 946–949, 1010
action of elastase on, 948
amino acid composition, 920, 946–947
elasticity, 948
electron micrograph studies on, 947–948
mucopolysaccharide in, 949
occurrence, 946
stability, 948
subunits in, 949
Elastoidin, 944–945
heat contraction of, 944
as basis for classification of teleost fishes, 945
sources, 911
Electron microscopy,
of bacterial flagella, 887–888
of collagens, 910, 912–914, 933
of elastin, 947–948
of fibrin, 905, 906
of silk fibroin, 849–850
of skeletal muscle, 1027–1034
Embden-Meyerhof cycle, 192
Embden-Meyerhof-Parnas-Cori cycle, 129
Energy-transfer agents,
in cells, 130
Enterokinase,
activity, pH optima, 1101
trypsinogen activation and, 1190, 1192
Enterotoxin, staphylococcal, as cause of food poisoning, 361

Enzyme-substrate complex,
properties, 132
Enzyme(s), see also Oxidizing enzymes, Plant enzymes, Proteolytic enzymes and individual compounds
action on wool, 873
active center of, 128
activity, cellular, 1294
sulfhydryl groups and, 963
adaptation, in bacteria, 1285
nucleic acids and, 1243
inhibitors of, 1285
protein synthesis and, 1244
adaptive, synthesis of, 1285
in animal tissues, 1283–1285
variations of, 1284ff.
antibodies and, 784, 798–799, 813, 1159–1160
antigens and, 844
associated with myosin, 964–969
as catalysts of glutamine metabolism, 1216
of glutamyl transfer, 1219
of peptide synthesis, 1221, 1222–1223
cathepsin-like, of rabbit muscle, metals as activators of, 1148
condensing, see Synthetase B
in egg white, 474
glutathione-splitting, of kidney, 1136
inactivation of FSH by, 601
of growth hormone by, 612
of human chorionic gonadotropins by, 620
of insulin by, 647–648
of pregnant mare serum gonadotropin by, 623
in milk, 428–432, see also individual enzymes
in muscle, glycolytic, 1051
in juice of, 1007
mitochondrial, 1050, 1051
oxidizing, see Oxidizing enzymes
from a particulate kidney fraction, 1134–1135
plant, see Plant enzymes
plasma, 732–733
protein synthesis and, 1252, 1291–1292, 1293, 1294
proteolytic, see Proteolytic enzymes
respiratory, see also Oxidizing enzymes and individual members
groups of, 125
in striated muscle, 955
specificity, homogeneity and range of, 1109
Epidermin, 868, 885–887
amino acid composition, 886
β-configuration, 848
cross β-form, 886, 887
thermal contraction and, 905
infrared spectrum, 887
solubility, 886
stability, 886
structure, 886
subunits of, 906
thermal contraction, 886, 905
x-ray diffraction pattern, 886–887
Equine encephalomyelitis virus, 91
Erythroblastosis foetalis,
placental permeability and, 738
Erythrocruorin, 281, 294
Erythrocytes,
agglutination by influenza virus, 88
inhibitor of, in ovomucin, 472
by ricin, 359
human, metal-activated proteases of, 1148
incorporation of amino acids into proteins of, 1260
lysis by hemolysin, 784
by phosphatidase A, 365
mammalian, hemoglobin content, 281
proteins, life of, 1273
turnover of, 1264
reversible reaction between antibody and, 843
Escherichia coli,
enzymatic adaptation in, nucleic acids and, 1243
protein turnover in, determination, 1292
transforming factor from, 56
Esterase(s), see also individual members
activity, determination, 1100
Estrogens, 596
action, 597
source, 597
Ethanolamine,
occurrence in plants, 563

Ethionine,
action on cellular organization, 1296
on pancreatic tissue, 1296
incorporation into proteins, catheptic action and, 1294
DL-Ethionine,
effect on formation of liver catalase, 1284
of tryptophan peroxidase, 1284
Ethionine-$C_2{}^{14}H_5$,
uptake into proteins, 1286
Eukeratins, 885
Excelsin, 487, 501
effect of solvents on, 494
elementary analysis, 491
molecular constants, 503
molecular weight, in urea solution, 507
water of crystallization, 503
Exopeptidases,
metal-activated, 1106
activity, pH optimum, 1102
Eye,
effect of bee venom on, 356
of dehydrogenase inhibitor of snake venom on, 378
of ricin on, 359
lens, autoantigenicity of proteins of, 759

F

FAD, see Flavin adenine dinucleotide
FMN, see Flavin mononucleotide and Riboflavin 5-phosphate
FSH, see Follicle stimulating hormone, of anterior pituitary
Factor(s),
antibacterial, of milk, 432
bacterial, transformation, 55–57
responsible for amino acid uptake into tissue proteins, 1258–1259
skin, formation of FLS fibrils and, 928
skin sensitivity, of tuberculin, 352
spreading, see Spreading factors, Hyaluronidase
stimulating growth of tissue cultures, 1238, 1240
incorporation of amino acids into proteins of tissue preparations, 1238–1240
Fatty acyl coenzyme A dehydrogenase, 254–257
copper in, 278
isolation, 255
purification, 224
Ferns,
amino acid composition, 541
Fetuin, 737–739
composition, 737
nature of, 737
physicochemical properties, 740
source, 737
Fibrin, 882–884, see also Blood, clotting compounds
amino acid composition, 877
applications, 884
conversion of fibrinogen to, 878–882
thrombin and, 878
end groups in, 879
fine structure, 882
gel, protein content, 882
structure, 882–883, 884
α-keratin structure, 868
subunits of, 905, 906
x-ray diffraction pattern, 876
Fibrinogen, 668, 725–730, 739, 876–882, see also Blood, clotting compounds
amino acid composition, 877
bovine, end groups in, 878, 1197
species differences in, 879
clotting by thrombin, 729–730
conversion to fibrin, 878–882, 1197
intermediates, 881–882
liberation of peptides during, 879
substances effecting, 881
effect of plasmin on, 728
half life, 1267
isolation, 745–746
molecular size and shape, 726, 728, 877, 878
molecular weight, 726, 728, 877
myosin and, 877
physicochemical properties, 728, 740
in plasma, 725, 876
preparation, 726
solubility, 727
stability, 726, 729
subunits in, 905
synthesis, in perfused livers, 1258
turnover curves for, 1266–1267

"unmasked," 880, 882
Fibrinolysin, see also Plasmin
bacterial, 382
Fibrinopeptides, 879, 1197
amino acid composition, 879, 1197
molecular weight, 879
Fibroin, silk, 849–859
action of chymotrypsin and trypsin on, 1187
amino acid composition, 850–851
denatured, configuration, 858
effect of inorganic salts on, 857
electron microscope studies on, 849–850
fibroinogen as precursor of, 854
heterogeneity of, 851, 854
isolation, 849
reaction with cupric ethylenediamine, 857
"renaturation," 857–858
configuration and, 858
Fibroinogen, 855
globular nature, 856
as precursor of fibroin, 854
Ficin,
as catalyst of replacement reactions, 1134, 1135, 1141
homogeneity, 1109
isolation, 1069
resolution of amino acids by, 1108
substrates for, synthetic, 1107
substrate specificity, 1133
Fishes,
venoms of, 353
Flagella,
algal, 889
bacterial, 887–889
amino acid composition, 888
electron microscope studies on, 887, 888
isolation, 887
α-keratin structure, 868
molecular dissociation, 887
molecular size and shape, 887
molecular structure, 887
as motile organs, 889
protein nature, 887
x-ray diffraction pattern, 887–888
Flavin adenine dinucleotide,
absorption spectra, 219
determination, 220
identification, 220
structure, 218
Flavin mononucleotide, see also Riboflavin 5-phosphate
determination, 220
identification, 220
structure, 218
Flavin nucleotide coenzymes,
isoalloxazine nucleus in, 216
Flavoenzymes, 216–259, see also individual enzymes
action mechanism, 222
chemical properties, 225
composition, 222
inhibition by riboflavin, 221
metabolic functions, 220
oxidation-reduction, 132, 221
prosthetic group, 221
purification, 223
specificity, 220
Flavoproteins, 216–259, see also Flavoenzymes
flavin nucleotide components of, 216
holoenzymes, 220
prosthetic groups of, metal ions in, 278
role in reoxidation of reduced pyridine nucleotides, 218
Fluoromethemoglobin, 315
Folic acid,
effect on uptake of labeled glycine by tissue homogenates, 1239
Follicle-stimulating hormone, of anterior pituitary, 598, 600–602
action of pepsin on, 1186
activity, 597, 598
essential groups for, 601
chemical constitution, 597
composition, 601
inactivation, 601
isolation, 600–601
molecular weight, 601
physicochemical properties, 601
species variations in, 602–603
Formaldehyde,
action on collagen, 923
effect on keratin, 874–875
on toxins, 348
reaction with proteins, 767

Formic acid,
effect on keratin, 875
on synthetic polypeptides, 890
FSL fibrils, 927ff.
structure, 936
Fumaric hydrogenase, 258

G

GH, see Growth hormone, of anterior hormone
GSH, see Glutathione
Galactose,
in human chorionic gonadotropin, 620
Galactozymase,
formation, nucleic acids and, 58
Gas gangrene,
effect on muscle, 383
Gastric juice,
catheptic enzyme in, 1103, 1127
proteolytic enzymes of, 1060
Gelatin, 923–925
cellular agglutination and, 757
collagen and, 923, 933
excretion, 762
gelling, mechanism of, 924–925
gels, properties, 925
hydroxylysine from, 533
nonantigenicity, 758, 760, 762
production of antigenic, 761, 763
sols, amino acid composition, 923
molecular weight, 924
properties of, 923–924
Genes,
role of nucleic acids in self-duplication of, 521
Gila monster, Mexican, venom of, 353
Gliadin, 498
amino acid composition, 288, 331, 333, 499, 511
elementary analysis, 491
glutamine in, 572
molecular constants, 496
Globin, 288–292
denatured, 288
reaction with hemin, 289
formation, 292
"hemaffin" group of, 331
horse blood, amino acid content, 290–291
interactions between heme and, 328–335
nature of, 330ff.
stable bond between globin and heme, 285–286
reaction with mesoheme and its dimethyl ester, 331
stability, 288
turnover rate, 292, 1264
α-Globulin(s),
carbohydrate content, 678, 737
α_1-Globulin(s), 741
concentration in plasma, 740
α_2-Globulin(s),
carbohydrate content, 741
concentration in plasma, 740
α_3-Globulin(s), 739
β-Globulin(s),
of barley, amino acid composition, 511, 512
carbohydrate content, 737
concentration in plasma, 740
β_1-Globulin, 699
β_2-Globulin, 711
composition, 712
γ-Globulin(s),
antibodies and, 710, 711, 713, 714, 782, 786
anticomplementary activity, 843
carbohydrate content, 737
cellular agglutination and, 757
composition, 722
denaturation, 722
human composition, 792
molecular weight, 720
intracellular distribution, 671
iodinated, turnover, 1277
isolation, 717ff., 744ff., 749, 750–751
in lymphocytes, 810
physical and chemical properties, 720–725
plasma, 708
antibody function, 710ff.
composition, 712
serum, action of proteolytic enzymes on, 1175
immune lactoglobulin and, 433
solubility, 720

stability, 721
stabilizers for, 721–722
terminal amino acid sequence in, 714
therapeutical application, 742, 743
γ_1-Globulin(s), 711
composition, 712
isolation, 718
γ_2-Globulin(s),
composition, 792
isolation, 718
plasma antibodies and, 741
physicochemical properties, 740
ϕ-Globulins,
concentration in plasma, 740
T-globulin(s), 739, see also γ_1-Globulins
composition, 792
Globulin X, 957
Globulin(s), 439
antibodies as, 782, 786, 790
serological difference between normal and, 786
"anti-hemophilic," 379, 740, 742
in cereals, 493
isolation, 496–497
molecular constants, 496
circulation, 1271
of coconut, 499
of dicotyledons, 500ff.
components, 503–504
molecular constants, 503
physicochemical analysis, 503–507
in seeds, 500
extraction, 502
in egg white, 473–474
horse serum, antigenicity and tyrosyl residues, 764
immune, of colostrum, origin of, 433
of cow milk, amino acid composition, 427
serum γ-globulin and, 433
formation, site of, 673
structure of, 1288
iron-binding, 698–699
placental transmission, 1272
of pumpkin, amino acid composition, 511
synthesis, in perfused liver, 1258
Glucagon, see Hyperglycemic-glycenolytic factor, of pancreas
Glucosamine,
in fetuin, 737
in mucoproteins, 735
in thyrotropic hormone, 607
Glucose,
biological oxidation, 129
effect on incorporation of labeled amino acids into proteins of tissue preparations, 1239
Glucose dehydrogenase,
liver, 183–185
assay, 184–185
inhibition, 185
purification, 184
specificity for prosthetic group, 133
Glucose oxidase(s), 257–258
isolation, 257
of molds, action mechanism, 222
prosthetic group of, 133, 221
purification, 224, 257
substrate specificity, 133, 258
Glucose 6-phosphate dehydrogenases, 185
of animal tissues, assay, 186
purification, 186
substrate specificity, 133
yeast, assay, 187
purification, 186
specificity for prosthetic group, 133
stability, 187
Glucosides, cyanogenetic,
occurrence in plants, 560
L-Glutamic acid,
incorporation into proteins of *Staph. aureus*, concentration and, 1237
Glutamic acid,
γ-amide, see Glutamine
labeled, utilization for protein synthesis in living tissues, 1261
in myosin, 1022
occurrence in plants, 561
in bulk proteins of, 534, 535, 536, 537, 538
role in nitrogen and protein metabolism of, 525, 578
in protein biosynthesis, 582
in transamination, 578
L-Glutamic dehydrogenase,
beef liver, purification, 169
reductive amination of α-iminoglutarate by, 134

Glutamic dehydrogenases, 187–190
of higher plants, 190
of mammalian tissues, action mechanism, 189
assay, 187
physical constants, 188
purification, 188
specificity, 189
stability, 189
of microorganisms, 190
Glutaminase(s),
action, 1217–1218
hydrolytic, 1216
Glutamine,
as bacterial growth factor, 569
biosynthesis, 570–571
enzymatic, 526
in gliadin, 572
metabolism, enzymes catalyzing, 1216
occurrence in plants, 559, 560ff.
metabolic relation to asparagine, 567–573
nitrogen metabolism and, 524
physiological role, 524, 525–526, 572
protein metabolism and, 566, 578
protein synthesis and, 568, 570, 582
structure, 569
physiological behavior and, 569–570
synthesis, *in vitro*, 1217–1219
activators, 1218
inhibitors, 1218
mechanism, 1218
in tobacco mosaic virus, 572
Glutamine peptides,
role in transpeptidation, 578
Glutamine synthetase, 1216
glutamyltransferase and, 1218
γ-Glutamyl peptides,
role in protein synthesis, 587–588
γ-Glutamyl transfer, 1219–1220
mechanism, 1219
from peptide derivatives, 1220
Glutamyl transferase, 571, 1216
bacterial, 1219
plant, 1219
Glutamylcysteine,
as precursor of glutathione, 1219
synthesis, 1215
Glutamylcysteinylglycyl peptide, see Glutathione
Glutathione,
enzymatic replacement of cysteinylglycine moiety of, 1136
in muscle, 956
occurrence, 559
synthesis, *in vitro*, 1211–1217
ATP and, 1213ff.
activators, 1213
incorporation of labeled amino acids from peptides, 1241
inhibitors, 1213, 1214
mechanism, 1211, 1217
nucleotides and, 1213, 1214
precursors, 1215
requirements for, 1211ff.
Glutelins,
in cereals, 493
in dicotyledonous seeds, 500, 502
effect of germination on, 510
extraction, solvents for, 490
formation in wheat grains, 509
role in seeds, 499
Gluten,
in cereals, 495
wheat, nature of, 492, 498
physicochemical investigations on, 492, 498
Glutenin,
in ripening barley, 508
Glyceraldehyde 3-phosphate,
oxidation to 1,3-diphosphoglyceric acid, 129, 166
Glyceraldehyde 3-phosphate dehydrogenase,
See Glyceraldehyde phosphate dehydrogenase
D-Glyceraldehyde 3-phosphate dehydrogenase, see Glyceraldehyde phosphate dehydrogenase
Glyceraldehyde phosphate dehydrogenases, 126, 192–202
action mechanism, 192, 201, 202
from rabbit muscle, activity, 197, 198
amino acid composition, 195
complexes with DPN, 167
crystallization, 196
cysteine and, 198
DPN in, 199, 200

glutathione and, 198, 199
inactivation, 198
isolation, 196
physical properties, 196
purification, 169, 196
specificity, 126
stability, 197
yeast, activation, 196
amino acid composition, 195
distribution, 193
fractionation, 193, 194
isolation, 193, 194
preparation, 193
purification, 169
specificity, 126, 195
stability, 195
α-Glycerophosphate dehydrogenases, 190–192
occurrence, 190
rabbit muscle, purification, 169, 191
specificity, 191
Glycine,
acylation, coenzyme A and, 1211
cellular, hippuric acid synthesis and, 1208–1209
in collagen, 873, 921
incorporation into plasma proteins, 1268
in myosin, 1022
occurrence in plants, 561
Glycine-1-C^{14},
incorporation into desoxypentose nucleohistone, 1245
into proteins of tissue cultures, 1257
into proteins of tissue homogenates, 1236
folic acid deficiency and, 1239
into proteins of tissue preparations,
concentration and, 1237
effect of insulin on, 1239
folic acid deficiency and, 1239
into type III pneumococcus antibody, 1254, 1255
Glycine-N^{15},
from peptides, incorporation into hippuric acid, 1241
Glycine oxidase, of pig kidney, 259
Glycinin,
homogeneity, 506
molecular constants, 503
occurrence, 503
Glycoprotein(s),
acid (M1 of Winzler and Mehl), 735–736
composition, 736
isolation, 735–736
plasma, composition, 734
Glycylbetaine,
occurrence in plants, 573
Gofman's plasma proteins, 707
composition, 702
function, 707
Gonadotropic hormones, of anterior pituitary, 568–604, 618–623, see also individual hormones
carbohydrate content, 623
human chorionic gonadotropin and, 620–621
occurrence, 618
Gonadotropin, chorionic,
human, 596, 618–621
action, 597
biological properties, 620
chemical nature, 597
composition, 620
inactivation, 620
isolation, 618–619
molecular weight, 619
pituitary gonadotropins and, 620–621
solubility, 620
source, 597
stability, 620
pregnant mare serum, 596, 621–623
activity, 621
composition, 622
glycoprotein nature, 622
inactivation, 622, 623
purification, 621–622
source, 621
stability, 622
Gorgonin, 945
tyrosine, 945
Gramineae, see also Cereals
bulk proteins of, amino acid composition of, 534, 536
globulins of, 494
physicochemical investigations, 492
seed, 495

Grasses,
bulk proteins of, amino acid composition, 539
seed globulins of, 495
Growth,
effect of amino acid analogs on, 1286
Growth factor,
plant, in coconut milk, 529
effect on protein synthesis, 530
Growth hormone, of anterior pituitary, 598, 604, 607–613
active center, 613
activity, 597, 608, 609
assay, 608
essential groups for, 609, 612
aggregation of, 610
amino acid composition, 611–612
chemical nature, 597
classification, 598
inactivation, 609, 612
isolation, 608–609
molecular weight, 610, 611
factors affecting, 610
nitrogen distribution in, 611–612
physicochemical properties, 609
purification, 608
sedimentation behavior, 610
sources, 597, 608
structure, 612
Guanidine,
effect on myosin, 962
Guanine (2-amino-6-hydroxypurine),
in nucleic acids, 10
Gum acacia,
cellular agglutination and, 757
retention in body, 762

H

HCG, see Gonadotropin chorionic, human
HGF, see Hyperglycemic-glycogenolytic factor, of pancreas
Hair, see also Wool
configuration, 861
effect of stretching on, 861
fractionation, 872
histology, 860
keratin, see Keratin(s), hair
mechanical properties, 861
stretching, 861–863
cross-linking and, 862–863
thermodynamic studies, 863
supercontraction, 861
Haptens, see also Antigens
bivalent, reaction with antibody, 831
specificity, antigenic determinant and, 769
isomerism and, 773–774
polar groups and, 773
terminal groups and, 774
Heart malic dehydrogenase,
coenzyme III and, 165
Helicorubin, 343
Hematin, 314
as prosthetic group of methemoglobin, 314
Hematoporphyrin, 285
Heme(s), 284
conversion to hemochromogens, 329
formation, 286–287
interaction between globin and, 328–335
essential groups for, 331ff.
nature of, 330ff.
stable bond between globin and heme, 285–286
stability, 329
Hemerythrin, 341–342
absorption spectrum, 341, 342
iron content, 341
isolation, 341
isoelectric point, 342
occurrence, 280, 341
oxygen capacity, 341
oxygenation, 341, 342
prosthetic group, 342
Hemin(s), 284
paramagnetic susceptibility, 329
Hemochromogens, 329
configuration, 329
formation, 329, 330
hemoglobin and, 330
magnetic properties, 330, 332, 334
Hemocuprein, ceruloplasmin and, 700
Hemocyanin(s), 336–341
absorption spectrum, 340
amino acid composition, 338
antigenicity of, 759, 760
Busycon, molecular weight, 817

copper content, 337
isoelectric points, 339
isolation, 337
molecular weights, 340, 698, 777
occurrence, 280, 336ff.
oxygen capacity, 338
oxygenation, 338, 339, 341
prosthetic group, attempted isolation of, 341
species differences, 337
sulfur content, 337
valency, 777
Hemoglobin(s), 281–335
absorption spectra, 295–299
action of cathepsin C on, pH and, 1103
of pepsin on, pH and, 1103
of trypsin on, 1172
activity, catalytic, 322–323
adult, 319
fetal and, 318, 319, 320, 321
antibodies to, species differences in, 780
in blood diseases, 321
in blood of the newborn, 319, 321
bond types in, and its derivatives, 334
carbon monoxide capacity, 309
carbonylation, mechanism, 333
composition, species variations in, 317
crystal structure, species specificity in, 283, 780
denaturation, 289, 290
denatured proteases hydrolyzing, 1068
derivatives, see also under Oxyhemoglobin, Carbonylhemoglobin, Methemoglobin
absorption spectra, 295–299, 33
ferrous iron-containing, 312–31
magnetic properties, 334
dissociation, 675
distribution, 280
of sulfur in, 291
electrochemical properties, 306–308
enzymatic digestion, 286, 291–292
extracellular, of invertebrates, see also Erythrocruorin
molecular weight, 280
fetal, 319, 781
adult and, 318, 319, 320, 321
stability, 318
heme content, 286
hemochromogens and, 330
horse, antigenicity, 759, 760
human, 319–322
differing properties, 781
number of, 781
"inactive," identity with choleglobin, 317
intracellular, molecular weight, 280
oxygen affinity, 305
of invertebrates, 294
isoelectric point, 293
molecular weight, 293
iron content, 292
magnetic properties, 330
effect of oxygenation on, 333
mammalian, in erythrocytes, 281
molecular configuration, 324, 334–335
molecular size and shape, 292–295, 675
species differences in, 295
molecular weight, 292–293, 294, 698
origin and, 292ff., 296
muscle, see Myoglobin
oxygen affinity, 299, 300, 304
oxygen capacity, 299
as oxygen carrier, 328, 335
oxygenated, see Oxyhemoglobin
oxygenation, 299–306
effect on crystal shape, 303
intermediates, 303
kinetics, 301–302
mechanism of, 333
physiological function, 299
plant, see Leghemoglobin
preparation, 281–282
properties, species variation in, 317ff.
prosthetic group, 283–288
quantitative determination, 327–328
relation between affinities and absorption spectra, 311
solubility, species differences in, 282–283
specificity, 317–322
species variations in, 317
stability, 306, 1109
structure, 332
Hemoglobin A, see Hemoglobin, adult
Hemoglobin C, 322, 781

Hemoglobin D, 322
Hemoglobin F, see Hemoglobin, fetal
Hemoglobin S,
in sickle-cell anemia, 321–322, 781
Hemolymph,
copper content, 338
oxygen capacity, 338
α-Hemolysin,
of streptococci, 368
β-Hemolysin,
of streptococci, 368
γ-Hemolysin,
of streptococci, 368–369
Hemolysin(s), 364–371
action of, 364
of *Cl. septicum*, 369
lysis of erythrocytes by, 784
oxygen-labile, 370–371
bacterial, 370–371
biological activity, 371
inhibition by cholesterol, 370
properties, 370–371
serological relationships between, 371
of cobra venom, 371
oxygen-stable, 364–370, see also individual hemolysins
phosphatidases in, 364–367
of snake venoms, 365ff.
nature of, 365, 366
species variations in, 366
staphylococcal, 367–370
rabbit anti-sheep, 793
molecular weight, 793
physical properties, 794
reaction with erythrocytes,
effect of sodium chloride on, 820
Hemolysis,
catalytic nature of immune, 843
Hemophilia,
treatment with *Vipera russellii* venom, 381
Hemovanadin, 343
Heparin,
action of, 665
Hexosamine,
in follicle-stimulating hormone, 601
in glycoproteins, 734, 736
in human chorionic gonadotropin, 620
in immune proteins, 712, 792
in mucoproteins, 734
in pregnant mare serum gonadotropin, 622
Hexose,
in follicle-stimulating hormone, 601
in glycoproteins, 736
in immune proteins, 712, 735
in mucoproteins, 735
in pregnant mare serum gonadotropin, 622
in thyrotropic hormone, 607
Hippocastanin,
molecular constants, 503
occurrence, 503
Hippuric acid,
synthesis *in vitro*, 1205–1209
cellular glycine concentration and, 1208–1209
free energy of, 1201
incorporation of labeled glycine from peptides, 1241
inhibitors of, 1205, 1206
intermediates, 1207, 1208
possible mechanism of, 1207ff.
Histamine,
occurrence in plants, 563, 564
in venoms,
of bee, 357
role in lethal action of, 375
Histidase,
activity, effect of pH on, 1101
Histidine,
in chlorocruorin, 336
cobalt complex, combination with oxygen, 306
in globin, 288, 331, 332
in hemocyanins, 338
incorporation into plasma proteins, 1268
in myosin, 1022, 1023
occurrence in plants, 560, 561
L-Histidine-N^{15}-imidazole,
uptake by protein of duck reticulocytes, 1230
Histone(s),
composition, 41, 42
distribution, 42
isolation, 36, 42
linkage between, and desoxyribonucleic acid in cellular nucleus, 49

Holoenzymes, 220–225
  reversible separation into components, 128
Homoserine,
  occurrence in plants, 559
Hordein,
  effect of germination on, 510
  elementary analysis, 491
  molecular constants, 496
  in ripening barley, 508
Hordenine,
  occurrence in plants, 560
Hormones, see also individual members
  classification, 596
  effect on incorporation of labeled amino acids into proteins of tissue preparations, 1239
  gonadotropic, see Gonadotropic hormones, of anterior pituitary and individual hormones
  metabolic, of anterior pituitary, 598, 604
  phenolic, 596, 597
  pituitary, see Pituitary hormones and under individual hormones
  plasma, 733
  protein, see Protein hormones
  sources, 596
  steroid, see Steroid hormones
  thyroid, see Thyroid, hormones of, and individual compounds
Hyaluronic acid, 921
Hyaluronidase(s),
  bacterial, 385
  effect on tendons, 922
Hydrocarbons,
  carcinogenic, antibodies for, 775
Hydrogen ions,
  effect on enzyme activity, 1100–1105
Hydroxamic acids,
  formation, 571
Hydroxy acids,
  aliphatic, in actin, 1023
β-Hydroxy acyl coenzyme A dehydrogenases, 202
  beef liver, purification, 170
  coenzyme III and, 165
Hydroxyapatite,
  in bones, 846
β-Hydroxybutyric dehydrogenase,
  muscle, purification, 168
Hydroxylamine,
  physiological role, 573
  role in nitrogen metabolism in plants, 525
    in protein metabolism, 558
    in reactivation of DFP inhibited enzymes, 1158
Hydroxylysine,
  in collagen, 921
  as proteolysis product, 533
5-Hydroxymethylcytosine, 104–105
  in desoxyribonucleic acid of bacteriophages, 104
Hydroxyproline,
  in collagen, 912, 920, 921
    species differences, 937
    structural significance, 936
  in elastin, 920
  occurrence in plants, 558, 562
    in bulk proteins of, 521
  toxic effects, 583, 593
Hydroxytyramine,
  occurrence in plants, 564
Hyoscyamus mosaic virus, 84
Hyperglycemic-glycogenolytic factor, of pancreas, 649–650
  activity, 650
  composition, 649–650
  insulin and, 649
  isolation, 649
  unit of, 650
Hypertensinogen,
  isolation, 747
"Hypnotoxin," 353

I

ICSH, see Interstitial cell-stimulating hormone, of anterior pituitary
Ichthulinic acid, 485
Ichthylepidin,
  sources, 911
Ichthyocol, 944
  fibrils from acid extracts of, 929
  sources, 911
Imino acids,
  occurrence in plants, 558–559
α-Iminoglutarate,
  reductive amination, 134
Immunity,
  mechanism, proteins and, 756

β-Indoleacetic acid,
occurrence in plants, 560, 573
Indoleacetonitrile,
occurrence in plants, 546, 560, 573
Influenza virus, 61, 87–89
action, 88, 89
agglutination of red blood cells by, 88
inhibition, 472
components, 89
groups of, 87
"incomplete," 89
inhibition by mucopolysaccharides, 89
properties, 88–89
purification, 87–88
size, 84
Inosine triphosphate,
action of adenosinetriphosphatase on, 966
Insects,
metamorphosis, hormonal control, 733
viruses of, 91–93
Insulin, 596, 634–650
action of proteolytic enzymes on, 1183, 1184, 1187, 1188
activity, 597
antigenic, 760
assay of, 635, 637
effect of removal of alanine on, 1188
of asparagine on, 1188
essential groups for, 646–649
international unit, 635
beef, A component of, 639
chemical composition, 638
species differences in, 760
crystalline, 635–636
zinc in, 635–636, 639
effect on incorporation of labeled amino acids into proteins of tissue preparations, 1239
fibrils of, 637
glucagon, 649
homogeneity, 637
inactivation, 647
isolation, 635–636
molecular dissociation, 733
molecular weight, 639–641
oxidized, action of trypsin on A and B chains of, 1119, 1125, 1128, 1184, 1185, 1187
fractions of, 642–645
structure, 643, 644
physicochemical properties, 636–638
plasteins from, 1143
polymerization, 640
removal of terminal alanine from, 1225
resynthesis, from proteolysis products, 1144
reversible aggregation, 908
solubility, 636
sources, 597, 645
species differences in, 645–646
stability, 636
structure, 639, 643, 645
terminal residues in, 641–642
Interstitial cell-stimulating hormone, of anterior pituitary, 598, 599–600
activity, 597, 598
essential groups for, 600
chemical nature, 597
inactivation, 600
isolation, 599
molecular weight, 599
species differences in, 599–600
Invertebrates,
smooth muscles of, 954
Iodine,
labeled, in protein turnover studies, 1271, 1272
in thyroglobulin, 652
in thyroid, 650, 657
in thyroxine, 655
Iodogorgoic acid (diiodotyrosine), as protein hydrolytic product, 533
Iron,
chlorophyll formation and, 593
effect on production of diphtheria toxin, 362–363
in hemerythrin, 341
occurrence in enzymes, 278
reaction with conalbumin, 460
Irradiation,
effect on proteolytic enzymes, 1093
Isinglass, 944
Isoagglutinin(s),
human, physical properties, 794
isolation, 744ff., 749, 750, 751–752
Isoagglutins, of human blood groups, 788
Isoamylamine,
occurrence in plants, 564

Isobutylamine,
occurrence in plants, 564
Isocitric dehydrogenases, 203–207
DPN specific, isolation, 206
occurrence, 204, 206
requirement for adenosine 5-phosphate, 206
stability, 207
TPN specific, action mechanism, 203, 204, 205
homogeneity, 206
occurrence, 204
preparation, 205
Isohemagglutinins,
human formation, 788
physicochemical properties, 740
purification, 716, 808
therapeutical application, 742, 743
Isoleucine,
occurrence in plants, 560
toxic effects, 583, 593
Isomers,
serological differentiation between, 773–774, 774–775

K

α-Keratin, 848
configuration, 870
density, 897
polypeptide chain, configuration, 896–905
Astbury model, 897–899
helical structures, 899, 901–904
ribbon structures, 899–900
x-ray diffraction pattern, 902
wide angle, 865ff.
β-Keratin, 848
density, 897
polypeptide chain of, configuration, 892–896
x-ray diffraction pattern, 892
wide angle, 865, 866, 868
Keratin(s), see also Hair
amino acid composition, 872–873
chemical properties, 872–875
classification, 885
conjugated, preparation, 771
cross-linking reactions, 874–875
effect of formic acid on, 875
elasticity, 863
feather, configuration, 896
subunits, 909
x-ray diffraction pattern, 869, 870, 896
hair, 860
infrared spectra, 871–872
myosin and, 1024, 1025–1027
occurrence, 847
solubilization, 875
mechanism, 876
spectroscopic studies on, 870–872
structure, 892–909
subunits of, 906, 907
supercontracted, 904–905
x-ray diffraction pattern, 904
α-β-transformation, 863
mechanism of, 863, 896
wool, 860
action of alkali on, 874
action of papain on, 1186
amino acid composition, 847
polypeptide chain, configuration of, 848
stability of, 873
cystine content and, 873–874
x-ray diffraction patterns, 848, 859, 863, 864–870
Keratin–myosin–epidermin–fibrinogen group, 848, see also individual members
in animal tissues, 859
configuration, α-helix, 901
elasticity, 896
formation, 909
x-ray pattern, 859
Keto acids, see also individual compounds
metabolic transformations of α-, 265
physiological role, 525, 577
α-Ketoglutaric acid,
protein metabolism and, 578
role in protein synthesis, 525
α-Ketoglutaric dehydrogenases,
purification, 168
α-Ketoglutaric oxidase(s),
coenzyme III and, 165
pig heart, 275
reaction pattern of, 266

Kidney,
acetylation of S-benzylhomocysteine in, 1209
acylases of, 1134
enzyme from particulate fraction of, 1134
function of, capillary porosities and, 674–676
glutathione-splitting enzyme in, 1136
swine, isolation of peptidases from, 1068, 1069
tissue, hippuric acid synthesis *in vitro* in, 1205
species differences in ability of, 1206
Krebs tricarboxylic acid cycle, 129, 579, 580
effect of light on, 581

L

LH, see Lactogenic hormone, of anterior pituitary,
LTPP, see Lipothiamide pyrophosphate and Lipothiamine pyrophosphate
Labiateae,
bulk proteins, amino acid composition of, 536
Laccase(s), 135, 157–160
ceruloplasmin and, 701
inactivation, 140
of mammalian plasma, activity, 159
effect on ascorbic acid, 159
isolation, 158
purification, 158
of plant tissues, 157
activity, 157
oxidation of ascorbic acid by, 157
pigment from, 158
purification, 157
role of copper in, 139
specificity, 139
Lactalbumin,
action of carboxypeptidase on, 1184, 1188
α-Lactalbumin, 417
molecular weight, 417, 418
preparation, 418–419
Lactase,
in cow milk, 428, 430
Lactation,
effect on composition of milk proteins, 393–395
"Lactenin," 432
Lactic dehydrogenase(s), 126, 207–210
of beef heart, activity, 209
assay, 209
preparation, 208
purified, 170
of Jensen sarcoma, 210
purified, 170
of rat liver, 210
purified, 170
of rat muscle, 126, 210
purified, 170
of yeast, 126
Lactobacillus arabinosus,
incorporation of foreign amino acids into proteins of, 1295
Lactogenic hormone, of anterior pituitary, 598, 602–604
action, 597, 598
activity, assay of, 602
essential groups for, 603
amino acid composition, 604
chemical nature, 597
inactivation, 603
international unit, 602
isoelectric point, 603
isolation, 603
molecular weight, 603
properties, species differences in, 603
source, 597
β-Lactoglobulin, 417, 418, 419–427
action of proteolytic enzymes on, 1114, 1173, 1176–1177, 1184, 1188
amino acid composition, 427
components of, 424–425
crystals, density, 420, 422
water content, 420–421
denaturation, 425–427
effect on optical rotation, 1189
of pressure on, 1114
of temperature on, 426
on tryptic digestion, 1172, 1173
rate of, 425, 426
density, 406–407
dissociation curves, 423, 424
molecular weight, 419, 420
preparation, 418, 419

reaction with small molecules, 425
solubility, 419, 1177
preparation and, 423
terminal amino groups in, 428
Lactoglobulins,
bovine, immune, composition of, 792
Lactoperoxidase,
of cow milk, 428, 429–430
components, 430
iron content, 430
isolation, 429
molecular weight, 429
Lanthionine,
from wool keratin, 874
Leaves,
nitrogen metabolism in, factors affecting, 575–576
proteins of, 542–550
amino acid composition, 550
biosynthesis, 556ff.
development and, 531–532
hormonal regulation, 532
regions, 531
template hypothesis, 556
virus and, 556
bulk, composition, 533
species differences in, 533
chloroplast, 543, 544, 548–550, 556
amino acid composition, 549
homogeneity, 549
species differences in, 549
types of, 549
classification, 543
cytoplasmic, 543, 544–548, 549–550, 556
amino acid composition, 549
auxin activity, 545, 546
effect of thiouracil on, 555
enzyme activity, 546
isolation, 545
nucleoprotein nature of fraction I, 547
physicochemical properties, 545–546
species differences, 547
virus protein and, 552–553
effect of tobacco mosaic virus infection on, 551–552
extraction, 543–544
inclusions in leave cells, 542–543
vacuolar, 543, 544, 550
Lecithin,
activation of snake venom hemolysins by, 365
occurrence in plants, 560
Lecithinases, see Phosphatidases
Lectins, 789, 806, 814
Legcholeglobin, 322
Leghemoglobin, in root nodules of legumes, 322
Legumes,
hemoglobin in root nodules of, 322
nitrogenous compounds in, 565
Legumin,
amino acid composition, 511
arachin and, 506
effect of germination on, 510
elementary analysis, 491
formation in pea seeds, 509
molecular constants, 503
occurrence, 503, 504
soybean globulin and, 506
Leguminoseae,
bulk protein of, amino acid composition of, 533, 534, 536
seed proteins, physical properties, species differences in, 504
Leucenol,
occurrence in plants, 559, 565
identity with mimosine, 565
structure, 565
Leucinaminopeptidase, see also Cathepsin III
isolation, 1068
Leucine,
daily allowance for man, 512
incorporation into visceral plasma proteins, 1268
occurrence in plants, 560, 561
L-Leucine,
copolymer, x-ray diffraction pattern, 890
L-Leucine-1-C^{14},
incorporation into proteins of reticulocytes, RNA and, 1243
DL-Leucine-2-C^{14},
incorporation into proteins of tissue homogenates, 1234
L-Leucine-N^{15},

uptake into proteins of liver tissues, 1259
Leucoflavins, 217
Leuconostoc mesenteroides, utilization of peptides by, 1241
Leucoriboflavin, 217
Light,
effect on respiration, 581, 582
Lima beans,
antibodies to blood group A antigen in, 789
isolation, 809
trypsin inhibitor in, 1163
isolation, 1164–1165
Lipase,
in cow milk, 428, 430
lactation and, 430
Lipides,
effect on solubility of seed proteins, 490
transport by plasma proteins, 706, 707
α-Lipoic acid,
structure, 261
in thiaminoproteins, 261
α-Lipoproteins,
in plasma, 706–707
cholesterol content, 707
components, 707
α_1-Lipoprotein, 707
composition, 702
physicochemical properties, 740
α_2-Lipoprotein,
physicochemical properties, 740
β-Lipoprotein,
intracellular distribution, 671
physicochemical properties, 740
β_1-Lipoprotein, 704–706
antibodies to, 706
composition, 702, 705
molecular size and shape, 705
molecular weight, 738
physiological function, 706
in plasma, 706
stability, 705
structure, 706
Lipoproteins,
in egg yolk, 474
plasma, 701–708
composition, 702
isolation, 744ff., 749, 750
Lipothiamide pyrosphosphate, 261, 269
Lipothiamine, 132
Lipothiamine pyrophosphate, 269
Lipovitellenin, 478
composition, 475
isolation, 478
solubility, 478
Lipovitellin, 476–477
composition, 475
isolation, 475, 476, 477
solubility, 477
Lipoxidase(s), 276–278
of soybeans, purification, 276
Liver,
acetylating system of pigeon, 1210
acetylation of *S*-benzylhomocysteine in, 1209
formation of plasma proteins and, 809
of serum albumin in, 673
homogenates, as catalyst of replacement reactions, 1141
protein, half life, 1274
synthesis of *p*-aminohippuric acid in, 1207
coenzyme A and, 1207, 1208
tissue, hippuric acid synthesis *in vitro* in, 1206–1207
species differences in ability of, 1206
Livetin, 480–481
amino acid composition, 477, 479
homogeneity, 481
isolation, 480–481
Luteotropin, see Lactogenic hormone, of anterior pituitary
Lymph, 674
protein concentration in, 669
protein loss from, 669
Lymphatic system,
passage of plasma protein into, 1270–1271
Lymphocytes,
antibody formation in, 810, 811
γ-globulin in, 810
adrenocortical hormones and, 810
Lysine,
in chlorocruorin, 336
daily allowance for man, 512
in hemocyanins, 338

incorporation into plasma proteins, 1268
incorporation into proteins of liver preparations, 1245ff.
optical specificity of, 1245
systems responsible for, 1245–1246
in myosin, 1022
occurrence in plants, 560, 561
reaction with phenylisocyanate, 770
DL-Lysine-2-C^{14},
incorporation into desoxypentose nucleohistone, 1245
Lysine-6-C^{14},
uptake into plasma albumin and globulin, 1260
Lysocephalin,
hemolytic activity, 365
Lysolecithin, 365
formation, 384
hemolytic activity, 365
toxic effects, 384
Lysozyme, 439, 462–465
activity, enzymatic nature of, 474
essential groups, 465
components of, 464
enzymatic activity of, 464
composition, 463
isolation, 440, 443
molecular weight, 465
physical constants, 449
physical properties, 464–465
preparation, 462
structure, 463

M

Magnesium,
in G-actin, 978
activation of actomyosin by, 967–968
of muscular ATPase by, 1004–1005
effect on nitrogen metabolism in plants, 591
Malic dehydrogenases, 211–214
decarboxylating, 212
homogeneity, 213
isolation, 212
purification, 212, 213
stability, 212
nondecarboxylating, occurrence, 211
purified, 170, 211
Mammals,
milk of, composition, 390
Mammary gland,
synthesis of casein in, 433
Manganese,
effect on nitrogen metabolism in plants, 592
Marsh factor, 1041
role in muscle physiology, 1053–1054
Melanin pigments,
formation by phenol oxidases, 137
Mercaptalbumin,
formation, 681, 689
Mercurials,
excretion, serum albumin and, 697
Mercury,
reaction with serum albumin, 681
H-Meromyosin,
ATPase activity of, 976
amino acid composition, 1020
physical constants, 976
L-Meromyosin,
amino acid composition, 1020
globulin nature, 976
physical constants, 976
Meromyosins, 976–977
amino acid composition, 1024
formation, 976, 977, 1190, 1198
molecular weight, 977, 1198
in myosin, 977
properties, 1198
Mesoheme,
reaction of, and its dimethylester with globin, 331
Mesoporphyrin, 285
Metabolites,
biological oxidation, 128
proteins as transporters of, 698–708
Metals,
binding of, by conalbumin, 459–461
groups involved, 460
effect on proteolytic enzymes, 1148–1150
occurrence in enzymes, 278
Methanol,
effect on α-chymotrypsin-catalyzed hydrolysis, 1122

Methemoglobin, 313–316
absorption spectrum, 297, 315, 334
from carbon monoxide hemoglobin, 310
conversion of oxyhemoglobin to, 284
denaturation, 314
heat of, 290
derivatives, 315, 316
toxicity of, 316
dissociation constant, 315
formation, 313
magnetic properties, 333
absorption spectrum and, 334
prosthetic group, 314
Methionine,
in actin, 1023
daily allowance for man, 512
low food value of soybeans and, 1170
occurrence in plants, 561
relation to cystine in plant bulk proteins, 539–540
DL-Methionine-2-C^{14},
incorporation into proteins of tissue cultures, inhibitory effect of analogs on, 1256
DL-Methionine-S^{35},
incorporation into proteins of pituitary, 1239
into proteins of tissue cultures, 1256
into proteins of tissue preparations, 1229–1230
concentration and, 1236
effect of homogenization on, 1232
γ-Methylbutylamine,
occurrence in plants, 564
5-Methylcytosine, 7, 11
in nucleic acids, 11
species differences in content of, 54, 55
γ-Methyleneglutamic acid,
occurrence in plants, 559
amide of, see γ-Methyleneglutamide
γ-Methyleneglutamide,
occurrence in plants, 559, 560, 567
enzymatic decarboxylation, 564
structure, 567
N-Methyltyramine,
occurrence in plants, 564
Metmyoglobin, 326
Michaelis complex, definition, 1111
Microorganisms,
utilization of asparagine by, 570
of glutamine by, 570
Microsomes,
incorporation of amino acids into proteins of, 1234, 1263
nucleic acids in, 1243, 1263
structure proteins in, 950
utilization of amino acids for protein synthesis by, 1263
Milk,
antibacterial factor of, 432
casein synthesis in, 1288–1289
mechanism, 1289
cholinesterase content, blood cholinesterase of offspring and, 390
coagulation by rennet, 409, 419
ions essential for, 410
composition, 390
development of young animals and, 389
species differences in, 390
enzymes in, 428–432
globulin of, see also Milk proteins
preparation, 411
proteins of, see Milk proteins
"regression," 434
blood serum and, 433
colostrum and, 433–434
serum albumin in, 434
Milk proteins, 389–434, see also individual compounds
amino acid composition, 427, 428
biological importance, 389–390
composition, lactation and, 393–395
copper in, 432
distribution in milk, 391–393
effect of lactation cycle on, 393–395
estimation, 391ff.
electrophoretic, 392
ultracentrifugal, 392–393
properties, 400–408
relationship to serum proteins, 433–434
separation, 395–400
as source of essential amino acids, 428
surface active, 432
Millet,
protein content, 493

Mimosine,
identity with leucenol, 565
Minerals,
nutritive supply, nitrogen metabolism in barley leaves and, 576–577
and protein biosynthesis in plants, 588–594
Mint,
bulk proteins of, amino acid composition, 534–535, 536, 538
Mitochondria,
activity, factors affecting, 1294
incorporation of labeled amino acids into proteins of, 1234, 1235, 1262
as site of enzyme activity, 548
structure proteins in, 950
uptake of azoproteins by, 1279
Molds,
glucose oxidase of, 133, 221, 222
Molybdenum,
dietary, xanthine oxidase and, 230
effect on nitrogen metabolism in plants, 592
occurrence in enzymes, 278
Monocotyledons,
cytoplasmic protein in leaves of, 547, 548
proteins of, composition, 494–495
physicochemical analysis, 495–499
seed, amino acid composition, 510ff.
Monophenolase, 135
Mucoid,
in egg yolk, 482
Mucopolysaccharides,
collagen and, 926, 927, 942
in elastin, 949
inhibition of influenza B virus by, 89
α_1-Mucoprotein,
physicochemical properties, 740
Mucoproteins,
in human urine, 676
plasma, composition, 734
electrophoretic pattern of, 736
reaction with procollagen, 928
stability, 735
Mucosa,
duodenal, as source of secretin, 597
Muscle,
cardiac, 955–956
hormonal control of, 956
contraction, chemistry, 957
mechanism, 1027
models of, 1040–1050
mechanical behavior, 1044–1045
thread and fiber models, 1043–1044
newest concepts of, 1051–1053
role of ATP in, 1042, 1043, 1044, 1045, 1054
of myokinase in, 1054
thermodynamics of, 1046–1050
enzymes in, 1050–1051
glycolytic, 1051
mitochondrial, 1050, 1051
fibers, behavior of minced, 1040–1043
response to ATP, 1041
composition, 1028
types of, 1034
granules, 1002–1004
adenosine triphosphatase of, 1004–1005
distribution, 1002
isolation, 1003
physiological significance, 1003
proteins of, estimation, 1010
types of, 1002
histology, 954
isolation of actin from, 979ff.
juice, enzymes in, 1007
proteins of, 1050
lactic acid formation in, iodoacetate and, 963
molluskan, paramyosin in, 1034, 1038
nonprotein components, 956
particulate components, 1002–1005, see also Muscle, granules and Myofibrils
physiology, 952–954
proteins, 956–1055, see also Actin, Actomyosin, Myosin, Tropomyosin
of adult, 1010ff.
of developing, 1013ff.
effect of atrophy on, 1015–1018
of contraction on, 1008
of fatigue on, 1008
electron microscope studies of, 1027–1034
enzyme components of, 1006
estimation, 1010–1018

extracellular, 1010
extractability, 1005–1010
factors affecting, 1005–1006
intracellular, 1010
isolation, 956–957
physical constants, 961
prosthetic groups of, 1019
structure, 1018–1024
x-ray diffraction patterns, 1024–1027, 1034–1038
relaxation, factors affecting, 1053–1054
role of ATP in, 1044
skeletal, see also Muscle, striated
electron microscope studies, 1027–1034
fractionation of protein nitrogen in, 1012
structure proteins of, 957–1001
smooth, 954–955
hormonal control of, 956
of invertebrates, 954
striated, 955, see also Muscle, skeletal
contraction, nerve impulse and, 956
myoglobin in, 955
particulate bodies in, 955
structure, 955
types of, 954–956
ultrastructure, 1027–1032, 1036–1037
F-actin and, 1032
x-ray diffraction pattern, 903
small-angle, 1034–1038
Mushrooms,
phenol oxidases in, 150–157
Mussels,
paramyosin in, 955
Mustard gas,
effect on proteolytic enzymes, 1154
reaction with proteins, 770
effect on immunological behavior, 769–770
Mutase, 178, 179
Mycobacterium tuberculosis,
α,ϵ-diaminopimelic acid from, 533
Myelin sheath,
structure proteins in, 950
Myeloma, multiple,
antibody production in patients with, 711
Myoalbumin, 1006
nature of, 1018
Myofibrils, 1004
actin content, 978
adenosinetriphosphatase of, 1004, 1005
components of, 957
composition, 1004
nucleotide content of, 978
preparation, 1004
size and shape, 1004
structure, 1028, 1029
Myogen, 956
heterogeneity, 957
Myogen A, see Aldolase
Myoglobin(s), 281, 323–327
absorption spectrum, 325
amino acid composition, 326
carbon monoxide affinity, 325
heme groups in, 334
iron content, 324
isolation, 324
liberation in crush syndrome, 1016
magnetic properties of, and its derivatives, 334
molecular configuration, 324
molecular weight, 305, 325
oxidation products, 326
oxygen capacity, 325
prosthetic group of, 324
species specificity, 326
stability, 326
in striated muscle, 955
Myokinase,
activity, in myosin, 969
role in muscle contraction, 1054
Myosin A, 958, 959
action of ATP on, 977
electron microscope studies, 986
sulfhydryl content, 964
Myosin B, 959
action of ATP on, 977
Myosin T, 959
homogeneity, 960
sulfhydryl content, 964
viscosity response to ATP, 960
Myosin α, 959
Myosin β, 959
Myosin γ, 959
L-Myosin, 959

S-Myosin,
  identity with actomyosin, 959
Myosin(s), 957–958, 971–972
  action of trypsin on, 976–977
  activation by calcium, 967, 968
  ATPase activity of, 964–969
    groups essential for, 976
    specificity, 966–967
    sulfhydryl groups and, 964
  amino acid composition, 972, 1018, 1019, 1020, 1021
  bacterial flagella and, 888
  characterization, by physical methods, 969–972
  combination with actin, groups essential for, 976
  composition, 1019
  β-configuration, 848
  crystalline, 971, 972
    sedimentation constant, 972
  deaminase activity of, 968
  denaturation, 962
  effect of guanidine on, 962
    of trypsin on, 966
    of urea on, 962, 973
  fibrinogen and, 877
  formation of meromyosins from, 1190, 1198
  globulin character, 958
  isoelectric point, 962
  isolation, 958–959
  keratin and, 868, 1024, 1025–1027
  molecular size and shape, 972
  molecular weight, 972
  myokinase activity, 969
  physicochemical properties, 960–962
  polymerization, 972
  preparation, 980–981
  solubility, 960, 961, 962
  structure, 972–976, 977, 1023
  subunits of, 906, 973–976
    molecular weight, 974, 975
    size of, 974
  sulfhydryl groups of, 963–964
    ATPase activity and, 963
    denaturation and, 963, 964
  titration data, 1022
  tropomyosin and, 1022
  turnover rates, 1274
  viscosity, effect of ATP on, 960
  x-ray diffraction pattern, 971, 972

N

NMN, see Nicotinamide mononucleotide
Nanograms, 347
β-Naphthoquinone sulfonate,
  reaction with proteins, effect on immunological behavior, 769
Nerve axon, structure proteins of, 950
Neurospora crassa,
  glutamyl transferase in, 1219
Neurotoxin(s),
  of *Clostridium botulinum*, 372–373
    types of, 372
  of *Crotalus terrificus* venom,
    identity with hemolytic phosphatidase A, 357
  properties of purified, 346
  of snake venoms, 374–377, see also individual substances
Newborn,
  hemoglobins in blood of, 319, 321
Nicotinamide,
  derivatives, 160, 161, 164
  physiological role in plants, 573
Nicotinamide mononucleotide,
  coenzyme activity, 165
Nicotinamide riboside,
  coenzyme activity, 165
Nicotine,
  occurrence in plants, 560
Nitric oxide hemoglobin, 312
  absorption spectrum, 296
Nitrogen,
  labeled, in study of relations between tobacco mosaic virus and leaf proteins, 563
  of plants,
    non-protein, protein metabolism and, 557–574
    over-all economy, 515
    ratio of protein to non-protein, in cells, 528–530
Nitrogen compounds,
  in plants,
    amino acid composition of soluble and protein, 583–585
    in growing points, 529–530

inorganic, 558
organic, 558–560
soluble, of leaves, tobacco mosaic virus and, 555–557
Nitrogen metabolism,
in plants, ammonia and, 525
calcium and, 590–591
copper and, 593
effect of sulfur on, 589–590
hydroxylamine and, 525
iron and, 593
magnesium and, 591
manganese and, 592
mineral nutrition and, 575, 576–577
molybdenum and, 592
photosynthesis and, 581–582
potassium and, 590, 591
respiration and, 521, 578–581
vitamins B and, 573
Noradrenaline, 596
action, 597
chemical constitution, 597
source, 597
Notatin, 257, see also Glucose oxidases
Nucleic acid(s), see also individual compounds
analogs, incorporation into nucleic acids, 1287
analysis, 14–20
biological role, 3, 52
chemical properties, 10–28
composition, 10–12
nucleosides in, 11
nucleotides in, 11
purines in, 10
pyrimidines in, 11
sugar components, 10
degradation products, 12, 14
distribution, 8–9
in microsomes, 1263
in myofibril, 1004
number of, 7
occurrence, 3
origin, composition and, 6
peptidase activity of, 588
physical properties, 28–34
protein synthesis and, 57–60, 1242, 1244
role in reproduction of genes, chromosomes, and nuclei, 521
structure, 6–7
synthesis, activators of, 1243
bacitracin as inhibitor of, 1243
thymus, 7
of tobacco mosaic virus, 67
types of, 8
in viruses, 10
wheat germ, 5
composition, 5–6
yeast, 5, 6, 8
"Nuclein," 4, 5
composition, 4
nature of, 5
Nucleoproteins, 1–59
cytoplasmic, 46–49
fish sperm, chemical composition, 45
isolation, 35, 40
isolation, 36–44
nature of, 45–49
relation to structure of cell nucleus, 49–52
separation, 38
types of, 36
Nucleosides, see also individual compounds
in nucleic acids, 11–12
Nucleotides, see also individual compounds
in actin, 978
cyclic, structure, 23–24
glutathione synthesis and, 1213, 1214
isolation, 12–14
in myofibrils, 978
nomenclature, 12
in nucleic acids, 12
structure, 14
Nucleotropomyosin, 49, 995, 996
Nucleus,
cellular, constituents, 39
desoxyribonucleic acid in, biological function, 52–55
combination between histone and, 49
isolation, 3
nucleic acid synthesis and, 59
protein constituents other than nucleoproteins, 43–44
protein synthesis and, 59
ribonucleic acid in, 9
structure, nucleoproteins and, 49–52

role of nucleic acids in self-duplication of, 521
Nutrition,
animal, essential amino acids for, 515

O

Oats,
protein content, 493
seed globulins of, 495
Ocytocin, 596, see also Oxytocin
Ophioadenosinetriphosphatase,
of snake venoms, shock-production by, 384
Ophio-L-amino acid oxidases,
see L-Amino acid oxidases, snake venom
Ophiotoxin, 355
Opsonins, 788
Ornithine,
synthesis, 1209
Ornithine-citrulline-arginine cycle,
in plants, 566–567
Ornithuric acids,
synthesis, 1209
Ovalbumin, 439, 443–455
action of proteolytic enzymes on, 1173, 1174, 1184, 1185–1186, 1188
biosynthesis, 1288
carbohydrate content, 445–446
components, 449
composition, 445ff.
amino acid, 447
conversion to plakalbumin, 454, 880, 1189, 1190
enzyme effecting, 1070
denaturation, 451–454
effect on physical properties, 453
of phosphatase on, 450–451
enzymatic degradation, 1248, 1254
by *Bacillus subtilis* enzyme, 454, 455
essential groups in, 452, 453
homogeneity, 448–451
immunological behavior, chemical alteration and, 77
isolation, 440, 441, 442, 443–444
molecular weight, 445, 448, 777
peptic hydrolysis, 1101
phosphorus content, 450
physical properties, 446–448, 449
plakalbumin and, 455
reaction with antibody, essential groups for, 816
rate of, 817
stability, 454
structuie, 446
sulfhydryl groups in, 452
synthesis, in oviduct tissue, 1248
precursors, 1248
titration curve, 446–448
valency, 777
Oviduct,
protein biosynthesis in, 588, 1248
Ovokeratin, 885
source, 911
Ovomucin, 471–473
composition, 473
glycoprotein nature, 471, 474
inhibitor for agglutination of erythrocytes by influenza virus in, 472
isolation, 442
preparation, 471–472
Ovomucoid, 439, 465–469
activity, antigenic, 770
antitryptic, 469
essential groups for, 468
carbohydrate content, 466–467, 1166
composition, 466–467
amino acid, 467, 1167
homogeneity, 467, 468
identity with trypsin inhibitor, 466
isolation, 440, 441, 442
molecular weight, 468
physical constants, 449
physical properties, 467–469
preparation, 465–466
properties, 1161
purification, 1165
reaction with trypsin, 1167
effect of chemical modification on, 1152–1153
stability, 468
Ovomucoid-β, see Ovomucin
Oxalacetic acid,
role in protein synthesis, 525
Oxidases,
action mechanism, 127, 128
copper in, 593

Oxidations,
  biological, generation of "energy-rich" phosphate compounds by, 130
    mechanisms of, 128–131
Oxidizing enzymes, 123–278, see also under individual enzymes
  action mechanism, 127, 131
  classification, 126
  isolation, 126
  nomenclature, 127
  properties, 127
  prosthetic groups of, 218
  purification, 126
  specificity, 133
Oximes, protein metabolism in plants and, 572
Oximinosuccinic acid,
  occurrence in plants, 572
Oxygen,
  combination with cobalt complexes of amino acids, 306
    with hemoglobin, 299–306
Oxyhemocyanin, 338
  absorption spectrum, 339
Oxyhemoglobin,
  absorption spectrum, 296, 297, 298
  carbon monoxide hemoglobin and, 309
  crystal structure, species specificity in, 283
  denaturation, 289
  deoxygenation, 299, 300
    kinetics, 305
  dissociation curve, 325
  magnetic properties, 333
  preparation, 281, 282
  reaction with carbon monoxide, 310
  stability, 283, 306
    species variations in, 318
Oxymyoglobin,
  dissociation curve, 325
Oxytocin, 597
  activity, 623, 632
  chemical nature, 597, 630
  composition, 629, 631
  isolation, 628–629
  molecular weight, 629
  occurrence, 597
  partial separation from vasopressin, 626–628
  structure, 630, 660
  synthesis, 660

P

PAH, see *p*-Aminohippuric acid
PMSG, see Gonadotropin, pregnant mare serum
Pancreas,
  biological activity of, 634
  effect of ethionine on, 1296
    of virus infection on, 1296
  elastase in, 948
  hyperglycemic-glycogenolytic factor of, 649–650
  juice, hydrolysis of peptides by, 1106
  proteolytic enzymes of, 1060
  as source of secretin, 597
  trypsin inhibitor in, 1163
    preparation, 1163–1164
Papain,
  action on wool keratin, 1186
  activators, 1145, 1147
    mode of action, 1145–1146
  activity, 1101, 1109
    clotting, 729
    factors affecting, 1103
    pH optimum, 1101, 1103, 1104
  amino acid composition, 1083, 1084
  as catalyst of replacement reactions, 1134, 1135, 1139, 1140
  effect on thyroglobulin, 653, 1176
  homogeneity, 1082, 1109
  immunological behavior, 1159
  inhibitors of, 1153
  isoelectric point, 1073, 1082, 1084
  molecular weight, 1072, 1081, 1082, 1084
  oxidized, structure, 1147
  physicochemical properties, 1081–1082
  preparation, 1069
  resolution of amino acids by, 1108
  specificity, 1133
    optical, 1140, 1141
  stability, 1092
  structure, 1083
  substrates for, synthetic, 1107, 1133
  synthesis of anilides by, 1135
Paracrinkle virus, 119, 120
Parahematins, formation, 329, 332

Paramecin, 120, 121–122
  "killer" factor (kappa) and, 121
Paramecium,
  "killer" factor in, 62
Paramecium aurelia,
  "killer" factor (kappa) of, 120–121
    nature of, 121
    paramecin and, 121
    transmission, 121
Paramyosin, 1034, 1038–1040
  in molluskan muscles, 1034, 1038
  in mussels, 955
Pathogens, latent, 62
Peanut plant,
  γ-methyleneglutamide in, 559, 567
Peanuts,
  albumins of, 506
    molecular weights, 504
    sulfur content, 506
  globulin of, 492, 499, 505, see also Arachin, Conarachin
    molecular constants, 503
  proteins of, see also individual compounds
    sarcosine from, 533
Peas,
  albumins of, amino acid composition, 511
    molecular weights, 504
  proteins of, see also individual compounds
    effect of germination on, 510
      of ripening on, 509
  seed globulins, 504
    molecular constants, 503
Pectin-tyrosyl-gelatin,
  toxicity, 387
Pectin-tyrosyl-globulin,
  toxicity, 387
Penicillium,
  protease from, 1070
Penicillium kinase,
  trypsinogen activation and, 1192
Pepsin, 1126–1128
  action on A and B chains of oxidized insulin, 1128
    of detergents on, 1094
    on follicle-stimulating hormone, 1186
    on β-lactoglobulin, 1173
    on ovalbumin, 1173
    on proteins, 1127, 1128
    of thioglycolic acid on, 1083
    of tyrosinase on, 1152
  activity, 1101
    assay of, 1095
    effect of acetylation on, 1087, 1127, 1151–1152
      of iodination on, 1151, 1152
      of irradiation on, 1093
      of pH on, 1086, 1102–1103, 1126, 1127
      of pressure on, 1094
    exopeptidase, 1128
  amino acid composition, 1083, 1084
  autolysis, 1086
  conversion of pepsinogen to, 1190, 1191–1192
      mechanism, 1191
  crystalline, action on diphtheria antitoxin, pH and, 1102, 1103
      on hemoglobin, pH and, 1103
    homogeneity, 1103
  denaturation, effect of concentration on, 1088
    factors affecting, 1088
    kinetics, 1086–1087
  effect on antibodies, 798
  homogeneity, 1071, 1088, 1175
  immunological behavior, 1159
  isoelectric point, 1073, 1083, 1084
  molecular weight, 1072, 1084
  nature of, 1128
  phosphorus content, 1083
  preparation, 1064, 1065
  reactivation of denatured, 1088
  solubility behavior, 1071
  sources, 1064, 1065
  stability, 1086–1088, 1094, 1148
    effect of pH on, 1086–1088
  substrates for, synthetic, 1102, 1103, 1107, 1127
      hydrolysis of, 1126
        temperature and, 1112
Pepsin A, 1103
Pepsin inhibitor, 1190, 1191, 1192
  isolation, 1064–1065
  molecular weight, 1192
  properties, 1162
  structure, 1192

Pepsinogen,
conversion to pepsin, 880, 1190, 1191–1192
mechanism, 1191
isoelectric point, 1071, 1073
isolation, 1064, 1065
molecular weight, 1071, 1072
solubility, 1071
sources, 1064, 1065
stability, 1088
Peptidases,
activation by metal ions, 1148
activity, determination in presence of amino acid oxidase, 1099
as catalyst of replacement reactions, 1135, 1136
complexes with metal ions, 1149
in egg white, 474
in egg yolk, 481
isolation from swine kidney, 1068
microbial, 1069
in serum, origin of, 672
Peptide bonds,
hydrolysis, enzymatic, 1058–1059
interaction between, 1201
synthesis, 1199ff.
adenosine triphosphate and, 1224
from amino acid derivatives, 1208
CoA and, 1210, 1224
energy requirement, 1291
energy transfer, 1205
free energy changes in, 1202
by γ-glutamyl transfer, 1220
in response to toxins, 1201
thermodynamics of, in closed biological systems, 1201–1204
in open biological systems, 1204–1205
by transamidation, 1221
cosubstrate effect, 1222
mechanism, 1224
thermodynamics, 1223–1224
Peptide hormones, 597
γ-Peptides,
conversion to α-peptides, 1220
Peptides,
antigenic specificity, 774
effect on bacterial growth, 1241
hydrolysis, by pancreatic juice, 1106
of proteolytic enzymes, heats of, 1116
as intermediates in protein biosynthesis, 1241
liberation of, during conversion of fibrinogen to fibrin, 880
metabolism, 1242
occurrence in plants, 559
as products of proteolysis, 1173–1174
protein synthesis and, 587–588, 1291
synthesis, 526, 1222–1223, 1242
energy supply in, 1223
enzymes as catalysts of, 1222–1223
by transpeptidation, 1060
Peroxidases,
effect on diphtheria toxin, 348
Phagocytosis,
of particulate antigens, 671
Phalloidin, 357
polypeptide nature, 358
structure, 358
Phenol oxidase, 135, 148–157
action mechanism, 138
of animal tissues, 149
functions in invertebrates, 137
inactivation of, 140
melanin formation and, 137
from mushrooms, 150–157
cultivated, action mechanism, 152, 153
copper content, 155
enzymatic properties, 155
homogeneity, 155
inhibition, 151
isolation, 151
purification, 151, 153
wild, activity, 151
isolation, 150
purification, 151
potato (Irish), inhibition, 150
purification, 149
specificity, 150
potato (sweet), 149
role of copper in, 139
Phenolase, 135
Phenyl isocyanate,
reaction with amino acids, 770
effect on immunological behavior, 770
Phenylalanine,
daily allowance for man, 512

incorporation into proteins of *Staphylococcus aureus*, 1237
in myosin, 1022
occurrence in plants, 560
DL-Phenylalanine-1-C^{14},
incorporation into desoxypentose nucleohistone, 1245
DL-β-Phenylalanine,
copolymer of, x-ray diffraction pattern, 890
DL-Phenylalanine-3-C^{14},
incorporation into proteins of tissue cultures, 1256
of tissue homogenates, 1234
of tissue preparations, insulin and, 1239
Phenylglycine,
occurrence in plants, 559
Phosphatase,
acid, carcinoma of the prostate and serum, 733
alkaline, in animal tissues, 1283
of cow milk, 428, 431
stability, 431
effect on ovalbumin, 450–451
in egg yolk, 481
Phosphate compounds,
"energy-rich," 130
protein biosynthesis and, 575
Phosphatidase A,
in animal venoms, 354, 357, 365
hemolytic, in bee venom, 357
identity with crotoxin, 357
lysis of erythrocytes by, 365
in snake venoms, 365
Phosphatidase C, 366–367
clostridia toxins and, 366
hemolytic activity of, 367
Phosphatidases, see also individual members
activity of, 364
bacterial, 366–370
of snake venoms, 364–366
Phosphatides,
occurrence in plants, 560
Phosphodiesterase, of snake venoms,
action on desoxyribonucleic acids, 27–28
on ribonucleic acids, 25–26
6-Phosphogluconic acid dehydrogenases, 214
Phosphoglyceraldehyde dehydrogenase,
in muscle proteins, 1006
Phosphopeptones,
isolation from casein, 408–409
Phosphorus,
in actin, 978
deficiency, in plants, nitrogen metabolism and, 576, 577
effect on nitrogen metabolism and protein synthesis in plants, 589
in egg yolk proteins, 475, 476, 478–480, 482–483
in pepsin, 1083
Phosphorus compounds,
organic, effect on proteolytic enzymes, 1154–1159
mechanism, 1155, 1157
inhibition of cholinesterase by, 1154
Phosphorylases,
in muscle proteins, 1006
Phosphorylations,
oxidative, 131
Phosvitin, 478–480
amino acid composition, 477, 479
composition, 480
isolation, 478–479
molecular weight, 479
solubility, 479
Photosynthesis,
nitrogen metabolism and, 577, 581–582
Phytotoxins, 357–360
Picograms, 347
Pinnaglobin, 343
Pipecolic acid,
occurrence in plants, 559, 560, 562, 565, 582
origin, 565
possible conversion to lysine, 566
structure, 565
Piperidinic acid, 564
Pituitary,
anterior, hormones of, 596–618, see also individual hormones
activity, 597
chemical nature, 597
classification, 598
number of, 598
sources, 597

extracts, isolation of proteases from, 1068
incorporation of labeled amino acids into, 1239
posterior, biological activity, 632ff.
unitary hormone theory of, 632
extracts,
activity, international unit of, 625
preparation, 625–626
hormones of, 597, 623–634, see also individual hormones
activity, 623
assay of, 625
molecular size, 634
molecular weight, 628
polypeptide nature, 628
protein, 632–633
essential groups in, 634
separation, 624
Placenta,
anatomy, species differences in, 737–738
human,
as source of human chorionic gonadotropin, 597
permeability of, 737, 738
erythroblastosis foetalis and, 738
role in transfer of antibodies from human mother to infant, 413
transmission of globulins by, 1272
Plakalbumin, 454–455
formation, 454, 1189, 1248
immunological behavior, 766–767
molecular weight, 455
ovalbumin and, 455, 1189
Plants,
amide synthesis in, 570–572
ATP and, 571
amino acids in, 527
in growing points of, 529–530
role in protein synthesis, 527
cells of, protein and nonprotein nitrogen in, 528–530
protein synthesis in, 528ff.
factors influencing, 529, 530
site of, 557
proteins as constituents of surface membranes of, 517
ribonucleic acids in, 8
enzymes, activators of, 1145
activity, sulfhydryl groups and, 1145
stability, 1092–1093
growth, protein synthesis and, 531
metabolic activities, protein metabolism and, 577
metabolic relations of asparagine and glutamine in, 567–573
nitrogen compounds of,
in growing points of, 529–530
nonprotein, chromatographic studies on, 561–563
decarboxylation products of, 563–564
nitrogen fractions, amino acid composition of soluble and protein, 583–585
factors affecting protein and soluble, 574–588
nitrogen metabolism in, 523–528
ammonia and, 525
boron and, 592–593
calcium and, 590
hydroxylamine and, 525
iron and, 593
magnesium and, 591
potassium and, 590, 591
respiration and, 521
sulfur and, 589–590
ornithine-citrulline-arginine cycle in, 566–567
over-all nitrogen economy in, 515
oxidative metabolism, terminal oxidase systems and, 574–575
polysaccharides in structure of, 846
protein metabolism in, 513–594
during germination, 509
non-protein nitrogen and, 557–574
oximes and, 572
during ripening, 508
proteins in, 528–532
biosynthesis, 555
bulk, 532ff., 541
amino acid composition, 532–542
species variations in, 534
analysis of, 533
relation of cystine to methionine in, 539–540
deposition of storage, 518, 520
protein synthesis in, 528ff.
cell growth and, 520–521

glutamine and, 568
respiration, nitrogen metabolism and, 521
unidentified ninhydrin-reactive substances in, 560, 561, 562–563, 567
viruses of, see Viruses, plant
zinc deficiency in, 593
Plasma,
antibodies in, distribution in fetal, 738
formation in cells, 811
γ_2-globulins and, 741
physicochemical properties, 720–725
blood, 666
composition, constancy of, 666
ionic, 667
proteins, see Plasma, proteins
cells of, antibody formation in, 811
copper components, 700
enzymes in, 732–733
fetal, antibody distribution in, 738
lack of γ-globulins in bovine, 719
fibrinogen content, 725
fractionation, 741–753
γ-globulin content, 708
hormones of, 733
human, fibrinogen content, 876
fractions of, 711, 744ff.
lipoproteins of, 701–708
composition, 702
β_1-lipoprotein of, 706
proteins, see also Antibodies and individual proteins
amino acid composition, 723
antigenicity, 760
antiinfective activities, 708–725
biological significance, 671
cellular, 670–671, 672
diagnostic significance, 672
methods for study of, 670–671
difference between maternal and fetal, 738
disappearance from circulation, 1278–1279
effect of immunological response on, 1280
distribution, 669–672
electrophoretic analysis, 739
liver and formation of, 809
metabolism of, 672–673
molecular weight, 667, 676
net production, 1246–1247
origin of, 664
osmotic properties, 676
passage into lymphatic system, 1270–1271
from peritoneal cavity into circulation, 1272
permeability of cells for, 1271–1272
cellular protein metabolism and, 1272
physicochemical properties, 667, 740, 741
circulatory function and, 667–678
physiological functions, 743
protein biosynthesis and, 1288
reservoirs of, 669–670
sites of formation, 673
species differences in, 739
synthesis *in vitro*, 1257
amino acid requirements, 1259
rates of, 1257
therapeutical applications, 742, 743
tissue proteins and, 1291
transferrin content, 699
turnover, 1275–1278
animal size and, 1278
basal metabolism and, 1278
rates, 1268–1278, 1295
species differences in, 1276
of unknown function, 733–781
uptake of labeled amino acids by, 1260
S_{20} components of, 738
zinc components, 700
Plasmagenes, 62, 119
plant viruses and, 119
Plasmin,
effect on fibrinogen, 728
isolation, 382
proteolytic activity, 731, 732
Plasminogen, 731–732
activation, 732
isolation, 744, 749, 750–751
solubility, 732
stability, 732
Plasmosin, 38
Plasteins,
formation, 1225
by proteolytic enzymes, 1143
molecular weight, 1143

Platypus, Australian, venom of, 353
Pneumococcus (type III),
  transforming factors from, 56, 57
Pneumococcus polysaccharides,
  antigenicity, 761
  isolation of horse antibody to, 719
  reaction with antibodies, rate of, 817
Poisons,
  antigenic, see Toxins
Poly-L-alanine,
  infrared spectrum of silk and, 854
  x-ray diffraction pattern, 890, 891
Poly-α-amino acids, see also individual compounds
  synthesis, 889
Polycarbobenzoxy-DL-lysine,
  infrared absorption spectrum, 892
Polyglutamic acid,
  bacterial, 949–950
    sources, 949
    structure, 949–950
Polyglycine,
  configuration, 895
  synthetic, infrared spectrum of silk and, 854
Poly-DL-isoleucine,
  x-ray diffraction pattern, 890
Polylysine,
  tryptic hydrolysis, 1125, 1126
Poly-γ-methyl-L-glutamate,
  x-ray diffraction pattern, 890
Poly-γ-methyl-L-glutamic acid,
  configuration, α-helix, 901
Polypeptide chains,
  configuration, 893–896
    in collagen, 932, 936–942
  liberation of amino acids from C-terminal position of, 1109
Polypeptide(s),
  synthetic, 889–892
    action of formic acid on, 890
    configuration, 868, 870, 892, 899
      α-helix, 901
    infrared spectra, 871, 891
    resemblance to k-m-e-f group of proteins, 890, 891
    x-ray diffraction pattern, 890–891
  toxic, in bee venom, 357
Polyphenol oxidase, 135
  biological role, 575, 578
Poly-DL-phenylalanine,
  infrared absorption spectrum, 892
Polysaccharides,
  in collagen, 912, 921, 922
  in fibrin, 877
  in fibrinogen, 877
  pneumococcal, 719, 761, 817
  in structure of microorganisms and plants, 846
Polystyrene, sulfonated,
  antigenicity, 761
Pomelin,
  molecular constants, 503
  occurrence, 503
Porcupine quill,
  configuration, 848, 870, 901
  infrared spectra, 870, 871
  x-ray diffraction pattern, 867, 868, 869
Porphine, 283
Porphyrins,
  absorption spectrum, 298
  formation, 286–287
Potassium,
  nitrogen metabolism in plants and, 576, 577, 580, 590, 591
Potato tuber,
  amino acid composition of nonprotein and protein fractions of, 584
  bulk protein of, amino acid composition of, 535, 536
  nonprotein nitrogen compounds in, 561ff.
Potato X virus, 71
  size, 84
Potato Y virus, 84
Precipitation reaction, serological, 822–839, see also Antibody-antigen reactions
  composition of precipitate, 824–826, 834
  determination of precipitate, 822
  effect of molecular weight of antigen on, 825
  flocculation rates, 826–828
    mathematical theory, 827
    optima, 826–827
      neutralization and, 828
  linear relation between antibody-antigen ratio in precipitate, 833ff., 835
  mechanism of, 828–839

solubility of precipitate, 826
theories of, 828ff.
alternation theory, 829, 831
Bordet's nonspecific, 828, 831
lattice theory, 829, 830
zones in, 822–824
Precipitins, see Antibodies, precipitating
Pregnancy,
biological test, 618
Probacteriophages, 118
Procarboxypeptidase,
activation to carboxypeptidase, 1066, 1067
conversion to carboxypeptidase, 1190
stability, 1061
Procollagen,
isolation, 926
properties, 926
reaction with mucoproteins, 928
reprecipitated, FLS fibrils of, 927–928
structure, 936
Prolamins,
amino acid composition, 512
in cereals, 493
molecular constants, 496
in dicotyledonous seeds, 500, 502
extraction, solvents for, 490
formation in wheat grains, 509
molecular weights, 496
role in seeds, 499
sedimentation behavior, 497
Proline,
in collagen, 873, 912, 920
species differences, 937
structural significance, 936, 937
in elastin, 946
occurrence in plants, 560
in bulk proteins of, 521
in spongin, 936
Progesterone, 596
action, 597, 599
source, 597
Prolactin, see Lactogenic hormone, of anterior pituitary
Prosthetic groups,
activation of substrate and, 131
definition, 128
of flavoenzymes, 221
of oxidizing enzymes, 218
reoxidation, 132
Protamines, 4, 35, 39–42
antigenicity, 758, 760
composition, 40, 41
distribution, 39ff., 41
isolation, 40
Proteans, formation, 506
Proteases, see also Proteinases
action on proteins, 1171
activity, determination in presence of peptidases, 1098
pH optima, 1101
of *Aspergillus oryzae*, isolation, 1070
of *Bacillus subtilis*, action on insulin, 1188
isolation, 1070
bacterial, see also Proteases, microbial and streptococcal
activation by metal ions, 1148
effect of organic phosphates on, 1154
specificity, 1134
as catalysts of replacement reactions, 1135, 1138
of cow milk, 428, 429
separation from α-casein, 42
digestive, 1060
preparation, 1060–1064
isolation from pituitary extracts, 1068
microbial, 1069–1079, see also Proteases, bacterial and streptococcal
pancreatic, preparation, 1061–1064, see also individual enzymes
penicillium, 1070
plant, isolation, 1069
streptococcal, activators, 1134
activity, pH optimum of crystalline, 1102
isolation from group A streptococcus filtrates, 1070
specificity, 1134
substrate specificity, homogeneity and, 1109
substrates for, synthetic, 1107
X-Protein, see β_1-Lipoprotein
Protein hormones, 595–661, see also individual compounds
action, 597
sources, 597
Proteinases, 1106, see also Proteases
pancreatic, specificity, 1106

Proteins,
acetylation, in *Clostridium Kluyverii*, 1210
action of carboxypeptidase on, 1183
of detergents on, 1094
of pepsin on, 1127, 1128
of proteolytic enzymes on, 1171–1198, see also Proteolysis
of soaps on, 1094
amides in, 572–573
anaphylactic sensitization by, 351
pathology of, 351
antibody formation to animal's own, 709, 758, 759
to artificially conjugated, 805
antigenic, 757, 758
antigenic determinants in, 774, 776
nature of, 763–765, 779
number of per molecule, 777
structure of, 760
aromatic amino acids and, 760–761
combination with antibody, 763
effects of chemical alteration on, 765–770
of hemolytic group A streptococci, 352–353
molecular size and, 760
production of, 761
requirements for, 759–763
specificity, functional, 779
species differences in, 779–781
of *Bacillus anthracis*, 353
basic, in animal sperm, 43
biosynthesis, 1224, see also Proteins, synthesis *in vitro*
effect of amino acid analogs on, 1286
on turnover curve, 1267
energy requirement, 1291
enzymes and, 1244, 1252, 1291–1292
nucleic acids and, 57–60, 1242
peptides as intermediates, 1241
in plants, activators of, 579, 580
amide synthesis and, 571
amino acids and, 526, 527, 582–583
energy-rich phosphate and, 587
entropy change during, 586–587
glutamine and, 568
growth and, 531–532
from inorganic N by plants, 515
keto acids and, 525
in leaves, viruses and, 555–557
mineral nutrition and, 594
phosphate and, 575, 589
polyphenol oxidase and, 575
regions of, 531
role of carbohydrates in, 527–528
of peptides in, 587
of sugar in, 579, 582
suggested mechanism, 585–586
sulfur and, 590
terminal oxidase systems and, 574–575
requirements for, 1288–1290
sites of, 812
desoxyribonucleic acid content of, 812
of specific, 1246–1255
steps in, 1253
theories of, 1249–1253
derived from immunological investigations, 1278
peptide theory, 1249
pluperfect concept of, 1286
by reversal of proteolytic reaction, 1224
template theory of, 517–518, 587–588, 812, 1249ff.
viruses and, 554ff.
breakdown, relation to protein turnover, 1293
carbohydrate-containing, 737
electrophoretic separation, 737
cellular, 671
in plasma, 672
self-duplicating, 57
chemical alteration,
effect on antigenic specificity, 765–770
complexes with chlorophyll, 548, 549
molecular weight, 549
structure, 549
conjugated, immunological behavior, 770–773
preparation, 771ff.
as constituents of surface membranes of plant cells, 517
denaturation, degrees of, 766
effect on optical rotation, 1177, 1180
on proteolysis, 1172

denatured, action of proteolytic enzymes on, 1176
immunological behavior, 766
deposition of storage, in plants, 518
diazotized, immunological behavior, 769
effect of ultraviolet radiation on, 1093
egg, see Egg proteins and individual compounds
enzymatic hydrolysis, effect of temperature on, 1113
fibrous, see also Proteins, structural, and individual compounds
chemical behavior, 846–847
subunits, configuration of, 907
globular, aggregation to fibers, 907–908
configuration, 859
diameter, 517
physiological activity, 516
subunits, configuration, 901–908
halogenated, essential grouping in antigenic determinants of, 768
immunological behavior, 768
heterologous, catabolism, 1278–1280
disappearance from circulation, 1279
half lives, 1278–1279
immune, see also Antibodies, γ-Globulins
composition, 712, 792
stability, 720–722
incorporation of foreign amino acids into, 1286–1288, 1295
of labeled amino acids into, of tissue preparations, activation of, 1236–1240
interstitial, 663ff., see also Plasma, proteins
iodinated, action of enzymes on, 1273
half life, 1277–1278
thyroactive, 657–660
preparation, 657
labeled, with I^{131}, species variations in turnover of, 1277
with isotopic chromium, use in biochemical studies, 1272
mechanism of immunity and, 756
metabolically active, of plants, 516–517
seed proteins and, 516, 517
metabolism, intracellular plasma proteins and, 671
milk, see Milk proteins and individual compounds
muscle, see Muscle, proteins and Proteins, structural
nerve, 885
optical rotation, denaturation and, 1177, 1180
pancreatic, isolation, 1061
plant, see also Proteins, biosynthesis in plants
bulk, 532ff.
amino acid composition of, 520, 532–542
analysis of, 533
chromatographic, 535–539
phylogenetic significance, 540–542
metabolism of, 513–594
amino acids and, 525, 578
ammonia and, 525
carbon dioxide and, 575, 579, 580
glutamine and, 578
γ-ketoglutaric acid and, 578
metabolic activities and, 577
oximes and, 572
methods used in study of, 521–523
plant viruses as, 550–557
plasma, see Plasma, proteins of, and individual compounds
poisonous, of Castor bean, 757
reaction with formaldehyde, 767
respiratory, 279–344, see also individual proteins
occurrence, 279–281
species differences in, 280
resynthesis, by proteolytic enzymes, 1143–1144
role in virulence of pathogenic bacteria, 352–353
seed, see under Seed proteins
serum, relationship to milk proteins, 433–434
in skeletal structure of animals, 846
spore, 487
storage, dietary factors effecting, 1289
structural, 845–1055, see also Muscle, Proteins, fibrous and individual members
biogenesis, 847
of cellular elements, 950
classification, 847–848

keratin-myosin-epidermin-fibrinogen group of, 848, 859–909
sulfhydryl groups in, 963
effect of denaturation on, 963
synthesis, *in vitro*, 1227–1245, 1258
in tissue cultures, nucleic acids and, 1244
in tissue homogenates, 1234–1235
in tissue preparations, 1227ff.
activators of, 1227
experimental complications, 1227–1229
inhibitors of, 1227
use of labeled amino acids in, 1229–1245
tissue, plasma proteins and, 1291
toxic, 345–387, see also Toxins, Venoms, and individual toxic proteins
distribution, 353–361
factors affecting toxicity of, 347–350
modes of action, 361–387
properties, 361ff.
immunological, 350–353
as transporters of metabolites, 698–708
as trypsin inhibitors, 1160–1171
tryptic hydrolysis, metal ions and, 1172
of tuberculin, 352
turnover, concepts of, 1280–1282, 1291
applicability, 1292–1296
dynamic state hypothesis, 1293, 1296
no turnover hypothesis, 1293, 1295–1296
curves, 1263ff.
factors affecting, 1267
types of, 1263–1265
data derived from *in vivo* experiments, 1263–1283
determination, 1274, 1276–1277
in vitro, 1267–1269
rates of, 1290–1291
factors affecting, 1282–1283
metabolic rate and, 1290–1291
tissue enzymes and, 1292
variations in, as explanation of blood types, 680
tissue grafting and, 680
Proteolysis,
limited, 1183ff.
reaction products of, ill-defined, 1183–1187
well-defined, 1189–1198
mechanism, 1175–1183
hypothetical scheme for, 1182
preliminary denaturation in, 1179–1180
products of, 1173–1175
theories of, 1173
Proteolytic enzymes, 1057–1198, see also individual enzymes
action, reversal of, 1224
activators, mode of action, 1145
active center of, 1059
number of, 1059, 1123–1124
reaction with structural analogs of synthetic substrates, 1059
activity, determination, 1095–1100
effect of pH on, 1100–1105
biological importance, 1291
chemical composition, 1082–1085
classification, 1106
effect of alcohols on, 1095
on antibodies, 790–791
of chemical modification on, 1151–1159
on cytochrome c, 1186–1187
of denaturing agents on, 1094–1095
on horse globin, 1186
on insulin, 1183
of mustard gas on, 1154
of organic phosphates on, 1154–1159
mechanism, 1155, 1157
on ovalbumin, 1186
on protein hydrolyzates, 1143–1144
on proteins, 1171–1198
intermediate products, 1174–1175
formation of plasteins by, 114
of gastric juice, 1060
immunological behavior, 1159
inhibition, 1059
by redox and sulfhydryl reagents, 1145
inhibitors of, 1144, 1059
intracellular, 1067–1069, see also Cathepsins
isoelectric points, 1073
molecular weights, 1073
pancreatic, 1060

physicochemical properties, 1070–1082
preparation, 1060–1070
proteolytic coefficient, definition, 1109
reaction with collagen, 922
resolution of amino acids by, 1108
role in action of snake venoms, 383
in peptide synthesis by transpeptidation, 1060
specificity, 1106–1135
applications of, 1109
optical, 1106, 1108
stability, 1086–1095
effect of irradiation on, 1093
of pH and temperature on, 1086–1093
of pressure on, 1094
substrates for, synthetic, 1059, 1106, 1107
applications of, 1108
hydrolysis, inhibition of, 1111, 1116
kinetics of, 1109–1116
temperature and, 1111–1115
structural analogs of, reaction with active center, 1059
transpeptidations catalyzed by, 1135–1143
Proteus vulgaris,
flagella, 887
glutamyl transferase in, 1219
Prothrombin, 730–731
coagulase factor and, 380
conversion to thrombin, 730, 731
possible mechanism, 731
trypsin inhibitors and, 1171
homogeneity, 730–731
isolation, 744ff., 749, 750, 751–752
physicochemical properties, 731, 740
in plasma, 730
purification, 730
Protins, 969
Protogen A, see α-Lipoic acid
Protoheme, 284
life span, 287
as prosthetic group of hemoglobin, 283
of myoglobin, 324
stability, 284
Protohemin,
catalytic activity of, 322–323
identity with α-hemin, 285
preparation, 285
stability, 285
Protoporphyrin, 287
formation, 285
Protoporphyrin IX,
ferrous complex of, 283, 284
structure, 283, 284
Protyrosinase, 149
Pseudokeratins, 885
Pseudomonas tabaci,
α,ε-diaminopimelic acid from toxin of, 533
Pumpkin,
globulin of, amino acid composition, 511
Purines,
in nucleic acids, 10
occurrence in plants, 560
physiological role, 573
Putrescine,
occurrence in plants, 559, 564
origin, 563
Pyrazolones,
antibodies to, 775
Pyridine nucleotide enzymes, 165–216, see also under individual enzymes
chemical properties, 168
mode of action, 165
oxidation-reduction, 132
purified, 169
specificity, 133
Pyridine nucleotides,
chemical stability, 162
oxidation-reduction potential, 162
reduced, reoxidation, 162, 218
ultraviolet absorption spectra, 163
Pyridine nucleotide transhydrogenase, 215
Pyridinoprotein enzymes, see Pyridine nucleotide enzymes
Pyridinoproteins,
see also under Pyridine nucleotide enzymes, 160–216
apoenzymes of, 165
purification, 168
pyridine nucleotide components of, 160
Pyridoxal,
occurrence in plants, physiological role 573

Pyrimidines,
  in nucleic acids, 10
  occurrence in plants, 560
    physiological role, 573
Pyruvate oxidation factor,
  see α-Lipoic acid
Pyruvic acid oxidase,
  pigeon breast muscle, 272–274
    identity with "diacetyl mutase," 274
    purification, 168, 273
Pyruvic dehydrogenases,
  purification, 168

R

RNA, see Ribonucleic acid
Rabbit papilloma virus (Shope), 61, 90–91
  isolation, 90
  properties, 90–91
  size, 84
Rattlesnake venom, see Crotalin neurotoxin, Crotoxin
Reagins, see Antibodies, allergic
Renal cortex,
  reticulin in connective tissue of, 943
Rennet,
  coagulation of casein by, 409–410
    of milk by, 409, 410
  preparation of crystalline, 409
Rennin,
  activity, pH optimum, 1101
  amino acid composition, 1084
  isoelectric point, 1082, 1084
  molecular weight, 1072, 1082, 1084
  preparation, 1065
Replacement reactions,
  catalyzed by enzymes, see Transpeptidations
Respiration,
  Krebs carboxylic acid cycle, of, 579, 580
    effect of light on, 581
  in plants, nitrogen metabolism and, 578–581
Reticulin, 931, 943, 1010
  amino acid composition, 943
  collagen and, 943
  in renal cortex, 943
  sources, 911
Reticulo-endothelial system,
  formation of antibodies in, 810ff.
Reticulocytes,
  incorporation of labeled amino acids into proteins of, inhibitors of, 1240
Retinene,
  from vitamin A by enzymatic oxidation, 178
Rhodanase,
  in animal tissues, 1283
Rib grass virus,
  amino acid composition, 66
Riboflavin,
  absorption spectra, 219
  derivatives, flavin nucleotide coenzymes and, 216
  enzymatic reduction, 217
  inhibition of flavoenzymes by, 221
  occurrence in plants, 560
    physiological role, 573
  reduction by xanthine oxidase, 228
Riboflavin 5-phosphate, 218, see also Flavin mononucleotide
  structure, 219
Ribonuclease,
  action on ribonucleic acids, 20–24
  of crotalin venoms, 384
  pancreatic, stability, 20
Ribonucleic acid(s),
  association with cytoplasmic particles, 46–48
  in bacteriophages, 103, 104
  complex formation with tropomyosin, 49
  composition, 17
  degradation, 15–19
    enzymatic, 20ff.
  formation, nucleus and, 59
  role in protein biosynthesis, 586, 1243
  structure, derived from degradation, 20–28
D-Ribose, 6
  in nucleic acids, 10
Rice,
  protein content, 493
Ricin, 357, 358–360
  agglutination of erythrocytes by, 359
  components, 358
  effect of proteolytic enzymes on, 358, 359

enzymatic activity, 359
homogeneity, 358
lethal dose, 346, 358
molecular weight, 346
physical properties, 346
toxic effects, 358, 359
toxicity, species variations in, 349
Ricinen,
occurrence in plants, 561
Rigor mortis,
cause of, 1043
Robin, 357
occurrence, 358
Rye,
protein content, 493
seed globulins of, 495

S

SLS fibrils, 929, 930
structure, 936
Saccharomyces cerevisiae,
adaptation to maltose, effect on free amino acids, 1285
Salmine,
composition, 40
molecular weight, 42
tryptic hydrolysis, 1125, 1126
Salmonella typhi,
polymolecular antigenic toxins, of, 386
Salyrgan,
action on F-actin, 1046
inhibition of ATPase by, 1046
Saponin,
oxygen-labile bacterial hemolysins and, 371
Sarcoplasm,
biological significance, 1050
Sarcosine,
in muscle, 956
occurrence in plants, 573
from peanut protein, 533
Sarcosomes, 1002, see also Muscle, granules
Scarlet fever toxin,
streptococcal, erythrogenic skin reaction due to, 361
Schardinger enzyme,
see Xanthine oxidase, milk
Scorpions,
venom of, 354
Secretin, 596, 597
Seed globulins, see also Seed proteins
denaturation, 506–507
depolymerization by urea, 507
stability, 506–507
Seed proteins, 487–512, 516, 517, see also Seed globulins and individual compounds
amino acid composition, 510–512
of cereals, effect of germination on, 509–510
of ripening on, 508–509
molecular constants, 496
characterization, 490–494
by chemical analysis, 490–491
by physicochemical methods, 491–494
extraction, 489–490
solvents for, 489
metabolism during germination, 509–510
during ripening, 508–509
nitrogen content, 490, 491
solubility, lipides and, 490
Seeds,
essential amino acids in, 512
nitrogen metabolism in, 518–519
proteins of, see Seed globulins, Seed proteins, and individual compounds
role of glutenin in, 499
of prolamin in, 499
Sericin, 849, 855
amino acid composition, 849
Serine,
in collagen, 921
occurrence in plants, 561
in bulk proteins of, 534, 536, 537, 538
DL-Serine-3-C^{14},
incorporation into proteins of tissue homogenates, 1234
stimulating factors, 1236
Seromucoid (Rimington), 734, 735
Serum,
agglutinating, inhibitory substances in, 822
albumin, see Albumin, serum
antitoxic, human, types of, 787

blood, metal-binding β-globulin of human, 459
trypsin inhibitor in, 1163
mucoproteins of, 735
"non-immune" bactericidal substance in, 724
rabbit, amount of antibodies in, 809
trypsin inhibitor in, 1163, 1165–1166
Shiga neurotoxin,
lethal dose, 346
molecular weight, 346
properties of purified, 346
toxicity, species variations in, 349
Shigella shigae,
polymolecular antigenic toxins of, 374, 385, 386
Siderophilin, see Transferrin
Silk,
amino acid composition, 847
antibodies for, 763
antigenic determinants in, 763
biosynthesis, 855–856
chemical properties, 856–859
β-configuration, 848
effect of alkali treatment on, 858
of nitration on, 858
fiber, configuration, 853–854
inhomogeneity of, 854
fibroin, see Fibroin, silk
formation, from labeled amino acids, 1246
isolation of fibroin from, 849
polypeptide chain, configuration, 851–855, 894, 895
spectroscopic studies on, 854–855
tensile strength, 856
ultraviolet spectrum, effect of alkali treatment on, 855
Silkworms,
jaundice of, 91
Skin,
factor, formation of FLS fibrils and, 928
as source of epidermin, 885
Skin sensitivity factor,
of tuberculin, 352
Snake venoms, 354–356, see also individual compounds
activity, coagulant, 381
cytolytic, substances responsible for, 383
composition, 355
glands expelling, 355
hemolysins of, activation by lecithin, 365
of cobra venom, 371
hydrolysis products, 365
lethal doses, 355
phosphodiesterase of, action on nucleic acids, 24–25, 27–28
shock-producing substances in, 384
toxic effects of, 356
Snakes,
poisonous, 354–355
immunity to own venoms, 356
Soaps,
action on proteins, 1094
Sodium dodecyl sulfate,
action on pepsin, 1094
Sodium fluoroacetate,
lethal dose, 347
Somatotropin, see Growth hormone, of anterior pituitary
Sorghum, protein content, 493
Southern bean mosaic virus, 77–78
properties, 78
purification, 77–78
Soybeans,
globulins of, 506, see also Glycinin
legumin and, 506
low food value, 1170
methionine and, 1170
proteins of, see also individual compounds
extraction, 502
number of, 506
trypsin inhibitor in, 1163
isolation, 1164
Sperm,
animal, proteins other than nucleoproteins in, 43ff.
fish, nucleoproteins of,
chemical composition, 45
isolation, 35, 40
Spiders,
venoms of, 354
Spirographis spallanzanii,
chlorocruorin in hemolymph of, 335
Spleen,
cathepsins of, 1067
formation of antibodies and, 811

Split protein,
 definition, 128
Spongin, 945
 proline content, 936
 sources, 911
 tyrosine content, 945
Spreading factors, 384–385, see also Hyaluronidase
 nature of, 385
 occurrence, 385
Stachydrin,
 occurrence in plants, 561, 573
Staphylocoagulase, 379–380
 action of, 380
 antigenicity, 380
Staphylococci,
 hemolysins of, 367–369
  α-hemolysin, 368
  β-hemolysin, 368
  γ-hemolysin, 368–369
Staphylococcus aureus,
 protein synthesis in, nucleic acid content of cells and, 1243
  inhibitors of, 1243
 toxin of, 360
Staphylococcus muscae phages,
 lytic cycle, 107
Steroid hormones, 596, 597, see also individual members
 action, 597
 sources, 597
Streptococci,
 hemolytic, group A, antigenic proteins of, 352–353
  isolation of protease and precursor from, 1070
Streptococcus pyogenes,
 toxin of, 360
Streptokinase, 382
 action, 382
Streptolysin S, 369–370
 composition, 369
Strychnine,
 antibodies for, 775
 substrate specificity, 133
Succinic dehydrogenase,
 inhibition by animal venoms, 378
Sugar,
 role in protein biosyntheses, 579, 582
Sulfanilamide,
 acetylation, 1210
  ATP breakdown and, 1210–1211
  coenzyme A and, 1211
Sulfhemoglobin, 299, 312–313
Sulfhydryl compounds,
 as activators of proteolytic enzymes, 1145, 1146
Sulfonamides,
 antibodies for, 775
Sulfur,
 in bacteriophages, 104, 105
 effect on nitrogen metabolism in plants, 590
 in hemocyanins, 337–338
 in posterior pituitary hormones, 628
 in serum albumin, 686, 687
Synthetase A, 1217
Synthetase B, 1215
 action, 1215
 isolation from yeast, 1217

T

TMV, see Tobacco mosaic virus
TPN, see Triphosphopyridine nucleotide
TSH, see Thyrotropic hormone, of anterior pituitary
T_2 bacteriophage,
 size, 84
T_3 bacteriophage,
 size, 84
Takadiastase,
 inactivation of follicle-stimulating hormone by, 601
Tanning agents, 922–923
 action of, 923
Taurine, 564
 in muscle, 956
Tendons,
 effect of hyaluronidase on, 922
Tetanospasmin, see Tetanus neurotoxin
Tetanus neurotoxin, 373–374
 lethal dose, 346
 molecular weight, 346
 properties of purified, 346
 toxicity, 347
  species variations in, 349
 toxoidation, 348

Tetrahymena,
incorporation of foreign amino acids into proteins of, 1295–1296
Tetrahymena geleii,
growth, inhibition by 8-azaguanine, 1287
3,5,3′,5′-Tetraiodo-L-thyronine, see Thyroxine
1,3,5,8-Tetramethyl-2-formyl-4-vinylporphine-6,7-dipropionic acid,
identity with Spirographis porphyrin, 336
Tetronerythrin, 344
Thiamine,
occurrence in plants, 560
Thiamine enzymes,
prosthetic groups of, 260, 261
Thiamine phosphates, 260, 261
Thiamine pyrophosphate,
see Diphosphothiamine
Thiaminoproteins, 259–276
isolation, 264
mechanism of substrate activation, 267
metabolism and, 261, 262
mode of action, 263
purified, 270
6-Thioctic acid, see α-Lipoic acid
Thioglycolic acid,
action on pepsin, 1083
Thiopyrimidines,
inhibition of tobacco mosaic formation by, 554
Thiouracil,
effect on cytoplasmic protein, 555
inhibition of tobacco mosaic virus formation by, 554
Threonine,
in collagen, 921
occurrence in plants, 561
in bulk proteins of, 534, 536, 537, 538
role in tobacco mosaic virus activity, 1188
Threonine-1,2-C^{14},
incorporation into proteins of tissue homogenates, 1236
Thrombin,
action, 878
clotting activity, 729
enzyme nature, 728, 878
formation, calcium and, 665
therapeutical application, 742
Thromboplastin, action of, 379
Thymine (5-methyluracil),
in nucleic acids, 11
Thymus gland,
nuclei, nucleoproteins of, 45
Thyroglobulin, 596, 597, 651–653
active moiety of, 650
antigenic determinants in, 771
composition, 653
effect of goiter on, 652
species differences in, 652
denaturation, 653
papain and, 653, 1176
3,5-diiodotyrosine in, 656
homogeneity, 652
iodine content, 652
molecular weight, 653, 777
nature of, 597
physicochemical data, 652
purification, 651–652
sources, 597, 650
thyroxine in, 656
valency, 777
Thyroid,
activity, iodine and, 657
triiodothyronine and, 657
hormones, 650–660, see also Thyroxine
assay of, 650–651
iodinated compounds in, 656
protein of, hydrolytic products, 533
Thyrotropic hormone, of anterior pituitary, 597, 598, 605–607
activity, 597, 604–605
assay of, 605
criteria for, 605
composition, 607
effective dose, 606
molecular weight, 607
nature of, 597
physicochemical properties, 607
purification, 606–607
sources, 597, 605
Thyroxine, 596, 650, 654–657
activity, 597, 655
optical configuration and, 655–656
structural requirements for, 656
antibodies for, 775
antigenic determinants of thyroglobulin and, 771

chemical nature, 597
diiodotyrosine as precursor of, 656
formation, 658–659
iodine content, 655
isolation, 654
from iodinated casein, 657, 658
as protein hydrolytic product, 533
structure, 654
synthesis, 654–655
Tissues,
animal, enzyme concentration in, 1283–1285
variations of, 1284
collagens in mammalian and other vertebrate, 943–945
genesis, collagen regeneration and, 930
grafting, protein variations and, 680
incorporation of amino acids into proteins of, 1226–1245
effect of homogenization on, 1232, 1234–1235
proteins of, plasma proteins and, 1291
turnover, 1274–1278, 1291
determination, 1263–1274
Tobacco mosaic virus, 61, 64–71
activity, effect of size and shape on, 557
of threonine removal on, 1188
amino acid composition, 66, 572
strain differences in, 66
antibody, molecular size, 796
biosynthesis, 555
inhibitors of, 554–555
soluble nitrogen compounds of leaves and, 555–557
concentration in host cell, 550
effect on tobacco leaf proteins, 551ff.
use of labeled N in studies of, 553–554
of ultrasonic radiation on, 556
on immunological behavior, 770
infective particle, protein nature, 61
molecular weight, 71, 550
nucleic acid of, 67
nucleoprotein nature, 550
phosphorus content, 61
properties, chemical, 65–67
physical, 67–71
purification, 64–65
stability, 64, 67
strains of, 64
subunits, configuration, 908
Tobacco necrosis viruses, 78–82
chemical composition, 80
properties, 80–82
Rothamsted strains, 80, 82
size, 84
Tobacco ring-spot virus, 82
Tomato bushy stunt virus, 61, 71–73
molecular weight, 73
properties, chemical, 72
physical, 72–73
purification, 72
size, 84
α-Toxin, of *Clostridium welchii* (*perfringens*),
phosphatidase nature, 347
κ-Toxin, of *Cl. welchii*,
enzyme nature, 347
Toxins, 345–387, 781–782, see also individual toxins
activity, essential groups, 348
antibodies to, 350
bacterial, 360–361
collagenase activity, 1134
nature of, 360
polymolecular, of gram-negative bacteria, 385–387
identity with O-antigens, 385, 386
immunology of, 386–387
nature of, 385, 386
role in infectious diseases, 360
conversion to toxoids, 348
distribution of, 353–361
enzymatic, 347
immunological behavior, 350–353, 781–782
neutralization, 820–821
flocculation optima and, 828
plant, see Phytotoxins
purified, properties of, 346–347
rattlesnake, 781
toxicity, species variations in, 348–349
Toxoids, 340
antibodies to, 350
formation from toxins, 348
immunological behavior, 767
Transamidation, 1221–1222
effect of side chain of replacement agent on, 1140

peptide synthesis by, 1222–1223
enzymes catalyzing, 1223, 1224
Transamination,
formation of amino acids by, 525
role of glutamic acid in, 578
Transferrin, 698–699
conalbumin and, 698
molecular weight, 699
in plasma proteins, 699
properties, 699
physiochemical, 740
Transformation factors, bacterial, 55–57
desoxyribonucleic acid and, 55, 56
Transfusions, blood, protein loss following, 669
Transpeptidations,
catalyzed by proteolytic enzymes, 1135–1143
peptide synthesis by, proteolytic enzymes and, 1060
role of glutamine peptides in, 578
types of, 1135
Trees,
nitrogen metabolism in budding, 519
Tributyrinase,
in egg white, 474
in egg yolk, 481
Triglutamine,
in *Pelvetia*, 559
Trigonelline,
occurrence in plants, 561, 573
3,5,3′-Triiodothyronine, 656–657
biological activity, 657, 695
occurrence in thyroid, 657
from tyrosine, 659
Triosephosphate dehydrogenase,
sulfhydryl groups of, 963, 964
Triphosphopyridine nucleotide,
isolation, 163
structure, 163, 164, 165
Triphosphopyridine nucleotide cytochrome reductases, 250–252
isolation, 251
purified, 224
specificity, 251
Triphosphopyridine nucleotide specific isocitric dehydrogenases, see Isocitric dehydrogenases, triphosphopyridine nucleotide specific
"Triticonucleic acid," 5–6
Tropomyosin, 48, 957, 994–1001
amino acid composition, 997, 999, 1018, 1019, 1020–1021
complex formation with ribonucleic acid, 49
composition, 1019
crystalline, 996
depolymerization, 998
globulin-like behavior, 997
isolation, 980–981
light scattering, 1001
molecular size and shape, 998, 999, 1000, 1001
molecular weight, 999, 1000, 1001
myosin and, 997, 1022
physicochemical properties, 961, 994–997
polarization of fluorescence, 1001
polymerization, 995
preparation, 995
stability, 997
structure, 868, 973, 997, 999–1000, 1001
subunits, 999
viscosity, 997, 999
Trypsin, 1125–1126
acetylation, 1153
action of detergents on, 1095
on clupein, 1178
on hemolysin, 1172
on insulin, 1125, 1183, 1185
on β-lactoglobulin, 1176–1181
effect of temperature on, 1176
on myosin, 966, 976–977
on silk fibroin, 1187
of tyrosinase on, 1152
activators of, 1149–1150
mode of action, 1150
active centers of, 1126
activity, 1101, 1109
determination, 1095
effect of chemical modification on, 1152–1153
of irradiation on, 1093
of organic phosphorus compounds on, 1154, 1155, 1156
of pressure on, 1094
esterase, 426
effect of trypsin inhibitors on, 1169
pH optimum, 1079, 1101, 1105
autolysis, 1090, 1091

as catalyst of replacement reactions, 1138
conversion of trypsinogen to, 1190, 1192–1193
enzymes effecting, 1190–1192
mechanism, 1193
dialkylphosphoryl derivative, physicochemical properties, 1079–1080
electrophoretic behavior, 1079
hydrolysis of β-lactoglobulin by, effect of pressure on, 1114
immunological behavior, 1159
inactivation by diisopropyl fluorophosphate, 1126
of follicle-stimulating hormone by, 601
inhibitors, see Trypsin inhibitors
isoelectric point, 1073, 1079
molecular weight, 1072, 1080, 1081, 1126
physicochemical properties, 1078–1081
plasteins from, 1143
preparation, 1061, 1062
purification, 1063
reaction with ovomucoid, chemical modification and, 1152–1153
with protein inhibitors, 1167–1171
mechanism, 1168–1170
sedimentation behavior, 1080
specificity, 1125
structural requirements, 1125, 1126
stability, 1079, 1090–1092, 1148
stabilizers, 1091, 1092
structure, 1083
substrates for, synthetic, 1107, 1112
hydrolysis of, 1125–1126
inhibition of, 1111
pressure and, 1114
temperature and, 1112
Trypsin inhibitor(s), 1150, 1153, see also Ovomucoid
Ascaris, biological importance, properties, 1166
colostrum, 414–416
function of, 415–416, 1171
molecular weight, 415
pancreatic trypsin inhibitor and, 415, 416
preparation, 1164
properties, 1161, 1166
purification, 415
egg white, 466, see also Ovomucoid
lima bean, amino acid composition, 1167
isolation, 1164–1165
properties, 1162
reaction with trypsin, 1167
mode of action, 1150
pancreatic, 1163
amino acid composition, 1167
biological importance, 1171
difference between, and colostrum, 415, 416
molecular weight, 415
preparation, 1163–1164
properties, 1161, 1166
reaction with trypsin, 1168
protein, 1160–1171
biological importance, 1170–1171
properties, 1166–1167
reaction with trypsin, 1167–1171
mechanism of, 1168–1170
serum, biological importance, 1171
properties, 1166
sources, 1163
soybean, isolation, 1164
properties, 1161, 1166
reaction with trypsin, 1167, 1168
Trypsinogen,
amino acid composition, 1083, 1084
conversion to trypsin, 1190, 1192–1193, 1197
immunological behavior, 1159
isoelectric point, 1073, 1074, 1083, 1084
isolation, 1061, 1062, 1063
molecular weight, 1072, 1084
physicochemical properties, 1078
purification, 1063
stability, 1090
Tryptamine,
occurrence in plants, 563
Tryptophan,
in hemocyanins, 338
in plants, 560, 563
in serum albumin, 686, 687
Tryptophan peroxidase,
in animal tissues, 1283
variations in, 1284

formation, inhibitor of, 1284
turnover of cytoplasmic liver protein and, 1284
Tuberculin,
components, 352
sensitivity to, 352
Tuberculinic acid, 7
Tulip,
bulb scales of, amino acid composition of bulk protein of, 535, 536
Tumors,
incorporation of amino acids into tissues proteins of, 1226, 1231
Turnip,
bulk protein of, amino acid composition, 537, 538
Turnip yellow mosaic virus, 74–77
molecular weight, 76
properties, chemical, 76–77
physical, 74–76
protein components of, 74–75
antigenic behavior of, 77
purification, 74
ribonucleic acids in, 54
size, 84
transmission, 74
Typhoid 0 agglutinin, see β_2-Globulin
Tyramine,
occurrence in plants, 563
Tyrosinamide,
derivatives, as substrates for chymotrypsin, 1104
Tyrosinase, 135
action on proteolytic enzymes, 1152
potato, oxido-reduction, 132
purification, 149
specificity, 149
Tyrosine,
in actin, 1023
biological activity of growth hormone and, 609
in gorgonin, 945
in hemocyanins, 338
occurrence in plants, 560, 561
in spongin, 945
DL-Tyrosine-N^{15},
uptake into proteins of living tissues, 1259
L-Tyrosyl-L-lysyl-L-glutamyl-L-tyrosine,
action of proteolytic enzymes on, 1131

U

Uracil (2,6-dihydroxypurine),
in ribonucleic acids, 10
Urea,
denaturation of β-lactoglobulin by, 1180
depolymerization of seed globulins by, 507
effect on myosin, 962, 973
on proteases, 1094
on serum albumin, 685
occurrence in plants, 566, 567
possible origin, 566
Urease,
activity, effect of antibody on, 351, 784
occurrence in plants, 566
Urine,
excretion, posterior pituitary hormones and, 623
mucoprotein in human, 676
of pregnancy, human chorionic gonadotropin in, 618
Uroporphyrin, 287
Uterus,
effect of posterior pituitary hormones on, 623

V

Vaccinia virus, 60, 61, 85–87
composition, 86
properties, 86–87
purification, 85–86
sedimentation constant, 85
size, 84, 85
Valine,
daily allowance for man, 512
occurrence in plants, 560
DL-Valine-2-C^{14},
incorporation into proteins of tissue homogenates, 1236
Vascular system,
maintenance of vascular volume, 673–698
Vasopressin, 596, 597
activity, 597, 624, 632
antidiuretic, assay of, 625
composition, 630–631
isolation, 630
nature of, 597

partial separation from oxytocin, 626–628
source, 597
structure, 661
synthesis, 661
Venoms,
action, anticoagulant, 382
role of histamine in lethal, 375
animal, distribution, 353–357
inhibition of succinic dehydrogenase by, 378
bee, see Bee venom
color of, flavin adenine dinucleotide and, 383
role of zinc in, 379
snake, see Snake venoms
spreading factors in, 385
Verdoglobins, 317
formation, 317
Verdoheme, 316
structure, 317
Versene, 1133
activation of papain by, 1147
Vicilin, 504
amino acid composition, 511
conarachin and, 506
effect of germination on, 510
formation in pea seeds, 509
molecular constants, 503
occurrence, 503
Vipera russelli,
venom of, treatment of hemophilia with, 381
Viperidae, 355
Virus(es), 60–122, see also individual viruses
animal, 85–93
effect on protein synthesis of host cell, 550
genetics, 114
nucleic acids and genetic specificity of, 55, 58
infection, effect on pancreatic tissue in mice, 1296
insect, 91–93
capsular, 93
polyhedral, desoxy ribonucleic acids in, 10, 62
isolation of polyhedral bodies, 85
nature of, 92
types of, 92–93
latent, 118
neutralization of, 820–881
plant, 62–85, 540, 541, see also individual plant viruses
inhibitors of protein synthesis in, 554–555
as proteins, 550–557
purification, 62–64
resembling plasmagenes, 119–120
ribonucleoprotein nature of, 10, 61, 62
reproduction, DNA and, 1244
Vitamin A,
oxidation by alcohol dehydrogenase, 178
Vitamins, see also individual members
occurrence in plants, 560
physiological role, 573
Vitellenin, 478
amino acid composition, 475
Vitellin, 476–477
amino acid composition, 475
enzymatic degradation, 480
isolation, 475, 476
phosphoprotein nature, 476
solubility, 477
Vitellinic acid, 480
Vitrosin, 943–944
fibrils of, 944
occurrence, 911, 943

W

Wassermann antibody (syphilitic reagin),
human, electrophoretic mobility, 796
molecular weight, 791
physical properties, 794
Wassermann test, for syphilis, 841
Wheat,
proteins of, 493
formation in grains, 508–509
seed globulins of, 495
Whey,
preparation, 411
proteins of, 392, 411–428, see also Lactalbumin, Lactoglobulin
albumin fraction, 416–426
components, 417
homogeneity, 417

effect of lactation on, 395
globulin fraction, 411–412
species differences in, 395, 411
Wool,
action of enzymes on, 873
fiber, heterogeneity of, 860
protofibrils in, 905
histology, 860
keratin of, see Keratin(s), wool
mechanical properties, 861

X

X-ray diffraction patterns,
of bacterial flagella, 887–888
of collagens, 848, 914–919, 934, 935
of copolymer of DL-β-phenylalanine, 890
of epidermin, 886–887
of fibrin, 876
of keratin–myosin–epidermin–fibrinogen group, 859
of keratins, 863, 864–870, 892, 896, 902, 904
of muscle proteins, 1024–1027, 1034–1038
of myosins, 971, 972
of poly-L-alanine, 890, 891
of poly-DL-isoleucine, 890
of poly-γ-methyl-L-glutamate, 890
of porcupine quill, 867, 868, 869
of silk fiber, 851, 852–853
of synthetic polypeptides, 890–891
Xanthine,
occurrence in plants, 560
Xanthine oxidase, 124, 220
in animal tissues, 1283
variations of, 1284
milk, 226–231, 428, 430–431
activity, 230
chemical properties, 226
components, 431
homogeneity, 227
inhibition, 231
isoelectric point, 430
molecular weight, 430
molybdenum content, 230
purification, 224, 226, 430, 431
specificity, 230
pig liver, 231
prosthetic group in, 221
substrate specificity, 132
synthesis, *in vitro*, 1254
Xanthoproteins,
immunological behavior, 768

Y

Yeast,
brewer's, amino acid composition of bulk protein of, 537, 538
Yeast alcohol dehydrogenase,
coenzyme III and, 165
Yellow enzyme,
new, isolation, 249
old, 186, 217, 220, 247, 248
isolation, 247
prosthetic group of, 219
purification, 224
Yellow pigment, 216

Z

Zein,
amino acid composition, 511
molecular constants, 496
Zinc,
in crystalline insulin, 636, 639
in plasma, 700
role in venoms, 355, 379
Zwischenferment, see Glucose 6-phosphate dehydrogenase
Zymogens, see also individual compounds
activation of, possible mechanism, 1196–1197

Volume II part A—pages 1 to 661
Volume II part B—pages 663 to 1296